Elementary School Mathematics

Teaching Developmentally

JOHN A. VAN DE WALLE
Virginia Commonwealth University

Longman
New York & London

Elementary School Mathematics: Teaching Developmentally

Longman, 10 Bank Street, White Plains, N.Y. 10606

Associated companies:
Longman Group Ltd., London
Longman Cheshire Pty., Melbourne
Longman Paul Pty., Auckland
Copp Clark Pitman, Toronto

Senior editor: Naomi Silverman
Production editor: Dee Amir Josephson
Cover design/Cover illustration: Robin Hessel-Hoffmann
Production supervisor: Joanne Jay

Library of Congress Cataloging-in-Publication Data

Van de Walle, John A.
Elementary school mathematics : teaching developmentally : a resource book for teachers / by John A. Van de Walle.
p. cm.
Bibliography: p.
Includes index.
ISBN 0-8013-0203-X
1. Mathematics—Study and teaching (Elementary) I. Title.
QA135.5.V34 1990
372.7—dc20 89-12506
CIP

7 8 9 10-HC-95949392

CONTENTS

PREFACE

> Education in any discipline helps students learn to think, but education also must help students take responsibility for their thoughts. While this objective applies to all subjects, it is particularly apt in mathematics education because mathematics is an area in which even young children can solve a problem and have confidence that the solution is correct—not because the teacher says it is, but because its inner logic is so clear.
>
> *Everybody Counts: A Report to the Nation on the Future of Mathematics Education*
> (National Research Council, 1989)

This book is about the challenging and rewarding task of helping children develop ideas and relationships about mathematics. The methods and activities that you will find throughout the book are designed to get children mentally involved in the construction of those ideas and relationships. Children (and adults) do not learn mathematics by remembering rules or mastering mechanical skills. They use the ideas they have to invent new ones and modify the old. The challenge is to create that clear inner logic, not master mindless rules.

AN OVERVIEW

Let me give you a brief guide to the way the book is written and some of the things you can expect to find within it.

In Chapter 1 you are given an opportunity to reflect on what it might mean to teach mathematics and to suggest some of the important factors that will shape mathematics education into the next century.

Chapter 2 describes the general philosophy behind the subtitle, "Teaching Developmentally." That approach is based on the nature of mathematics knowledge, some basic understanding of how children actually construct that knowledge, and finally, what this says to us about how to help children develop that knowledge.

Problem solving has correctly become a major focus of the mathematics curriculum. In Chapter 3, meaningful learning of mathematics is shown to be a problem-solving process regardless of the particular content. Problem solving as a topic within the curriculum is also discussed.

Activities, Learning, and Children

Chapters 4 through 19 each address a different area of the elementary mathematics curriculum. The most important feature of these chapters is the activities suggested throughout each one. While each activity is identified with a number, I have woven most of them directly into the text; you will discover many activities among the figures, too. It will be very difficult to read the book and skip over the activities as you go, or at least that is my intention.

As a mathematics student, I was frequently told to read a mathematics book with a pencil and paper. The point was that to get involved with and therefore understand the mathematics, I had to wrestle with it firsthand; try it out, reason along with the author, fill in the holes and gaps, make it my own. This book is not about mathematics but about *children learning mathematics*. It is children's learning or construction of mathematics that you need to get involved with, and try to understand as you read. Rather than read this book with a pencil, you should read it with the materials (counters, blocks, graph paper, calculators, etc.) suggested along the way.

A brief story may explain. In one of my undergraduate classes we were exploring a second-grade activity that starts with children using tens and ones pieces to show 35. My students placed materials on the desk as shown here.

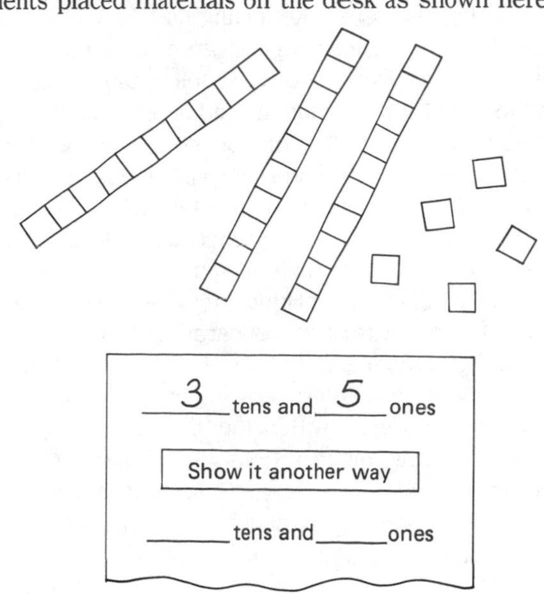

The task was: "Can you find another way to show this number?" Immediately they picked up one of the strips and began to replace it with 10 ones pieces, making 2 tens and 15 ones. I could tell this was being done with little thought on their part, because the activity presented *them* with no problem. They were not thinking about how a child would know to do this or why to do it, so I stopped them. "Suppose students did not know what to do here?" Several responded that they would tell the children to put a ten piece back, get 10 ones, and write down 2 tens and 15 ones. With this direction, there is little doubt that their students would be able to complete the exercise. But my students were focusing on completing the activity; on getting students to do the task. They were not focusing on how the children might be thinking; why they might not know what to do. The real purpose of the task was not to exchange a strip for 10 little squares; it was to help children develop the idea that there are 35 little squares in all, some in strips of 10 and some not, and that they do not have to be in three groups of 10, but can be grouped in other ways and it will be the same amount. Telling the children what to do is doing their thinking for them. It suggests all that is required is to follow blindly the superior knowledge of the teacher. It does not help children construct the inner logic that makes mathematics so clear and so much fun to learn.

Try, then, not only to do some of the activities as you read along, but work hard at doing them from the perspective of the child. Avoid using your adult knowledge that makes many elementary-level tasks seem trivial. The goal of each activity is not to "be able to do the activity," or to "get the answer," but to construct ideas. Activities are designed to cause children to think. As you read, try to figure out how children might think about it and how an activity will affect their thinking. Children and adults do not think alike. In Appendix B a guide for relfective thought will help you get the most from reading this book. Even if your instructor does not make it an assignment, you should read Appendix B after reading Chapter 2. Reflecting on how children learn from activities is the best way to grow as a teacher.

In its own way, each of the last four chapters deserved to be in the Chapter 4 position instead of position 20, 21, 22, or 23. You may want to read one or more of these chapters early, depending on your interests and needs. In Chapter 20, the role of calculators and computers in mathematics is discussed in some depth, even though calculators are in the activities of almost every chapter and computers in quite a few. Chapter 21 discusses planning lessons, suggestions for classroom use of materials, cooperative learning groups, homework, and the role of the basal textbook. Chapter 22 discusses assessment, with special emphasis on diagnosis and how to listen to children within the classroom. Chapter 23 explores the special considerations that should be given to children with special needs. Separate sections cover children with cognitive processing difficulties (learning disabilities), mentally handicapped children, and gifted children.

Features

This book was not written with a lot of "gadgets" or special sections. At the end of each chapter you will find a discussion and exploration section. The purposes of this section are varied. You may wish to read through these questions and suggested activities just to jog your thinking about what was in the chapter. Some suggestions may help you to examine a student textbook or textbook series, to observe a classroom lesson, or to try some activities with students. Your instructor might use these ideas to help structure an assignment. Look on them as "food for thought."

Each chapter ends with a list of suggested readings. These are usually short articles, chapters, or occasionally books that have been written for classroom teachers. No such listing could be "complete" in any way, but these represent items that I have found useful or thought-provoking. Some may have been referenced in the text. (All works cited in the text are found in the references at the end of the book.) In addition to articles and books, I have occasionally included teacher resource materials relevant to the chapter topic. These may be activity books for children, teacher's guides, activity cards, or even kits of materials.

In Appendix C you will find a collection of black-line masters and some brief directions for making important materials. Suggestions for their use are found throughout the book. You are encouraged to duplicate these pages to make materials or to use in activities. I hope that they will prove to be a practical resource for you.

A Few Notes to Instructors

Hey, I know this is a long book! One reviewer suggested I begin to look for some four-semester methods courses.

I chose to write a book that talks to teachers and provides them with enough information about how children learn and about conceptual activities for helping them learn so that they can use this book as a resource as well as a text. I hope that you look on this as a luxury. It allows you to make choices and provide emphasis on topics or activities that are most important to you. Those topics that you choose not to cover in class or activities you select not to explore are described in sufficient detail that they can be studied independently. At the very least, your students will carry those ideas and discussions with them if they see this book as a resource instead of as just a text. There is plenty from which to choose.

The activities throughout the text have been numbered for your convenience (note that some figures serve as activities). With the use of the numbers you can identify activities that you would especially like your students to explore or be familiar with. I encourage you to have your students utilize the four-step chart in Appendix B that will guide students in reflecting on an activity. The four steps provide a built-in method for very flexible out of class assignments. Students can be assigned one or more activities "for reflection" according to the guide. Their individual or group reflections, especially step 4, can be shared in class or written up briefly to be turned in. Different students might be assigned different related activities to discuss in class. Use the guide with your class for activities you conduct in class.

The general constructivist approach that is described in Chapter 2 and reflected in the guide I have just discussed is firmly grounded in the theories of Piaget, Skemp, Dienes, and Bruner. At the same time, very little specific reference to these men or their theory is made. This book is about helping children construct mathematics, not about learning theories. Should you wish to emphasize Piaget to a greater extent than I have, especially his theory of assimilation and accommodation, you will find no contradictions within the text.

As mentioned already, the last four chapters are essentially independent of the content chapters. Many instructors would like their students to know at the start of the course about some of the issues around lesson planning, classroom management, cooperative learning groups, and the use of textbooks. This would be most important if your students are in a practicum setting while in your class. My personal preference is to have my students read this information on their own during the first few weeks of the semester. Similar comments can be made about the material in the other chapters at the end.

Several of the content chapters had difficulty finding a natural home in the numeric sequence. Chapter 4 ("Instruction with Process Problems") and Chapter 17 ("Classification and Patterning") both deal with the teaching of processes related to problem solving and can be explored at any time or as an ongoing project. The classification and patterning activities should not be viewed as "prenumber" experiences as they are in some texts, but rather as a piece of the broader strand of problem solving. In my own classes, I do attribute and patterning activities the first week. They are fun, easy, and set the stage for getting children to think.

Several logical groupings of chapters should probably be kept intact, such as measurement and geometry (Chapters 15 and 16) or place value, whole number computation, and mental computation and estimation (Chapters 8, 9, and 10). With those obvious exceptions of groupings, you should be able to move around in the text in a manner that suits your needs or to assign some chapters to be read independently.

Acknowledgments

Much of the credit for this textbook must go to the mathematics educators who took time from their own professional endeavors and took great care in offering comments on drafts of the manuscript. Each provided many, many helpful suggestions and insights that served to substantially improve the quality of the book. My most sincere and heartfelt thanks is given to John Dossey (Illinois State University), Bob Gilbert (Florida International University), Warren Crown (Rutgers), and Steven Willoughby (University of Arizona). These four each reviewed and commented on the entire manuscript. I am also deeply indebted to Arthur Baroody (University of Illinois-Champaign) and James Bruni (Herbert H. Lehman College, CUNY), who reviewed significant portions of the manuscript.

Thanks is also extended to Kathy Maitland (Virginia Commonwealth University), who took the photographs of the computer screens.

Two people have played a significant part in my own professional development, and each can be found in the spirit with which this book is written. My thanks for a lifetime goes to Harold "Bud" Trimble, professor emeritus at The Ohio State University, to whom I owe the fact that I am a mathematics educator. Dr. Trimble guided me into the world of education, showed me the thrill of working with children, and demonstrated faith and confidence that a young college mathematics teacher could work with elementary teachers and children. Richard Shumway, also of Ohio State, provided a challenge to succeed and much of the help with which to do so. Dick once told me I should be teaching teachers not how to teach but "how to *learn* to teach." This book represents, in part, an answer to his challenge.

Finally, and most importantly, this book would never have been finished if it were not for the constant love, support, encouragement, and patience far beyond the ordinary that my wife has given without reservation or complaint during the long months of writing as well as during our lives together. A professional in her own right, she has had to take a greater share of our daily tasks and endured long nights and weekends alone while I worked on the book. With all my love, thank you, Sharon.

CHAPTER 1

TEACHING MATHEMATICS: REFLECTIONS AND DIRECTIONS

Thinking About Teaching Mathematics

BEGINNING WITH YOUR PERCEPTIONS

"Teaching mathematics."

What kinds of images and emotions does that simple phrase bring to your mind? Consider first the *mathematics* part. What do you think mathematics is all about? What is mathematics in the elementary school? Pause right now and reflect on your own ideas about the topic of mathematics. What is it? How does it make you feel? What does it mean to "do mathematics?" Where do calculators and computers fit in? What parts of the subject seem to you to be most important? Write down three or four of your strongest thoughts about mathematics. Compare your thoughts with those of others.

Next focus on the *teaching* part. Someday soon you will find yourself in front of a class of students, or perhaps you are already teaching. Your goal is for children to learn mathematics. What general ideas will guide the way you will teach mathematics? Do you think your ideas are influenced by your view of what mathematics is? Do children learn mathematics differently than other topics? How can you make it interesting and enjoyable? If mathematics is not exactly your favorite subject, do you think it had anything to do with the way you were taught? How can you help children like the subject more than you do?

These are hard questions. They do not have simple, unique answers that everyone will agree on. But the nature of the subject of mathematics and the way that students learn it can and should influence how we teach it to children.

TWO EXAMPLES

In the following hypothetical (but not unrealistic) examples of teaching mathematics, two teachers are guided in their approach by the way they perceive the topic and by the way they think it is learned. The topic is comparing fractions: given two fractions, how do you tell which is greater? It was chosen because you should easily identify it as part of the familiar curriculum—something you no doubt learned in the fifth grade, the approximate grade level of these examples. See what you think about the two approaches.

Mr. C's Approach

Mr. C focused his lesson on presentation of content. He began by showing the class three categories of fraction pairs as in Figure 1.1. For the first type of category, he explained how the larger numerator means more pieces. In the second case he showed them that the smaller denominator means larger pieces (so the smaller denominator is the larger fraction). In the third case, he suggested that the easiest thing to do is to get common denominators and then the first rule could be used. Several pairs of fractions were then put on the board. The class identified which category each pair belonged in and applied the corresponding rule. A worksheet or text page was assigned in which students were to select the greater fraction for 20 different pairs. Starred problems required the students to put four fractions in order from least to most.

Ms. L's Approach

For Ms. L, the same lesson was designed somewhat differently. Instead of an explanation, she began the lesson with a question. It essentially was a problem, since the students had not been taught how to do exercises like this before. The fractions $\frac{2}{3}$ and $\frac{1}{4}$ were put on the board, and the students were asked to select the fraction which was larger. However, before any answers were accepted, she put the students in groups of three and gave out pie-piece models for fractions. Each team had to answer the question and come up with at least two explanations for their answer. They could use the pie pieces if they wished. After 5 minutes, different groups were asked to share their ideas with the class. As many ideas as possible were considered.

Following this exercise, the groups were given a worksheet with eight pairs of fractions. Representatives of each category in Figure 1.1 were included but not identified. The teams were required to select the greater fraction in each pair

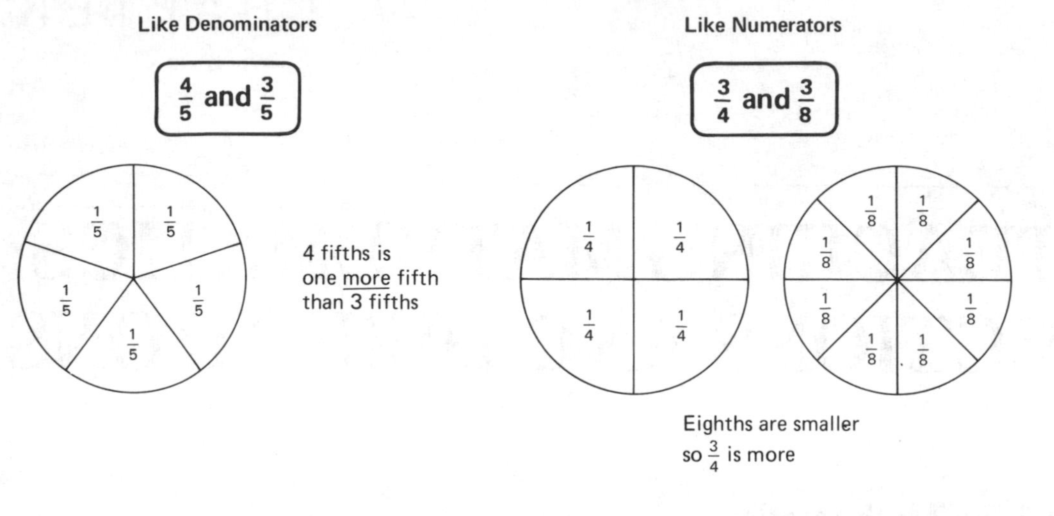

Figure 1.1
Categories of fraction pairs.

and provide some meaningful explanation for their choice. They were encouraged to draw pictures or refer to their pie pieces. While they were working, Ms. L circulated around the room, listening to the student interaction, providing hints to some teams, and additional challenges to others.

Contrasts

Both teachers presented good lessons. Mr. C apparently gained a bit more ground than Ms. L because his students were already practicing a basic skill and were using rules that Ms. L had not yet explained. The students in her class were encouraged to devise their own rules. Besides those presented by Mr. C, Ms. L's students also noted that sometimes one fraction is more than a half and the other is less. At other times they could choose the larger because one was clearly closer to 1 (7/8 is closer to 1 than 4/5 is). In fact, many more relationships about fractions, numerator, and denominator meanings were discussed and used in her class. Mr. C's students quickly dropped the use of the pictures for fractions and concentrated on deciding which rule to use.

Both teachers would have papers to examine in planning the next day. By grading the papers, Mr. C would know which students used the rules correctly and which did not. For those who did not do well, he would probably try a more careful explanation of the content. However, he has few clues concerning why students missed problems. For example, for a pair requiring finding a common denominator, the difficulty could be with not understanding equivalent fraction concepts. In that case a student might well wonder why a different fraction (9/12 instead of 3/4) would help at all. Another, but very different difficulty, might be a lack of skill in finding common denominators, or perhaps only a weakness with factors and multiples. Subsequent lessons would likely be variations of the first one. Careful articulation and practice of the rules would be the theme.

Ms. L would have the informal information gathered during the lesson as well as the reasons the students wrote on the worksheet. This feedback would provide a window to her students' thinking. Based on this information, her second lesson might very well involve grouping students so that those with the best ideas and concepts would be paired with the average students and challenged with a similar task as the first day. For both variety and understanding, a different model, perhaps paper folding, might be used. Those who demonstrated the most trouble quite likely also demonstrated concept deficiencies. She might work with that group on basic meanings of fractions while the rest of the class works independently.

Effectiveness

Both classes could easily end up with a similar degree of proficiency on this particular task. By that measure, both teachers would be successful.

What would you have done? Which method is closest to the way you were taught? Which class would you learn best in? In which class would you rather be?

If you had a Mr. C for each grade K to 8, would your perception of mathematics be different than your perception if you had been taught by teachers like Ms. L? Consider the following two lists. Which statements would students from the two classes most likely choose?

Mathematics is

Numbers and getting answers

Work

A lot of rules

Adding and multiplying and things like that

Confusing and difficult to remember

Always either right or wrong

Making me nervous (I never was very good at mathematics)

OK except for the word problems

Mathematics is

Full of relationships about numbers and shapes

Neat because you can always figure it out if you forget something

A lot more than just computation

Working with graphs, and shapes, and measuring, and patterns

Fun—it's like solving a bunch of interesting puzzles

A subject where you can discover new ideas

Learning to think through a situation logically

Solving problems you've never seen before

The approach taken by Mr. C might be characterized as *content-oriented,* that of Ms. L as *learner-oriented.* One is concerned a bit more with skills and methods of getting answers, while the emphasis of the other is to help her students confront ideas and reason them through. The spirit of this text is in keeping with a learner-oriented approach to teaching mathematics. It is based on a belief that it is important, beneficial, and more productive in the long run for students to talk about mathematics, to confront problems and struggle with them, to use materials to help them think and explore ideas and relationships, to make errors and profit from them, and to learn to reason logically. A major consideration that is reflected in the methods suggested throughout the book is the manner in which children learn mathematics.

Directions in the Elementary Mathematics Curriculum

The manner in which we teach children is only one of the things that affect what happens in the mathematics classroom. The curriculum, the content to be taught, is another major factor to be considered. Many influences affect and change the content of elementary school mathematics. Some understanding of these influences can provide us with a professional perspective on the changing curriculum and thereby help us react to change appropriately—perhaps even to influence change.

SOCIETAL DEMANDS

The very nature of the society we live in has a real impact on what is taught in schools. We no longer live in an industrial economy but one that is dominated by information and services. Such an economy requires a smaller and smaller percentage of unskilled labor and an increasing number of highly skilled and well-trained personnel. The rapid changes in our society brought about by technology make it difficult to train people specifically for jobs that may change or not exist 10 years later. Even those jobs that are not directly in scientific areas are impacted by technological changes that require workers to learn to adapt to new situations.

Public education is highly political and responsive to a vocal constituency, especially parents. In many instances this influence has been in a different direction than that of professional mathematics educators. A case in point was the "back to basics" movement of the 1970s. As college entrance scores declined, the public clamor rose for a return to the basic skills that they associated with mathematics from their childhood. That mathematics was dominated by computational skills. Unfortunately, those skills were not and are not adequate.

More recent explanations of problems with U.S. schools have been very explicit in noting that our students do reasonably well with lower-level skills such as computation. Tests have consistently pointed to the failure of U.S. students in areas of conceptual knowledge, reasoning, and applications. Between 1973 and 1986, there have been four National Assessments of Educational Progress (NAEP), testing programs which sampled the proficiency of the nation's 9-, 13-, and 17-year-old students. In reviewing trends of all four assessments, Dossey, Mullis, Lindquist, & Chambers (1988, p. 10) came to the following conclusion: "While average performance has improved since 1978, the gains have been confined primarily to lower-order skills. The highest level of performance attained by any substantial proportion of students in 1986 reflects only moderately complex skills and understandings. Most students, even at age 17, do not possess the breadth and depth of mathematics proficiency needed for advanced study in secondary school mathematics."

Comparisons of U.S. students' performance in mathematics with other nations, especially Japan and the Soviet Union, point to a similar need to move our curriculum toward conceptual understanding and reasoning skills. Results of the Second International Mathematics Study, conducted in 1981–82, indicated that average Japanese students outperform even the top 5% of U.S. students enrolled in college preparatory programs (McKnight, et al., 1987.)

Other reports have also sounded serious warnings concerning the nature of our nation's curriculum, the overemphasis on low-level skills at the expense of mathematical thinking, and failure of students to achieve. Among these have been *A Nation at Risk* (National Commission on Excellence in Education, 1983), and *Everybody Counts* (National Research Council, 1989). The latter report notes that, "Communication has created a world economy in which working smarter is more important than merely working harder. Jobs that contribute to this world economy require workers who are mentally fit—workers who are prepared to absorb new ideas, to adapt to change, to cope with ambiguity, to perceive patterns, and to solve unconventional problems" (p. 1). These needs, not the routine requirements of calculations, speak to the importance of mathematics today. For the most part, calculators and computers do the routine. We must train children to think. At the present time, while students in the United States are learning computational skills, they are not mastering enough mathematics to sustain our present technologically based society.

Schools are slowly responding to the needs and concerns raised by these reports. The result is an increased emphasis in the mathematics curriculum on logical reasoning, problem solving, and spatial or geometric reasoning. These are generic skills that can be used in a wide variety of situations needed in a changing, technological, information society. Beyond these examples of mathematical reasoning, new topics are being added to the curriculum or are receiving increased attention. Measurement, probability, and statistics are the most significant examples. Conceptual development is also given a higher value.

THE NCTM STANDARDS FOR SCHOOL MATHEMATICS

In 1989, the National Council of Teachers of Mathematics (NCTM) released a major document entitled *Curriculum and Evaluation Standards for School Mathematics* (NCTM, 1989). The NCTM report is the first time ever that the council has attempted to establish national expectations for school mathematics. Released only after a full year of gathering reactions to a draft of the document, the *Standards* were designed to establish a broad framework that might guide reform in school mathematics into the 1990s. While stopping short of national curriculum, the *Standards* describe in broad terms the mathematics content that NCTM believes to be appropriate for schools. The document describes the focus the curriculum should take for each content area as well as examples of appropriate learning activities that suggest the intended spirit of instruction. The *Standards* are divided into four major sections: K–4, 5–8, 9–12, and evaluation. The latter discusses appropriate strategies and areas of focus for evaluating both curriculum as well as student outcomes.

A general direction for mathematics suggested by the *Standards* document can be found in the introduction under the heading *New Goals for Students*. There it states:

> Educational goals for students must reflect the importance of mathematical literacy. Toward this end, the *Standards* K–12 articulate five general goals for all students: (1) that they learn to value mathematics, (2) that they become confident in their ability to do mathematics, (3) that they become mathematical problem solvers, (4) that they learn to communicate mathematically, and (5) that they learn to reason mathematically. These goals imply that students should be exposed to numerous and varied interrelated experiences that encourage them to value the mathematical enterprise, to develop mathematical habits of mind, and to understand and appreciate the role of mathematics in human affairs; that they should be encouraged to explore, to guess, and even to make and correct errors so that they gain confidence in their ability to solve complex problems; that they should read, write, and discuss mathematics; and that they should conjecture, test, and build arguments about a conjecture's validity (p.5).

Many of the specific recommendations within the *Standards* are already being implemented by school systems. Others will require teacher education and changes in long-standing beliefs about traditional curriculum. The *Standards* strongly emphasize the importance of problem solving. They make a special effort to broaden the meaning of computation to include estimation, mental arithmetic, and the use of calculators. Calculators are recommended for use at every grade level. A major theme of the recommendations is an increased emphasis on conceptual development and the development of mathematical reasoning.

It is safe to say that the *Standards* document represents the viewpoint of a consensus of mathematics educators and that it has wide support in general from professional educators. Like any set of suggested "standards," NCTM's document will certainly attract its share of criticism. However, NCTM has committed unprecedented resources to the *Standards* project and anticipates that it will have a significant influence on the direction of school mathematics through the 1990s.

Appendix A contains two charts taken from the *Standards* that outline the relative changes in content emphasis that the document suggests. Even a casual review of these charts can provide a snapshot of suggested changes from the curriculum of the 1980s. As you read different chapters of this text, you may find it interesting to examine the corresponding topic in the charts.

TECHNOLOGY

The calculator and the computer have had two kinds of effects on mathematics curriculum. First, both have drastically reduced the importance of pencil-and-paper computational skills. Second, these technologies have provided new ways to teach a variety of important ideas. You will find activities for both calculators and computers throughout this book. Contrary to a persistent belief held by many teachers and parents, there is, after hundreds of studies over the past 15 years, *no support* for the fear that calculators will have a negative effect on basic skills or concepts. In fact, learning in general and concept development in particular seem to be enhanced by the presence of calculators. Chapter 20 presents a complete discussion of calculators, computers, and their use in mathematics instruction. It is sufficient at this point to note that technology has had and will continue to have an impact on school mathematics by changing what is teachable and by improving methods of instruction.

TESTING AND EVALUATION

New directions and content in mathematics have begun to make test designers look hard at how tests are constructed and what should be permitted. Issues include whether to permit the use of calculators on tests, how to test mental arithmetic, estimation, and problem-solving skills that are difficult to test in the usual multiple-choice format, and how to use materials or pictures that some students may not be familiar with.

How children are evaluated and on what content they are tested has always had a significant impact on what is actually taught in classrooms. Standardized testing programs frequently lag behind the changes suggested by recommending groups or the curriculum changes implemented by local school systems. The classroom teacher feels pressure to teach the content on the standarized test even when that content is in conflict with the suggested or prescribed curriculum. If, for example, a new curriculum suggests deemphasis on division with decimal divisors, but the test includes them, what should the teacher do? When major or even minor changes in content occur, there are always going to be discrepancies for some time between the content of textbooks, tests, and school system curriculum guides. Such gaps between testing and the curriculum are keenly felt by teachers. The result is frequently several different curriculums: the "prescribed" curriculum, the "tested" curriculum, and the "taught" curriculum.

An Invitation to Learn

This book is about helping children learn mathematics. Mathematics educators are learning more and more about how children learn mathematics. We begin with a discussion of how children learn or "construct" mathematics and build from that point. A child-centered, developmental approach to teaching is rewarding and challenging.

New directions in mathematics, while perhaps not under your control, have opened the world of mathematics to a wide range of exciting investigations. Elementary mathematics can no longer be equated with mundane computational skills.

Teaching mathematics is an exciting venture. Perhaps the most exciting part is that you can, and will, learn and grow along with your students.

For Discussion and Exploration

1. Get a copy of the NCTM *Curriculum and evaluation standards for school mathematics* and select one recommendation that you find especially important. Compare the content of the recommendation with a current basal textbook for a particular grade level. How do the two compare? If there are differences, why do you think they exist?
2. Discuss how technology, especially computers, has influenced mathematics instruction and curriculum in your local school district. For input, talk with teachers or curriculum specialists or visit a school and see how computers affect instruction.
3. What proportion (percentage) of time is currently being spent in schools on pencil-and-paper computation? Make a guess for several different grade levels. How much time do you think is spent on other forms of computation (estimation, mental computation, calculators)? How much time is spent on problem solving? Compare your estimates with a textbook series and/or discuss these areas with a classroom teacher or curriculum specialist. Are the emphases you observe appropriate? What changes need to be made?
4. How does curriculum get changed? Select any of the following roles: teacher, district supervisor, state supervisor. Select a specific change that you believe should be made in curriculum or instruction in mathematics. How could you implement that change? What factors would make change difficult? How do you think change happens?
5. Get a copy of the very easy to read report, *Everybody Counts* (National Research Council, 1989). Select any section of that report that is of interest to you and share your thoughts in a brief report. *Everybody Counts* should be "must reading" for everyone who is even remotely concerned about children, mathematics, and the future of this nation.

Suggested Readings

Brownell, W. A. (1987). Arithmetic teacher classic: Meaning and skill—Maintaining the balance. *Arithmetic Teacher, 34* (8), 18–25.

Lindquist, M. M. (1984). The elementary school mathematics curriculum: Issues for today. *The Elementary School Journal, 84*, 595–608.

Lindquist, M. M. (1989). It's time to change. In P. R. Trafton (Ed.), *New directions for elementary school mathematics.* Reston, VA: National Council of Teachers of Mathematics.

National Council of Supervisors of Mathematics. (1989). Essential mathematics for the twenty-first century: The position of the National Council of Supervisors of Mathematics. *Arithmetic Teacher, 36* (9), 27–29.

National Council of Teachers of Mathematics: Commission on Standards for School Mathematics. (1989). *Curriculum and evaluation standards for school mathematics*. Reston, VA: The Council.

National Research Council. (1989). *Everybody counts: A report to the nation on the future of mathematics education.* Washington, DC: National Academy of Sciences.

Trafton, P. R. (Ed.). (1989). *New directions for elementary school mathematics.* Reston, VA: National Council of Teachers of Mathematics.

CHAPTER 2

LEARNING AND TEACHING MATHEMATICS: A DEVELOPMENTAL VIEW

Relational Understanding

In this chapter a framework is developed for *how children learn* mathematics and, correspondingly, how we can most effectively *help children learn* mathematics. First, however, it is important to ask *what* we want children to learn; what is our goal in a developmental approach to mathematics? As summed up in Figure 2.1, this goal consists of helping children develop

1. Conceptual knowledge of mathematics
2. Procedural knowledge of mathematics
3. Connections between conceptual and procedural knowledge

This three-part objective of teaching mathematics developmentally will be referred to throughout this book as *relational understanding:* an understanding of mathematics ideas (conceptual knowledge), a facility with the symbolism and methods of performing mathematics processes (procedural knowledge), and a clear connection between the methods and symbols and the corresponding concepts.*

By way of example, consider the following statement: "The fractions $3/4$ and $9/12$ are equivalent." A conceptual understanding of this statement includes knowing that each of the two fractions represents exactly the same quantity—that $3/4$ and $9/12$ are two symbols for the same number. A mechanical or procedural understanding involves knowing a process that can be used on one fraction to get the other, namely, multiply the top and bottom numbers of $3/4$ by 3 as shown here:

$$\frac{3}{4} = \frac{3 \times 3}{4 \times 3} = \frac{9}{12}$$

It is possible for a child to know and use this procedure to verify that the fractions are "equivalent" with absolutely no understanding at all of what that means conceptually. It is also possible for a child to understand that two fractions can stand for the same quantity and yet not know this symbolic procedure. These represent two types of knowledge about the same topic. In addition to the distinction between conceptual and procedural knowledge, consider how both could be known with or without knowing how the two ideas are related. The connection between conceptual and procedural knowledge must be a specific part of our goal in teaching mathematics.

CONCEPTUAL KNOWLEDGE OF MATHEMATICS

Conceptual knowledge of mathematics consists of relationships that are integrated with or connected to other mathematical ideas and concepts. There are two important ideas in this definition: mathematical concepts are relationships, and these relationships are integrated with other concepts.

Mathematical Concepts Are Relationships

Consider the block and the stick shown in Figure 2.2. We can describe several attributes of these objects that we can see or feel with our senses: they are hard, made of wood, have color; one rolls and the other does not; they make noise if dropped, and so on. In the figure, the block is *above* the stick. The stick is *longer than* the block. The two shapes are *different.* These properties of being *above, longer,* and *different* are not physically in the block nor are they in the stick. We may think we "see" them, but if we could see *above,* for example, then we could touch it or put our finger on it or point to it or in some way tell where it is. Where is *above* in the figure? You can see the color of the blocks, hear them drop, see

*Richard Skemp, a psychologist and mathematics educator, writes eloquently about "relational understanding" and "instrumental understanding" (1978, 1979). In Skemp's terms, instrumental understanding is "rules without reasons" while relational understanding consists of knowing "both what to do and why" (1978, p. 9). These two terms, *relational* and *instrumental understanding,* are used freely in this text.

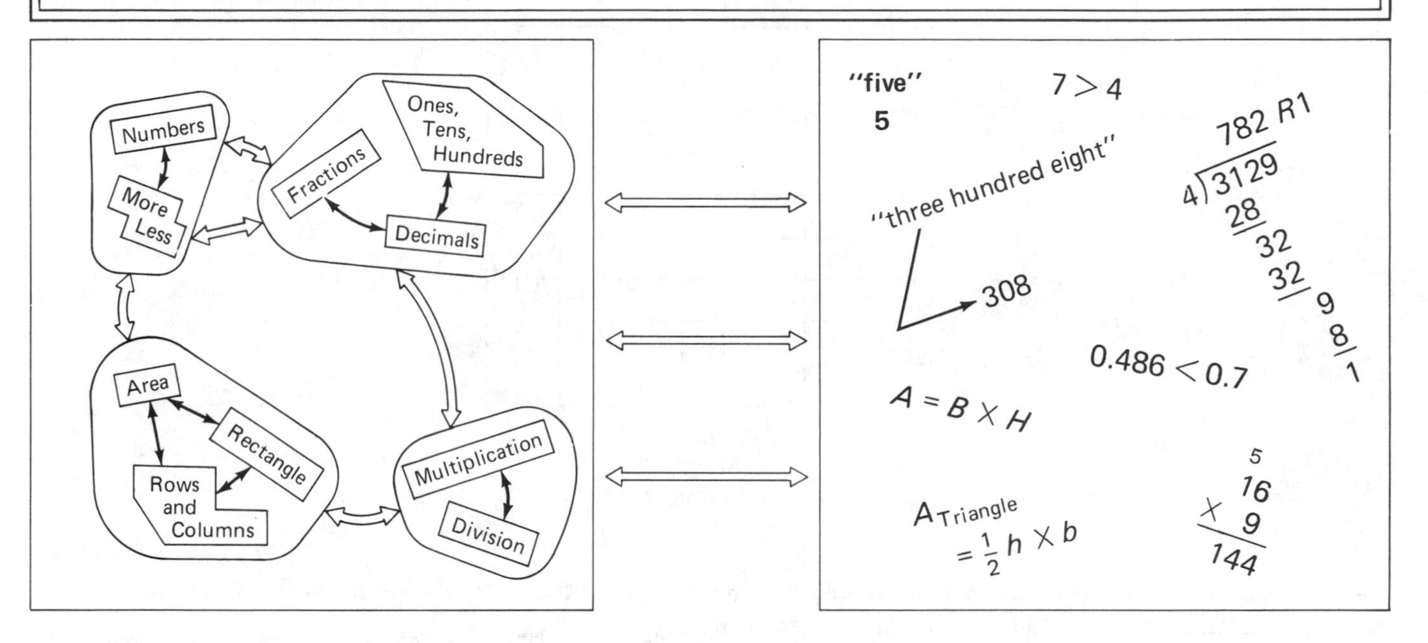

Figure 2.1
Relational Understanding: 1) Well-integrated conceptual knowledge, 2) Well-developed procedural knowledge, and 3) Clearly developed connections between concepts and procedures.

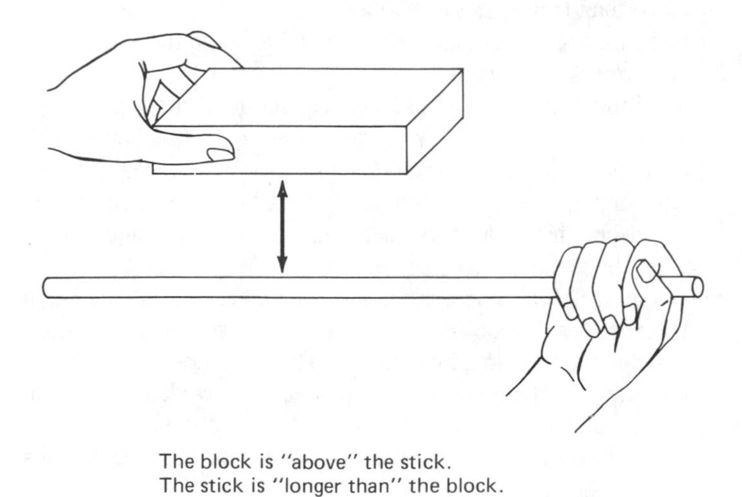

Figure 2.2

them roll, feel their hardness. But our senses cannot locate *above*. What kind of thing is *above?* It is not a physical thing but a *relationship*. It is a relationship between the current position of the block to that of the stick and that relationship is a creation of our own mind. We did not abstract the relationship *from* the block or stick because it is not *in* them to be abstracted. We formed a relationship between the pieces on our own. If the pieces are moved, they do not change, but the relationship between them changes. However, the relationships called *above* (and also *longer, harder,* and *different*) remain as concepts in our minds. Those ideas are not bound to the materials.

Piaget called these relationships *logico-mathematical* concepts and distinguished them from physical concepts and social concepts (Labinowicz, 1985).* In elementary mathematics, ideas such as seven, rectangle, ones/tens/hundreds (as in place value), sum, product, equivalent, ratio, and negative are all examples of ideas that are relationships. If a child identifies the long wooden stick in Figure 2.3 as a "ten" and

*A *physical concept* is one for which there are examples in the physical world that can be seen or experienced with the senses. Shoe is a physical concept. We can see examples of shoes and abstract from these examples a concept of shoe. Colors, sounds, and smells are all examples of physical concepts. *Social concepts* include the social conventions that have been designed by people. These include the words we use to name things, proper names, the various symbols used to represent things, and the fact that Christmas is on December 25.

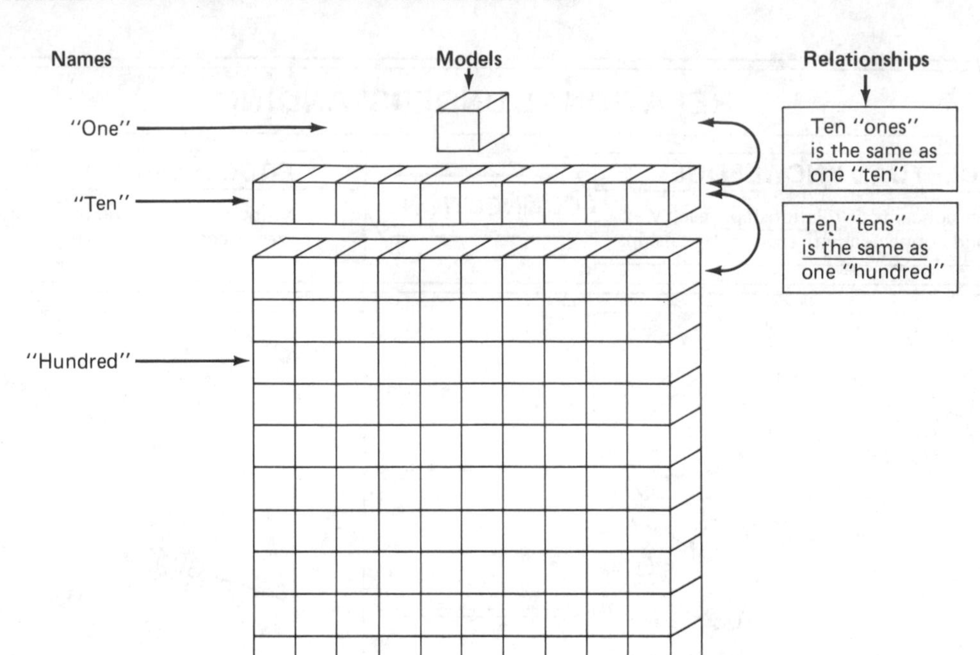

Figure 2.3
Names and materials are not the same as the relationships between them.

the little cube as a "one," that does not necessarily mean he has a mathematics concept of "a ten" as *being the same as 10 ones.* That he recognizes the stick and can name it is very similar to recognizing and naming a chair. The relationship between the stick and 10 small cubes is a relationship between those physical pieces. That relationship is not in the stick nor in the little cube, but must be created within the child's mind. When this relationship is completely constructed, the physical sticks will no longer be necessary for the child, just as you understand the concept of *above* without seeing any physical objects in that relationship.

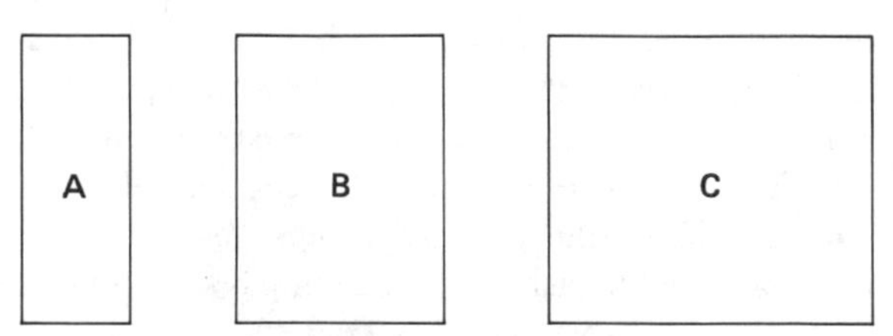

Figure 2.4
Three shapes—different relationships.

We refer to the shape labeled A (Figure 2.4) as a rectangle. But, if we call shape B "one" or a "whole," then we might refer to shape A as "one half." The idea of "half" *is* the relationship between shapes A and B that we constructed in our minds. If we let shape C be the whole, A becomes "one fourth." It is still the same rectangle. The concepts of *"half"* and *"fourth"* are not in shape A but are relationships. We construct these relationships. The rectangles help us "see" (construct mentally) the relationships.

Conceptual Knowledge Is Integrated

As new relationships are constructed, they are integrated with other relationships or mathematical concepts in the mind. The preschool child begins with very primitive concepts such as *one* and *more than.* With these and through counting, other number concepts are created. Eventually the concept of *seven* is developed and connected to the concept of *more;* seven is more than three. If it is also integrated with the idea of "one more than six" and "one less than eight," then the knowledge of seven is more complete.

Students who are first learning about fractions use their existing ideas about sharing to make equal-size parts. The concept of *fraction* is a relationship between parts and wholes. Connecting the idea of number of parts to counting words (three and third, four and fourth) makes it easier to understand why these particular names for fraction words are used. Counting is again connected to fraction ideas as students begin to count the fractional parts: one fourth, two fourths, three fourths, four fourths, five fourths. These concepts will later be integrated with and used to help develop new ideas, including equivalence of fractions and decimal numeration.

Other examples of connected concepts in mathematics include the following:

Addition and subtraction facts are related to each other and to concepts of numbers. Knowing $5 + 9 = 14$ is helpful in knowing $14 - 9 = 5$. It is also useful to connect these ideas to the concept of 14 as 10 and 4.

In geometry, the formulas for rectangles, parallelograms, trapezoids, and triangles are all integrated with the single idea that area can be measured by multiplying the base times the height, and that idea is bound up with an understanding of area units and multiplication concepts.

The concept of ratio can be connected to the geometric concept of similarity.

PROCEDURAL KNOWLEDGE OF MATHEMATICS

Procedural knowledge is knowledge of the symbolism that is used to represent mathematics and the rules and the procedures that one uses in "doing" mathematics tasks (Hiebert & Lefevre, 1986).

Knowledge of symbols means that expressions such as $(9 - 5) \times 2 = 8$ are recognizable and can be read correctly. It does not imply any understanding of what is read. The concepts or relationships that the symbols represent are distinct from the procedural knowledge.

The procedures that are used in mathematics are characterized by their step-by-step nature. "To add two three-digit numbers you first add the numbers in the right-hand column. If the answer is 10 or more, put the 1 above the second column and write the other digit under the first column. Proceed in a similar manner for the next two columns in order." The mechanical, step-by-step rule can be learned without any justification for any of the steps or for the order in which they are done. A computer program is a suitable analogy to a mathematical procedure. Computers do exactly as they have been programmed, and they do so one step at a time. When the program has been loaded into the computer, we might say that the computer has "knowledge" of the rule or procedure and can perform it.

$$\frac{3}{4} + \frac{1}{4} = \frac{4}{8} = \frac{1}{2}$$

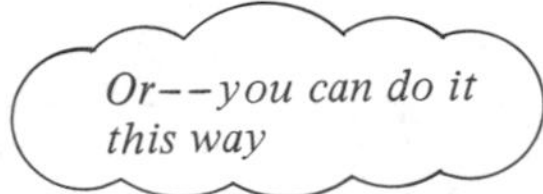

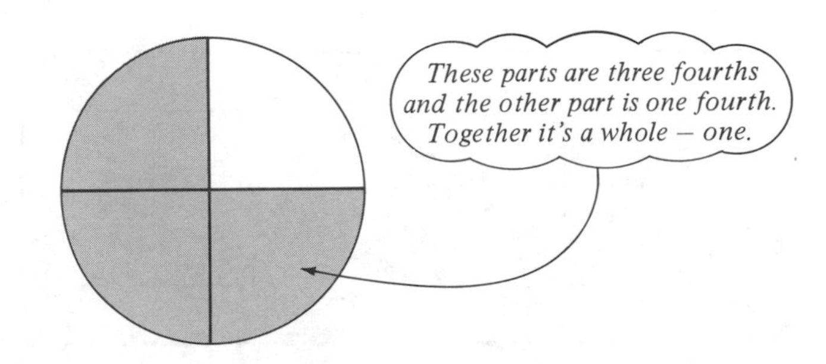

Figure 2.5
Fifth-grade child gives two explanations and answers for the same problem. (Erlwanger, 1975)

CONNECTIONS BETWEEN CONCEPTUAL AND PROCEDURAL KNOWLEDGE

To have *connections between conceptual and procedural knowledge* means that the rules and processes of procedural knowledge have a conceptual basis or meaningful rationale and that the symbolism used represents the appropriate concepts. When we do a mathematical procedure, the steps make sense and we understand why we are doing them in that way. Without a conceptual basis for procedures, rote learning must be used to master them. When rules or procedures are well grounded in our conceptual knowledge, we can explain not only what we are doing but why.

Erlwanger (1975) tells of an interview with a bright fifth-grade child that is a classic example of conceptual and procedural knowledge taught without clear connections. When asked to add $\frac{3}{4} + \frac{1}{4}$, the child wrote $\frac{4}{8}$ and reduced that to $\frac{1}{2}$. The student explained that you add the tops and the bottoms, noting that was how he had been taught (probably a confusion with the *rule* for multiplication). The child also demonstrated by drawing and shading parts of a circle that the answer is a whole (Figure 2.5). He then wrote $\frac{3}{4} + \frac{1}{4} = 1$. The child was not bothered by giving two different answers to the same problem. In his mind, the context (either symbols or pie pieces) was partly responsible for the meaning given to the problem. He was not aware of the error in his symbolic version. Since each context or method was different, he was content that these were different answers. Without connections, the two domains of knowledge do not necessarily interact and can actually be quite distinct from the vantage of the child.

INSTRUMENTAL UNDERSTANDING

Having procedural knowledge without connections to concepts or a conceptual rationale can be referred to using Skemp's term, *instrumental understanding* (1978). Figure 2.6 illustrates two ways that procedural knowledge might exist without being connected to a conceptual foundation. Either the corresponding conceptual knowledge has not been constructed and is not present for the child, or it may exist but be totally unrelated to the procedural knowledge.

Many children know and can use the procedural rule for dividing two fractions: "Invert the divisor and multiply the numerators and denominators." However, of these children who possess this procedural knowledge, many cannot explain what $\frac{3}{4} \div \frac{1}{2}$ means. They cannot make up a story situation for which the computation would be appropriate. They cannot tell you why $1\frac{1}{2}$ is a reasonable answer to the problem, nor can they use fractional pie pieces to show why the process works. These children have procedural knowledge without any conceptual underpinning. They have an instrumental understanding.

On the Fourth National Assessment of Educational Prog-

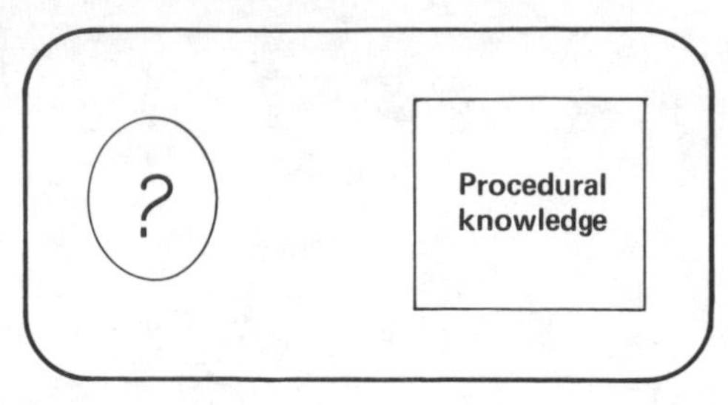

No conceptual basis

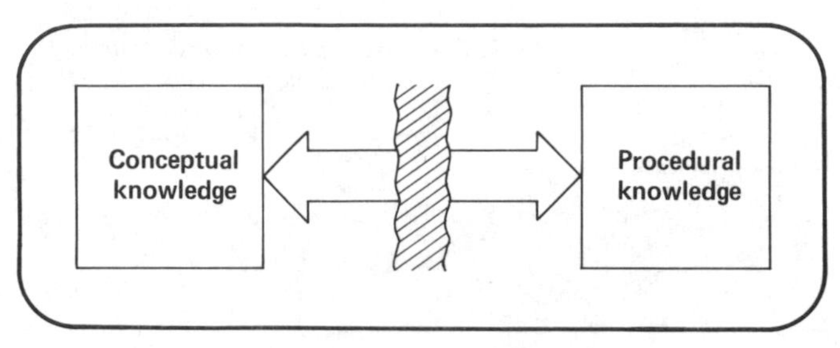

Concepts and procedures not connected

Figure 2.6
Two ways to have only instrumental understanding.

"Before you borrowed you had 56 and after you borrowed you had this much (looped). Did you have more before you borrowed, or after you borrowed, or was it the same?" (Cauley, 1988)

Figure 2.7

ress, roughly 80% of seventh-grade students were able to correctly express $5\frac{1}{4}$ as $\frac{21}{4}$. However, when asked to select the meaning of $5\frac{1}{4}$ ($5 + \frac{1}{4}$, $5 - \frac{1}{4}$, $5 \times \frac{1}{4}$, or $5 \div \frac{1}{4}$), fewer than half of the children chose the correct expression. Apparently many children are using procedures with fractions without an understanding of the concepts behind them (Kouba, et al., 1988a).

At the second- and third-grade levels, many children are able to subtract with pencil and paper, but are unable to explain the meaning of the little numbers that they write when they "borrow." They also do not understand that the number written after regrouping is the same quantity as before (Figure 2.7). These same children are able to use sticks and bundles of 10 sticks to do the same subtraction (Cauley, 1988). They seem to have both conceptual and procedural knowledge of the same process, but do not have these ideas connected. Their procedural knowledge is instrumental.

Instrumental understanding of a whole array of rules and procedures is all too common in schools. Too many children are mindlessly and mechanically pushing pencils across paper, sometimes correctly but frequently not. The following are examples of rules or procedures that are frequently learned instrumentally, without connection with a conceptual basis:

Turn the second fraction upside down and multiply.

In division, after you subtract bring down the next number.

Subtract the two numbers to do $7 + \square = 12$.

Area is length × width.

To change decimals to percents, move the decimal point two places to the right.

In percent, divide to get the "of number." (20 is 35% *of* what?)

Line the ruler up with one end of the object and read the number on the ruler at the other end.

The key words "in all" in a word problem mean "to add."

BENEFITS OF RELATIONAL UNDERSTANDING

As you will discover, to teach for relational understanding requires a lot of work and effort. It takes time to develop concepts. Materials have to be made, and children must be taught how to use them. More talking goes on. You must rely less on the textbook and more on class discussions. There are many important reasons why the effort is not only worthwhile but essential for quality instruction.

Relational understanding is

1. Intrinsically rewarding
2. Easier to remember and retained longer
3. An aid to learning new concepts and procedures
4. Directly related to general problem-solving skills
5. "Organic," leading to self-directed learning of new ideas
6. A major factor in preventing math anxiety

Intrinsically Rewarding

Nearly all people, and certainly children, enjoy learning, adding, and relating new ideas to existing ones. Children who learn by rote for the sake of a test, to please a parent or teacher, or out of fear of failure find the process of learning distasteful. External rewards such as an extra recess or a star on a chart may be effective in the short run, but do nothing to encourage a love of the subject when the rewards are removed. When mathematics is learned relationally, the meaningful learning feels good and can often be just plain fun.

Easier to Remember

There is substantial evidence to support the claim that meaningful material is more easily retrieved or remembered than nonmeaningful material. "Meaningful" implies well-established connections to other ideas. The more connections or associations people have made with a piece of information, the easier it is to retrieve from memory. An isolated idea is difficult to recall. Furthermore, the more associations you have with an idea, the longer an idea will stay in memory. A large percentage of school time is devoted to reteaching and review. If ideas were learned relationally instead of instrumentally, it is likely that much less time would be spent in review.

Aid to Learning New Concepts and Procedures

An idea that is fully understood in mathematics can be extended to more easily learn a new idea. An understanding of whole number place value aids in the development of decimal concepts. An understanding of the four operations and how they can be illustrated on the number line can help with the understanding of integers. The better that children understand concepts of numbers, the easier it is for them to master basic addition and subtraction facts.

Mathematics is a highly organized and logical subject. Understanding in one area makes learning a new, related idea much easier. Without the integrated nature of conceptual understanding and well-formed connections with procedures, each new idea that students learn can be completely isolated from existing ideas, with the result being rote learning.

Enhances Problem-Solving Abilities

Between 1973 and 1986, the NAEP gathered data on the mathematics proficiency of the nation's 9-, 13-, and 17-year-olds. A consistent trend was a significantly lower level of performance in both problem solving and concepts than in traditional computational skills (Dossey, et al., 1988). While the results are largely a reflection of the emphasis on basic skills in American curriculum, they also point out that skills developed in isolation are not very useful when it comes to solving problems and thinking. Problem solving requires both procedural and conceptual knowledge. Both are much more useful to the problem solver when intertwined and connected (Silver, 1986).

Leads to Self-Directed Learning

The term *organic* is used by Skemp (1978) to denote this searching and growth quality of relational understanding. In his explanation, Skemp notes that when knowledge or gaining knowledge is found to be pleasurable, people who have had that experience of pleasure are likely to seek new ideas on their own. Jerome Bruner's (1963) concept of "going beyond the information given" is very much the same idea.

Helps Prevent Math Anxiety

Mathematics anxiety involves definite fear and avoidance behavior. It is also self-destructive in that the more one fears and avoids mathematics, the more one is reinforced in the belief that he or she is inadequate. When students learn mathematics relationally, the resulting knowledge is not foreign or strange. There is no reason to fear or be in awe of mathematics. Relational understanding engenders a belief system that says mathematics is not a mysterious world of unfathomable ideas that only smart people can learn. On the contrary, relational understanding builds self-confidence.

There is no mathematics in the elementary school curriculum that cannot be learned meaningfully or relationally by children with average intelligence. Children should be taught in such a way that all of the mathematics they learn makes good sense *to the children,* not just to the teacher. To do otherwise is to negate virtually all of the advantages listed in this section.

Constructing Knowledge

AN EXAMPLE OF LEARNING

Listed below is a string of numbers. Before reading further, try spending about one minute memorizing it so that you can repeat it either verbally or in writing.

2581114172023

How did you approach the task? Many separate the list into smaller chunks; for example, 258–111–417–2023. The four or five chunks are easier to remember than the entire string of 13 separate numbers. If you tried this method or one similar, how long do you think you will remember the list? an hour? a day? If you practice, especially if you say it aloud, invent a sing-song type cadence, and perhaps write it down 40 or 50 times, your memory would probably be improved. In fact, most people know several strings of numbers that they have learned in just that way. Phone numbers, social security numbers, and auto license tags are just a few examples.

If you think your mastery of the string of numbers is probably weak and that you will likely forget it soon, look again at the numbers. However, this time look for some kind of pattern or rule. Try it now!

In a group of adults given this same memory task, one woman had the string "mastered" in less than 20 seconds and was quite confident that she would recall the string two or three weeks later with no practice. She pointed out that the list starts with 2 and then 3 is added to each successive number: 2, 5, 8, 11, 14, 17, 20, 23. She commented that it was because she was in a mathematics class that convinced her there must be some logic or pattern involved, so she looked for one from the outset. Had that been the number on her credit card, she guessed she would have used one of the methods mentioned earlier and not even looked for the pattern.

What can we learn from this example? First, the idea of adding 3 each time is not visible in that string of numbers. It is a relationship of "3 more than" between certain numbers, and you had to construct that relationship.

It is significant that a disposition to look for a pattern or a relationship played a key role. Did you see a pattern at first? Would you have seen a pattern if the same task were presented in a reading textbook or a novel? When you were told that there was a pattern, was it then easier to find?

Once the +3 relationship is observed, the string is very easy to recall. This has little to do with the quantity of material to be mastered. Relationships within the new material are integrated with your existing ideas (patterns, addition, number, and a 3-more-than relationship). The connections provide stability.

Finally, there is a positive feeling of satisfaction at having accomplished the task so easily when it seemed a bit formidable at the outset. That feeling is greater for those who discover the pattern on their own than for those who are shown the pattern.

A COGNITIVE THEORY OF LEARNING

In a *cognitive theory of learning* (Baroody, 1987), the learner is actively engaged in the learning process. To understand a new idea involves making connections between old ideas and new ones. "How does this fit with what I already know? How can I understand this in the face of my current understanding of this idea?" When there are more connections and relationships, the new material is more completely understood and better remembered. Mastery of a new idea is measured in terms of its connectedness to existing ideas rather than the strength of a stimulus-response bond or the amount of practice given to the idea. The list of numbers in the preceding section was integrated with existing ideas, concepts, or knowledge base. The ideas of pattern, number, addition, and the 3-more-than relationship were all useful in learning the list.

By contrast, in what Baroody calls "absorption theory," the learner is viewed as a blank slate, a passive receptor of knowledge. With this view of learning, teaching would be seen as planning a careful sequence of content, communicating this content to the children, and providing for practice using the new concepts or procedures. The more children would practice, the more stable would be the new ideas. These new ideas and skills are simply added on, not connected or integrated with existing concepts. An absorption theory of learning is essentially a behavioristic view, found in the theories of Thorndike and Skinner.

All theories are just that—theories. If, however, a theory of learning is found to be useful in effectively helping us to be better teachers, then that theory is worthy of consideration. A cognitive theory heightens our awareness as teachers that we must help children integrate new ideas with their existing knowledge. We should help them develop a web of connected ideas rather than isolated bits of mechanically learned rules. A cognitive theory of learning provides the basis for a developmental approach to teaching.

REFLECTIVE THOUGHT AND THE CONSTRUCTION OF KNOWLEDGE

In a cognitive theory of learning, the learner must play an active role rather than a passive one in the learning process. When new material is presented, the desired integration or connections with existing ideas do not happen automatically, but require active, reflective thought by the learner. The woman who discovered the pattern in the numbers was actively searching for a way to give meaning to what appeared at first to be little more than a random string of numbers.

The existing ideas in a learner's mind are more than simply the net effect of learning. These ideas are an organizational framework for new ideas; they are the tools with which new knowledge is formed or constructed. The construction of new ideas takes place in an active way. The meaning that a new idea has is given to it by the learner as he or she reflects on it, acts on it mentally, thinks about it, and connects or relates the new information to already learned information. To construct an idea is to give meaning to it, and to know it in a very personal way using knowledge already possessed as tools.

The idea of learners actively constructing ideas through reflective thought is particularly important when we consider the nature of mathematical concepts. Since mathematics concepts are relationships and as such do not exist outside our minds, there is no way that we can simply absorb them passively or even abstract them from things that we are shown. When learning a physical concept such as *color,* or *car,* or *smooth,* we can see or touch physical examples of those concepts. We learn to sort examples of the concept from nonexamples. To teach young children what a kangaroo is, you might show them pictures of kangaroos and pictures of animals that are not kangaroos. While children must actively make the distinctions between what is and what is not a kangaroo, the teacher can be sure that when children look at a picture of a kangaroo (or an example of whatever physical concept is being taught), they at least visually see an example of that concept. In contrast, with relationships or mathematical concepts, the children may be looking at counters or base ten models or fraction pie pieces, but there can be no assurance that any individual child is "seeing" or constructing the desired relationships. If the child is not reflecting on the relationships involved and thereby constructing them internally, the concept is not being seen—only the models are.

A CLASSROOM EXAMPLE

Consider a third-grade child who has made a quite common error in subtraction, as shown in Figure 2.8. The child was presented with a situation that was partly familiar and partly

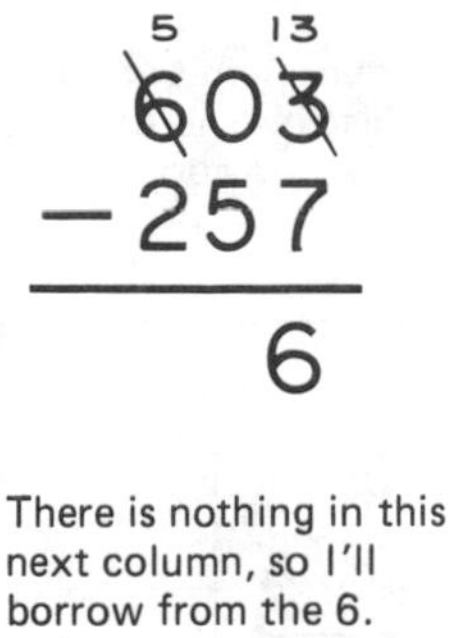

```
 5  13
 6 0 3
-2 5 7
------
     6
```

There is nothing in this next column, so I'll borrow from the 6.

Figure 2.8

not. What was familiar was that the problem appeared on a mathematics worksheet, it was subtraction, and the class had been doing subtraction with borrowing. This context narrowed the choice of ways to give meaning to the situation, just as the mathematics class helped the woman with the number string. But this problem was a little different from the child's existing ideas. The 7 was more than the 3 in the top. She knew she should borrow from the next column. But the next column contained a 0. She could not take 1 from 0. That part was different. The child decided that "the next column" must mean the next one that has something in it. She therefore borrowed from the 6 and ignored the 0. This child gave her own meaning to the rule "borrow from the next column."

There is some evidence that children will resist a change in their thought patterns or ideas in favor of relying on existing ways of knowing a situation. The child in this situation did not change her idea structure. Rather, she knew, or constructed the problem in the same way as those that did not involve a zero in the top number. Her understanding of the relationships between the values of the ones, tens, and hundreds column was either incomplete, not being utilized, or perhaps not present at all.

The example is not at all unusual. There are a variety of quite probable explanations for the way she interpreted the problem. Perhaps in her haste she did not reflect carefully enough on the situation and did not search her collection of concepts completely. If she knew about the nature of ones, tens, and hundreds, perhaps she may have utilized that knowledge to construct the situation differently. Another possibility is that the needed conceptual knowledge of place value just was not there to use.

Ginsburg (1977), a developmental psychologist interested in mathematics education, believes that children rarely give random responses. Their answers tend to make sense in terms of their perspective, or in terms of the knowledge they are using to understand the situation. The connection or integration of new ideas with existing knowledge is a key principle of learning. In many instances with children, their existing knowledge is incomplete or inaccurate, or perhaps knowledge we assume they have is not even present. In these situations, a new idea is likely to be misunderstood or learned differently than we expect. The example just discussed is such a situation. Not only is conceptual knowledge important for children to have as they learn new ideas, but it is also important that they reflect on that knowledge. As teachers, we must try to be aware of the knowledge that different children in our class possess, and provide methods of encouraging active, reflective thought. A challenging task.

Helping Children Construct Mathematical Concepts

To this point we have considered the nature of relational understanding of mathematics and a cognitive-constructivist view of learning. These ideas can guide us in our efforts to help children learn mathematics. Three important teaching principles emerge:

1. Use models to help children form relationships.
2. Promote reflective thought.
3. Listen to children to be aware of the ideas they are constructing.

USE MODELS FOR MATHEMATICS CONCEPTS

A *model* for a mathematical concept refers to any objects or pictures that can help a student construct or understand that concept. Since mathematics concepts are relationships, we cannot "show" students the concept directly. Rather, we provide materials that can exhibit the relationship which is the concept. In Figure 2.9 common examples of models are illustrated for a variety of mathematical concepts. Consider each of the concepts and the corresponding model. Try to separate the actual model from the relationship which the model exhibits.

For the examples in Figure 2.9:

The concept of *six* is a relationship between sets that match the words *one, two, three, four, five, six.* Changing a set of counters by adding one changes the relationship. The difference between the set of 6 and the set of 7 is the relationship "one more than."

The concept of *hundred* is not in the larger square but in the relationship of that square to the strip (ten) and the little square (one).

The concept of *length* could not be developed without making comparisons of the length attribute ("longness") of different objects. The length measure of an object is a relationship of the length of the object to the length of the unit.

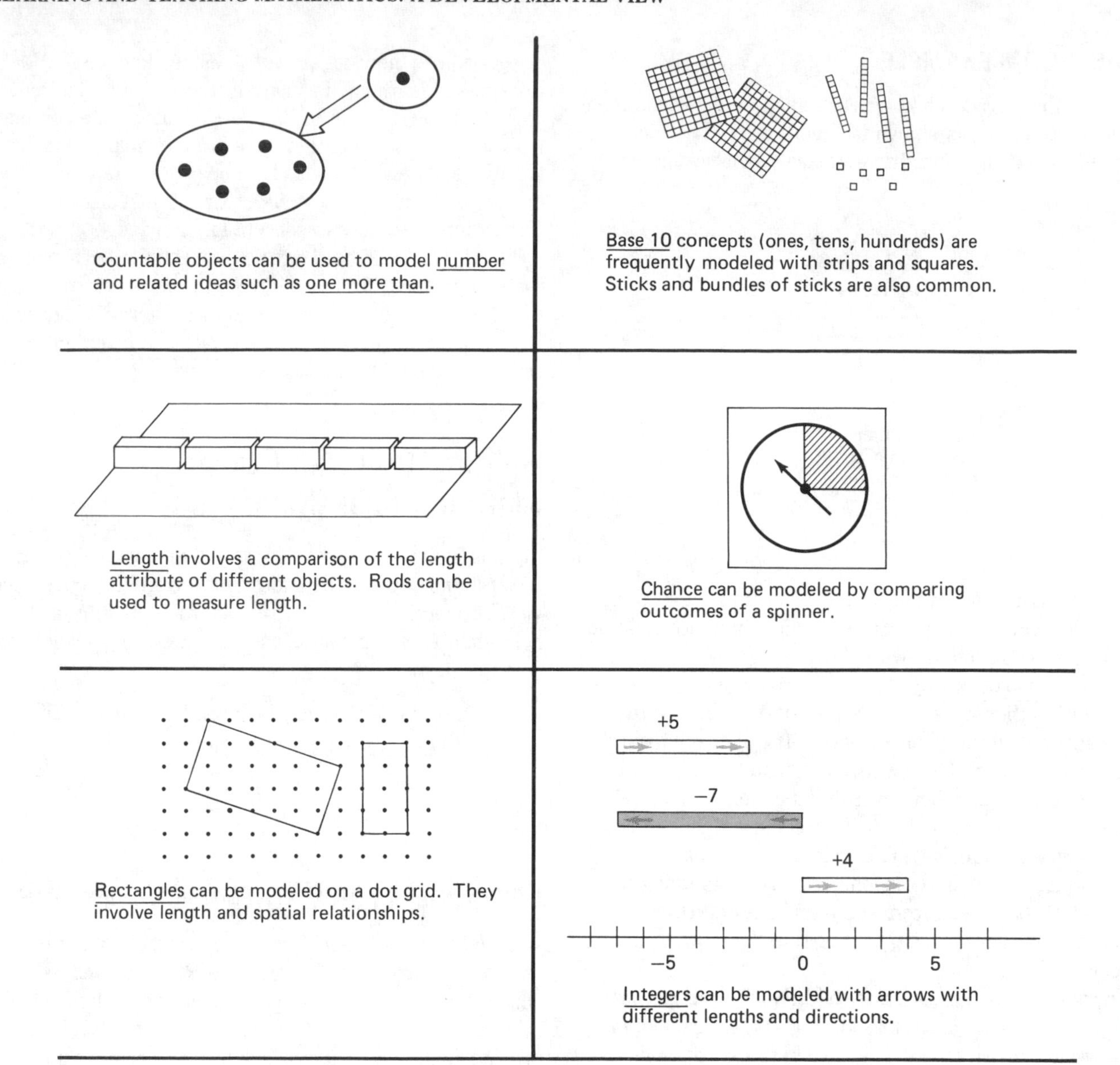

Figure 2.9
Examples of models to illustrate mathematics concepts.

The concept of *rectangle* is a combination of spatial and length relationships. The dot paper can illustrate the relationship of the opposite sides being of equal length and the sides meeting at right angles.

Chance is a relationship between the frequency of an event happening compared with all possible outcomes. The spinners can be used to create relative frequencies. These can be predicted by observing relationships of sectors of the spinner. Note how chance and probability are integrated with ideas of fractions and ratio.

The concept of a negative integer is based on the relationship "is the opposite of." Negative quantities only exist in relationship to positive quantities. Arrows on the number line are not themselves negative quantities but model the "opposite of" relationship in terms of length and direction.

A model is only effective if the students actually construct the desired relationship. That construction cannot be forced. If a model consists of physical materials that can actually be moved (relationships changed) or drawings that students make, the relationships exhibited have a better chance of being "seen." But the teacher should never take that seeing for granted. When children manipulate something (change, move, count, compare, draw, measure) there is a better chance that they will have to think about how and why they are doing that particular action. Firsthand physical interaction with something is simply a better thinking tool than passively observing it.

The ultimate test of a model's effectiveness is what happens inside the student's mind. Being attractive or nice to feel does not in itself make a model a good one. Simple pictures of models, like those found in textbooks, would be quite adequate if children constructed the desired concepts simply by looking at them. However, the important reflective thought that is so vital in the construction of mathematics concepts is much more likely to occur when models are being manipulated.

PROMOTE REFLECTIVE THOUGHT

Reflective thought requires that our students be actively involved in the learning process. In addition to the use of models, there are two things we can do to encourage reflective, active involvement: conduct lessons in a problem-solving atmosphere and encourage self-validation of student ideas.

Use a Problem-Solving Atmosphere

Students should be encouraged to test, challenge, discuss, explain, show, represent, validate, disprove, conjecture, and develop ideas, make decisions, make predictions, argue logically, and make mistakes. In such an environment students not only create relationships but become mathematical thinkers and problem solvers.

When content is presented in a prepackaged, ready-to-learn, simple and organized manner, it is the presenter or teacher who is creating relationships and ideas. The students become passive absorbers of preformed ideas. The result may be that ideas are learned superficially or even by rote, and are not integrated with other concepts. Such an approach to teaching is based on an absorption view of learning rather than on a cognitive view.

The nature of a problem-solving environment is discussed more completely in Chapter 3.

Encourage Self-Validation

The teacher or textbook need not and should not be the standard by which an idea or process is judged correct. Ideas suggested by a student or group for evaluation ("Is this right?") can be turned back to the students:

Why do you think that's right?

That seems like a good idea. How could we tell?

Can you show me with those blocks why that makes sense?

That looks like a good idea, but it's not the same as Sara's group. Why don't you get together and see what you did differently?

When teachers make a regular habit of responding to both right and wrong answers this way, interesting things begin to happen in the classroom. Students learn that they are capable of figuring things out. With a little thought, ideas begin to make sense. Self-esteem improves. Ideas become connected as students struggle to explain them to peers and teachers.

LISTEN TO CHILDREN

Perhaps it goes without saying that we should listen to our students and pay attention to what they are thinking, yet it is an easy maxim to forget. How children are forming the ideas you are trying to teach is of paramount importance. Encourage children to talk about the mathematics they are doing. Learn to ask questions that require explanations rather than simple one-word responses. Challenge students' ideas, both right and wrong, so that they will have to reflect on their responses and defend them. Besides assisting children in reflecting on their own actions, talking with children will provide some notion of how a child is constructing the ideas we think we have presented.

Listening goes beyond listening with your ears. Listening means looking at errors that students make on worksheets and quizzes and watching how students use materials. Look for evidence of misconcepts, ideas that have been constructed incorrectly. Remember that not all children will necessarily construct the same concept just because they have been given the same class experience. In Chapter 22, an entire section is devoted to ways of listening to children.

Helping Children Learn Procedural Knowledge Connected with Concepts

Conceptual knowledge is only one of three parts of the goal of relational understanding. Procedural knowledge must also be developed and must be carefully connected to a conceptual base. The following four maxims or principles are suggested to guide the design of instructional strategies that focus on connections between conceptual and procedural knowledge.

1. Develop conceptual knowledge before procedural knowledge.
2. Emphasize the use of language.
3. Use models as a link between concepts and symbolism.
4. Avoid premature symbolism.

DEVELOP CONCEPTUAL KNOWLEDGE BEFORE PROCEDURAL KNOWLEDGE

Children should learn that symbolism is simply a way to record or represent the ideas that they have already experienced, discussed, and understood. To begin with symbolic rules and procedures and expect that children will later attach meanings to these rules is both wishful thinking and a disservice to students. Written procedures should evolve as methods of doing or recording meaningful processes and ideas.

Some procedures, such as measuring lengths with a ruler,

are not symbolic. It still makes sense to develop the meaning first. Lining up the ruler at one end of the length and reading the ruler at the other end is a simple example of a procedure. It is quite possible to learn this procedure correctly and not have any idea what the result means or indicates. Would it make sense to have students running around the room measuring objects if they did not understand such concepts as length, unit of length, or how units are used to design rulers? How did the numbers get on the ruler? What do these answers mean?

Many procedures are developed as a result of making a conceptual activity with models more automatic and less cumbersome. Working with counters or number lines or base ten models is slow and deliberate. Students need to reflect on each step as they proceed. The more they practice, the less they need to rely on the conceptual ideas. The procedure becomes more automatic, and the models are gradually replaced with a more efficient symbolic mode. However, if required, students should be able to recall the concepts that the symbols represent.

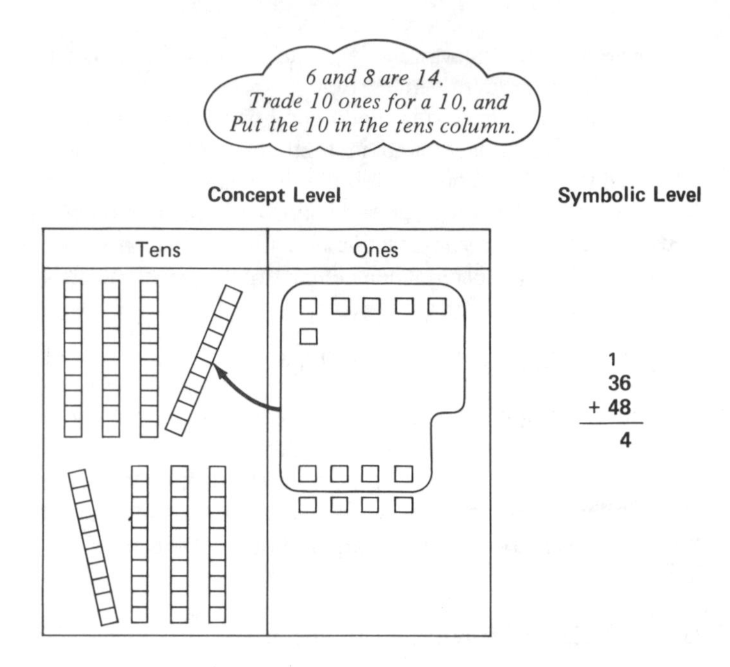

Figure 2.10
The same oral language is used in both conceptual and symbolic modes.

EMPHASIZE THE USE OF LANGUAGE

When children or adults talk, they generally use ideas that make sense to them. When discussing a new experience, the talking helps relate the new experience to their existing ideas. Using a similar approach, we should encourage children to talk about and express their ideas orally or in writing during early conceptual development activities. Later, when the instruction shifts to procedural knowledge, the same language and ways of expressing ideas that were used earlier can be continued and repeated as symbolic rules and procedures are introduced to represent those ideas. The words used, especially when they are the same or similar in both conceptual and procedural activities, is a powerful linkage (Figure 2.10).

Hamrick found that children who were able to easily talk about addition and subtraction ideas prior to the introduction of symbolism connected symbolism readily with these ideas. Students who were not able to verbalize ideas had more difficulty with symbolism. Delaying the introduction of symbolism until these children were verbally "ready" proved to be beneficial in connecting ideas with symbols (Hamrick, 1979).

Robert Wirtz (1978) noted that by about fourth grade, children can be roughly separated into two major groups. He called them "The I Can Do Its" and the "I Can't Do Its." He went on to say that the major distinguishing factor between the two groups is that the "I Can Do Its" know how to talk to themselves. When stumped with a procedural or problem-solving situation, the child who can talk about the ideas involved has a much better chance of retrieving the previously established connections required to make sense of the task.

USE MODELS AS CONNECTING LINKS

Models should always be a part of introducing symbolism or developing procedural knowledge. As noted earlier, models play a major role in helping children develop or construct conceptual knowledge. Since the models can actually be seen, pointed to, arranged and moved, the teacher is able to help children learn to write numbers and symbols and to perform symbolic operations that are directly linked to the models. Therefore, if models are used effectively, they are a major linkage between concepts and symbolism; a sort of intermediary between the relationships in children's minds and the symbols on paper (Figure 2.11).

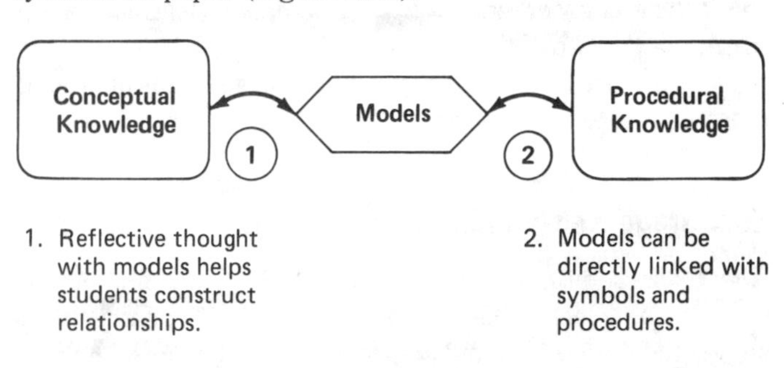

Figure 2.11
If models are used for both purposes they help students develop connections between conceptual and procedural knowledge.

In order for the model linkage to be effective, the first part of the connection, that with conceptual knowledge, must be quite clear (Carpenter, 1986). If the connection with concepts is well made, then encoding and decoding activities can form the second part of the linkage.

In general, *encoding* means to translate something meaningful into code. *Decoding* means just the opposite—to translate symbolism or code into something meaningful. In the context of mathematics, encoding means to write down the numbers and/or other symbols that represent a meaningful activity. Decoding begins with a symbolic expression or a symbolic task, the meaning of which is then demonstrated with the use of models (Post, 1980; Van de Walle, 1983).

Examples of Encoding

Figure 2.12 shows a simple concept activity for addition. Nine counters are put on a two-part mat. One student separates the nine counters, placing some in one section of the mat and the rest in the other. Before doing any writing, two students working together read the combination aloud: "Two and seven is nine." Then this combination is encoded as they write an addition equation. The activity continues by rearranging the counters to show different combinations with a new equation for each.

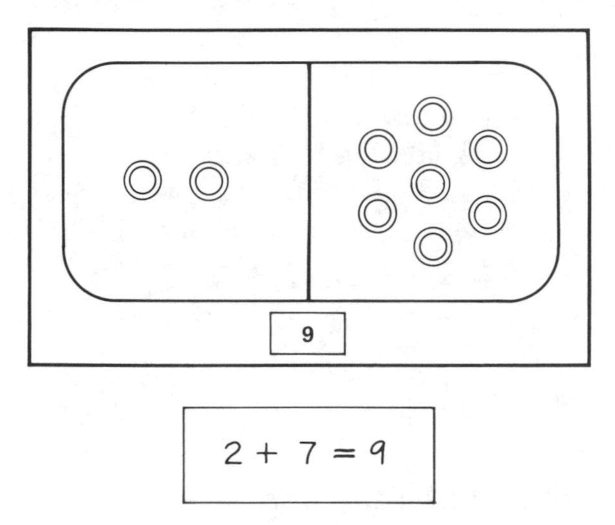

Figure 2.12
Encoding in addition.

In Figure 2.13 children represent each fraction with pie pieces and then find an equivalent way to show both fractions using all the same size pieces. When the equivalent models are found, a new equation is written down (encoded) along with the sum. The focus of the otherwise symbolic exercise is kept on the connection between the materials and the corresponding symbolism. Similar activities could be assigned as homework, with the students being required to draw pictures of the pie pieces and/or write short explanations of what they did to arrive at the symbolic results.

Notice that encoding is recording something that is being done conceptually, usually with models. Simply writing down answers is not the same thing.

Examples of Decoding

Suppose that students in the second grade have used base ten materials (ones, tens, and hundreds models) to do addition with regrouping. Later they work in pairs doing an addition problem with pencil and paper. They then take turns using their familiar base ten materials to "teach" each other how the problem was worked. The same type of decoding can be done in a larger group, with one student providing an explanation with models for an exercise presented by the teacher. It is fun to purposely make an error in a computation and have students use models to explain where you made the mistake.

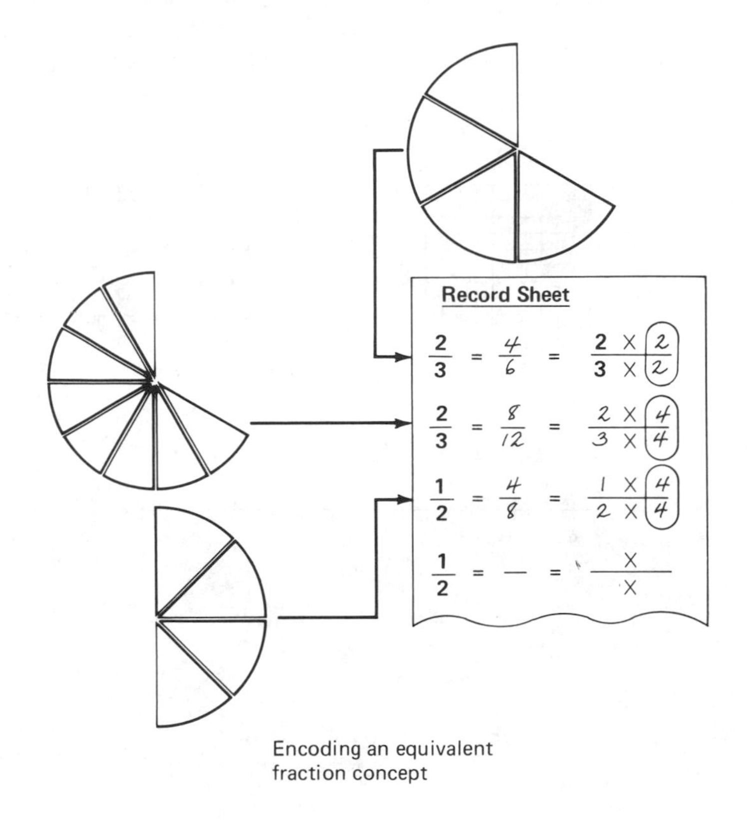

Figure 2.13
Encoding an equivalent fraction concept.

In the upper grades there is the advantage of students being more adept with simple drawings and writing brief verbal explanations. For example, students can be asked to select the larger of two decimals or to put three decimals in order from least to most. Having done this, they can be told to draw representations of the decimals by shading in a 10×10 grid or by sketching a number line as in Figure 2.14. They might also be asked to write a sentence explaining what they did.

AVOID PREMATURE SYMBOLISM

As a rough guideline, the first 50 to 60 percent of the total time spent on a topic should be devoted to concept development and making connections with procedural knowledge. This is a unit or chapter perspective, not a lesson-by-lesson guideline. This large proportion of time given to conceptual and connecting activity is highly unusual in most classrooms.

Most basal textbooks are written with both symbolic and

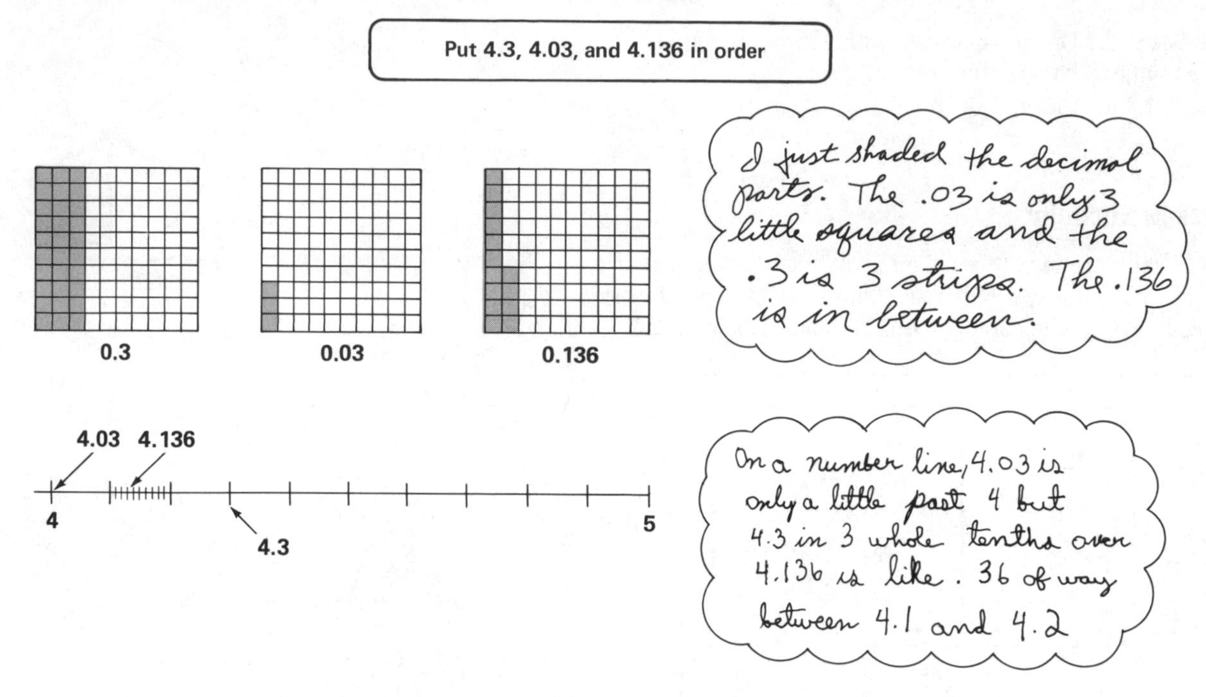

Figure 2.14
Decoding the meaning of decimals.

conceptual material in nearly every lesson. However, the development of a new concept almost always is a matter of days or even weeks. That means that the developmental teacher should make modifications in the way concepts are presented in books. The good activities and conceptual ideas that are usually found in the pupil books and especially in the teacher's guide should be used and, where appropriate, additional activities added. However, just because there are symbolic exercises in the text lesson does not mean they must be done that day. To use a textbook while teaching developmentally means to view the text as a resource rather than as a series of lesson plans.

The following lines from Roach Van Allen provide food for thought:

What I can do, I can think about.
What I can think about I can talk about.
What I can say, I can write.
What I can write, I can read. The words remind me of what I did, thought, and said.
I can read what I can write and what other people can write for me to read.

(Cited in Labinowicz, 1980, p. 176)

Although written about a language-experience approach to reading, these thoughts make compelling common sense when applied to mathematics. Consider what it would mean if a child trying to do symbolic mathematics could not "do" it conceptually. Would he be able to think about it? Would he be able to write about it (use mathematical symbols and rules)? Children should only be asked to use symbolism for ideas they have explored, reflected on, and discussed—what they have done, thought about, and talked about.

Hallmarks of Teaching Developmentally

In this chapter we have explored the nature of the mathematics we want children to learn (relational understanding) and have examined a theory of how children develop this knowledge (a cognitive-constructivist theory). These ideas provide a framework and a guiding philosophy for how to teach mathematics developmentally. Principal characteristics of this approach are summarized here.

Teaching mathematics developmentally—

Is child-oriented rather than content-oriented.

It is listening to children to gain their perspective.

It is realizing that children, not teachers or books, give meaning to ideas and procedures.

Is based on a cognitive-constructivist view of learning.

It is recognizing the role and value of manipulative models in helping children form conceptual relationships.

It is understanding that existing ideas give meaning to new ones.

It is encouraging children to talk about concepts and relationships.

It is emphasizing the interconnections among concepts.

Emphasizes meaningful connections of concepts with symbols and procedures.

It is using manipulative models as a major tool to create

linkages between conceptual and procedural knowledge.

It is capitalizing on children's oral language in the promotion of relational understanding.

It is being careful to see that conceptual knowledge is developed prior to the introduction of symbolism.

It is avoiding an overemphasis on mindless drill.

For Discussion and Exploration

1. Discuss the meaning and validity of the following statement: Children see what they understand rather than understand what they see.
2. Read the first chapter of Labinowicz's *Learning from children* (1985) or the first chapter of *Children's mathematical thinking* (Baroody, 1987). Relate the ideas of these authors to those in this chapter.
3. Visit an elementary school classroom to observe several mathematics lessons over a period of a few days. Discuss how it appears that children are learning. Is the general approach based more on a cognitive theory of learning or on an absorption theory? Take special note of the use of physical or even picture models. How and for what apparent purpose are these being used? How are students encouraged to discuss and articulate ideas?
4. Examine the teacher's edition of a current popular basal textbook. Select any chapter and explore the development within that chapter in terms of concept development and the introduction and use of symbolism. Consider the use of models, opportunity for encoding or decoding activities, or other means of connecting concepts and procedures. Select activities from throughout the chapter that could be used as good developmental or concept activities. Use the text to outline a concept development lesson and a different lesson connecting concepts and procedures.

Suggested Readings

Baroody, A. J. (1987). *A guide to teaching basic mathematics in the primary grades.* Boston: Allyn and Bacon.

Baroody, A. J. (1987). *Children's mathematical thinking: A developmental framework for preschool, primary, and special education teachers.* New York: Teachers College Press.

Burton, L. (1984). Mathematical thinking: The struggle for meaning. *Journal for Research in Mathematics Education, 15,* 35–49.

Byers, V., & Herscovics, N. (1977). Understanding school mathematics. *Mathematics Teaching, 81,* 24.

Davis, R. B. (1984). *Learning mathematics: The cognitive science approach to mathematics education.* Norwood, NJ: Ablex.

Driscol, M. J. (1980). Meaning in elementary school mathematics. In *Research within reach: Elementary school mathematics.* St. Louis, MO: CEMREL, Inc.

Erlwanger, M. (1973). Benny's concept of rules and answers in IPI mathematics. *Journal of Children's Mathematical Behavior, 1,* 7–26.

Ginsburg, H. P. (1989). *Children's arithmetic: How they learn it and how you teach it* (2nd ed.). Austin, TX: PRO-ED.

Hamrick, K. B. (1980). Are we introducing mathematical symbols too soon? *Arithmetic Teacher, 28*(3), 14–15.

Hiebert, J. (1984). Children's mathematics learning: The struggle to link form and understanding. *The Elementary School Journal, 84,* 497–513.

Hiebert, J. (1989). The struggle to link written symbols with understandings: An update. *Arithmetic Teacher, 36*(7), 38–44.

Labinowicz, E. (1985). *Learning from children: New beginnings for teaching numerical thinking.* Menlo Park, CA: Addison-Wesley.

Labinowicz, E. (1980). *The Piaget primer: Thinking, learning, teaching.* Menlo Park, CA: Addison-Wesley.

Juraschek, W. (1983). Piaget and middle school mathematics. *School Science and Mathematics, 83,* 5–13.

Post, T. R. (1988). Some notes on the nature of mathematics learning. In T. R. Post (Ed.), *Teaching mathematics in grades K–8: Research based methods.* Boston: Allyn and Bacon.

Skemp, R. (1978). Relational understanding and instrumental understanding. *Arithmetic Teacher, 26*(3), 9–15.

Stigler, J. W. (1988). The use of verbal explanation in Japanese and American classrooms. *Arithmetic Teacher, 36*(2), 27–29.

Suydam, M. N. (1986). Manipulative materials and achievement. *Arithmetic Teacher, 33*(7), 3–20.

Van de Walle, J. A. (1983). Focus on the connections between concepts and symbolism. *Focus on Learning Problems in Mathematics, 5*(1), 5–13.

Weaver, J. F. (1987). What research says: The learning of mathematics. *School Science and Mathematics, 87,* 66–69.

CHAPTER 3

MATHEMATICS AND PROBLEM SOLVING

What Is a Problem?

There have been several definitions of what a problem actually is. Two are offered here.

> A *problem* is a doubtful or difficult question; a matter of inquiry, discussion, or thought; a question that exercises the mind.

This definition is from the *Oxford English Dictionary*. The following one appears frequently in some form or another in much mathematics education literature. It is taken from Charles and Lester (1982, p. 5).

A *problem* is a task for which:

1. The person confronting it *wants or needs to find a solution.*
2. The person has *no readily available procedure* for finding the solution.
3. The person must *make an attempt* to find a solution.

Both definitions rule out the computational drill-type exercises that are commonly referred to in mathematics classes as problems. These are not problems in the sense intended here because there is a readily available method for getting the answer.

The Charles and Lester definition is more explicit in pointing out three essential features of a problem: *desire, blockage,* and *effort.* Reflection will show that these features are also implicit in the first definition.

Problem Solving and the Curriculum

In 1976 the National Council of Supervisors of Mathematics offered a list of 10 basic skill areas (NCSM, 1977). The list was updated in 1988 as "Essential mathematics for the 21st century" (NCSM, 1988). Both papers state that "Learning to solve problems is the principal reason for studying mathematics." In 1980 NCTM published *An agenda for action* in which it states, "Problem solving must be the focus of school mathematics in the 1980s." As noted in Chapter 1, NCTM placed a heavy emphasis in the *Standards* on both problem solving and reasoning mathematically.

Those recommendations span more than a decade. The mathematics education community and, more recently, the public at large continue to push for a greater, almost pervasive emphasis on problem solving. In fact, the 1980s have witnessed a significantly greater emphasis on problem solving in textbooks and most curriculums around the country. And yet, as we enter the 1990s, the goal to which these recommendations collectively point, the improvement of students' problem-solving and reasoning skills, is yet to be achieved. It is important that we as teachers understand the full meaning and intent of these recommendations. Just how can problem solving be such a pervasive part of school mathematics? How do these recommendations affect what we teach? Do they also affect how we teach?

Growth and learning in mathematics can be enhanced in an atmosphere of inquiry, investigation, analysis of mathematical situations, and problem solving, because it is through such an atmosphere that students are actively engaged in the construction of ideas. Problem-solving approaches can be applied to most any content area in mathematics and contribute to more effective learning. It is in this sense that learning mathematics or creating new mathematics ideas is problem solving.

While mathematics learning is enhanced by a problem-solving approach, specific strategies and skills useful in solving problems is an objective in itself. Helping students develop their own problem-solving abilities as lifelong skills is a way of providing them with the tools of mathematical thinking, not just to solve problems but to continue to learn mathematics.

This chapter addresses both of these aspects of mathematics as problem solving: an atmosphere for learning mathematics and processes for solving problems. An understanding of mathematics as problem solving does affect how we teach and what we teach.

A Problem-Solving Atmosphere for Learning Mathematics

A constructivist view of learning suggests a problem-solving atmosphere. Since problem solving is never a passive activity, students who are actively engaged in finding a solution to a task posed in any area of mathematics are already engaged in reflective thought. The opportunity for new relationships to be formed and integrated with familiar, existing ideas or concepts is maximized. The following are hallmark features of a problem-solving atmosphere:

1. Goals or tasks for exploration
2. A spirit of inquiry
3. A frequent use of models
4. Encouragement of self-validation
5. Verbal expression and group work

GOALS OR TASKS FOR EXPLORATION

A problem-solving environment includes provisions for problems or tasks for students to explore, work on, and solve. The following are examples of tasks that can be presented to students.

Consider the rectangle in Figure 3.1. How big is it? What are some ways to decide?

Pick three numbers less than 10. By adding or subtracting using only these numbers, what numbers from 1 to 20 can you make? Try to find three numbers that will let you make the most numbers from 1 to 20.

I have 20 cubes in this bag. Some are red, some are yellow, and the rest are blue. You can take out one cube at a time and then put it back. How could we figure out how many cubes of each color are in the bag?

On a sheet of square-centimeter grid paper, draw a rectangle that has 24 squares inside and write down how far it is around the rectangle. Now try to draw three other shapes that have the same distance around. Are these shapes all the same size?

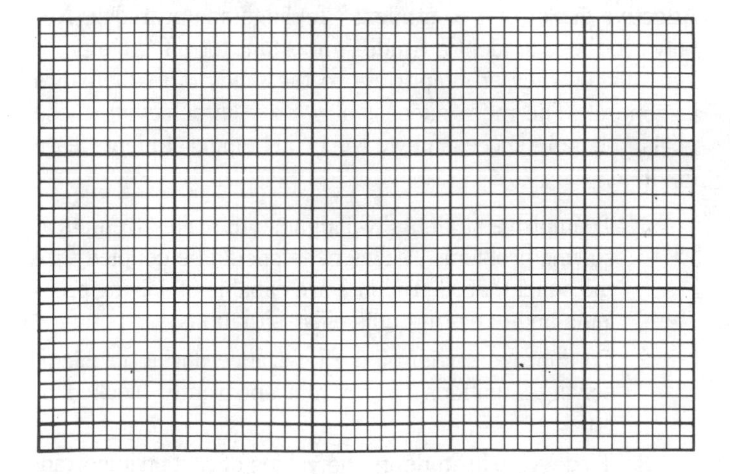

Figure 3.1
How big is this rectangle? (A premultiplication task.)

Each of these tasks can be worked on in small groups, in pairs, or individually. Each involves some specific goal but includes no well-defined way of approaching it. Students can explore tasks like these and begin to develop an assortment of ideas and relationships that can then be applied to the content we want them to understand. From the first task students can develop the relationships required to understand the procedure for two-digit by two-digit multiplication. While exploring the third task, students begin to construct ideas about probabilities, predicting outcomes, and estimating probabilities through large numbers of trials. In the second example, a concept of number can be improved as students attempt to find ways to "make" certain numbers with combinations of other numbers. The contrast between size and distance around provided by the last task sets the stage for a more directed discussion of area and perimeter.

Certainly not every lesson has to begin with a problem-solving situation, and perhaps not every concept and procedure can be drawn from an exploration such as those that have been discussed. However, the values of a problem-solving atmosphere are hard to ignore.

SPIRIT OF INQUIRY

When tasks without solutions are regularly presented, we can help students realize that being perplexed or confused is natural and OK. At the same time, we want to help students learn that they themselves can work their way out of this confusion. To promote this spirit of inquiry, teachers can incorporate the following approaches in their instruction:

1. Provide hints and direction rather than solutions.
2. Encourage risk taking.
3. Listen to and accept ideas—avoid censorship.

Provide Hints and Direction Rather Than Solutions

If students, blocked in an approach to a problem or task, are simply shown how to proceed, a sense of lacking self-worth can easily begin to develop. In contrast, if hints and encouragement are provided instead of solutions, students find their own paths to solutions, connect them to their own ideas, and develop a sense of confidence in solving problems on their own.

Encourage Risk Taking

Most students have some ideas concerning a problem, but many lack the confidence to share them with peers or teachers for fear of being wrong. Teachers can help overcome such insecurities by praising the efforts of *all* students who volunteer an idea, not just those who are correct or exceptionally creative. As students begin to see that saying or doing something is more rewarded than doing nothing, they will begin to take risks and venture forth with their ideas rather than fear being wrong. They will quite frequently be surprised to find out how much they are able to contribute.

Listen to and Accept Ideas—Avoid Censorship

When students present an idea to the class or group, immediate teacher evaluation can cause increased fear of taking risks. Moreover, it demonstrates to students that it is the teacher who really has the answers and not the students. By being an active listener and accepting the student's comments, the teacher becomes part of the inquiry and not the judge or evaluator. Phrases such as "That's a good idea. Who else has an idea?" or "That's good thinking. What do the rest of you think?" can easily become habit.

FREQUENT USE OF MODELS

Physical materials give students something to think with, something to do. They encourage a trial-and-error approach, with the feedback coming from the materials. Of the four example tasks presented, three have a model of some type included. The suggestion to use base ten pieces for the first task makes that problem even more approachable than the drawing. For the second task, young students could make a bar of connecting plastic cubes for each of their three numbers.

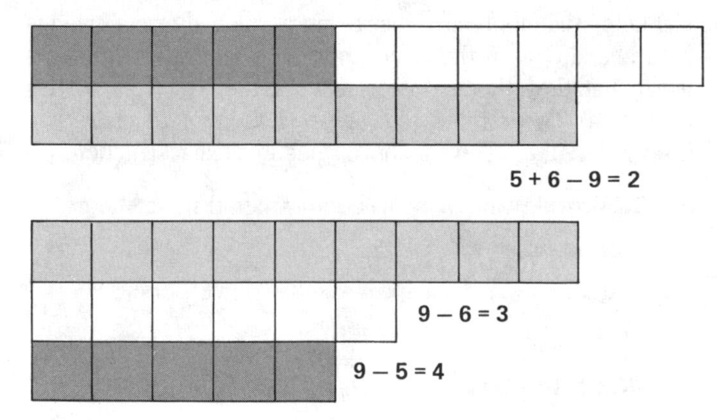

Figure 3.2
Using bars of blocks to find combinations.

As shown in Figure 3.2, the bars can then be used to "make" different lengths.

SELF-VALIDATION

The notion of self-validation was introduced in Chapter 2, but is worth reiterating here. Mathematics classes should not be seen by students as a game of testing answers against the teacher's or those in the back of the book. Teachers should regularly turn the question "Is this right?" directly back to the student: "Well, it certainly is a good idea. How could we tell?" Reasoning to the validity of their own answers conveys to students that mathematics makes sense and is not mysterious. When students learn that the teacher thinks they are capable of appropriate judgments, they begin to conclude that it must be possible to make sense of these ideas. The result can be increased confidence, a sense of satisfaction and an interest in further exploration.

VERBAL EXPRESSION AND GROUP WORK

A problem-solving environment is not a silent one in which students diligently work quietly at their seats. In order to explore, test ideas, and create relationships, young children need to formulate their thoughts in words. Explaining, questioning, showing, and making observations verbally forces us to formulate thoughts and connect ideas together. Even hearing themselves talk is frequently a way that students discover errors or observe new relationships in their own thoughts. Responding to classmates involves some evaluation of other students' ideas in relationship to their own. Children are frequently better at explaining ideas to a classmate than is their teacher, who is not operating at the same conceptual level, not using the same relationships, and probably not using the same vocabulary. Having small groups work on problem-solving tasks is an excellent way to maximize the amount of verbal interaction that can take place at one time in the classroom.

Problem Solving as a Curriculum Objective

GOALS OF A PROBLEM-SOLVING–ORIENTED CURRICULUM

Listed here are three sets of objectives which help to define a comprehensive problem-solving curriculum.

Affective Goals

Affective goals refer to students' feelings and attitudes. Not only are these student attitudes important for effective problem solving, they are important to the study of mathematics in general. The more successful we are in achieving these goals, the more successful we will be in helping students learn mathematics.

1. To improve students' willingness to make attempts at solving problems. Willingness should be coupled with the belief that all students are capable of solving problems and exploring unfamiliar situations.
2. To improve students' perseverance during problem-solving situations even when apparently blocked or stuck.
3. To develop in students the values of certain important problem-solving behaviors, such as using a systematic approach, perseverance, and alternative strategies.
4. To provide successful experiences in problem solving so that students will gain personal confidence and esteem in their abilities to do mathematics.

5. To establish that to know and do mathematics is to examine and explore unfamiliar problems and situations rather than master rules and formulas invented by others. Thus to learn to be a problem solver is a major reason for studying mathematics.

Process Goals

The term *process goal* or *process objective* is frequently used when discussing the teaching of problem solving per se. These goals are not concepts or skills in the traditional sense but processes—the mental thought patterns and strategies, also known as *heuristics,* that are used in the act of problem solving. These processes are used when solving all types of problems, including those that lead to growth in other areas of mathematics. In this broader sense, teaching to the process goals of problem solving is teaching students how to learn mathematics.

1. To improve students' abilities to analyze unfamiliar problem situations and clearly identify the goal of a problem, including the use of models, the identification of needed information, and the formulation of key questions.
2. To help students develop a collection of problem-solving strategies or heuristics that are useful in a variety of problem-solving settings.
3. To improve students' ability to select and use strategies appropriately.
4. To help students learn to monitor their own problem-solving progress.
5. To help students learn to evaluate their own solutions to problems.
6. To help students go beyond the solution to a problem and to search for generalizations and/or formulate new explorations.

Related Skills

In addition to the broadly stated affective and process goals, a lengthy list of specific skills is frequently included in the goals of a problem-solving program. These include the ability to complete a chart or table, to make and interpret graphs, to estimate an answer, to create and extend patterns, to identify unnecessary or missing information, to write an equation for a given situation, and to make up a problem for given data.

PROBLEMS AS TEACHING TOOLS

Problem solving is relative to the person solving the problem (Schoenfeld, 1985; Charles & Lester, 1982). A situation that is a problem for one person may not be a problem for another. If the task is not a problem for the student for whom it is posed (an easy exercise, no interest or effort, blockage too insurmountable), then the mental processes which are the object of problem solving cannot and will not be exercised.

Problems then are like teaching tools that we can use to help students develop problem-solving processes. The problem in the classroom can be thought of analogously as an overhead projector. The overhead is a tool that we use to place material in clear view of the class to see, discuss, and so forth. The problem is a tool that we use to place strategies or processes in "clear view" of the class. In problem-solving instruction, the class works on problems and reflects on, discusses, labels, modifies, refines, and so on, the processes of solving the problem. What was done to understand the problem? How did different groups solve it? How can the solution be evaluated and the problem extended? The focus is on the methods, processes, or strategies—the heuristics of problem solving—rather than on the problem itself.

TYPES OF PROBLEMS AND USES

Pursuing the "tool" analogy, we can select different types of problems to serve different teaching objectives. There is a wide variety of problem types, and each can be used to help develop different processes, concepts, or skills.

One-Step and Multistep Translation Problems

Sometimes referred to as simple story problems, *translation problems* present a "real" situation which requires the application of the appropriate arithmetic operations. In a multistep problem, one or more operations must be done in advance to provide data for the final computation. The following two problems are examples of one-step and two-step translation problems, respectively:

> Each crate of oranges weighs approximately 38 pounds. Vito's Van Service is to pick up 24 crates of oranges. How many pounds will the shipment weigh?

> The local candy store purchased candy in cartons of 12 boxes per carton. The price paid for one carton was $42.50. Each box contained 8 candy bars which the store was going to sell individually. What was the candy store's cost for each candy bar?

These problems are called translation problems, because we solve them by making translations between three different languages. The problem situation is presented in oral or written words, a *word language.* This must be translated to a *symbolic language* or computational form. Frequently a third language, that of *models,* is used as an intermediary to assist in the translation to symbols. The model, which may be a drawing or actual materials such as counters, helps clarify which operation is required. When the computation is complete, the result is translated back to the context of the original problem. Figure 3.3 illustrates these translations for the first problem.

Students' ability to solve simple and complex translation problems is closely related to their understanding of the mean-

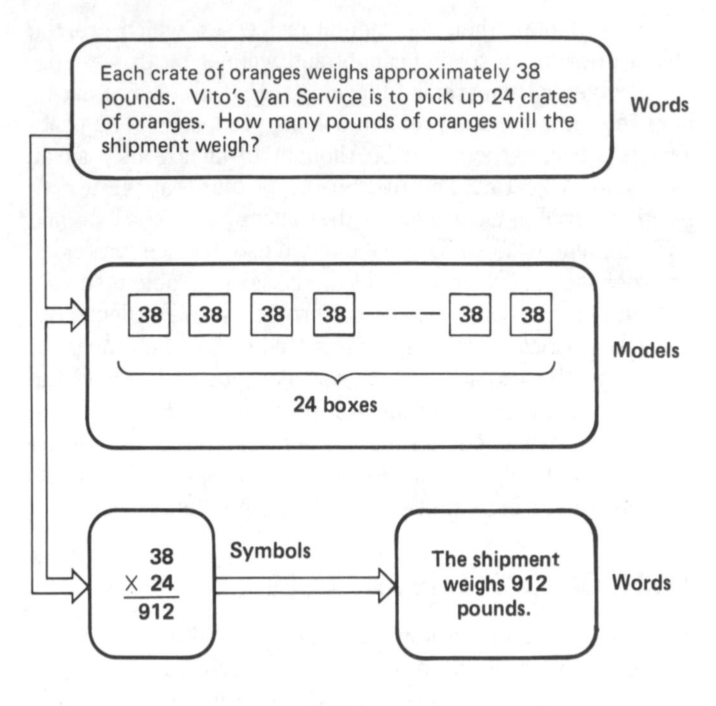

Figure 3.3
Translations in Word Problems.

ings of the operations. In fact, for years, "word problems" were found at the ends of the chapters where an operation was taught. They were also found in chapters on computation so that students could see some application for the skills they were learning.

Translation problems (one-step and multistep word problems) should now be seen as only one small part of a total program of problem solving. They should be encountered every week of the year regardless of what chapter the class is working on.

Translation problems are addressed again in Chapter 5 on the meanings of the operations.

Modifications of Translation Problems

Translation problems can be modified to provide students an opportunity to develop certain heuristic strategies. Listed here are examples of some frequently suggested modifications and the purpose each serves:

Translation problems with extra or superfluous data.

> George and Bernie each bought 3 bags of marbles. The marbles cost 69 cents a bag. There were 25 marbles in a bag. How many marbles did the boys buy?

Problems with too much data help students with the strategy of identifying essential and nonessential information in a problem. Students can be presented with such problems, and a discussion can be conducted around what data are important for solving the problem and why the data are important. The actual solution of the problem may not even be sought.

Translation problems with insufficient data.

> Mrs. Saunders gave Annette $10.00 to spend at the circus. She bought cotton candy that cost $1.25 and went on 3 rides. Admission to the circus was $1.50. How much money did she have left over?

Problems with insufficient information help students with the strategy of identifying required additional data. Many real-world problems require us to seek out additional information. When data is omitted from problems, the discussion can focus on what else is required to solve the problem. A good idea is to mix these problems with regular translation problems and problems with too much data.

Translation problems without questions.

> Brenda is reading a new storybook that has 327 pages. Today she stopped reading on page 120. She reads 20 pages in a day.

These situations allow students to formulate relationships among the given information and decide for what questions the given data can supply the answer. A similar idea is simply to present a collection of data (measurements, a menu or catalog, price lists, etc.) and let students make up problems that can be answered with that information. Research has demonstrated that this is a significant tool in helping students to solve translation problems.

Problems with too little or too much information are now commonly found in basal textbooks. However, the limited exposure to this type of problem provided by textbooks is probably not adequate to develop the skills of identifying appropriate data for a problem.

Process Problems

Process problems or "nonstandard" problems require the use of one or more general heuristics rather than a straightforward application of a computational procedure. Many process problems require no computation at all. The kinds of processes or strategies that might be used include (but are certainly not limited to) guess-and-check, look for a pattern, draw a picture, or try a simpler related problem. The following problem has become a "classic" in elementary mathematics education literature as an example of a nonstandard or process problem. It is appropriate for fourth- or fifth-grade children.

> One day Farmer Brown was counting his pigs and chickens. He noticed that the pigs and chickens together had a total of 60 legs. If Farmer Brown had 22 animals (pigs and chickens), how many of each did he have?

The pigs-and-chickens problem exemplifies a number of properties of process problems. For the fourth- or fifth-grade child there is no prescribed or clearly defined procedure for getting the answer (although there might be for an algebra student). Straightforward computation with the numbers in the problem probably will not produce the answer. Different students may use or attempt different solution strategies. For this particular problem a guess-and-check approach is one common solution strategy, but making a table or drawing a picture are other possibilities used by many children. Figure 3.4 illustrates these different solutions.

The principal purpose of process problems in the classroom is to help students develop various strategies for solving problems. A class that has just been working on the pigs-and-chickens problem might have a discussion about how different groups solved or tried to solve the problem. After evaluating the solution, a major focus of the discussion would be on the strategies used to solve the problem. For example, a student

Guess-and-Check Method

22 animals, so try 11 each.

11 pigs = 44 legs
11 chickens = 22 legs
66 total

That's too many – need fewer pigs.
Try 9 pigs. That leaves 13 chickens.

9 X 4 = 36
13 X 2 = 26
62 legs

Still too many.
Try 8 pigs and 14 chickens.

8 X 4 = 32
14 X 2 = 28
60 That's it!

Draw a Picture Method

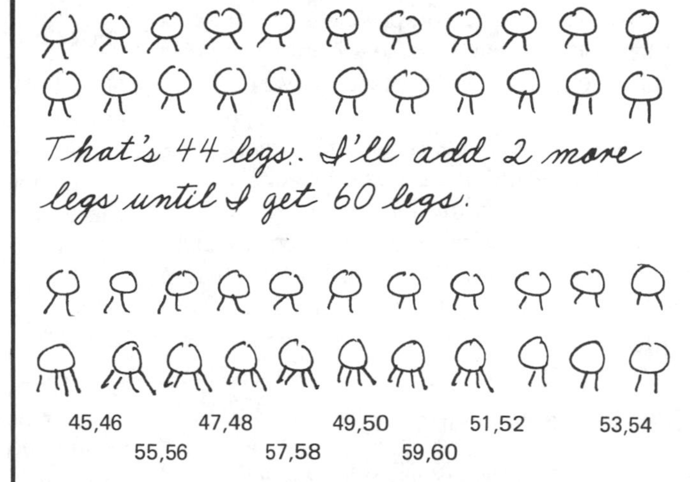

Pigs	Chickens	Pig Legs	Chicken Legs	Total
1	21	4	42	46
2	20	8	40	48
3	19	12	38	50
4	18	16	36	52
5	17	20	34	54
6	16			56
7	15			58
8	14			60
9	13			
10	12			

← Hey. These go up by 2. I can extend this column to 60.

← Ans: 8 pigs & 14 chickens

Figure 3.4
Three ways to solve the "pigs-and-chickens" problem.

might say, "We tried 11 pigs and 11 chickens first because that was sort of in the middle. But that worked out to too many legs. So then we figured it must be less pigs and more chickens." The student thinking is completely focused on the specific aspects of the problem. We could then respond, "I can see how you were thinking. You used a *guess-and-check* approach. You made a first guess that you thought might be close. And then for this problem, you were able to see if your guess really worked. You made a guess, and then you checked it." We would have helped these students articulate the strategy in more general terms and put a label on it for future reference.

In the solution of a process problem, the strategies or heuristics are placed in front of the class. Without the problem, the students cannot "see" the strategy. We select problems or the textbook presents problems that have a high likelihood of being solved by the particular strategy that is the objective for that lesson. The problem, after being solved by the students, has, like an overhead projector, placed the strategy "in view" of the class for discussion. If some students use other strategies as well, they can also be discussed as we desire or as time permits.

Helping students develop heuristics through process problems is a real art and requires some experience. Chapter 4 is entirely devoted to teaching children strategies with process problems. In Chapter 17, classification activities and an entire range of patterning activities are discussed. These activities should also be viewed as a form of process problems. They meet many of the objectives of a problem-solving curriculum.

Applied or Real-World Problems

The following is an example of an applied problem:

> The refreshment committee for the fall dance is trying to decide what to charge for drinks. What should drinks cost?

Problems in this category are usually realistic, or nearly so. As in the refreshment problem, they generally require students to formulate additional related questions. (What size? What beverage(s)? What will they cost us? Do we need ice?) They rarely have unique answers, and different groups might need to defend their rationale along with their solution. Frequently the applied problem is an opportunity for students to bring together different mathematics skills and concepts and to integrate mathematics with other areas such as social studies or science. Groups may work on gathering data and preparing a solution over a period of several days or even weeks. Probability, graphing, and statistics concepts can frequently be incorporated into these projects as students prepare their presentations.

The applied problem is an opportunity for students to see mathematics in realistic situations, to pull ideas together and see how different ideas can interact, to see the usefulness or practical side of mathematics, and to integrate mathematics with other content areas. Frequently such problems serve as enrichment projects or extensions of studies in other content areas.

A few additional examples are provided here to give a broader idea of their potential.

> In what pattern should the grass on the school lawn be mowed?
>
> The church is building a new parking lot. Where should the entrances and exits be, and how should the lines be painted on the pavement?
>
> Which of these 5 brands of paper towels is the "best buy?"
>
> How many pens and pencils should the school buy for the full year?
>
> How should the company design a box for its new candy? shape? method of assembly? size? cost?

Puzzle Problems

Puzzle problems usually require an unusual way of thinking about a situation, special insight, a play on words, or just plain luck. These problems abound in books of recreational mathematics, magazines, and newspapers. The following is an example:

> How can 4 pennies be placed in 3 cups so that each cup has a different number of pennies and no cup is empty?
>
> (A solution can be found at the end of the chapter.)

The solution to this problem requires very little mathematics or even mathematical thinking, but rather a twist of meaning from the obvious. Many people love these puzzles and thrive on the challenge to find a new way of thinking about a situation. Others are frustrated or totally uninterested. For these latter people, puzzles such as these are really not problems according to the definitions provided earlier. (Why?)

The puzzle problem does have a place in the classroom. They should be provided as extras or options for fun and pleasure for those who enjoy them. Students should never be made to feel stupid or inadequate because a fellow student solved a puzzle problem that to them was totally elusive. For those students who are challenged by puzzles, puzzle problems provide an opportunity to be rewarded for perseverance. They learn the value of backing off of a problem and finding new and creative approaches.

FACTORS AFFECTING PROBLEM-SOLVING ABILITY

Charles and Lester (1982) point out three sets of interacting factors that play a part in an individual's general problem-solving ability: affective factors, experience factors, and cognitive factors.

Affective factors include such things as willingness, self-confidence, stress and anxiety, tolerance for ambiguity, perseverance, interest in solving problems or in the particular problem situation, motivation of various types, such as desire for success or the need to please a teacher, and others.

Experience factors include things such as age and exposure to the context of a particular problem. Experience also includes exposure to and prior use of specific problem-solving strategies. An old maxim of some merit states: "In order to learn to solve problems, one must solve problems."

Cognitive factors include knowledge of specific mathematical content, ability to reason in a logical manner, ability to use spatial reasoning when appropriate, short- and long-term memory capability, computational skills (including estimation), and similar skills.

It would be wrong to conclude that good problem solvers just *are*. Though innate characteristics may color or affect some of the factors relating to problem-solving behavior, a careful reflection on these characteristics will show that most can be strongly influenced by instruction. This has been proven to be true not only for cognitive characteristics but for affective factors as well. Not all students will become expert problem solvers, but all students can be helped to improve their problem-solving abilities.

For Discussion and Exploration

1. Select a topic for a specific grade level. Design (or discuss) an approach to this topic that has an exploratory or problem-solving spirit. How effective would this approach be? Compare it with the approach to that topic found in a basal textbook.
2. How much time should be devoted to problem solving compared with other topics? Consider the average five-day week with 1 hour per day allotted to mathematics. Make a case for your position on problem solving. What are the competing issues here?
3. In the NCTM *Standards* document, the third standard for both the K–4, and 5–8 section is "Mathematics as Reasoning." Read these standards and discuss what you think the recommendations mean in various problem-solving situations. What do they mean in the contexts of other mathematics topics? How are these *Standards* suggestions being met in schools today?
4. What kinds of problem-solving activity are included in today's curriculum as defined by current basal textbooks? Examine a current book at a grade level in which you are interested, to see how much problem solving is there and of what type it is. How much problem solving is being conducted in classrooms in your vicinity? Should there be more? less? What topics need a change of emphasis in order to allow for more problem solving?

(*Solution to the puzzle problem on page 26)

Put 1 penny in each of 2 cups and 2 in the third. Now put the cup with 2 pennies inside 1 of the cups with 1 penny. That cup now holds 3 pennies, not 1.

Suggested Readings

Burns, M. (1976). *The book of think.* Boston: Little, Brown.

Burns, M. (1977). *The good time math event book.* Palo Alto, CA: Creative Publications.

Burns, M. (1987). Teaching the basics through problem solving. *Arithmetic Teacher, 35*(1), 24–25.

Charles, R. (1985). The role of problem solving. *Arithmetic Teacher, 32*(6), 48–50.

Charles, R., & Lester, F. (1982). *Teaching problem solving: What, why & how.* Palo Alto, CA: Dale Seymour.

Cobb, P., & Merkel, G. (1989). Thinking strategies: Teaching arithmetic through problem solving. In P. R. Trafton (Ed.), *New directions for elementary school mathematics.* Reston, VA: National Council of Teachers of Mathematics.

Garofalo, J. (1987). Metacognition and school mathematics. *Arithmetic Teacher 34*(9), 22–23.

Krulik, S., & Rudnick, J. A. (1988). *Problem solving: A handbook for elementary school teachers.* Boston: Allyn and Bacon.

Lenchner, G. (1983). *Creative problem solving in school mathematics.* Boston: Houghton Mifflin.

Lesh, R., & Zawojewski, J. S. (1988). Problem solving. In T. R. Post (Ed.), *Teaching mathematics in grades K–8: Research based methods.* Boston: Allyn and Bacon.

Lindquist, M. M. (1977). Problem solving with five easy pieces. *Arithmetic Teacher, 29*(6), 7–10.

Locket, A. (1988). Learning through problem solving. *Mathematics Teaching, 124,* 6–7.

Rosenbaum, L., Behounek, K., Brown, L., & Burcalow, V. (1989). Step into problem solving with cooperative learning. *Arithmetic Teacher, 36(7),* 7–11.

Schroeder, T. L., & Lester, F. K., Jr. (1989). Developing understanding in mathematics via problem solving. In P. R. Trafton (Ed.), *New directions for elementary school mathematics.* Reston, VA: National Council of Teachers of Mathematics.

Silver, E. A., & Thompson, A. G. (1984). Research perspectives on problem solving in elementary school. *The Elementary School Journal, 84,* 529–545.

Slesnick, T. (1984). Problem solving: Some thoughts and activities. *Arithmetic Teacher, 31*(7), 41–43.

Smith, B. (1985). Towards a problem solving school: 2 being challenged by Tom. *Mathematics Teaching, 112,* 2–7.

Van de Walle, J. A., & Thompson, C. S. (1985). Promoting mathematical thinking. *Arithmetic Teacher, 32*(6), 6–12.

Walter, M. (1987). Generating problems from almost anything: 1. *Mathematics Teaching, 120,* 3–7.

Walter, M. (1987). Generating problems from almost anything: 2. *Mathematics Teaching, 121,* 2–6.

Watson, A. (1986). Opening up. *Mathematics Teaching, 115,* 16–18.

Williams, H., Brown, H., & Rowland, S. Frameworks. *Mathematics Teaching, 109,* 2–9.

CHAPTER 4

INSTRUCTION WITH PROCESS PROBLEMS

Process problems were described in Chapter 3 as an instructional tool for helping children develop general strategies or heuristics for solving problems. The use of these tools and the careful nurturing of the young problem solver is an exciting and challenging form of teaching.

Strategies for Solving Problems

A LOOK AT THE STRATEGIES

This section will acquaint you with those problem-solving strategies or heuristics that are commonly taught in the elementary school or are found frequently in resource materials. Each of the nine strategies is explained briefly, followed by a problem solved using the strategy. When looking at these solutions, realize that different people would work these problems differently, even if they were using the same general strategy. The solution is there to illustrate the particular strategy. A second problem is then suggested that can be solved with a similar approach. If you are just learning these strategies, it would be profitable for you to try to work the second problem using the suggested strategy. Sometimes you may feel that a different approach may well be possible or even easier for you. That's fine. However, in order to learn the strategies, you should make an attempt to use the given approach. Frequently you will find yourself using more than one strategy. That is certainly acceptable. Strategies are often used in combination. These problems represent a variety of difficulty levels, but most have been chosen from those suitable for grades 3 to 8. Problems which lend themselves to the same strategies exist for both lower and higher grades.

Guess-and-Check

The guess-and-check strategy involves making a guess at the solution based on an estimation of a reasonable answer, or even just a blind "shot in the dark." However, the strategy is only useful in problems where you can tell if an answer is correct by checking against the conditions of the problem. You can make a guess at any problem, but with many you would not be able to tell if your guess was correct. Further, in problems where guess-and-check is most useful, a check of an incorrect guess should give you information about how to make a better guess. Therefore, the strategy is most likely profitable if you sense that successive guesses are moving you closer and closer to the correct solution.

Example:

Larry and Pete play marbles almost daily. Since Larry is the better player, he agrees that, when he wins, Pete pays him 5 marbles, but if Pete wins, Larry will pay Pete 8 marbles. In one month the boys played 26 games, and they each ended with just as many marbles as when they began. How many games did Larry win? (Solution given in Figure 4.1.)

1st guess:
What if both won the same:
That would be 13 each.
26 − 13 = 13
Larry wins: 13 × 5 = 65 marbles
Pete wins: 13 × 8 = 104
Too many wins for Pete.

2nd guess:
Try Larry wins 15. That means Pete wins 11.
26 − 15 = 11
Larry: 15 × 5 = 75
Pete: 11 × 8 = 88 — still too many

3rd guess:
Try 17 for Larry.
26 − 17 = 9
Larry: 17 × 5 = 85
Pete: 9 × 8 = 72
Now Pete doesn't have enough.

4th guess:
Tried 15 and 17. It must be in between. Try 16 – that's 10 for Pete.
Larry: 16 × 5 = 80
Pete: 10 × 8 = 80
That's it! Larry 16, Pete 10.

Figure 4.1

Try using a guess-and-check method with this problem:

> Ross collects lizards, beetles, and worms. He has more worms than lizards and beetles together. There are 12 creatures in all, with a total of 26 legs. If lizards have 4 legs and beetles have 6, how many lizards does Ross have?

Draw a Picture, Act It Out, Use Models

Even the best mathematicians and thinkers draw pictures to help them think about the relationships in a problem. Pictures here refer to any sort of sketch that helps organize the data and relationships involved in the problem. Just because a problem is about elephants does not mean we should draw pictures of elephants. Simple dots or X's will do fine. Lines, loops (for sets), dots (for things), or geometric drawings (circles, rectangles, triangles, etc.) are a few of the kinds of things we can draw easily to help solve problems. Think about sketching the ideas and relationships and avoid nonessential elements of the problem.

In some problems the action in the problem can actually be carried out by children doing what the problem describes and gathering data as they go. Other problems lend themselves to using counters, blocks, or geometric models to help in thinking about the problem. Using models and acting out a problem can be looked on as variations of the draw-a-picture strategy, although many will list these as separate approaches. With drawing, models, and actions, the idea is to get something visual in front of us to help determine the necessary relationships.

Drawing a picture is a strategy that permeates mathematical thinking, not just the solution of process problems. It is frequently used in connection with other strategies.

Example:

> Jane wanted a board cut into 8 equal pieces. The lumber company charges 60 cents for cutting a board into 4 equal pieces. How much will it charge for cutting Jane's board? (Solution given in Figure 4.2.)

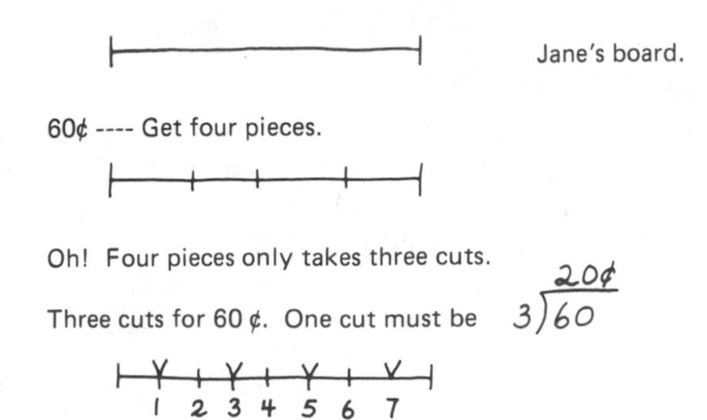

Figure 4.2

Use a drawing to help with this problem:

> Terri's new computer center is in 3 sections that sit side by side: a desk, a printer stand, and a set of drawers. The desk is as long as the stand and drawers together. The stand, which is 16 inches wide is ⅓ the width of the drawers. How wide is the entire unit?

Look For a Pattern

Mathematics is filled with patterns. With practice we can learn to expect a pattern to exist in certain situations. Patterns may occur in problems where there is a progression of circumstances or in geometric problems in which there are successive variations or elements added to a figure in a regular manner. Sometimes the pattern is not apparent to problem solvers because they do not have (or do not think of) the appropriate relationships that form the pattern. Looking for a pattern involves arranging the elements of the problem in such a way that the pattern will emerge, and then bringing to bear other ideas that will make the pattern "visible." The process of observing and extending patterns is so basic to mathematical thinking that a major portion of Chapter 17 is devoted to this topic.

Example:

> Carlos is building stairsteps with toy blocks as shown. This set of stairs has 4 steps. How many blocks will he need to have a stairs with 20 steps?

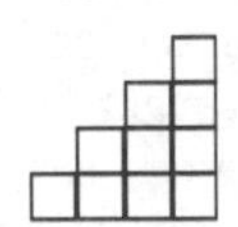

> (Solution given in Figure 4.3.)

Find a pattern to help with this one:

> A number of small towns set up their own telephone system. Between each pair of towns they stretched a main phone cable. In all they put up 28 separate cables connecting all of the towns. How many towns were there?

Make a Table or Chart

The make-a-table strategy is useful when there is a series of numbers in a problem or where the answer could be found in a list if it were available. Charts or tables consist of rows or columns that list important variables in the problem. Usually one row of the table begins at some natural starting place and progresses in an orderly, numeric manner; for example: 1, 2, 3, 4, . . . or 5, 10, 15, ... The other parts of the table are filled in accordingly. At some point along the way the table should yield the answer. To use this strategy, you need to learn what helpful things might go in the chart or table; what should the headings be?

The make-a-table strategy is frequently used in conjunction with another strategy. Drawing a picture can be used to generate the elements in a table. As the table is constructed, we can look for a pattern. In the stairstep problem just discussed, a

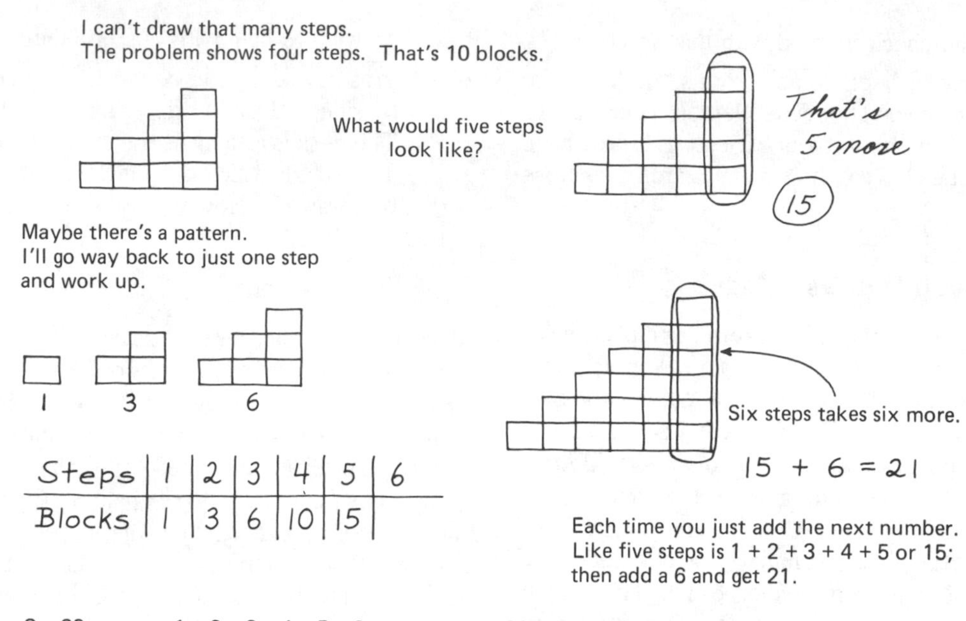

Figure 4.3

table was constructed showing the number of blocks in each set of stairs. The table helped in seeing the pattern.

Example:

> Mark turned on the radio at 6:00 A.M. and heard the temperature was −13 degrees. That night he heard that the temperature had risen 3 degrees each hour until 3 P.M., when it was at its highest for the day. What was the high temperature that day? (Solution given in Figure 4.4.)

I bet I could list each hour and then figure out what the temperature is at that time.

Time	6	7	8	9	10	11	12	1	2	3
Temp	−13	−10	−7	−4	−1	2				

Three degrees up from minus 13 is minus 10, . . . and then minus 7, . . .

Figure 4.4

Make a table to solve this problem:

> Carlotta read that 2 of every 5 people have blue eyes. She decided to use this idea to predict how many students in her class of 32 would have blue eyes. What would she predict?

Make an Organized List

Sometimes referred to as systematic counting, the strategy of making an organized list is useful in problems that require some method of counting things or making certain that every possibility has been covered. It may appear on the surface that simple counting of the different ways or situations would be impossible. To make the situation manageable for counting, we can try to devise an organizational scheme that permits listing all of the elements in some way. If the scheme or organized list is adequate, not only is it usually easier to count, but we will be able to tell that we have included all things that need counting and none were counted twice.

An organized list is also useful when you are looking for one or more of something from an unruly collection of things and you are not sure which combination you want or how many there may be. If you can construct an organized list, all of the cases you are searching for will show up.

Example:

> Pablo has 4 three-cent stamps and 3 four-cent stamps. What different amounts of postage can he make with these stamps? (Solution given in Figure 4.5.)

Now you try the next one:

> Suppose that you are making bike license plates, using 4 letters and numbers on each. The first position must be a letter, and the last 3 places must be numbers. The only letters to be used are A, B, and C, and the only numbers are 1 through 3. Numbers may be used twice but not 3 times on one plate. How many license plates can be made?

Work Backwards

The work-backwards strategy is sometimes useful when a series of events takes place and we know the result but need to determine the start condition. If the events are a matter of arithmetic operations such as earning or spending money, the

If he uses all of the stamps that's the most. Now if I start taking away 3¢ stamps one at a time, - - - -

Fours	Threes	$	
3	4	24	(−3)
3	3	21	−3
3	2	18	−3
3	1	15	−3
3	0	(12)	

4 × 3 = 12   3 × 4 = 12 = 24

That's all the ways to do it with 3 fours. I can do the same thing with 2 fours and 1 four and no fours.

2	4	20
2	3	17
2	2	14
2	1	11
2	0	8

1	4	16
1	3	13
1	2	10
1	1	7
1	0	4

0	4	(12) X
0	3	9
0	2	6
0	1	3
0	0	(0) ?

That's 20 ways altogether. WHOOPS! Twelve cents happened twice! Should I count that? Problem says how many different amounts. Those aren't different. When he gets zero, he isn't using any stamps. Don't count zero. That leaves 18.

Figure 4.5

task is to reverse these operations. In other problems a series of unknown events results in some condition, and the task is to find out what that sequence of events is. If not all steps are obvious, occasionally the last one is relatively clear. From the last step you can then get to the next to last, and so on.

Example:

Two players play a game of Nim in which they take turns playing 1, 2, or 3 square tiles on this strip.

The play starts at one end, and players add their tiles adjacent to those already played. The person who plays the last tile wins. If you start, how can you be assured of winning? (Solution given in Figure 4.6.)

I win if he leaves me with any of these.

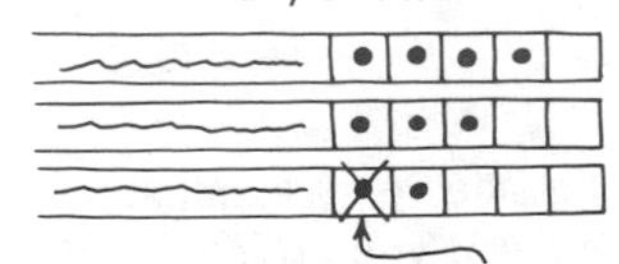

I'll get one of these if I play here and leave four spaces.

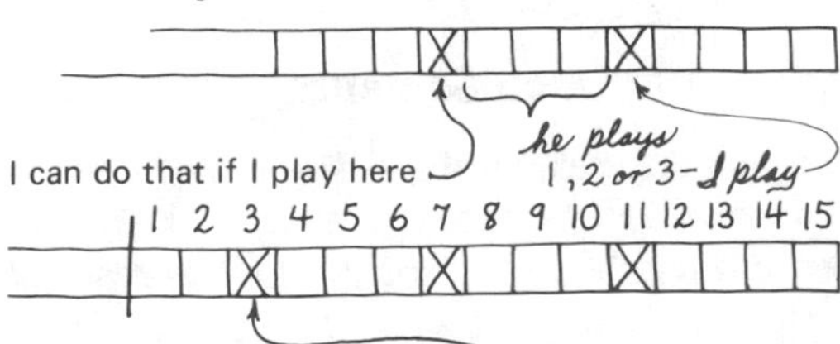

That means I have to play here. But hey! I can do that if I go first. I just play three counters on the first turn. Then I play to fill space 7 and then space 11. Got him!

Figure 4.6

Working backwards is useful in this next problem:

Molly, Max, and Buz earned $150 raking leaves, but they each earned a different amount. They agreed to share equally. Since Molly had the most, she took half of her money and shared that part equally with Max and Buz. But then Buz had too much, so he gave $10 each to Molly and Max. Finally, Max gave Molly $2, and they all had the same amount. How much did each earn originally?

Logical Reasoning

Perhaps this strategy should be called "using if-then reasoning" or "eliminating cases." The problems for which this strategy is useful involve establishing connections or relationships. We can use the strategy when a series of if-then situations can be considered that helps to narrow the number of relationships involved or in some way moves us closer to the solution.

A chart is commonly used along with the strategy to help keep track of possible connections. The entries in the chart stand for all possible pairings, and the strategy is to eliminate those that do not fit the conditions of the problem. If the problem can be solved by this method, the use of logic will permit ruling out those cases that are not solutions.

Another type of logic problem involves relationships among sets. For these problems it may be helpful to draw a set picture that illustrates the relationships among the elements in the problem. The use of "logical reasoning" is a rather broad strategy compared with many of the others discussed here. As described, it is combined with the strategies of make a chart or draw a picture.

Example:

Tom, Dick, and Harry work in a bank. One is the manager, one is the cashier, and one is the teller. The teller, who was an only child, earns the least. Harry, who married Tom's sister, earns more than the manager. What job does each one have? (Solution given in Figure 4.7.)

This is a bit different but still requires if-then thinking:

Three cards from an ordinary deck of cards are placed face down in a row. To the right of a king is a queen or two. To the left of a queen is a queen or two. To the left of a heart is a spade or two. To the right of a spade is a spade or two. Name the three cards.

Try a Simpler Problem

At times we encounter problems that seem to be overwhelming due to their complexity and/or large numbers. When this happens, it may be useful to try the same problem with much smaller numbers or fewer conditions. By working the easier problem, we hope that one of two things will happen: (1) Solving the easier problem may help us see a way of solving the original problem. (2) There is a chance that working a

They could each be one of these three things — I'll put that in a chart. Now I can X out a space if the man can't have that job.

	Manager	Cashier	Teller
Tom		X	X
Dick		X	
Harry	X	yes	X

The teller is an only child — therefore he isn't Tom — Tom has a sister.

If Harry earns more than the manager, then he isn't the manager. Put an X in there.

And Harry can't be the teller because the teller earns the least, and Harry earns more than someone — that means he isn't least. That's only one spot left for Harry. He has to be the cashier.

If Harry is the cashier, then Dick and Tom get X's there.

I can finish the rest of this easy.

Figure 4.7

Wow! I'll never figure this out for 20 players.

Maybe I could do it for three or even four players. I can draw a little picture for that.

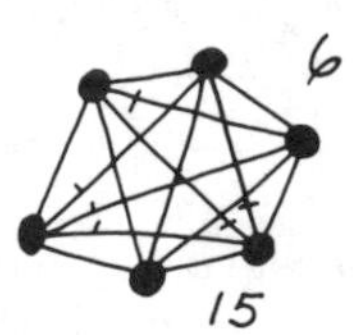

Let's see — 7 is going to really be a mess! Maybe I can look at these numbers —

3 → 6 → 10 → 15 → Add three, then four, then five . . .
That's just like the stairsteps!

$$1 + 2 + 3 + 4 + 5 + 6 + \cdots + 20$$

Do I add up to 20? No, I think only 19. It's always one less. Two players is 1, then three is 1 + 2, and four is 1 + 2 + 3. Yes — just add the numbers up to 19.

Figure 4.8

series of simpler problems, starting with the easiest possible and systematically increasing the difficulty, may lead to a pattern or generalization that will solve the original problem.

Example:

In a club checker tournament, each player played every team exactly once. If there were 20 players, how many games were played? (Solution in Figure 4.8.)

In addition to trying a simpler problem, drawing a picture or using physical models will help with this one:

Suppose you have 36 identical rectangular tiles. If you are to put them together to make larger rectangles in such a way that short sides are matched only with short sides and long sides only with long sides, in how many different ways can you do it?*

Write an Equation or Open Sentence

Sometimes the numeric relationships involved in a problem can be written down in terms of equations or inequalities. Young children can use boxes or triangles to represent un-

*The original problem involved matchboxes with three different faces. The question is how many larger boxlike structures can be built if like faces are always paired up. In this case, one condition that could be reduced is to change from three dimensions to two dimensions. (Mason, Burton, & Stacey, 1982)

known numbers in their equations. In either case, such equations or inequalities are referred to as *open sentences,* since until a value replaces the boxes or variables they are neither true nor false. They are "open." If there is only one equation required, usually straightforward computation will solve the problem. If there are two or more conditions in the problem, several equations or sentences involving two or more unknowns may be required. Even children in the fourth grade can then use a guess-and-check procedure to find a solution, while older children can use simple algebraic techniques. Writing down an equation to represent numeric relationships in a problem is an algebraic skill that is practiced in the seventh and eighth grades.

Example:

> Harvey's Gift Shop had a sale on figurines and posters. A figurine and a poster together cost $8. Arthur spent $51 dollars buying 7 figurines and 2 posters. How much did each cost? (Solution given in Figure 4.9.)

Now try an equation approach on this one:

> Abe and Zack live in opposite directions from the town. The distance from Abe's house to Zack's house is 29 miles. Abe is 5 miles closer to town than Zack is. How far does each live from town?

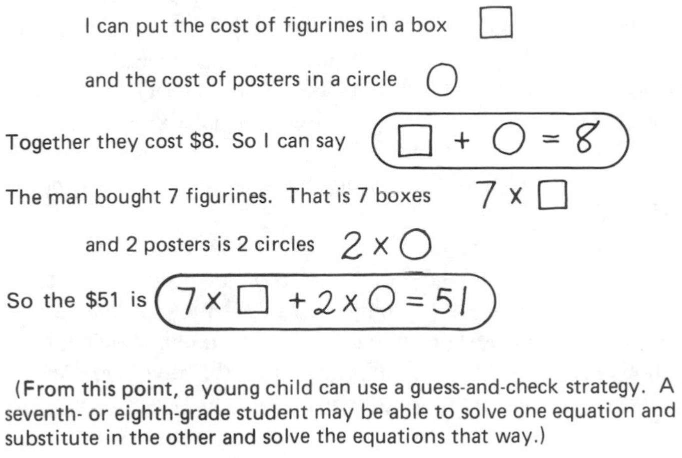

MIXED PROBLEMS FOR YOU TO WORK ON

Recall that part of what makes a problem a problem is not having a well-defined method of getting at the answer—blockage. In the problems you just worked through, a strategy was strongly suggested, so in some sense the blockage was reduced or perhaps even eliminated. The purpose of being so directed was to expose you to specific strategies and to label or name them for you.

Now that you have seen some strategies for solving problems and had a chance to try them a bit, here are some more problems to work on. By removing the strong clue concerning which strategy to use, the feature of problem solving that requires the solver to search for a method of solution is now included. It is possible to use each of the nine strategies at least once while solving these problems, and you will likely use many in combination. It is also possible that in *your* approach to these problems, there will be some strategies you do not use at all. A good idea is to write down all of your thinking and first attempts. Even write down the name of the strategy or strategies you think you want to try. If you change your approach midstream, make a note of that too. Then compare your notes, ideas, and methods with others.

No answers are given. Part of good problem solving is to decide if your answer makes sense and to convince yourself that you have actually solved the problem. In the real world there are no answer books.

I can put the cost of figurines in a box □

and the cost of posters in a circle ○

Together they cost $8. So I can say □ + ○ = 8

The man bought 7 figurines. That is 7 boxes 7 × □

and 2 posters is 2 circles 2 × ○

So the $51 is 7 × □ + 2 × ○ = 51

(From this point, a young child can use a guess-and-check strategy. A seventh- or eighth-grade student may be able to solve one equation and substitute in the other and solve the equations that way.)

Figure 4.9

1. How many rectangles are in this drawing?

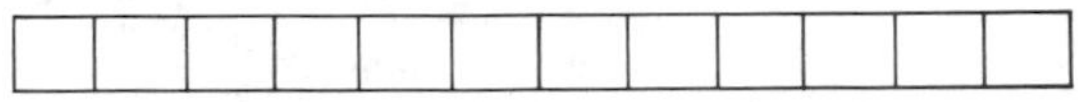

2. You have 50 boxes to deliver to a series of stops. At the first stop you are to leave 1 box. At the second stop you are to leave 2 more boxes than the first. At the third stop you are to leave 2 more than at the last stop, and so on. At what stop will you not have enough boxes to make a delivery?
3. How many different ways can you make 27 cents, using pennies, nickels, dimes, or quarters?
4. Robyn and Lois swam toward each other at the same speed from opposite ends of the pool. They passed each other after 12 seconds. They continued swimming at the same speed and lost no time in turning. In how many seconds will they pass each other again?
5. A game board has 9 spaces in a row with 4 black and 4 white markers arranged as shown in Figure 4.10. The black markers can only move to the right by sliding to an empty space or by making a jump over a single marker to an empty space. The rules for white markers are the same, but they move to the left. What is the minimum number of moves required to interchange the black and white markers?

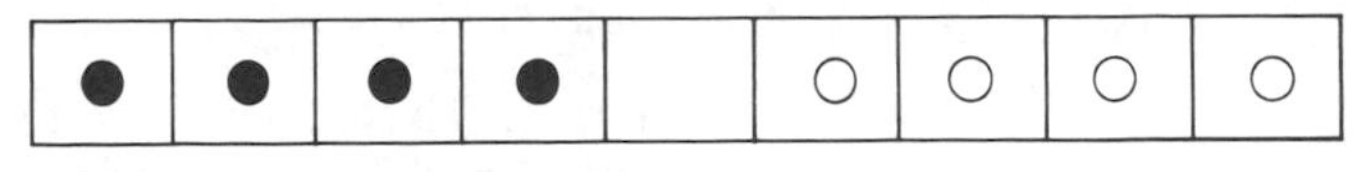

● Can slide or jump to the right.

○ Can slide or jump to the left.

Figure 4.10
Interchange ● and ○

6. (This is an oldie.) A sock drawer has 20 blue socks and only 10 brown socks. They are all mixed up. You reach into the drawer in the dark and pull out socks one at a time. How many should you take out before

you are certain to have 2 socks of the same color? (What if the drawer also had 15 green socks?)

7. If you walk on the lines in Figure 4.11, it is 8 blocks from A to B, always moving down or toward the right. How many different 8-block paths are there?

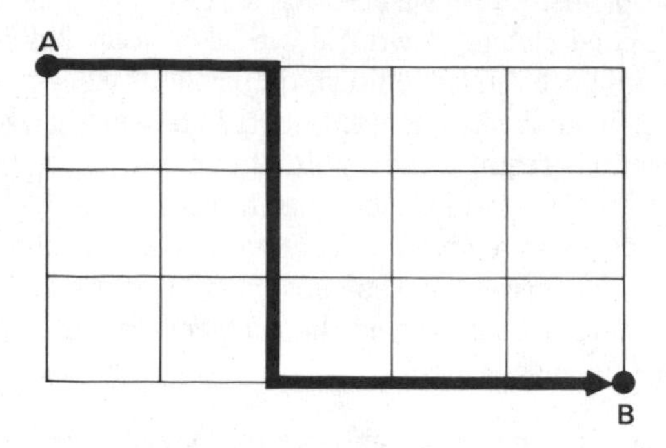

Figure 4.11
One 8-block path.

8. The seventh-grade class had a jack-o'-lantern design contest. All of the faces with teeth were smiling faces. Half of the faces that had triangles in them were smiling. Among the 40 smiling faces there were just as many with teeth as with triangles. None of the 20 faces with triangles had teeth. How many smiling faces had neither teeth nor triangles?
9. A baker has perfected a magic bread dough which, when placed in the oven, doubles in size every minute. He has further determined a special measure of this dough that will exactly fill the oven in 30 minutes. If the baker puts 2 special measures in the oven, when will the oven be full?
10. Diana and Debbie are saving cereal box tops to get a free record album. Diana has 9 more box tops than Debbie. Together they have 35. How many do they each have?
11. Three times my age and 5 is the same as 77 less my age. How old am I?
12. A man buys a necklace for $6, sells it for $7, buys it back for $8, and sells it again for $9. How much does the man make or lose in all of this buying and selling?
13. The dart board has three rings. The bullseye is worth 11 points, the middle ring is worth 7 points, and the outer ring is worth 3 points. Marshal threw 5 darts and all hit the target. His score was 43. Where might the darts have landed?
14. Eggbert had a basket of eggs which he was taking to market. On the way he met a poor lady in need of food. He gave her half of all of his eggs plus half of an egg and traveled on. Later he met a poor man in need of food. He gave the man half of his eggs plus half of an egg and traveled on to market. There he sold the 10 eggs he had remaining. No eggs were broken at any time. How many eggs did he have to begin with?
15. Stewart has a collection of baseball cards. If he puts them in piles of 2 he has 1 left over. He also has 1 left over if he puts them in piles of 3 or 4. In piles of 7 he has none left over. What is the smallest number of cards he could have?
16. Two to the 100th power (that is $2 \times 2 \times 2 \times \cdots \times 2$ with 100 twos) is a very, very large number. If you did calculate it, what would be the digit in the ones place?
17. The product of three consecutive numbers is 32,736. What are the numbers? (Use a calculator!)

Teaching Heuristics with Process Problems

THE TEACHER VIEWPONT

While process problems are a tool for teaching problem-solving strategies or heuristics, the teacher remains the most significant factor in helping students grow in their development of strategies and abilities to solve problems. This is a difficult and artful task but one that is fun and rewarding.

The teacher should have an appropriate perspective of three things. First, the processes of problem solving evolve and develop slowly over time. Students *do* grow in their understanding and ability to use strategies. They *do* grow in their positive attitudes toward problem solving, their willingness to persevere, and their sense of self-confidence. However, both the skills and the attitudes grow over months and years, not weeks.

Second, the teacher should have in mind an ultimate goal of developing in students an understanding of a general, overall approach to problem solving. This general scheme involves the stages of thought that a person goes through as they solve problems. George Polya (1957), widely referenced as the foremost leader in developing problem-solving strategies, is responsible for a general four-step approach: (1) understand the problem, (2) devise a plan, (3) carry out the plan, and (4) look back. But explaining this general strategy to students is woefully insufficient in helping them develop such a global approach.

Third, teaching heuristic reasoning is more a matter of encouraging and guiding students' thought processes than it is a matter of directly teaching them. Activities for conducting problem-solving lessons should be well planned, should model good problem-solving behavior, consider the development and experiences of the students, and aim in an indirect manner at a general problem-solving model. But students must become actively engaged in actually solving problems before they can "see" various ways to solve problems.

METACOGNITION AND PROBLEM SOLVING

"*Metacognition* refers to the *knowledge* and *control* one has of one's cognitive functioning, that is, what one knows about one's cognitive performance and how one regulates one's cognitive actions during performance" (Garofalo, 1987, p. 22). Metacognition is thinking about our own thinking. How metacognition affects learning and, more specifically, how it can affect mathematics learning and problem-solving proficiency has become a matter of interest to mathematics educators (Schoenfeld, 1988).

The knowledge part of metacognition includes an awareness of your own strengths and weaknesses, your tendencies, and your general methods of doing things. In the area of mathematics and problem solving, do drawings help you think about a problem? Do you tend to do something impulsively and examine it later, or do you more frequently ponder how to approach a task before beginning? It also includes your beliefs and values. Do you place a value on neatness? Do you believe that practice with a process can help you understand it and use it better?

The control part of metacognition involves decision making about your own actions. For example: "How should I approach this task? Should I take a direct route or an indirect route? I'm usually more successful if I can brainstorm first, so I'll just put down a lot of ideas to get me going. This time I'll be a bit more careful about keeping track of details and computations since I know I tend to be sloppy." Notice that decision making is related to what you know about yourself. Choices may best be made in terms of an accurate understanding of your strengths and weaknesses.

Garofalo suggests that teachers should spend more time helping children be aware of their own thinking—to actually help students develop metacognitive awareness. In problem solving this includes discussions and questions that focus on thought processes and decision making. What kinds of errors do you make a lot? How can you avoid that? How did you know you were on the right track (or wrong one)? What kind of thinking seems to work best to get going on a problem. Is it a good idea to talk about a problem before you get started? Why do you think so?

We are only beginning to learn how to help children be more reflective in their own thinking. However, simply training students to mindlessly follow a patterned behavior, especially in the area of problem solving, will not very likely help students become effective problem-solving adults.

DEVELOPING A GENERAL SCHEME FOR PROBLEM SOLVING

Variations of George Polya's four-step approach have become standard models for problem-solving behavior. Polya's four steps (1957) do not direct or describe how to teach problem solving, but they identify the goals of problem-solving behavior that students should construct. In Figure 4.12, a problem is solved along with the solver's thought processes provided at each step. As you read the following discussion of Polya's four steps, refer to the solutions of this problem as an example. Even better, start with an unfamiliar problem yourself and go through it following the general scheme described. Have a "metacognitive experience."

Understanding the Problem

Many students (and adults) are overwhelmed when they first see an unfamiliar situation or problem. A first step in problem solving is to calmly examine the information in a problem. Articulate or write down all relevant information. Decide what information is important and what seems unimportant. Examine the conditions of the problem that will have an impact on the problem. Be very clear about what is being asked for in the problem. It may be helpful to reformulate some of the information, perhaps make lists of knowns and unknowns, to draw pictures, charts, or diagrams. Begin to think of what similar situations you have experienced that may be like this problem or contribute to its solution. Write these down or test to see if they really are similar situations.

Besides the obvious benefit of being clear about what a problem is asking, the very act of going through this understanding phase is calming. It gets the problem solver doing something productive without having to decide what to do. Thinking about a problem is a very active and involved process. To say to someone, "First be sure you understand the problem," does not communicate the kinds of things noted in the last paragraph. Understanding a problem fully goes a long way toward solving it, and gets the problem solver mentally involved in the problem. The amount of time and effort spent in this first phase is a major difference between good problem solvers and poor problem solvers. Good problem solvers really get involved with a problem and understand it before they waste any time trying to solve it. Poor solvers tend to barge forward with little thought.

Devising or Selecting a Plan

Once the problem is well understood, the experienced problem solver begins to change his or her focus of thought. Rather than getting information and ideas *from* the problem, thoughts turn to what can be *brought to* the problem. "Does this problem lend itself to special cases? Is a drawing or chart a useful approach? Have I ever done anything similar to this before? How was that solved? Could I make a guess and if so would that help? Could I test a solution? Perhaps the problem could be restated in such a way to give me a better clue to the solution, or perhaps there is a special case of the problem that can be solved quite easily." And so on.

For a novice problem solver who has not developed a collection of heuristics, the selection of a plan may be reaching into a nearly empty bag of tricks. Even the experienced problem solver may try several unproductive approaches and begin to wonder if in fact he or she has a method of solving the problem. Students in school must be helped to develop a set

Problem: How many squares on an 8 X 8 checkerboard?

Understanding the Problem

Seems simple. There are 64 little squares. There must be something else. What other squares could there be? The board itself is a square. What about a 2 X 2 square? There could be all sorts of sizes?

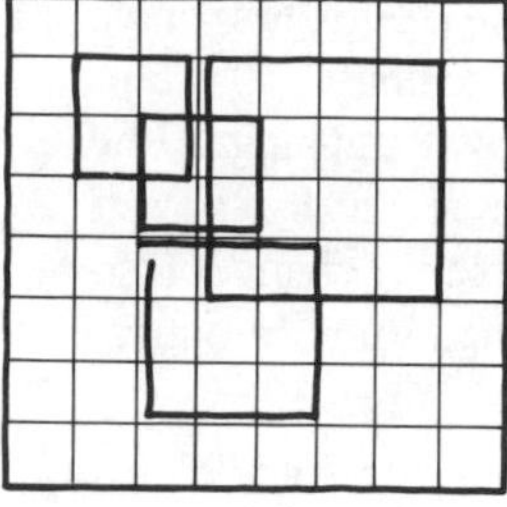

There are 1-squares, 2-squares, 3-squares, . . . , 8-squares (biggest). But they could also overlap. Wow! I need to find out how many there are of each size and add these up. I probably could tackle each of those separately. Need to remember that they could overlap. There really is no way I can guess what the answer is.

Devising or Selecting a Plan

I think this seems like a very visual problem, so I'll try to draw a picture. In my picture I can draw all of the squares of one type and then count them, and then I can draw another to count the next size, etc. Maybe I will see some sort of pattern to this as I go and I really won't have to draw them all. But to see a pattern, I better count them in some type of order.

Carrying Out the Plan

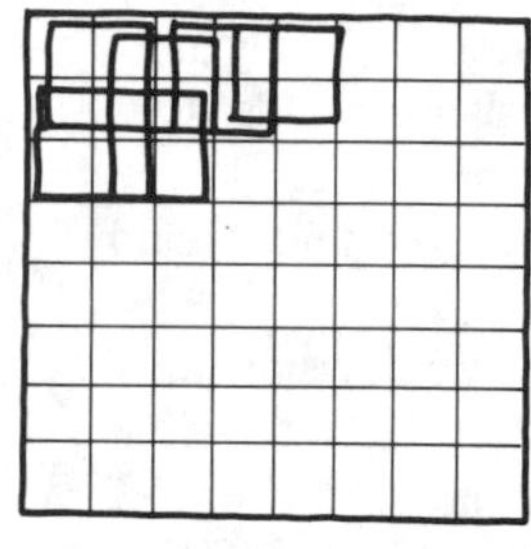

This is getting very messy. I can't really see what I have drawn (sense blockage). There must be some system that I can use to help keep track. Instead of drawing pictures, maybe it would help to cut out a little square and sort of slide it around on the board.

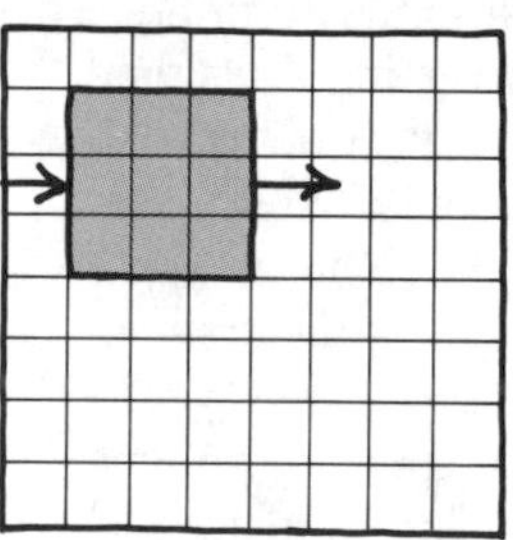

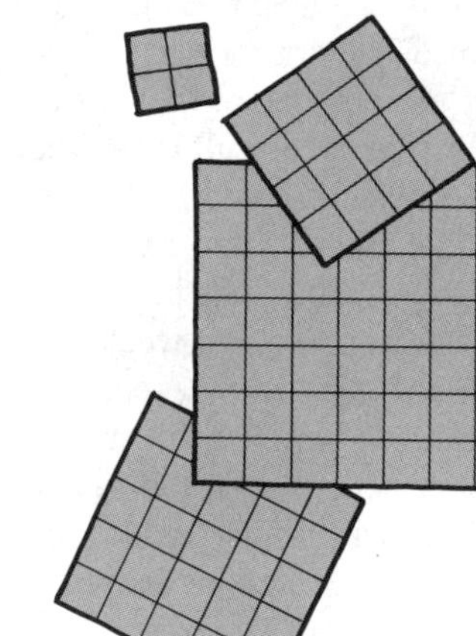

There are clearly 64 little squares. The 2 X 2 paper square can slide across the top of the board in seven different places. But then I can move it down one row and get seven different squares. And slide down again, and again. There are seven rows that I can slide that one on, each has seven squares. That's 64 + 49. Try the 3 X 3 square. It slides across in six places. Go down. Six rows. 6 times 6 squares. Now 64 and 49 and 36. Hey. Looks like 8 X 8 then 7 X 7 then 6 X 6. Check out a 4 X 4 square. Yes! Five places, five rows. 5 X 5.

That's the pattern. Get the calculator and add these up.

$64 + 49 + 36 + 25 + 16 + 9 + 4 + 1 = \underline{204}$

Looking Back

Does 204 squares sound right? I never would have guessed that so I can't tell. Did I miss any? I counted squares from 1 X 1 to 8 X 8. That is all the sizes. There is no other way to draw squares. Any square I can draw would be counted in my scheme. Yes. I believe I can trust that.

This was sort of a draw a picture, systematic count, look for a pattern approach. The picture gave me an idea for how to count them. But the pattern was so clear I didn't need to count them all. I could have started with a smaller checkerboard. A 6 X 6 board wouldn't have been much easier. But I could sure solve it for a 2 X 2 board almost without trying. Five squares—4 and 1. Wonder if that would have helped.

(In fact, the smaller-problem approach can be pursued, and another similar pattern discovered).

Hey, I can do this for any size board. Just add up all the square numbers up to $n \times n$. (There is a formula for this, and discovery or use of it depends on the background of the solver). What else? Rectangles! Could I count them? They come in different sizes and can be positioned in different ways. (This, too, is solvable in a similar manner.) What if the checkerboard wasn't square? Maybe 6 X 12 or something. What If I used a grid of equilateral triangles? Could I count triangles using a similar approach? What other grids and shapes are there?

(Each of the conjectures above leads to a solvable problem, perhaps a bit more difficult, but approached using similar techniques).

Figure 4.12
Problem: How many squares on an 8 × 8 checkerboard?

of solution strategies as part of their repertoire. In addition, they must learn to recognize when a particular situation calls for this or that approach.

Carrying Out the Plan

Carrying out the plan is partly a matter of following through with the approach selected, being careful of each step along the way. The more sophisticated aspect of carrying out the plan is the self-monitoring of your own progress and the methods you are using, a metacognitive activity. You may have been working feverishly on an approach for a long time, producing lots of pencil markings, charts, drawings, calculations, and so on. But are you making any progress? Or, a particular approach may very well have run head on into a dead end. You must learn to recognize this and decide if the blockage is due to the approach or some other factor, such as overlooking a condition or an incorrect assumption or incorrect data recording. When blockage midway in a solution process occurs, some recycling of the general scheme is called for. Sometimes it requires returning to the original problem and the understanding stage. At other times it is the approach to the problem that requires rethinking. In such cases a return to the general process of devising a plan is called for to search for a new way of approaching the problem.

Looking Back

When a solution to a problem is found, the problem-solving process is not over. Three significant looking-back activities should always be considered:

1. Look at the answer.
2. Look at the solution process.
3. Look at the problem itself.

In the real world, problems are never "solved" when an answer is found. There is no answer book or teacher to verify that the answer is correct. Consequently some effort must be made to be sure that the answer arrived at is indeed a solution. The method of doing this varies with the problem. Does the answer seem to be reasonable? Is it in the "ballpark" of what was expected? Could there be other answers? Is there a way to verify the answer by checking it against all of the conditions? Are there any contradictions between the answer and the conditions of the problem? Sometimes the answer can only be accepted or rejected by looking at the process used to arrive at the answer. Was the logic used appropriate? Were the calculations correct?

How was the problem solved? This question sometimes helps validate the solution as just discussed. However, it is an important step in itself. Could it have been solved a different way? an easier way? What was the method used to solve it? The last question helps identify or label the process to make it more readily used in future problems.

Finally, consider in retrospect the problem itself. Now that the problem is solved, are there other questions similar to or related to this problem that can be answered? If the numbers were bigger or different somehow, could you still solve the problem? Is there a general case that can be solved because of having solved this special case? What features of the problem might be changed to create a similar problem that may be solvable, interesting, challenging, or be more useful? Return to the problem of the staircase built of blocks (p. 30). Suppose you wanted 100 steps. Would you have to add the numbers 1 to 100? (See Figure 4.13 for one idea about this.)

If two sets of stairs the same size are put together, a rectangle is formed that is one block longer on one side than on the other. If the two stairs each have 20 steps, the rectangle will be a 20 X 21 rectangle. The number of blocks in one set of stairs will be half of 20 X 21. We can use this argument to demonstrate that the sum of the numbers $1 + 2 + 3 + 4 + \ldots + n$ is $n(n + 1)/2$.

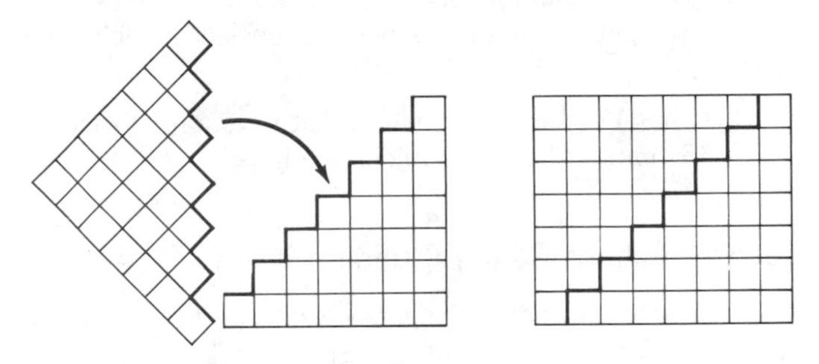

Figure 4.13
Two sets of seven stairs make an 8 × 7 rectangle; the number of cubes in one set is ½ (8 × 7).

Suppose the steps were built with two types of bricks, like those shown in Figure 4.14? In business, engineering, architecture, art, science, law, and other fields, this last looking back at the problem is what makes a problem solver truly valuable to the employer. It is a habit of looking for new worlds to conquer, new ideas to explore, new applications, new knowledge.

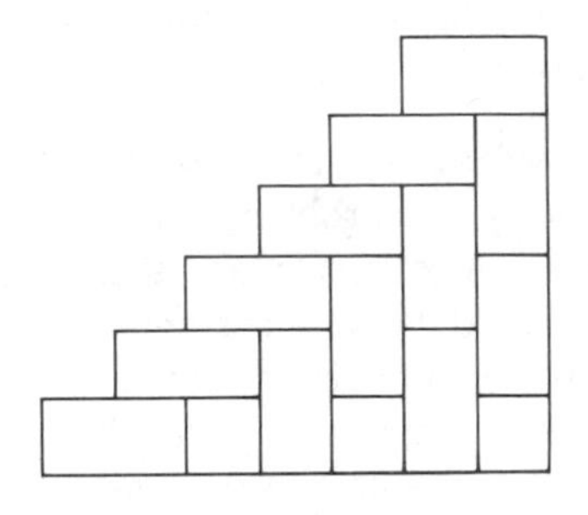

Figure 4.14
Stairs built with 2 × 1 rectangles and squares.

THE TEACHER'S ROLE IN PROCESS PROBLEM-SOLVING INSTRUCTION

What should the teacher do during a problem-solving session with a class? The broad goals of such lessons are to help students make progress toward the general scheme we have described and to construct, identify, and apply the various problem-solving strategies.

The teaching plan described here is modeled after the three-part approach described, developed, and tested by Charles and Lester (1982). It is also very nicely carried out in the resource books, *Problem solving experiences in mathematics* (Charles et al., 1985). The plan focuses on teaching actions *before, during,* and *after* the time students are solving a problem. The three-part approach is an excellent scheme for a teacher at any level to prepare a lesson plan for problem solving.

Problems can be presented in any appropriate manner: read orally, prepared on an overhead or chalkboard, distributed on paper, or taken from books. It is generally a good idea to have students work in cooperative groups of three or four to maximize interaction and discussion.

For the purpose of the discussion, consider the following problem:

> Lisa has 12 coins in her purse that have a total value of 45 cents. What coins might she have?

Teacher Actions Before Students Begin

There are some standard things you can expect to do with all problems. These include reading the problem, having students restate the problem in their own words, asking for identification of what is known and what is being asked for. Besides these things, good planning includes preparation of some specific questions related to the problem.

> How many coins does Lisa have?
>
> How much are they worth in all?
>
> What different kinds of coins are there?
>
> How much is a quarter (and other coins) worth?
>
> Could she have a half dollar?
>
> What is one way to have 45 cents that you know of? How many coins?

With each problem the first questions should cover all of the important data in the problem, and the questions can be answered directly. Any vocabulary that may be confusing should be clarified. Make few assumptions about students' knowledge. Help students clarify any implicit assumptions in the problem. For example, in the checkerboard problem in Figure 4.12, the squares can be of all sizes and can overlap, but that is not clearly in the problem.

Some questions can begin to give hints and are appropriate for beginning problem solvers at this time. They can also be held back as hints for individual groups later on. Could there be pennies? Could there be less than five? Could there be six?

The purpose in this *before* stage is to model the type of thinking and problem-understanding behavior that some day they will do on their own. There is no reason to hide assumptions, use vague wording, double meanings, or to in any way be tricky. That would discourage students and defeat the purpose.

Teacher Actions While Students Are Working

The next thing is to get your students working. Exactly what you plan to do at this time depends on the prior experiences of your students, the strategy that you want to draw out of the lesson, and whether they have experienced that strategy before. For a totally novice class, it may be wise to simply set them to work and prepare a series of hints and suggestions to give to individual groups. Another option is to guide a full class discussion toward a particular solution strategy, and then set the groups to work with the hope that many will follow up on your lead.

Whatever your specific approach, it should never communicate to students that there is some predetermined "right" way to do the problem. Accept ahead of time that some students will solve problems in ways you had not planned on and in ways that had not even occurred to you. If your goal was to use the problem to teach the organized list strategy, then plan your hints and suggestions to steer students toward that strategy. However, do not be upset that several groups found the guess-and-check heuristic to be more useful to them. In the final analysis, if your class solves one problem using three different methods or strategies, that will be more valuable to their development than 30 problems all solved exactly as you prescribe.

Suppose your intention in using Lisa's coin problem is to work on the organized list strategy. The first thing you should do in preparation is to solve the problem yourself according to this strategy and reflect on the kinds of decisions you made and in what order. Then think about things that could go astray or cause difficulty. You might prepare the following hints and questions:

> Would it help to find *all* of the ways to make 45 cents? Could you make a list of them?
>
> What is the fewest number of coins that could possibly be used?
>
> What is the greatest number of quarters you could have?
>
> What would you have left to worry about after one quarter? (20 cents)
>
> Perhaps it would help to organize the different ways in some kind of chart. What would you need to keep track of?

Notice that these questions aim at the strategy of finding all the ways to make 45 cents, starting with the least number of coins. If a group decided to start with all pennies and work from that direction, the strategy is exactly the same but with different specifics. Both should be encouraged. If you have thought about that ahead of time, you will be prepared to help that group without letting them think they are on the wrong track. Never force your method on students, especially if theirs is just as effective.

For this example, there is another listing strategy that you may think is more useful. It focuses on the number of coins and not the amounts.

How many coins does Lisa have? What if they were all pennies?

What are some other ways to have 12 coins?

All pennies makes only 12 cents. Could the 12 coins be pennies and nickels? How many different ways can you have pennies and nickels and have just 12 coins?

Do you need to consider all possible numbers of pennies? (No. Only 0 and 5 are possibilities.)

You have tried pennies and nickels. What coin should you add to the list now? What combinations can you get that have 12 coins?

You should try to work the problem ahead of time with as many different approaches as you can that seem to be probable. If you have only one approach in mind, you may communicate to students that your approach is "the correct" one. Students with that perception will have difficulty building their self-confidence in their own abilities to solve problems.

In planning, it is a good idea to list as many hints, questions, and suggestions as you can think of. Write them down as they come to you and then examine the list. Put the hints and questions in an order that goes from least guided to most guided. You may not need them all for all groups. You want the students to do as much thinking on their own as possible. The more help you give them, the more you are solving the problem for them. On the other hand, the kind of reflective thinking required to produce just the right hint or guidance is difficult to do while walking from group to group if you have not thought through some possible guidance ahead of time.

During the actual lesson be an active listener. If a group is searching for a place to begin, some of the first hints in your list may be appropriate. If a group has started on a strategy but needs some help in using it effectively, you need to decide if intervention and assistance is more helpful or if this is a time to let students learn from their struggles. When students are working hard but clearly going in the wrong direction, try to praise their work and ask questions that will help them discover their own misdirection. You will frequently encounter approaches you have not thought of. They may be more inventive and valuable than what you had planned. Do not be too quick to guide students in another direction.

Besides helping those groups that need guidance, you must also be prepared for those groups or individuals that zoom directly to the solution ahead of the rest of the class. First, require that they have carefully checked their reasoning and their answer. Encourage them to be ready to explain their approach. But then what? Be prepared to offer an extension or a related challenge. Simply giving the group another problem can be construed as punishment for being effective and clever. Rather, your praise of their excellent work can include a challenge to go beyond the original problem.

That is super work! Are you sure there is only one solution? (There are two solutions to the coin problem, but the class need not know that.)

I wonder if this group could figure out what other amounts you could have with exactly 12 coins. Could you change the amount from 45 cents to something else so that there is only one answer to this problem?

Teacher Actions After the Problem Is Solved

After a sufficient time return to a full class discussion. Ask several groups to give their answers and then discuss with the class how they can decide if the answer is correct. You want the students to learn to validate or assess their own answers. Do not be an answer book!

In the example of 45 cents and 12 coins, it is relatively straightforward to check that the solution meets both conditions. But is it the only solution? That question can turn the discussion back to the processes used.

While students were working, you will have been able to determine the approaches used by different groups. Select one or two groups to explain how they did the problem. (Each group should always have both a recorder and a spokesperson.) As different solutions are described, help the class identify the strategies used. As noted earlier, there are at least two approaches to this example problem that use the organized list strategy. The same problem can also be done with a guess-and-check approach. If a group has done it that way and you have time to discuss it, then there is an opportunity to both see another strategy and communicate that there is no mystical "right" way to solve a problem. The guess-and-check method is less likely to turn up two solutions, but there is no way anyone could know that before doing the problem. At the very least, a quick acknowledgment of a group's different approach will keep them from thinking they were wrong in what they did.

If a strategy used is a new one for the class, then it should be labeled and added to your classroom list of strategies. A chart or bulletin board with a growing list of useful approaches will be valuable as your problem-solving program continues.

Finally, discuss the problem itself. What else is known, or what else can be done now that we have solved the problem? Encourage students to look for generalizations if appropriate. The coin example does not lend itself to a generalization in the same way as the checkerboard problem.

Here are some ideas that you might consider when looking for a new but related problem (Charles & Lester, 1982):

Can the wanted and given information be reversed?

What if the context were changed? (Suppose that instead of money, Lisa had boxes of candy of different sizes.)

Can other conditions be added, changed or removed? (What if Lisa had "more than 12 coins" and/or "less than 75 cents?")

If the numbers were just changed, would the problem be any easier or harder? Could we make one up and solve it easily now?

Are there any problems that seemed like this one somehow? (The coin problem has the same structure as the pigs-and-chickens problem.)

Can the problem be generalized? (Sometimes you can get to this by working toward larger and larger numbers.)

Classroom Environment for Problem Solving

DEVELOPING POSITIVE STUDENT ATTITUDES

If students do not demonstrate a willingness and interest in solving problems, it is very difficult to teach them problem-solving processes. Willingness to solve problems, even enjoyment, should be one of the highest priorities in your problem-solving program. What can be done to develop that type of positive attitude?

Build in Success

Nothing succeeds like success. This is a cliché but very true in this context. In the beginning of the year, plan problems that you are confident your students can solve. The success should clearly be your students', and not due to your careful guidance. Avoid creating a false success that depends on you showing the way at every step and curve. Students will quickly see that you are the one that solves the problems and without you, they will believe, they are helpless. Rather, if problems are solved through their own efforts, you can correctly heap praise on their outstanding good work. Even as much as two months of relatively simple problems that did nothing to move the students forward in their thinking skills would be more worthwhile than frustrating students with weak self-concepts about problem solving.

Praise Efforts and Risk Taking

Students need to hear frequently that they are "good thinkers" capable of good, productive thought. When students volunteer ideas, listen carefully and actively to the idea and give credit for the thinking and the risk that children take by venturing to speak out. The more that students are encouraged, the more effort they will expend to receive that praise. Conversely, if weak or incorrect ideas are put down or ignored, the children who ventured forth with those ideas will think twice before ever trying again. After some time the reluctance turns into a belief that says "I'm no good! I can't think of good ideas."

Use Nonevaluative Responses

Develop a response form that is nonevaluative, which can be used all of the time for both good and weak ideas. For example, "That's a *good* idea. Who else has an idea?" As another example, "That's good thinking. I can tell you are really thinking about this situation." Even the child who spurts out without thinking can be rewarded: "I always want to hear your good ideas, Cindy. Be sure you've thought through what you want to say first." Be sure to return to this child and reward the more reflective thought.

Listen to Many Students

Avoid ending a discussion with the first correct answer. As you make nonevaluative responses, you will find many children repeating the same idea. Were they just copying a known leader? Perhaps, but more likely they were busy thinking and did not even hear what had already been said by those who were a bit faster. If 10 hands go up, 10 children may have been doing good thinking. If you forget to listen to lots of children, lots of children do not hear your praise.

Provide Special Successes for Special Children

Not all children will develop the same problem-solving abilities. Those who are slower or not as strong need success also. One way to provide their success is to involve them in groups with strong and supportive children. In group settings all children can be made to feel the success of the group work. Slower students can also be quietly given special hints that move them toward success.

Better students need special successes also. Be especially prepared with a super challenge for these students. Be careful, however, to be discreet.

COOPERATIVE GROUP WORK

Problem solving is an activity that is strongly improved when students work together. From the teacher standpoint it also makes interacting, helping, and evaluating all much more manageable.

In problem-solving groups there should always be a recorder and a presenter. Make it clear that you are interested in group solutions, not four solutions from one group. If students think that they each have to turn in a paper, what frequently happens is a focus change from productive thought to copying what everyone else has written. Individual recording also leads to independent work because there is no need to share ideas. The recorder should be the one who is writing down the ideas agreed upon by the group. This forces the group to interact and agree on what they are doing. A presenter can be designated ahead, or you can explain that any group member may be called upon. This is a way of telling the group members that they are all responsible for the group's work.

CALCULATORS DURING PROBLEM SOLVING

At every grade level, calculators should be a standard tool available to all students during any problem-solving activity. Since computational skill development is a completely separate issue from problem-solving objectives, there is no reason for not including their use. More importantly, however, calculators can serve to enhance problem solving. Computations are made so easily that numbers do not need to be artificially small. Trial and error and pattern analysis are promoted when students can make multiple computations that, without the calculator, would be excessively tedious. In short, the calculator allows the mind to be more productively occupied.

The use of the calculator during problem solving has other benefits as well. Students need to learn when to use the calculator, when to estimate, and when to use mental computation techniques. In group activities, different students will select different forms of computation. Over time students will begin to learn when it is most important to select the calculator and when other forms are important. The need to record results will also teach another skill related to calculators—recording intermediate and end results.

THE TEACHER AND THE PROBLEM-SOLVING CLIMATE

There is no more important ingredient in developing student problem-solving abilities than the teacher. The teacher sets the tone, the spirit, the climate for the classroom.

Enthusiasm

A teacher's enthusiasm for a topic is easily detected by students. If the teacher is "into" problem solving, is clearly enjoying the activity, is spirited in discussions with students, then students will begin to share in that enthusiasm. This can be as simple as excitement over finding a new problem to solve or a bit of exuberant behavior when someone has demonstrated a good idea.

Values

If problem solving is to be the most important part of the mathematics curriculum, then your students should have every indication of that value system. Your value system around problem solving is reflected in the amount of time devoted to it, compared with other topics, and its position relative to other work in the day and week. "We *will* make time for our problem-solving lesson today. Fraction practice will just have to wait." A comment like that says something quite different than, "If we have time after going over our fraction review, we will have a problem-solving lesson."

Perhaps more important than time and emphasis is evaluation. Students need to know that the problem-solving work is being evaluated, monitored, and made a significant part of the regular grading scheme. Parents should also be kept informed about your problem-solving program. Let them know of the part it has in grades, the time devoted to it in class, and the importance that you believe thinking skills have for their children.

Challenge

Problem solving should be fun and serious business at the same time. One way to bridge this gap is to approach problem solving as a challenge. It is important and fun to conquer challenges. Monotonous work is rarely important and never fun.

Model Problem-Solving Behavior

Let your students know that you, too, are still growing as a problem solver. Occasionally do problems with them that you have never seen before. With older children, share problems that you have found in magazines or books that you find challenging and interesting. Become a problem solver yourself, and let your students know that it is important to you.

Assessment of Problem Solving in the Classroom

Classroom assessment of problem solving is a complex problem that will require some effort, especially at first. Simply getting the correct answer to a problem is not sufficient as evidence of good problem-solving skills. Some students get correct answers but use incorrect reasoning, and others use excellent strategies and simply make careless errors. The objectives of problem solving include the thinking that goes on in all stages of the problem-solving process. Problem-solving ability cannot be adequately equated with getting answers.

Before deciding how to evaluate problem solving, take some time and decide exactly what in your program should be evaluated. This list will likely change with different grade levels. It will probably also change throughout the year. For example, attitudes toward problem solving are important at all times but are more likely a major objective in the beginning of the year. The use of a particular strategy may be of special concern at one point in the year, while the ability to select from a number of strategies may be more relevant toward the end of the year.

There are a variety of ways to evaluate problem solving. You will need to work with them for a while in order to find a mix that serves your needs. Different evaluation techniques serve different purposes.

DECIDING WHAT TO EVALUATE

The skills and attitudes listed in this section can all be evaluated or assessed in one or more ways (Charles, Lester, & O'Daffer, 1987). Few would attempt to measure all of these all of the time. However, it is important to review your program periodically to be sure it is appropriately focused.

Affective Considerations:

Demonstrates a willingness to try to solve problems

Demonstrates a self-confidence in problem-solving abilities

Perseveres in attempts to solve problems even when progress is slow

Enjoys solving problems

Believes he or she is capable of being a problem solver

Works cooperatively in groups while solving problems

Has pride in problem-solving abilities

Understands the Problem:

Is able to determine the relevant information in a problem

Can clearly state what is being asked for

Seeks clarification of ambiguous statements or unfamiliar vocabulary

Writes down important information for clarity

Identifies subproblems or tasks before beginning work

Draws pictures or uses models to help clarify a problem

Makes estimates of solutions when appropriate

Strategy Knowledge and Selection:

Given a strategy to use for a problem, can use it appropriately

Frequently uses different strategies

Given a problem, can determine an appropriate strategy to use

Is able to use the _______ strategy(ies)

Working on Problems:

Is careful and accurate in his or her work

Is able to self-monitor progress

Knows when stuck and is time to choose another approach

Keeps careful records of work

States answers in complete sentences or meaningful phrases

Looking Back:

Can assess answers in terms of the problem and the solution process

Remembers to assess solutions before quitting or going on

Regularly goes beyond the problem given (looks for generalizations, related, or extended problems)

Can identify problem-solving strategies that were used

WHEN TO EVALUATE

During class, students solve problems in interactive groups, brainstorm, doodle, experiment with different approaches, and share ideas. Traditional test situations with the pressure of grades are not conducive to the kind of problem-solving behavior we encourage during instruction. Therefore, some problem-solving performance should be evaluated under non-test conditions. This may mean that assessment is carried out during the regular class times that students are solving problems, including those times that they are working in cooperative groups.

There are test formats that can be used in more traditional test settings. However, in these tests, students are not solving problems but reflecting on various parts of problem solving. The pressure of solving a problem from beginning to end is removed from the testing situation.

You certainly cannot evaluate every child during every problem-solving session, nor should you try to do that. However, it is also difficult to adequately assess all children in one session. This means that some evaluation should be carried out in most problem-solving sessions. One or two groups of students can be assessed on certain objectives in one period. Over a short time, data on that objective will have been collected for the entire class.

CLASSROOM EVALUATION METHODS

Charles, et al. (1987) describe four different techniques for evaluating problem solving: observing and questioning, self-assessment data, holistic scoring, and multiple-choice and completion tests. Each of these methods will be described briefly here.

Observing and Questioning

The idea behind observation and questioning methods is to observe students as they solve problems. This can be done during regular full class or group problem-solving lessons, and it can also be done in more structured situations, such as a one-on-one interview.

Full class observations are best if a checklist with specific criteria has been prepared. The checklists can vary from time to time to match specific objectives. They should be relatively short and simple so that specific items can be kept in mind during class time and then filled in accurately as soon after a class session as possible. Figure 4.15 shows three possible forms suggested by Charles et al. The first is a narrative, the second is a checklist, and the third is a rating scale. Any of these formats could be utilized with equal effectiveness, and modified to suit the specific objectives of your current program.

It is certainly a good idea to let your class know that you will be using these observation techniques; tell them exactly what the criteria are. In this way you clearly communicate that you value the entire problem-solving activity. Attitude

Problem-solving Observation Comment Card

Student Sue Trent Date 10/5

Comments:

Knows how and when to look for a pattern.
Knows that a table will help her find a pattern.
Keeps trying even when she has trouble finding a solution.
Needs to be reminded to check her solutions.

Problem-solving Observation Checklist

Student ______ Date ______

_____ 1. Likes to solve problems
_____ 2. Works cooperatively with others in the group
_____ 3. Contributes ideas to group problem solving
_____ 4. Perseveres—sticks with a problem
_____ 5. Tries to understand what a problem is about
_____ 6. Can deal with data in solving problems
_____ 7. Thinks about which strategies might help
_____ 8. Is flexible—tries different strategies if needed
_____ 9. Checks solutions
_____ 10. Can describe or analyze a solution

Problem-solving Observation Rating Scale

Student ______ Date ______

	Frequently	Sometimes	Never
1. Selects appropriate solution strategies	____	____	____
2. Accurately implements solution strategies	____	____	____
3. Tries a different solution strategy when stuck (without help from the teacher)	____	____	____
4. Approaches problems in a systematic manner (clarifies the question, identifies needed data, plans, solves, and checks)	____	____	____
5. Shows a willingness to try problems			
6. Demonstrates self-confidence	____	____	____
7. Perseveres in problem-solving attempts	____	____	____

Figure 4.15
R. Charles, F. Lester, and P. O'Daffer (1987). How to Evaluate Progress in Problem Solving. *Reston, VA: National Council of Teachers of Mathematics. Reprinted by permission.*

and perseverance items should be given special attention in discussions with your students. Keep dated checklists in student folders. Share the checklists with parents as you keep them informed of your problem-solving curriculum.

For more detailed evaluation, such as for diagnostic purposes, you may decide to use an interview or small-group approach. In these settings, the student or group is encouraged and frequently reminded to "think out loud." After explaining that you want to find out how the students are thinking, present a problem for them to work. Have the student read the problem. From that point on, you should try to ask only questions which help you see what is going on in their thinking process. Avoid questions that lead the student in how to solve the problem. Examples of appropriate questions are:

What did you do after you read the problem?

Can you explain what is being asked? What do you know and understand about the problem?

What strategy do you think you will use? Why?
(if stuck) What are you thinking now?
Did you change your approach? Why?
Are you sure that is the answer? Why do you think so?
What else can you tell me about this solution? Do you think you could solve it in a different way?

Those questions do not represent a complete interview but should provide some suggestions. Important points that you wish to assess should be planned ahead of time and written down. With practice, you will improve your technique. The interview approach is time-consuming, but allows you to be more careful and to probe more deeply into what a student or group of students is thinking.

Self-Assessment Data

Self-assessment refers to students doing some form of reporting to you. This can take the form of writing down what they are thinking as they solve problems, talking into a tape recorder, or answering open-ended questions you have prepared concerning processes or attitudes. A more common and easier-to-use form of self-assessment is a rating scale which students fill out. These rating scales are most common for assessing attitudes. Scales and self-report forms depend on student assessment and judgment, not yours. As such, they should only be used in conjunction with other techniques. They are interesting in that they give the student's view in contrast to yours.

Holistic Scoring

By holistic scoring Charles and his co-workers mean a systematic method of assigning a score to the whole problem-solving process or specific aspects or phases of the process. The scoring could be done during an observation or problem-solving session. More commonly, this form of scoring is based on the written work done by students while solving a problem. If an entire class is asked to work independently on a problem and write down all of their work, it is possible to use a holistic approach to evaluate each student on the same problem.

The scoring system shown in Figure 4.16 is a method of assigning points to different phases of the problem-solving process. While the results give good information, especially if used repeatedly over time, summing the scores and making student-to-student comparisons may be misleading. A score of 1,1,2 would mean something quite different from a score of 2,2,0.

Instead of assigning points to individual portions of the process, a single score can be assigned to the entire process. This requires listing ahead of time a series of very specific criteria.

The biggest drawback to holistic methods is the need to infer student processes from student written work. The student's thinking or approach to a problem is not always clear from the written work. This is especially true for the understanding portion of a problem and the looking-back phase. With younger children an attempt should be made to gather the data or related information through observation.

Analytic Scoring Scale	
Understanding the Problem	0: Complete misunderstanding of the problem 1: Part of the problem misunderstood or misinterpreted 2: Complete understanding of the problem
Planning a Solution	0: No attempt, or totally inappropriate plan 1: Partially correct plan based on part of the problem being interpreted correctly 2: Plan could have led to a correct solution if implemented properly
Getting an Answer	0: No answer, or wrong answer based on an inappropriate plan 1: Copying error; computational error; partial answer for a problem with multiple answers 2: Correct answer and correct label for the answer

Figure 4.16
R. Charles, F. Lester, and P. O'Daffer (1987). How to Evaluate Progress in Problem Solving. *Reston, VA: National Council of Teachers of Mathematics. Reprinted by permission.*

Multiple-Choice and Completion Methods

Problem-solving tests are beginning to be developed which can be administered easily to large groups of students such as an entire school or school system. With a multiple-choice format it now appears possible to gather reasonably reliable information about students' ability to understand a problem, use a particular strategy, select a strategy, evaluate an answer, and so forth. For example, to assess problem understanding, the teacher can give a problem and provide several statements which might describe what the problem is asking for. The task is to pick the statement that best describes the problem question. Strategies can be matched to problems, different ways of evaluating an answer can be chosen for a particular problem, and incomplete solutions can be determined to be "most likely to produce a solution."

Completion tests are more open-ended, requiring students to list given data, state the question the problem asks, name a strategy that is appropriate, and so on.

Of these two approaches, the multiple-choice format is the most difficult for the classroom teacher to construct. To be valid, the alternative choices must be very carefully chosen, and questions must be cleverly selected. Prepared tests of this sort are likely to become more and more common, however, as textbook publishers and testing companies begin to meet the demand for problem-solving evaluation. The completion test, in contrast, is quite easy to construct. Simply think of specific questions that could be answered in an interview. The major drawbacks are the need for students to be able to write responses and the need to give credit for partially correct answers.

SUMMARY

Problem solving is a complex topic and nontrivial to evaluate. In fact, the lack of clearly defined evaluation methods is partly responsible for the slow rise of problem solving in the curriculum. Testing companies, textbook publishers, and state and local school systems are finally beginning to create instruments for use in the classroom. However, it remains the teacher's responsibility to assess student progress, and it is unlikely that any single prepared test will be completely adequate for all classroom needs. To do a good job of assessing problem solving for both diagnostic and grading purposes, the teacher is going to have to select from available tests and use a variety of techniques.

For Discussion and Exploration

1. Review the list of suggested problem-solving strategies (sometimes called heuristics) and discuss whether they are appropriate for primary-grade students. Are other strategies more appropriate? Which ones are not? Find problems that have been suggested for primary students and see what strategies can be taught with those problems.
2. What would you do if you had a class do a process problem and no group used the strategy that was your objective for the day?
3. Review the affective objectives for problem solving. How do you stack up on these objectives? What have been the major factors in your experiences that caused these attitudes toward problem solving?
4. Examine a commercially available problem-solving program. Two possibilities are *Problem solving experiences in mathematics* (Charles et al., 1985) and *The problem solver: Activities for learning problem solving strategies* (Goodnow, Hoogeboom, Moretti, Stephens, & Scanlin, 1987). How do these programs fit your understanding of the needs of problem-solving instruction?
5. Try some problem-solving instruction in a classroom. Remember that there will be big differences in how children work with problems, depending on the amount of such experiences they have had.
6. Examine a basal textbook series for problem solving. What types of problems are provided? How frequently are process problems provided? What strategies are introduced? What form of problem-solving evaluation is provided? It will very likely be necessary to examine the teacher editions of the books to find everything you want to know.
7. How much effort should be given to problem-solving evaluation by the teacher in the classroom? What if no evaluation were done? Could there be too much time, effort, or pressure placed on evaluation? What are the positive benefits of evaluation? What are the negative features?

Suggested Readings

Billington, J., & Evans, P. Levels of knowing 2: "The handshake." *Mathematics Teaching, 120,* 12–19.

Brannan, R. (1983). An instructional approach to problem solving. In L. G. Shufelt (Ed.), *The agenda in action.* Reston, VA: National Council of Teachers of Mathematics.

Brown, S. I., & Walter, M. I. (1983). *The art of problem posing.* Hillsdale, NJ: Lawrence Erlbaum.

Charles, R., et al. (1985). *Problem-solving experiences in mathematics (grades 1 to 8).* Menlo Park, CA: Addison-Wesley.

Charles, R., & Lester, F. (1982). *Teaching problem solving: What, why & how.* Palto Alto, CA: Dale Seymour.

Charles, R., Lester, F., & O'Daffer, P. (1987). *How to evaluate progress in problem solving.* Reston, VA: National Council of Teachers of Mathematics.

Dolan, D. T., & Williamson, J. (1983). *Teaching problem-solving strategies.* Menlo Park, CA: Addison-Wesley.

Fisher, L. (1982). *Super problems.* Palo Alto, CA: Dale Seymour.

Goodnow, J., Hoogeboom, S., Moretti, G., Stephens, M., & Scanlin, A. (1987) *The problem solver series (grades 1 to 8).* Palo Alto, CA: Creative Publications.

House, P. A., Wallace, M. L., & Johnson, M. A. (1983). Problem solving as a focus: How? When? Whose responsibility? In L. G. Shufelt (Ed.), *The agenda in action.* Reston, VA: National Council of Teachers of Mathematics.

Mason, J., Burton, L., & Stacey, K. (1982). *Thinking mathematically.* London: Addison-Wesley.

Meyer, C., & Sallee, T. (1983). *Make it simpler: A practical guide to problem solving in mathematics.* Menlo Park, CA: Addison-Wesley.

Ohio Department of Education. (1980). *Problem solving . . . a basic mathematics goal: A resource for problem solving.* Columbus, OH.

Ohio Department of Education. (1980). *Problem solving . . . a basic mathematics goal: Becoming a better problem solver.* Columbus, OH.

Otis, M. J., & Offerman, T. R. (1988). How *do* you assess problem solving? *Arithmetic Teacher, 35*(8), 49–51.

Seymour, D. (1984). *Problem parade.* Palo Alto, CA: Dale Seymour.

Szetela, W. (1986). The checkerboard problem extended, extended, extended, ... *School Science and Mathematics, 86,* 205–222.

CHAPTER 5

THE DEVELOPMENT OF NUMBER CONCEPTS AND RELATIONS

Number is a complex and multifaceted concept. A rich understanding of number involves many different ideas, relationships, and skills: more-and-less relationships, counting and what that tells, special relationships with other numbers, recognition of special patterned arrangements of objects without counting (such as on dice or dominoes), combinations of numbers in a part-part-whole context, and so on. It is important that we provide children with an equally rich assortment of activities that will help them construct these many ideas of number.

The Beginnings of Number Concepts

Parents help children count their fingers, toys, people at the table, and other small sets of objects. Questions concerning who has more or are there enough are part of the daily life of children as early as two to three years of age. Considerable evidence has been gathered indicating that these very young children have some form of understanding of number and counting (Gelman & Gallistel, 1978; Gelman & Meck, 1986; Baroody, 1987; Fuson & Hall, 1983; Ginsburg, 1977).

A developmental approach to number will capitalize on the simple beginning ideas about counting and number that children have before entering kindergarten. It will help them construct new ideas about number to expand, refine, and enhance the concepts that have been developing in the preschool years.

COUNTING

The Development of Counting Skills

Counting involves at least two separate skills. First, one must be able to produce the standard list of counting words in order: "One, two, three, four, ..." Second, one must be able to connect this sequence in a one-to-one manner with the items in the set being counted. Each item must get one and only one count, as shown in Figure 5.1a.

These two skills involve no conceptual knowledge in and of themselves. At least the first 12 or so number words in the counting sequence must be learned by rote. (After 12, there is a rule or pattern that helps.) The procedure of saying one count for each object must also be practiced. Preschool and kindergarten children frequently make errors in either or both of these skills, as illustrated in Figure 5.1b, c, d. Oral counting drills and guided activities in which sets are counted are necessary so that children develop these skills.

Experience and guidance are the major factors in the development of these counting skills. Many children come to kindergarten able to count sets to 10 or beyond. At the same time, children from impoverished backgrounds may require considerable practice to make up their experience deficit. The size of the set is also a factor related to success in counting. Obviously longer number strings require more practice to learn. The first 12 counts involve no pattern or repetition, and many do not recognize a pattern in the teens.

Counting a set of objects that can be moved as they are counted is easier than counting objects that cannot be moved or touched. Counting a set that is "ordered" in some way such as in a string of dots or other pattern is easier than counting a randomly displayed set. There seems to be little advantage in making counting tasks difficult. Therefore, children still learning the skills of counting, that is, matching oral number words with objects, should be given sets of blocks or counters that they can move or pictures of sets that are arranged for easy counting.

Activity 5.1

Play "Catch the Mistake." Have children watch you (or a puppet) count a set of objects and ask them to catch you making a mistake. They should then explain the error. Include mistakes such as counting objects twice, missing an object, use of an incorrect counting sequence, pointing to or moving objects faster or slower than the count.

Meaning Attached to Counting

There is a difference between being able to count (as just described) and knowing what counting tells. When we count a set, the last number word used is the name of the "many-

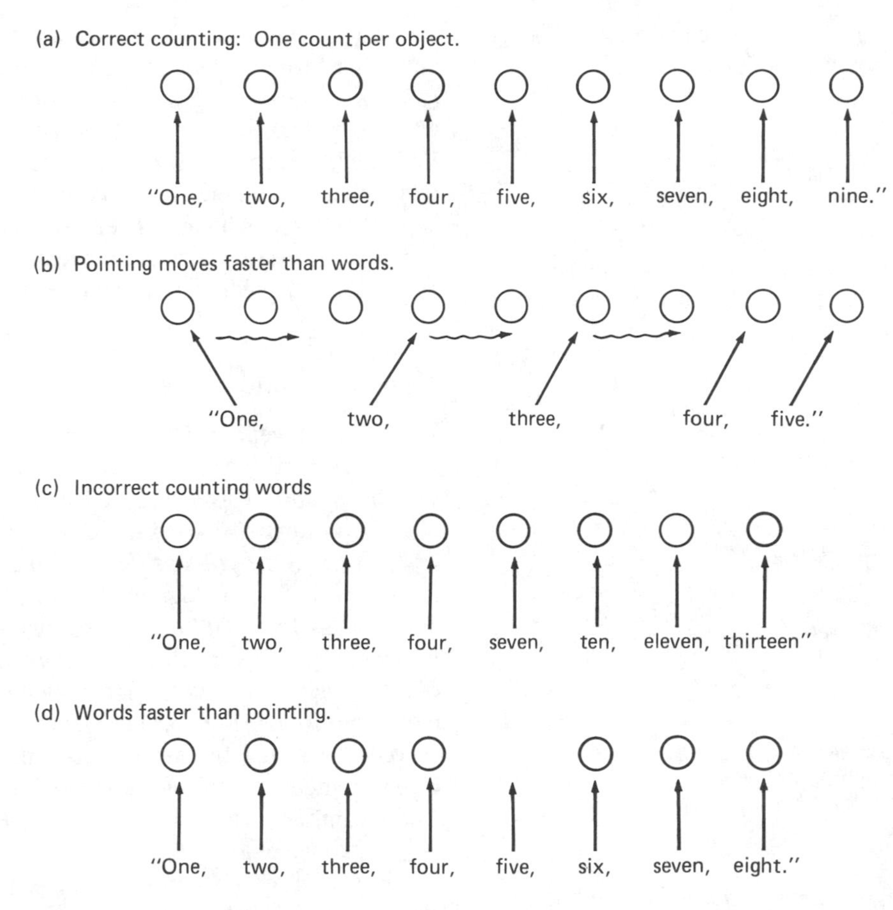

Figure 5.1
Examples of counting and common counting errors.

ness" of the set or the *cardinality* of the set. While very young children seem to have some concept of quantity or manyness, they do not immediately connect this concept with the act of counting. When children understand that the last count word names the quantity of the set, they are said to have the cardinality principle. While experience is a major factor, most children by age 4½ have made this connection (Fuson & Hall, 1983).

To determine if children have the cardinality rule, they can be given a set of objects and asked "How many?" If they repeat or emphasize the last count, then it can be inferred that they have switched meaning from a count use of the last word to a cardinality use of the last word. For example, "One, two, three, four, five, *six.* (There are six.)" If this cardinal usage is not clear, the "How many?" question can be repeated. If the child then announces the total without counting it is clear that he or she is using the cardinal meaning of the counting word. However, a recount of the entire set would indicate that the question "How many?" was interpreted by the child as a command to count rather than a request for the quantity in the set.

In the classroom you can help students develop a deeper sense of cardinality through counting activities such as the following:

Activity 5.2

Count several sets with the same number of objects and discuss how they are alike and different.

Activity 5.3

Have students make (count out) sets that have a specified amount.

Activity 5.4

Have students count a set. Then rearrange the set and again ask "How many now?" If they see no need to count over, the connection is probably made. If they choose to count again, discuss why they think the answer is the same.

Activity 5.5

Give students a number of different sets and have them find pairs that have the same amount. Sets can be prepared on cards with different arrangements and different objects. A large number of such cards with sets ranging from 4 to 12 objects provides a good activity for four or five children at a table or on the floor.

Activity 5.6

Have students make sets of objects that have the same amount as a given set. The cards used in the previous activity could be used. Next to each card the children can make a set of counters with the same amount (Figure 5.2).

Figure 5.2
An early number activity.
Make a set that has as many.

Once counting becomes a meaningful activity, it becomes a major tool in the development of other relationships on number. Meaningful counting is only the beginning of number concept development, not the end goal.

THE RELATIONSHIPS OF MORE, LESS, AND SAME

The concepts of *more, less,* and *same* are basic relationships contributing to the overall concept of *number.* The beginnings of these ideas are developed before children enter school. An entering kindergarten child can almost always choose the set which is *more* if presented with two sets which are quite obviously different in number. In fact, Baroody states "A child unable to use 'more' in this intuitive manner is at considerable educational risk" (1987, p. 29). Classroom activities should help children build on this basic notion and refine it.

How Children Make Comparisons of Quantities

For a long time it was popular to teach children to make comparisons between two sets by a process of one-to-one matching. Objects in one set are carefully matched in some way (paired physically, lines drawn, connected with string, etc.) with the objects in the other. The set with leftover objects clearly has more. If the two sets match, they are the same. These one-to-one matching activities became popular due to the strong influence of Piaget during the 1960s and 1970s. According to Piaget, an understanding of number depends on an understanding of one-to-one correspondence.

Researchers have found that almost all children use counting rather than matching as the basis of comparison. Children can *learn* to use matching, but it does not seem to be the natural method that they use intuitively (Fuson, Secada, & Hall, 1983; Baroody, 1987). In fact, children who are not taught to use matching are in no way less able to make comparisons than those who are. There seems to be no real reason to force children to engage in matching activities if they can successfully use counting to make comparisons.

More, Less, and Same

While the concept of *less* is very similar to the concept of *more,* the word *less* proves to be more difficult for children than does *more.* A possible explanation is that children have many opportunities to use the word *more* and have very limited exposure to the word *less.* Given two unequal sets, selecting the set with more is logically the same as not selecting the set that has less. To help children with the concept of *less,* frequently pair it with the word *more* and make a conscious effort to ask "which is less" questions as well as "which is more" questions. For example, suppose that your class has correctly selected the set that has more from two that are given. Immediately follow with the question, "Which is less?" The unfamiliar term and concept can be connected with the better-known idea.

For all three concepts (more, less, and same) children should construct sets using counters as well as make comparisons or choices between two given sets. The activities described here include both types.

Activity 5.7

At a work station or table provide about eight cards with sets of 4 to 12 objects, a set of small counters or blocks, and some word cards labeled MORE, LESS, and SAME. Next to each card students make three collections of counters: a set that is more, one that is less, and one that is the same. The appropriate labels are placed on the sets (Figure 5.3).

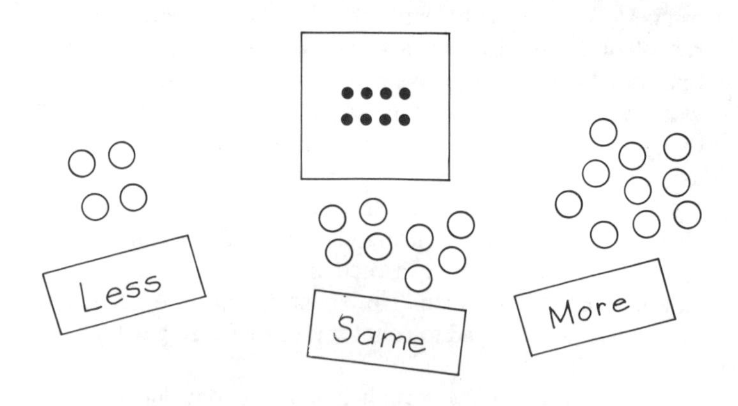

Figure 5.3
Making sets that are more, less, and same.

Activity 5.8

As a seatwork variation, a worksheet such as that shown in Figure 5.4, can be provided. On a given day students can make sets which are less than the given set (or more, or the same). After making the set with small counters, they can use a crayon to make a dot for each counter.

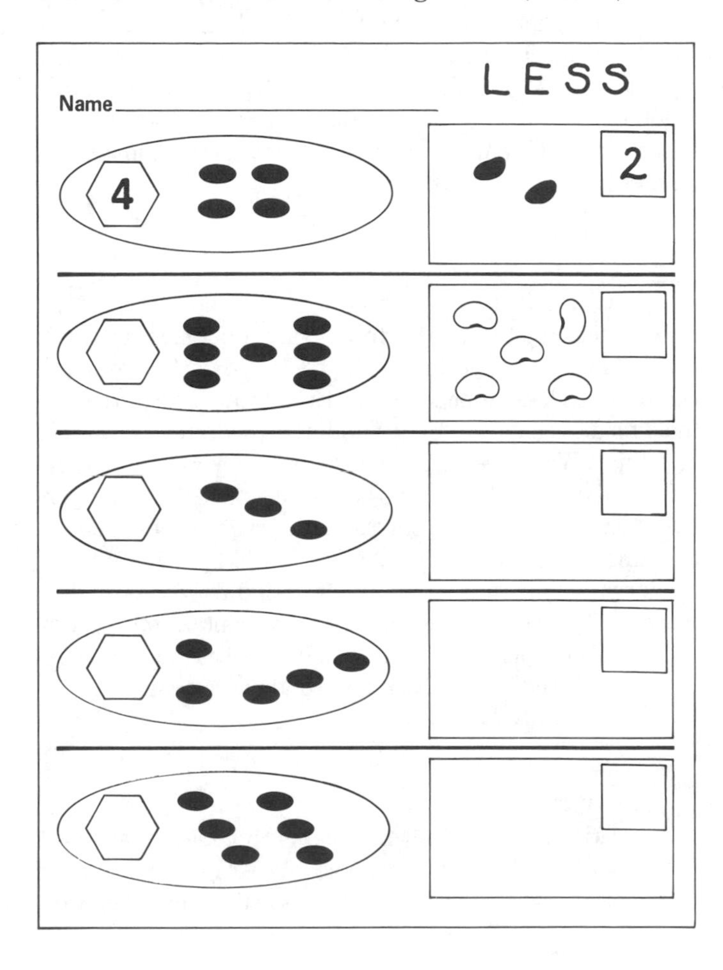

Figure 5.4
A more, less, same worksheet that can be used with counters.

Activity 5.9

The familiar game of war can be played with any cards that have sets on them, or you can use standard playing cards with the face cards removed. To stress the idea of less, periodically play the game where the winner is the person with less rather than more. Another variation is to spin a more/less spinner on each turn. The spinner face is divided in two parts labeled MORE and LESS. If the spinner shows LESS, the player with less wins, and vice versa (Figure 5.5).

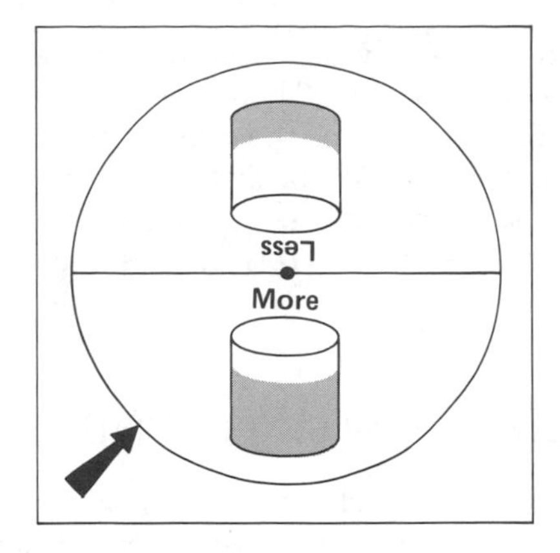

Figure 5.5
A more/less spinner.

Notice that these activities are done with sets and counters but not with numerals. Numerals can be added for connecting level activities. However, to be conceptual activities, the sets must be included.

ANOTHER LOOK AT READINESS ACTIVITIES

For many years it has been popular to have kindergarten children engage in three types of "prenumber" or readiness activities: seriation, classification, and matching. In a *seriation* or ordering activity, materials are arranged from shortest to longest or smallest to largest. *Classification* activities involve students in making sets in which the objects within the sets are alike according to some property. *One-to-one matching* activities (described earlier) involve making set comparisons without counting.

There has also been great concern over the Piagetian conservation of number tasks. In these tasks, a child agrees that two paired-up sets of 7 to 10 counters have the same number of objects. The experimenter then rearranges one of the sets, usually spreading the counters out. The child is then asked if the sets are still the same or is one set more. Many four- and five-year-old children will respond that the now spread-out set is more. These children are said to be "nonconservers of number." By age six very few children remain nonconservers, and virtually all children eventually conserve number.

The argument has been that these readiness activities, designed to develop logical operations, are prerequisites to number concept development. Furthermore, it is argued that counting and other number activities may be useless or even harmful if done with nonconservers.

More recently, a number of researchers have come to the conclusion that these readiness activities are not, in fact, requirements for the construction of number concepts (Ginsburg, 1977; Baroody, 1987; Fuson & Hall, 1983; Clements, 1984; Gelman & Meck, 1986). Clements, for example, compared students taught the logical operations with students taught only a carefully designed counting program. He found that for number concepts, students in a counting program not only outperformed the students taught logical operations but equaled them on the very same logical tasks the other group had been specifically taught. There are a number of explanations for students' failures on the conservation task (see, for example, Fuson & Hall, 1983). Suffice it to say that there is no evidence that nonconservers cannot count meaningfully or that they cannot learn new relationships through counting activities.

Procedural Knowledge of Numbers

Some procedural knowledge of numbers has already been discussed, namely knowledge of the standard counting words in order (oral counting) and the skill of using this counting sequence to accurately count a set. Other procedural skills related to number concepts are the ability to count on and count back, the ability to read and write numerals, and the use of symbols to represent the relationships of more, less, and same (>, <, =).

NUMERAL WRITING AND RECOGNITION

Helping children to read and write single-digit numerals is similar to teaching them to read and write letters of the alphabet. Traditionally, instruction has involved various forms of repetitious practice. Children trace over pages of numerals, repeatedly write the numbers from 0 to 10, make the numerals from clay, trace them in sand, write them on the chalkboard or in the air, and so on. The principle has been that numeral form is a rote activity that can best be learned through practice.

Baroody (1987) argues convincingly that while some numeral writing practice is clearly necessary, much time can be saved by taking a meaningful approach. Such an approach involves directly focusing on the defining characteristics of each numeral and on a motor plan for forming the numerals. The characteristics of the numerals should be articulated and discussed. For example, 1, 4, and 7 are made up of straight lines, while 2 and 5 have a straight and a curve. Similar numerals should be taught together in order to focus on the properties which are necessary to distinguish them from each other. The numerals 6 and 9, for example, are distinguished from all other numerals by a closed loop on a stick. The loop for the 6 is at the bottom and the loop for the 9 is at the top. These features distinguish one from the other.

COUNTING ACTIVITIES

Rote Counting

Oral counting that does not attempt to enumerate a set is simply an exercise to practice the number sequence. While the forward sequence is relatively familiar to most young children, counting on and counting back are difficult skills for many. Frequent short practice drills are recommended.

Activity 5.10

Correlate rhythmic counting with some form of movement. For example, in a count-to-7 exercise, children might stretch their hands above their heads and clap on each count up to 7. Then in the same rhythm and without slowing down, lean over and clap each count, again from 1 to 7, down near their knees. A rhythmic count is useful to keep everyone together. Use a xylophone, triangle chime, or small bell to strike out a rhythm. Count to different target numbers on different days.

Activity 5.11

Counting up to and back from a target number in a rhythmic fashion is an important counting exercise. For example, line up five children and five chairs in front of the class. As the whole class counts from 1 to 5 the children sit down one at a time. When the target number, 5, is reached, it is repeated, the child who sat on 5 now stands, and the count goes back to 1. As the count goes back, the children stand up one at a time, and so on. "1, 2, 3, 4, 5, 5, 4, 3, 2, 1, 1, 2, ..." Kindergarten and first-grade children find exercises such as this quite fun and challenging. Any movement (clapping, turning around, jumping jacks, etc.) can be used as the count goes up and back and up and back in a rhythmic manner.

Activity 5.12

A variation of the last exercise is to count up and back *between* two numbers. For example, start with 4 and count to 11 and back to 4, etc. Keep a rhythm as in the other exercises.

Activity 5.13

The calculator provides an excellent counting exercise for young children because it includes seeing the numerals as the count is made. Have each child press [+] 1 [=] [=] [=] [=] [=]. The display will go from 1 to 5 with each [=] press. The count should also be made out loud in a rhythm as in the other exercises. To start over, press the clear key and repeat. Counting up and back is also possible, but the end numbers will not be repeated. The following illustrates the key presses and what the children would say in rhythm:*

3 [+] 1 [=] [=] [=] [=]   [−] 1 [=] [=] [=] [=]   [+] 1 [=] [=]. . .
"3 plus 1, 4, 5, 6, 7, minus 1, 6, 5, 4, 3, plus 1, 4, 5, . . ."

*To do activity 5.13 requires that the calculator have an automatic constant feature for addition and subtraction. This feature automatically stores the last operation of addition and subtraction. For example, if you press 3 [+] 2 [=], the display shows 5. A subsequent press of the [=] will add 2 to the display and result in 7. If you press 4 [=], the 2 will be added to the 4. Subtraction works in a similar manner.

Counting Sets

Children need practice applying their counting skills to counting sets. Many useful games exist in which children count a set and then make another set of the same amount. Usually a die is rolled or a card drawn that shows a set of objects. The player then counts the corresponding number of counters or moves a marker around a track on a board.

Activity 5.14

"Fill Up" can be played by two or three players. Each player has a board with spaces marked and a collection of counters to place in the spaces. In turn each child draws a card with a set of dots. He or she then counts out that many counters. When the correct amount has been counted (to the side of the board), the counters are placed on the board, one in each space. The first to fill up their board is the winner. The number of spaces on the board, the arrangement, and the number of dots on the cards can all be varied to suit the ability level of the students. Figure 5.6 shows a board and some cards that could be used.

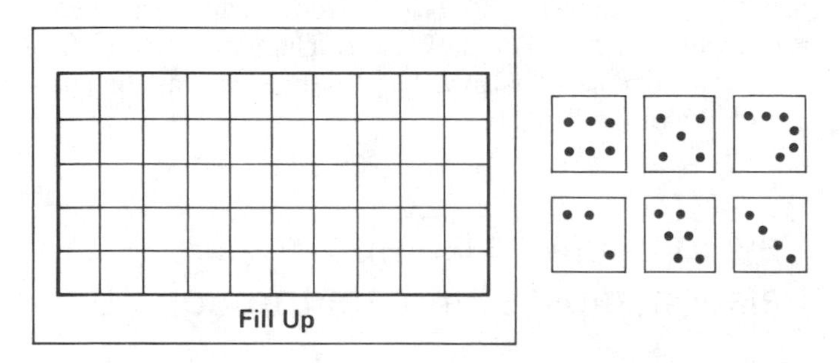

Figure 5.6
The game "Fill Up": Take turns. Draw a card. Put that many counters on your board.

Perhaps the most common textbook exercises are those which match sets to numerals. The children are given pictured sets and asked to write or match the number that tells how many in each set. Alternatively, they may be given a number and told to make or draw a set with that many objects. Set-to-numeral matchings practice only the skills of counting sets and numeral recognition or writing. They are connecting-level activities in that the cardinality concept is being connected to the numeral which represents the concept.

Any counting activity can be converted from a concept to a connecting-level activity by including numerals. For example, the "Fill Up" game can be played by drawing numeral cards or rolling a number cube with numerals instead of or along with dots.

Activity 5.15

A board game can be designed with a track that has a numeral on each space. The children roll dice or draw a dot card and move to the next space that shows that number.

Counting on rhythmically is quite different from counting on from a given set. Here are several counting on activities.

Activity 5.16

Give each child a collection of 10 or 12 small counters which they line up left to right on their desks. Tell them to count four counters and push them under their left hands (Figure 5.7). Then say, "Point to your hand. How many are there?" (4) "So let's count like this: F-o-u-r (pointing to their hand), 5, 6, ..." Repeat with other numbers under the hand. The same activity can be done on the overhead projector. Display the counters, count off some, and cover them. Ask how many are covered and then collectively count starting with the amount covered.

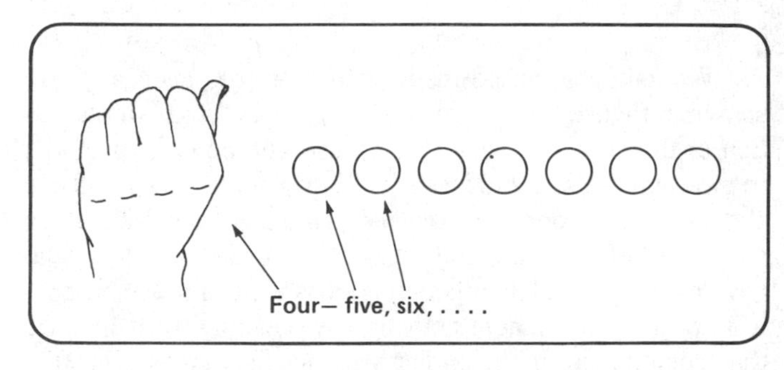

Figure 5.7
Counting on: Hide four. Count, starting with those hidden.

Activity 5.17

Line up some children in front of the room. Count off some of them from one end and have them sit. Then have the class count all of the children beginning with the seated group as a single count and count on from there.

In each of these counting-on exercises, the children first count the group. You stop them, ask how many are there, and then count using a count-on procedure. The children can understand that they are counting on to enumerate the set because they in fact counted the first group. A more abstract version has the first set previously counted.

Activity 5.18

A box has a numeral on it indicating the number of counters in the box. You then drop from one to three additional counters in the box, one at a time, so all can see. Ask how many are now in the box.

Activity 5.19

On the overhead, cover some counters with a card before the light is turned on and place additional counters to the side of the card, as in Figure 5.8.

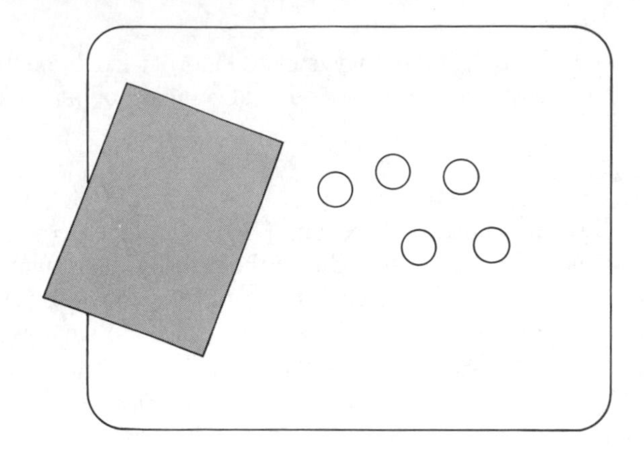

Figure 5.8
Four under the card. How many all together? Let's count.

Worksheets can be made with examples such as those shown in Figure 5.9. Notice the progression from an illustration of the above activities to the use of a numeral and dots to two numerals. Leutzinger (1979) recommends that there should be considerable emphasis on these activities using counters before using the worksheet mode. Children who have not integrated their counting skills with number concepts will do these worksheet activities by counting from 1 rather than counting on. In the earlier activities they understand that the first amount is already counted and need not be recounted.

Activity 5.20

At the most conceptual level have the children count out a number of objects, say eight, and cover them all. Next have them take out one counter and ask "How many covered?" Continue taking out one at a time and asking for the amount that remain covered.

At the next level, show a covered amount and some counters to the side. Tell the students how many you started with and have them count back to find out how many are still covered.

Finally, this can be done in a worksheet mode as in Figure 5.10. Here the numeral tells the total amount.

RELATIONSHIP SYMBOLISM

The symbols for *is less than* (<) and *is greater than* (>) traditionally cause children difficulty. Given two quantities or numbers side by side, children can be taught to insert the correct symbol between the two. To assist in this, liken the two symbols to a mouth (an alligator or a hungry duck) and explain that the mouth always turns to the larger quantity. Another idea is to point out that one side of the symbol is big (the open end) and the other side is small (the point). Reading sentences such as $5 < 8$ is considerably more difficult because the students are not first picking which is larger but reading a symbol, and they are not used to saying "is less than" but rather identifying the amount which is less or more. What is most important is to note that none of the corresponding concepts depends on this symbolism.

The equal sign (=) is more important, since it is used much more frequently in the early grades. However, it really does not need to be introduced until addition is introduced. At that time, and from then on, it is important to stress that the meaning of the equal sign is *is the same as*. What is on one side of the equal sign is a representation of the same quantity that is on the other. The notion of a two-pan balance is a useful analogy for children (Van de Walle & Thompson, 1981). The quantity on one side "balances" or "is the same" as the quantity on the other.

Development of Number Relationships: Numbers through 10

Set-to-numeral matchings and other counting activities are vitally important to number development. However, they only focus on the basic meaning of number and on accurate counting skills. Once children have acquired a concept of manyness or cardinality and can adequately use their counting skills in meaningful ways, little more is to be gained from the kinds of counting activities that have been described so far. More relationships must be created in order for children to have what is called "number sense" or a flexible concept of number not completely tied to counting.

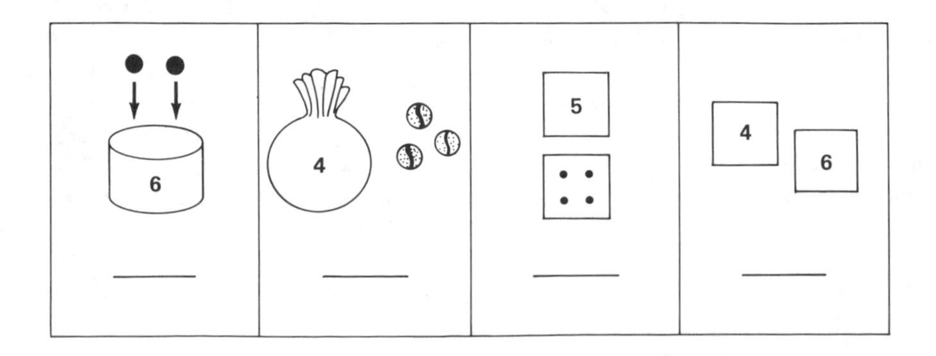

Figure 5.9
A progression of counting-on activities for worksheets.

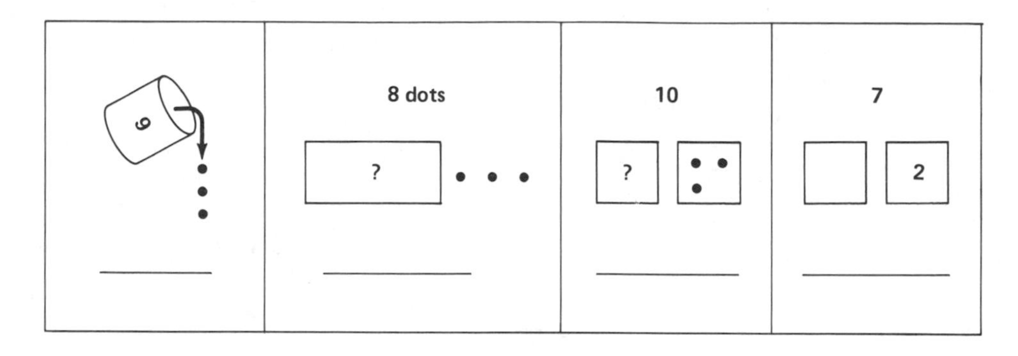

Figure 5.10
A progression of counting back activities for worksheets.

TYPES OF NUMBER RELATIONSHIPS

Patterned Set Relationships

When people see the dots on dice or on a domino, they do not have to count them to tell how many are there. Adult familiarity with these patterns is something that is known about numbers, a spatial arrangement that is connected to the concept of that number. For some numbers, several patterns may be recognized quickly. It is possible that different arrangements are connected to different ideas about number. For example, these three patterns each convey something different about 5.

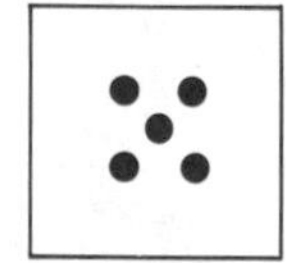
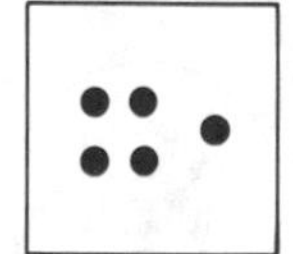

The first may just be a recognizable spatial pattern. The second may be related to the notion that 5 is 1 more than 4. The last helps with the idea that 5 is composed of 3 and 2. For the numbers 3 to 10, each pattern that children recognize without counting can add to their overall concept of number.

Relationships to Other Numbers

In counting activities, children are not focusing on how one number differs from another. A very important set of relationships for number concerns how each number is compared with or related to other numbers. Some of these relationships are especially important.

One and Two More, One and Two Less. While it is good to know global more and less relationships, it is quite useful to know that 7, for example, is *1 more than* 6 and *2 less than* 9.

The 5 and 10 Benchmarks. How each of the numbers between 1 and 10 is related to the numbers 5 and 10 is an especially useful relationship. The importance of 10 in our place value system is clear. The use of both 5 and 10 as "anchors" for the other numbers also makes mental computation and basic addition and subtraction facts much easier.

Part-Part-Whole Relations

Due to the "straight-line" sequential nature of counting, the concept of number constructed from counting does not promote thinking about sets or number in terms of component parts. Children who have counted seven objects have not been helped to think about that set as a set of 5 with a set of 2 or a set of 3 with a set of 4. These ideas are known as part-part-whole relationships. A noted researcher in children's number concepts, Lauren Resnick, states:

> Probably the major conceptual achievement of the early school years is the interpretation of numbers in terms of part and whole relationships. With the application of a Part-Whole schema to quantity, it becomes possible for children to think about numbers as compositions of other numbers. This enrichment of number understanding permits forms of mathematical problem solving and interpretation that are not available to younger children. (1983, p. 114)

ACTIVITIES FOR PATTERNED SETS

Make a set of dot plates as shown in Figure 5.11. Luncheon-size paper plates work well. Dots can be made with a marker, or use the stick-on dots that can be found in stationery stores. Note that some patterns are combinations of two smaller patterns or a pattern with one or two additional dots. These should be made in two colors as indicated. Keep the patterns compact. If the dots are spread out, the patterns are hard to see.

Activity 5.21

To introduce the patterns, provide each student with about 10 counters and a piece of construction paper as a "mat." Hold up a dot plate for about 3 seconds. "How many dots did you see? How did you see it? Make the pattern you saw using the counters on the mat." Spend some time discussing the configuration of the pattern and how many dots. Do this with a few new patterns each day.

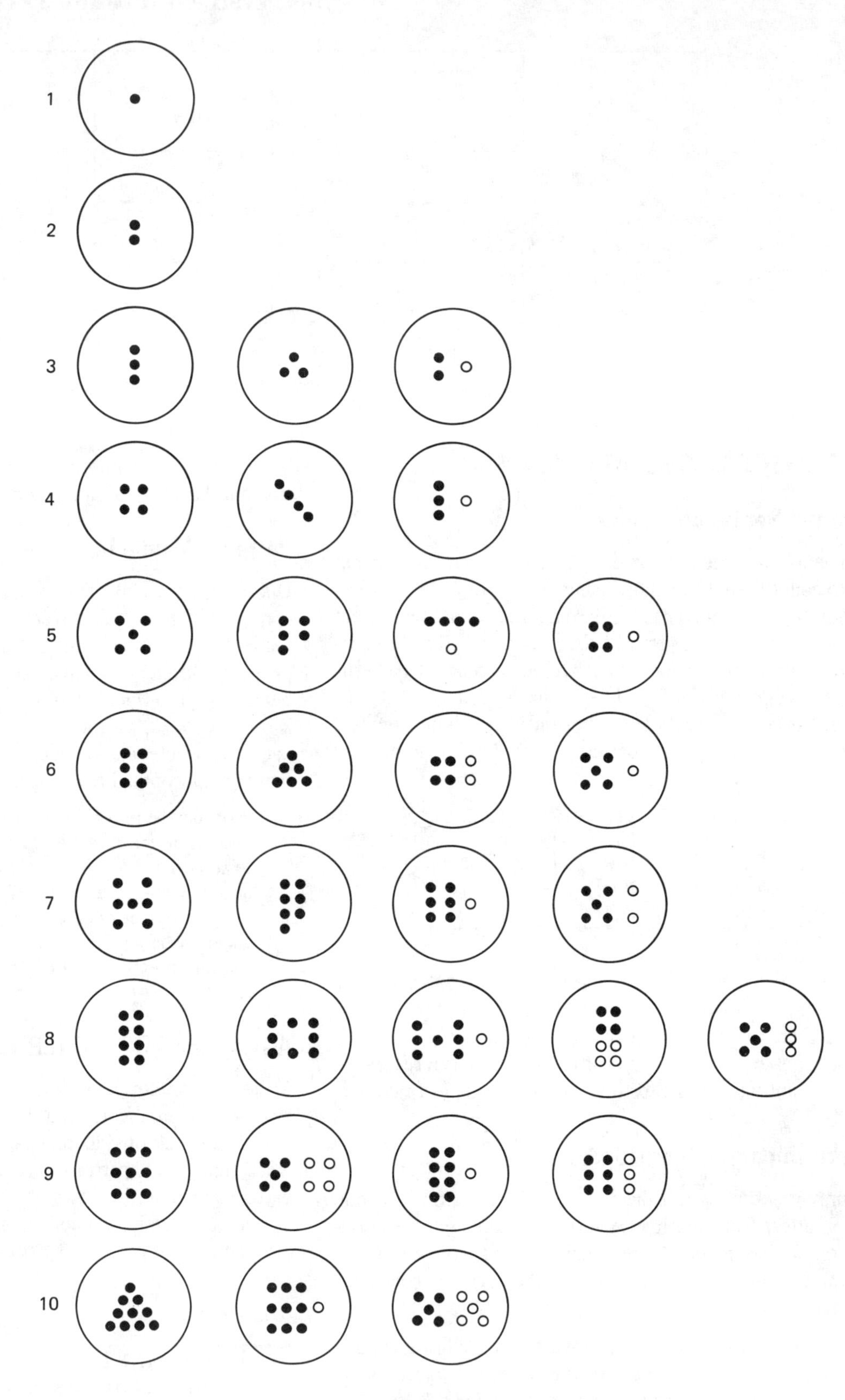

Figure 5.11
A useful collection of dot patterns for "dot plates."

Activity 5.22

Hold up a dot plate for only 1 to 3 seconds. "How many? How did you see it?" Children like to see how quickly they can recognize and say how many dots. Include lots of easy patterns and a few of those with more dots as you build their confidence. Students can also flash the dot plates to each other as a workstation activity.

Activity 5.23

Make a set of dominoes out of poster board and put a dot pattern on each end. The dominoes can be about 5 centimeters by 10 centimeters. The same patterns can appear on lots of dominoes with different pairs of patterns making up each one. Let the children play dominoes in the regular way, matching up the ends. As a speed activity, spread out all of the dominoes and see how fast they can play all of the dominoes or play until no more can be played. Regular dominoes could also be used, but there are not as many patterns.

Activity 5.24

Make dot patterns on acetate and use the overhead projector instead of flashing plates.

The dot plates and patterned sets in general can easily be used in many other activities, as we will see. However, the instant recognition activities with the plates are exciting and can be done in 5 minutes at any time of day or between lessons. There is value in using them at any primary grade level and at any time of year.

ACTIVITIES FOR ONE AND TWO MORE (AND LESS)

The first activities in this section involve no numerals. Children construct or find a set that is one (two) more than or less than a given set.

Activity 5.25

Use the dot pattern dominoes or a standard set and play "one-less-than" dominoes. Play in the usual way, but instead of matching ends a new domino can be added if it has an end that is 1 less than the end on the board. A similar game can be played for 2 less or 1 (2) more.

(Subsequent activities in this section can be done with any of the four relationships, but each will be described for only one.)

Activity 5.26

Using the dot plates, flash the plates and have the students say 1 more than the amount on the plate.

Activity 5.27

With the dot plates or any set of dot cards, have students construct a set of counters that is 2 more than the set. Similarly, spread out 8 to 10 dot cards and find another card for each that is two less than the card shown. (Omit the one and two card for two less than, etc.)

Activity 5.28

Using a worksheet like that in Figure 5.4, students can make a set that is 1 more than the given set. To make a record, after making a set with counters, they line them up and make a dot for each counter. Notice that numerals can easily be added to the worksheet.

In activities where children find a set or make a set, they can add a numeral card (a small card with a number written on it) to all of the sets involved. They can also be encouraged to take turns reading a *number sentence* to their partner. If, for example, a set has been made that is 2 more than a set of 4, the child can "read" this by saying the number sentence "Two more than four is six."

The following activities involve numerals and sets together or just numerals. These are especially important in late first grade through the third grade.

Activity 5.29

Provide each child with six to eight number cards about the size of index cards. (Children can cut up paper and make their own.) On the cards put the numbers you will need for the day (e.g., 5 through 10). Now flash a dot plate and have students hold up the card that is 1 more than the plate. Nothing is said out loud, and all students respond. Similarly, you can hold up a numeral card or say a number orally, and they respond with their numeral cards.

Activity 5.30

Duplicate a sheet of response cards with dot sets. Eight "cards" on one sheet is about right. Let the students cut these out and use them in the same way as in activity 5.29. This time, flash or say a number and the students respond with the appropriate dot set.

Activity 5.31

Worksheets can be made with a numeral given, and the students can make a set that is 2 less. Alternatively, the worksheet can show the set, and the students can write the number that is 2 less.

Activity 5.32

Add more dominoes to your set. The new ones can have numerals on either or both ends instead of dot patterns. Play the same games as before.

Many activities that have simple student responses to simple input, such as those that have been described, can be made into machine games. Draw a funny-looking "machine" on the board. It requires an input hopper and an output chute, as shown in Figure 5.12. Tell the students what the machine does. For example, "This is a magic ONE-MORE-THAN machine. It takes in a number up here and spits out a number that is one more."

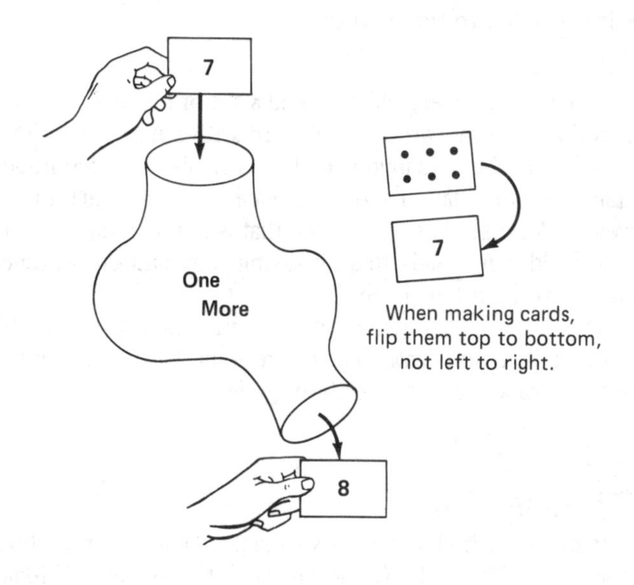

Figure 5.12
A one-more-than machine drawn on the chalkboard. Machines like this can be used for a variety of teacher-directed drills.

One way to "operate" a machine is to prepare some cards with the input on one side and the output on the reverse. Hold the card over the input hopper, and as you slide it down the board toward the output the children call out what will come out. Just as the card gets to the chute, flip it over to confirm. The input side of the cards can be dot patterns or numerals. Machine activities can also be done without cards by just announcing the input numbers and thereby making a quick and easy drill.

Activity 5.33

The calculator is every child's personal input-output machine. For example, press [+] 2 [=] to store plus 2. Now press any number and predict the two-more-than result. For example, press 5. Two more is 7. Press [=] to confirm. Since the +2 is stored, the process can be repeated over and over.

ACTIVITIES FOR FIVE AND TEN BENCHMARKS

The Ten-Frame

A ten-frame, originally developed by Robert Wirtz (1974), is a 2×5 array in which counters or dots are placed to represent numbers. The numbers 1 to 5 are shown on the top line, left to right. For numbers 6 to 10, additional counters are placed left to right in the second row (Figure 5.13). The ten-frame can be an egg carton with two end sections cut off or covered, or it can simply be a drawing on a sheet of construction paper. A classroom set of ten-frames should be easy to have. Activities can be conducted with the full class or can become workstation activities for small groups of students.

Before doing any activities, show children how to "show numbers" on the ten-frame. It is only by adhering to the notion of filling up the top row first and then adding more to the second row that the relationships to 5 and 10 become apparent.

Activity 5.34

"Crazy Mixed-Up Numbers" is a great introductory game adapted from *Mathematics their way* (Baratta-Lorton, 1976). All children make their ten-frame show the same number. The teacher then calls out numbers between 0 and 10. After each number, the children examine their ten-frames and decide how many more counters need to be added ("plus") or removed ("minus"). They then call out plus (or minus) whatever amount is appropriate. If, for example, the frames showed six and the teacher called out "four," the children would respond "minus two!" and then change their ten-frames accordingly. The activity continues with a random list of numbers. Children can play this game independently by preparing strips of about 15 "crazy mixed-up numbers." One child plays "teacher," and the rest use the ten-frames. Children like to make up their own number strips.

Activity 5.35

Give students a number in some form other than a ten-frame and have them show that number on their ten-frames. Show them a dot pattern from the dot plates, hold up fingers, show some counters on the overhead, hold up a numeral card, or simply say a number out loud. These activities could also be done independently at a workstation by providing the number cards or sets along with the ten-frames and counters. Worksheets could be prepared for independent seatwork, as shown in Figure 5.14. The children would use an actual ten-frame and then record the result on the paper.

Ten-frame cards provide another useful variation. Make cards from poster board about the size of a small index card and draw a ten-frame on each. Dots are drawn in the frames. A set of 20 cards consists of a 0 card, a 10 card, and two each of the numbers 1 to 9.

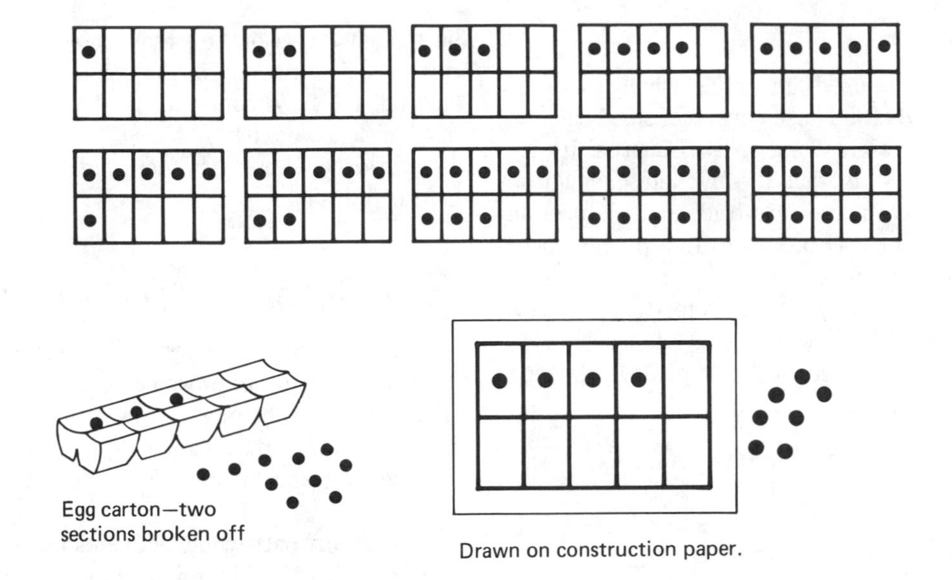

Figure 5.13
Ten-frames.

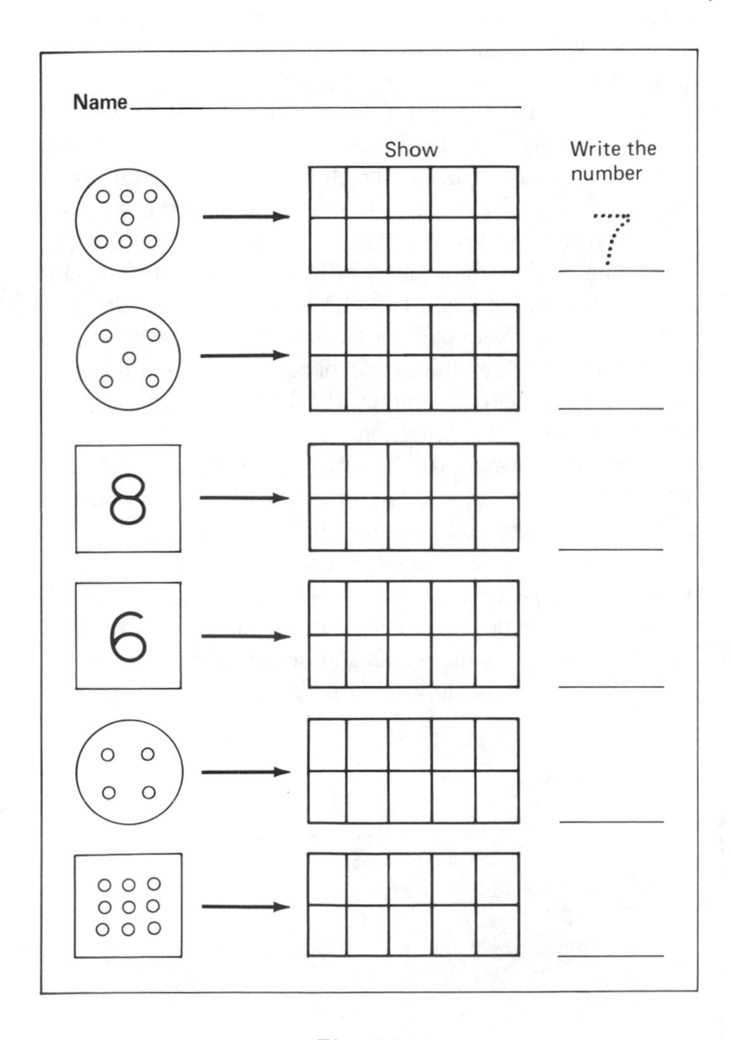

Figure 5.14
A ten-frame record sheet. Children can use a large ten-frame with counters and then draw dots in the ten-frames.

Activity 5.36

Flash ten-frame cards to the class or group and see how fast they can tell how many dots are shown. This activity is fast-paced, takes only a few minutes, can be done at any time, and is a lot of fun if you encourage speed.

Important variations of the previous activity include

Saying the number of spaces on the card instead of the number of dots.

Saying one more than the number of dots (or two more, and also less than).

Saying the "ten fact." For example: "Six and four make ten."

After students have become familiar with the ten-frame, simply having a large blank ten-frame drawn on the board can profitably influence children's thinking about five and ten as they do number activities. Try looking at a ten-frame while doing the following two activities.

Activity 5.37

In "Five and" the teacher calls out numbers between 5 and 10. The children respond "Five and ______" using the appropriate number. For example, if you say, "Eight!" the children respond "Five and three."

Activity 5.38

In "Make Ten" the children respond to a number called by calling how many more are needed to make 10. This is most effective with numbers between 5 and 10.

Ten-Frame Alternatives

In Japan, number "tiles" are used almost exclusively to represent numbers (Hatano, 1982). These consist of small squares and unmarked strips that are as long as five squares. It may be, as some have suggested, that the five tiles should be marked to show the five squares. Both ideas are shown in Figure 5.15(a). Tiles can easily be made from poster board using squares of about one inch.

Another variation is also borrowed from Robert Wirtz. He glued beans onto popsicle sticks or tongue depressors, as shown in Figure 5.15(b). These could also be made by drawing dots on strips of poster board.

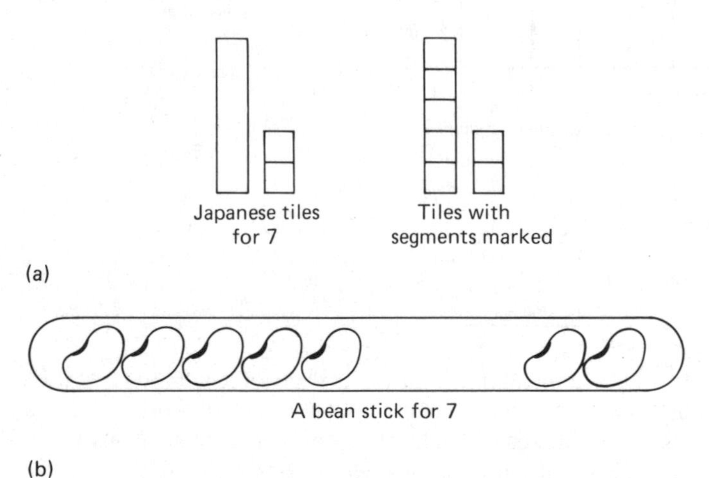

Figure 5.15
Alternatives to ten-frames.

The basic idea with ten-frames, tiles, and bean sticks is to encourage children to think about numbers in relation to five and ten. Most of the activities described for ten-frames can be modified to be done with tiles or bean sticks, and certainly other activities can be devised in the same spirit. The textbooks developed by Wirtz for Curriculum Development Associates include many creative, thought-provoking activities for both ten-frames and Wirtz's bean sticks.

Once again the calculator provides a symbolic approach. If [+] 5 is stored, then a press of any key from 0 to 5 can be thought of as putting that many counters in the second row. As a variation, if the minus sign on the display is ignored, the child can store [−] 10, and then a press of any number followed by [=] shows how much more to get to 10. Children should try to predict the result each time before pressing [=].

ACTIVITIES FOR PART-PART-WHOLE

In part-part-whole activities, children either construct a specified quantity in two or more parts, or they separate a specified amount into two or more parts. After much experience with these activities, they can orally provide a missing part. (I have seven counters in all. Two are showing. How many are covered?)

Most part-part-whole activities focus on a single number at a time. Thus a child might spend 15 to 20 minutes on the number 5. Generally, kindergarten children can begin these activities working on the numbers 4 or 5. There is little value or need to do part-part-whole activities for 1, 2, or 3. Children can begin with numbers from 4 to 6 and work with numbers up to 10 or 12 as their concepts develop.

Another important feature of part-part-whole activities is to get children to say the combinations out loud. Not only does this reinforce the combination, but it also serves to focus attention. Part-part-whole activities can be encoded as addition equations and missing part activities encoded as subtraction. This is illustrated in Figure 5.16. Addition and subtraction equations can likewise be decoded or explained by children in terms of part-part-whole models. Thus there is a real linkage between this aspect of number concept development, the meaning of addition and subtraction, and the mastery of addition and subtraction facts.

In most part-part-whole activities, the children make eight or more sets, each illustrating the prescribed number in two or more parts. When complete, all of the sets are still available. The teacher can ask individual students to read a number sentence to go with each set. They may also write an equation (encode) for each set if the child is ready. With these basic ideas, as many different formats and materials as can be thought of are appropriate.

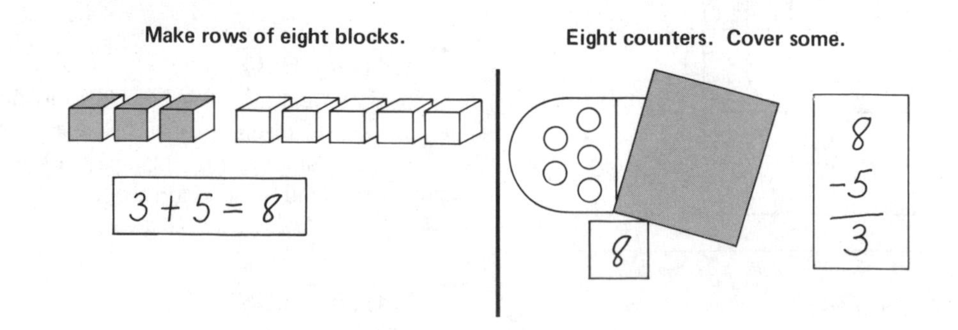

Figure 5.16
Encoding part-part-whole activities.

Figure 5.17
Two-color materials for building parts of six.

Activity 5.39

The ideas here are illustrated in Figure 5.17.

Make sets with "two-color counters" such as lima beans spray-painted on one side or picture mat board cut into small squares.

Make bars of Unifix cubes or other plastic cubes that interlock. Make each bar with two colors.

Color rows of squares on 1-inch grid paper.

Make combinations with "dot strips," which are simply strips of poster board with dots on them. Make lots of strips with from 1 to 4 dots and some strips with 5 to 10 dots. The strips are only as long as the row of dots that is on them. (Punch holes in the dots to make a set for the overhead projector.)

Make combinations with "two-column" strips. These are cut from tagboard ruled in 1-inch squares. Except for the single square, all pieces are cut from two columns of squares. Odd numbers will have an "odd" square on one end. (Punch holes in center of the squares to make a set for the overhead projector.)

Activity 5.40

Have students make designs with different materials. Each design has the same prescribed number of objects. When talking with the children about their designs, ask how they can see their design in two parts or in three parts.

Make arrangements of wooden cubes.

Make designs with pattern blocks. It is a good idea to use only one or two shapes at a time.

Make designs with flat toothpicks. These can be dipped in white glue for a permanent record on small squares of construction paper.

Make designs with touching squares or touching triangles. Cut a large supply of small squares or triangles out of construction paper. These can also be pasted down.

It is both fun and useful to challenge children to see their designs in different ways producing different number combinations. In Figure 5.18, decide how children might be looking at the designs to get the combinations listed under each.

Missing-part activities require two people, or they can be done in a teacher-directed manner. Again, children work with only one number at a time. One person hides one of the two parts, and the other person tries to tell what is hidden. If there is hesitation or the student does not know the hidden part, it is immediately shown. The focus is on learning and thinking, not on testing and anxiety.

Activity 5.41

Place all counters under a margarine tub or cardboard. Pull some out. "How many are covered?"

Activity 5.42

For each number 4 to 10, make missing-part cards that show all possible parts. For the number 8 there would be nine cards showing from zero to eight dots. Students try to say (or write) a number sentence that includes the part hidden under the flap (Figure 5.19, p. 61).

Activity 5.43

Play "I wish I had." For example, hold out a bar of Unifix cubes, a dot strip, a two-column strip, or a dot plate showing seven or less. Say "I wish I had seven." The children respond

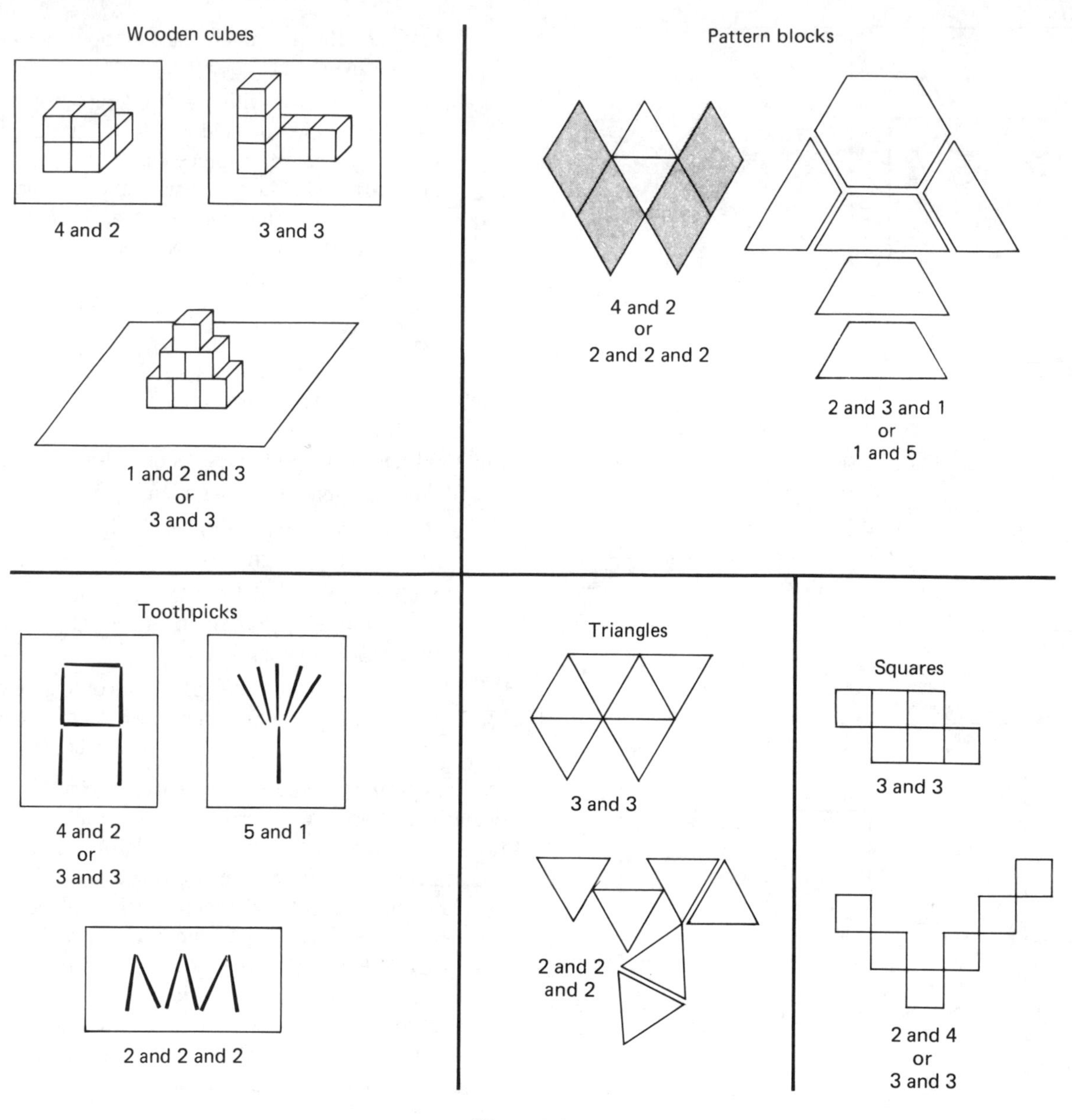

Figure 5.18
Designs for six.

with the part that is needed to make seven. Counting on can be used to check. The game can either focus on a single whole, or the "I wish I had" number can change each time. At the symbolic level, the game can be played orally or by holding up numeral cards.

The following are also missing-part activities but are at the symbolic level.

Activity 5.44

Draw a machine on the board similar to the one in Figure 5.12. This time the machine is a "parts-of-8" machine (or any number). A number "goes in" and the students say the other part. If 3 goes in a parts-of-8 machine, a 5 comes out.

Activity 5.45

On the calculator store the whole number by pressing [−] 8 [=], for example. (Ignore the minus sign.) Now if any number from 0 to 8 is pressed followed by [=], the display shows the other part. Children should try to say the other part before they press [=].

MORE ACTIVITIES FOR NUMBER RELATIONSHIPS

Many good number development activities involve more than one of the relationships discussed so far.

A special set of small dot cards, about 5 cm square can be easily made using the black-line masters in Appendix C (Figure 5.20). These cards include single-dot patterns, combinations of

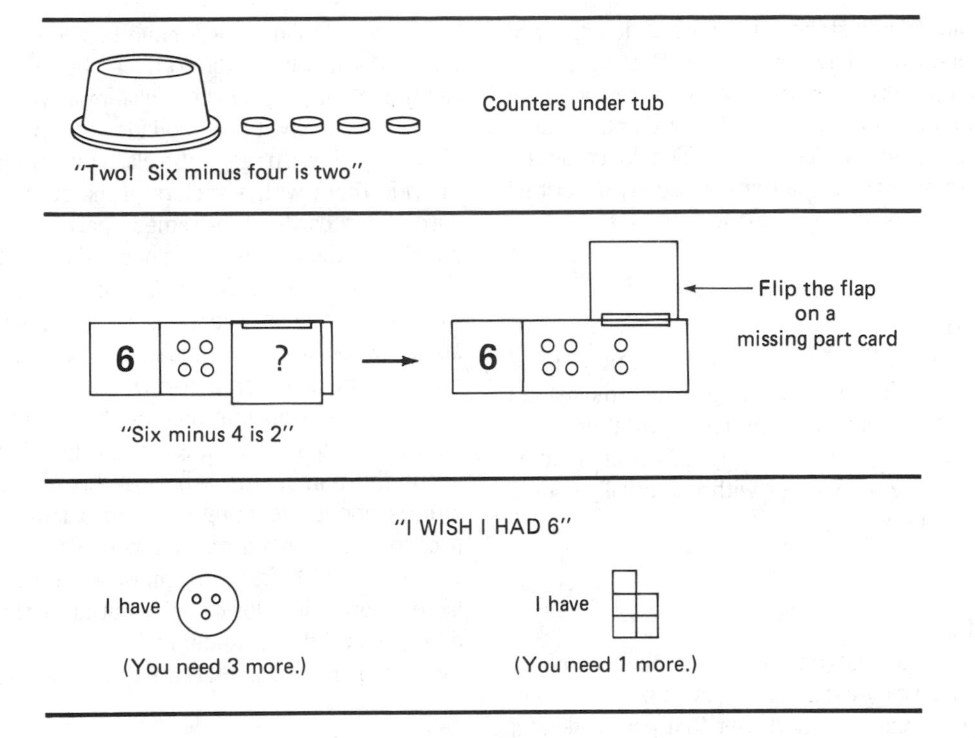

Figure 5.19
Missing-part activities.

Figure 5.20
Dot cards.

dot patterns, ten-frames with standard as well as nonstandard configurations of dots. Some have one or two dots outside of the ten-frame. When children use these cards for almost any activity that involves number concepts, the cards make them think about numbers in many different ways. The dot cards add another dimension to many of the activities already described and can be used effectively in the following activities.

Activity 5.46

The game of "Double War" (Kamii, 1985) is played like war, but on each play both players turn up two cards instead of one. The winner is the one with the larger total number. Children playing the game can use many different number relationships to determine the winner without actually finding the total number of dots.

Activity 5.47

Spread the dot cards out on the table. Select one at random and see how quickly a card that has as many (or one more, or one less, etc.) can be found. Set that pair aside and repeat until no more cards can be paired.

Activity 5.48

Make a long row of cards from one up to nine and then start back again to one and then up, and so on. This is especially effective with the dot cards described in Figure 5.20.

Activity 5.49

Play "Difference War." Besides dealing out the cards to the two players as in regular war, prepare a pile of about 50 counters. On each play the players turn over their cards as usual. The player with the greater number of dots wins as many counters from the pile as the difference between the two cards. The players keep their cards. The game is over when the counter pile runs out. The player with the most counters wins the game.

Activity 5.50

Select a number between 5 and 10 and find combinations of two or three cards that total that number. When students have found at least 10 combinations, one student can then turn face down one card in each group. The next challenge is to name the card that was turned down.

Chapter 18 discusses how to make bar graphs with children in grades K to 2. Once a simple bar graph is made, take a few minutes to ask as many number relationship questions as is appropriate for the graph. Refer to Figure 5.21 for an example.

At first children will have considerable trouble with the questions involving differences, but repeated exposure to them in a bar graph format will improve their understanding.

Occasionally it is a good idea to have a "What can we say about six" day. Arrange the class into cooperative groups and provide them with a variety of materials: counters, numeral cards, dot cards, ten-frames, part-whole mats, and other suitable materials. It is not necessary that every group have the same materials. The group task is to come up with at least five different ideas, questions, pictures, ways to show or think about, word stories, comparisons, or whatever that have to do with the number 6 (or whatever number has been selected). After sufficient time has been given, solicit ideas from various groups. Encourage drawings that can be displayed about the number on a bulletin board. Help students write stories about the number as in a language-experience approach. ("Six birds are on a fence. Two are bluebirds and four are redbirds.") Early attempts will be simple, and you will have to provide a lot of help, but after this activity has been done several days, students will begin to invent new ideas of their own. Encourage creativity and effort.

Relationships for Numbers 10 to 20

Even though young children daily experience numbers up to 20 and beyond, it should not be assumed that they will automatically extend the rich set of relationships they have developed on smaller numbers to the numbers beyond 10. And yet these numbers play a big part in many simple counting activities, in basic facts, and in much of what we do with mental arithmetic. Relationships on these numbers are just as important as relationships on the numbers through 10.

EXTENSION OF COUNTING ACTIVITIES

Counting activities are still important vehicles for early conceptualization of numbers. All of the counting activities discussed earlier in this chapter should be extended to counts of at least 20. Special consideration should be given to the oral pattern in the teen numbers to help children relate this extended sequence to what they already know.

Counting-on and counting-back activities are especially useful within the teens. Counting in both directions with the calculator is also helpful in establishing patterns for the teens.

A PRE-PLACE-VALUE RELATIONSHIP WITH TEN

A seemingly logical approach to the teen numbers is to view them in terms of tens and ones. However, a place-value approach may not be developmentally appropriate. A place-value understanding of the number 16 would imply that the digit 1 stands for 1 ten and the 6 for 6 ones. To count 10 objects as a single entity (1 ten) and to represent 10 with the digit 1 in the second place are two ideas that lie at the very heart of place-value numeration and base ten concepts. These

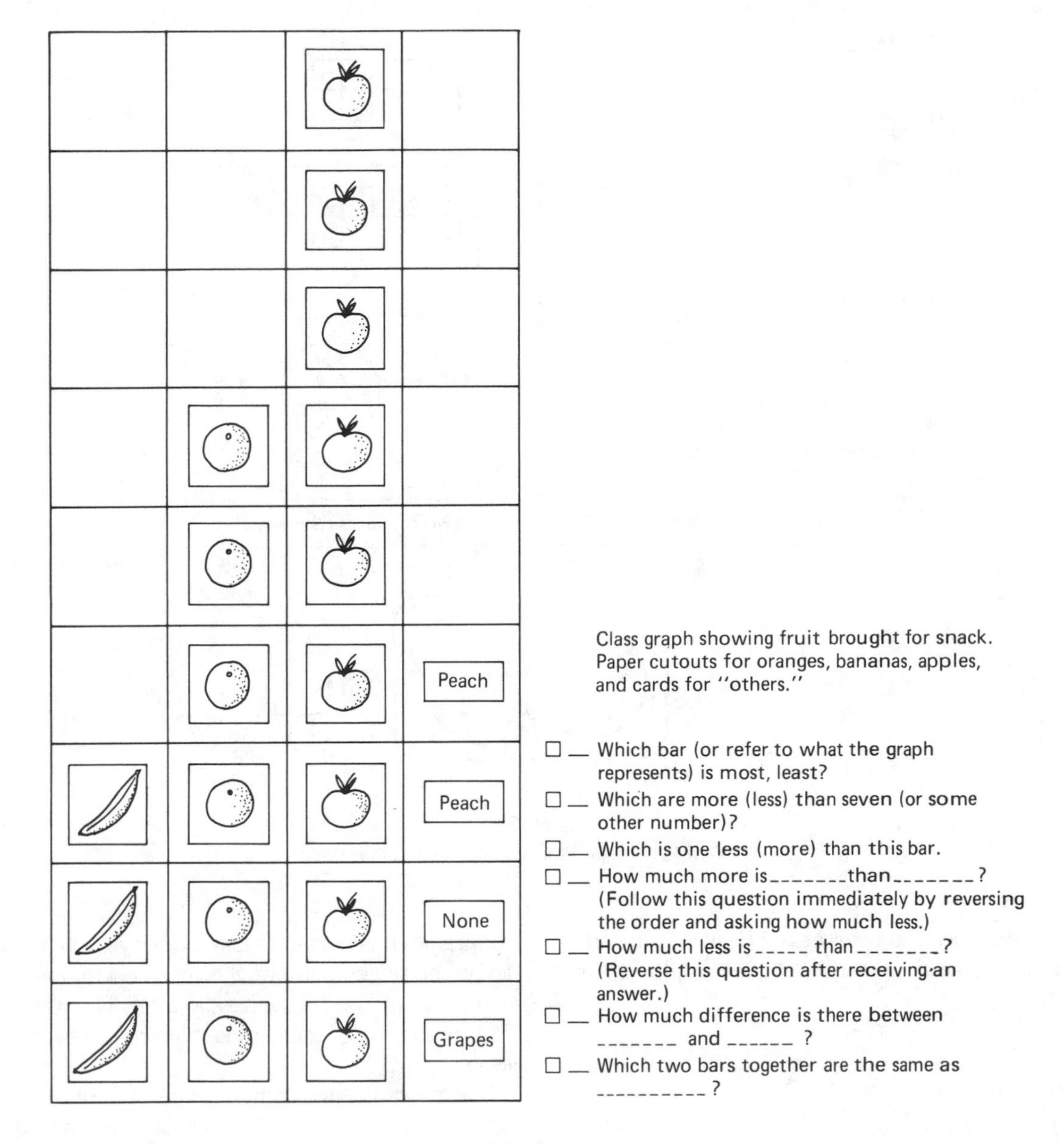

Figure 5.21
Comparisons in a bar graph.

base ten concepts are very difficult for young children to construct.

While a place-value approach may not be appropriate, the number 10 can and should play a significant role in the conceptualization of the teens. Children should learn that one very special part-part-whole relationship for the numbers greater than 10 makes them very easy to think about, namely, those combinations where one of the parts is 10. Both orally and with physical materials, children can learn that 16 is 10 and 6 and that 2 and 10 is 12. In this form, 10 is treated in exactly the same way as the non-10 part and is not counted as 1 ten, as a single entity. Figure 5.22 illustrates the contrast between the place-value construct for 16 and a pre-place-value or part-part-whole construct. The latter is simply an extension of the way children have been constructing numbers. The place-value approach requires sophisticated new ideas. It is probably best to initially develop base ten place-value concepts in the context of numbers from 20 to 100 where the oral names all follow the same pattern and where there is more than 1 ten for the children to discuss.

The following activities involve this very special part-part-whole understanding for the teens.

Activity 5.51

Use a simple two-part mat and have children count out 10 counters onto one side. Next have them put five counters on the other side. Together count all of the counters by ones. Chorus the combination: "Ten and five is fifteen." Turn the

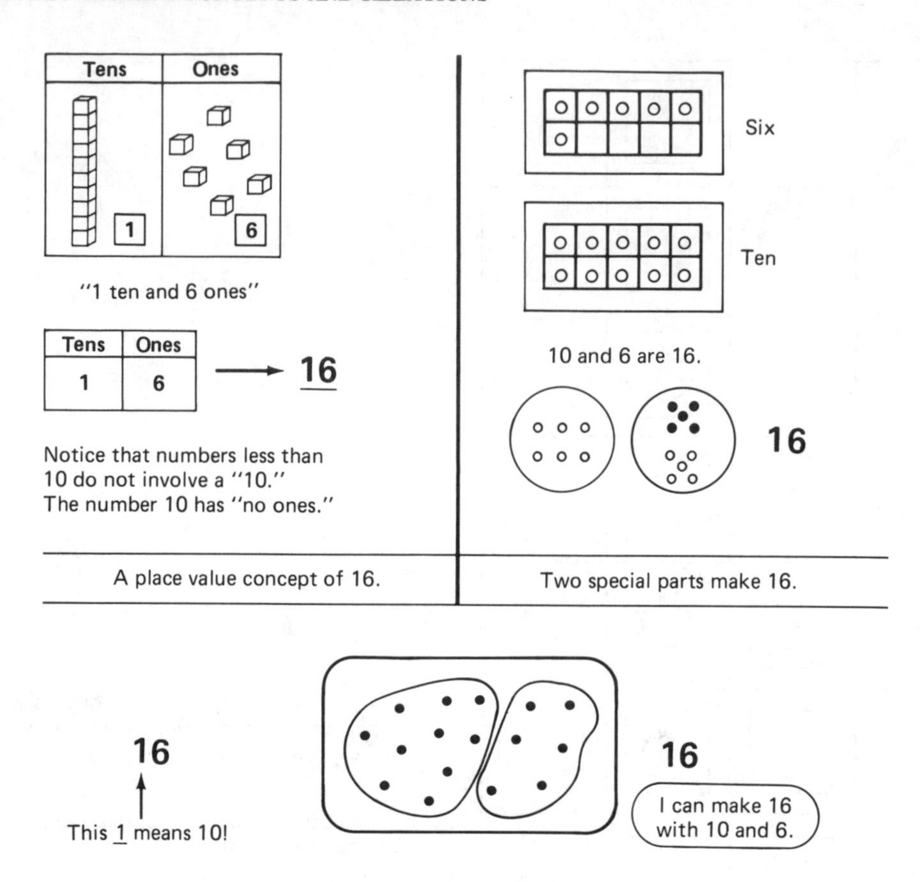

Figure 5.22
Understanding 16 with and without place-value concepts.

mat around: "Five and ten is fifteen." Repeat with other numbers in a random order but without changing the 10 side of the mat.

Activity 5.52

Place 10 counters on the overhead and have children count together. In a separate pile place various amounts from one to nine and count those. Ask, "How many all together?" Or, with 10 counters on the overhead ask, "How can I make seventeen?"

Activity 5.53

Extend the "Crazy Mixed-Up Numbers" activity (No. 34) to include numbers to 20. Provide each child with two ten-frames drawn on a construction paper mat. One ten-frame is placed under the other. For numbers greater than 10, the new rule is that one ten-frame must be completely filled. Periodically say a number sentence for a number represented. For numbers less than 10, one part is always 5: "Five and three make eight." For numbers greater than 10, one part is always 10: "Ten and seven make seventeen."

Frequently represent numbers between 10 and 20 using one ten-frame filled and additional counters or dots not in a ten-frame. It is not necessary to always have the frame to the left of the single counters. The oral description is simply "Ten and ______" or "______ and ten." The place-value notion of 1 ten is avoided until place-value concepts are firmly established.

Besides the ten-frame, some patterned sets are useful for numbers greater than 10. Those that can be quickly recognized show either two or three groups of five and additional dots. Once discussed, these can be added to instant-recognition activities as done before.

EXTENDING MORE AND LESS RELATIONSHIPS

The relationships of one more than, two more than, one less than, and two less than are important for all numbers. However, these ideas are built on or connected to the same concepts for numbers less than 10. The fact that 17 is 1 less than 18 is connected to the idea that 7 is 1 less than 8. Children may need help in making this connection.

Activity 5.54

On the overhead show seven counters and ask what is two more, or one less, etc. Now add a filled ten-frame to the display (or 10 in any pattern) and repeat the questions. Pair up questions by covering and uncovering the ten-frame as illustrated in Figure 5.23.

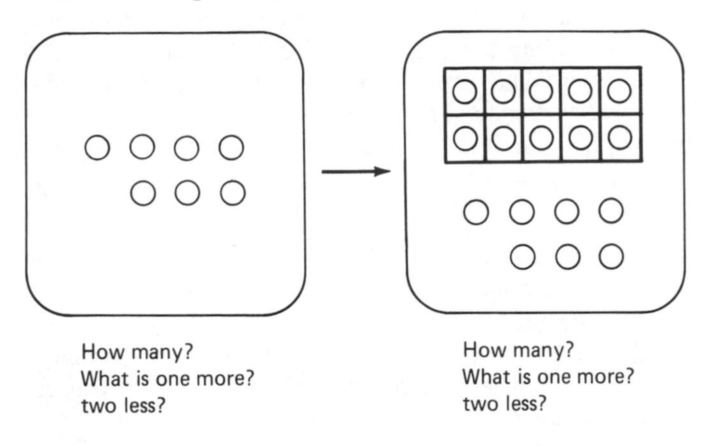

Figure 5.23
Extending relationships to the teens.

DOUBLE AND NEAR-DOUBLE RELATIONSHIPS

The use of doubles (double 6 is 12) and near-doubles (13 is double 6 and 1 more) is generally considered a strategy for memorizing basic addition facts. There is no reason why children should not begin to develop these relationships long before they are concerned with memorizing basic facts. Doubles and near-doubles are simply special cases of the general part-part-whole construct.

Activity 5.55

Relate the doubles to special images. Thornton (1982) helped first graders connect doubles to these visual ideas:

Double 3 is the bug double: three legs on each side.

Double 4 is the spider double: four legs on each side.

Double 5 is the hand double: two hands.

Double 6 is the egg carton double: two rows of six eggs.

Double 7 is the two-week double: two weeks on the calendar.

Double 8 is the crayon double: two rows of eight crayons in a box.

Double 9 is the 18-wheeler double: two sides, nine wheels on each side.

Children should draw pictures or make posters for each double.

Activity 5.56

Make large dot dominoes for the doubles. Use the easiest dot pattern for the numbers 2 through 10 and make it the same on both ends of the domino. Use these in instant-recognition activities. Stress the use of the word *double*, as in "Double seven is fourteen."

Activity 5.57

Use the double dominoes, but ask for one more than the double.

Activity 5.58

Play "What Is the Double Number?" Give a double number orally and ask students to tell what double it is. "What is fourteen?" (Double 7!) When students can do this well, use any number up to 20. "What is seventeen?" (Double 8 and 1 more.)

Activity 5.59

On the calculator store the "double maker" (2 ☒ ⊟). Now a press of any digit followed by ⊟ will produce the double. Children can work in pairs or individually to try to beat the calculator.

Number Relationships and Addition and Subtraction

In the past, most curriculum materials moved children directly from counting and numeral writing activities to addition and subtraction and mastery of basic facts. Children were placed in the position of needing answers to questions such as 3 + 5 = ☐ without any skills or concepts except those of counting. There is a growing awareness that children need more time and experiences to develop the broad range of relationships for number that go beyond those developed via counting. Many of these ideas are closely related to addition and subtraction concepts and also to addition and subtraction fact mastery. The thrust of this chapter is that children should be given significant amounts of time in kindergarten through second grade to develop number relationships. In fact, many children in the third through sixth grades can benefit from these same number relationship activities.

The transition from teaching number relationships to teaching addition and subtraction need not be abrupt. Part-part-whole activities can be encoded or written down as addition equations. Missing-part activities can be encoded as subtraction equations. The ten-frame and dot card or dot plate formats can easily be incorporated directly into early addition and subtraction activities.

Number relationships play an enormous role in helping children master basic facts. Most classes spend days and weeks doing pages of facts with sums up to 10. If instead, the time were spent on number relationships, then writing the answers to these exercises would simply be a matter of writing what they already know. When the relationships have not been developed, the children are left with only counting to get answers. The resulting emphasis on counting is simply not necessary.

Figure 5.24 shows a basic addition chart. Of the 100 addition facts, 55 involve sums of 10 or less. These can potentially be an immediate consequence of number relationships. Twelve more facts are doubles or nearly doubles. The remaining facts are directly related to the ten and some more concept of the teens. For example, to think about 5 + 9, a child can think "Nine and one more (from the five) makes ten and four more is fourteen." A child without the ten benchmark and the special ten and four concept of fourteen must either count on or memorize. In fact, so many children do not possess these relationships that it has become popular to teach them as "strategies" for learning the basic facts. (See Chapter 7.)

	0	1	2	3	4	5	6	7	8	9
0	0	1	2	3	4	5	6	7	8	9
1	1	2	3	4	5	6	7	8	9	10
2	2	3	4	5	6	7	8	9	10	11
3	3	4	5	6	7	8	9	10	11	12
4	4	5	6	7	8	9	10	11	12	13
5	5	6	7	8	9	10	11	12	13	14
6	6	7	8	9	10	11	12	13	14	15
7	7	8	9	10	11	12	13	14	15	16
8	8	9	10	11	12	13	14	15	16	17
9	9	10	11	12	13	14	15	16	17	18

55 combinations of numbers 10 and less

12 more doubles and nearly doubles

24 special relationships involving 10

Figure 5.24
Addition fact chart.

Subtraction is also related to how children conceptualize number. In particular, the relationship between addition and subtraction is very much the same as the relationship between part-part-whole and missing-part concepts. More will be said about this in Chapters 6 and 7.

For Discussion and Exploration

1. Examine a textbook series for grades K–2. Compare the treatment of counting and number concept development with that presented in this chapter. What ideas are stressed? What ideas are missed altogether? If you were teaching in one of these grades, how would you plan your number concept development program? What part would the text play?
2. Explore number concepts of some children in the first or second grade. Consider checking their understanding of counting on, counting back, and their use of the relationships of one (two) more and less. Try using dot plates and ten-frame cards to see how children relate to these representations of numbers.
3. Discuss how much time out of the year can and should be spent on the development of number relationships in K, 1, or 2.
4. Many teachers in grades above the second find that their children do not possess the number relationships discussed in this chapter but rely heavily on counting. Given the pressures of other content at these grades, how much effort should be made to remediate these number concept deficiencies?

Suggested Readings

Baratta-Lorton, M. (1976). *Mathematics their way.* Menlo Park, CA: Addison-Wesley.

Baratta-Lorton, M. (1979). *Workjobs II.* Menlo Park, CA: Addison-Wesley.

Clements, D. H., & Callahan, L. G. (1983). Number or pre-number foundational experiences for young children: Must we choose? *Arithmetic Teacher, 31*(3), 34–37.

Coombs, B., & Harcourt, L. (1986). *Explorations 1.* Don Mills, Ontario: Addison-Wesley.

Fuson, K. C. (1989). *Children's counting and concepts of number.* New York: Springer-Verlag.

Kelly, B., Wortzman, R., Cornwall, J., Kennedy, N., Maher, A., & Nimigon, B. *Mathquest one: Teacher's edition.* Don Mills, Ontario: Addison-Wesley.

Kroll, D. L., & Yabe, T. (1987). A Japanese educator's perspective on teaching mathematics in the elementary school. *Arithmetic Teacher, 35*(2), 36–43.

Leutzinger, L. P., & Bertheau, M. (1989). Making sense of numbers. In P. R. Trafton (Ed.), *New directions for elementary school mathematics.* Reston, VA: National Council of Teachers of Mathematics.

Liedtke, W. (1983). Young children—small numbers: making numbers come alive. *Arithmetic Teacher, 31*(1), 34–36.

Thompson, C. S., & Rathmell, E. C. (Eds.). (1989). Number sense. Special issue of *Arithmetic Teacher, 36*(6).

Van de Walle, J. A. (1988). The early development of number relations. *Arithmetic Teacher, 35*(6), 15–21, 32.

Van de Walle, J. A., & Thompson, C. S. (1981). A poster-board balance helps write equations. *Arithmetic Teacher, 28*(9), 4–8.

Wirtz, R. (1974). *Mathematics for everyone.* Washington, DC: Curriculum Development Associates.

CHAPTER 6

DEVELOPING MEANINGS FOR THE OPERATIONS

Constructing Meanings for Operations

This chapter is about helping children connect different meanings, interpretations, and relationships to the four operations so that they can effectively use those operations in other settings, both in and out of the world of mathematics.

MEANINGS VERSUS FACT MASTERY OR COMPUTATION

It is important to separate the goal of operation meaning from mastery of basic facts and from computation. Many children can respond quickly to $4 \times 9 = \square$ or $12 - 8 = \square$ or can compute with large numbers, but have little understanding of when to multiply or subtract or how to use these operations in different settings. Students' poor performance in selecting the correct operation for standard, one-step word problems, even though demonstrating computation skills with the very same numbers and operations, is testimony to the fact that these are separate issues and objectives for instruction.

It should likewise be clear that the activities of this chapter are not designed to produce basic fact mastery or computational skill. Conceptual knowledge of the operations is related to and enhances those procedural skills but will not produce them. Unfortunately, the distinction between fact mastery and operation meaning is almost nonexistent in the standard elementary basal textbook.

GENERAL STRATEGIES FOR ALL OPERATIONS

For all four of the operations, models and word problems are the two basic tools the teacher has to help students develop operation concepts.

Basic Meanings Developed with Models

Models for the operations include countable objects (counters), arrays or things in rows and columns, and number lines, rods, or segmented strips. Relationships (groupings, actions such as joining or partitioning, set and subset relations, etc.) are formed with these models by moving or arranging the materials and/or drawing lines, arrows, loops, and so on, to indicate various relationships. In general, children are directed to do something with the models and should verbalize the relationships they make. For example, children may be given a set of 24 counters to be put into equal subsets in as many ways as possible. In a group, children might share results: "I made my counters in eight piles of three." The teacher may ask different children to share results with the class or discuss findings individually or in small groups.

Encoding and Decoding

Operations must be intimately connected to symbols (+, −, ×, ÷). Initially children learn to say the relationships with the use of the operation names. "Eight *times* three is twenty-four." But almost immediately the idea and language is connected to the corresponding symbols or equation: $8 \times 3 = 24$. In the example, the children could write a multiplication equation for each different arrangement of equal subsets they found for 24 counters. The connection is with the entire equation, not just answers. This is encoding the concept into the symbols that represent it. The concept of the operation is the focus of the activity, not answers or procedures; the writing or encoding connects that idea with the written form.

The reverse activity is decoding. Children might be given a completed equation such as $7 \times 5 = 35$, and asked to use a familiar model to show what the equation means and why it makes sense. They could also explain verbally why the model explains the meaning of the equation.

The Role of Word Problems

The word problem provides an opportunity for examining a much more diverse set of meanings for each operation. Models without word problems are useful for examining basic structures or meanings of the operations. When both problems and models are used, a much more comprehensive understanding of each operation can be developed.

With a word problem the relationship is in the problem.

Children analyze the problem with the help of models and thereby construct the relationship. For example:

> Marjorie's mother made bags of 3 candies each for all of the children at Marjorie's birthday party. There were 8 children at the party. How many pieces of candy did Marjorie's mother use?

Children could be given grid paper, counters, paper to draw sets on, or number lines to help them think about and solve the problem. It is not a matter of telling children to multiply, but rather having them look for a relationship in a verbal situation. Word problems go beyond the basic relationships provided by the models. They provide a context for the numbers involved. Phrases such as "three for each," "three children," "3 miles per hour," or "three times as many," have more meanings than can be found in counters alone.

When word problems are presented at the beginning of the development of an operation, children tend to focus attention on the structure of the problem (Carpenter & Moser, 1982; Campbell, 1984). In discussing solutions, children will discuss how they thought about the problem and how they used their models to help analyze it. "I made rows with three squares. One row stands for one person at the party. There were eight people and so I made eight rows. Then I could tell to multiply to find the number of squares." If the problem comes after the concepts are supposedly developed, children focus on trying to choose the correct operation. "I multiplied the numbers together. Is that right?" There may or may not be analysis.

Translations: Models, Words, Symbols

It is useful to think of models, word problems, and symbolic equations as three separate languages. Each language can be used to express the relationships involved in one of the operations. The model and word problem languages more clearly illustrate the relationships involved, while the equation is a convention used to stand for ideas. Given these three languages, a powerful approach to helping children develop operation meaning is to have them make translations from one language to another. The translations from words to models to symbols and also the translations between models and symbols (encoding and decoding) have already been discussed. Once children develop a familiarity with an assortment of models and drawings, and they gain exposure to a wide variety of meanings through word problems, they can begin to make translations from any one language to the other two (Figure 6.1). These translation activities can be done in short 10-minute periods every week, all year long.

In a translation exercise students are provided with an expression in one of the three languages: a model (usually in the form of a drawing), a word problem, or an equation. They are then asked to come up with expressions in each of the other two "languages" that represent the same relationships. Since different cooperative groups or individuals will devise different ideas, the results provide a wonderful source of

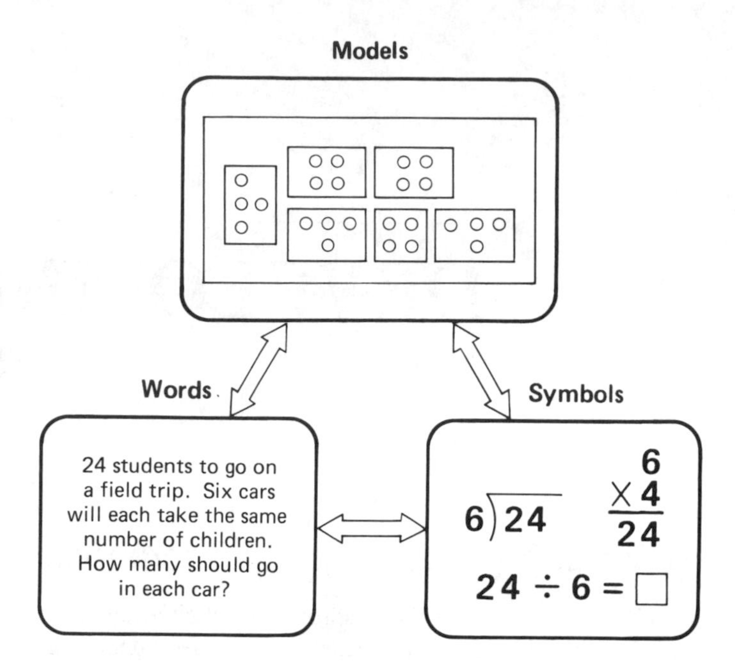

Figure 6.1
Translations can be made from any one of three languages to the other two.

discussion. In Figure 6.2, several examples illustrate what might occur.

Addition and Subtraction Concepts

Addition and subtraction concepts are very closely related. Both can be derived from the same basic relationships between sets: either a part-part-whole relationship or a comparison relationship. For each of these there are several models that children can effectively use in the classroom to learn about addition and subtraction.

As you read through the following discussion, notice that the meanings of addition and subtraction are relationships between sets and have little or nothing to do with action. When these relationships are imbedded in word stories, some will involve actions and some will not.

PART-PART-WHOLE MEANINGS

The part-part-whole concept as discussed in Chapter 5 is a way of understanding that a quantity or *whole* can be composed of two or more separate quantities or *parts*. As illustrated in Figure 6.3, this idea can be modeled with sets, in which case the whole is made of the exact same materials (dots, counters) as the individual parts, or a length model can show the parts as equivalent to the whole. Each of these two can be used to show more than two parts. All materials that can be used to model a part-part-whole meaning of addition or subtraction are simply variations of these two ideas.

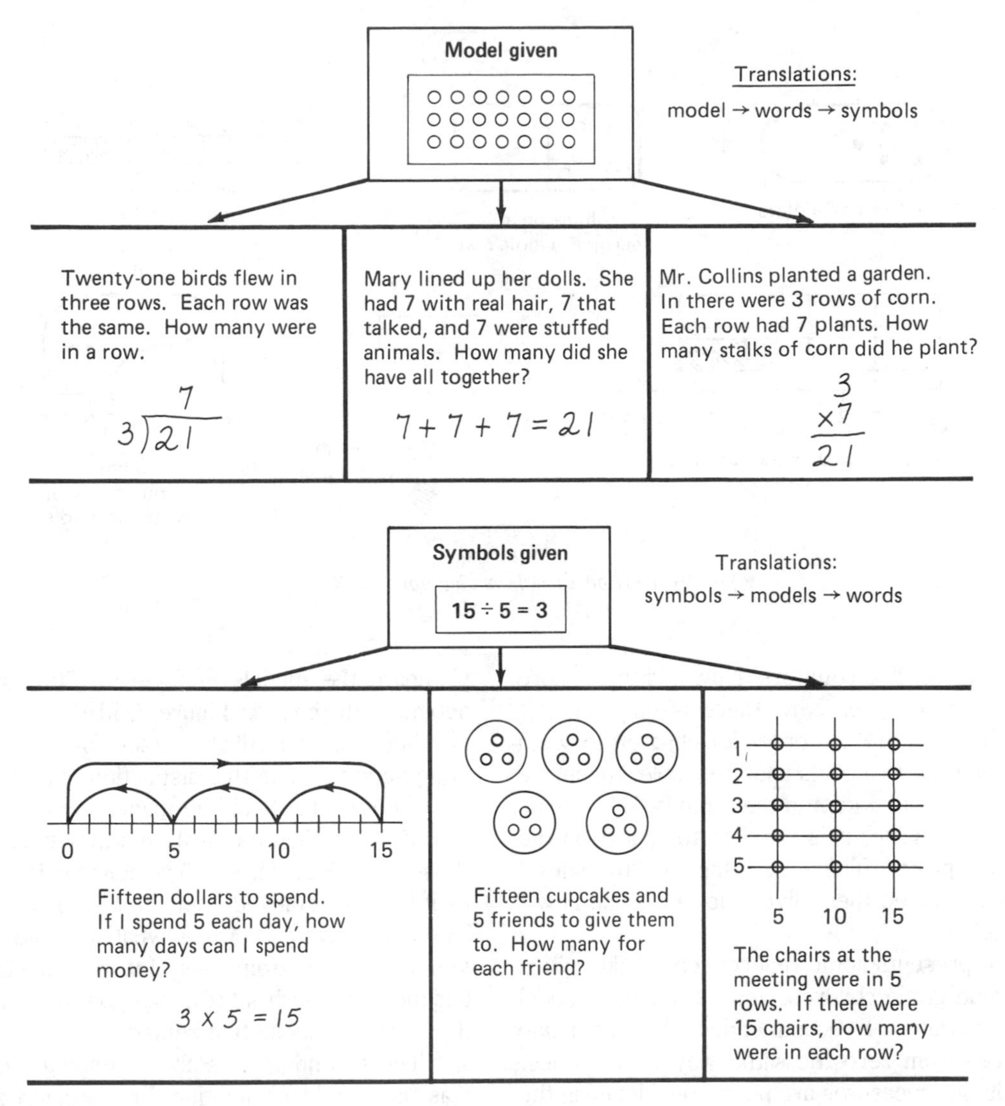

Figure 6.2
Two translation tasks; varied results.

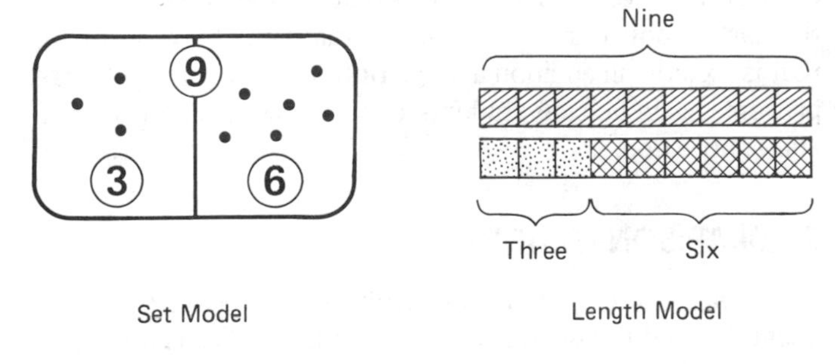

Figure 6.3
Two models for part-part-whole.

Part-Part-Whole Addition Concepts

When the parts of a set are known, *addition* is used to name the whole in terms of the parts. Some researchers have made a distinction between addition situations that involve a joining or "put-together" action and those that are static having no action. For example, if you place five red counters on a mat and then add or join three blue counters with them the result is a set with two parts, five and three. If you have a mat with five counters on one side and three on the other, then that also is a set made of two parts, five and three, but there was no action or joining. Children will probably not make the distinction between action and nonaction, but it does lead to different types of word story problems, as will be seen.

Examine the different variations of part-part-whole models for addition shown in Figure 6.4. Each can be encoded as $5 + 3 = 8$. Some of these are the result of a definite put-together or joining action, some are not. It really makes no difference. What is important is to notice that in every example both of the parts are distinct—even after two parts are joined. If counters are used, then the two parts should be kept in separate piles, kept in separate sections of a mat, or be two distinct colors. To illustrate the value of keeping the parts

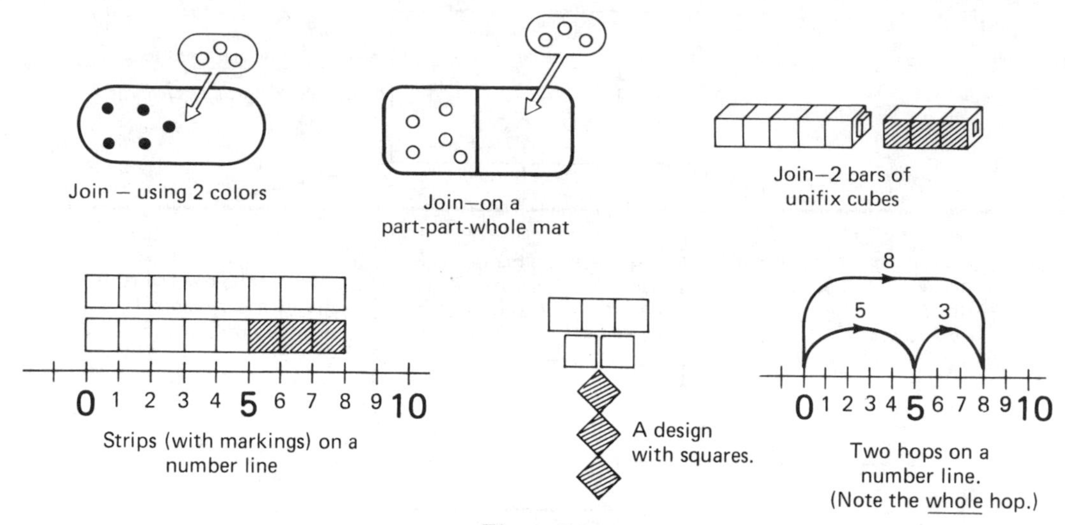

Figure 6.4
Part-part-whole models for 5 + 3.

distinct, try this. Set out five counters. Now add three more of the same type to the pile. Now there is only one set remaining, the whole set of 8. In order for children to see a relationship between the two parts and the whole, the image of the five and three must be kept in their minds. By keeping the two sets distinguishable, it is possible to reflect on the action after it takes place. "These red chips are the ones I started with. Then I added these five blue ones, and now I have eight all together."

A number line presents some real conceptual difficulties for first- and second-grade children, and its use as a model at that level is generally not recommended. A number line measures distances from zero the same way a ruler does. The concepts of length measures are poorly developed in the early grades. Children focus on the dots or numerals on a number line instead of the spaces (distances or unit lengths). They think of numbers in terms of sets and objects, not lengths. However, if strips or rods are placed over a number line, or if arrows (hops) are drawn for each number in an exercise, the length concept is more clearly illustrated. To model the part-part-whole concept of three plus five, start by placing a bar or drawing an arrow from 0 to 3, indicating, "This much is three." Do not point to the dot for 3, saying "This is three." Strips with units of length marked on them are a good alternative and may be thought of as a readiness for the number line.

Part-Part-Whole Subtraction Concepts

In the part-part-whole model, when the whole and one of the parts are known *subtraction* names the other part. This definition is in agreement with the drastically overused language of "take-away." If you start with a whole set of 9 and remove a set of 4, the two sets that you know are the sets of 9 and 4. The expression $9 - 4$, read "nine *minus* four," names the five remaining. Therefore, nine minus four *is* five. Compare the models in Figure 6.5(a), showing a remove action, with those in Figure 6.5(b), showing only a missing part but no action. All are models for $9 - 4 = 5$. Again, there is no need to make this distinction with children.

Two points should be made about the models in Figure 6.5. First, in Figure 6.5(b), you may have a tendency to say those are models for $9 - 5$ rather than $9 - 4$ because it looks as if 5 were removed. But the *known* part is what you see. The expression $9 - 4$ tells what the unknown part is. When you take 4 away from a set of 9, you also know the whole (9) and the part, which is 4 (the four you counted). The expression $9 - 4$ names those that remain.

The second point is about keeping the parts distinct, as was discussed with addition. In take-away situations, there is a tendency to focus only on those left. They become the "answer." Those that were removed are simply shoved aside or returned to the supply box. As with addition, this prevents children from reflecting on the situation and reversing it. If the part removed in a $9 - 4 = 5$ situation is then returned, that is exactly an addition action. Both parts and the total are present, again helping students connect the ideas of addition and subtraction.

COMPARISON MEANINGS

The *comparison* relationship involves two distinct sets or quantities and the difference between them. We can refer to these as the smaller set, the larger set, and the *difference* amount. Several ways of modeling the difference relationship are shown in Figure 6.6.

Comparison Addition Concepts

If the smaller of two sets and the difference between them are known, then *addition* tells how many are in the larger set. For example, if you know that your stack of cubes is three

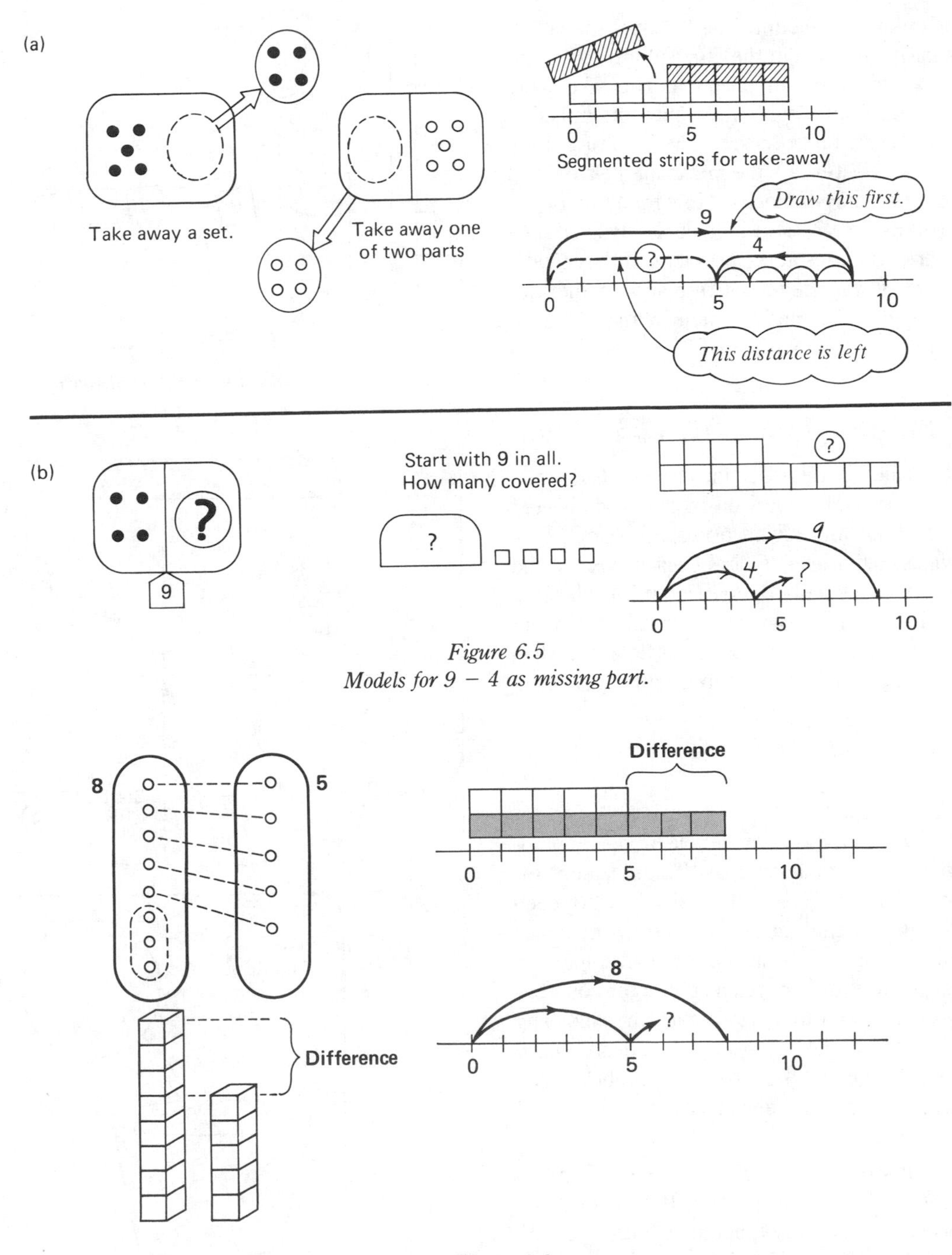

Figure 6.5
Models for 9 − 4 as missing part.

Figure 6.6
The difference between 8 and 5 is 3.

cubes shorter than the one behind the wall and your stack has five, then 5 + 3 tells how big the larger stack is.

For some reason, this relationship is almost never used to illustrate a meaning for addition in schools. However, the relationship does exist in the real world, even the world of children. If Suzy is taunting Tommy by saying, "I've got three more pennies than you have," and Tommy looks at his five pennies, then Tommy is exactly in this situation. He knows the smaller of two sets and the difference amount. Addition tells how many Suzy has.

While you may not find the comparison model for addition in your second- or third-grade texts, you may find story problems that involve that relationship.

Comparison Subtraction Concepts

A comparison meaning for subtraction occurs in two situations: (1) The larger of the two sets and the difference is known and the smaller set is unknown. (2) Both of the sets are known and the difference is unknown. In both cases subtraction names the unknown amount. The "what is the difference" situation is fairly common in the curriculum, but is not given anywhere near the same emphasis as the take-away concept for subtraction. As a result, when students drilled in take-away confront a word story such as the following, they experience difficulty. "Yesterday the mailman delivered 8 pieces of mail and today only 5 pieces. How many more did he bring yesterday than today?" The difficulty is involved with the fact that there are 13 things in the problem. If they model this with counters, they will need a set of 8 and another set of 5. Now they have two sets, and for many it makes sense at this point to add rather than subtract.

Figure 6.7
Writing about what you do.

ADDITION AND SUBTRACTION ACTIVITIES

In the activities described in this section, children use variations on part-part-whole and comparison models and connect these models with addition and subtraction symbolism. Many of the part-part-whole activities are direct extensions of the part-part-whole activities described in Chapter 5 to develop number concepts. From the viewpoint of number concept development, addition and subtraction are little more than an elaboration of part-part-whole concepts with symbolism added.

Activity 6.1

Provide children with any part-part-whole model for number and have them use the model to make "plus names" for a specified amount (refer to Figures 5.17 and 5.18). For each combination for the number that they make, they write a plus equation or an addition "number sentence." For example, if a child makes a design for eight, or separates eight counters into two parts, he or she can then write the corresponding equation. The equation will match the way the children have learned to read their designs. If the design or combination is read, "Four and five is nine," the equation is $4 + 5 = 9$.

The popular games of *Mathematics their way* (Baratta-Lorton, 1976), *Workjobs II* (Baratta-Lorton, 1979), and many excellent activities in the *Explorations* books (Coombs & Harcourt, 1986) are similar to the activity just described. These books add a wealth of creativity to an otherwise simple activity.

As another variation on these activities, have two children work together with a model. One child makes a set in two parts using any amount no higher than some value you set. The other child says the addition fact out loud. Both children can then write the equation on paper or use numeral cards to write the equation as in Figure 6.7.

To model the joining action, you will have to provide some direction, either verbally or in the form of a worksheet, since the whole or total amounts will vary with each exercise. Notice in Figure 6.8 that students write complete equations, not just answers.

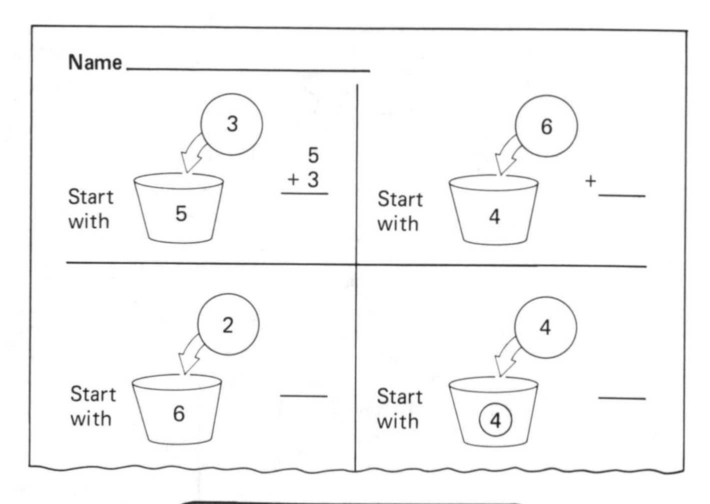

Figure 6.8
Worksheet and "mat" for join addition.

Decoding exercises can be done by providing a completed addition equation in either horizontal or vertical form and having students use the materials provided to illustrate the meaning. You can do this as a class activity, with the teacher writing equations on the board, or as worksheet exercises like the one shown in Figure 6.9.

Having children model and write subtraction sentences is done in much the same way as for addition.

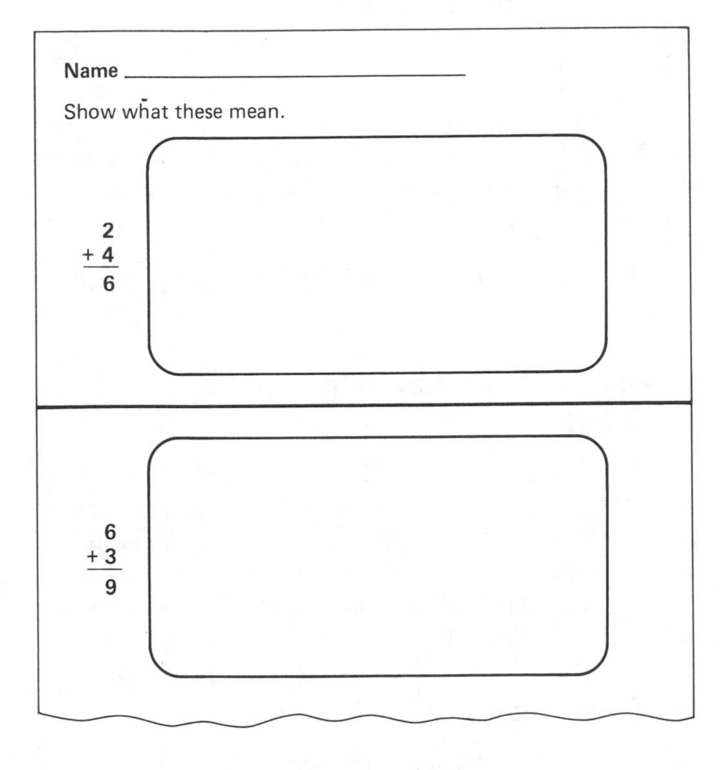

Figure 6.9
Decoding worksheet.

Activity 6.2

The "Take-Apart" game (Thompson & Van de Walle, 1984) is played by two children using a fixed total of counters. One child places all of the counters under his left hand and then removes some, placing them to the right. The second child says, "You started with eight, you took away three, and there are five left." The amount left can be uncovered as the child gets to that point in the sentence. The subtraction equation can then be written by both children, and the roles reversed.

Activity 6.3

A missing-part game can be played very similarly. A fixed number of counters are placed on a part-part-whole mat. One child separates the counters while holding a piece of tagboard to shield what is being done. Then she can place the tagboard over one of the parts, revealing the other. The second child can say the subtraction sentence, "Eight minus three (the visible part) is five (the covered part)." The covered part can be revealed if necessary for the child to say how many are there. Both the subtraction and the addition equation can then be written.

Activity 6.4

On the overhead projector you can use counters to model part-part-whole subtraction and have the class say the equations out loud and then write them.

When modeling subtraction as a part-part-whole relationship, it is a good idea to have children write the subtraction equation that is modeled first and then, since the two parts (one removed and one remaining) are still visible, write at least one addition equation for the same model. This clearly relates the two concepts. The part-part-whole addition worksheet in Figure 6.10 can be used in this way. Counters are

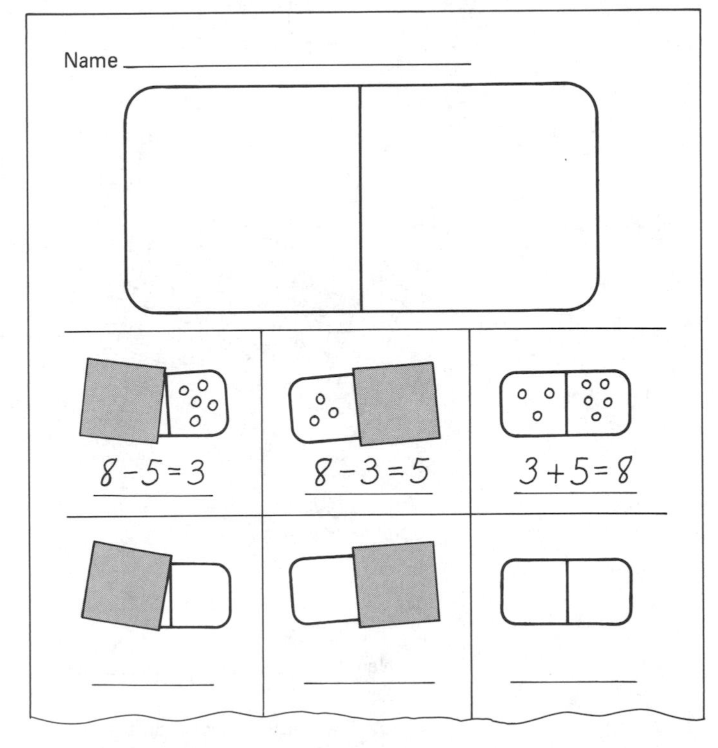

Figure 6.10
A missing-part subtraction worksheet.

used in the space at the top of the page. Below the drawings on the lower part of the page, children can draw dots for the counters and write corresponding equations. Several new textbook series are beginning to include pages similar to this one for first and second grade. Children use counters directly on the pages to help with the meaning of the activity.

Activity 6.5

The "Compare" game is played by two players as a way to connect the comparison relationship to subtraction symbolism (Thompson & Van de Walle, 1984). A fixed number of blocks or bar of Unifix cubes is lined up on the left side of a mat. One child places a second bar, no longer than the first, on the right side of the mat as in Figure 6.11. The other child says, "The difference between eight and three is five." If necessary the bars are placed side by side to make the comparison. A challenge version is played by placing the first bar under the mat out of sight. Players know how many are hidden and try to make the comparison mentally. A subtraction equation is written on paper or with the aid of numeral cards.

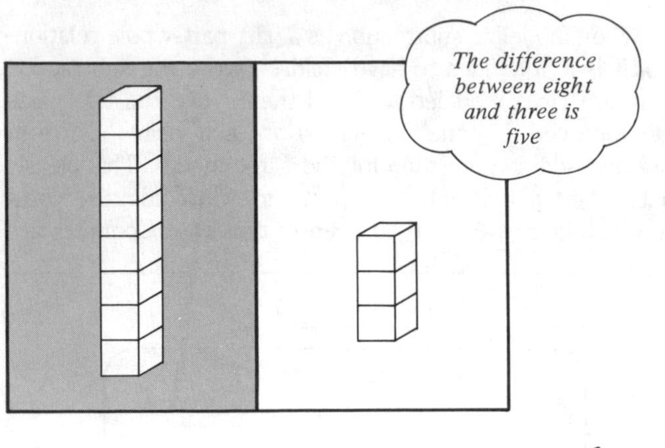

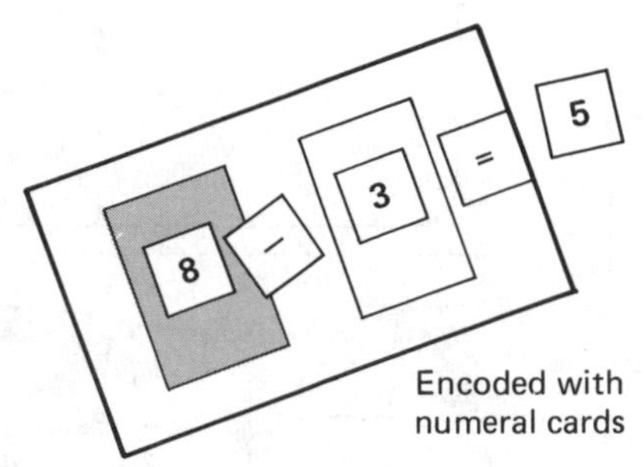

Figure 6.11
The compare game.

Activity 6.6

Teachers can easily create worksheets or activities that tell the child what to do with a model and have them record the activity and encode it. Figure 6.12 illustrates a comparison activity on number lines. Children would have strips or Cuisenaire rods of different lengths corresponding to the number line. They record what they do with the strips as "hops" on the number line.

PROPERTIES THAT ARE IMPORTANT FOR CHILDREN

There are a few properties or ideas that deserve some attention because they are helpful to children. The names of the properties are not important for children, but the ideas are. This is not the complete list of addition and subtraction properties you may have learned, but only those that have a direct impact on children's learning.

The Order Property in Addition

Put simply, the *order* (or *commutative*) property says that it makes no difference in which order two numbers are added.

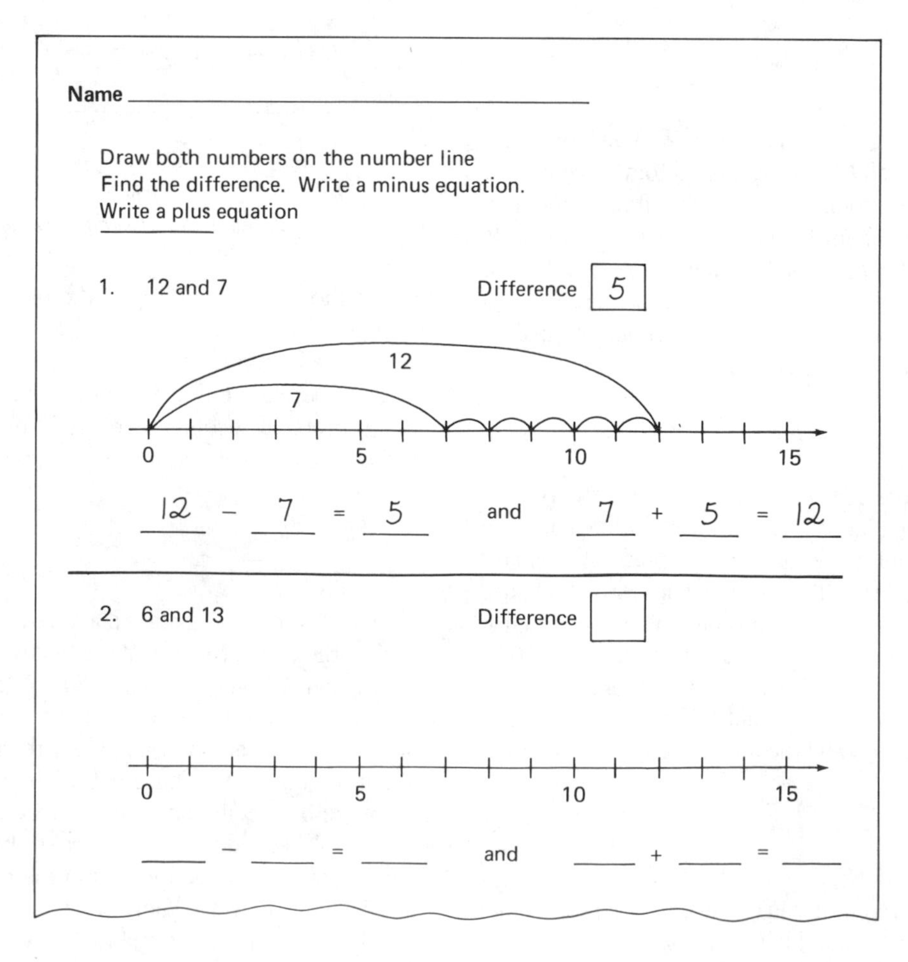

Figure 6.12
A number line worksheet for the comparison concept of subtraction.

Most children find little difficulty with this idea. Since it is quite useful in problem solving and for mastering basic facts, there is value in spending some time helping children construct the order relationship.

Activity 6.7

Show a set on a part-part-whole mat on the overhead projector. Ask the children to say the addition name for how much is shown, for example "eight plus four." Now, draw careful attention to the mat as you turn it completely around. Again request the addition name, "Four plus eight." Discuss which is more, eight plus four or four plus eight. Repeat for a few other examples, and eventually have your students discuss the rule that the order makes no difference. Notice that the rule is not arbitrary, but comes from the students and the model.

Activity 6.8

Provide a sheet with addition combinations all over the page. The activity is to circle names for the same number with the same color crayon (Figure 6.13). Be sure to include combinations in reverse order. When discussing the answers, take some time to point out these special pairs.

Zero in Addition and Subtraction

Many children have difficulty with addition and subtraction when zero is involved. There may be many reasons for this. Zero as a number is probably not understood as early as other numbers. There is an intuitive idea that addition "makes numbers bigger" and subtraction "makes numbers smaller." To help with this problem, some effort should be made to model addition and subtraction involving zero without simply providing arbitrary sounding "rules." Figure 6.14 illustrates several ideas.

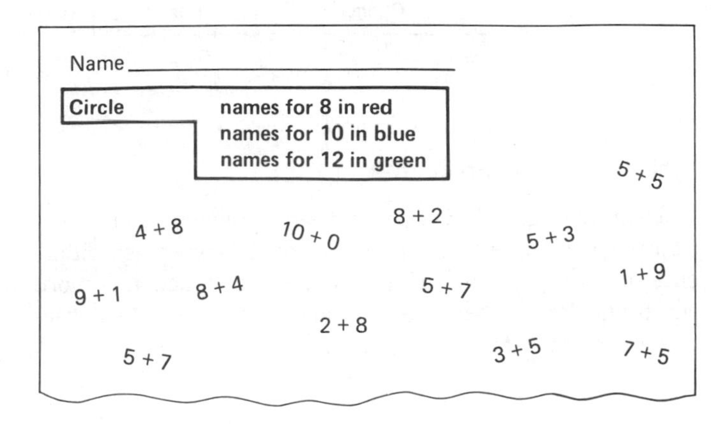

Figure 6.13
Emphasize the order property for addition.

WORD PROBLEMS FOR ADDITION AND SUBTRACTION

While models provide straightforward, basic concepts of addition and subtraction, word problems involve a wider variety of relationships and semantic differences. Researchers have separated problems into categories based on the kind of relationships involved: *join and separate problems, part-part-whole problems,* and *compare problems* (Carpenter & Moser, 1983; Thompson & Hendrickson, 1986).

Join and Separate Problems

For each of the two actions of join and separate, there are three quantities involved: the *initial* quantity, the *change* quantity (amount joined or removed), and the *resulting* quantity. Any one of the three quantities can be the unknown. An example of each of the six possibilities (three for join, three for separate) is given here.

Join: Result unknown

Sandra had 8 pennies. George gave her 4 more. How many pennies does Sandra have altogether?

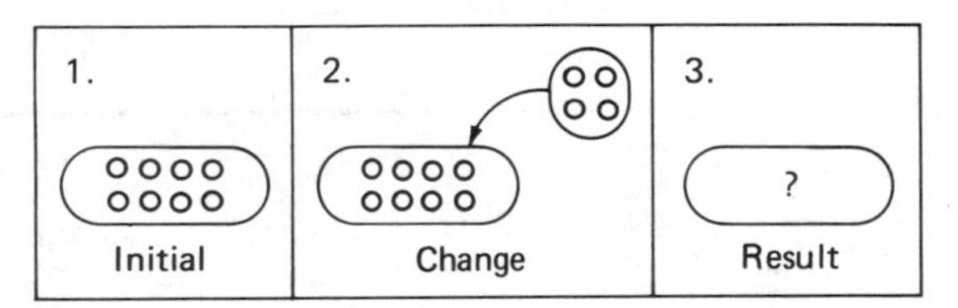

Join: Initial unknown.

Sandra had some pennies. George gave her 4 more. Now Sandra has 12 pennies. How many pennies did Sandra have to begin with?

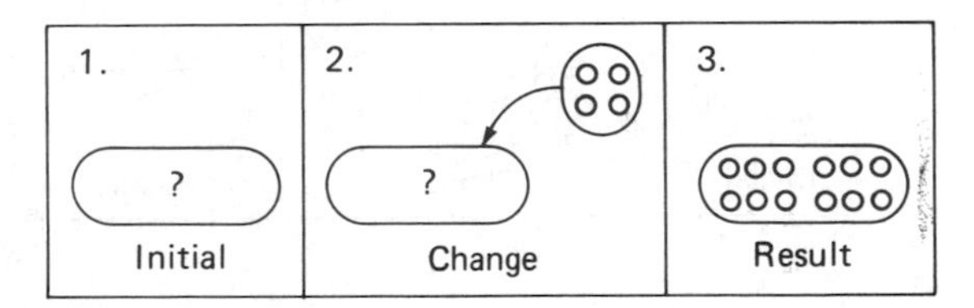

Join: Change unknown.

Sandra had 8 pennies. George gave her some more. Now Sandra has 12 pennies. How many did George give her?

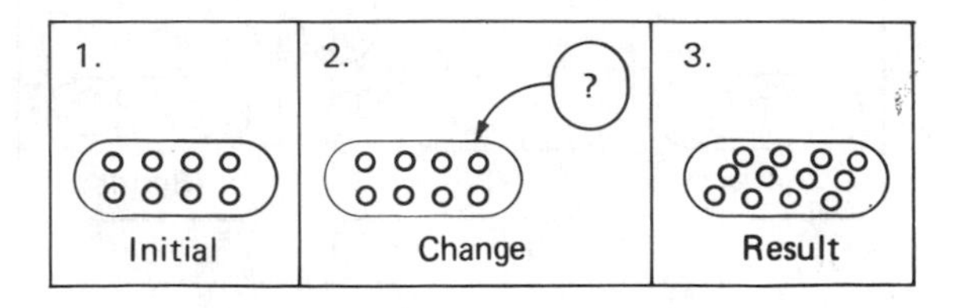

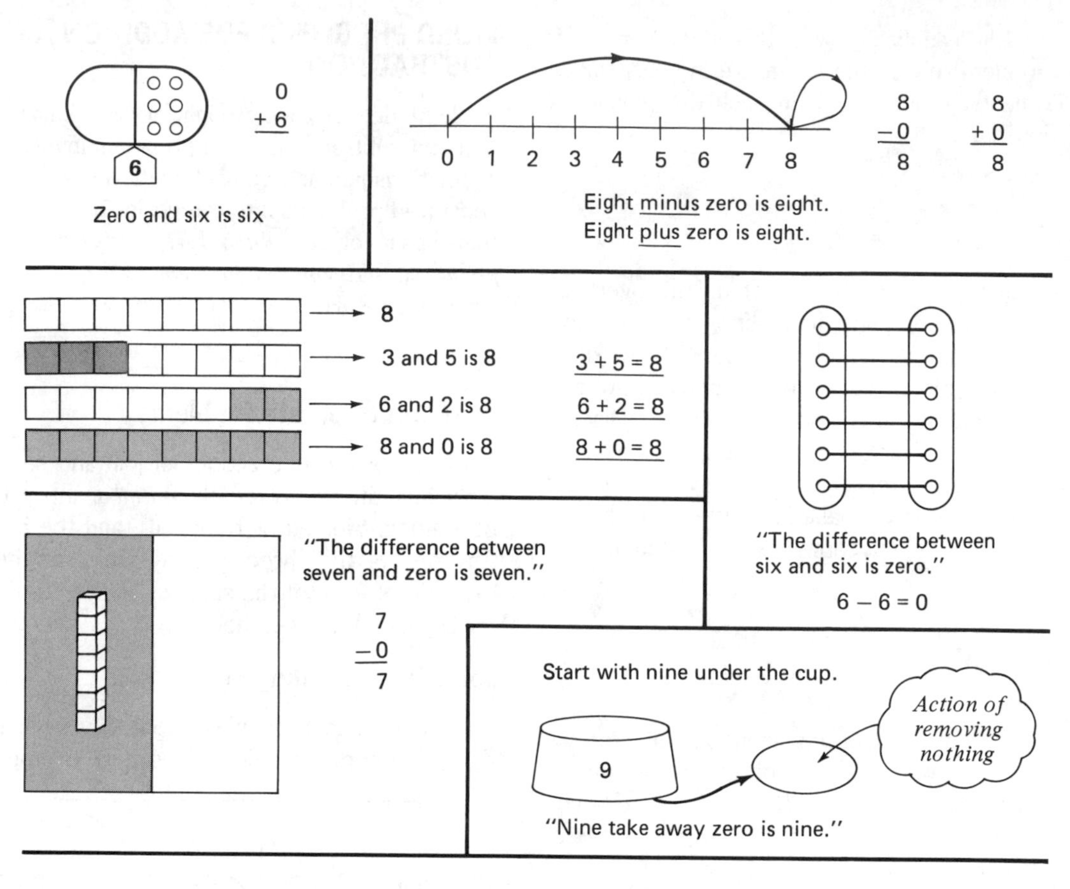

Figure 6.14
Zero in addition and subtraction.

Separate: Result unknown.

Sandra had 12 pennies. She gave 4 pennies to George. How many pennies does Sandra have now?

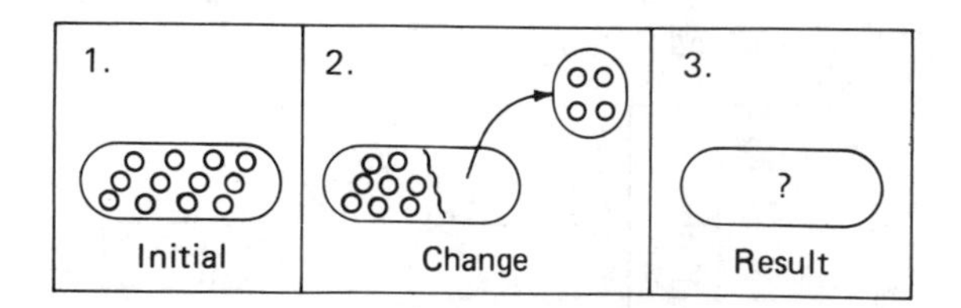

Separate: Change unknown.

Sandra had 12 pennies. She gave some to George. Now Sandra has 8 pennies. How many did she give to George?

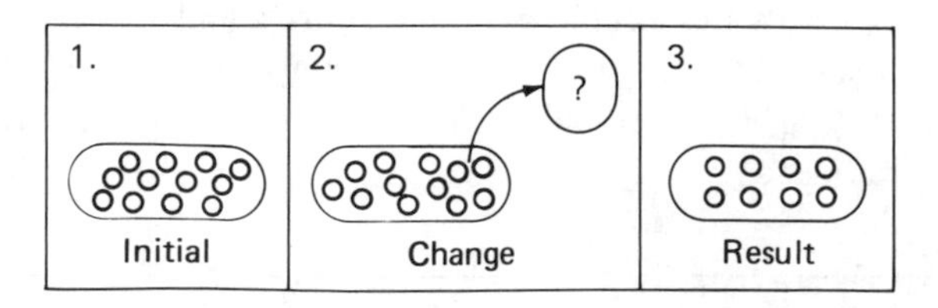

Separate: Initial unknown.

Sandra had some pennies. She gave 4 to George. Now Sandra has 8 pennies left. How many pennies did Sandra have to begin with?

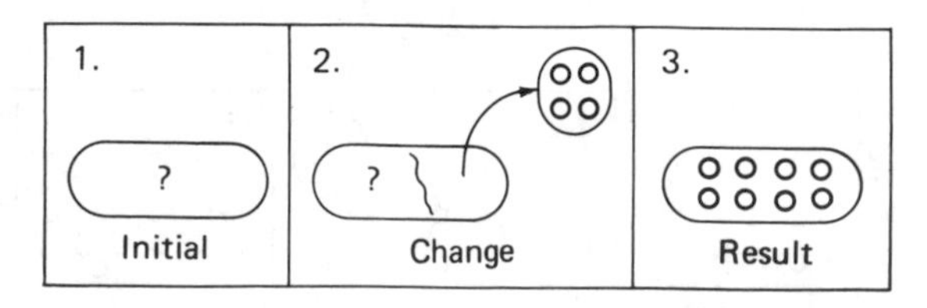

Part-Part-Whole Problems

While the difference between action and no action is not necessarily obvious to children when using models, there is a difference in story problems. Here the absence of action is more significant. Remember, the meaning of addition and subtraction is a relationship, not an action.

Part-part-whole: Whole unknown.

George has 4 pennies and 8 nickels. How many coins does he have?

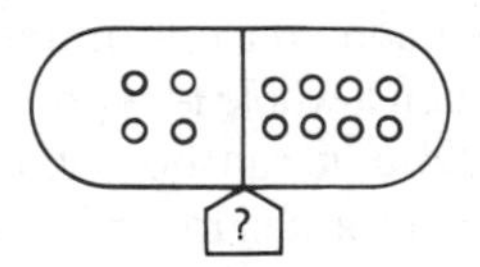

Part-part-whole: Part unknown.

George has 12 coins. Eight of his coins are pennies, and the rest are nickels. How many nickels does George have?

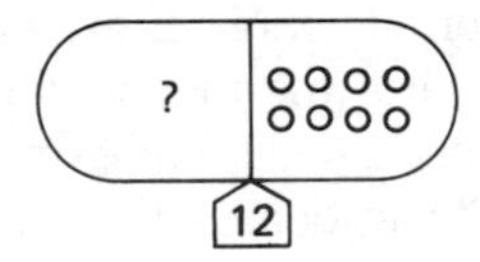

Compare Problems

There are three types of compare problems corresponding to which quantity is unknown (smaller, larger, or difference). For each of these, two examples are given: one in terms of more, the other in terms of less.

Compare: Difference unknown.

George has 12 pennies and Sandra has 8 pennies. How many more pennies does George have than Sandra?

George has 12 pennies. Sandra has 8 pennies. How many fewer pennies does Sandra have than George?

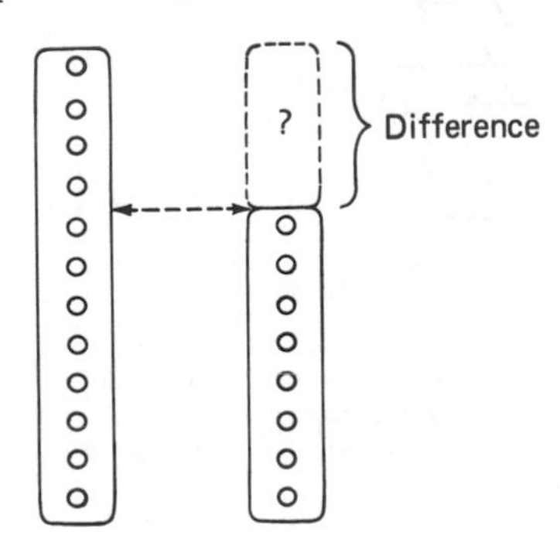

Compare: Larger unknown.

George has 4 more pennies than Sandra. Sandra has 8 pennies. How many pennies does George have?

Sandra has 4 fewer pennies than George. Sandra has 8 pennies. How many pennies does George have?

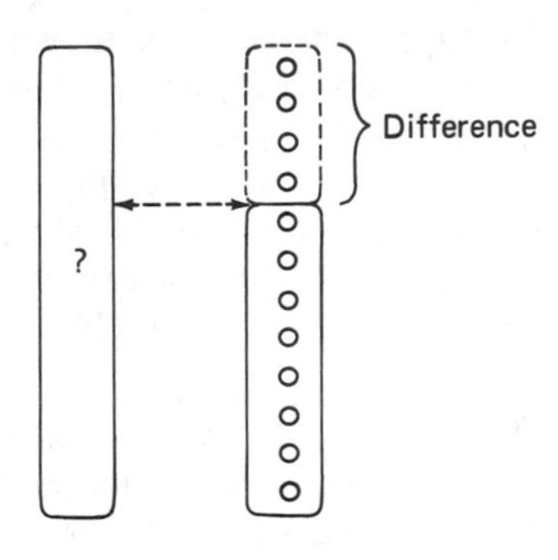

Compare: Smaller unknown.

Sandra has 4 fewer (less) pennies than George. George has 12 pennies. How many pennies does Sandra have?

George has 4 more pennies than Sandra. George has 12 pennies. How many pennies does Sandra have?

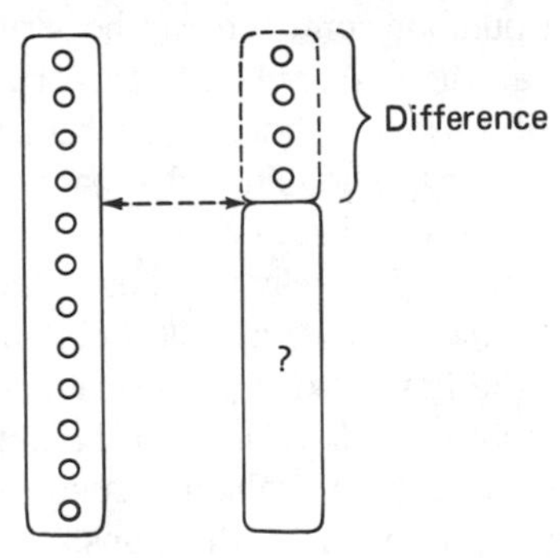

Using Addition and Subtraction Word Problems in the Classroom

The important goal here is not just getting answers, but to analyze relationships. Provide children with counters, strips, Unifix, or other blocks. Present the problems to children orally. If children work in pairs or small groups, they can discuss their ideas out loud.

In the early grades, children solve word problems with their models and afterward write an equation *that tells what they did.* Equations can be "written" with numeral cards so that changes can be made easily. The answer to a problem may be any number in the equation. Thompson and Hendrickson (1986) suggest that after the sentence is written, children should circle or identify the number which tells the answer. In this way a variety of correct equations are possible for the same word problem. Discussions will help children make connections between addition and subtraction in different forms. For example, 8 + [4] = 12 and 12 − 8 = [4] are both likely equations for the first compare problem on page 78.

Obviously you will want to use small numbers with children in grades K to 2, but small numbers are just as useful with older children. Some of these problems are unfamiliar and difficult even for fifth graders and above. Many older children have learned that word problems mean only "select the operation." With smaller quantities they can be required to draw pictures of models to explain their reasoning.

Another approach for the upper grades is to use problems with numbers and quantities which are appropriate for them, but then have them change all numbers to whole numbers less than 20. After modeling and solving the new problem, they can translate the process to the original problem. For example:

> Tania has been working at the corner grocery store after school. This month she made $137.75. That was $28.60 more than she made in the first month. How much did she make the first month?

Small-Number Version:

> This month Tania made $15. That is $6 more than she made in the first month. How much did she make in the first month?

The small-number version can be modeled with a simple drawing, and a number sentence can be written for it. The same process can then be applied to the original problem.

Not all of the various addition and subtraction word problems are equally easy. The join and separate problems with the initial set unknown and compare problems with the reference set unknown are generally the most difficult (Thompson & Hendrickson, 1986). However, all problems should be explored, even in the first grade. As children are exposed to a richer and more varied collection of problem types, they can begin to use these different problem structures as they make up their own word problems in the translation exercises described earlier in the chapter.

Multiplication Concepts

REPEATED ADDITION

By far the most common concept of multiplication is that of *repeated addition.* When a whole is represented in two or more equal parts, *multiplication* can be used to name the whole amount. It is important for children to connect the new multiplication concept with their already existing knowledge of addition. For example, the addition expression $4 + 4 + 4 + 4 + 4 + 4$ and the expression 6×4 represent the same relationship. The 6 and 4 are referred to as the first and second *factors,* and the total, 24, is the *product.* Notice that each factor represents a different idea. The first factor tells *how many sets,* while the second factor represents the *number in each set.*

Sets and strips or number lines are models for multiplication, since they are also the models for addition. A model not generally used for addition, but extremely important and widely used for multiplication, is the array. An *array* is any arrangement of things in rows and columns, such as a rectangle of square tiles or blocks.

In order to make clear the connection to addition, early encoding exercises should also include writing an addition sentence that represents the same model. A variety of models is shown in Figure 6.15 and encoded as both addition and multiplication. Notice that the products are not included; only addition and multiplication "names" are written. This is to avoid the tedious counting of large sets. The purpose of these

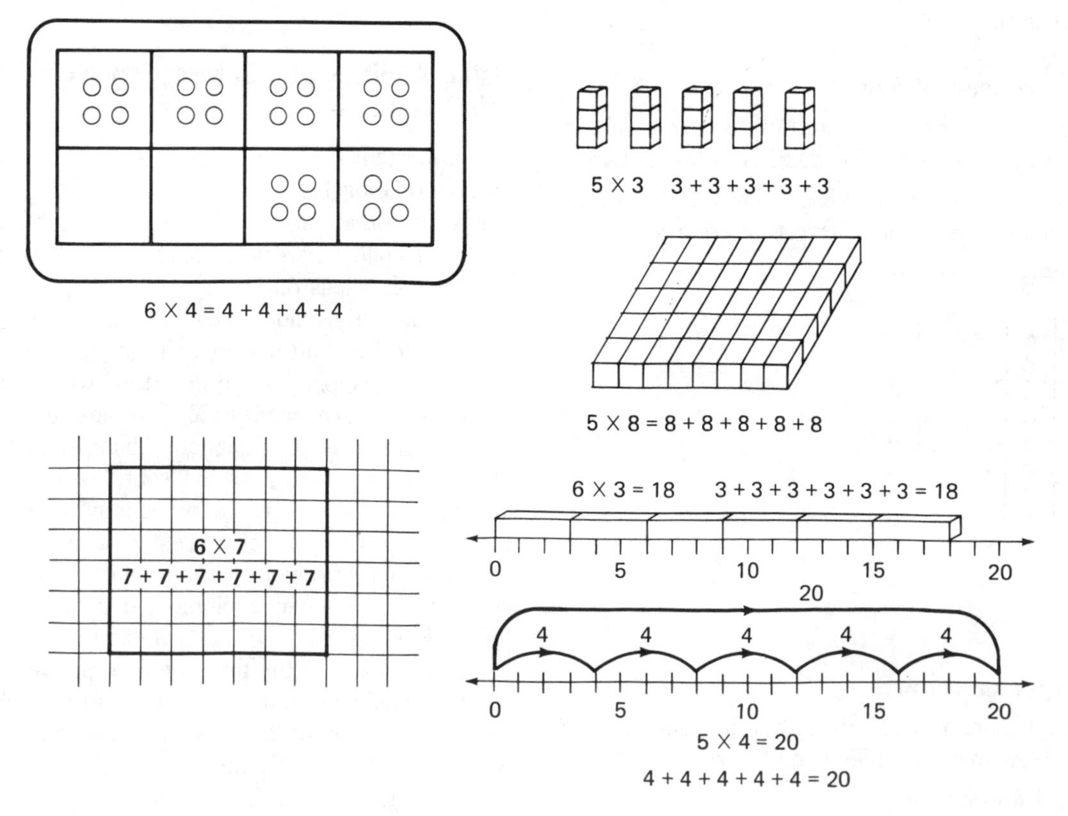

Figure 6.15
Encoded models for repeated addition.

encoding activities is to associate the symbolic knowledge of multiplication with the concept. Mastering the basic facts is a later objective. Another approach is to write one sentence that expresses both concepts at once; for example $9 + 9 + 9 + 9 = 4 \times 9$.

COMBINATIONS (CARTESIAN PRODUCTS)

A very different concept of multiplication involves only two sets. Multiplication can be used to express the number of possible pairings between the elements of two sets. (This is also known as the *Cartesian product* concept of multiplication.) In elementary and middle school, this concept is usually modeled in terms of some real context. Figure 6.16 illustrates several ways to model the combination concept. The array model for the combination concept helps illustrate that the two concepts of multiplication are not as different as they first appear.

The need to count possible combinations or outcomes, such as in simple probability experiments, suggests that an increased emphasis on this concept may be appropriate. At present, this concept is given only limited attention, mostly in the upper grades.

AREA

Consider the rectangle shown in Figure 6.17 filled in with square units. The multiplication 4×7 tells how many squares there are. Compare that with the task of using the length-times-width formula for area of a rectangle. Now we are multiplying two *lengths* to get an *area.* Neither factor is the size of a set or a number of sets but a length. Furthermore, the product is an area, a different unit from either factor. While there is a clear connection between the array of squares and the length $\times$ length $=$ area relationship, the latter is frequently considered as a separate meaning of multiplication. In fact, the area concept for multiplication will be used extensively in Chapter 9 to develop computation procedures for multiplication.

MULTIPLICATION ACTIVITIES

Multiplication is usually first introduced in second grade and developed with more emphasis in the third and fourth grades. The general approach of encoding models is the same as with addition and subtraction. As noted earlier, multiplication and addition "names" can be used to avoid tedious counting and still explore large products. Fact mastery is a later objective.

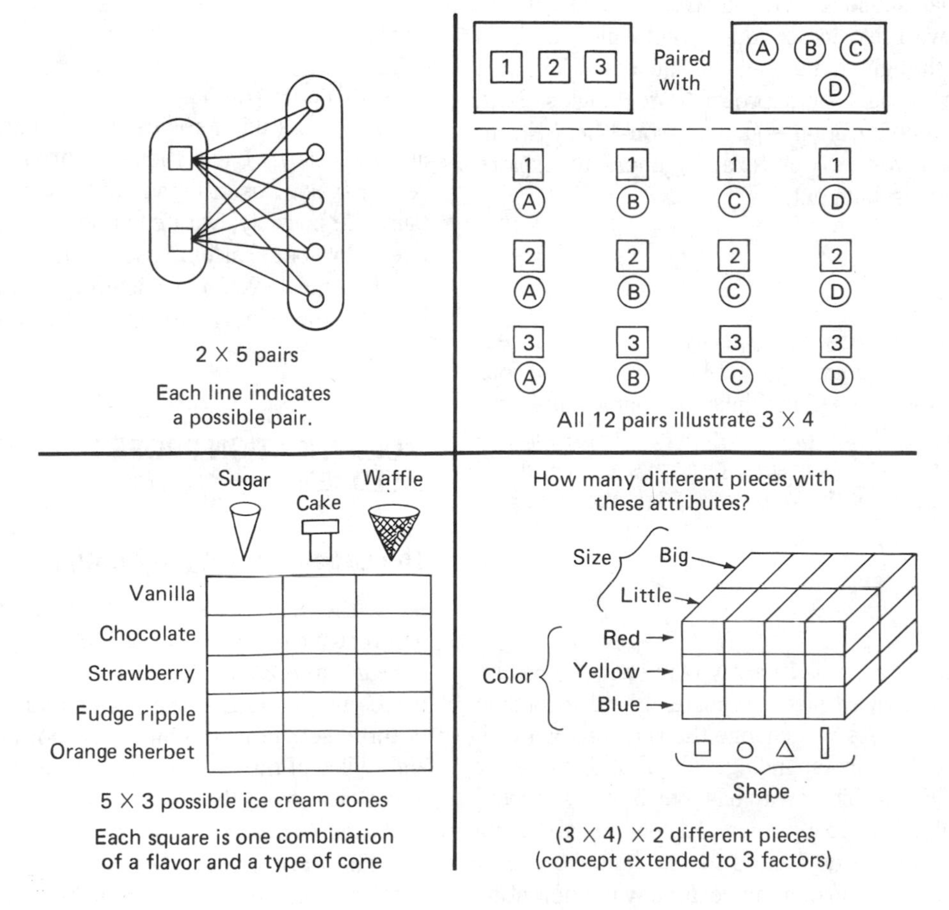

Figure 6.16
Models for the combination concept of multiplication.

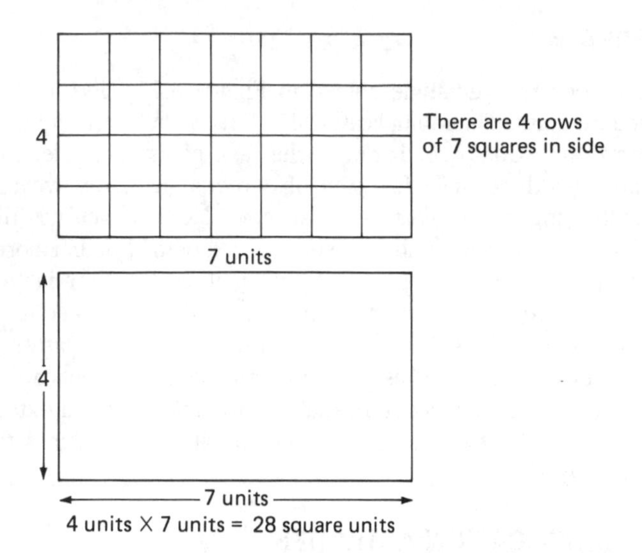

Figure 6.17

Activity 6.9

Start with a number that has several factors; for example, 12, 18, 24, 30, or 36. Have students use a model to find multiplication expressions for this number. With counters, students attempt to find ways to separate the counters into equal subsets. With arrays, students try to build rectangles with blocks or draw rectangles on grid paper that have the given number of squares. For each such arrangement of sets or appropriate rectangles, both an addition and a multiplication equation should be written. Students could be given two or three numbers to explore for an independent assignment.

Activity 6.10

A more directive exercise is to specify the number of sets and the size of each. For example, "Make six rows of four squares in each." Addition and multiplication names are written for each instance. When physical models are used, students can draw pictures of their sets, or arrays as a record of their activity. Examples of these exercises are shown in Figure 6.18.

Activity 6.11

A transparency of a 10 × 10 array (Appendix C) can be used with a large L-shaped piece of poster board to outline arrays of various sizes. As you change the array shown with the L, students say the corresponding multiplication name for that array (Figure 6.19). The array has four 5 × 5 sections, making it easy to see how many rows and columns are being shown. The array can be duplicated for students to use at their seats to give them an instant model for any multiplication combination up to 10 × 10.

Activity 6.12

Student-made "fold-ups" are another variation of the array. Cut paper lengthwise into strips about 5 or 6 cm wide. Have students fold up the strips from one end in folds about 1 cm wide. Fold back one fold and write 0 × 6 on the roll. Roll down another fold. Write 1 × 6 on the fold and make a row of six evenly spaced dots on the paper, as shown in Figure 6.20. Unroll another fold, write 2 × 6 and make a second row of dots exactly under the first. Continue to unfold and make rows of dots until there are nine rows for 9 × 6. Students can make fold-ups for each factor 2 through 9. The fold-up helps to emphasize the repeated-addition concept and can be used later to help with basic fact memory. Turn it sideways to see that 4 × 6, for example, is the same as 6 × 4.

Activity 6.13

The calculator is a good way to relate multiplication to addition. Students can be told to find various products on the calculator without using the ⊠ key. For example, 6 × 4 can be found by pressing ⊞ 4 ⊟ ⊟ ⊟ ⊟ ⊟ ⊟. Let students use the same technique to add up sets of counters they have made. Students will want to "confirm" these results with the use of the ⊠ key.

Activity 6.14

Provide students with small squares of paper, four red, and six yellow. Label these A through D and A through F, respectively. Pose the task of finding out how many different pairs are possible, with each pair having one red square and one yellow one. Encourage students to work together, record results, and look for a pattern. This is a good example of a problem-solving approach to the combination concept of multiplication.

MULTIPLICATION PROPERTIES IMPORTANT TO CHILDREN

The Order Property in Multiplication

Since the two factors in a multiplication expression carry different meanings, it is not intuitively obvious that 3 × 8 is the same as 8 × 3 or that, in general, the order of the numbers makes no difference (the commutative property). A picture of three sets of eight objects cannot immediately be seen as eight piles of three objects. Eight hops of 3 land at 24, but it is not clear from that model that three hops of 8 will land at the same point.

The array, on the other hand, is quite powerful in illustrating the order property, as is shown in Figure 6.21. Children should draw or build arrays and use them to demonstrate why

Draw 3 rows 8 in each — Write X and + names	Draw 6 rows 4 in each — Write X and + names
○○○○○○○○ ○○○○○○○○ ○○○○○○○○ 3 X 8 8 + 8 + 8	

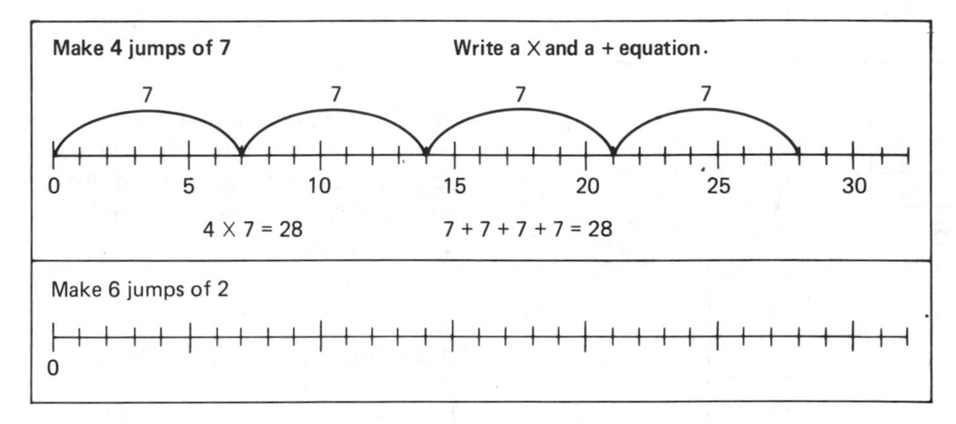

Make 9 sets of 3. Draw a picture. Write X and + names.

Figure 6.18
Three different examples of how students model and encode multiplication as directed by written exercises.

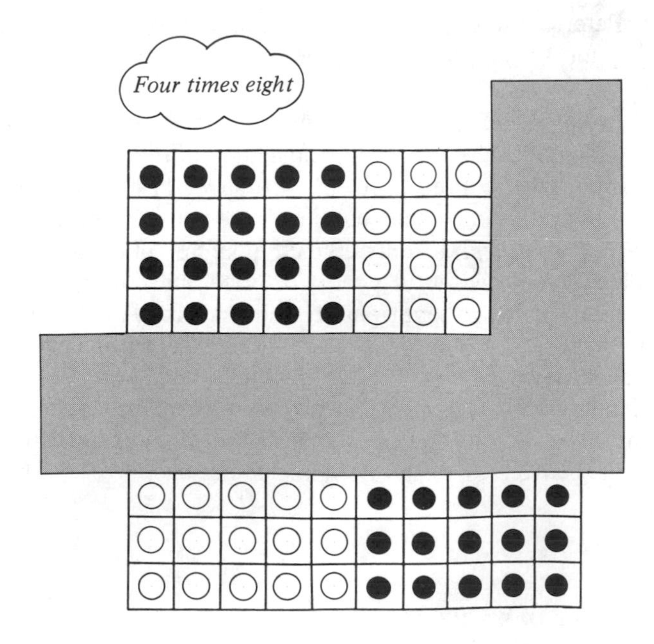

Figure 6.19
A 10 × 10 array can be used to quickly show models for 1 × 1 to 10 × 10.

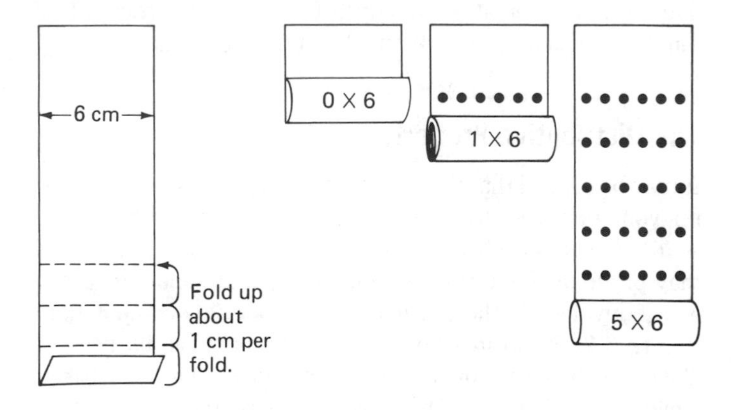

Figure 6.20
A fold-up model for multiples of six.

each array represents two different multiplications with the same product.

The Role of Zero and One in Multiplication

Factors of 0 and, to a lesser extent, 1 can cause difficulty for children. They are somewhat extreme cases. In one second-grade textbook (Bolster, et al., 1988) a train pulling four cars

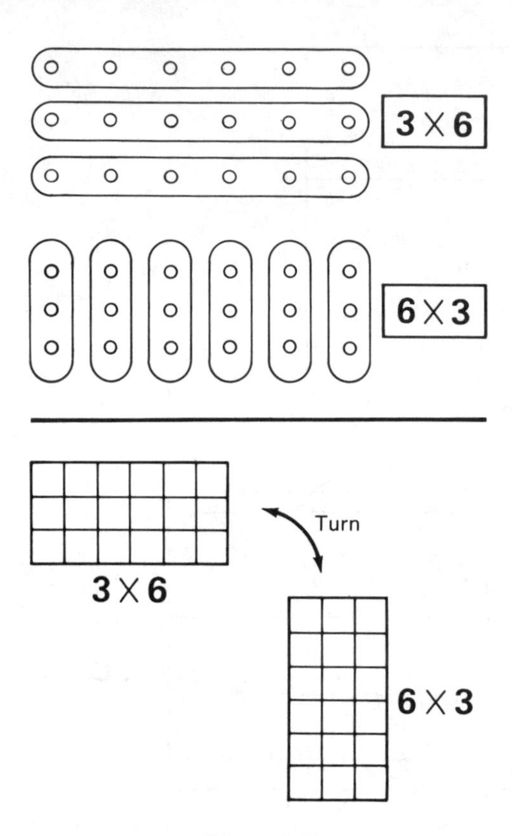

Figure 6.21
Two ways an array can be used to illustrate the order property for multiplication.

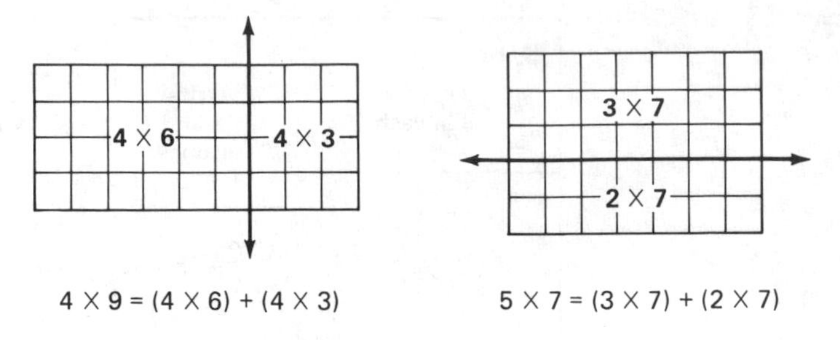

Figure 6.22
Models for the distributive property.

of six people models 4 × 6. The train pulling six empty cars models 6 × 0, while a train with no cars of 6 models 0 × 6. Factors of 1 can be interpreted similarly. On the number line, five hops of 0 land at 0. No hops of 5 is the same result. It is fun to discuss factors of 0 and 1 in terms of an array.

The Distributive Property

It may be argued that the distributive property is not essential for young children to know. In the formal form, $a \times (b + c) = a \times b + a \times c$, that may be true. But the concept involved may prove useful in relating one basic fact to another, and it is also involved in the development of two-digit computation. Figure 6.22 illustrates how the array model can be used to illustrate that a product can be broken up into two parts. Children could be asked to draw specific rectangles and "slice them" into two parts and write the corresponding equations.

WORD PROBLEMS FOR MULTIPLICATION

A scheme for categorizing both multiplication and division problems together (as was done for addition and subtraction) is possible (Hendrickson, 1986). For clarity, however, types of multiplication problems are discussed here, and related division problems are discussed later in the chapter.

Three types of multiplication word problems can be identified. The first two are ultimately related to or modeled by the repeated-addition concept and the third involves combinations or Cartesian products. Word problems for the area concept are not included, since those problems are essentially measurement tasks.

Rate Times a Quantity

In each of the following, one of the numbers is a *rate* or comparison of some amount to one set or one item. Rate examples include the "number in each set," the "cost per unit item," or the "speed per unit time." The second number is an actual *quantity*. It is the number of sets or the number of units to which the rate applies.

Mark has 4 bags of apples. There are 6 apples in each bag. How many apples does Mark have altogether?

Apples cost 7 cents each. Jill bought 4 apples. How much did they cost in all?

Peter can walk 4 miles per hour. If he walks at that rate for 3 hours, how far will he have walked?

In Figure 6.23, each problem is modeled with sets, an array, and a number line. An examination of the models will show how these three problems are alike. In the modeled form, the rate can be seen as the number in each set or the size of the set, and the quantity is the number of the sets. It is difficult for children to deal with price and time rates in problems. Helping them translate these ideas to models can help them relate these ideas to the more basic equal-addition concept they have developed for multiplication. According to Quintero (1986), children's difficulty with multiplication comes from understanding the situations in the problems and the various number uses, and not in selecting the correct operation.

Multiples of a Quantity

In these problems only one number is an actual quantity and stands as a reference set. The other number is a *multiplier* that indicates how many copies of the reference are in the

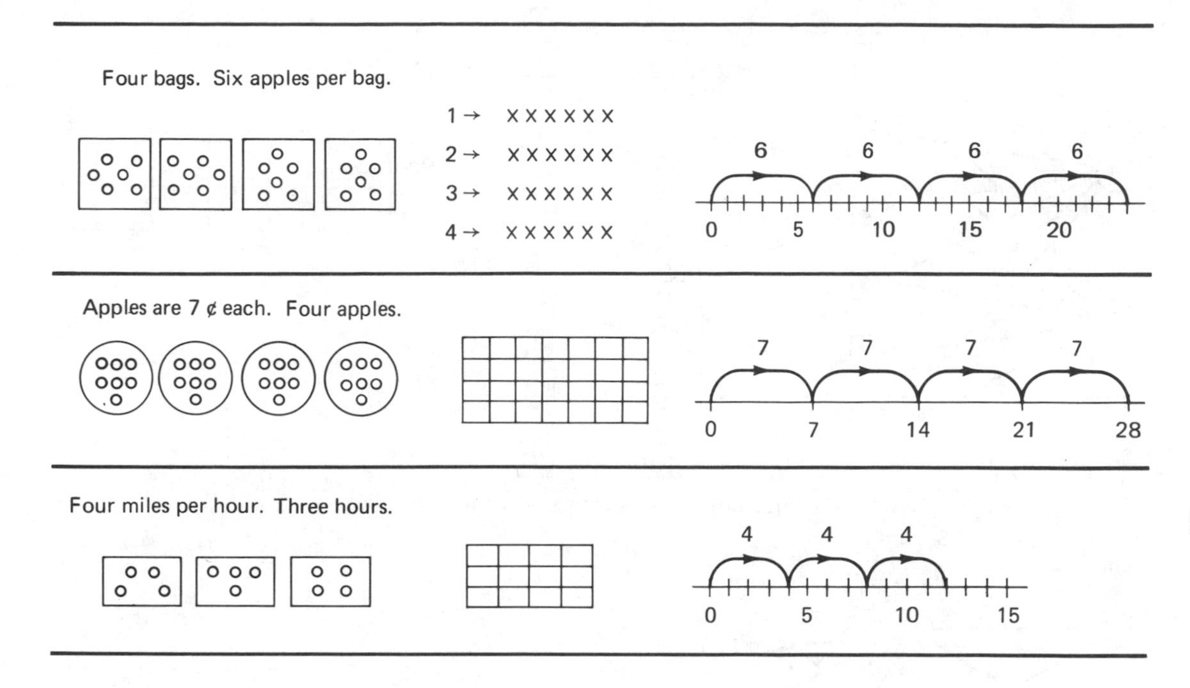

Figure 6.23
Three models of three rate times quantity problems.

total or product. These problems can be interpreted using any of the repeated-addition models.

Jill picked 6 apples. Mark picked 4 times as many apples as Jill. How many apples did Mark pick?

Mark now has 4 times as many dollars saved as he had last year. Last year he had $6. How many dollars does he have now?

Combination Problems (A Quantity Times a Quantity)

Sam Slick is really excited about his new clothing purchases. He bought 4 pairs of pants and 3 jackets, and they all can be mixed or matched. For how many days can Sam wear a different outfit if he wears one new pair of pants and one new jacket each day?

You want to make a set of attribute pieces that have 3 colors and 6 different shapes. If you want your set to have exactly 1 piece for every possible combination of shapes and colors, how many pieces will you need to make?

An experiment involves tossing a coin and rolling a die. How many different possible results or outcomes can this experiment have?

These problems reflect directly the combination concept of multiplication. There are different ways to model these problems: indicate pairings using lines, model all pairs directly, or use an array. Figure 6.24(a) shows how the coin and die experiment could be modeled all three ways.

Combination problems can have more than two quantities. For example, Sam may also have six different ties ($4 \times 3 \times 6$); the attribute pieces could include two sizes and have one, two, or three holes in each piece ($3 \times 6 \times 2 \times 3$); and the experiment may include two coins instead of one ($2 \times 2 \times 6$). Modeling combinations from three or more sets is possible but quickly becomes tedious. [Figure 6.24(b)]. However, in the upper grades it is worthwhile exploring such combinations with small numbers to see that this concept of multiplication can be extended to any number of sets. As probability and finite mathematics become more and more common in high school, early experiences with combination problems become more important.

Using Multiplication Word Problems in the Classroom

It is easy to become more interested in getting answers than in modeling problems. An emphasis on "choosing the operation" may be due to the fact that with multiplication and division large numbers appear to make modeling tedious or seemingly impossible. This is very unfortunate, since modeling is just as important for understanding relationships with these operations as it is for addition and subtraction. One way around the tedious counting of large quantities is to use the calculator to determine products. The main value in the word problem exercise is the relationship found in building the model, not in getting the answer.

In upper grades the numbers may in fact be hard to model. Instead we can use exactly the same approach as was suggested for addition and subtraction. Substitute small whole

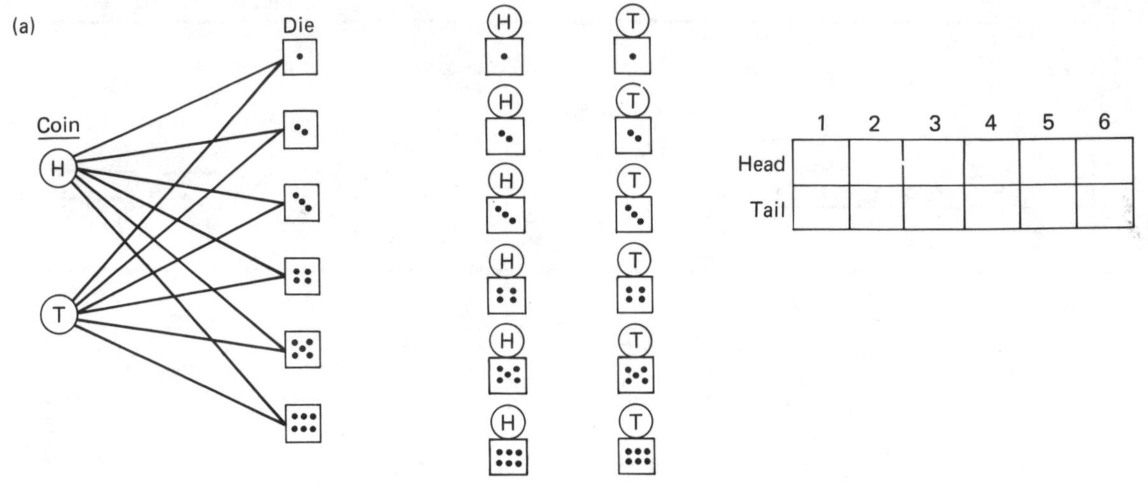

Three models for outcomes of 1 coin and 1 die.

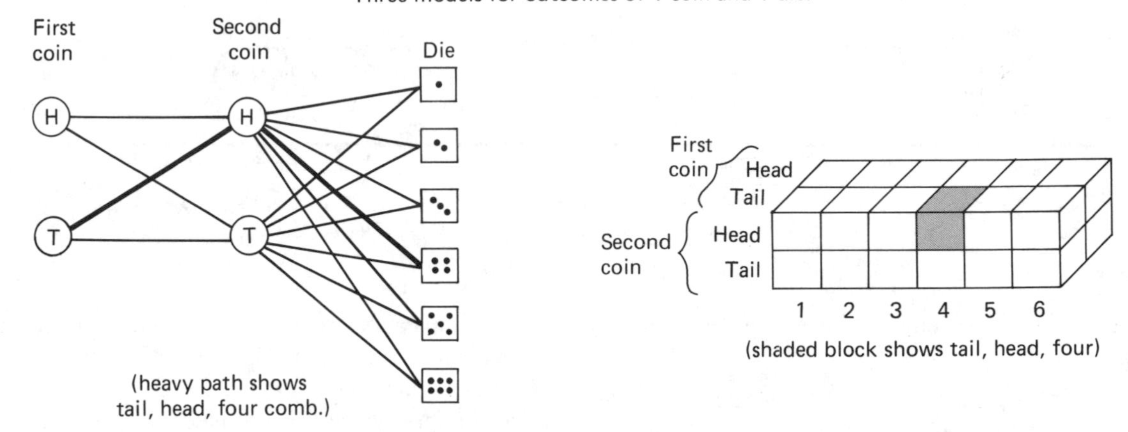

Two possible models for outcomes of two coins and one die.

Figure 6.24
Two possible models for outcomes of 2 coins and 1 die. Modeling outcomes of simple experiments.

numbers for the hard-to-model numbers (money, fractions, measurements) in the problems. The results are frequently humorous, adding some fun to the class. The value of being able to model problems with small numbers and then apply ideas to large numbers will become apparent very quickly. A second idea is to use pictures of sets or number line drawings with large numbers shown symbolically as in Figure 6.25.

Students should be exposed to all types of multiplication problems and should use all types of models. Do not make the mistake of presenting only multiplication problems and only when you are discussing multiplication. Word problem exercises should include all operations that have been covered. They also should be part of the weekly problem-solving program all year long.

Division Concepts

Corresponding to the repeated-addition concept of multiplication, there are two different concepts of division, depending on which factor is unknown. The models that we can use to illustrate division concepts are exactly the same as those for multiplication. In fact, when a division is modeled, the result always looks like a multiplication model. A division concept related to the combination concept is possible but unusual, and is not discussed here.

FAIR SHARING OR PARTITION

If a quantity is to be separated evenly into a given number of subsets, then *division* expresses the number in each subset. For example, if 24 counters are to be separated into six equal piles, the expression $24 \div 6$ is used to tell *how many are in each subset.* This meaning of division is referred to as the *partition* concept or *fair sharing.*

Besides counters, partitioning can be modeled with a number line or an array. On a number line, the distance from 0 to 24 could be partitioned or separated into six equal parts. The size of each hop or part corresponds to the number of things in each set. If 24 items are put in 6 rows, $24 \div 6$ refers to the length or number of things in each row (Figure 6.26).

In Figure 6.26, notice that after the partitioning is completed for each model, a multiplication relationship is apparent. The equation 24 ÷ 6 = □ is clearly related to 6 × □ = 24. The *size of the set* is what is unknown.

REPEATED SUBTRACTION OR MEASUREMENT

If a quantity is to be measured out into sets of a specified size, then *division* expresses the number of such sets that can be made. For example, if the 24 counters in the previous example were to be arranged in as many sets of 6 as possible, then the expression 24 ÷ 6 tells *how many sets of 6* can be made. This meaning of division is referred to as *measurement* or *repeated subtraction.* (The equal-size sets are "subtracted" from the total.)

To model the measurement concept of division on a number line the total distance from 0 to 24 is "measured out" in jumps or lengths of 6. Alternatively, the lengths could be subtracted by starting at 24 and working backward. If the 24 counters were to be arranged in rows of 6, division tells how many such rows are required. All three models for measurement division are shown in Figure 6.27.

As before, after the sets of 6 are made, the models look like a multiplication situation. For the measurement concept of division, each resulting model illustrates [4] × 6 = 24. That is, the measurement concept for 24 ÷ 6 = □ is related to □ × 6 = 24. The *number of sets* is unknown, with the size of the sets and the total amount given.

NOTATION, LANGUAGE, AND REMAINDERS

Unfortunately there are two commonly used symbols for division: 24 ÷ 6 and 6)24. The second form is the computational form. It would probably not exist if there were no pencil-and-paper computational procedure that made use of it. (The other three operations use a vertical form for computation.)

These two forms cause some troubles worth noting. First, the order of the numbers is reversed. Therefore, to read them both as "twenty-four divided by six," one is read left to right and the other right to left. To compound this difficulty is the meaningless but traditional phrase, "four *goes into* twenty-four." This expression probably originated with the computational form in conjunction with the question "How many fours are in twenty-four?" The "goes into" terminology is so ingrained in our society that most adults and teachers

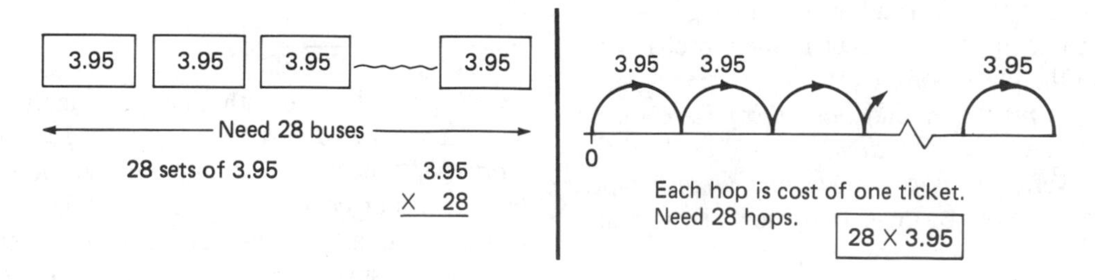

Figure 6.25
Modeling big numbers and decimals.

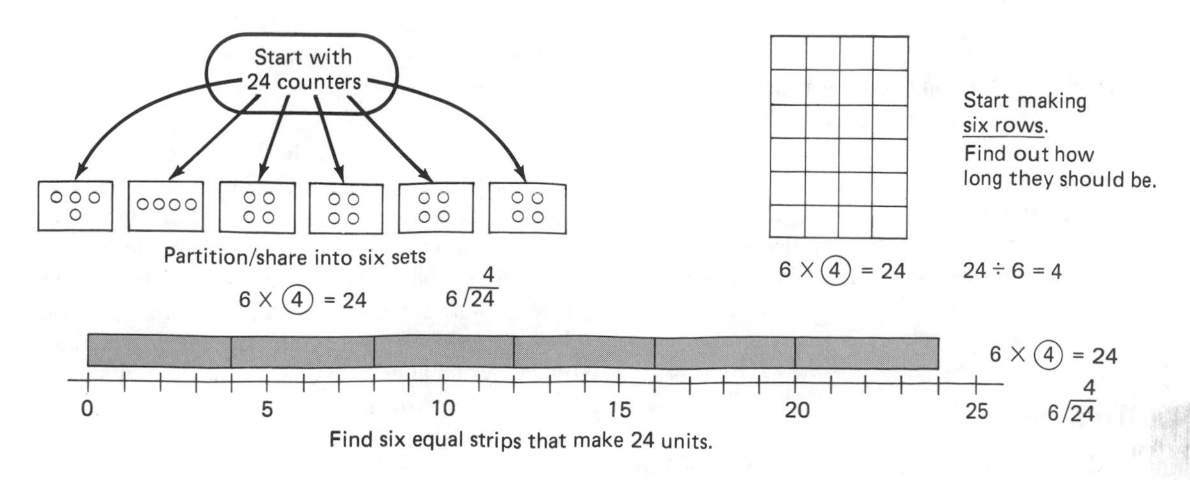

Figure 6.26
Three models of the partition concept *for 24 ÷ 6.*

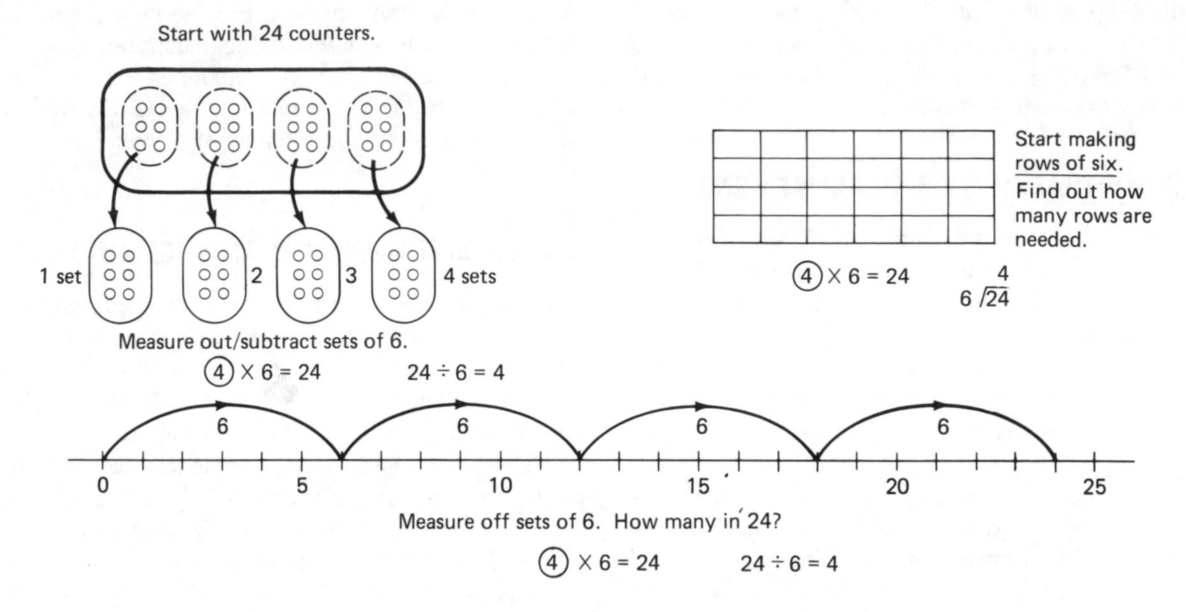

Figure 6.27
Three models of the measurement concept *for 24 ÷ 6.*

continue to use it as if it carried some clear meaning. The phrase has not been in student textbooks for many years. Perhaps it would help if we realize that "goes into" is language that is connected to *our* tradition and understanding, not children's. If "goes into" is in your vernacular, and it probably is, try not to use it in the classroom.

While the terms *multiplier* and *multiplicand* are rarely used anymore, the terms *divisor, dividend,* and *quotient* are still around. However, the words *factor* and *product* more clearly connect the concept of division to multiplication and can be used instead, as shown here.

$$\begin{array}{r} \text{factor} \\ \text{factor}\overline{)\text{product}} \end{array} \qquad \begin{array}{r} \text{quotient} \\ \text{divisor}\overline{)\text{dividend}} \end{array}$$

More often than not, division does not result in a simple whole number result. That is, problems such as 33 ÷ 6 = □ are more common than 30 ÷ 6 = □. In the absence of a word problem context, *remainders* can only be dealt with in two ways: remain a quantity left over or be partitioned into fractions. In Figure 6.28, 11 ÷ 4 = □ is modeled both ways. Notice that with whole numbers there is no way to express the remainder using the "÷" notation.

DIVISION ACTIVITIES

The partition and measurement concepts of division are the most important. When modeling and encoding these concepts, children should write both division and multiplication equations to emphasize the connection between the operations.

Activity 6.15

Provide children with an ample supply of small counters and some way to sort them into groups. Small paper cups work well. Specify a total number and the number of sets to form (partition concept) or the size of the sets (measurement concept). As a teacher-directed activity, students can work together in small groups and report the results orally. You can show students how these actions can be recorded as division equations. "You separated thirty beans into five cups. You got 6 beans in each cup. This is how we can write that." Have students write 30 ÷ 6 and/or $6\overline{)30}$. Your students should then be able to give the corresponding multiplication equations for this arrangement.

Be sure to examine both concepts in the same lesson. That is, after partitioning the beans into six cups, you might follow with separating the same 30 beans into cups of six.

The same activity just described can be done by having the children build arrays with square tiles or blocks, or by having them draw arrays on centimeter grid paper. Present the exercises by specifying how many squares are to be in the array and how many rows. The students use their models to determine the number in each row (partition). Alternatively, give them the number of squares in each row and let them determine the number of rows (measurement). They should write both multiplication and division equations for each example.

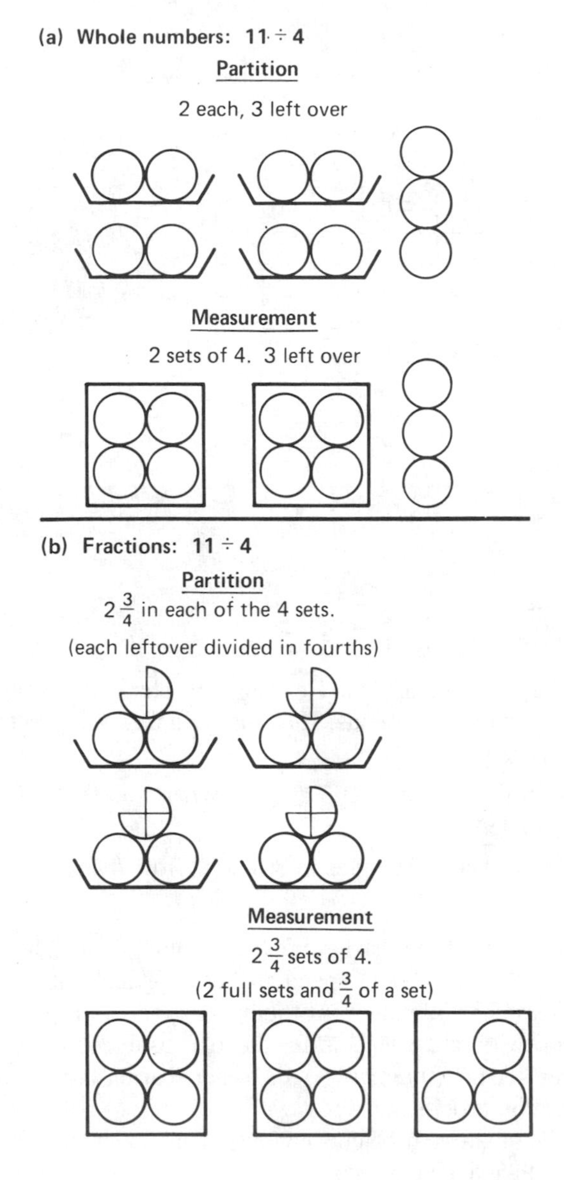

Figure 6.28
Remainders: Whole numbers and fractions.

Activity 6.16

A meter stick makes a good number line for division. Use Cuisenaire rods for the "arrows" or hops. Strips of poster board cut into lengths of 1, 2, 3, . . . , 10 cm can be used instead of the rods. Students are given a total length and either the number of rods (partition) or the length of the rods to use (measurement). When the other factor is determined, they write the corresponding division and multiplication equations.

Activity 6.17

Plastic connecting cubes such as Unifix are a useful model for grouping large sets into smaller amounts. For example, students can figure out how to use the cubes to show 36 in sets of 9 (or in 9 sets). When completed, the cubes can be arranged in separate bars (a set model), the bars can be placed side by side into an array, or if made of different colors, can be connected into one long bar like a number line (Figure 6.29).

Activity 6.18

Have children work in groups to find different methods of using the calculator to solve division exercises *without using the divide key*. For partition situations, a trial-and-error approach can be used. For 45 ÷ 6, press 6 [×] and a trial quotient and then [=] . If the trial is too large or too small, just enter a different try and press [=] . The factor of 6 is retained. For measurement, the method of counting backwards from the total or forward from zero (storing a [−] constant and pressing [=]) can be used. These calculator methods are just as conceptual as using physical models, since they clearly model the meanings of the operation. (To clearly understand how this is done, try it out.)

Introductory activities should first be done with totals or products that are multiples of the divisor; that is, the results should come out "even." Very soon, however, do similar activities with totals that produce remainders, such as 35 in four sets.

WHY NOT DIVISION BY ZERO?

Some children are simply told, "Division by zero is not allowed." While this fact is quite important, it is not always understood by children. To avoid an arbitrary rule, pose problems to be modeled that involve zero. "Take thirty counters. How many sets of zero can be made?" Or, "Put twelve blocks in zero equal groups. How many in each group?" Even if they think they may have a result that makes sense, suggest that they write the corresponding multiplication equations as they have done for other exercises.

WORD PROBLEMS FOR DIVISION

For the purpose of simplicity, division problems can be reduced to two categories corresponding to the two most prevalent meanings of division: partition and measurement. However, recall that for multiplication word problems, there were two types that related to the basic model of repeated addition: rate times a quantity and multiples of a quantity. Therefore, within the measurement and partition categories discussed here, you will note problems that correspond to each of these multiplication types. The difference is worth noting and can help us provide our students with more variation in the types of word problems we give them.

Each problem given as an example involves the same numbers and contexts as the multiplication problems on page 84 to which they correspond. It may be helpful for you to

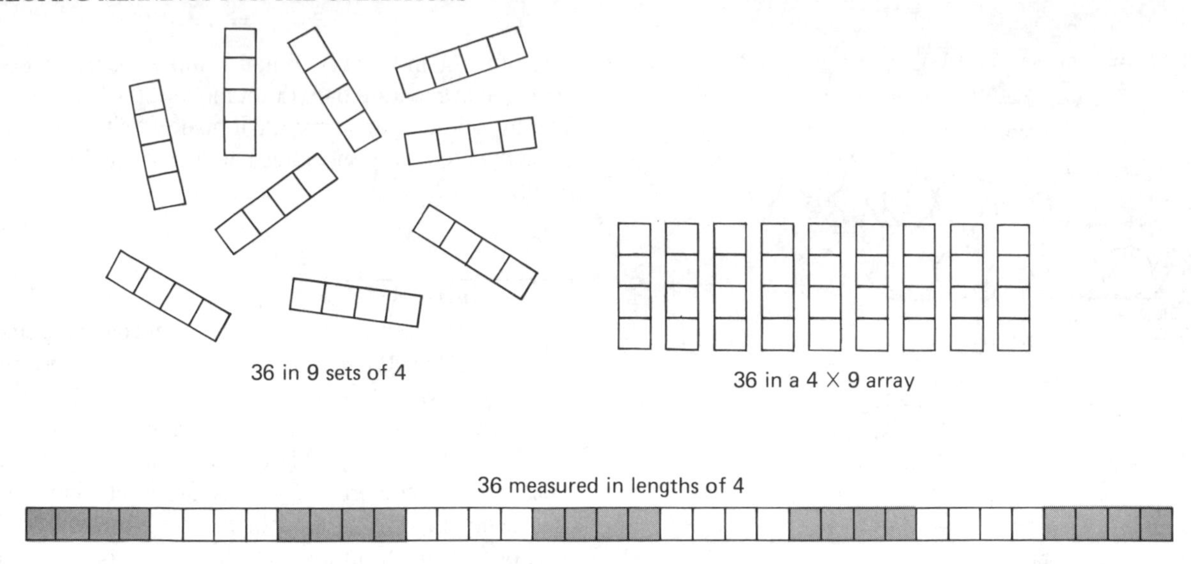

Figure 6.29
Unifix cubes in division activities.

make these comparisons so that you can strengthen your own relationship between multiplication and division. Another suggestion is to use counters, arrays, or number lines to model these problems. Then match the model with the corresponding multiplication example problem.

Measurement Problems

These first examples correspond to the rate-times-quantity multiplication problems. The quantity (number of sets) is the unknown, while the total and the rate (size of set) is given.

Mark has 24 apples. He put them into bags of 6 apples each. How many bags did Mark use?

Jill bought apples for 7 cents a piece. The total cost of her apples was 35 cents. How many apples did Jill buy?

Peter walked 12 miles at a rate of 4 miles per hour. How many hours did it take Peter to walk the 12 miles.

The next two examples are modeled in a similar manner but correspond to the multiple-times-a-quantity problems. Here the multiplier is unknown. See if you can see how these are the same and how they are a little different from the examples just given.

Mark picked 24 apples and Jill picked only 6. How many times as many apples did Mark pick than Jill did?

This year Mark saved $24. Last year he saved $6. How many times as much money did he save this year over last year?

Partition Problems

These first partition problems correspond to the rate-times-quantity problems with the rate (size of set) the unknown and the quantity (number of sets) and total given.

Mark has 24 apples. He wants to share them equally among his 4 friends. How many apples will each friend receive?

Jill paid 35 cents for 5 apples. What was the cost of 1 apple?

Peter walked 12 miles in 3 hours. How many miles per hour (how fast) did Peter walk?

The second and third examples are more difficult because price ratios and rates of speed are more difficult for children to understand and model. At the upper grades, progress can be made with rate problems like these by using models to help children see how they are like the easier sharing and repeated-subtraction problems.

The next two problems involve an unknown quantity (set size) with a given multiplier of the set and a total. Again, you are encouraged to see how these problems are alike and different from the examples just given and to relate them to the multiplication examples.

Mark picked 24 apples. He picked 4 times as many apples as Jill. How many apples did Jill pick?

This year Mark saved 4 times as much money as he did last year. If he saved $24 this year, how much did he save last year?

Using Division Word Problems in the Classroom

It is not important for students to be able to label these problems as measurement or partition. A good class activity is to have students represent the same problem using different models. The more flexibility that students develop modeling these one-step translation problems, the better will be their concepts of the operations.

By the time that division is introduced, regular weekly exercises with simple translation problems should involve all four operations. Do not announce, "Today we will do a division problem." Simply present a problem, have students model the problem and write one or more equations to go with the model. The more general translation activities from model to word problem can now become quite varied and interesting. One model can easily provoke word problems for two or more different operations and an interesting discussion can follow.

The suggestions of using small numbers in lieu of realistic larger numbers and the use of drawings, such as those in Figure 6.25, are again recommended for division problems.

Remainders in Word Problems

Earlier (Figure 6.28) it was noted that when division problems do not come out evenly, the remainder is either "left over" or can be partitioned to form a fraction. In real contexts, remainders sometimes have three additional effects on answers: (1) the remainder is discarded, leaving a smaller whole number answer, (2) the remainder can "force" the answer to the next highest whole number, and (3) the answer is rounded to the nearest whole number for an approximate result. The following problems illustrate all five possibilities.

You have 30 pieces of candy to share fairly with 7 children. How many will each receive? Ans.: 4 and 2 left over. (left over)

Each jar holds 8 ounces of liquid. If there are 46 ounces in the pitcher, how many jars will that be? Ans.: 5 and 6/8 jars. (partitioned as a fraction)

The rope is 25 feet long. How many 7-foot jump ropes can be made? Ans.: 3. (discarded)

The ferry can hold 8 cars. How many trips will it have to make to carry 25 cars across the river? Ans.: 4. (forced to next whole number)

Six children are planning to share a bag of 50 pieces of bubble gum. About how many will each get? Ans.: about 8. (rounded, approximate result)

Students should not just think of remainders as "R3" or left over. They should be put in context and dealt with accordingly.

Summary

An emphasis on concepts and understanding includes a clear understanding of the meanings of the four operations. In this chapter, each concept was developed in two different ways: through the use of models and through the use of word problems or simple translation problems. The models provide basic relationships that are easy for children to work with. The word problems extend these relationships to provide both realistic contexts and shades of meanings. The process of making translations from models to symbols and from words to symbols is a good way to help children give meaning to the operations.

By incorporating the translation problems into the weekly problem-solving program, you can find the time to develop a variety of meanings for the operations. At the same time, the process of making translations and modeling is good general problem-solving behavior. As noted in Chapter 3, translation problems belong simultaneously in two parts of the curriculum, problem solving and meanings of the operations. When the two areas are blended, both are enhanced.

For Discussion and Exploration

1. Examine a basal textbook at one or more grade levels. Identity how, and in what chapters, the meanings for the operations are developed. Discuss the relative focus on meanings of the operations with models, one-step translation problems, and mastery of basic facts.
2. Look up the *Arithmetic Teacher* article, "Verbal multiplication and division problems: Some difficulties and some solutions" (Hendrickson, 1986) and compare his classifications of multiplication and division problems with those presented here. Which problems match? Which were not included in this presentation?
3. Consider these problems that involve a rate times a rate and therefore constitute a type of problem different from those discussed.

   Mark put 6 candies in each bag and gave 3 bags to each of his friends. How many pieces of candy did each friend get?

   There are 8 ounces of cereal in a box. The store sells 3 boxes of cereal for $2. How many ounces do you get for $2.

   Discuss ways that these problems might be modeled.
4. Find some two-step translation problems for various grade levels. Discuss how these problems might be modeled using the same techniques as covered in this chapter. How would you conduct a lesson around one of these two-step problems?

Suggested Readings

Burns, M. (1989). Teaching for understanding: A focus on multiplication. In P. R. Trafton (Ed.), *New directions for elementary school mathematics.* Reston, VA: National Council of Teachers of Mathematics.

Katterns, B., & Carr, K. (1986). Talking with young children about multiplication. *Arithmetic Teacher, 33*(8), 18–21.

Mahlios, J. (1988). Word problems: Do I add or subtract? *Arithmetic Teacher, 36*(3), 48–52.

Quintero, A. H. (1985). Conceptual understanding of multiplication: Problems involving combination. *Arithmetic Teacher, 33*(3), 36–39.

Rathmell, E. C., & Huinker, D. M. (1989). Using "Part-whole" language to help children represent and solve word problems. In P. R. Trafton (Ed.), *New directions for elementary school mathematics.* Reston, VA: National Council of Teachers of Mathematics.

Sowder, L., Threadgill-Sowder, J., Moyer, M. B., & Moyer, J. C. (1986). Diagnosing a student's understanding of operation. *Arithmetic Teacher, 33*(9), 22–25.

Stuart, M., & Bestgen, B. (1982). Productive pieces: Exploring multiplication on the overhead. *Arithmetic Teacher, 29*(5), 22–23.

Suydam, M. N. (1985). Improving multiplication skills. *Arithmetic Teacher, 32*(7), 52.

Talton, C. F. (1988). Let's solve the problem before we find the answer. *Arithmetic Teacher, 36*(1), 40–45.

Thompson, C. S., & Hendrickson, A. D. (1986). Verbal addition and subtraction problems: Some difficulties and some solutions. *Arithmetic Teacher, 33*(7), 21–25.

Thompson, C. S., & Van de Walle, J. A. (1984). Modeling subtraction situations. *Arithmetic Teacher, 32*(2), 8–12.

Weiland, L. (1985). Matching instruction to children's thinking about division. *Arithmetic Teacher, 33*(4), 34–35.

CHAPTER 7

HELPING CHILDREN MASTER THE BASIC FACTS

Basic Fact Overview

Basic facts for addition and multiplication refer to those combinations where both addends or both factors are less than 10. Subtraction and division facts correspond to addition and multiplication facts. Thus, $15 - 8 = 7$ is a subtraction fact since both parts or addends are less than 10.

GOALS

Mastery of a basic fact means that a child can give a quick response (less than 3 seconds) without resorting to a less efficient means, such as counting. Work toward mastery of addition and subtraction facts begins in the first grade. Most books include all addition and subtraction facts for mastery in the second grade, although much additional drill is usually required in grade 3 and even higher. Multiplication and division facts are generally a target for mastery in the third grade. Additional practice is almost always required in grades 4 and 5. Unfortunately, there are many children in grade 8 and above who do not have a complete command of the basic facts.

A DEVELOPMENTAL PERSPECTIVE

A developmental approach to basic fact mastery involves three components:

1. A strong development of number relationships and operation meanings before mastery activities
2. The introduction of *thinking strategies* to help children develop ways to use their conceptual ideas to master facts
3. An adequate time allotment for children to develop effective use of conceptual strategies and relationships to master facts.

The Role of Concepts and Relationships

Children who possess a rich collection of relationships for numbers up to 18 are well on their way to mastering addition and subtraction facts. Thus, the number concept activities in Chapter 5 are very much a part of a good addition and subtraction fact mastery program. At the same time, those children who have not developed ideas such as one more than, two less than, five and ten relationships, and an understanding of the teens as a set of 10 and some more (see Chapter 5) are at a significant disadvantage. A strong connection between addition and subtraction concepts permits children to use knowledge of addition in mastering subtraction. For example, $13 - 7 = 6$ because $6 + 7 = 13$. Mastery of multiplication facts is similarly enhanced by good concepts of number and the meaning of multiplication, especially the order property or commutative property.

Thinking Strategies

A *thinking strategy* is the use of specific number and/or operation relationships to help memorize a particular set of basic facts. For example, one way to think about $9 + 6$ is to capitalize on relationships with 10: 9 and 1 more (from 6) is 10, and 5 more is 15. Not all facts lend themselves to the same relationships, and many facts can be processed by using more than one conceptual strategy.

The general approach is to first help children understand and use a thought process that works well for a particular collection of facts. Facts that lend themselves to this strategy are clustered together for practice exercises. As children use a strategy over and over, the thought process becomes more and more automatic. Eventually the response can be made quickly without conscious use of the strategy.

When a new strategy is introduced, children should see it as something that makes good sense, a way to use ideas they already possess. Without a good background, basic fact strategies may simply be seen as the teacher's clever tricks that are nearly as difficult as the facts themselves. The use of strategies for fact mastery is not at all a new idea. Brownell and Chazal (1935) recognized that children use different strategies and thought processes on different facts. Since the mid-1970s there has been a strong interest among mathematics educators in the idea of directly teaching strategies to children (for example, Rathmell, 1978; Thornton & Noxon, 1977; Thorn-

ton and Toohey, 1984; Steinberg, 1985; Baroody, 1985; Fuson, 1984; Bley & Thornton, 1981; Thornton, Tucker, Dossey, & Bazik, 1983). Carol Thornton has especially contributed to creative instructional techniques for fact mastery, and many of the ideas in this chapter have been adapted from her work.

Required Time and Reflection

Teachers must realize that young children require time to fully develop fact strategies and make them their own. In the early stages of a new strategy, children should be encouraged to talk through the processes in their own words, use models or drawings, explain why their processes make sense, and discuss different ways to approach the same tasks.

If children are pushed prematurely to immediate recall, they will fall back on their inefficient but reliable method of counting rather than try to use the new strategy faster. Patience pays off in the long run.

ACTIVITIES FOR FACT STRATEGY PRACTICE

The nature of fact strategy exercises is similar from one operation and one strategy to the next.

Teacher-Directed Discussions

Initial work with a strategy begins with teacher-directed discussions. Models or drawings should be used to explain how a strategy works. Children also can use models at their desks as they begin to figure out a thought process. Students can give their own explanations for working through a strategy. One or two explanations are not nearly as effective as each child being given the opportunity to explain how to use a strategy for some fact.

Independent and Small-Group Exercises

It is a good idea to have as many as 10 different activities for each strategy or group of facts. File-folder or boxed activities can be used by children individually, in pairs or even in small groups. With a large number of activities available, children can work on the facts that they need the most help with and also have sufficient variation to hold their interest.

Flash cards are among the most useful approaches to fact strategy practice. Cards can be prepared with reminders, cues, or conceptual references that help children remember to use the strategy being practiced. Ordinary flash cards can be put into groups of facts that correspond to, and thus provide practice for, a single strategy.

Other activities involve the use of special dice made from wooden cubes, teacher-made spinners, matching activities of all sorts, and simple games. These activities are designed to practice, in a single group, those facts that can be answered with a particular strategy. Drawings, models, and labels are frequently used to remind the children of the thought process they are to be practicing.

In the following sections a few activities will be suggested for each strategy. Adapt a method used for one strategy to an activity for another. All that is necessary is to provide a little variety for the students.

Strategies for Addition Facts

The strategies for addition facts are directly related to one or more number relationships. In Chapter 5, numerous activities were suggested to develop these relationships. When the class is working on addition facts, the number relationship activities can and should be included with those described here. The teaching task is to help children connect these number relationships to the basic facts.

ONE-MORE-THAN, TWO-MORE-THAN FACTS

Each of the 36 facts in the chart in Figure 7.1 has at least one addend of 1 or 2. These facts are a direct application of the one-more-than and two-more-than relationships.

Activity 7.1

Ask "What is one more than seven?" As soon as you get a response, ask "What is one plus seven?" or hold up a 1 + 7 flash card. Be sure to connect the one-more-than-seven relationship to both 1 + 7 and 7 + 1.

Activity 7.2

Make several sets of "one-more-than" flash cards.

Activity 7.3

Make a die labeled +1, +2, +1, +2, ONE MORE, and TWO MORE. Use with another die labeled 4, 5, 6, 7, 8, and 9. After each roll of the dice, children should say the complete fact: "Four and two is six."

Activity 7.4

In a matching activity, children can begin with a number, match that with the one that is two more, and then connect that with the corresponding basic fact.

Activity 7.5

A lotto-type board can be made on a file folder. Small fact cards can be matched to the numbers on the board. The back of each fact card can have a small answer number to use as a check.

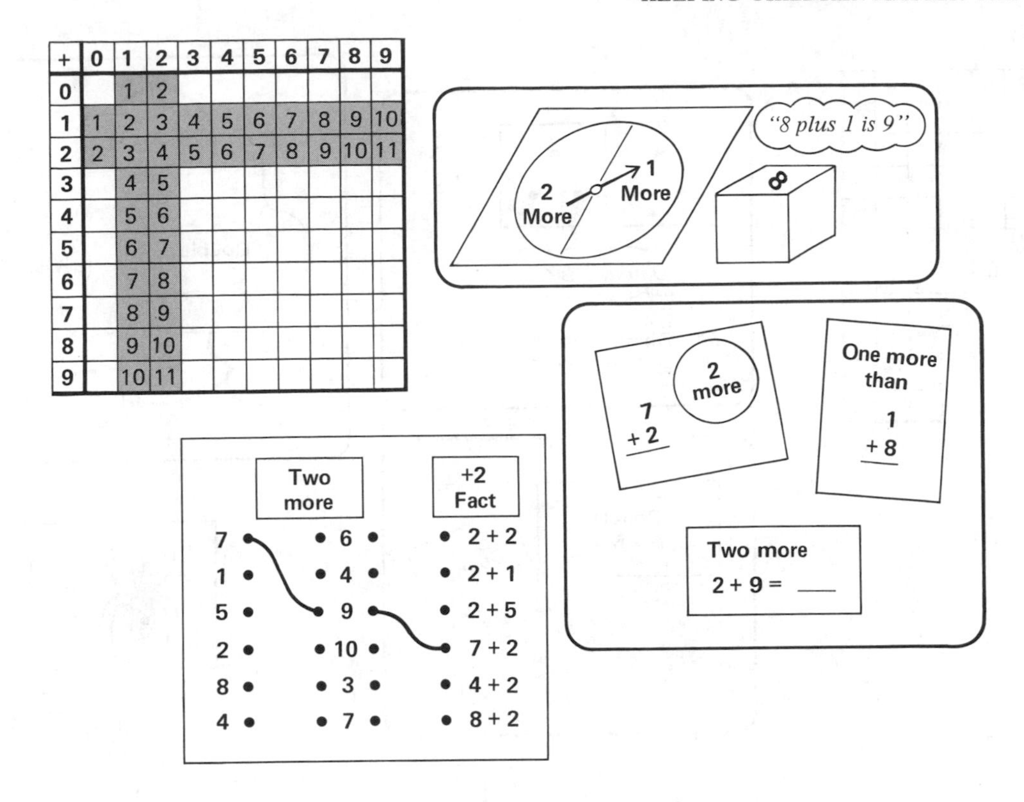

Figure 7.1
One-more and two-more facts.

FACTS WITH ZERO

Nineteen facts have a zero as one of the addends (Figure 7.2). While generally easy, some children overgeneralize the idea that answers to addition are bigger.

Activity 7.6

Write about 10 zero facts on the board, some with the zero first and some with the zero second. Discuss how all of these facts are alike. Have children use counters and a part-part-whole mat to model the facts at their seats.

Activity 7.7

Flash card and dice games should stress the concept of zero.

DOUBLES

There are only 10 doubles facts from 0 + 0 to 9 + 9, as shown in Figure 7.3. These 10 facts are relatively easy to learn and become a powerful way to learn the near-doubles (addends one apart). Some children use them as anchors for other facts as well.

+	0	1	2	3	4	5	6	7	8	9
0	0	1	2	3	4	5	6	7	8	9
1	1									
2	2									
3	3									
4	4									
5	5									
6	6									
7	7									
8	8									
9	9									

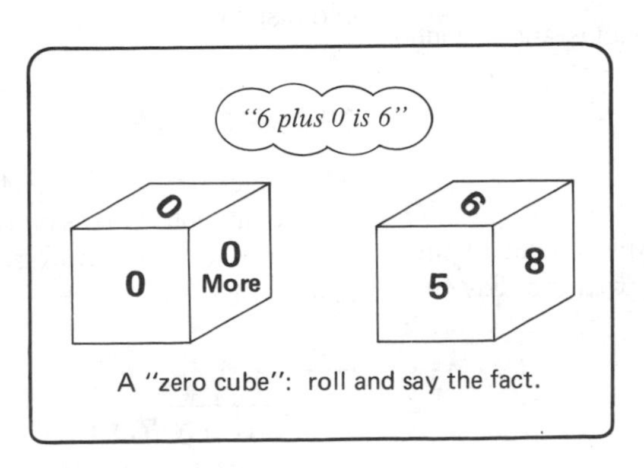

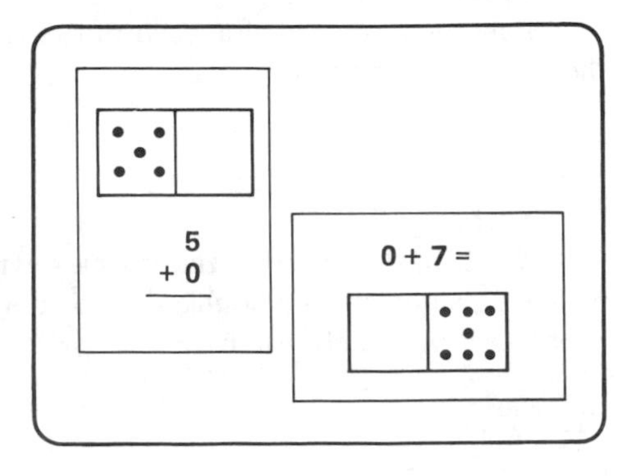

Figure 7.2
Facts with zero.

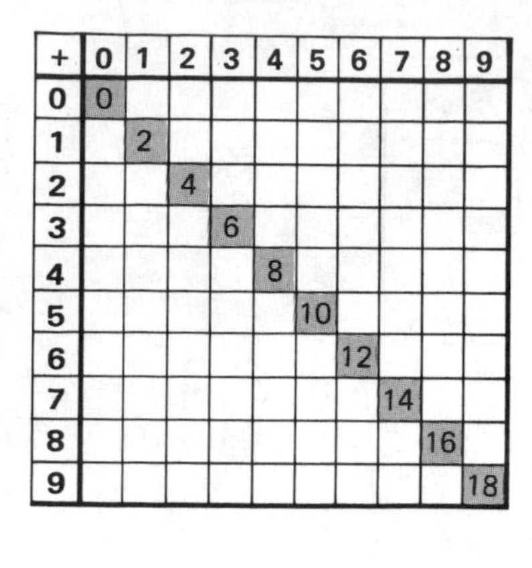

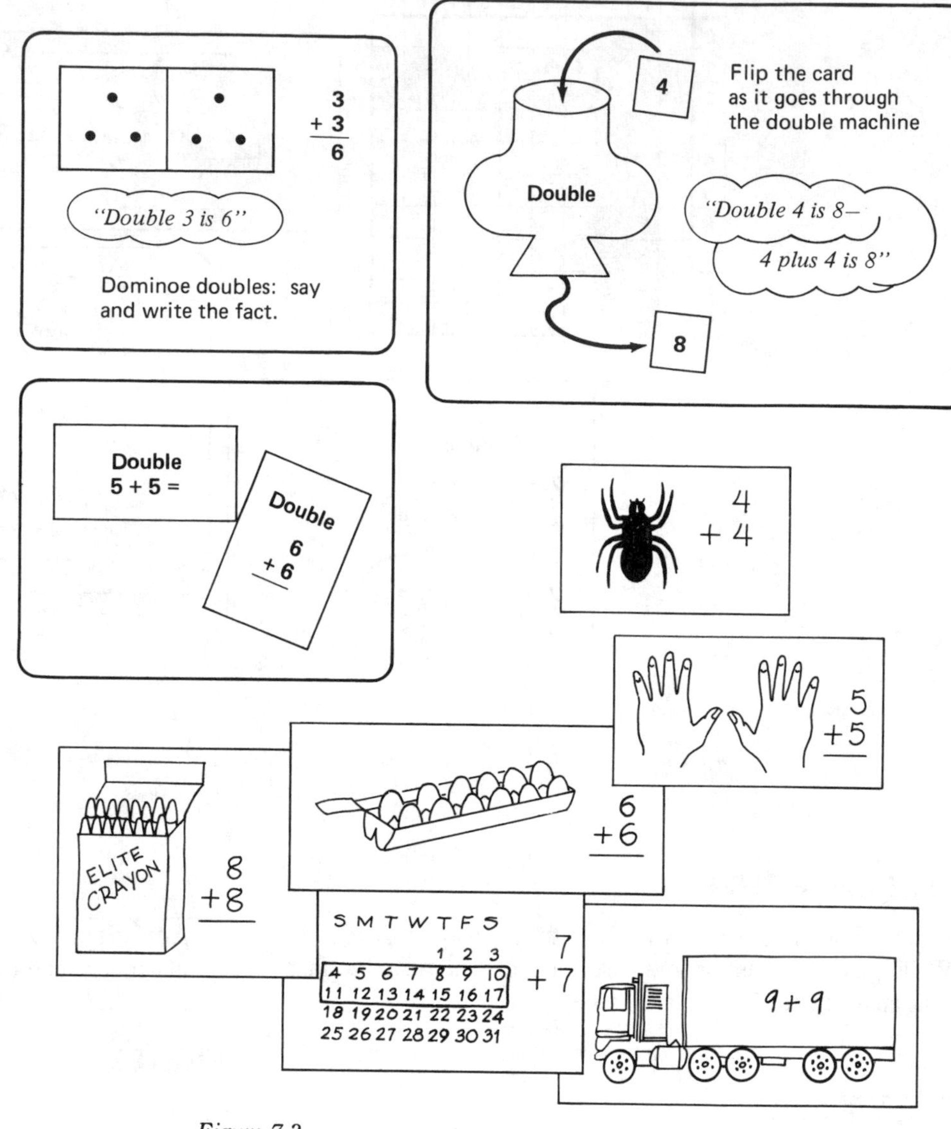

Figure 7.3
Doubles facts.

Activity 7.8

Make picture cards for each of the doubles and include the basic fact on the card.

Activity 7.9

Match flash cards for the doubles with pictures of the doubles images or with double-domino dot patterns. Say the doubt fact with each match.

Activity 7.10

Use the calculator and enter the "double maker" (2 ☒ ⊟). Let one child say, for example, "seven + seven." The child with the calculator should press 7, try to give the double (14) and then press 7 ⊟ to see the correct double on the display.

NEAR-DOUBLES

These facts are also called the *doubles plus 1* facts and include all combinations where one addend is one more than the other (Figure 7.4). The strategy is to double the smaller number and add 1.

Activity 7.11

After discussing the strategy with the class, write 10 or 15 near-doubles facts on the board. Use vertical and horizontal formats and vary which addend is the smaller. Quickly go

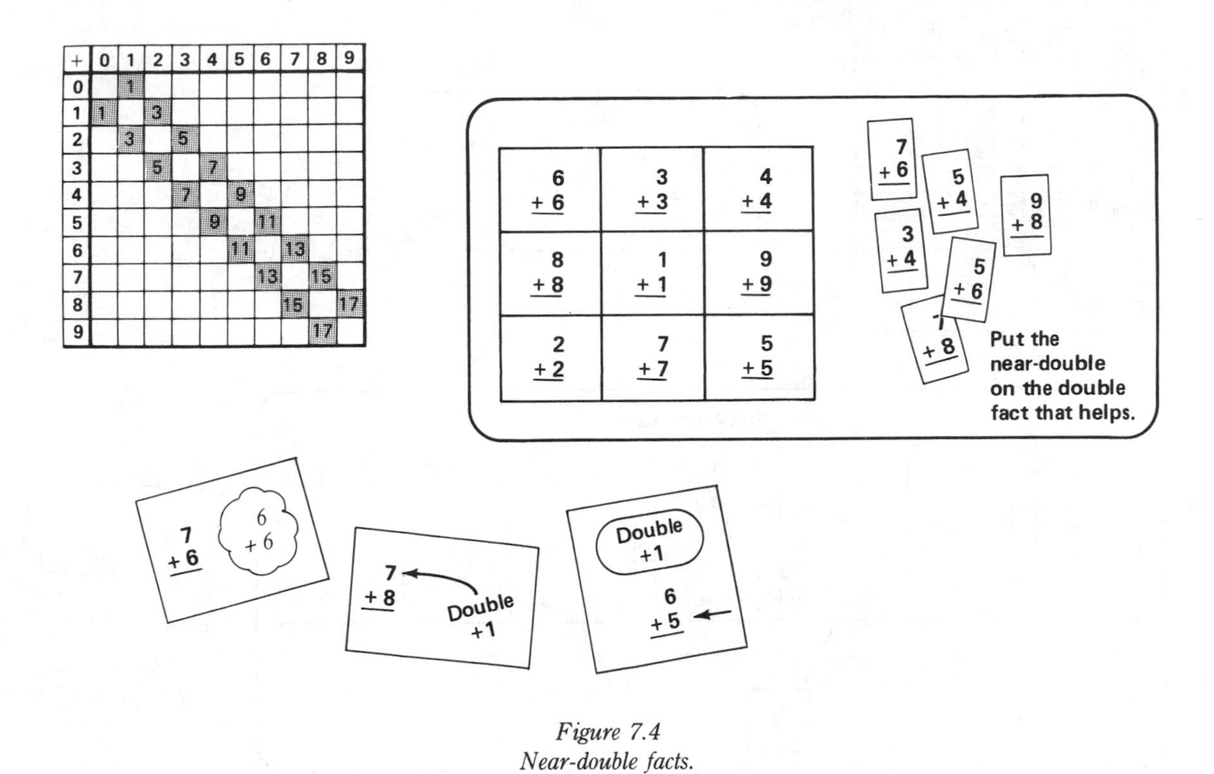

Figure 7.4
Near-double facts.

through the facts. First have students only identify which number to double. The next time through, have students name the double that will be used. The third time, have students say the double and then the near-double.

Activity 7.12

Use two sets of doubles cards and a complete set of near-doubles cards. Mix both sets and have students match them up, giving the answer for both facts.

Activity 7.13

Roll a single die with numerals or dot sets and say the complete double-plus-one fact. That is, for seven, students should say "seven plus eight is 15."

Activity 7.14

Make a matching sheet or a matching game board in which double-plus-one facts are matched with corresponding doubles facts.

FIVE AND TEN FACTS

This is really two categories of facts in one: those with a five as one addend and those with a sum of 10 (Figure 7.5). The five and ten benchmark relationships for numbers are so powerful that it makes sense to capitalize on them to help master the facts. The ten-frame and the five-bar (see figure) are important models related to this strategy.

Activity 7.15

On the board draw seven "cards" as shown in Figure 7.5. Select numbers from 5 to 9 and discuss what two cards can be used to make the given number. At least one of the two cards must be a five-bar. For each number have the class say an addition fact and write it on the board.

The last activity can be expanded to include target numbers up to 14. For numbers 10 to 14, the object is to find two numbers that make the target, using the cards in the set to model these numbers. For the target 13, 5 and 8 or 6 and 7 can be used, and each can be made with the five cards.

Activity 7.16

Make sets of five-bar cards for each child. After you say a number, the children use their cards to make the number and hold it up. They must use at least one five-bar for each number. Have them say the fact that goes with their combinations.

Activity 7.17

Make flash cards using both five-bars and ten-frames. (The facts with sums to 10 must be mixed in with the "five" facts; otherwise all answers would be 10.)

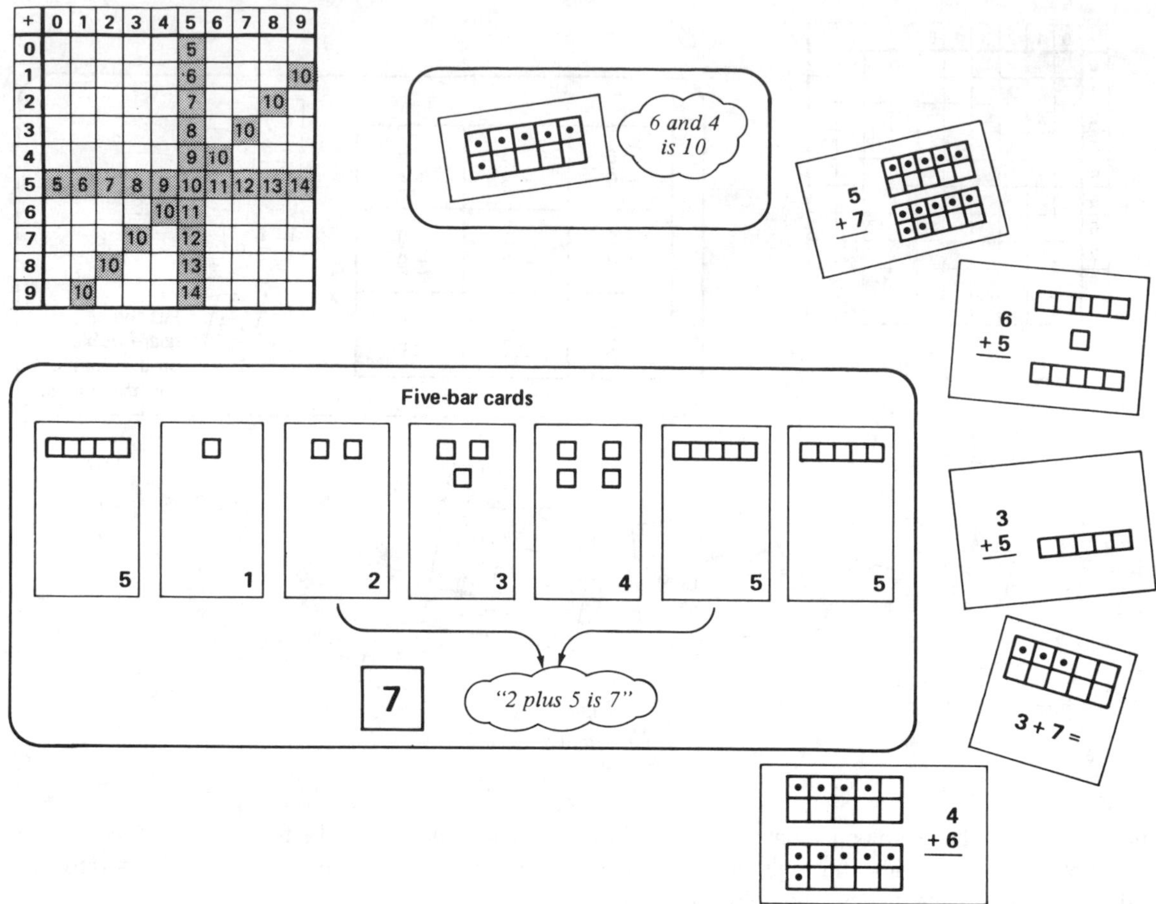

Figure 7.5
Five and ten facts.

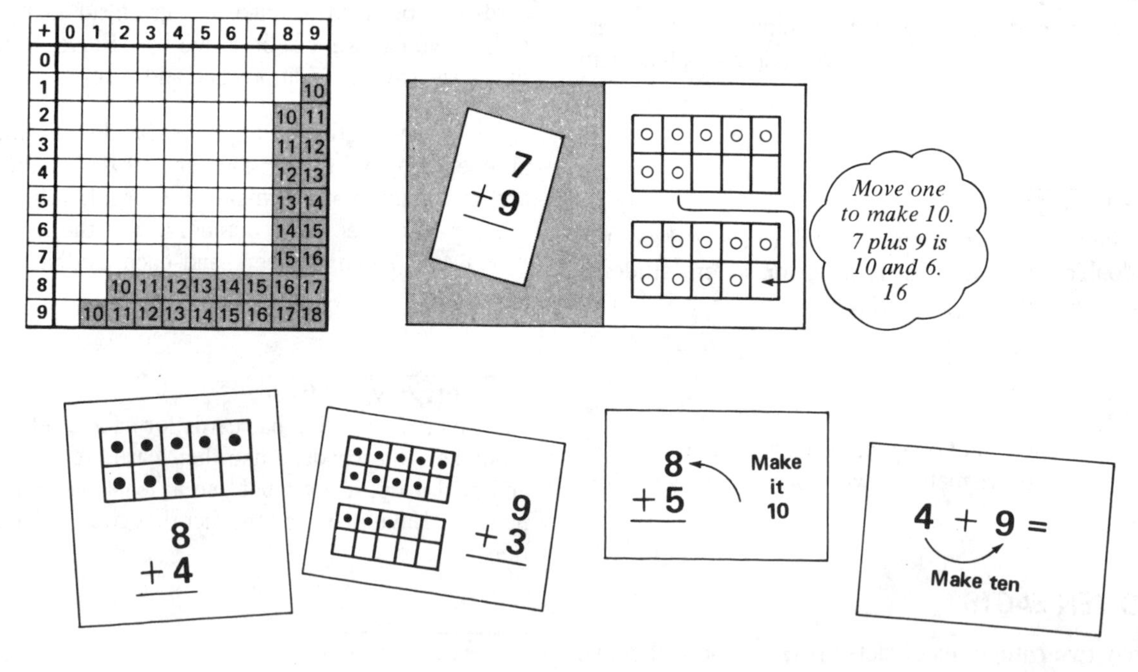

Figure 7.6
Make-ten facts.

Activity 7.18

Use the calculator to practice "plus five." Enter [+] 5 [=]. Try to beat the calculator as different single digits are pressed at random.

Activity 7.19

Hold up a ten-frame card and have children say the "ten fact." For a card with seven dots, the response would be "seven and three is ten."

Later, with a blank ten-frame drawn on the board as a reminder, say a number less than 10. If you say, "four," they say, "four and six is ten." This idea can be adapted into a variety of independent games or activities.

MAKE-TEN FACTS

These facts all have at least one addend of 8 or 9 (Figure 7.6). The strategy is to build onto the 8 or 9 up to 10 and then add on the rest. For 6 + 8: start with 8, then 2 more makes 10 and that leaves 4 more or 14.

Before using this strategy be sure that children have learned to think of the numbers 11 to 18 as 10 and some more. Many second- and even third-grade children have not constructed this relationship. (Refer to the section of Chapter 5 on "Relationships for Numbers 10 to 20," p. 63).

Activity 7.20

Give students a mat with two ten-frames. Flash cards are placed next to the ten-frames or a fact can be given orally. The students should first model each number in the two ten-frames and then decide on the easiest way to show (without counting) what the total is. The obvious choice is to move counters into the frame showing either 8 or 9. Get students to explain what they did. Have them explain that 9 + 6 is the same as 10 + 5.

Provide a lot of time with the last activity. The make-ten relationship is also the basis for many subtraction facts to be learned later on.

Activity 7.21

Make ten-frame cards on acetate. Show an 8 (or 9) card on the overhead. Place other cards beneath it one at a time. Suggest mentally "moving" two dots into the 8 ten-frame. Have students say the complete fact: Eight and four is twelve." The same activity can be done independently with regular ten-frame cards.

Activity 7.22

Make flash cards with either one or two ten-frames shown and also with reminders to "make ten" out of the 8 or the 9.

OTHER STRATEGIES

As summarized in Figure 7.7, the suggested strategies cover all but two facts, 6 + 3 and 7 + 4, and their commutative partners, 3 + 6 and 4 + 7. (Thornton calls these "turn-arounds.") Most teachers would be delighted for a method of mastering the other 96. There are strategies other than the ones presented so far. Most teachers tend to adopt their own "favorites."

Counting on is the most widely promoted of the strategies not included here. Counting on is applied for all facts with one addend of 1, 2, or 3. For the fact 3 + 8, the child starts with 8 and counts three counts: "9, 10, 11." Many textbooks and supplementary materials encourage this strategy. It was omitted from the preceding strategies because it is more procedural than conceptual, because it is frequently used where it is not efficient, as in the 5 + 7 fact, and because it is not required if all of the other strategies are developed. The count-on strategy is certainly an acceptable alternative, especially for students who have not constructed the one-more-than and two-more-than relationships. More than three counts, for a fact such as 7 + 4 or 7 + 5, is difficult without using fingers or counters.

The make-ten strategy is frequently extended to facts with a 7 as one of the addends. That is, 7 + 4 and 5 + 7 could be done by adding to the 7 to make 10.

The *doubles-plus-two* strategy is applicable to all facts with addends that are 2 apart and works the same as the near-

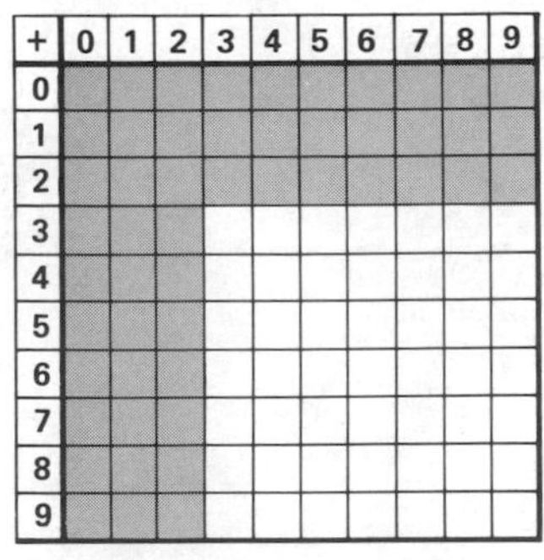

Zeroes, one-more and two-more facts

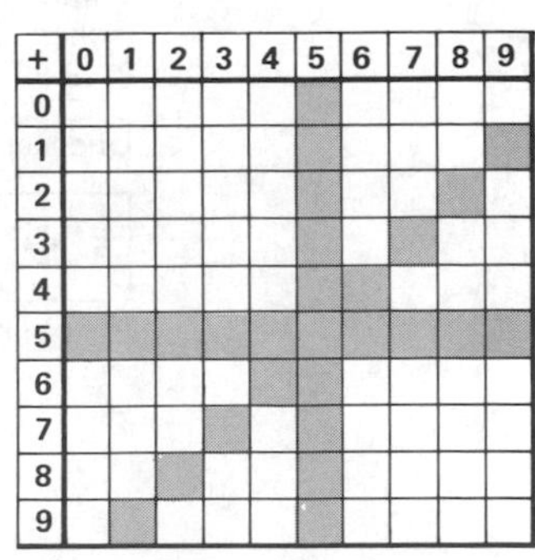

Five and ten facts

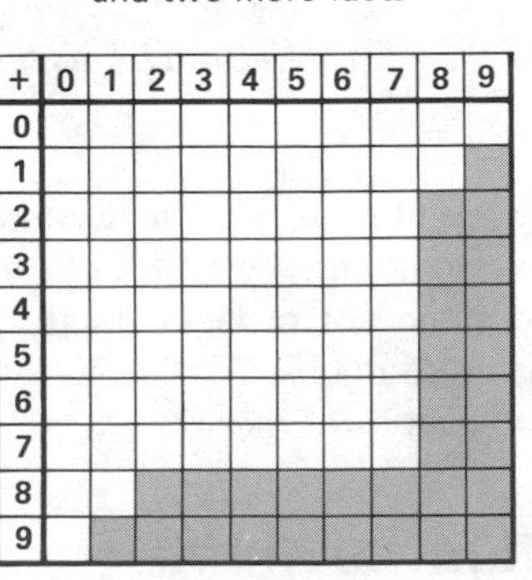

Make ten facts

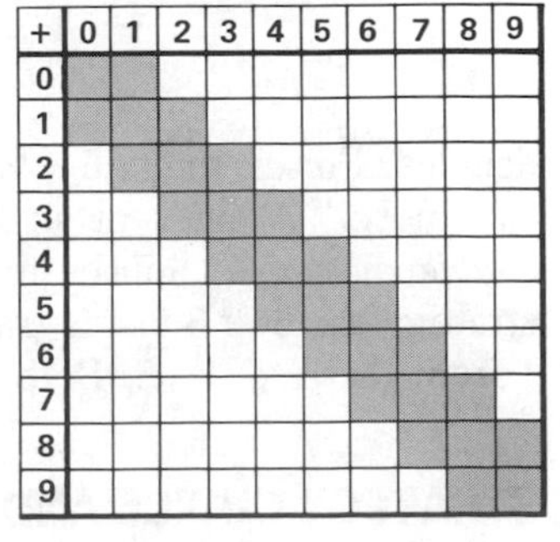

Doubles and near doubles

Figure 7.7
A summary of addition fact strategies.

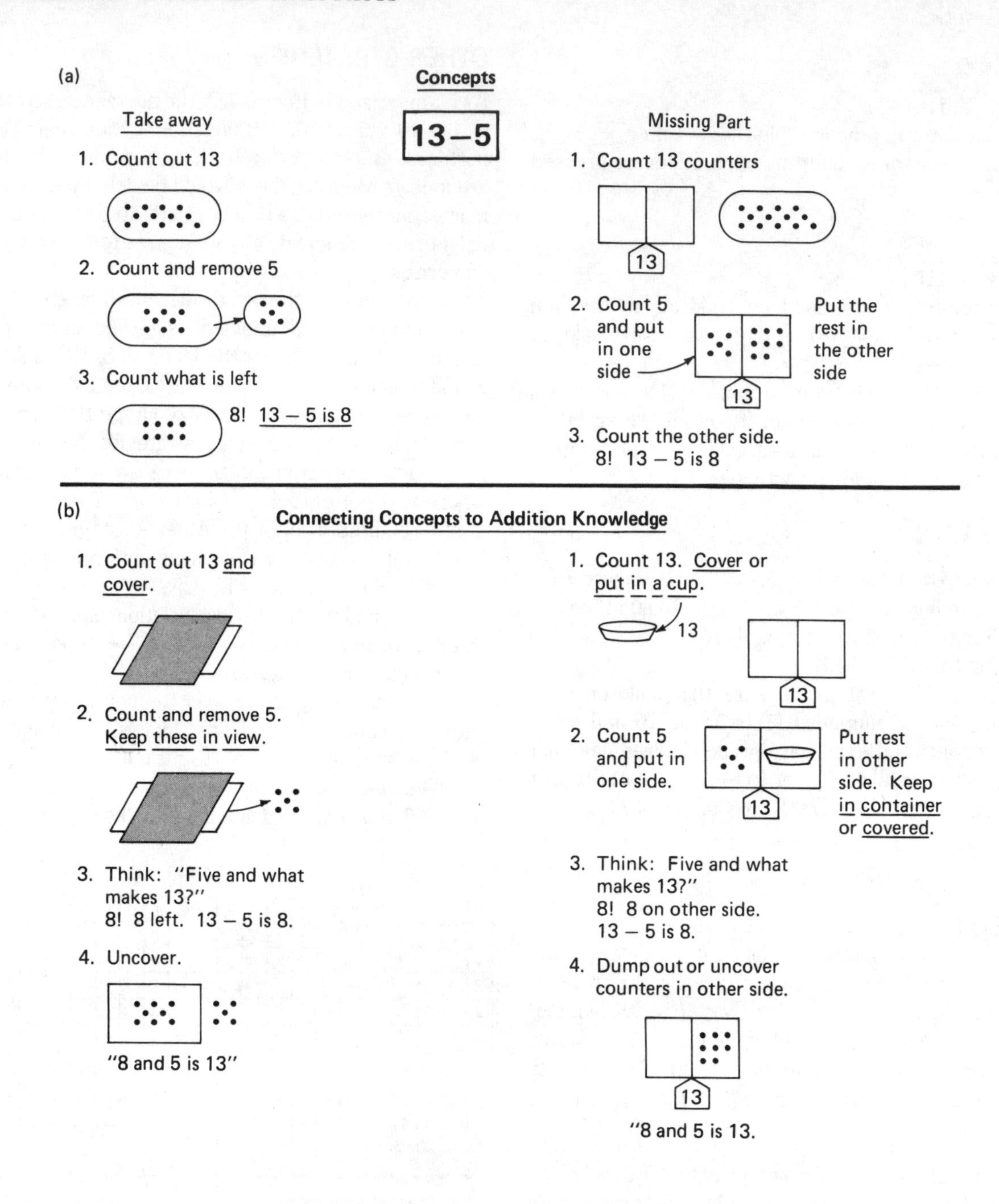

Figure 7.8
Using a think-addition model for subtraction.

doubles. For example, 6 + 8 is thought of as "double six and two more." A different strategy, sometimes called *make a double inbetween,* applies to the same set of facts. In that strategy the 6 + 8 fact is translated to double 7 by a shift of 1 from the 8 to the 6. Both strategies have merit.

STRATEGY CHOICE AND FINGER COUNTING

By now you have probably noticed that some facts belong in more than one strategy group. Children should be encouraged to use the strategy of their choice. If that strategy is to count on their fingers or some other inefficient process, it is quite likely that they have not developed sufficiently strong number relationships and/or have not worked with the more efficient strategy long enough. Children who count on their fingers are telling you that their concepts and strategies are not yet well developed. Spend more time working on relationships and exploring strategies conceptually rather than moving forward in a drill mode.

Strategies for Subtraction Facts

THE ADDITION-SUBTRACTION LINK

Difficulties

Subtraction facts prove to be more difficult than addition for children. Consider the fact 13 − 5. At a concept level, most children will count out 13 counters, count off 5 of the 13 and remove them, and finally count the remaining counters. There seems to be little evidence that children who do master subtraction facts are using this counting procedure to assist them. Unfortunately, many sixth, seventh, and eighth graders are still counting.

Modeling Subtraction as Think Addition

In Figure 7.8(a), the take-away and missing-part concepts for subtraction are illustrated, showing the three counts used to get answers. In Figure 7.8(b) the same actions are repeated, but the modeling encourages the child to think, "What goes with this part to make the total?" When done in this *think-addition* manner, the child uses known addition facts to produce the unknown quantity. If necessary, counting can be used to verify a response. When the unknown part is uncovered, the addition relationship is clearly evident and the addition fact can be verbalized or written.

SUMS TO TEN AND NUMBER RELATIONS

The think-addition approach is applied first to subtraction facts with sums of 10 or less since these 64 facts are generally introduced first. The basic idea is to cluster facts in correspondence to the strategies used for addition.

Facts with Zero

This set of facts includes those involving minus zero and those with a difference of zero (e.g., 7 − 0, 7 − 7).

Using the think-addition approach, model these facts on the overhead. You may be surprised to find some children confused by them, especially those involving subtracting zero.

Make flash cards as shown in Figure 7.9.

Figure 7.9
Flash cards for zero facts.

One-Less-Than, Two-Less-Than Facts

This group includes all facts with differences of 1 or 2 as well as those that involve −1 or −2 (e.g., 8 − 7, 8 − 6, 8 − 1, 8 − 2). The relationships of more than and less than must be connected.

Review the 1- and 2-less-than relationships using any of the ideas in Chapter 5. With these relationships in students' minds, model facts in this group on the overhead projector using the think-addition approach. Frequently follow one fact with its partner. That is, follow 9 − 2 with 9 − 7 and discuss how the two facts are alike. Then write the 7 + 2 = 9 fact. Help children see how the 2-more-than and 2-less-than relationships are involved in both facts.

Activity 7.23

Use the calculator to practice "minus two" and "minus one" facts. Press [−] 2 [=] to make a "minus two" machine. Make flash cards for all 36 facts in this group. Use the words "one less" or "two less" on all cards, as shown in Figure 7.10.

Activity 7.24

Have children draw a number card from numbers 2 to 10 (or spin a spinner or roll a die). Let them make a set of that many counters and cover them with a piece of tagboard as in

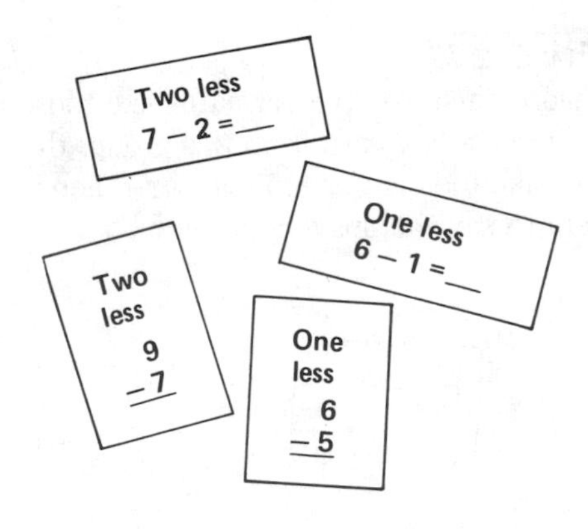

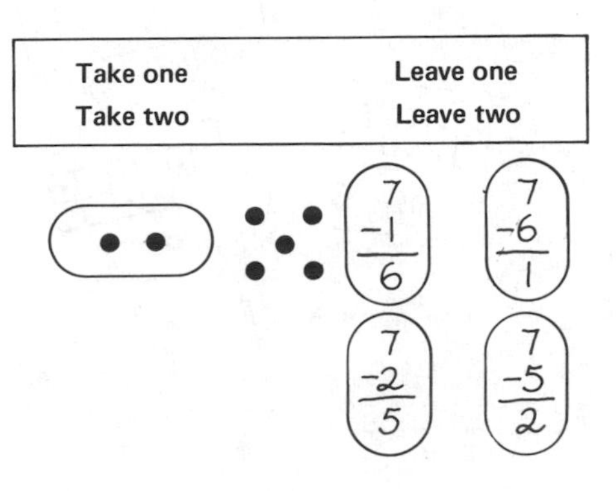

Figure 7.10
One-less and two-less.

the think-addition model. Then using the counters to model what they are doing, the children should write a subtraction fact for each of these four statements that you have written on a board or poster.

TAKE ONE   LEAVE ONE
TAKE TWO   LEAVE TWO

Five and Ten Facts

This group includes all facts with the first number of 10, those involving -5 and those with a difference of 5. Therefore, $10 - 7$, $8 - 5$, and $8 - 3$ are all in this group.

Activity 7.25

Make flash cards for all three types of facts in this group. All are shown in Figure 7.11. Use all the cards together. Similar drawings can be incorporated into other activities.

Activity 7.26

Show a ten-frame card and have students say a subtraction fact and then an addition fact. For a frame showing six dots, students would say, "Ten minus six is four," and then, "Six and four is ten."

Activity 7.27

Give children a set of five-bar cards like those in Figure 7.5. As they draw a fact card from this group, they first use the cards to make the sum or top number. Then it should be easy for them to say the answer to the fact.

Doubles and Near-Doubles

This group of facts corresponds to the addition facts of the same name. For most activities it will be useful to mix the doubles and near-doubles.

Activity 7.28

Review the addition doubles and the near-doubles. Then model the double and near-double subtraction facts on the overhead projector, using the think-addition approach. For each subtraction fact, ask if it was a double or a near-double, and have children say both facts. ("Near double! Nine minus four is five. Four plus five is nine.")

Activity 7.29

Write a large number of subtraction facts on a page. Make roughly a third of them doubles, a third near-doubles, and the other third other facts. Feel free to repeat facts several times. Have students circle with a crayon all of the doubles facts and then answer them. Next have them circle the facts that are near-doubles with a different crayon and answer them. The remaining facts can be left blank. Next to each subtraction fact, have them write the addition double or near-double that goes with it.

Activity 7.30

Make a matching worksheet or a matching game in which subtraction facts are paired with the double or near-double that helps answer it.

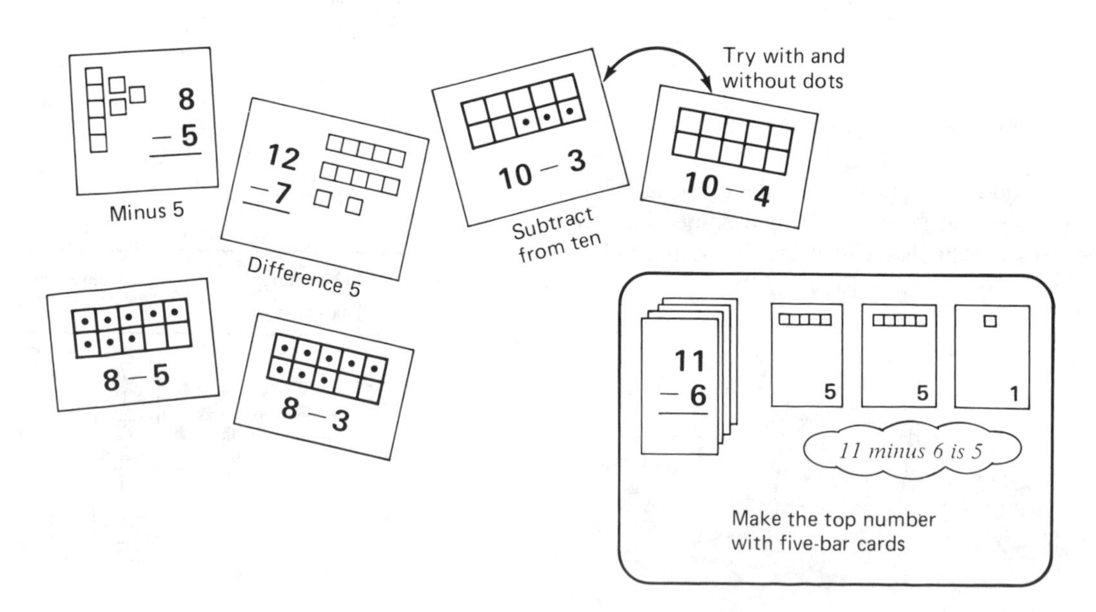

Figure 7.11
Five and ten subtraction facts.

THE 36 "HARD" SUBTRACTION FACTS: SUMS GREATER THAN 10

Many teachers find that the 36 subtraction facts with sums greater than 10 (Figure 7.12) tend to give children the most difficulty. Three overlapping strategy groups can be used to master these.

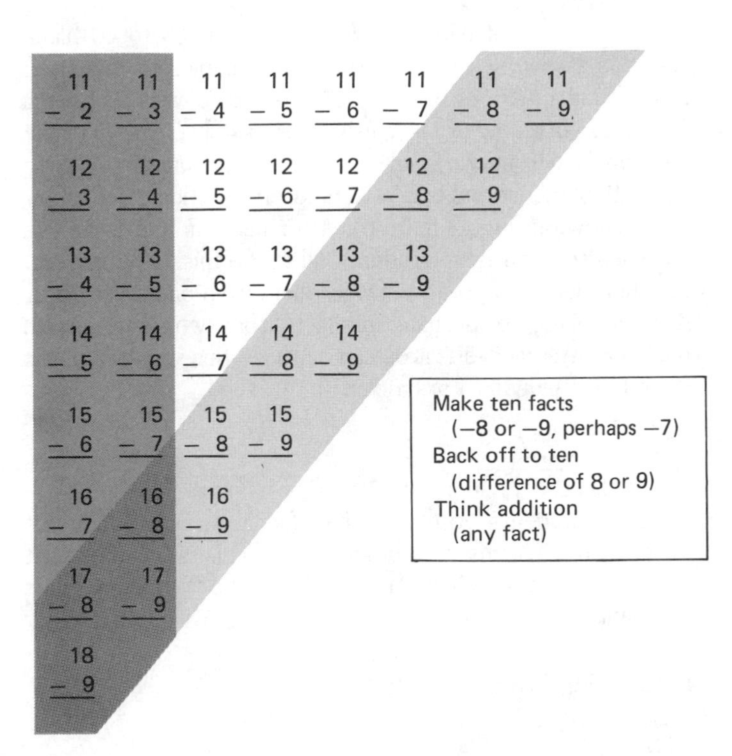

Figure 7.12
The 36 "hard" subtraction facts.

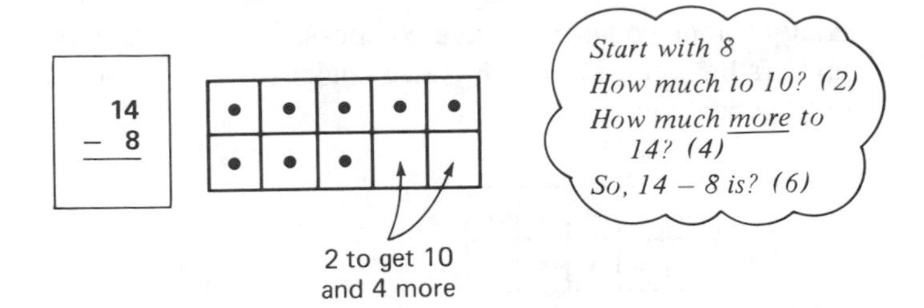

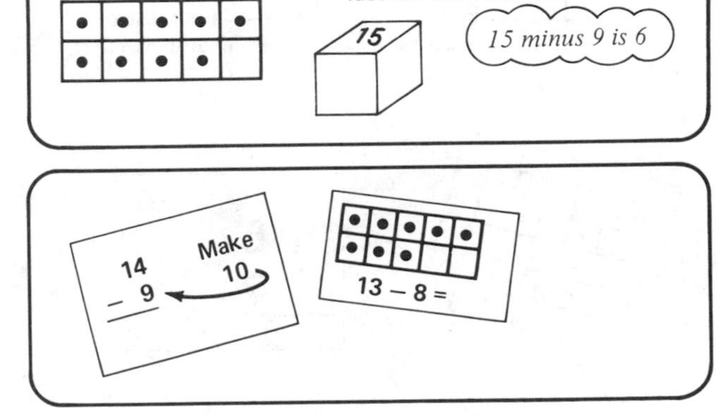

Figure 7.13
Make-ten subtraction facts.

Make-Ten Facts

This group includes those facts where the part or subtracted number is either 8 or 9. Examples are 13 − 9 and 15 − 8. The strategy, illustrated in Figure 7.13, is very similar to the make-ten strategy for addition.

Activity 7.31

On the board or overhead draw a ten-frame with nine dots. Discuss how you can build numbers between 11 and 18, starting with the 9 in the ten-frame. Stress the idea of *one more to get to* 10 and then the rest of the number. Repeat for a ten-frame showing 8. Next, with either the 8 or 9 ten-frame in view, call out numbers from 11 to 18 and have students respond with the difference between that number and the one on the ten-frame. Later, use the same approach, but show fact cards to connect this idea with the symbolic subtraction fact.

Activity 7.32

Make a mat with two ten-frames on which students can place counters. On one of the ten-frames draw in nine counters. Provide students with flash cards for the "minus nine" facts. Have them use counters to build the sum starting with the nine dots already there. Use a frame with eight dots for the "minus eight" facts.

Back Off to 10

Here is one strategy that is really take-away and not think addition. It is useful for all of those facts that have a difference of 8 or 9, such as 15 − 6 or 13 − 5. For example, with 15 − 6 you start with the total or 15 and "back off" five. That gets you down to 10. Then take off 1 more to get 9. For 14 − 6, just back off 4 and then take off 2 more to get 8. Here we are working backward with 10 as a "bridge." In the make-ten strategy we start with the subtracted number and work *up* through 10. Taken together these facts correspond to all of the make-ten addition facts and use the same basic relationship.

Activity 7.33

Start with two ten-frames on the overhead, one filled completely and the other partially filled as in Figure 7.14. For 13, for example, discuss what is the easiest way to think about taking off four counters or five counters. Repeat with other numbers between 11 and 18. Have students write or say the corresponding fact.

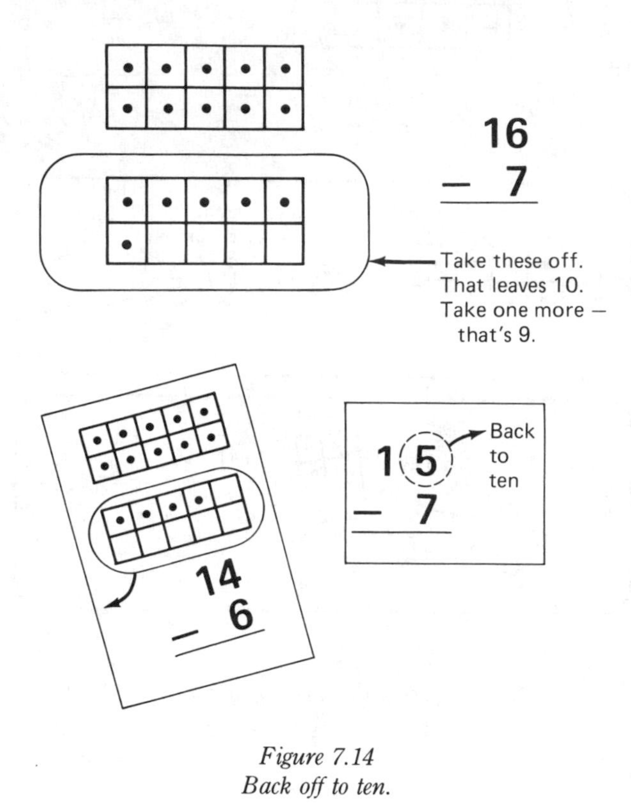

Figure 7.14
Back off to ten.

Extend the Think-Addition Strategies

The "five facts," doubles, or near-doubles include facts with sums greater than 10. The activities suggested earlier for sums to 10 can be extended to include these new facts. The think-addition model is a general approach that can include any fact.

MORE THINK-ADDITION ACTIVITIES

The activities in this section are think-addition activities that can be used with any of the 100 subtraction facts.

Missing-Number Cards

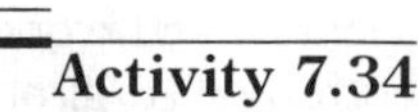

Activity 7.34

Show children, without explanation, families of numbers with the sum circled as in Figure 7.15(a). Ask why they think the numbers go together and why one number is circled. When this number family idea is fairly well understood, show some families with one number replaced by a question mark [Figure 7.15(b)] and ask what number is missing. When students understand this activity, explain that you have made some missing-number cards based on this idea. Each card has two of three numbers that go together in the same way. Sometimes the circled number is missing (the sum), and sometimes one of the other numbers is missing (a part). The cards can be made both vertically and horizontally with the sum appearing in different positions. The object is to name the missing number.

Activity 7.35

On a sheet of paper draw neatly about 35 to 40 blank cards as shown in Figure 7.16(a). Use copies of the blank form found in Appendix C to make a wide variety of drill exercises. In a row of 12 "cards," for example, put all of the combinations from two families with different numbers missing and with numbers and blanks in different positions. An example is shown in Figure 7.16(b). After filling in numbers, run the sheet off and have students fill in the missing numbers. Another idea is to group together facts from one strategy or number relation or perhaps mix facts from two strategies on one page. Actual flash cards can be made this way and put in packets for individual practice.

Activity 7.36

Have students write an addition fact and a subtraction fact to go with each missing-number card. This is an important step because many children are able to give the missing part in a family but do not connect this knowledge with subtraction.

Find a Helping Plus Fact

Activity 7.37

Select a group of subtraction facts that you wish to practice. Divide a sheet of paper into small "cards," about 10 or 12 to a sheet. For each subtraction fact, write the corresponding addition fact on one of the cards. Two subtraction facts can be related to each addition fact. Duplicate the sheet and have students cut the cards apart. Now write one of the subtraction facts on the board. Rather than call out answers, students find the addition fact that helps with the subtraction fact. On your signal, each student holds up the appropriate fact. For 12 − 4 or 12 − 8 the students would select 4 + 8.

Activity 7.38

The same idea of matching a subtraction fact with a helping-addition fact can be made into a matching card game or a matching worksheet.

Activity 7.39

Supply a list of facts and have students first write the helping fact and then answer both. Begin by giving a list of addition facts and have students write either subtraction fact

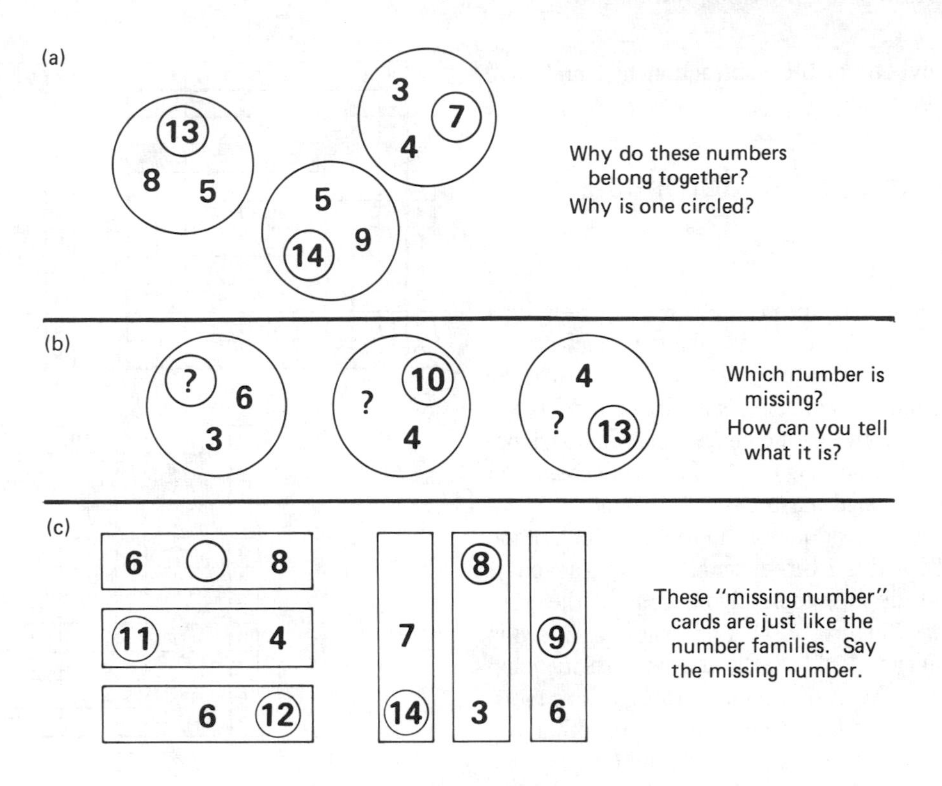

Figure 7.15
Introducing "missing-number" cards.

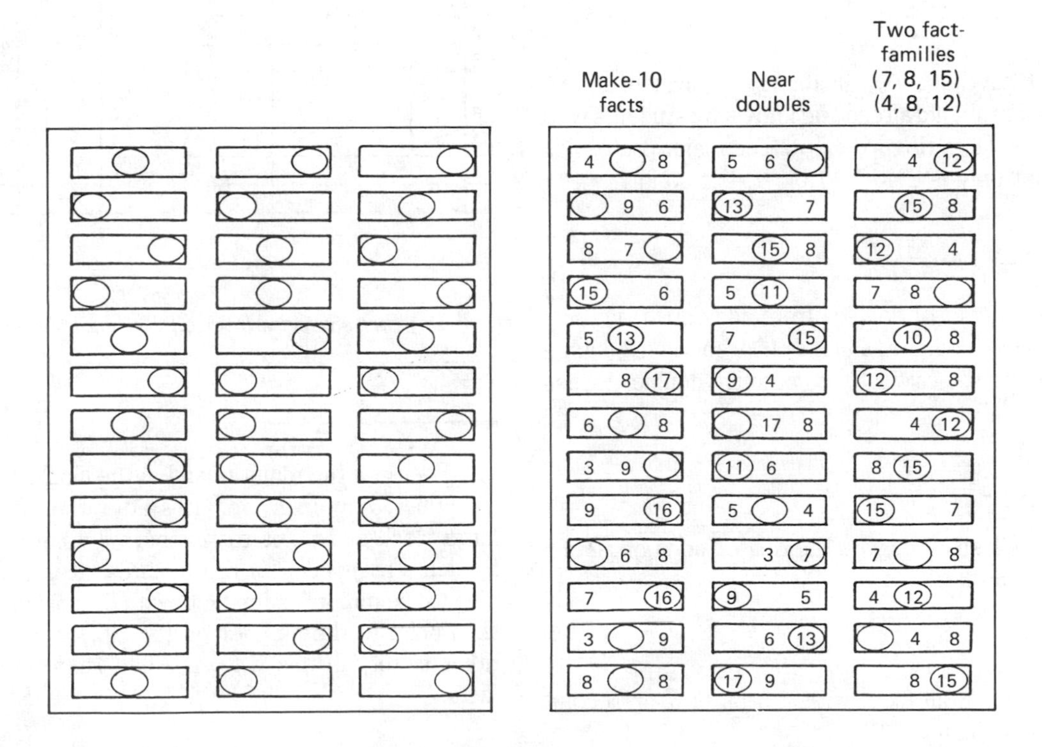

Figure 7.16
Missing-part worksheets. The blank form found in Appendix C can be used to fill in any sets of facts you wish to emphasize.

they wish. Later, give them the subtraction fact and have them write the addition fact.

Strategies for Multiplication Facts

OVERVIEW

Multiplication facts can also be mastered by relating new facts to existing knowledge. For example, the facts with a factor of 2 are related to the addition doubles. The fact 4 × 7 can be found from double 7 and then double again. While models are sometimes used, they are there for constructing relationships, not for getting answers. Counting the elements in six rows of eight will seldom help a child master the 6 × 8 fact.

Since the first and second factors in multiplication stand for different things (7 × 3 is 7 threes and 3 × 7 is 3 sevens), it is imperative that students completely understand the commutative property (see Figure 6.21). For example, 2 × 8 is related to the addition fact, double 8. But the same relationship also is applied to 8 × 2. Most of the fact strategies are more obvious with the factors in one order than in the other but turnaround facts should always be learned together.

Of the five groups or strategies discussed next, the first four strategies are generally easier and cover 75 of the 100 multiplication facts (Figure 7.17).

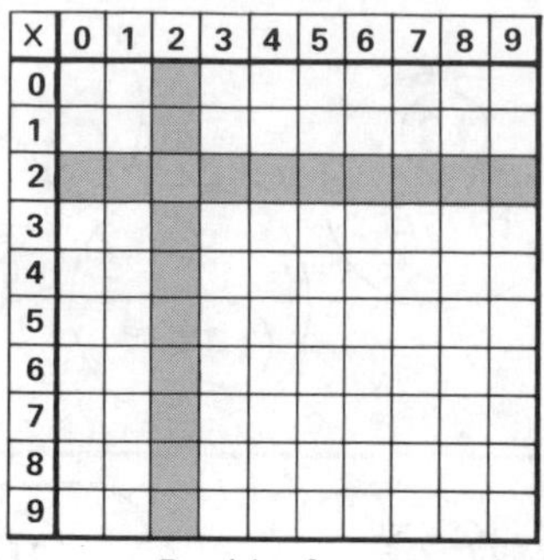
Doubles facts

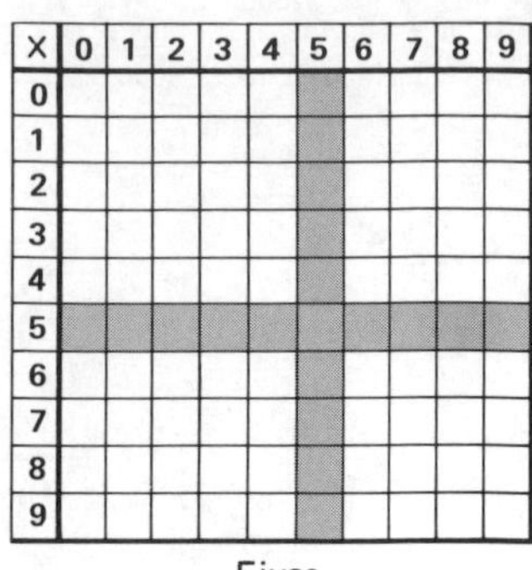
Fives

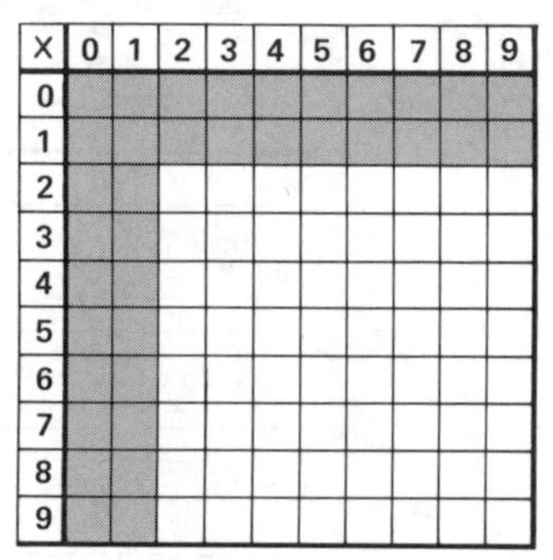
Zeroes and ones

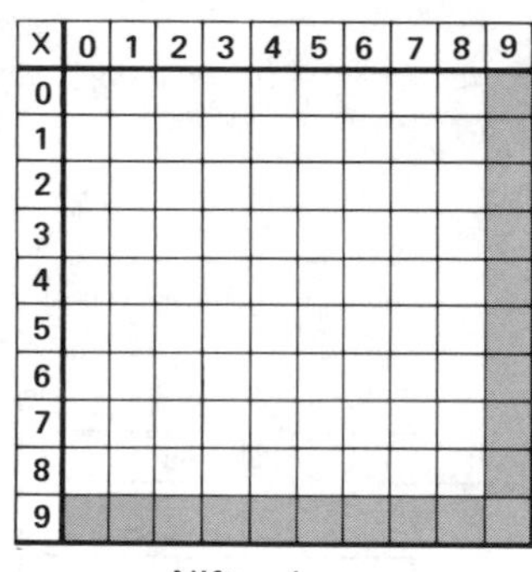
Nifty nines

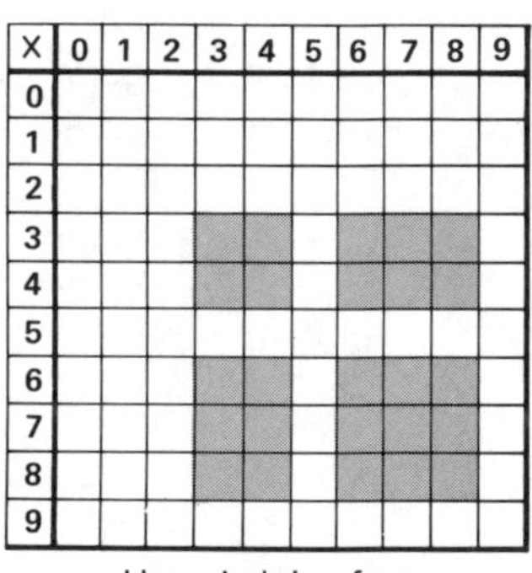
Use a helping fact

Figure 7.17
Five strategies cover all 100 multiplication facts.

DOUBLES

Those facts that have a 2 as a factor are equivalent to the addition doubles and should already be known by students who know their addition facts (Figure 7.18). The major problem is to realize that not only is 2 × 7 double 7, but so is 7 × 2.

Activity 7.40
Review the concept of doubles from addition. Play "Say the Double." You say a number and the children say the double of that number. Use the calculator to practice doubles (press 2 [×] [=]).

Activity 7.41
Make and use flash cards with the related addition fact or word "Double" as a cue.

FIVES FACTS

This group consists of all facts with a 5 as first or second factor, as shown in Figure 7.19.

Activity 7.42
Practice counting by fives to at least 45. Connect counting by fives with rows of five dots. Point out that six rows is a model for 6 × 5, eight rows is 8 × 5, and so on.

Activity 7.43
Focus on the minute hand of the clock. When it points to a number, how many minutes after the hour is it? Draw a large clock and point to numbers 1 to 9 in random order. Students respond with the minutes after. Now connect this idea to the multiplication facts with 5. Hold up a flash card and then point to the number on the clock corresponding to the other factor. In this way, the five facts become the "clock facts."

Activity 7.44
Include the clock idea on flash cards or to make matching activities.

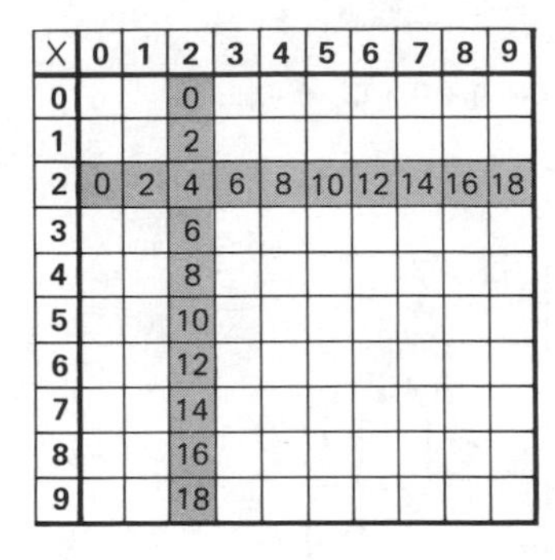

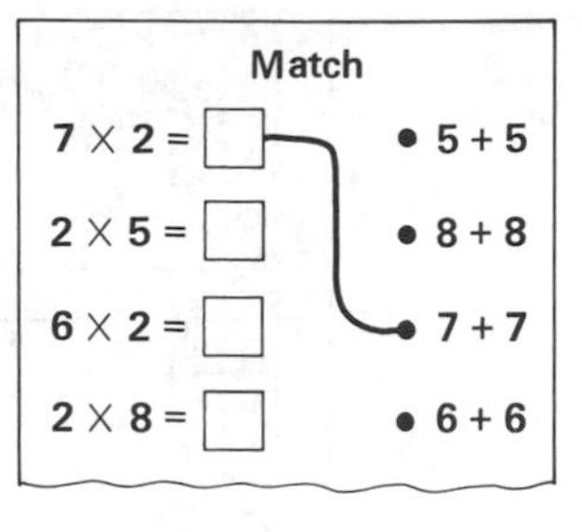

Figure 7.18
Multiplication doubles.

ZEROS AND ONES

Thirty-six facts have at least one factor that is either a 0 or a 1 (Figure 7.20). These facts, while apparently easy, tend to get confused with "rules" that some children learned for addition. The fact 6 + 0 stays the same, but 6 × 0 is always zero. The 1 + 4 fact is a one-more idea, but 1 × 4 stays the same. Make flash cards and games that reflect a conceptual approach to these facts. For zero, stress sets of nothing. Zero sets of 6 is difficult to conceptualize. One set of 6 for 1 × 6 is just as easy as 6 ones for 6 × 1.

NIFTY NINES

Those facts with a factor of 9 include the largest products, but can be among the easiest to learn (Figure 7.21). The table of nines facts includes some nice patterns that are fun to discover. Two of these patterns are useful for mastering the nines: (1) the tens digit of the product is always one less than the "other" factor (the one other than 9), and (2) the sum of the two digits in the product is always 9. So these two ideas can be used together to quickly get any nine fact. For 7 × 9: 1 less than 7 is 6, 6 and 3 make 9, so 63.

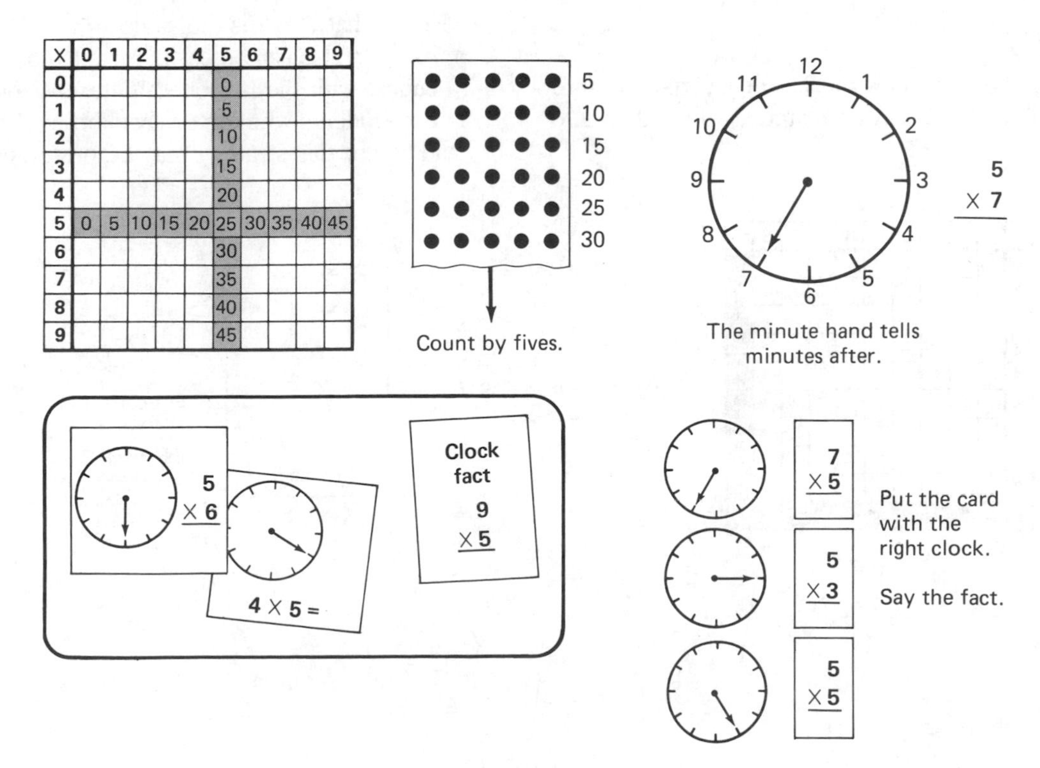

Figure 7.19
Fives facts.

Figure 7.20
Zeros and ones facts.

Activity 7.45

Practice quickly giving the other part of nine. If you say "six" the class answers "three."

Activity 7.46

Make and use flash cards like those shown in Figure 7.21 Notice that some cards practice only the one-less part of the rule that determines the tens digit, and others practice the sum-to-9 rule to determine the ones digit.

Activity 7.47

Either part of the nifty nines rule can be incorporated into a matching activity: 4 × 9 can be matched with either __6 or __3.

Activity 7.48

A "machine" like that shown in Figure 7.21 can be an independent activity or could be drawn on the board for full class practice.

A warning: While the nine's strategy can be quite successful, it also can cause confusion. Since two separate rules are involved and a conceptual basis is not apparent, children may confuse the two rules or attempt to apply the idea to other facts.

An alternative strategy for the nines is almost as easy to use. Notice that 7 × 9 is the same as 7 × 10 less one set of 7, or 70 − 7. This can be easily modeled by displaying rows of 10 cubes, with the last one a different color as in Figure 7.22. For students who can easily subtract 4 from 40, 5 from 50, and so on, this strategy may be preferable.

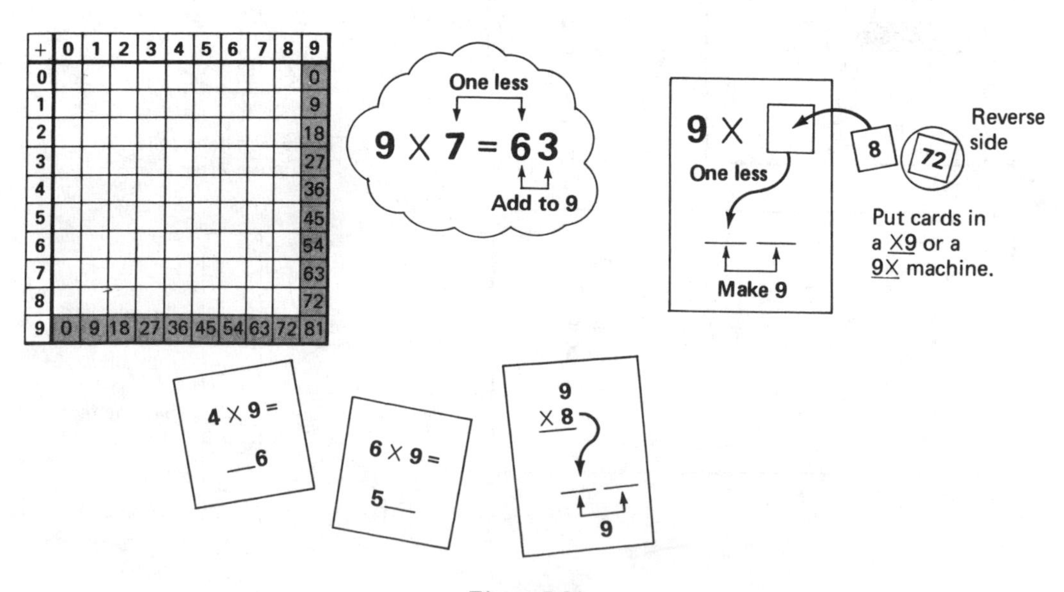

Figure 7.21
Nifty nines rule.

Figure 7.22
Another way to think of the nines.

USE A HELPING FACT

The chart in Figure 7.23 shows the remaining 25 multiplication facts. It is worth pointing out to children that there are actually only 15 facts remaining to master since 20 of them consist of 10 pairs of turnarounds.

X	0	1	2	3	4	5	6	7	8	9
0										
1										
2										
3				9	12		18	21	24	
4				12	16		24	28	32	
5										
6				18	24		36	42	48	
7				21	28		42	49	56	
8				24	32		48	56	64	
9										

Figure 7.23
Use a helping fact for these remaining facts.

These 25 facts can be learned by relating each to an already known fact or *helping* fact. For example, 3 × 8 is connected to 2 × 8 (double 8 and 8 more). The 6 × 7 fact can be related to either 5 × 7 (5 sevens and 7 more) or to 3 × 7 (double 3 × 7). The helping fact must be known, and the ability to do the mental addition must also be there.

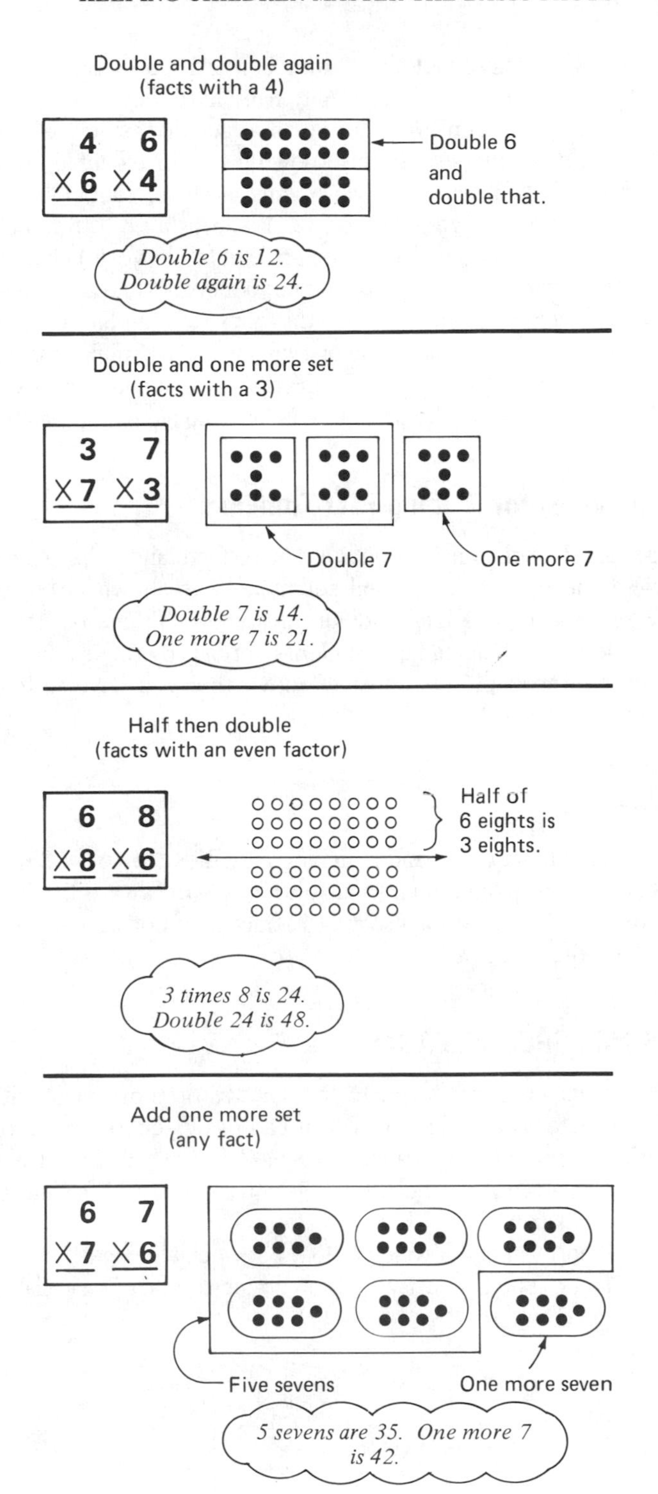

Figure 7.24
Finding a helping fact.

Finding a Helping Fact

How you can find a helping fact that is useful varies with different facts and sometimes depends on which factor you focus. Figure 7.24 illustrates models for four overlapping groups of facts and the thought process associated with each.

The *double and double again* approach is applicable to all facts with a 4 as one of the factors. Remind children that the idea works when the 4 is the second factor as well as the first. For 4 × 8, double 16 is also a difficult fact. Help children with this by noting, for example, that 15 + 15 is 30, 16 + 16 is two more, or 32. Adding 16 + 16 on paper defeats the purpose.

Double and one more set is a way to think of facts with one factor of 3. With an array or a set picture, the double part can be circled, and it is clear that there is one more set. Two facts in this group involve difficult mental additions.

If either factor is even, a *half then double* approach can be used. Select the even factor and cut it in half. If the smaller fact is known, that product is doubled to get the new fact. For 6 × 7, half of 6 is 3. Three times 7 is 21. Double 21 is 42.

For 8 × 7, the double of 28 may be hard, but it remains an effective approach to that traditionally hard fact.

Many children prefer to go to a fact that is "close" and then *add one more set* to this known fact. For example, think of 6 × 7 as 6 sevens. Five sevens is close; that's 35. Six sevens is one more seven or 42. When using 5 × 8 to help with 6 × 8, the set language "*six* eights" is very helpful in remembering to add 8 more and not 6 more. While admittedly difficult, many children do use this approach, and it becomes the best way to think of one or two particularly difficult facts. "What is seven times eight? Oh, that's forty-nine and seven more—fifty-six." The process can become almost automatic.

Activities for Helping-Fact Thinking

Model the relationships between hard facts and helping facts with the use of arrays and set models as shown in Figure 7.24. Spend time having different students "think out loud" as they use a helping fact. Students verbalizing these relationships is probably the most effective tool you have to help students with these facts.

Activity 7.49

Use flash cards and matching activities (Figure 7.25) to help children remember to use a helper. Be aware that one student's preferred approach to a fact may not be the same as another student's.

SOME SPECIAL FACTS

Many fact programs isolate the *squares* or those facts with both factors the same. Children can draw square arrays on grid paper to see why a fact such as 7 × 7 is called a square. This isolation and labeling of the squares may be helpful to some children.

Thornton and Toohey (1984) use special images for certain facts. The 3 × 3 fact is the "tic-tac-toe fact," associated with a tic-tac-toe board. The 3 × 7 and 7 × 3 facts are the "three week facts." The most unusual is the association of the numeral 7 with a gold miners pick so that the 7 × 7 fact becomes the "forty-niners fact."

Division Facts and "Near Facts"

DIVISION FACTS ARE NOT REALLY FACTS

Children are likely to use the corresponding multiplication fact to think of a division fact. If we are trying to think of 36 ÷ 9, we tend to think, "Nine times what is thirty-six." Most of us have memorized 42 ÷ 6, not as a separate and unrelated fact, but one closely tied to 6 × 7. (Would it not be wonderful if subtraction were so closely related to addition?)

It is an interesting question to ask, "When children are working on a page of division facts, are they practicing division or multiplication?"

DIVISION "NEAR FACTS"

Exercises such as 50 ÷ 6 are much more prevalent in computations and even in real situations than division "facts" or division without remainders. To determine the answer to 50 ÷ 6, most people run through a short sequence of the multiplication facts, comparing each product to 50: "Six times seven (low), six times eight (close), six times nine (too high). Must be eight. That's forty-eight and two left over." This process can and should be practiced. That is, children should be able to do problems with one-digit divisors and one-digit answers plus remainders and do them mentally.

Activity 7.50

To practice "near facts," try a "Price Is Right" type of exercise. As illustrated in Figure 7.26, the idea is to find the one-digit factor that makes the product as close to the target

Figure 7.25
Practice thinking of a helping fact.

without going over. Help children develop the process of going through the multiplication facts as described. This can be a drill with the full class by preparing a list for the overhead, or it can be a worksheet or flash card activity.

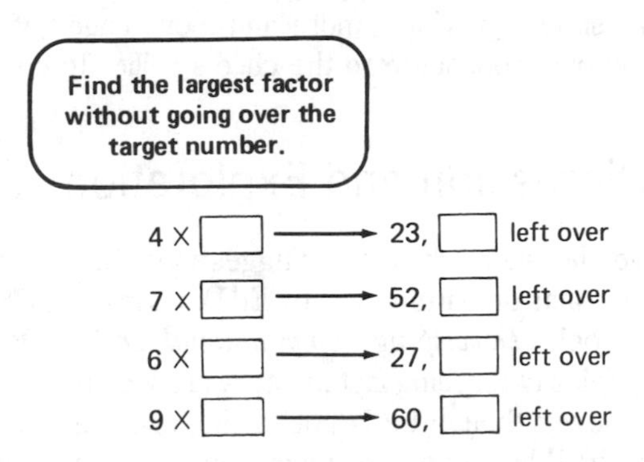

Figure 7.26
A "Price Is Right" exercise for division.

Effective Fact Drill

THE USE OF STRATEGIES

The major thrust of this chapter has been the development of effective thinking strategies or relationships to help children master the basic facts. Now that you have seen these strategies and understand the idea of grouping facts accordingly, the following sequence is suggested for designing your thinking strategy program.

1. For each strategy, be sure that prerequisite relationships and concepts are already developed.
2. Teach the strategy. Spend ample time with the class discussing how a particular strategy works. Remember that it will not be as clear to them as it is to you. Have children use models and verbalize the process to help develop the strategy or approach.
3. Group facts according to the strategy and begin drills. Children should be well aware of what strategy group they are working on.
4. Introduce another strategy or group of facts as the previous strategy is entering a drill stage.
5. Do strategy sorting activities.

In a *strategy sorting* activity, children are given a collection of facts from two or more strategy groups. Their first task is to identify strategies for the facts. This can be done in several ways.

Activity 7.51

On a worksheet, have students circle the facts that belong to a specified strategy they have been working on and then answer only those facts. The same approach can be used for two or even three strategies on one sheet.

Activity 7.52

Mix ordinary flash cards from two or more strategies into a single packet. Prepare simple little pictures or labels for the strategies that are in the packet. For each card, students first decide which of the labeled strategies they will use for that fact, place it in the appropriate pile, and then answer the fact.

Activity 7.53

Orally give facts to the class or group and first have them select and name a strategy they will use to answer the fact.

Strategy sorting is a crucial part of your fact strategy program. Children will generally use a strategy consistently when they know that all facts belong to one approach. When taken out of this mindset, as in a mixed drill exercise, they forget to use the strategies and revert to inefficient counting. Strategy sorting is a way to remind children to think about using a strategy.

THE ROLE OF SPEED

Early in this chapter it was noted that if children are pressured too early to give fact answers in speed drills, they will revert to inefficient counting methods. Under pressure, children forget new ideas and fall back on the secure and familiar. Counting is secure and familiar. Strategies are new. Give strategy practice a lot of time so that the strategies become a more secure part of your children's thought patterns.

At the same time, speed is a significant factor in getting children to memorize instead of using less efficient counting procedures (Davis, 1978). Therefore, listen to your children. As they become proficient with a particular strategy, try to increase their speed. For example, work faster and faster through a small deck of flash cards all for a single strategy. After successful strategy sorting, increase pressure to go faster with more varied sets of facts.

A short speed drill is much better than a speed drill of 100 facts. Consider rows of 8 to 10 facts. See how fast students can write the facts for one row instead of the whole page. Check answers after each row to provide feedback. Subsequent rows can have many repetitions of facts that are being worked on. After 10 speed rows, children will not only get faster and feel better, they will have practiced a few new facts over and over.

IMMEDIATE FEEDBACK

The frequent suggestion of flash cards in this chapter is partly because this method provides individual students with immediate feedback on every fact. Not only do they know if they are right or wrong, but they see the correct answer after each response. The calculator is another method of providing quick feedback in those situations that lend themselves to calculator drill (adding or multiplying by the same number). There are also many computer software programs that drill facts with virtually all using some form of instant feedback. Certainly not every activity can provide immediate feedback, but it is a feature to look for in good fact drills. Note, for example, that a long worksheet graded overnight provides no feedback. The same worksheet broken into short segments can have a separate answer key for each section.

FACT REMEDIATION WITH UPPER-GRADE STUDENTS

Children who have not mastered their basic facts by the sixth or seventh grade are in need of something other than more practice. They have certainly seen and practiced those facts endless times over the past several school years. They need a new approach.

Provide Hope

Children who have experienced difficulty with fact mastery can begin to believe that they cannot learn facts or that they are doomed to finger counting forever. These children have either not seen or have not mastered efficient techniques such as the strategies described in this chapter. Let these children know that *you* will provide them with some new ideas to help them and *you will help them!* Take that burden on yourself and spare them the prospect of more defeat.

Build in Success

Begin with easy strategies and introduce only a few new facts at a time. Even with pure rote drill, repetitive practice with five facts in three days will provide more success than introducing 15 facts in a week. Success builds success! With strategies as an added assist, success comes even more quickly. Point out to children how one idea, one strategy, is all that is required to learn many facts. Use fact charts to show what set of facts you are working on. It is surprising how the chart quickly fills up with mastered facts.

Keep reviewing newly learned facts and those that were already known. This is success. It feels good, and failures are not as apparent.

Diagnose

Find out exactly which facts are mastered (less than 3-second response) and which need practice. For a fact not mastered, ask the student how he or she would figure it out. What thought processes are they using? Sometimes you can select strategies that build on ideas they already have. Find out about their concepts of the operations and their command of number relationships. It may be more important to work on number relationships or a think-addition concept of subtraction than to begin directly with subtraction facts. Check on their use of turnaround facts. If they know 3×7, do they also know 7×3? Many do not even know these facts are related or do not take advantage of that relationship.

Looking for answers to questions like these will permit you to tailor an effective drill program for the remedial student, whereas simply providing more and more pages of practice will likely only contribute to the child's ability to count.

For Discussion and Exploration

1. Read the short chapter, "Suggestions for Teaching the Basic Facts of Arithmetic," by Ed Davis in the 1978 NCTM Yearbook, *Developing computational skills.* Davis's 10 principles of teaching fact mastery are worthy of consideration even though he is not talking about a strategy approach. Which ones are most important? In view of the current chapter, are there any of the 10 you disagree with?
2. Try using a calculator as a basic fact drill device. One idea is to have someone say a fact and let the child try to press the keys and say the answer before pressing [=]. Soon, the calculator actually is seen as slow. Another approach uses the automatic constant features. To practice multiplication facts with a factor of 8, press [8] [×] [n] [=]. Then press any number followed by [=] to get the product. Again try to beat the calculator. The same approach works for all four operations. (Different calculators may operate differently.)
3. Explore a computer software program that drills basic facts. There are perhaps more of these programs than any other area of drill-and-practice software. Many utilize a variation of an arcade format in order to encourage speed. Very few if any have organized the facts around thinking strategies. Do you think these programs are effective? How would you utilize such a piece of software in a classroom with only one or two available computers?
4. One view of thinking strategies is that they are little more than a collection of tricks for kids to memorize. This view suggests that direct drill may be more effective and less confusing. Discuss the question, "Is teaching children thinking strategies for basic fact mastery in keeping with a developmental view of teaching mathematics?" For some perspective on the issue, you might want to consult *Learning from children* (Labinowicz, 1985). Chapter 5 presents a discussion of this issue, although it is highly biased in favor of the view in this book.
5. Examine a recently published second-, third-, or fourth-grade textbook and determine how thinking strategies for the basic facts have been developed. Compare what you find with the groupings of facts in this chapter. How would you use the text effectively in your program? (Note: Before 1988, only two textbook series, Harper and Row and Addison-Wesley, had an explicit program of fact mastery for basic facts. Other series used many of these same ideas but were less overt in their approach.)

Suggested Readings

Baroody, A. J. (1984). Children's difficulties in subtraction: Some causes and cures. *Arithmetic Teacher, 32*(3), 14–19.

Baroody, A. J. (1985). Mastery of the basic number combinations: Internalization of relationships or facts? *Journal for Research in Mathematics Education, 16,* 83–98.

Brownell, W. A., & Chazal, C. B. (1935). The effects of premature drill in third-grade arithmetic. *Journal of Educational Research, 29,* 17–28.

Davis, E. J. (1978). Suggestions for teaching the basic facts of arithmetic. In M. N. Suydam (Ed.), *Developing computational skills.* Reston, VA: National Council of Teachers of Mathematics.

Flexer, R. J. (1986). The power of five: the step before the power of ten. *Arithmetic Teacher, 34*(3), 5–9.

Grove, J. (1978). A pocket multiplier. *Arithmetic Teacher, 25*(6), 25.

Labinowicz, E. (1985). *Learning from children: New beginnings for teaching numerical thinking.* Menlo Park, CA: Addison-Wesley.

Lessen, E. I., & Cumblad, C. L. (1984). Alternatives for teaching multiplication facts. *Arithmetic Teacher, 31*(5), 46–48.

Rathmell, E. C. (1978). Using thinking strategies to teach the basic facts. In M. N. Suydam (Ed.), *Developing computational skills.* Reston, VA: National Council of Teachers of Mathematics.

Rightsel, P. S., & Thornton, C. A. (1985). 72 addition facts can be mastered by mid-grade 1. *Arithmetic Teacher, 33*(3), 8–10.

Steinberg, R. M. (1985). Instruction on derived facts strategies in addition and subtraction. *Journal for Research in Mathematics Education, 16,* 337–355.

Thornton, C. A., & Smith, P. (1988). Action research: Strategies for learning subtraction facts. *Arithmetic Teacher, 35*(8), 8–12.

Thornton, C. A., & Toohey, M. A. A matter of facts: (Addition, subtraction, multiplication, division). Palo Alto, CA: Creative Publications.

CHAPTER 8

WHOLE NUMBER PLACE-VALUE DEVELOPMENT

Relational Understanding of Place Value

EXTENDED COUNTING BY ONES

Most children count to 100 in the first grade by learning the verbal patterns for the counts beyond 10. They also learn to read and write numbers with two or three digits. These initial counting and numeral skills extend their one-more-than concept of counting to larger numbers. "Forty-seven" is, for the young child, an amount consisting of 47 single things and is only understood in terms of counting by ones. Even the numeral 47 is viewed simply as "the way you write forty-seven." No grouping or place value understanding can be inferred from a child's ability to count to 100 or beyond. But this counting-by-ones knowledge of large numbers is where the child *is* before base ten concepts develop. We need to connect new ideas with this initial understanding of number.

CONCEPTUAL KNOWLEDGE OF PLACE VALUE

Conceptual knowledge of place value is the understanding of the *base ten* convention in which single items are grouped in sets of ten, these sets or *tens* are grouped in sets of tens or *hundreds,* and so on.

Quantities in Terms of Groups and Singles

Glance briefly at the set of dots in the figure. Even if you did not know or use multiplication facts, you probably "know" how many dots are there: 4 sets of 6 and 3 more. Your

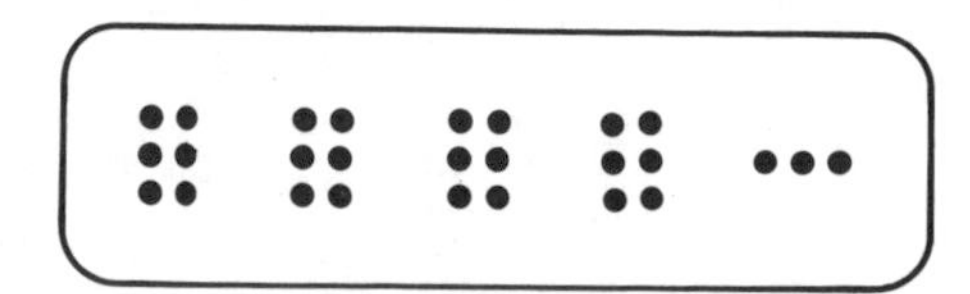

understanding of that quantity is different than the count-by-ones understanding of the first grader. The double count of four groups and three single dots is more complex than counting say four boys and three girls. While you counted the groups separately, you were at the same time aware that each group has six dots. You can connect that idea with counting all the dots by ones.

For a first-grade child, that integration of counting groups and singles with a counting-by-ones concept of number is nontrivial. Kamii (1985) talks in terms of a "two-level hierarchical structure" that requires constructing a single entity out of each group of 10 objects while understanding these groups as 10 singles.

Groups of Groups

Our base ten convention uses the same group size of 10 for all groupings. This leads to groups of groups. Ten groups of 10 are treated as a single entity: one group of 10 tens. Developmentally, this extension of the groups and singles concept is another conceptual hurdle. The single group called a *hundred* must be simultaneously understood as *one group,* 10 *tens,* and 100 *single items.*

The groups-of-groups process continues forever. Ten hundreds are grouped and called *thousands,* and so on. The desired goal is that each new grouping will at least be understood as 10 of the previous group and also as a set of singles.

Equivalent Representations for the Same Quantity

Sixty-two can be thought of as 6 tens and 2. Sixty-two can also be thought of as 5 tens and 12, or 3 tens and 32. Different groupings for the same amount can be called *equivalent representations* (Figure 8.1). That *6 tens and 4* can be the same as *5 tens and 14* is not at all obvious to children even when physical models are used (Ross, 1986; Cauley, 1988). The concept of equivalent representations is at the very heart of whole number computation procedures ("borrowing" and "carrying," for example).

PROCEDURAL KNOWLEDGE OF PLACE VALUE

Procedural knowledge can be roughly separated into *labels and names, counting skills,* and *symbolism* or the way we write base ten concepts using a positional system. These

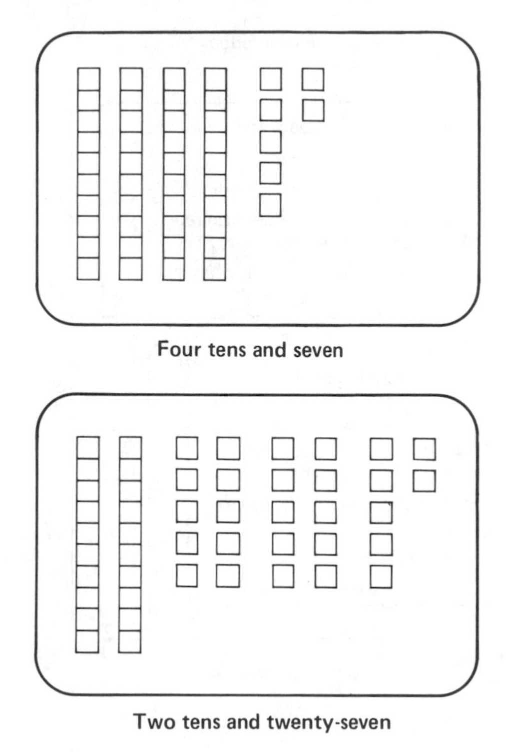

Figure 8.1
Two equivalent representations.

different aspects of procedural knowledge are quite interrelated with each other and with a conceptual knowledge of groups of 10.

Labels and Names

This category includes knowledge of the names given to groups and the standard names for numbers. Each group or group of groups has a name: *ten, hundred, thousand,* and so on. These group names can be used in phrases such as "two hundreds, no tens and six ones," a *place-value language* for the standard number name "two hundred six."

Group names are very strange for children because they are also counting words. In the phrase "ten ones makes one ten," the first "ten" refers to a count or a quantity of things: the number after 9. The second "ten" in the phrase is a singular noun that refers to a group. It is difficult for children to connect this word "ten" to two ideas at the same time. In a similar manner, the group names hundred, thousand, ten thousand, hundred thousand, million, and so forth, carry at least a dual role. One thousand is 10 hundreds, 100 tens, and a count-by-ones number.

The oral names for our numbers have a definite pattern, but they also have many exceptions and idiosyncrasies. The teens stress the ones first (*fif*teen) while the numbers 20 to 99 sound the tens part first. Twenty, thirty, and fifty might more easily have been two-ty, three-ty, and five-ty. Many big numbers are given dual names, such as three thousand five hundred and thirty-five hundred. Numbers with zeros, such as 307, are read with no mention of the tens.

Counting Skills

Counting by tens and ones, as in 10, 20, 30, 31, 32, 33, 34, is a closely related skill to that of counting by ones. As the group names hundred and thousand are added, counting skills must be elaborated accordingly.

Sequence of numbers refers to the related skill of saying the number that comes next or comes before a given number and is clearly related to place-value counting patterns.

Symbolism

To write numerals with two or more digits requires mastering the arbitrary convention of our place-value system in which the ones are recorded on the right with tens, hundreds, thousands, and so on, to the left. Reading numerals and writing numerals as one counts or hears them is intimately tied to the other procedural skills. Unfortunately, the procedural connections of reading and writing numerals and activities such as "naming the number in the hundreds place," are frequently overemphasized at the expense of connecting symbolism to concepts.

INTEGRATION OF CONCEPTUAL AND PROCEDURAL COMPONENTS

Figure 8.2 illustrates all of the components of place-value knowledge and the connections between them. The new conceptual knowledge of base ten relationships must be integrated with children's established knowledge of a count-by-ones concept of number. Counting bundles of 10 sticks and single sticks and never counting the entire collection by ones fails to address this fundamental connection. The procedural aspects of place value are also highly interconnected. But the most significant connections are between the procedural knowledge and concepts.

Spend some time with Figure 8.2. In terms of children working with numbers, think about the nature of the connections indicated by each arrow in the diagram. Especially consider the possibility and implications of a child possessing a particular feature of procedural knowledge without connections with concepts. Would it be possible?

Models for Base Ten Place-Value Numeration

The key instructional tool for developing the conceptual knowledge of place value and also for connecting these concepts to symbolism is the use of base ten models.

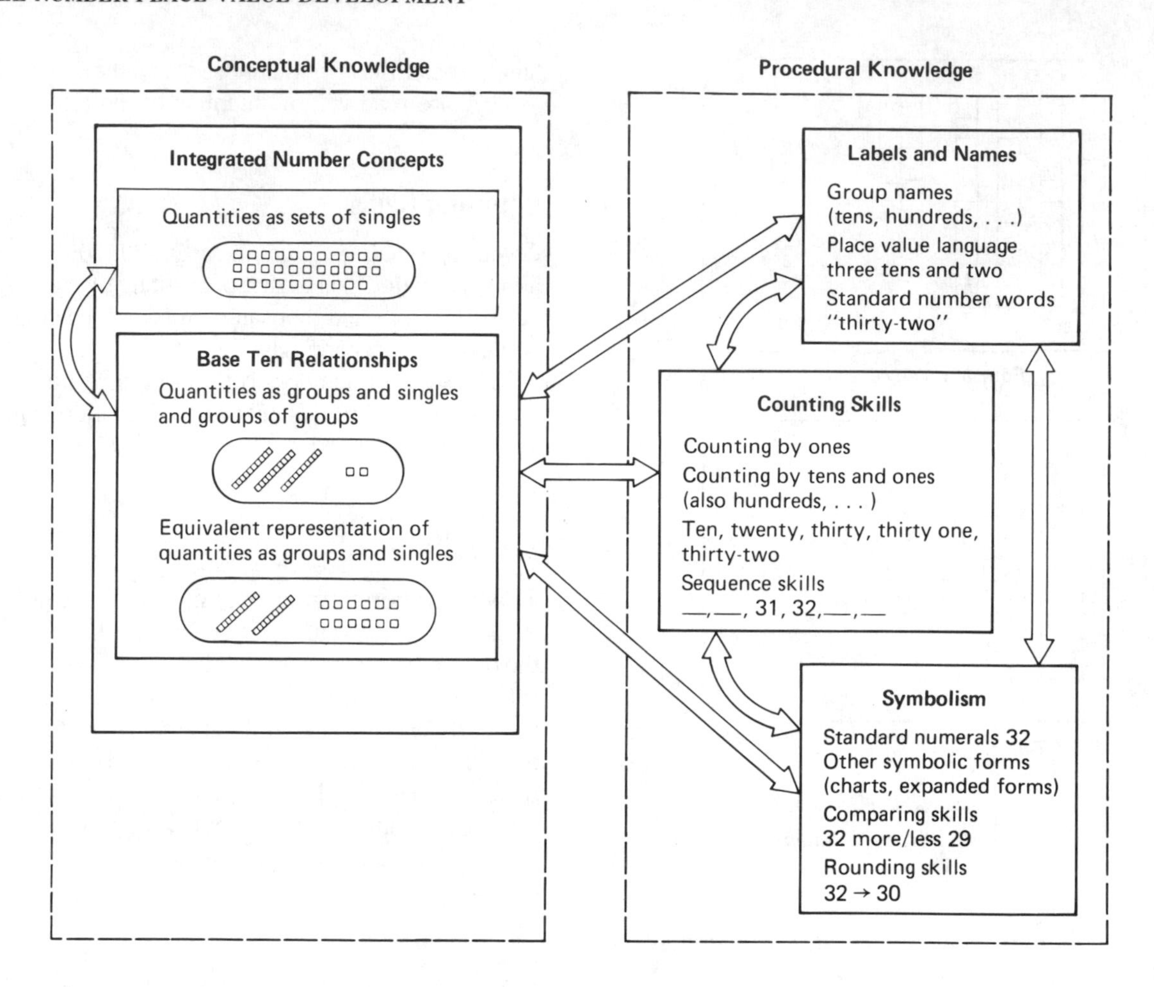

Figure 8.2
Relational understanding of place value.

BASE TEN MODELS SHOW A "TEN MAKES ONE" RELATIONSHIP

A good base ten model for ones, tens, and hundreds is *proportional.* That is, a *ten* model is physically 10 times larger than the model for a *one,* and a *hundred* model is 10 times larger than the ten model. Base ten models can be categorized as *groupable* and *pregrouped.*

Groupable Models

Models which most clearly illustrate the relationships between ones, tens, and hundreds are those for which the ten can actually be made or grouped from the singles. When you bundle 10 popsicle sticks, the bundle of 10 literally *is the same as* the 10 ones from which it was made. Examples of these "groupable" models are shown in Figure 8.3(a). These could also be called "put-together take-apart" models.

Of the groupable models, beans or counters in cups are the cheapest and easiest for children to use. (Paper or plastic portion cups can be purchased from restaurant supply houses.) The plastic Unifix cubes or the equivalent, are highly attractive and provide a good transition to pregrouped tens sticks. Bundles of popsicle sticks are perhaps the best known base ten model, but small hands have trouble with rubber bands and actually making the bundles. Hundreds are possible with most groupable materials, but are generally not practical for most activities in the classroom.

Pregrouped or Trading Models

At some point there is a need to easily represent hundreds. Therefore, models which are pregrouped must be introduced. As with all base ten models, the ten piece is physically equivalent to 10 ones, and a hundred piece equivalent to 10 tens [Figure 8.3(b)]. However, children cannot actually take them apart or put them together. When 10 single pieces are accumulated, they must be exchanged for or *traded* for a ten, and, likewise, tens must be traded for hundreds.

The chief advantage of these models is the ease of use and the efficient way they model numbers as large as 999. The disadvantage is the increased potential for children to use them without reflecting on the ten-to-one relationship. For example, in a ten-for-ones activity, children are told to trade 10 ones for a ten. It is quite possible for children to make this exchange without attending to the "tenness" of the piece they

(a) Groupable Base Ten Models

Counters and cups.
Ten single counters are placed in a cup.
Hundred: ten cups in a margarine tub.

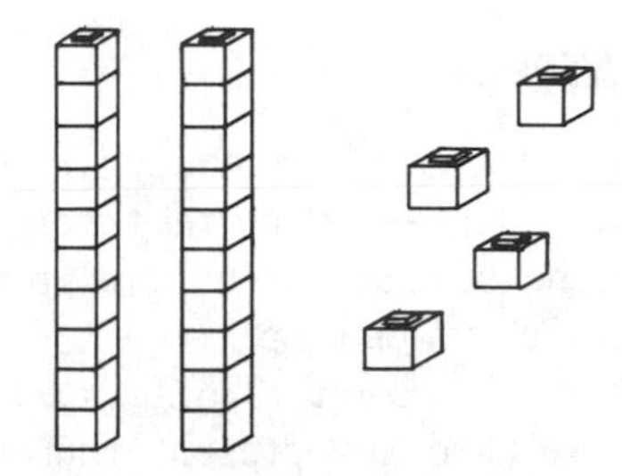

Unifix (or any interlocking cubes).
Ten single cubes form a bar of 10.
Hundreds: Ten bars on cardboard backing.

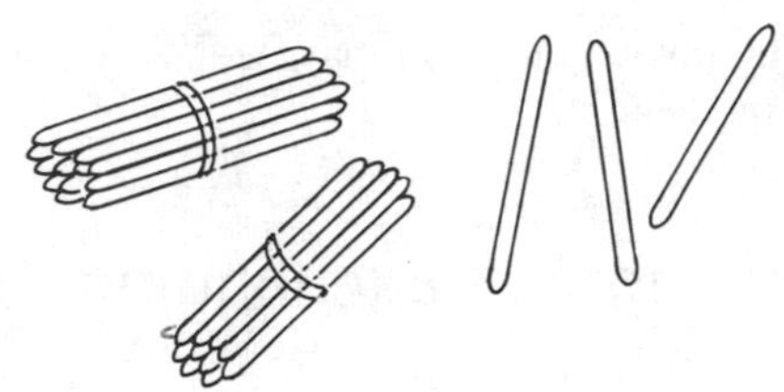

Bundles of sticks (wooden craft sticks, coffee stirrers)
(if bundles are left intact, these are a pregrouped model)
Hundreds: Ten bundles in a big bundle.

(b) Pregrouped Base Ten Models

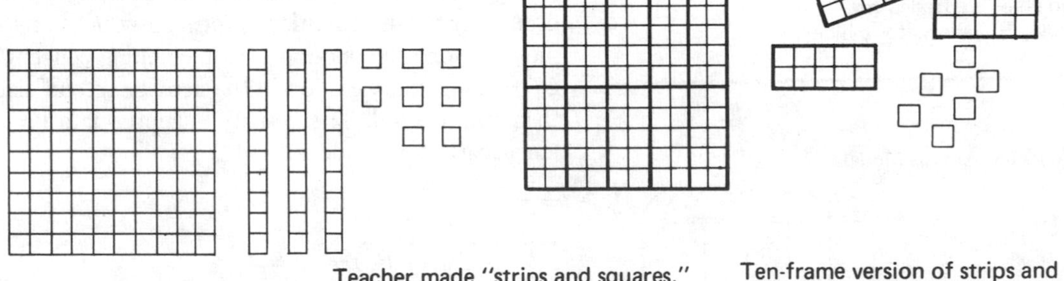

Teacher made "strips and squares."
Made from mount board and
poster board. See Appendix C.

Ten-frame version of strips and
squares. Made the same way but
ten is in arrangement of ten-frame
and hundred shows the tens.

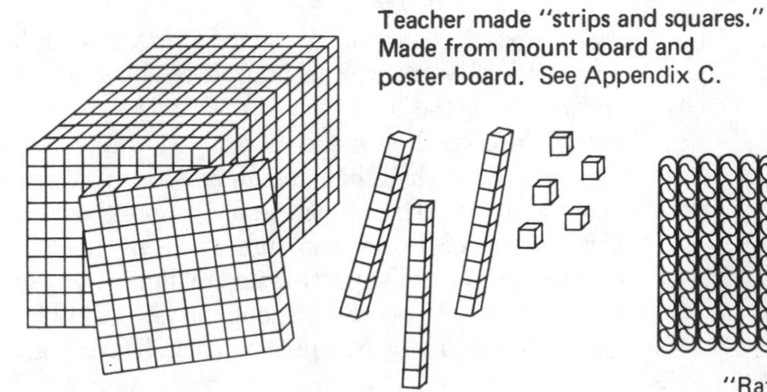

Wooden or plastic units, longs, flats and block.
Also known as Dienes blocks or base ten
blocks. Expensive, durable, easily handled,
only model with thousand.

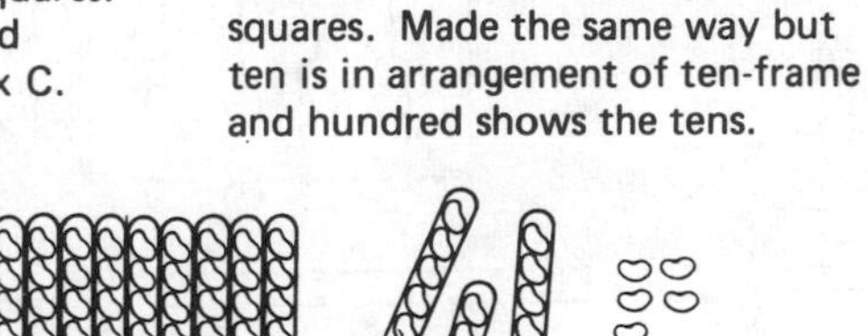

"Raft"

Bean sticks.
Beans glued to craft sticks.
Ten sticks in a raft is also
made from cardboard.

Directions and alternatives in Appendix C.

Figure 8.3
a] Groupable base ten models b] Pregrouped base ten models.

call a ten. While no model, including the groupable models, will guarantee that children are reflecting on the ten-to-one relationships in the materials, with trading pieces, we need to make an extra effort to see that children understand that a ten piece really is the same as 10 ones.

(In Appendix C there are directions and a black-line master for making base ten strips and squares.)

NONPROPORTIONAL MATERIALS

In this text, colored counters, abacuses, and money are not considered to model base ten ideas since the relationships must first exist in the mind of the child, and the materials play no part in helping to develop that relationship.

With *Chip Trading Materials* (Davidson, 1975), for example, different colored chips are used to represent different place values; red for ones, yellow for tens, and so on, as shown in Figure 8.4(a). Trades are made between pieces in a variety of excellent activities, all of which can be done with proportional base ten models. With the chips, the base can be changed arbitrarily by simply designating different exchange rates.

Somewhat similar to the colored-chip model are the various forms of the abacus, some of which are shown in Figure 8.4(b).

The use of money as a model involves the same issue as colored chips. Pennies, dimes, and dollars are frequently used by teachers and by textbooks as a place-value model. If the relationship that a dime is worth 10 pennies and that a dollar is worth 10 dimes is part of the children's existing understanding, then money will model base ten place value. On the other hand, if the teacher must explain or impose this relationship, then the trade rate between pennies and dimes is simply a "rule of the game" from the children's view. They can learn to obey the rule, but the model is not helping them see why 10 ones is the same as 1 ten. The relationship should be in the model.

The pivotal question may be, "Why not just use proportional models?"

(a)

Chip Trading Materials

Green	Blue	Yellow	Red

10 red chips are traded for a yellow, 10 yellows for a blue.

(b)

Assorted Abacus Models

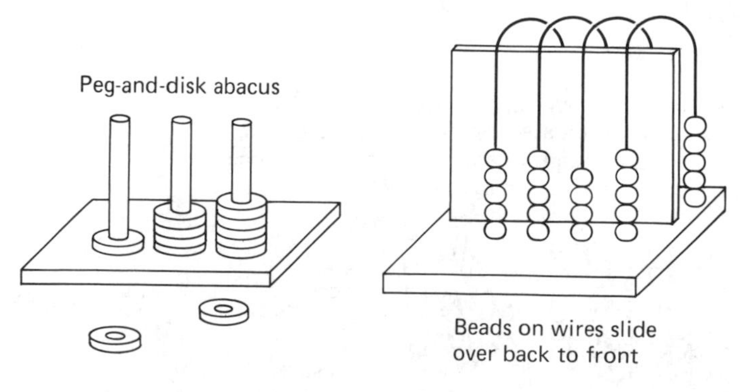

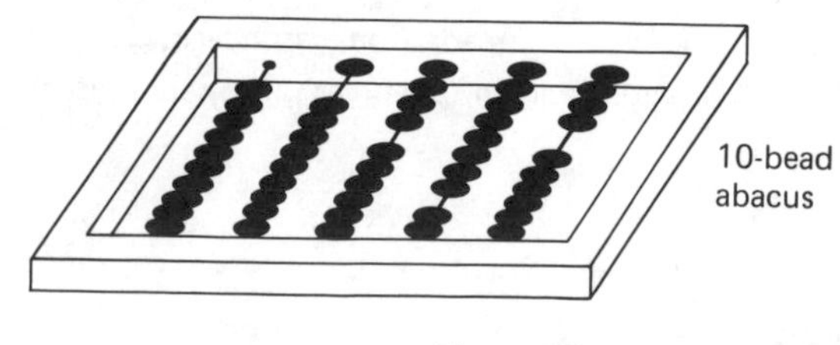

Figure 8.4
Nonproportional materials.

Grouping and Exchanging Activities

MAKING AND COUNTING GROUPS

Readiness Activity

The grouping activity described here is designed to help children think about quantities in terms of groups and singles: to move beyond the one-more-than counting concept. For ease of counting and variation in the activity, group sizes smaller than 10 are used. The activity is appropriate for children in kindergarten or grades 1 and 2.

Activity 8.1

Give each child a pile of counters. Any amount from about 15 to 40 will work well at first. It is not important that all children have the same amount, so you need not precount them. Have children make as many sets of six counters as they can. Each child then counts how many groups and how many leftovers. Have them orally report their amounts: "I have four piles of six and three." Using the same set of counters, repeat the exercise several times using group sizes from 3 to 10. A recording sheet can be made for recording "groups" and "leftovers" next to each different group size.

Worksheets can be developed similar to the one in Figure 8.5 so that the last activity can be done independently. Groups can be made in a variety of ways. For example, Unifix cubes can be made into bars; counters can be placed in cups, muffin tins, or baggies; craft sticks or coffee stirs can be banded or placed in cups; wooden cubes can be lined up or arranged in rectangles.

Name ______________

My set of unifix can be __

Bars of 3		Singles
_____	and	_____
Bars of 7		**Singles**
_____	and	_____
Bars of 5		**Singles**
_____	and	_____
Bars of 6		**Singles**
_____	and	_____
Bars of 10		**Singles**
_____	and	_____

Figure 8.5

Groups of 10

Activities with groups of 10 should be done exactly like the one just described with two exceptions. All group sizes are 10, and the language of "ten" as a group name is introduced. A significant aid both conceptually and manipulatively, is to use a ten-frame drawn on a half-sheet of construction paper to assist in counting out the groups. From the pile of counters, children fill the ten-frame. Then the counters are moved from the ten-frame to form a group. For children working at tables, a lot of ten-frame mats can be provided and the counters can remain directly on the mats.

In counting groups, initially use language such as "three bars of ten" or "six cups of ten." Gradually, this language can be abbreviated to "three tens." In this way, the use of the word "ten" as a group name is introduced naturally as a shortened form of more natural language that the children have been used to.

Surprisingly, there is no hurry to use the word "ones" for the leftover counters for quite a while. Language such as "four tens and seven" works very well. The terms *leftovers* or *singles* are also useful as in "five tens and two singles."

It is important for children to count their total collections by ones, either before or after grouping. Counting by ones is the child's way of thinking about the total quantity and helps connect that existing concept to the new idea of groups and singles. With the use of written number words, children can read and write these quantities without requiring the procedural knowledge of place value. Children can count "forty-seven" cubes but not be ready to connect that to "47." A chart can be made like the one in Figure 8.6 to help with spelling the number words.

Number Words		
eleven	ten -	one
twelve	twenty -	two
thirteen	thirty -	three
fourteen	forty -	four
fifteen	fifty -	five
sixteen	sixty -	six
seventeen	seventy -	seven
eighteen	eighty -	eight
nineteen	ninety -	nine

Figure 8.6
A chart can help children write number words.

Figure 8.7 shows some ideas that can be used for worksheets or task cards as variations on the grouping-by-tens activity. Note that quantities are given as sets of objects or as number words, encouraging children to count the collections by ones as well as by groups of tens and ones. These activities should always be done with manipulatives.

Equivalent Groupings

An important variation of the grouping-by-tens activities is to find more than one way to describe an amount using groups of 10. For example, when a set of 47 is found to be 4 tens and 7 singles, it can also be described as 2 tens and 27 singles or 3 tens and 17 singles. The recording suggestions of Figure 8.7 can be amended to allow for two or three different ways to record the same amount.

Once the idea of different groupings for an amount is presented, children can be given nonstandard amounts such as 2 tens and 15 and asked to use counters to find the number word (thirty-five) for these amounts.

By late second grade and above, equivalent groupings should be explored using pregrouped materials or ones, tens, and (later) hundreds pieces.

Activity 8.2

Make lots of different collections of place-value pieces in plastic bags. Most should include nonstandard collections such as 4 tens and 35 ones or 3 hundreds, 27 tens, and 18 ones. Students select a bag and record the number of each piece. Then they use the pieces and make trades to find the standard number name. This should be recorded as a number word and/or as a numeral. They can also be required to find at least one other representation of the same amount. When finished, they return the original collection of pieces, as they had recorded them, and trade the bag for another one.

After children have had sufficient experiences with pregrouped materials, both teacher and student can use a "stick-

Name ______________________

Bag of	Number word	
Toothpicks		Tens ☐ Singles ☐
Beans		Tens ☐ Singles ☐
Washers		Tens ☐ Singles ☐

Get this many.	Write the number word.
	______________ Tens _____ and Ones _____

Get forty-seven beans.

Fill up ten frames. Draw dots.

Tens _____ Extras _____

Loop sixty-two in groups of ten.

Tens _____ Ones _____

Figure 8.7
Number words and making groups of ten.

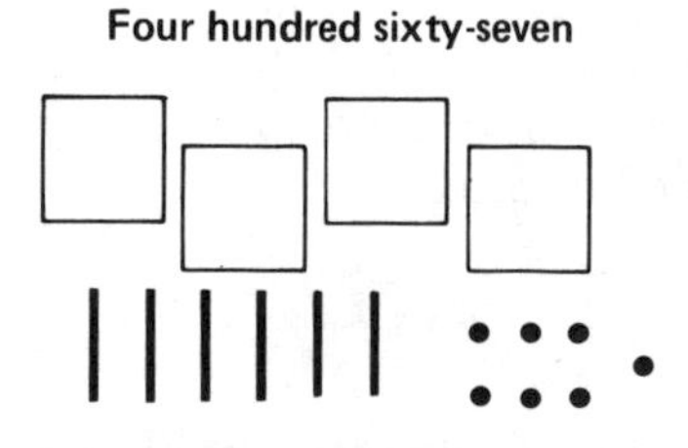

Figure 8.8
Dots, sticks, and squares to show ones, tens, and hundreds.

and-dot" notation for tens and ones. By third grade, children can use small squares for hundreds as in Figure 8.8. With this notation, it is easy to create activity sheets involving place-value representations. Use the drawings as a means of telling the children what pieces to get out of their own place-value kits, and as a way for children to record results. Soon, the children themselves can make stick-and-dot drawings in the activities as in Figure 8.9.

Activity 8.3

Base ten riddles can be presented either orally or in written form. Children should use base ten materials to solve them. The examples illustrate a variety of possibilities with different levels of difficulty.

I have 23 ones and 4 tens. Who am I?
I have 4 hundreds, 12 tens, and 6 ones. Who am I?
I have 30 ones and 3 hundreds. Who am I?
I am 45. I have 25 ones. How many tens do I have?
I am 341. I have 22 tens. How many hundreds do I have?
I have 13 tens, 2 hundreds, and 21 ones. Who am I?
If you put 3 more tens with me, I would be 115. Who am I?
I have 17 ones. I am between 40 and 50. Who am I?
I have 17 ones. I am between 40 and 50. How many tens do I have?

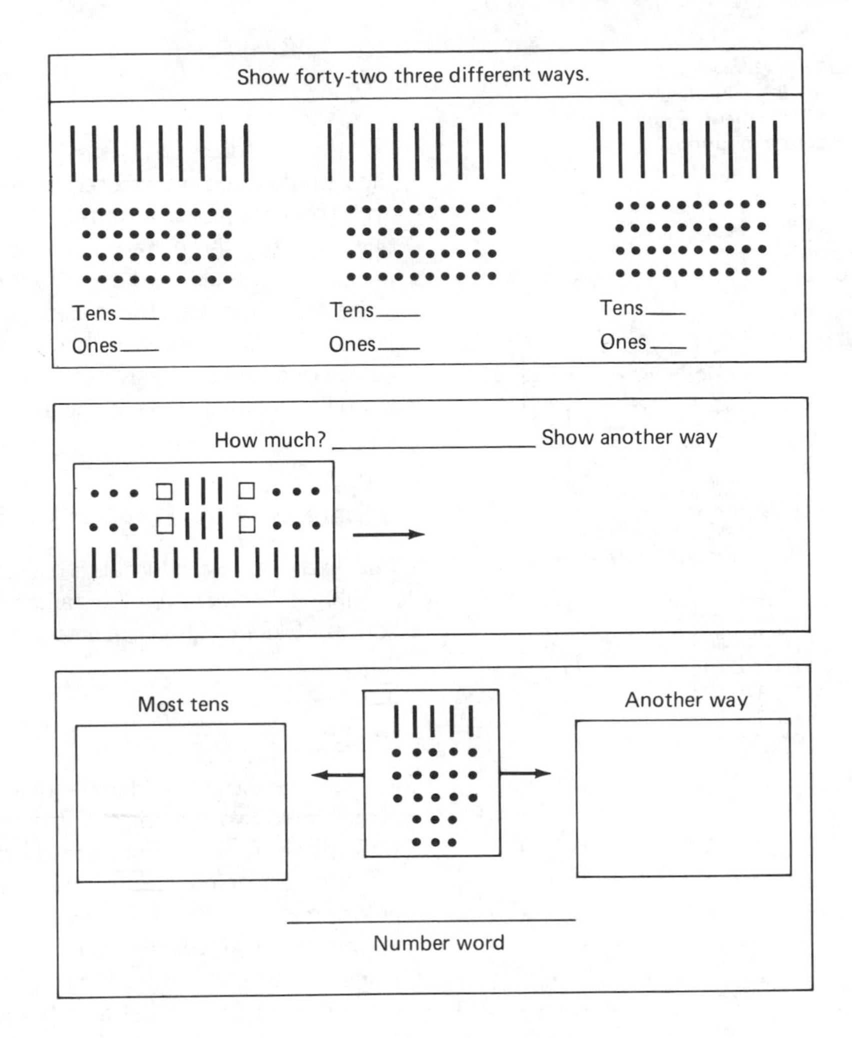

Figure 8.9
Equivalent representation exercises using stick-and-dot pictures.

SEQUENTIAL GROUPING OR TRADING

The initial goal of the activities in this section is to tie the grouping idea more closely to counting by ones. That is, groups of tens are made as we count. The activities introduce the right-to-left place-value convention for ones, tens, and later hundreds and also lay the groundwork for *regrouping* in computation ("borrowing" and "carrying").

There are two basic activities, counting forward (tens made from ones) and backward (ones made from tens). They can be done as early as first grade, but even fourth-, fifth-, and sixth-grade children with weak conceptual backgrounds will benefit from doing these activities with pregrouped place-value materials.

Materials

All of these activities are done on a place-value mat with either groupable or pregrouped models. *Place-value mats* are simple mats divided into two or three sections. They are easily made from construction paper, as shown in Figure 8.10. While there is no requirement to have anything printed on the mats, it is strongly recommended that two ten-frames be drawn in the ones place as shown. These can be reproduced on the paper used for the ones place before cementing to the mat. As children accumulate counters on the ten-frames, the number of counters there and the number needed to make a set of 10 is always clearly evident, eliminating the need for frequent and tedious counting (Thompson & Van de Walle, 1984). Most illustrations of place-value mats in this book will show two ten-frames even though that feature is strictly an option.

The Forward Game

The "game" is described first as a teacher-directed group activity using counters and cups. Any base ten material can be substituted and other variations are described later.

Mount half sheet of light-colored paper on construction paper (9 X 12). Use spray adhesive or rubber cement. Print ten-frame on paper before mounting.

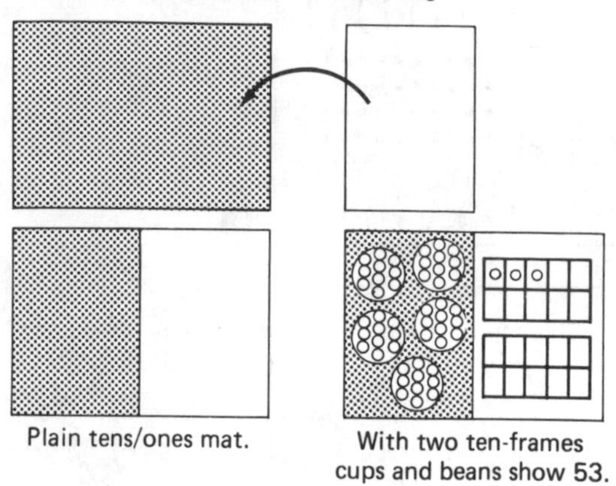

Strips and squares show 2 3 7 on three-place mat.

Figure 8.10
Place-value mats are easily made.

Activity 8.4

Each child has an empty place-value mat, a supply of counters, and small cups. Explain that each time you signal (snap fingers, ring a bell), they are to place one more counter on the ones or singles side of the mat. Whenever a ten-frame is filled (or there are 10 counters on the ones side), the class should call out in unison, "ten!" They then take their counters off of the ten-frame, put them in a cup, and place the cup of 10 on the left side or tens place of their mats. Periodically pause and have the class read their mats in unison. When you say "read your mat," children point to the tens side and then the ones side, saying what is there as they point: "Two tens and six." Occasionally have them say the usual name for that many counters: "twenty-six." Continue in this manner for 10 or 15 minutes or as long as seems reasonable for your class. If you stop at say, 4 tens and 5, you can write this on the board and begin there the next day. Go to 9 tens and 9 on a two-place mat.

The Backward Game

This activity is simply the reverse process from the forward game.

Activity 8.5

Have children place, say, 4 cups of 10 and 6 singles on their mats. Explain that each time you give the signal, they are to remove one counter from their boards instead of putting one on. When there are no more counters on the ones side, a cup of 10 should be dumped out onto the ones side and a ten-frame filled with the counters. Then a counter can be removed. Single counters should never be removed from cups, since cups are to always contain 10 counters. As before, periodically stop and have children read their mats.

Variations of the Forward and Backward Game

For first- or second-grade children, the variations below should only come after the games have been played as described. Older children can begin with one of these variations.

Activity 8.6

Change the amount to put on or take off. The first change should be to two at a time. In this way, the change to or from tens will always come out even. Then try putting on or taking off three at each signal. Always put all three on the board before making a ten. Use the second ten-frame when there are more than 10 counters on the ones side. In the backward game, nothing should be removed until all three can be removed. Figure 8.11 illustrates the sequence.

Activity 8.7

Use pregrouped materials. This is a significant switch for young children because now they *trade* 10 ones for a ten piece instead of grouping the singles into a ten. With pregrouped materials the games are referred to as "Forward Trading" and "Backward Trading" activities.

Activity 8.8

Play as a game. Each of two players has a mat. In turn, players roll a die to determine how many singles should be put on their mat (forward) or taken off of their mat (backward). In the forward version, the mats begin empty, and the first to reach a designated goal is the winner. In the backward game, both players start with a designated amount on the mats, and the winner is the first to clear his or her mat. In the forward game, the entire amount rolled should be placed on the mat before any groups or trades are made. In the backward game, nothing should be removed from the mat until the entire amount can be removed at once. That is, if a ten must be dumped (broken, traded), it should be done before any singles are removed.

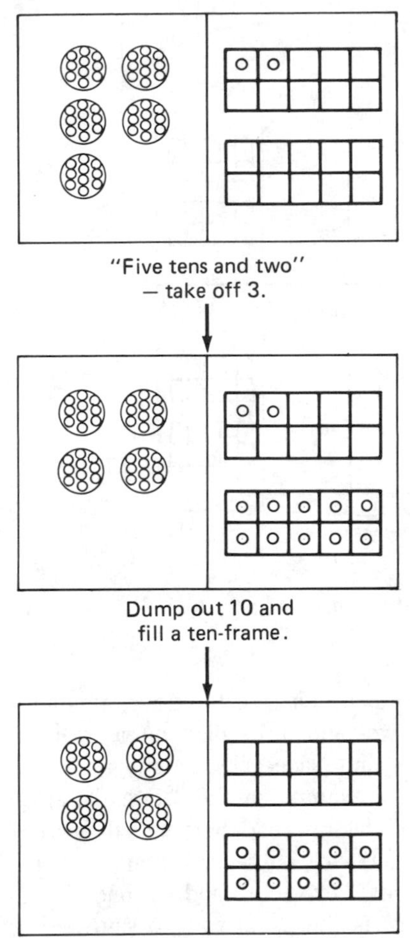

Figure 8.11
Backward sequential grouping. Taking 3 off each signal.

Activity 8.9

Use hundreds on a three-place mat. Trading can start at any point such as 3 hundreds, 5 tens, and 6. Ones can be added or removed in fixed amounts such as four at a time, or a die or spinner can be used to determine the amounts.

Activity 8.10

Add or remove both tens and ones on each move. In a teacher-directed format simply announce how many ones and tens to put on or remove. Vary the amounts each time. For independent activities and for the two-player game version, use two different dice, perhaps designating a red die for tens and a white die for ones. Cubes can be marked TENS and ONES, and numerals can be written on them to make dice. You may wish to use only the numbers 0, 1, 2, and 3 on the dice instead of 1 through 6. In forward games, both tens and ones should be added to the mats before any trades are made, as shown in Figure 8.12. In backward games, no partial amounts of ones or tens should be removed. You may wish to add the rule of always working first with the ones column and then the tens. This latter rule is totally arbitrary but does match the standard procedure for adding and subtracting with pencil and paper.

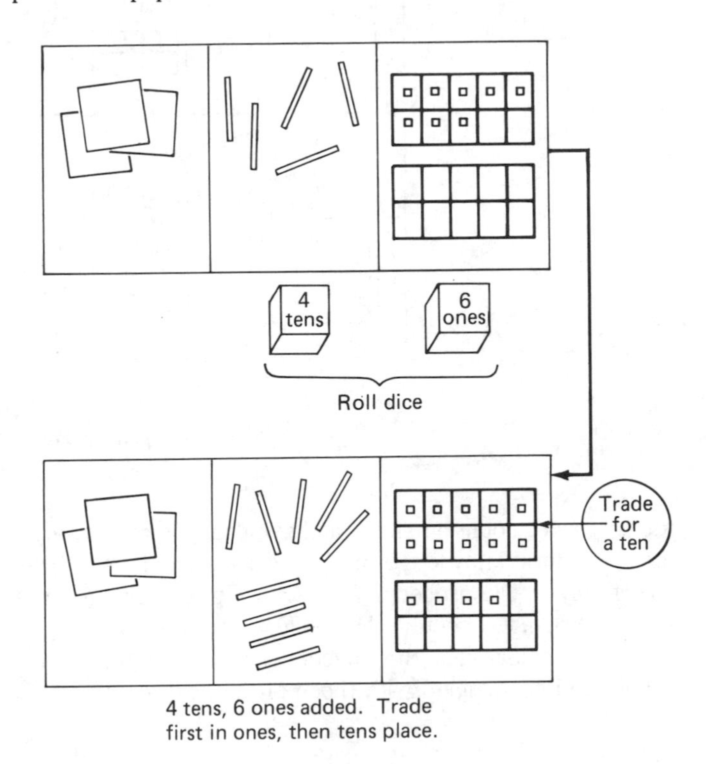

Figure 8.12
The forward trading game using 2 dice.

Activity 8.11

Worksheets can be made as in Figure 8.13 to designate the amounts to be added or removed. Children use materials as directed and record the results with the square, stick, and dot notation. This is especially useful to increase trades involving hundreds or to practice the double trade required in the backward game when there are no tens. The games in this form are simply a concept level of addition and subtraction with regrouping.

ALL THREE ACTIVITIES

Grouping activities, equivalent representation activities, and the sequential grouping or trading activities just described can be intermixed one with the other to provide variety and change of pace. It is not necessary for children to do or master one before the other. They have been presented separately here only for the sake of clarity.

Together, the three types of activities contribute to the development of all of the conceptual knowledge related to place value. Many of these activities include or can include

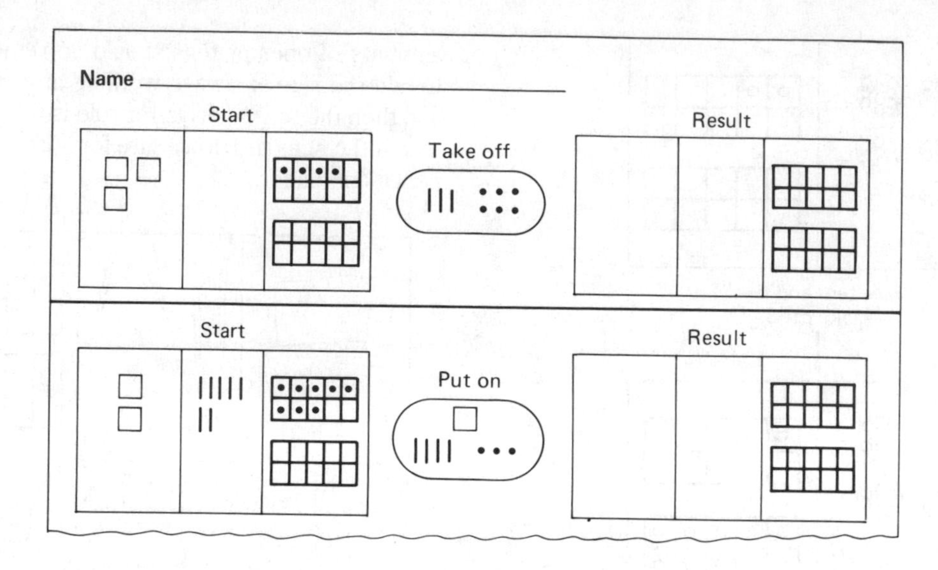

Figure 8.13
Students use place-value mats and models and record results.

the language, counting skills, and symbolism that are part of the procedural knowledge of place value. Activities directed specifically to the procedural aspects of place value are presented in the next section. Depending on grade level, experience, and objectives, the concept activities of this section should be intermingled with those of the next section.

Focus on Procedural Knowledge of Place Value

ORAL NAMES FOR NUMBERS

Standard Names versus Place-Value Names

The standard name of the collection in Figure 8.14 is "forty-seven." A more explicit terminology is "4 tens and 7 ones." This latter form will be referred to as *place-value language.*

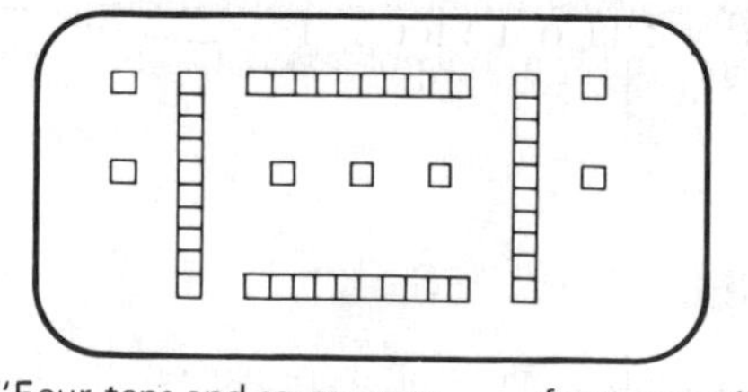

"Four tens and seven ones ––– forty-seven"

Figure 8.14
Mixed model of 47.

The more explicit place-value language is rarely misunderstood by children working with base ten materials and encourages thinking in terms of groups instead of a large pile of singles.

A general approach to the development of the standard names for two- and three-digit numbers is to juxtapose or pair standard and place-value languages frequently in lots of exercises. For example, in the sequential grouping or trading games, when children read their mats they are using the place-value name for what is there. When a mat has been read as "two hundreds, zero tens, and six ones," they can then be asked, "What is the usual way to say that?" Notice in this example how the place-value language is explicit about 0 tens, whereas in the standard name "two hundred six" the 0 tens are simply not mentioned. Place-value language provides a practical linkage between models and standard oral names for numbers.

Two-Digit Number Names

In first and second grade, children need to connect the base ten concepts with the oral number names they have used many times. They know the words, but have not thought of them in terms of tens and ones. The following sequence is suggested:

Start with the names "twenty," "thirty," "forty," . . . , "ninety."

Next do all names "twenty" through "ninety-nine."

Emphasize the teens as exceptions. Acknowledge that they are backwards and do not fit the patterns.

Almost always use base ten models while learning oral names. Use place-value language paired with standard language.

Activity 8.12

Use a 10 × 10 array of dots on the overhead projector. Cover up all but two rows [Figure 8.15(a)]. "How many tens? (2) Two tens is called *twenty.*" Have the class repeat. Sounds

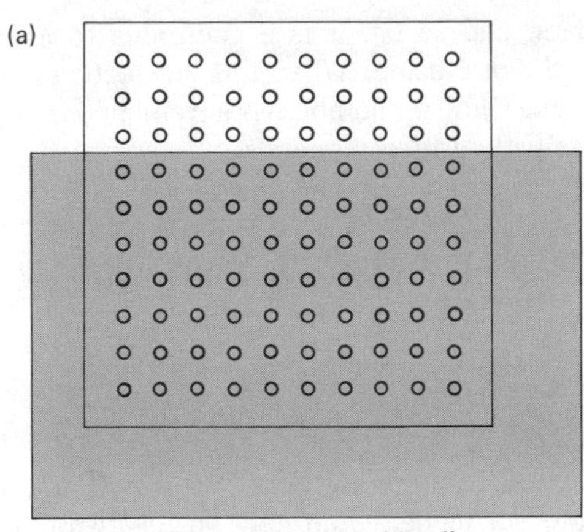

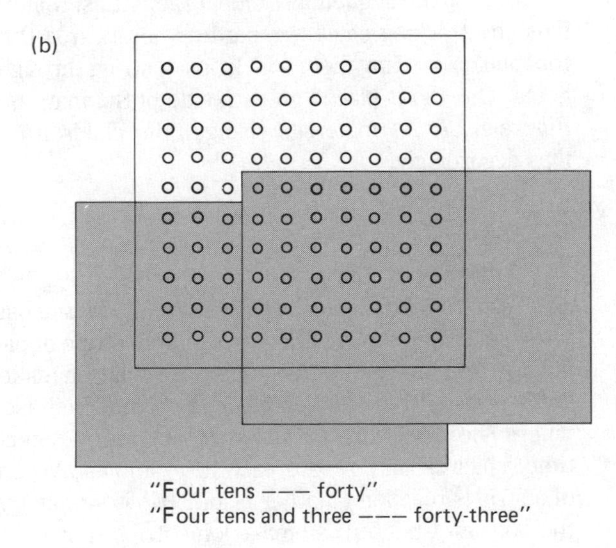

Figure 8.15
Ten-by-ten dot arrays are used to model sets of ten and singles.

a little like twin. Show another row. "Three tens is called *thir*ty. Four tens *for*ty. Five tens should have been *five*ty rather than *fif*ty. The names *six*ty, *seven*ty, *eigh*ty, and *nine*ty, all fit the pattern. Slide the cover up and down the array, asking how many tens and the name for that many.

Use the same 10 × 10 array to work on names for tens and ones. Show, for example, four full lines, "forty." Next, expose one dot in the fifth row. "Four tens and *one*. Forty-*one*." Add more dots one at a time. "Four tens and *two*. Forty-*two*." "Four tens and *three*. Forty-*three*" [Figure 8.15(b)]. When that pattern is established, repeat with other decades from twenty through ninety.

The basic approach described with the 10 × 10 dot picture should be repeated with other base ten models.

Activity 8.13

Show some tens pieces on the overhead. Ask how many tens. Ask for the usual name. Add a ten or remove a ten and repeat the questions. Next add some ones. Always have children give the place-value name and the standard name. Continue to make changes in the materials displayed by adding or removing 1 or 2 tens and by adding and removing ones. For this activity show the tens and ones pieces in different arrangements rather than the standard left-to-right order for tens and ones. The idea is to connect the names to the materials, not the order they are in.

Activity 8.14

Reverse the above activity by having children use place-value pieces at their desks. For example, you say, "make sixty-three." The children make the number with the models and then give the place-value name.

Activity 8.15

Ask your class: "How can you show thirty-seven fingers?" (This question is really fun if preceded by a series of questions asking for different ways to show 6 fingers, 8 fingers, and other amounts less than 10.) Soon children will figure out that four children are required. Line up four children and have three hold up 10 fingers and the last child, 7 fingers. Have the class count the fingers by tens and ones. Ask for other children to show different numbers of fingers. Emphasize the number of sets of 10 fingers and the single fingers (place-value language) and pair this with the standard language.

In all of the preceding activities, it is important to occasionally count an entire representation by ones. Remember that the count by ones is the young child's principal linkage with the concept of quantity. For example, suppose you have just had children use Unifix cubes to make 42. Try asking, "Do you think there really are forty-two blocks there?" Many children are not convinced, and the count by ones is very significant.

The language pattern for two-digit numbers is best developed and connected with models using numbers 20 and higher. That is where the emphasis should be. That is *not* to imply that we should hide the teens from children. Teens can and should appear in any of the previous activities, but should be noted as the exceptions to the verbal rules you are developing.

One approach to the teen numbers is to "back into them." Show, for example, 6 tens and 5 ones. Get the standard and place-value names from the children as before. Remove tens one at a time, each time asking for both names. The switch from 2 tens and 5 ("twenty-five") to 1 ten and 5 ("fifteen") is a dramatic demonstration of the backward names for the teens. Take that opportunity to point them out as exceptions. Count them by ones. Say them in both languages. Add tens to get to those numbers that are "nice." Return to a teen number. Continue to contrast the teens with the numbers 20 and above.

In Chapter 5, the teens were developed in activities that emphasized 10 and some more in a part-part-whole manner.

Fifteen is "ten and five" or "five and ten." However, 15 is not, at that early stage 1 ten and 5 ones. The concept developed early as simply 10 and 5 can be integrated with base ten concepts once they are firmed. The point is, use larger numbers to develop the base ten ideas.

Three-Digit Number Names

The approach to three-digit number names is essentially the same as for two-digit names. Show mixed arrangements of base ten materials. Have children give the place-value name and the standard names. Vary the arrangement from one example to the next by changing only one type of piece. That is, add or remove only ones, or only tens, or only hundreds.

Similarly, at their desks have children model numbers that you give to them orally using the standard names. By the time that children are ready for three-digit numbers, the two-digit number names, including the difficulties with the teens, are usually mastered. The major difficulty is with numbers involving no tens, such as 702. As noted earlier, the use of place-value language is quite helpful here. The 0 tens difficulty is more pronounced when writing numerals. Children frequently write 7002 for "seven hundred two." The emphasis on the meaning in the oral form will be a significant help.

Independent worksheet activities can be easily made by using the square, stick, and dot drawings for hundreds, tens, and ones. As shown in Figure 8.16, these can go from name to model or model to name.

WRITTEN NAMES FOR NUMBERS OR NUMERALS

Children both see and write two- and three-digit numerals well before they have them connected with concepts. While you cannot shield young children from room numbers, page numbers, calendars, and so on, it is a good idea to focus attention first on the oral name. When this connection is in place, the place-value name for numbers is a tremendous help in linking concepts with written numerals.

Connecting Symbols to Sequential Grouping and Trading

Activity 8.16

An idea borrowed from *Mathematics their way* (Baratta-Lorton, 1976) is to put numerals directly on the two-place mats during the sequential grouping activities. Make "numeral flips" by stacking small tagboard numeral cards 0 to 9 (0 on top) and connecting them with loops of string through punched holes. One flip is placed on each side of the mat, as shown in Figure 8.17. As materials change, the children change the flips accordingly.

Activity 8.17

Also from *Mathematics their way* is the idea of using place-value recording strips. These are simply strips of paper ruled in two columns (or three columns). As children make changes on their mats, they also record on the strips. In those versions where the same amount is either put on or taken off each time, children only record each new amount. When the end of a strip is reached, another blank strip is simply taped on at the bottom. Children can make long strips showing the count by ones or twos or threes from "00" to "99" and back [Figure 8.18(a)]. Children can work independently with their mats and their recording strips.

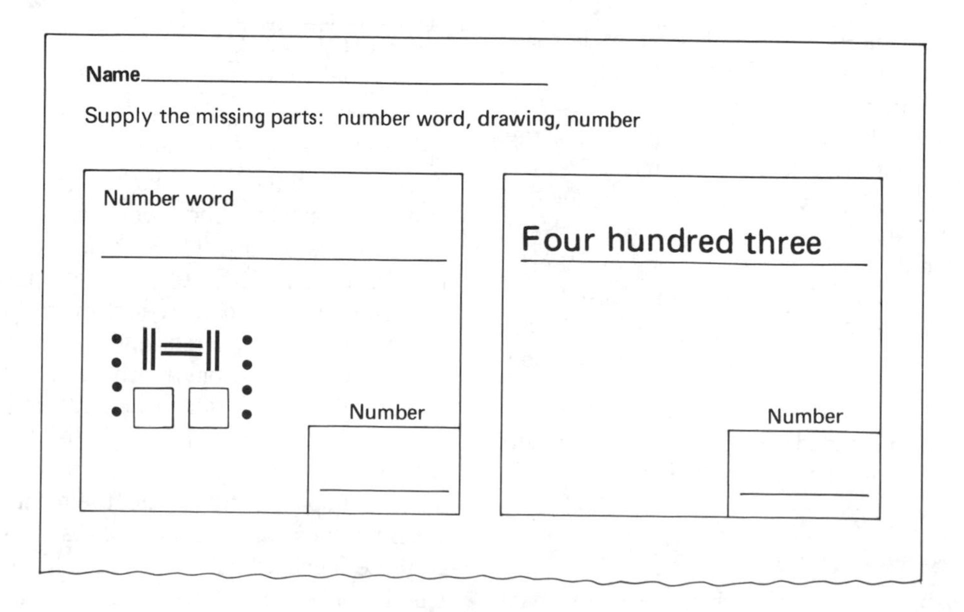

Figure 8.16
Connecting language, models, and numerals.

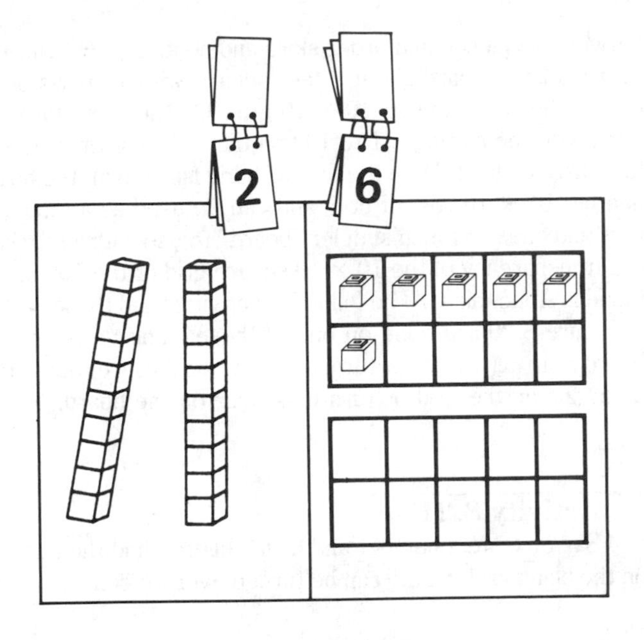

Figure 8.17
Flip cards connect model and numeral.

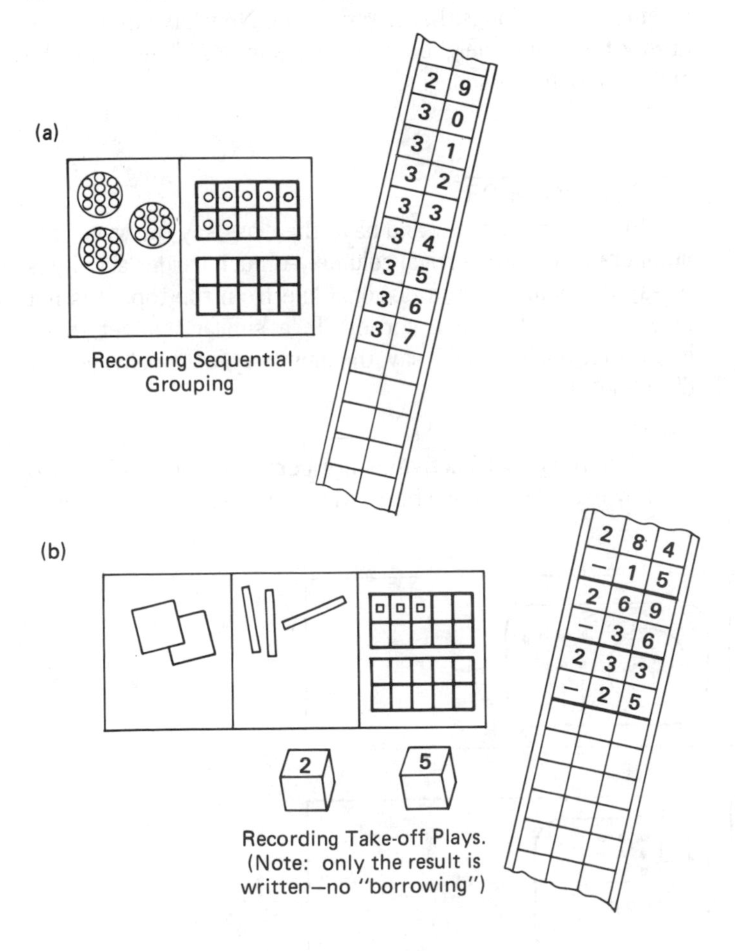

Figure 8.18
Using recording strips with grouping/trading activities.

In versions of sequential grouping where varying amounts are put on or removed, the strips can be used to indicate these change amounts as well. For example, if children were playing a forward-trading game in which they rolled a die for tens and a die for ones, they would record the amounts rolled and the result on the strips [Figure 8.18(b)]. This looks exactly like symbolic addition or subtraction, but is approached as recording the turns in the game.

Activity 8.18

Another way to connect symbols to the sequential grouping and trading activities is to use a calculator. For example, if students are putting 3 more on at each signal, they can store [+] 3 in the calculator and press [=] after each move. The calculator can then serve as a check to what they have done. A written record is still appropriate.

Other Connecting Activities

Try to conduct activities in which a model is shown and the students write the number, or vice versa. Children are aided in making the connections in these activities by use of the oral names for the numbers and by placing models onto a place-value mat. The place-value mat structures the left-to-right convention of hundreds, tens, and ones. When you show base ten materials for which you want students to write the numeral, arrange the materials in random order. This will ensure that children will attend to the left-to-right convention rather than simply copying the number for how many pieces are in each position.

Activity 8.19

Simple show-and-tell activities using any model can be done quite easily. For example, write a number on the board and have children show that many fingers (see earlier finger activity). Or have students model the number shown at their desks, using base ten materials. To reverse the activity, show base ten materials and have students write the corresponding numbers. For three-digit numbers, be sure to emphasize those numbers with no tens.

Almost any activity that has been suggested so far can be modified slightly to include a written form of the number involved. If the activity used number words, the students can also write the standard number. In grouping activities, if the students start with a large collection of singles, they can make the groups of 10 and then write the numbers. If they are given a collection of hundreds, tens, and ones including more than 9 ones and/or tens, they can not only make the trades physically, but they can record the resulting standard number symbolically.

Activity 8.20

Have students combine a written and a modeled number and write the standard name for the result. As illustrated in Figure 8.19, model and symbols activities may or may not involve grouping or trading. These activities can be done on worksheets or, with the use of models on the overhead, can be teacher-directed.

SEQUENCE AND COMPARISON SKILLS

Many children have difficulty filling in blanks in exercises such as the following:

____, ____, 59, ____, ____, 62
378, ____, ____, 381, ____
____, ____, ____, 241, 242

Two suggestions may help. First, relate this task to sequential-trading activities. Have children make a given number on their mats and use models to add one more or take one off to get the next numbers in the row. The second idea is to focus on the written patterns in the numeral sequence. These are especially easy to see if numbers are written one under another. Point out how the ones digit goes from 0 to 9 while the tens digit stays the same. Discuss especially the changes from 1 hundred to the next.

The calculator can provide excellent feedback for sequence activities. If, for example, a child is writing the numbers backward from 425 to 390, he or she could enter 425 [−] 1. Each time a number in the sequence is written, the [=] is pressed to confirm that it is correct.

Also related to conceptual knowledge is the ability to select the greater of two numbers. Rather than suggest rote rules such as "only look at the first digit," have children build or model each number in a pair and determine which is greater based on the model.

Many hundreds board activities are a good connection between sequence and order skills and concepts. A ***hundreds board*** is large board about 2 feet square with 10 rows of 10 nails or hooks in a neat, 10 × 10 grid. On the nails, tags are hung with the numbers 1 to 100 on them. Metal-rimmed key tags are excellent. With the tags all hung face down, the board is a big 10 × 10 row of dots and can be used as a model of tens and ones. A much simpler "board" to use with a full class is a transparency of the 10 × 10 square grid or the 100 beans found in Appendix C. The following activities and those shown in Figure 8.20 are done on one of these blank grids or on a board with all the tags turned down. You can either write numerals on the grid or turn tags over on the board.

Activity 8.21

Say or write a number and have children find the number on the blank grid. (A tag can be turned over for confirmation.)

Activity 8.22

Point to a square on the grid and have children decide what number belongs there. Write it in. Now have them write or say the eight "neighbors" of the number. These are the eight surrounding squares.

Activity 8.23

Start in the top row and have the class say (or write) the numbers in a given vertical column. A much harder activity is to say the numbers in a diagonal line from the top. It is not necessary to start in a corner. In a similar manner, start in the bottom row and say the numbers diagonally (either direction) upward.

Hundreds board activities are certainly not restricted to the few examples shown here. Most can also be converted to

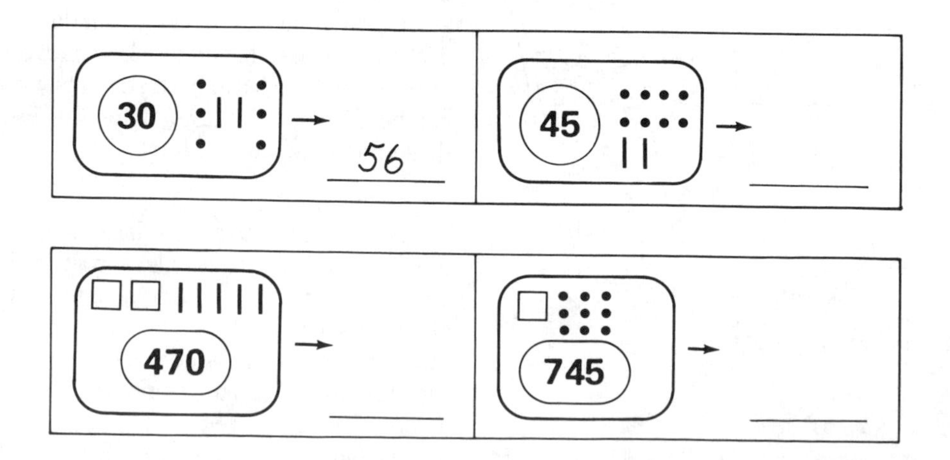

Figure 8.19
Combining models and numerals.

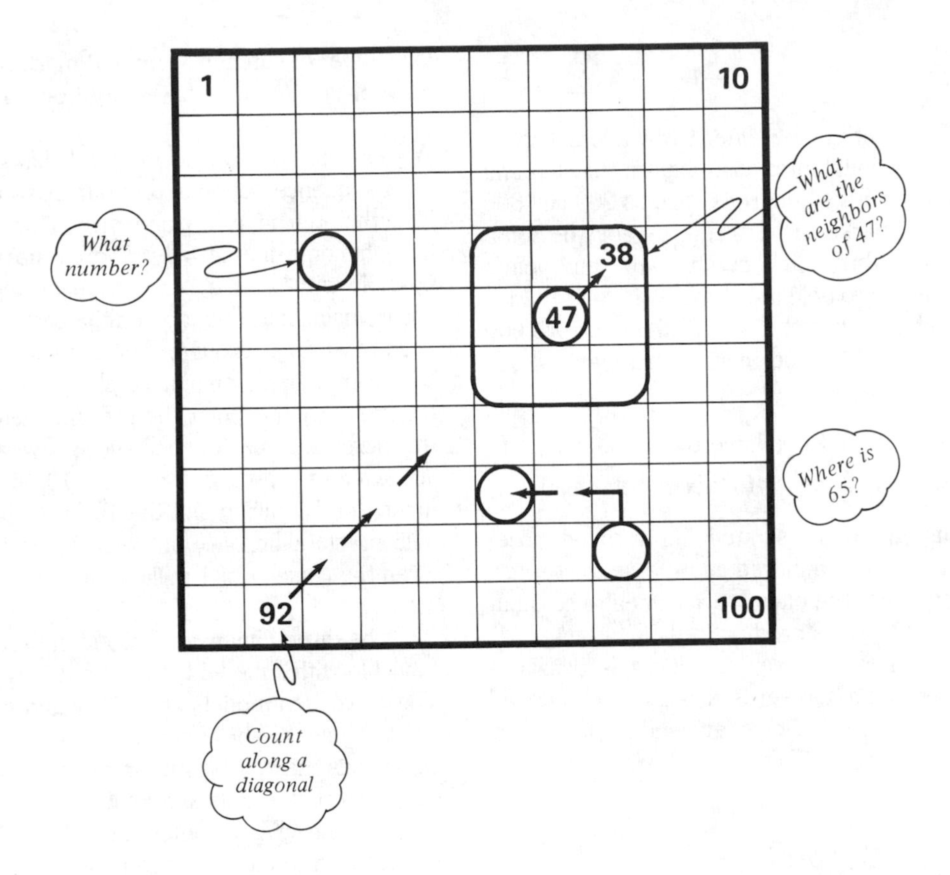

Figure 8.20
Numerals and number names on a hundreds board.

worksheet activities for independent work. Simply reproduce a 10 × 10 grid and supply directions.

Activity 8.24

The number line game is related to both sequence and order. On the board draw a long line and label the ends 0 and 100. Write down a secret number. On the line put a mark where your secret number would be. Children then try to guess your number. For each guess that is made, mark and label an appropriate point on the line. Each successive guess should be closer to the secret number, but you may very well be surprised.

For variations to the number line game, label the ends with any two multiples of 100. For example, they might be marked 700 and 800 or 300 and 600. After playing this way for a while, keep even the end points a secret and let children get their clues strictly from where you place their guesses.

ROUNDING NUMBERS

The skill of rounding numbers is closely related to an understanding of place value and is a significant tool in estimation skills. Too often, rounding is a mechanical activity that has no meaning. Consider the following two exercises and decide which is easier to think about:

A. What is 427 rounded to the nearest ten?

B. Is 427 closer to 420 or to 430?

Given exercises like exercise A, children are placed in a position of trying to recall a procedure: "What is the rule for rounding off?" Question B, on the other hand, asks not for a procedure but for a conceptual response. The distinction indicates a general approach to instruction in rounding. Eventually, children should be able to respond to both exercises, but developmental instruction should focus on the concept of rounding implicit in question B.

Suppose the number in questions A and B was 425. Question A implies there is a well-defined answer. Question B can be answered by noting that halfway is closer to neither. Students capable of understanding rounding are also able to handle an honest approach to rules that are completely arbitrary. Most schoolbooks round up whenever the number ends in 5. Many engineers use a rule involving even and odd numbers. That rule rounds up when the digit in question is odd and down when it is even. Thus, 435 is rounded to 440, but 425 is rounded to 420. Children should understand that there is no one "right" rule but that most groups of people adopt a convention in order to be consistent.

Activity 8.25

Play "Closer To" on a number line. Draw a long line on the board marked in 10 equal intervals. Explain that the end points are always consecutive hundreds such as 200 and 300 or 700 and 800. The marks in between are for the tens. Each time you name a three-digit number, the end points automatically are adjusted to hold the number. If you say 149, the ends are 100 and 200. If you say 612, the ends are 600 and 700. When this is understood, name a number and ask two questions:

Which end is it closer to? (Name the number.)

Which mark in between is it closer to? (Name the number.)

These two questions, in effect, ask students to round three-digit numbers to the nearest hundred and the nearest ten.

For four-digit numbers the ends are assumed to be thousands, and the between marks are hundreds.

Modeling a number with base ten pieces is usually a convincing way to see that a number is closer to one rounded value than another. It does not, however, lend itself to quick drills like the "Closer To" game just described.

Numbers Beyond 1000

EXTENDING THE THREE-DIGIT SYSTEM

Two important ideas developed for three-digit numbers should be carefully extended to larger numbers. First, the grouping idea should be generalized. That is, ten in *any position* makes a single thing in the next position, and vice versa. Second, the oral and written patterns for numbers in three digits is duplicated in a clever way for every three digits to the left. These two related ideas are not as easy for children to understand as adults seem to believe. Because models for large numbers are so difficult to have or picture, textbooks must deal with these ideas in a predominantly symbolic manner. That is not sufficient!

Activity 8.26

Have a "What comes next?" discussion with the base ten strips and squares. The unit or ones piece is a 1-cm square. The tens piece is a 10 × 1 strip. The hundreds piece is a square, 10 cm by 10 cm. What is next? Ten hundreds is called a *thousand.* What shape? It could be a strip made of 10 hundreds squares. Tape 10 hundreds together. What is next? (Reinforce the idea of "ten makes one" that has progressed to this point.) Ten one-thousand strips would make a square 1 meter (m) on a side. Draw one on butcher paper and rule off the 10 strips inside to illustrate the ten-makes-one idea. Continue. What is next? Ten ten-thousand squares would go together to make a strip. Draw this 10-m by 1-m strip on a long sheet of butcher paper and mark off the 10 squares that make it up. You may have to go out in the hall.

How far you want to extend this square, strip, square, strip sequence depends on your class and your needs. The idea that ten in one place makes one in the next can be brought home dramatically. There is no need to stop. It is quite possible with older children to make the next 10-m by 10-m square using masking tape on the cafeteria floor or with chalk lines on the playground. The next strip is 100 m by 10 m. This can be made on a large playground using kite string for the lines. By this point the payoff includes an appreciation of the increase in size of each successive amount as well as the *ten-makes-one* progression. The 100-m by 10-m strip is the model for 10 million and the 10-m by 10-m square models 1 million. The difference between 1 and 10 million is dramatic. Even the concept of 1 million tiny centimeter squares is dramatic.

The three-dimensional wooden base ten materials are all available with a model for thousands, which is a 10-cm cube. These wooden models are heavy and expensive, but having at least one to show and talk about is a good idea. Some companies make a plastic or cardboard version of the large cube, which is significantly cheaper.

Try the "What comes next?" discussion in the context of these three-dimensional models. (It is only necessary to actually have 10 ones, tens, and hundreds, and 1 thousand cube.) The first three shapes are distinct: a *cube,* a *long,* and a *flat.* What comes next? Stack 10 flats and they make a cube, same shape as the first one only 1000 times larger. What comes next? (See Figure 8.21.) Ten cubes makes another long. What comes next? Ten big longs makes a big flat. The first three shapes have now repeated. Ten big flats will make an even bigger cube, and the triplet of shapes begins again.

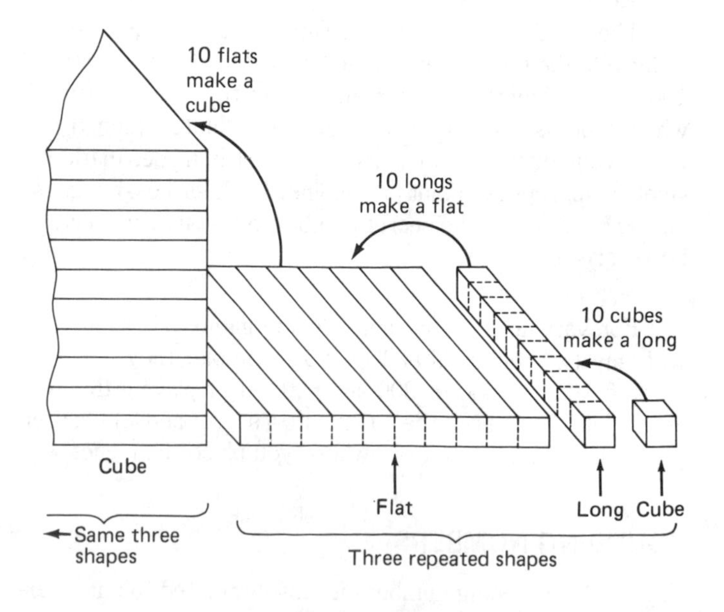

Figure 8.21
Every 3 places the shapes repeat.

A good discussion revolves around the metric dimensions of each successive cube. The million cube is 1 meter on an edge. The billion cube is 10 meters on an edge or about the size of a three-story building.

Each cube has a name. The first one is the unit cube, the next is a thousand, the next a million, then a billion, and so on. Each long is 10 cubes; 10 units, 10 thousands, 10 millions. Similarly, each flat shape is 100 cubes.

To read a number, first mark it off in triples from the right. The triples are then read, stopping at the end of each to name the unit (or cube shape) for that triple (Figure 8.22). Leading zeros are ignored. If students can learn to read numbers like 059 (fifty-nine) or 009 (nine), they should be able to read any number. To write a number, use the same scheme. If first mastered orally, the system is quite easy.

It is important for children to realize that the system does have a logical structure, is not totally arbitrary, and can be understood.

CONCEPTUALIZING LARGE NUMBERS

The ideas discussed in the previous section are only partially helpful in thinking about the actual quantities involved in very large numbers. For example, in extending the square, strip, square, strip sequence, some appreciation for the quantities thousand or of a hundred thousand is included. But it is hard for anyone to translate quantities of small squares into quantities of other items, distances, or time.

Creating Benchmarks for Special Big Numbers

In these activities numbers like 1000, 10,000, or even 1,000,000 are translated literally or imaginatively into something that is easy or fun to think about. Interesting quantities become lasting reference points or benchmarks for large numbers and thereby add meaning to numbers encountered in real life.

Activity 8.27

Collections: As a class or grade-level project, collect some type of object with the objective of reaching some specific quantity. Some examples: 1000 or 10,000 buttons, walnuts, old pencils, jar lids, pieces of junk mail. If you begin aiming for 100,000 or a million, be sure to think it through. One teacher spent nearly 10 years with her classes before amassing a million bottle caps. It takes a small dump trunk to hold that many!

Activity 8.28

Illustrations: Sometimes it is easier to create large amounts. For example, start a project where students draw 100 or 200 or even 500 dots on a sheet of paper. Each week different students contribute a specified number. Another idea is to cut up newspaper into pieces the same size as dollar bills to see what a large quantity would look like. Paper chain links can be constructed over time and hung down the hallways with special numbers marked. Let the school be aware of the ultimate goal.

Activity 8.29

Real and imagined distances: How long is a million baby steps? Other ideas that address length: toothpicks, dollar bills, or candy bars end to end; children holding hands in a line; blocks or bricks stacked up; children laying down head to toe. Real measures can also be used: feet, centimeters, meters.

Activity 8.30

Thinking of time: How long is 1000 seconds? How long is 1,000,000 seconds? a billion? How long would it take to count to 10,000 or 1,000,000? (To make the counts all the

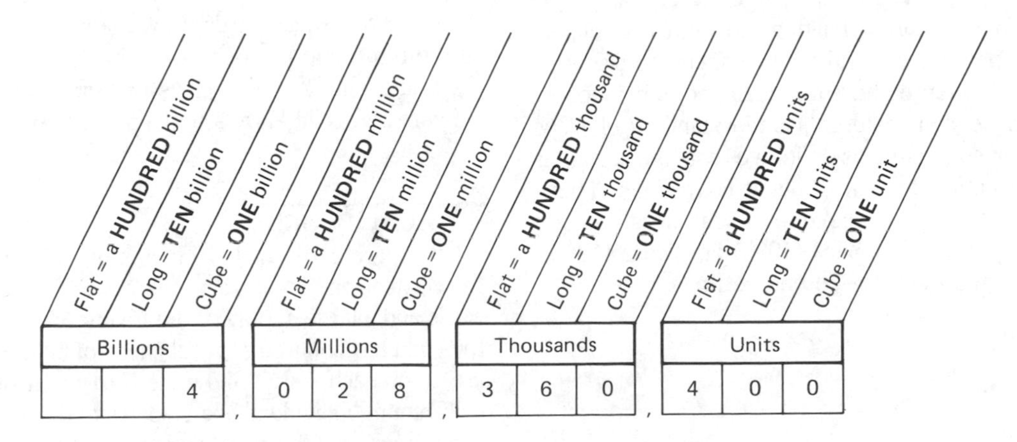

Figure 8.22
The triples system for naming large numbers.

same, use your calculator to do the counting. Just press the [=].) How long would it take to do some task like buttoning a button 1000 times?

Benchmarks from Easily Imagined Quantities

The foregoing activities aim at a specific number. The reverse idea is to select a large quantity and find some way to measure, count, or estimate it.

Activity 8.31

How many

Candy bars would cover the floor of your room?

Steps would an ant take to walk around the school building?

Grains of rice would fill the wastebasket?

Quarters could be stacked in one stack floor to ceiling?

Pennies can be laid side by side down an entire block?

Pieces of notebook paper to cover the gym floor?

Seconds have you (or the teacher) lived?

Big-number projects need not take up large amounts of class time. They can be explored over several weeks as take-home projects, group projects, or, perhaps best of all, translated into great schoolwide estimation contests.

Diagnosis of Place-Value Concepts

Assuming that the age level is appropriate, almost all children can be taught *to do* the activities described in this chapter. Even the most *meaningful* activities with place-value pieces can be performed without the child constructing the concepts involved. Going through the motions will not guarantee that concepts are formed. This is, of course, true of any mathematics activity. However, for the concept of place value, the distinction between conceptual thinking and surface-level performance of activities is frequently very difficult to discern.

The activities presented here are designed to help teachers look more closely at children's understanding of place value. They are not suggested as definitive tests but as means of obtaining information for the thoughtful teacher. These activities have been used by several researchers and are adapted primarily from Labinowicz (1985) and Ross (1986). The tasks are designed for one-on-one settings.

Activity 8.32

A variety of oral counting tasks provide insight into the counting sequence.

Count forward for me, starting at 77.

Count backwards, starting at 55.

Count by tens.

Count by tens starting at 34.

Count backwards by tens starting at 130.

In the tasks which follow, the manner in which the child responds is at least as important as the answers to the questions. For example, counting individual squares on tens pieces that are known to the child as "tens" will produce correct answers, but indicates the structure of tens is not being utilized.

Activity 8.33

Write the number 342. Have the child read the number. Next have him or her write the number that is *one more than* the number. Next ask for the number that is *ten more than* the number. Following the responses, you may wish to explore further with models.

Activity 8.34

Dump out 36 blocks. Ask the child to count the blocks and then have him or her write the number that tells how many are there. Circle the 6 in 36 and ask, "Does this part of your thirty-six have anything to do with how many blocks there are?" Then circle the 3 and repeat the question exactly. Do not give clues. (Ross, 1989, refers to this as the "digit-correspondence" task. Based on responses to the task, Ross has identified five distinct levels of understanding of place value.)

Activity 8.35

Dump out 47 counters and have the child count them. Next show the child at least 10 cards, each with a ten-frame drawn on it. (Spaces should be large enough to hold the counters used.) Ask, "If we wanted to put these counters in the spaces on these cards, how many cards could we fill up?" (If the ten-frame has been used in class to model sets of 10, use a different "frame" such as a 10-pin arrangement of circles. Be sure the child knows there are 10 spaces on each card.)

Activity 8.36

Prepare cards with 10 beans or other counters glued to the cards in an obvious arrangement of 10. Supply at least 10 cards and a large supply of the beans. After you are sure that the child has counted several cards of beans and knows there are 10 on each, ask, "Show me thirty-four beans." (Does the child count individual beans or use the cards of 10?) The activity can also be done with hundreds.

An understanding of equivalent representations, critical for understanding computation, can be assessed with the following activity.

Activity 8.37

Have the child represent a number using any base ten materials. Next trade one or more tens for ones (or hundreds for tens). "What number is this now?" or "Is there more now or less, or is it just the same?"

For the next two tasks use a "board" made from a half-sheet of posterboard, a box lid or some other cover, and pregrouped base ten materials.

Activity 8.38

Show a board with some base ten pieces covered and some showing. Tell the child how many pieces are hidden under the cover, and ask him to figure out how much is on the board all together [Figure 8.23(a)].

(a)

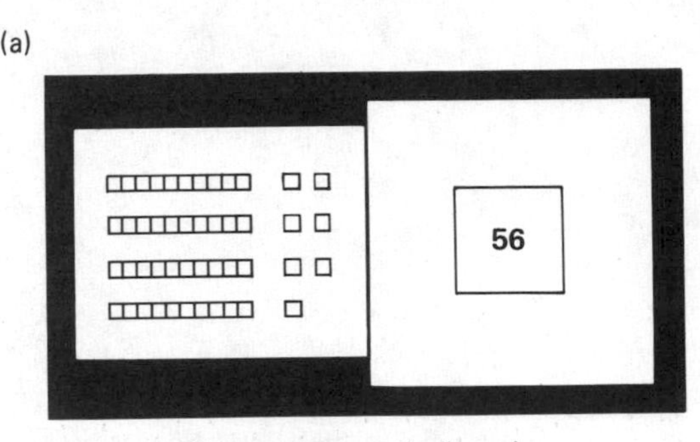

"How many on the board all together?"

(b)

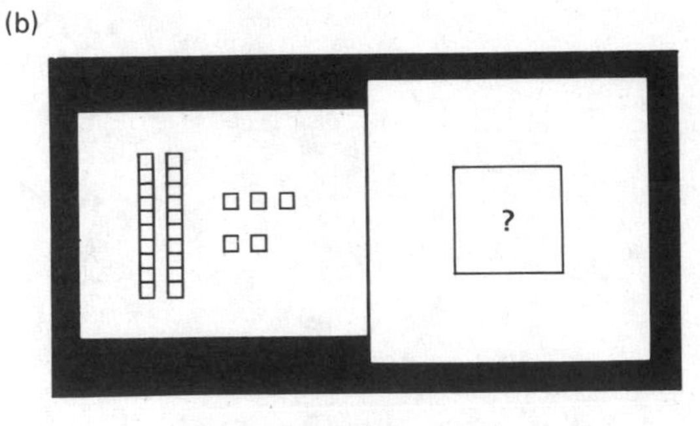

"There are 62 on the whole board. How many covered?"

Figure 8.23
Two diagnostic activities.

Activity 8.39

Show a board partially covered as before. Tell the child how many pieces are on the board altogether and ask how many are hidden [Figure 8.23(b)].

The last two tasks could also involve hundreds. The amounts that you tell the child could be given in written form instead of orally.

For Discussion and Exploration

1. Examine any textbook series starting in the first grade through the fourth grade for the place-value lessons provided. Compare the development with the elements of place-value understanding found in Figure 8.2. What ideas are developed well? How would you add to what is done on the student pages?
2. Visit a classroom in grades 2 to 5. Find out what base ten materials are used in the classroom and how they are used.
3. Conduct a big-number project in an upper-grade classroom. Design your project as a group effort. Before you actually begin, decide on some way to find out what concepts of numbers 1000 and larger the children in the class have.
4. Examine the excellent activities in the resource book *Picturing numeration* (Madell & Larkin, 1977). Pay special attention to those in which models and symbols are used in the same exercise.
5. Based on the suggestions in the last section and on the content found in the basal textbook, design a diagnostic interview for a child at a particular grade level and conduct the interview. (See Chapter 22 for interview techniques.) It is a good idea to take a friend to act as an observer, and/or use a tape recorder or video recorder to keep track of how the interview went. For a related discussion of children's place-value thinking and suggestions for preparing your interview, read Chapters 10 and 11 of *Learning from children* (Labinowicz, 1985).
6. The development of place value in the *Mathematics their way* program (Baratta-Lorton, 1976) includes several weeks of "readiness" in which the sequential grouping activities are conducted in bases 4, 5, and 6. For example, children make sets of five cubes into a Unifix bar or put groups of four beans in a cup. The groups are given humorous names such as "Bozos," so place-value mats are read "three Bozos and two." This means that children can make more groups in less time, avoid tedious counts to 10, and learn to use a word (e.g., "Bozos") to designate groups without the confusion of using a number word (ten) for a group name. Read the place-value chapter in that book or preferably find a first- or second-grade teacher who uses this approach. Discuss the pros and cons of this readiness activity.

Suggested Readings

Baratta-Lorton, M. (1976). *Mathematics their way.* Menlo Park, CA: Addison-Wesley.

Bickerton-Ross, L. (1988). A practical experience in problem solving: A "10 000" display. *Arithmetic Teacher, 36*(4), 14–15.

Creative Publications. (1986). *Hands on base ten blocks.* Palo Alto, CA: Creative Publications.

Kamii, C. (1986). Place value: An explanation of its difficulty and educational implications for the primary grades. *Journal of Research in Childhood Education, 1,* 75–86.

Kurtz, R. (1983). Teaching place value with the calculator. In L. G. Shufelt (Ed.), *The agenda in action.* Reston, VA: National Council of Teachers of Mathematics.

Labinowicz, E. (1985). *Learning from children: New beginnings for teaching numerical thinking.* Menlo Park, CA: Addison-Wesley.

Madell, R., & Larkin, E. (1977). *Picturing numeration.* Palo Alto, CA: Creative Publications.

Ross, S. H. (1989). Parts, wholes, and place value: A developmental perspective. *Arithmetic Teacher, 36*(6), 47–51.

Thompson, C. S., & Van de Walle, J. A. (1984). The power of 10. *Arithmetic Teacher, 32*(3), 6–11.

Van de Walle, J. A., & Thompson, C. S. (1981). Give bean sticks a new look. *Arithmetic Teacher, 28*(7), 6–12.

CHAPTER 9

PENCIL-AND-PAPER COMPUTATION WITH WHOLE NUMBERS

General Notes on Teaching Algorithms

COMPUTATIONAL PROCEDURES OR ALGORITHMS

The word *algorithm* refers to a rule or procedure for solving a problem. The whole number computational algorithms are the specific procedures used with pencil and paper to compute exact answers to arithmetic problems such as 489 + 367 or 56 × 39. In this country, with the possible exception of the long-division algorithm, most schools teach the same four algorithms, although others do exist. Therefore, this chapter concentrates on teaching what have become the "standard" algorithms for whole numbers.

It is important to remember that there are other algorithms for computation besides pencil-and-paper methods. Mental computation procedures are also algorithms. The procedure for using a calculator is an algorithm in the broadest sense. Within this chapter, the phrase "pencil-and-paper whole number computational algorithm" is generally shortened to "algorithm" for convenience.

The skillful use of an algorithm is very clearly in the area of procedural knowledge. To do something "algorithmically" implies a mindless or mechanical following of rules or procedures. What has become widely accepted is the value of developing algorithmic skills but with a very clear conceptual basis; that is, to connect the algorithms to concepts rather than to teach them as rules without reasons.

TOWARD A BROADER VIEW OF COMPUTATION

As we begin the 1990s, it is still true that pencil-and-paper computation holds a dominant place in the curriculum. There is, however, a long overdue trend toward viewing the algorithms with a different perspective. In the past, it was true that computational proficiency was a required skill in our society. We taught computation because it was necessary for everyday living. It was important for both students and adults to be able to compute sums of long columns of four- and five-digit numbers, to find products like 378 × 2496, and to do long divisions such as $71.8\overline{)5072.63}$. Today, computations with numbers as large as these are virtually never required of us, thanks to the readily available calculator.

An extreme reaction might be to delete pencil-and-paper computations from the curriculum. That would be a mistake. Some simple computations will always be done with pencil and paper when that is convenient. But perhaps more important than a utilitarian view of the algorithms is the effect that understanding them can have on other aspects of a more modern curriculum. A strong conceptual approach to the algorithms developed through the use of base ten models and discussion—and avoiding a premature rush toward mastery—can have a beneficial effect on computation taken more broadly.

Pencil-and-paper computation

Mental computation

Estimation

Use of calculators

All of these forms of computation, with whole numbers or decimals or fractions, can and should be interrelated. Understanding and use of one helps and contributes to the understanding and use of the others. All are tied to base ten place-value concepts. While many mental algorithms are different than those used with pencil and paper, conceptual knowledge of both processes is very much the same. For example, 42 × 8 mentally might be done this way: 8 forties is 320 and 16 more is 336. While not the same as the paper-and-pencil method, the use of tens and ones conceptually aids in the process. The better children understand the paper-and-pencil method, the easier will be the mental method. If, however, all that children learn is a rote skill (multiply the 8 by 2, write the 6 and carry the 1, . . .), the instrumental knowledge that results is useless in learning mental procedures. This calls for a heavy emphasis on meanings with pencil-and-paper algorithms with a reduced emphasis on skills.

A similar observation can be made concerning estimation. In the division $71.8\overline{)5072.63}$, a child who understands place-value concepts and the algorithm can see how that problem is the same as $7.18\overline{)507.263}$ and that it is close to $7\overline{)507}$. Since

$7\overline{)490}$ is 70, 70 is a reasonable estimate for the original problem. (If that example seemed difficult to you, reflect on the type of training you received in division. Did it prepare you to think that way?)

As you go through this chapter, the focus is clearly on a meaningful development of the algorithms. In the classroom, as you work with children doing pencil-and-paper computation exercises, it is not only appropriate but important to consider estimation and mental computation at the same time. For example, before children even begin a computation, have them write down a rough estimate of the result. "Is it going to be closer to 500 or 1000? Would 60 or 600 seem about right for this one? Let's try to get this one in our heads first." And then discuss different methods used. These discussions will almost always have a place-value orientation. Point out how estimation and mental computations are not the same as doing them on paper. These discussions can add to better place-value understanding, to an appreciation for different methods and approaches, to increased mental process skills, and to better understanding of the algorithms.

Even the calculator belongs in the pencil-and-paper computation program. For example, do a computation with the class using traditional approaches. Use the calculator and get the same result. "What would happen if we changed the problem from 32 × 8 to 34 × 8?" Try it with the calculator. What is the difference in the results? "Why is it 16 more? Can you explain that with the base ten pieces? What if we changed the 32 to 42?" The calculator is helping with the tedium, but the focus is still on the meaning of the algorithm, and the entire discussion is excellent background for mental computation skills.

Estimation, mental computation, and calculators all contribute to each other in a broader view of computational skill. There is a time to address each topic separately as they are addressed in this text. It is also appropriate to compare, contrast, integrate, and discuss these methods with your students.

PREREQUISITES FOR ALL ALGORITHMS

Concept of the Operation

Each algorithm is based on one meaning of the operation. That meaning should be familiar to children so that attention can focus on the algorithm. Children also must learn that the algorithm result applies to all meanings of the operation. For example, while long division is usually based on the partition concept, it can be used to solve problems involving the measurement concept.

Place-Value Concepts

Each of the algorithms depends conceptually on the place-value system of numeration to make it work. Most importantly, sequential grouping or trading activities with base ten materials should be totally familiar both conceptually and orally to any class being taught any computational algorithm. If your students do not have this knowledge and cannot talk about trading in a meaningful way, it will be very difficult to teach the algorithms in a meaningful manner. With a firm grasp of the concepts involved in trading or grouping, including familiarity with the language and the models, the algorithms can be taught in much less time and with much less remediation required later. The following discussion of each algorithm is based on the assumption that children have these prerequisites.

Textbooks and curriculum guides have come to refer to trading or grouping as *regrouping,* although that term seems to have no conceptual value for children. Language such as "trade," "make a group," or "break a ten," or whatever words were used when those activities were done, have the most promise of carrying any meaning for children. The most traditional terms of "borrowing," "carrying," and "bring down" do not at all relate to conceptual activities. It is strongly recommended that this older terminology not be used. It may be familiar and comfortable to adults, but not to children.

Basic Facts

In the initial stages of learning the algorithms, base ten models are used and it is possible to do activities without fact mastery. The focus of attention during these beginning stages of instruction is on the use of place-value concepts. If necessary, children can use counting methods with the base ten models.

However, the end goal of algorithm development is a reasonably efficient use of the procedure without models. This level requires fact mastery. Any children still counting on their fingers or referring to a multiplication fact chart will have their thoughts diverted from the procedure itself. Your goal should be to have all required basic facts mastered before your students begin computation at a totally symbolic level.

The Addition Algorithm

Some work with the addition algorithms usually begins in the last part of the first grade. The greatest emphasis is placed on both the addition and subtraction algorithms in the second and third grades. Especially as early as grade 1, and also in grade 2, the emphasis should be on an understanding of the procedure. Allow students to use base ten models for a long time to do addition and subtraction. In any discussion over what grade level addition (or subtraction) should be introduced, the pivotal consideration is the use of models. First graders, for example, have no need to be skillful with the addition algorithm using only pencil and paper.

CONCEPT-LEVEL INSTRUCTION

Unstructured Addition Activities

The idea of an unstructured activity is to have children solve a problem before they learn the standard algorithm. Students must then rely on their background knowledge and work together to find a way to arrive at an answer.

Activity 9.1

Give students a word problem similar to the following: Farmer Brown picked 36 apples from one tree and 48 more from another tree. How many did he pick altogether?

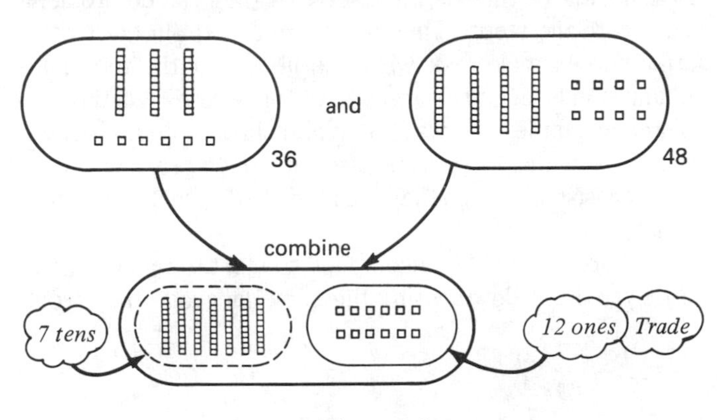

Figure 9.1
Gathering the big pieces first.

In this activity, once the suggestion of using base ten pieces is made, children will very likely gather the tens pieces before the ones and make a trade for a ten last (Figure 9.1). The method chosen is not important. Success with this activity demonstrates the principles of trading and more importantly communicates to the students that they actually know how to add two "big numbers." Children with hundreds models can combine numbers such as 348 and 276 in a similar way. Again, it is both natural and acceptable to begin combining the big pieces first.

All activities should involve trading from the very first example.

Structuring the Addition Algorithm

When children are comfortable combining two numbers with base ten models, the next step is to direct students to use the standard algorithm. Have students make two numbers on their place-value mats. Direct attention first to the ones column and decide if a trade is necessary (or if a group of 10 can be made.) If so, have the children make the trade. Repeat the process in the tens column. If you are working with hundreds, go on to the third column. Explain that from now on you want them to always begin working right to left, starting with the ones column. An example is shown in Figure 9.2. Children solve the problems with models and record only the answers. At the concept level, no symbolism is used to represent trades (carrying).

Continue at this level for several days or until children are able to add, using a right-to-left process, without any intervention or teacher direction. Your goal is to have children completely master the conceptual basis for the algorithm before attempting to connect it to a symbolic procedure.

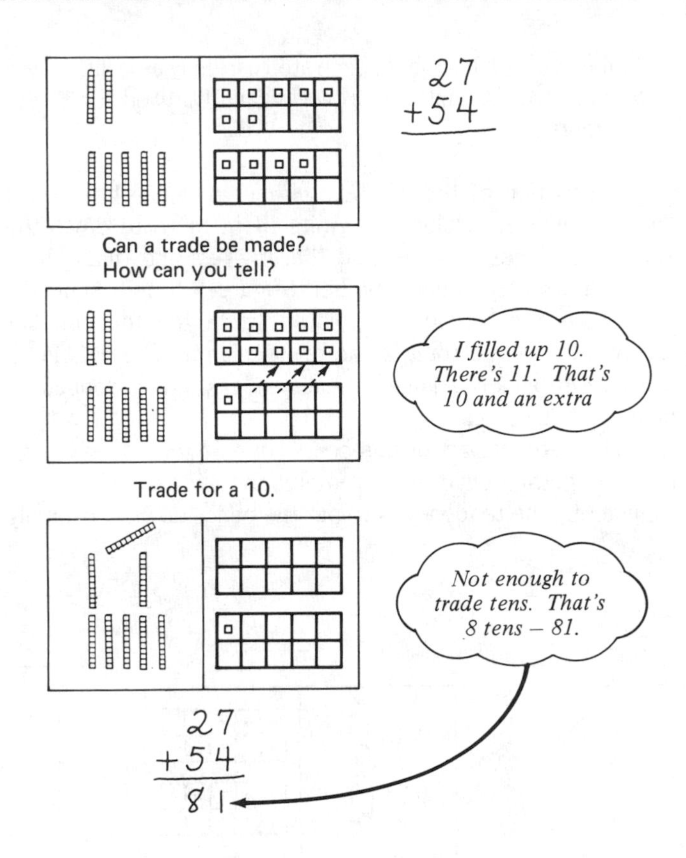

Figure 9.2
Working from right to left.

CONNECTING CONCEPTS TO SYMBOLISM

Activity 9.2

Reproduce pages with simple place-value charts similar to those shown in Figure 9.3. The charts will help young children record numerals in columns. The general idea is to have children record on these pages each step of the procedure they do with the base ten models *as it is done.* The first

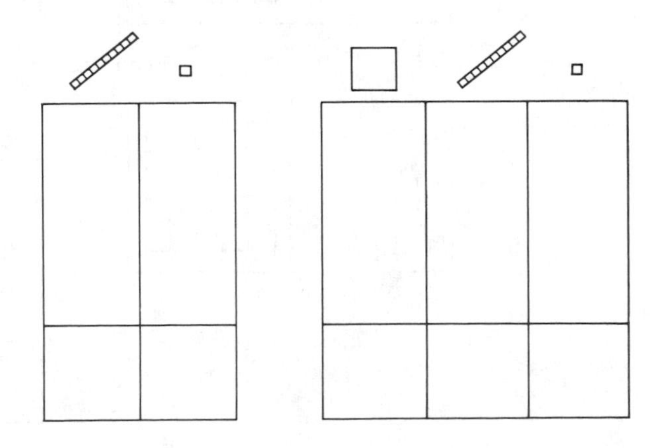

Figure 9.3
Blank recording charts are helpful. These are actual size.

few times you do this, guide each step carefully as is illustrated in Figure 9.4. A similar approach would be used for three-digit problems.

A variation of the recording sheet that has been used successfully with children permits them to write down the sum in the ones column, and thereby see and think about the total as a two-digit number (Figure 9.5 and Appendix C). When using models, children trade ten for one but rarely think of the total in the ones column. The bubble on the side provides a linkage that can easily be dropped or faded out later on.

The hardest part of this connecting or record-as-you-go phase is getting children to remember to write each step as they do it. The tendency is to put the pencil down and simply finish the task with the models. One suggestion is to have children work in pairs. One child is responsible for the models and the other for recording the steps as they are done. Children reverse roles with each problem.

To determine if children have adequately made the connection with the models, first see that they can do problems and record the steps. Then ask them to explain the written form. For example, ask what the little 1 at the top of the column stands for and also why they only wrote a 4 in the ones column when the total was 14. When children demonstrate a connection with concepts and also can do the procedure without any assistance, then they are ready to do problems without models.

Children should be encouraged to talk through each step as they write it down, using the same language they would

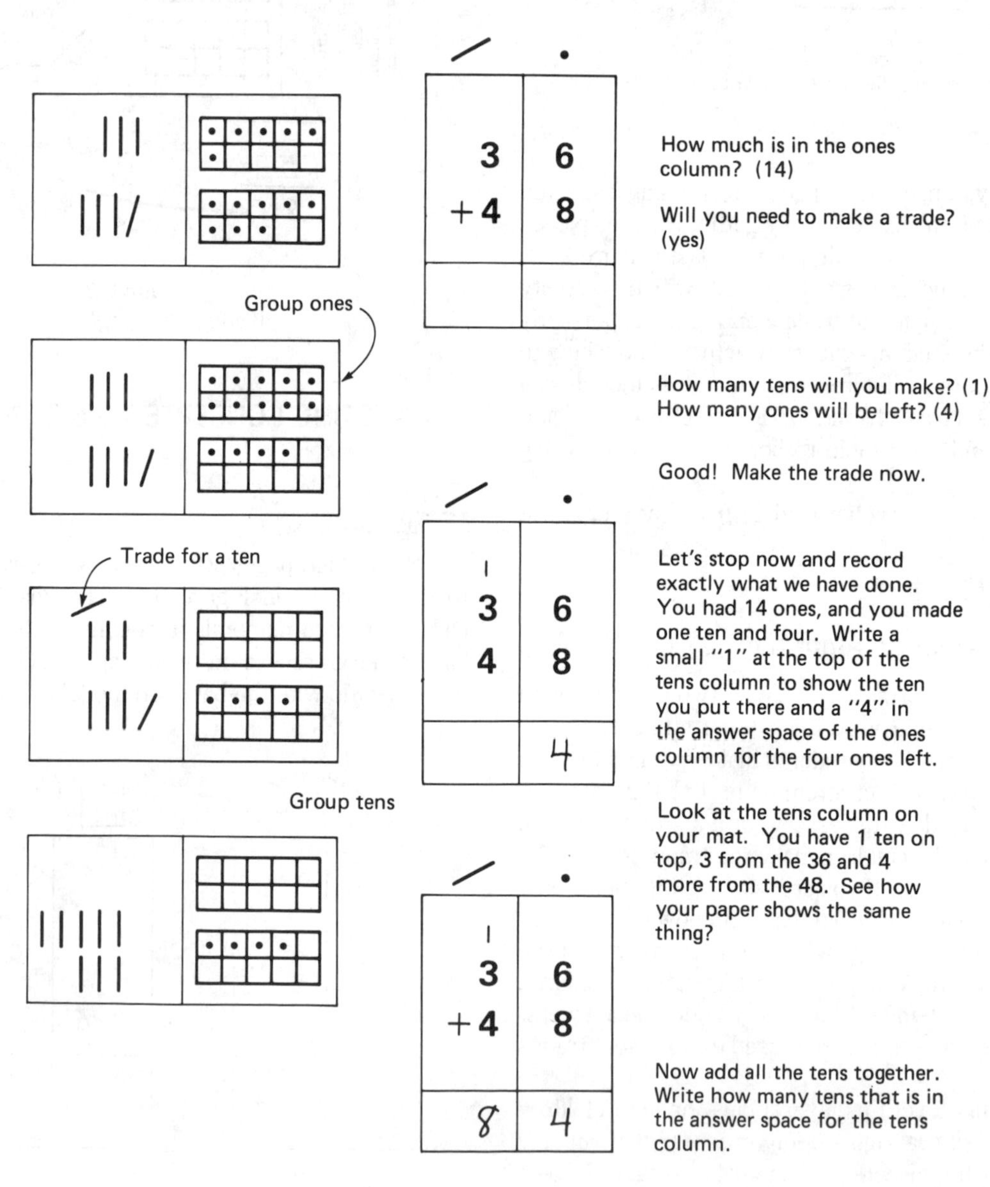

Figure 9.4
Help students record on paper each step that they do on their mats ***as they do it.***

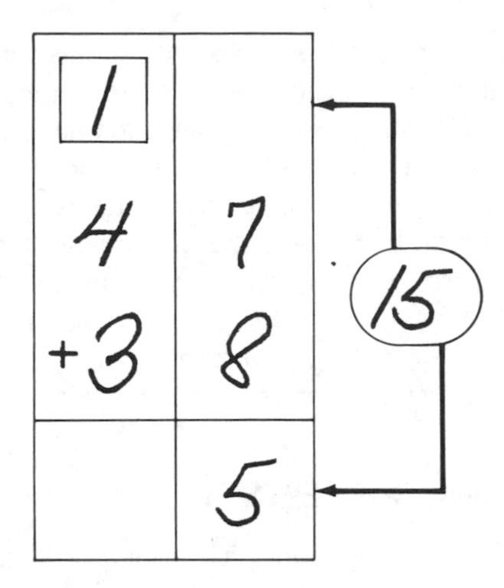

Figure 9.5
The bubble provides children with a place to write the sum in the ones column.

use if the base ten models were present. At this symbolic level, children should be ready to practice the algorithm in order to become more proficient.

Addition exercises with three or four addends need some special attention conceptually. These problems can require trading for 2 or 3 tens at one time. The ability to add more than two digits mentally is a multistep process that also is more difficult. Do not use the algorithm as the context for helping children add 9 + 8 + 5 mentally. This is a separate issue and should be dealt with individually.

The Subtraction Algorithm

The subtraction algorithm is usually taught immediately after the addition algorithm and with the same number of digits. Subtraction does prove to be harder for children. The general approach to teaching the subtraction algorithm is essentially the same as the addition algorithm.

CONCEPT-LEVEL INSTRUCTION

Unstructured Subtraction Activities

Activity 9.3

Use a word problem context to present a problem involving take-away. For example: There are 614 students in the school and 358 bought their lunch. How many did not buy lunch?

As with addition, do not provide children with a method but suggest that their base ten models may be useful. First- and second-grade children should begin with two-digit numbers. It would be a good idea for you to stop right now and use models to take 38 from 72. Do not try to mimic your pencil-and-paper method, just do what seems natural.

It is interesting that children's natural or spontaneous algorithms are not the same as those they are eventually required to master. In the exercise 72 − 38, they are likely to first remove 3 tens and then try to figure out how to take away 8 ones. They may get all 8 from a ten piece by trading it for 2 ones, or else take away the 2 existing ones and then trade a ten. They will very rarely trade a ten to create 12 ones, as in the standard algorithm. These methods are conceptually sound and should be permitted.

Structuring the Subtraction Algorithm

The goal is to guide children's natural tendencies and understanding to the standard procedure of working with the ones column first.

Activity 9.4

Start by having children model the top number in a subtraction problem on the top half of their place-value mats. A good method of dealing with the bottom number is to have children write each digit on a small piece of paper and place these near the bottom of their mats in the respective columns, as in Figure 9.6. The paper numbers serve as reminders of how much is to be removed from each column. Nothing should be removed from any column until the total amount can be taken off. (Recall that was the rule in the backward-trading game.) Also explain that they are to begin working with the ones column first as they did with addition. The only justification for this latter rule is that when done symbolically, it is a bit easier to work with the ones first and that is how "most people" do subtraction. The entire procedure is exactly the same as the backward-trading games when two dice were used. In fact, the paper numerals are simply a variation of the two dice as shown in Activity 8.10 on page 122.

Only the answers are written during this concept level. No indication of trades or "borrowing" is written at the top of the problem. Use a set of your own models to follow through the steps in Figure 9.6. Notice how the empty ten-frame is filled with ten ones when the backward trade is made. Without a frame, many children add onto the 5 ones already there and end up with only 10 ones.

Difficulties with Zeros

Exercises in which zeros are involved anywhere in the problem tend to cause special difficulties, especially in symbolic exercises without models. Enough attention should be given to these cases at the concept level to provide an experience base for working at the symbolic level.

A zero in the ones or tens place of the bottom number

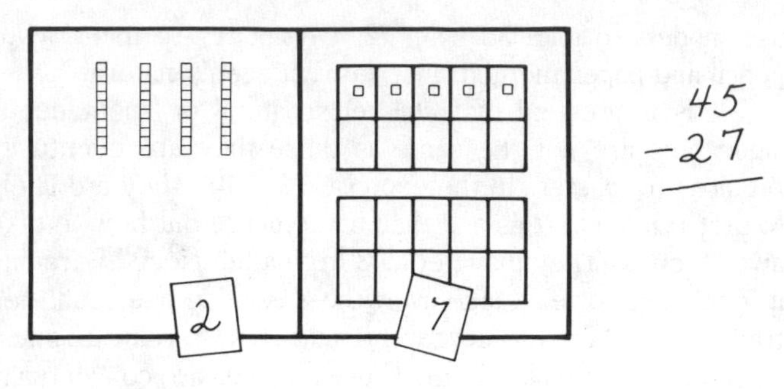

Not enough ones to take off seven.
Trade a ten for 10 ones.

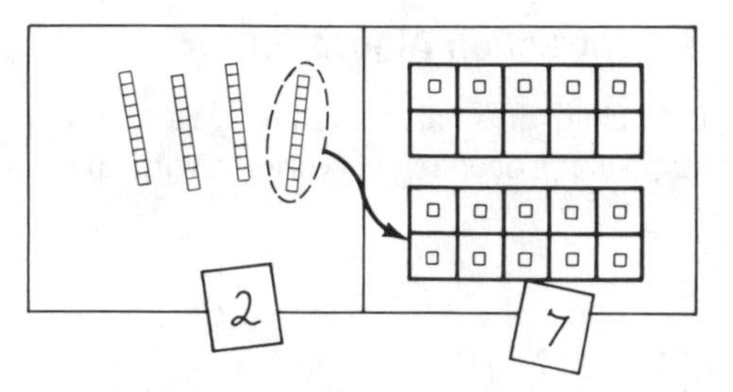

Now there are 15 ones. I
can take seven off easily.

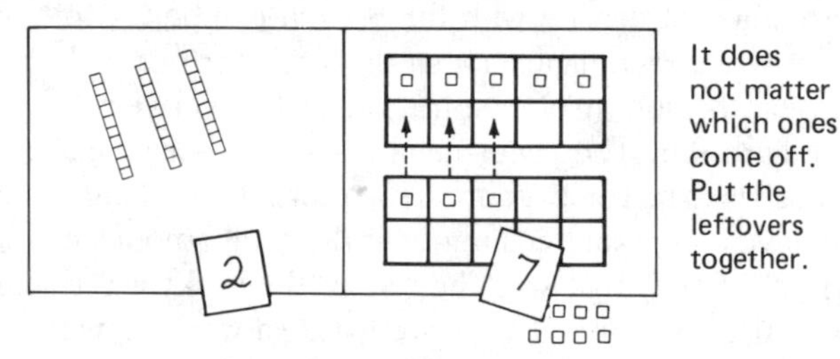

And now I can take
off 2 tens.

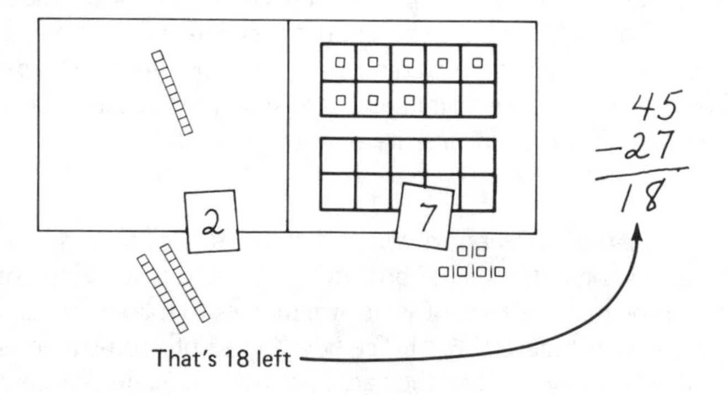

Figure 9.6
Two-place subtraction with models.

means there is nothing to take away and leaves some children wondering what they are supposed to do.

With a zero in the ones place of the top number there are no materials in that column which presents an unusual situation when done with models. Make a special effort to include problems such as 70 − 36 or 520 − 136 while children are working with models and before they begin learning how to record.

The most difficult zero case is the one with no tens in the top number, as in 403 − 138. With models, children must make a double trade, exchanging a hundreds piece for 10 tens and then one of the tens for 10 ones. With models it is relatively clear that the hundred should not be exchanged directly for ones. Symbolically, that is a very common error: "Take 1 from the 4 and put it with the 3 to make 13."

If you have not done so already, use base ten models to do these subtraction exercises with zeros.

CONNECTING CONCEPTS TO SYMBOLS

As with all of the algorithms, children should remain at the concept level using materials and recording only answers until they can comfortably do problems without assistance, and demonstrate an understanding. At that point, the process of recording each step as it is done can be introduced in the same way as was suggested for addition. The same recording sheets (Figure 9.3) can be used. Figure 9.7 shows a sequence for a three-digit subtraction problem, indicating the recording for each step.

When an exercise has been solved and recorded, children should soon be able to explain the meanings of all of the markings at the top of the problem in terms of base ten materials. This ability to decode is a signal for moving children on to a completely symbolic level.

The Multiplication Algorithm

The multiplication algorithm is probably the most difficult to teach with conceptual understanding. The relatively simple case with single-digit multipliers (bottom numbers) is usually introduced in the third grade. It is in the fourth and fifth grades, when the bottom number moves to two digits, that the conceptual difficulties begin.

In this chapter, one approach for helping children understand the general two-digit algorithm is carefully developed. A comparison of this sequence with one more commonly found in textbooks is also provided.

A PRELIMINARY EXPLORATION

Suppose for a moment that you have not learned how to multiply numbers when either one of them has two or more digits. That is, all you know are your basic multiplication facts. How could you go about finding out how many squares are in the rectangle in Figure 9.8 without counting each one individually? (It is strongly suggested that you actually do this exercise before continuing.)

There are many ways to approach the task. Note that the heavy lines on the grid are drawn every ten spaces in each direction. That means there are squares of 100 each. Can you find sets of tens that would be easy to count?

This task could reasonably be given to students in the

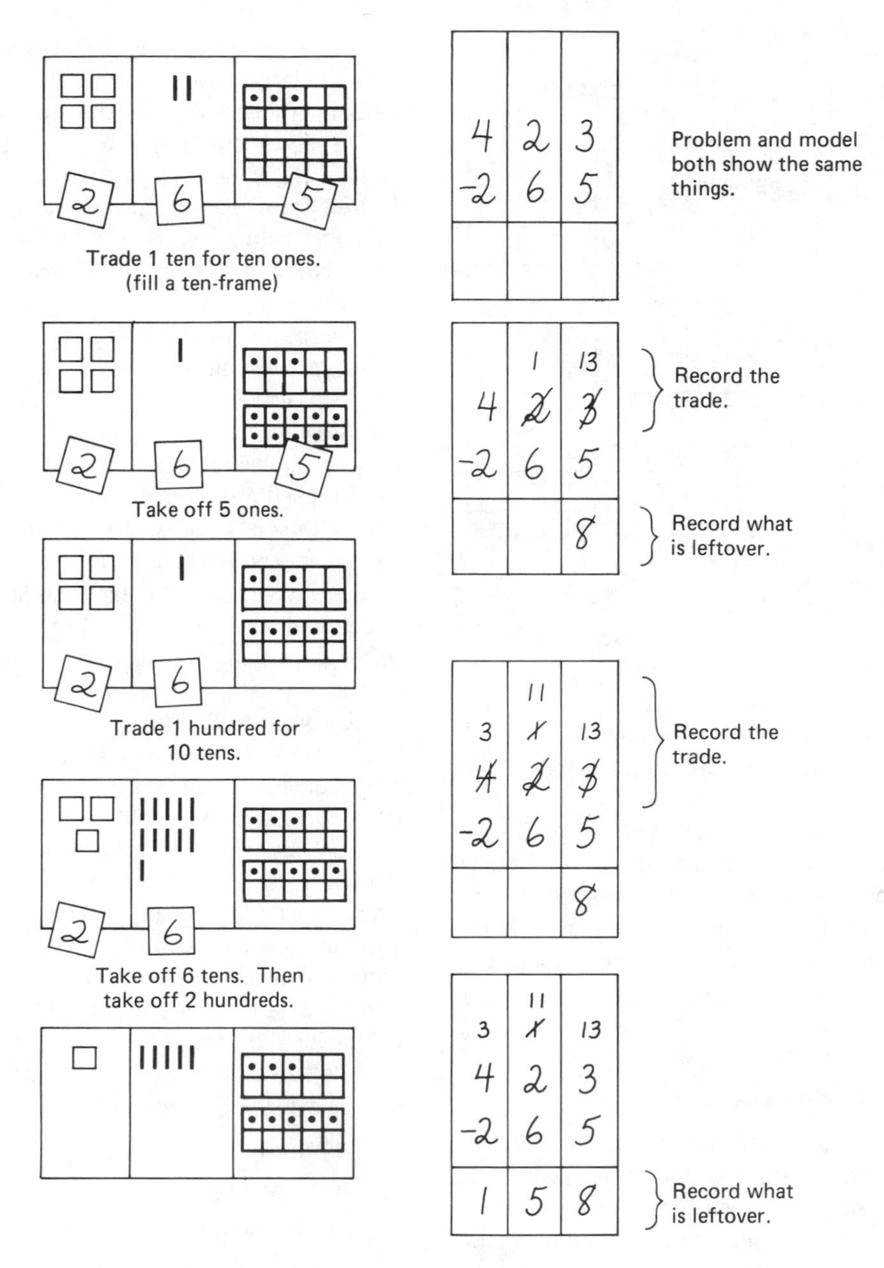

Figure 9.7
Help children record each step as they do it.

fourth or fifth grades as a problem-solving introduction to the multiplication algorithm. Rather than discuss the task more fully here, it is left for you to explore. The sections which follow should clarify one possible approach.

ONE-DIGIT MULTIPLIERS

The distinction between concept and connecting levels of instruction is not as distinct in this development as it is for addition and subtraction algorithms. Small ideas are developed one at a time using models. A strong base ten language component is also developed, and a method of recording is suggested. The language component is extremely important. It provides children with the principal means of connecting concepts and procedural knowledge.

Models

The activities suggested for the multiplication algorithm can be modeled by children in either of two modes. First, base ten models like strips and squares or the wooden base ten blocks can be used. Second, *base ten grid paper* with a heavier line drawn to show 10 × 10 squares can be used (Appendix C). Children can draw on this base ten grid paper to show base ten pieces as if they were ones, tens, and hundreds pieces. To ensure that the grid-paper version is meaningful,

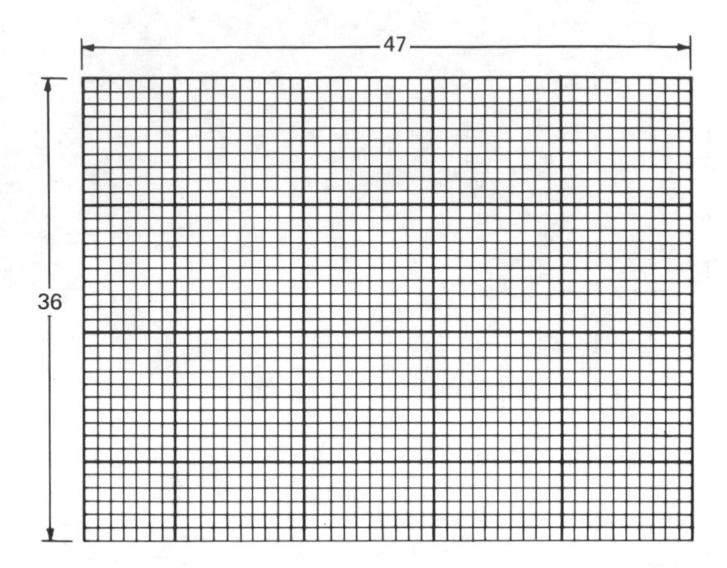

Figure 9.8
How big is this rectangle? (How many □'s are inside?)

do some modeling with ones, tens, and hundreds pieces. A rush to a symbolic mode with this algorithm can simply eradicate any benefits of working with models.

The arrangements of base ten materials can easily get larger than will conveniently fit on standard desktops. Groups can work at tables or on the floor to see a few examples with these pieces. You are strongly encouraged to use both strips and squares and base ten grid paper as you read through the development which follows.

Developing Base Ten Products

The exercises in this section involve filling in rectangles with base ten pieces. If the dimensions of the rectangles are in centimeters, then centimeter base ten strips and squares will fit the dimensions the way that you want. Wooden or plastic models all are in metric dimensions and are much easier to work with for these tasks.

It is probably better if you have students work together in small groups of two or three and give them predrawn rectangles. For dimensions longer than 20 cm you will need to use large sheets of paper or tagboard. One way to prepare these rectangles for your class is to cut one of each size you want from poster board and quickly trace around it with a felt marker.

Activity 9.5

Provide each group of children two rectangles with these dimensions:

Rectangle A: 3 ones by 5 ones (Note: 3 ones is 3 cm in length.)

Rectangle B: 3 ones by 5 tens

Next have them fill in the rectangles with base ten pieces, using the largest pieces that will fit as shown in Figure 9.9. "How many pieces are in the small rectangle?" (15) "What kind of pieces are they?" (ones) "How many pieces are in the long rectangle?" (15) "And what are they?" (tens) Have children "read" the dimensions of the rectangles and the area (amount filled inside) using base ten language, not centimeters: "Three ones by (times) 5 ones is 15 ones. Three ones by 5 tens is 15 tens."

The goal is to generalize two "base ten products:" *ones × ones are ones* and *ones × tens are tens.* Draw other rectangles and label in terms of tens and ones. Students should measure these with centimeter rulers. Have students read the length, width, and areas using the base ten language. Factors used are always less than 10. For example, 6 ones times 7 tens is 42 tens. The relationship between rectangle dimensions and the type of base ten pieces that can fill them is the essential concept behind this approach to the algorithm. Students should know why the product is tens or ones (because the rectangle is filled in with that type of piece). The language used in "reading the rectangle" will provide an important link with symbolism.

Rectangles should also be drawn on base ten grid paper instead of filling them in with strips and squares. Provide base ten paper and give the outside dimensions in terms of ones and tens. Students draw the corresponding rectangles on the grid. Each rectangle is "filled" with *all ones* or with *all tens.* Always "read the rectangle" with the class as in Figure 9.10. Another good idea is to have children shade in one or two "pieces" in each rectangle. Talk about how many base ten pieces would be inside if you had a full-size rectangle filled with physical models.

The language of base ten products can be drilled orally. The following series relates basic facts with base ten product language.

Eight times four is .... thirty-two
Eight ones times four tens is .... thirty-two tens
Eight ones times four ones is .... thirty-two ones

Recording Base Ten Products

When base ten product language is clearly understood, the next step is to learn to record the products in appropriate place-value columns. Worksheets can be prepared like those in Figure 9.11. The base ten factors are written to the left of the place-value columns and the products are written in the chart.

The natural place to write the product of 4 ones times 8 tens is in the tens column. Suggest that children write the two-digit products in the columns as shown in Figure 9.12. "What would you have to do if you had thirty-two tens all in one column?" (Make a trade) When the class determines that 32 tens will make 3 hundreds and 2 tens, have them cross out the 32 and show the 3 and 2 in the respective columns as shown. Continue with other factors using ones times ones and also ones times tens. Move quickly to recording a product

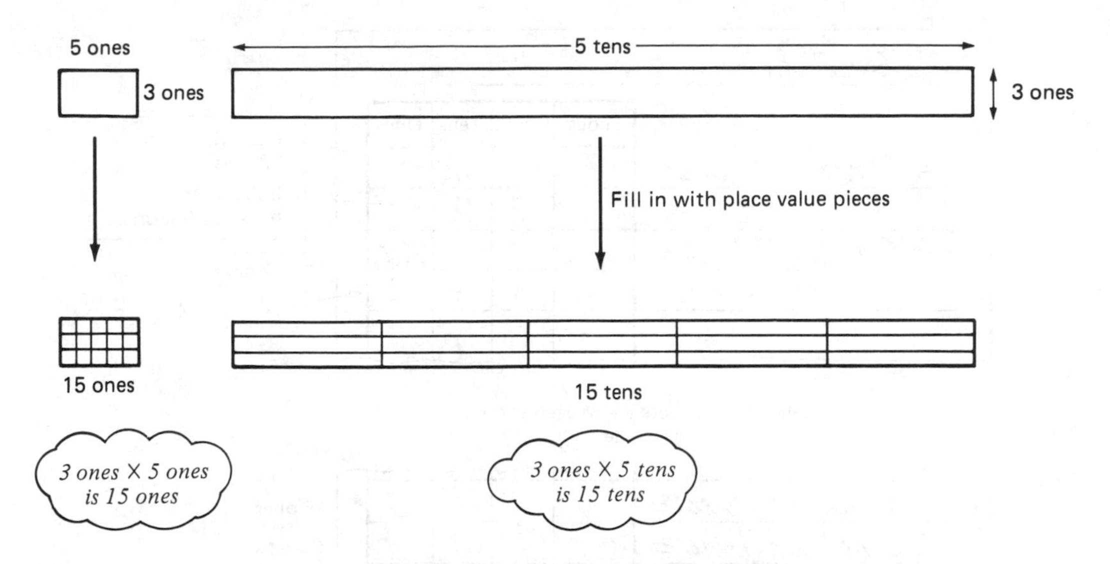

Figure 9.9
Fill in rectangles to develop base ten products.

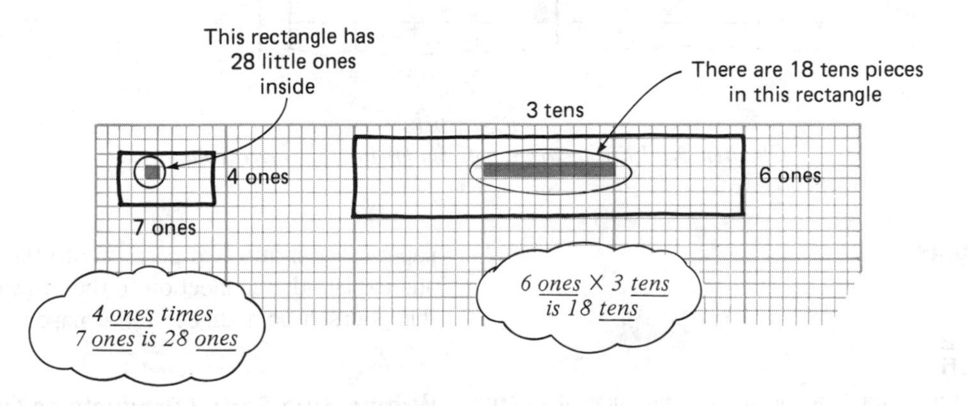

Figure 9.10
Draw and read rectangles on base ten grid paper.

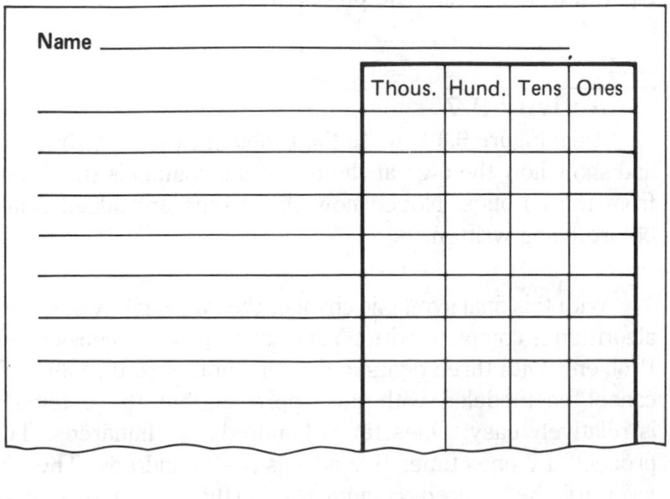

Name ______________________

	Thous.	Hund.	Tens	Ones

Figure 9.11
A place-value record sheet for recording base ten products.

with only one digit in a column and omitting the step of crossing out to show the trade. Students will discover that the last digit of the product will end in the named column.

Empty columns to the right are not filled with zeros. This makes the language agree with what is written. If 32 tens were written with a zero in the ones column, it would be read as three hundred twenty and not as 32 tens.

Understanding the recording of base ten products is the principal link between the procedural algorithm and conceptual knowledge. If you yourself do not have a good grasp of the links between filling in the rectangles, how that develops a base ten language, and how these products are recorded in a chart, it would be good to stop here and work through these last two sections again with your own models. Work with a friend and talk about all of these ideas in your own words. This will also demonstrate to you the value of learning an idea conceptually rather than just learning the rules.

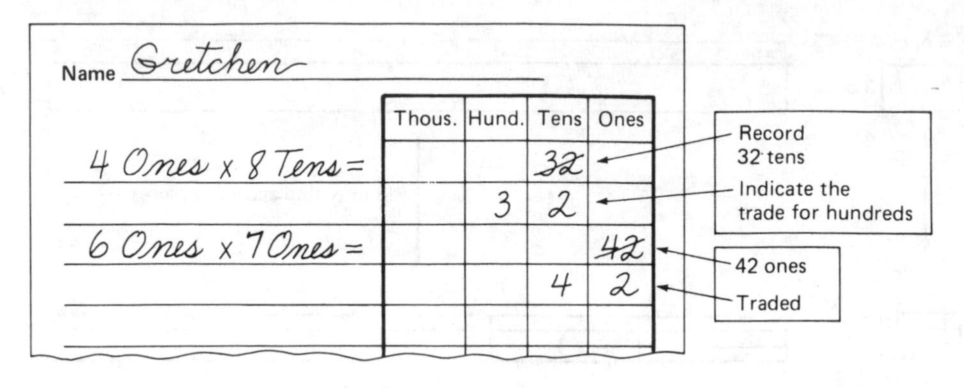

Eventually the products are written without showing a trade.

	Thous.	Hund.	Tens	Ones	
5 Ones x 3 Ones =			1	5	15 ones
7 Ones x 9 Tens =		6	3		63 tens

Figure 9.12
Children can learn to record base ten products in correct columns.

Two-Part Products

Activity 9.6

Give students a plain rectangle with a long side of 47 cm and a short side of 6 cm. How could this be filled in with base ten pieces?

As shown in Figure 9.13, this rectangle can be "sliced" or separated into two parts so that one part will be 6 ones by 7 ones or 42 ones and the other will be 6 ones by 4 tens or 24 tens. Have children read each part of the rectangle as before, using base ten language. Finally, since they already know how to record these products, show how the recording can be done in two lines beneath the problem as shown. The area of the entire rectangle is easily found by adding the two parts. Since the whole rectangle represents the product of 47 and 6, each section is referred to as a *partial product.*

Now go to the written record and repeat the problem over orally, pointing to each digit. "Six ones (point to 6) times seven ones (point to 7) is forty-two ones. Six ones (point) times four tens (point) is twenty-four tens." Notice that the base ten value of the factors is determined by which column the digits are in. Students have also learned how to record each partial product with only one digit per column, so they should already know where each product is to be recorded.

Students should soon be able to start with a problem such as 39 × 5, draw the appropriate rectangle on base ten grid paper, slice it into two parts, write the two partial products, and explain the connection to the drawing. Stop now and try this yourself on a sheet of grid paper.

Writing Two Partial Products as One

So far, the way the answer is written down requires writing two separate products which are then added. In the standard algorithm, this is all done on one line.

Activity 9.7

As in Figure 9.14, write the problem twice, side by side, and show how the digit at the top of the column is the 2 tens from the 24 ones. Notice how these tens are added orally before being written.

With this final symbolic change, the two-digit by one-digit algorithm is complete with no special cases to be considered. Problems with three digits in the top number (e.g., 639 × 7) cannot be modeled with this approach, but the extension is relatively easy. Ones times hundreds are hundreds. The product of 7 ones times 6 hundreds is 42 hundreds. The 2 is written in the hundreds column, the 4 in the thousands column. (Why?) Try the product 639 × 7 yourself. Say each of the three partial products in base ten language and record each separately. Then repeat the problem, combining the products in one line using a similar language to that of Figure 9.14.

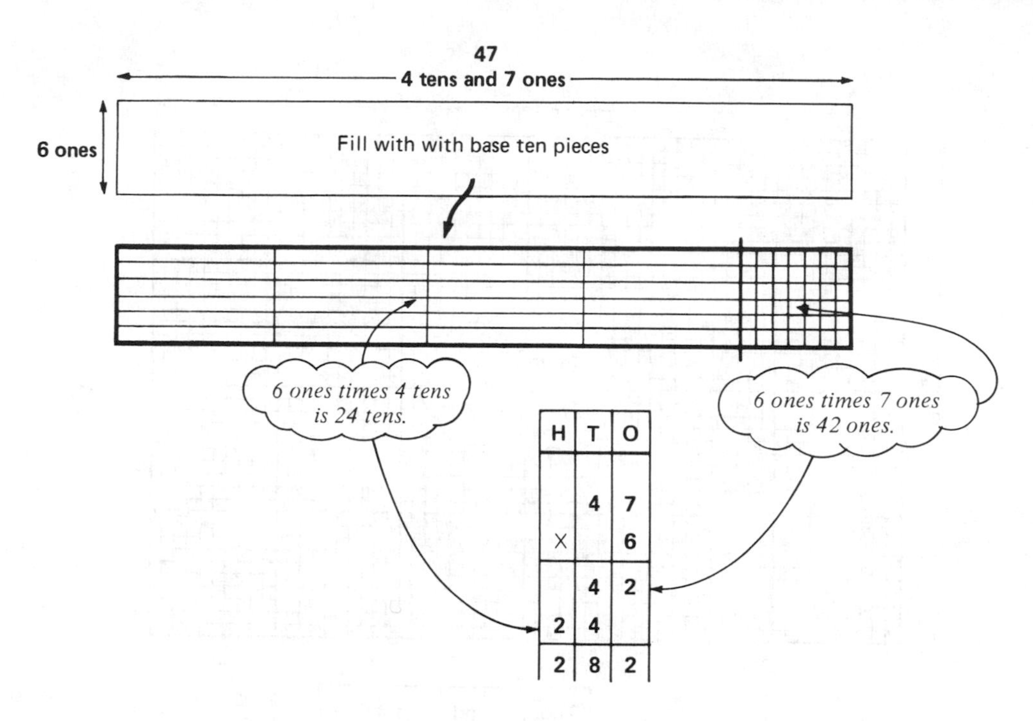

Figure 9.13
A model and a record of a two-digit by a one-digit product. Notice how the same language is used with the model and the symbolism.

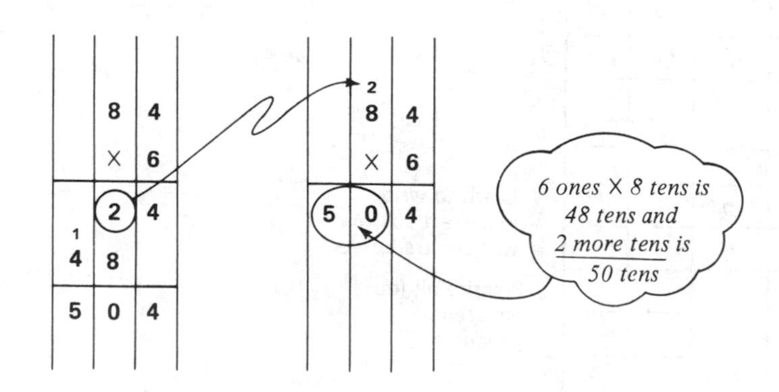

Figure 9.14
Two separate partial products can be added orally and the total written on one line in the standard form.

TWO-DIGIT MULTIPLIERS

If the algorithm for one-digit multipliers is completely understood, the extension to a two-digit multiplier is quite easy. Each of the next four sections is an exact parallel to the preceding development.

Developing Base Ten Products

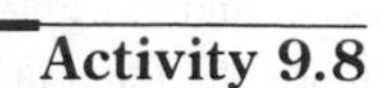

Activity 9.8
On large sheets of paper, provide students with a rectangle with dimensions 40 cm by 60 cm. Have them fill it in with base ten pieces. The obvious choice of pieces is hundreds, as in Figure 9.15(a). Read the rectangle as before: "Four tens by six tens is twenty-four hundreds." Generalize this new base ten product: *tens times tens are hundreds*. The other new base ten product is the turnaround or commutative partner of the old one: *tens times ones are tens*. Practice these ideas orally with specific numbers as before, and be sure that students can explain them conceptually. For example, "six tens times eight ones is forty-eight tens. Six tens times four tens is twenty-four hundreds."

Recording Base Ten Products

Go through exactly the same process as before in teaching children how to record the new base ten products. That is, permit or even encourage the recording of two digits in one column, and then make a trade symbolically as in Figure 9.15(b). Eventually children should be able to record all four base ten products with no more than one digit in each column.

Four-Part Products

Activity 9.9
Give students a rectangle (or have them draw one on base ten paper) with one side of 47 and the other side 36. How could this be filled in with base ten pieces?

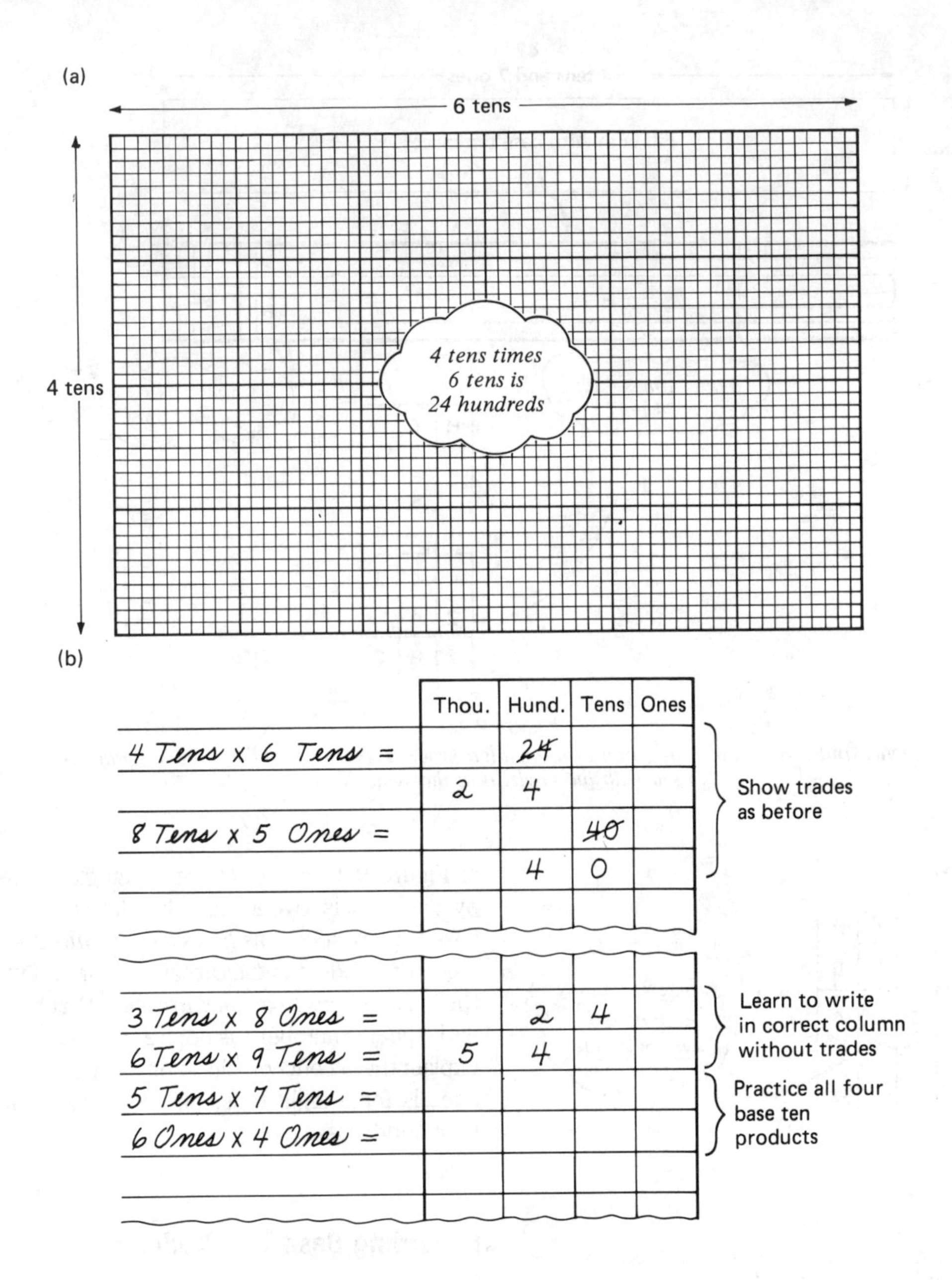

Figure 9.15
a) Tens × tens = hundreds. A new base ten product.
b) Children learn to record the new base ten products.

While many arrangements are possible, nearly everyone begins with hundreds placed in a corner, with a result similar to Figure 9.15(c).

When the rectangle is filled in, notice that there are four separate sections; one with hundreds, two with tens pieces, and one with singles or ones. Have students read each of these sections. As was done for the one-digit multiplier, record each section or partial product in place-value columns under the problem. The order in which the four sections or partial products are recorded is an arbitrary convention of the algorithm. The first two sections are the same as if the tens digit of the multiplier were not there. It is a good idea to record separately all four partial products, as shown in Figure 9.15(c).

Repeat the language that goes with the problem (reading the rectangles), but this time point to each digit in the problem. As before, the column each digit is in gives its base ten name and the exact same words can be used with the symbolism as were used with the rectangle. Students should be able to explain the connection of each partial product and each factor in the problem with the rectangle model.

You might wish to stop at this point and draw a rectangle that is 56 cm × 34 cm and fill it in with base ten pieces. Then

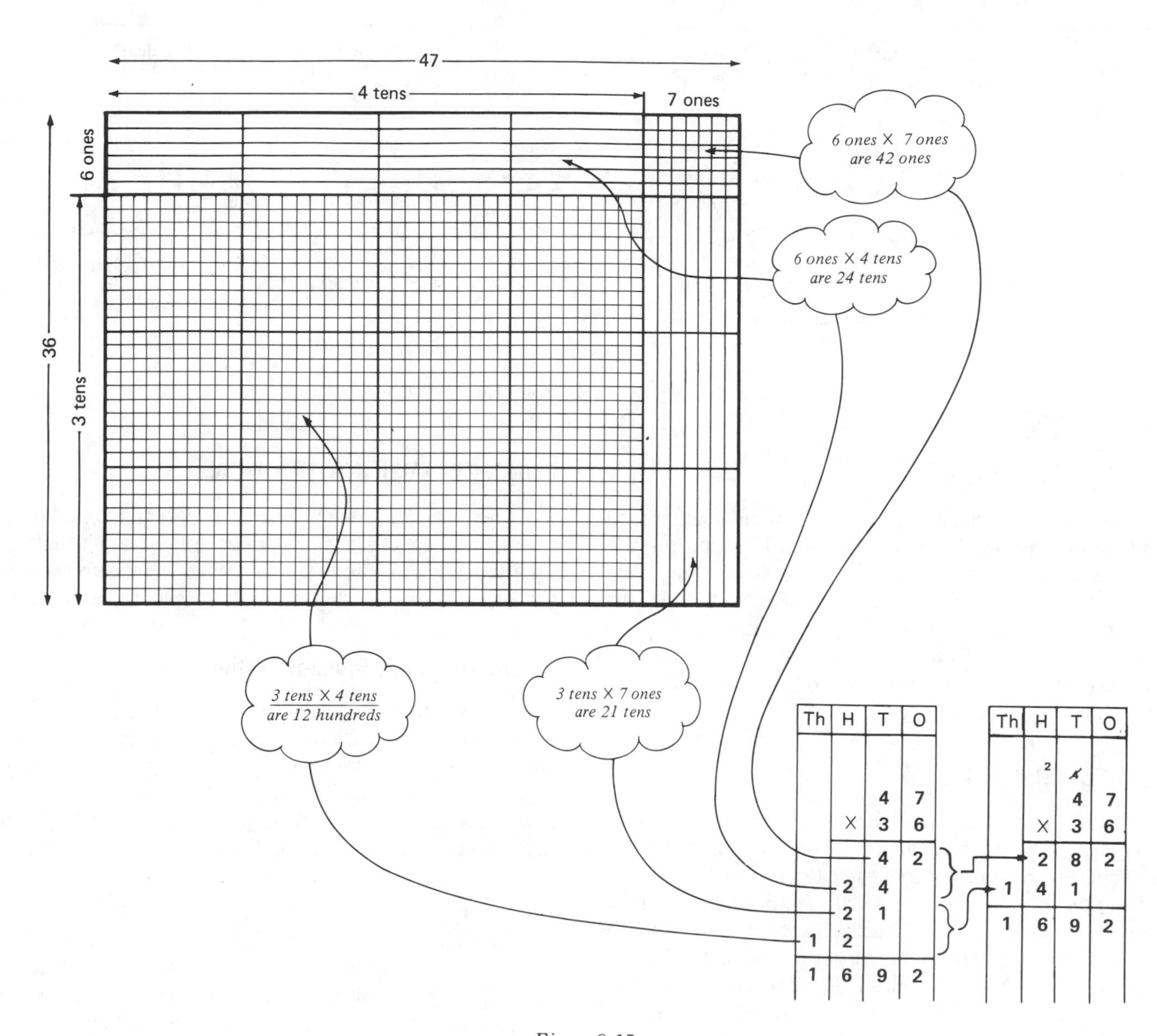

Figure 9.15
c) A 47 × 36 rectangle filled with base ten pieces.
Language connects the four sections of the rectangle to the four partial products.

read the four sections as you point to each. Record all four partial products in order, using the same oral language as you do so. Try another example drawn on a base ten grid.

Writing Four Partial Products as Two

The final step is to collapse the writing of all four partial products separately into two lines. The process is the same as before. Repeat a problem that has been completed. The second time write the left digit of each partial product at the top of the problem. It is a good habit to cross out these little numerals at the top of a problem as they are used. A whole variety of careless errors and bad habits result from a misuse or misunderstanding of the "trade digits."

ALTERNATIVE DEVELOPMENTS

The rectangle-area approach that has just been developed is consistent from one-digit to two-digit cases, can be modeled, can be generalized to more digits, and, perhaps most importantly, involves a strong conceptual language component that students can use in the absence of models. The difficulty lies in the time and tedium required and the need for at least drawing the models on base ten grid paper. Not every teacher is willing to invest that time and effort. The approaches in most basal textbooks tend to be much less developed, especially with two-digit multipliers. The sequence of problem types shown in Figure 9.16 is the one most generally followed. Explanations for the two-digit case require large drawings and lots of page space, which is just not possible in the student books.

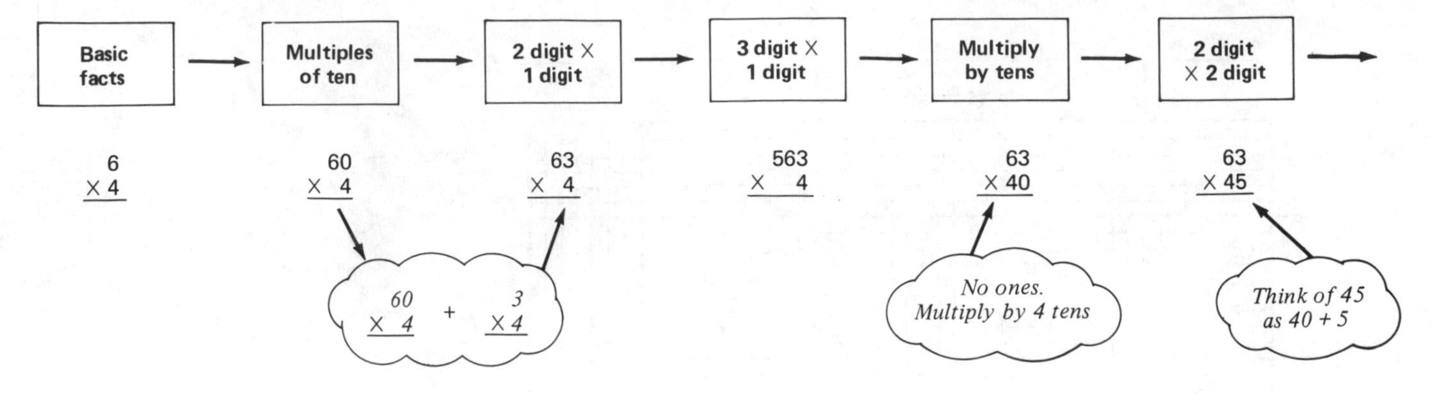

Figure 9.16
A fairly standard sequence of problem types for multiplication.

One-Digit Multipliers

A repeated-addition approach is popular for one-digit multipliers because it is easy to model and discuss. Notice in Figure 9.17 that the multiplier 3 tells *how many sets* of 2 tens and 4 ones. It is not read or interpreted as *3 ones.* Notice also that the arrangement of the base ten pieces can be made very similar to the rectangle-area approach. However, many texts will show pictures of separate sets of pieces rather than a rectangular arrangement.

Two-Digit Multipliers

Few fourth- and fifth-grade textbooks can afford to devote a lot of page space to development of the two-digit algorithm. The repeated-addition concept cannot reasonably be modeled and is not very helpful. (It is difficult to offer very many drawings of 38 sets of 46.) Usually, a special set of problems is practiced first where the multiplier is a multiple of 10; for example, 24 × 30 (see Figure 9.17). Different books offer different explanations for why the 72 is written to the left of the 0. It would be useful to examine some fourth- and fifth-grade textbooks to see how different books approach this topic.

The Division Algorithm

Division with one-digit is traditionally begun in the third grade, with two-digit divisors introduced in the fourth grade. A two-digit divisor substantially increases the difficulty of the task. Fortunately, there is a movement to totally remove three- and four-digit divisors from the curriculum. For example, the state of Michigan objectives restrict whole number divisors to two digits. Even these divisors are to be less than 30 except for multiples of 10, which may be as high as 90 (Michigan State Board of Education, 1988). This restriction allows students to focus on the rationale for the algorithm and avoid much of the stress caused by working with larger divisors. It also allows teachers to make better connections with estimation and mental computation methods.

CONCEPT-LEVEL INSTRUCTION

Most textbooks now teach a division algorithm based on the partition concept of division. During the 1970s a quite different algorithm was promoted that was based on repeated subtraction. This latter algorithm is not discussed in this chapter.

An Unstructured Division Activity

Before suggesting the following activity, have students solve one or two simple division exercises such as this: If 17 things are to be placed in 3 equal piles, how many will be in each and how many left over? Even if they can give an answer mentally, have them model the problem using counters to emphasize the concept and process of partitioning.

Activity 9.10

Immediately following the above task, have students figure out how many will be in each of four piles if there are 583 things to be shared equally. Let students try to solve this in any way they wish. The only restriction is that they be able to explain their procedure in a meaningful way.

It would be a good idea for you to stop and explore this task yourself. Would base ten models be helpful?

Structuring the Division Algorithm

Activity 9.11

Provide students with base ten pieces and six or seven pieces of paper about 6 inches square. (Construction paper cut in fourths is perfect.) Begin with a problem similar to the previous one, $4\overline{)583}$. Discuss the meaning of the task in base ten language. You have 5 hundreds pieces, 8 tens, and 3 ones to be shared evenly among four sets or piles. Place an emphasis on language that focuses on *5 things, 8 things,* and *3 things* rather than 583 single items. Have students model

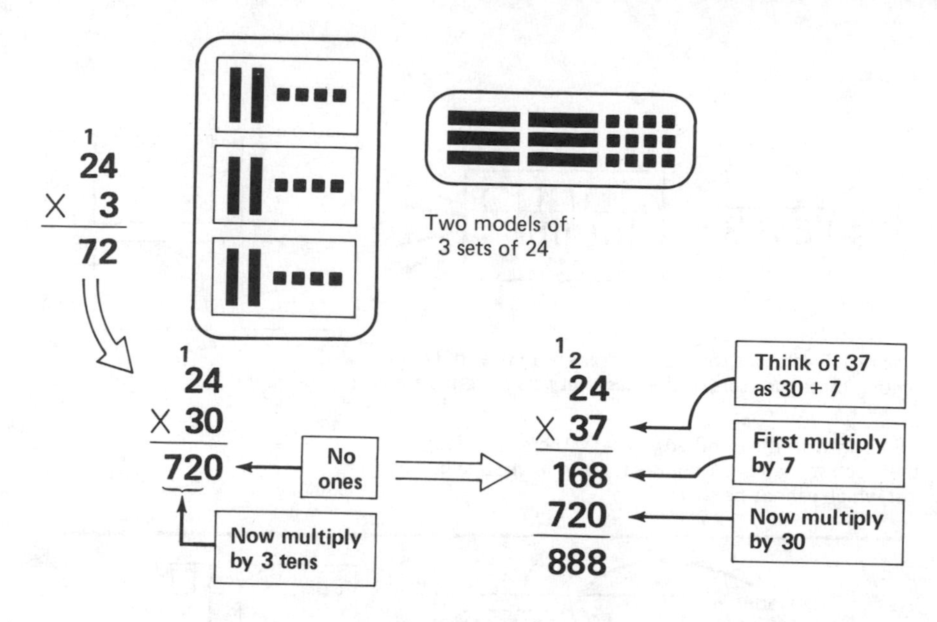

Figure 9.17
A traditional sequence to develop the 2-digit algorithm.

the quantity 583 with their base ten pieces, and set out four of their paper squares to represent the four sets. Explain that it is a good idea to always begin with largest pieces, since if they cannot be completely distributed evenly, they can be exchanged for smaller pieces. Figure 9.18 illustrates the complete solution to this task as students would do it with models.

It is very important to get students to talk this process out in their own words. Notice that with each place value, several distinct things happen:

The number of pieces available is considered, and a decision is made to see if there are enough to be distributed.

The pieces that each set can get are actually passed out to the sets—that is, placed on the paper mats.

Any leftover pieces are considered. Generally there will not be enough to distribute more to each set, but that possibility should be examined.

Any remaining pieces are traded for the next smaller size. The pieces received in trade are added to any that were already there. At this point the process begins over.

The sharing of individual pieces within each place value is the total conceptual knowledge basis for the long-division algorithm. Students should do exercises using only models until they can talk through that process (using their own words) in such a way that the sharing of each type of base ten piece and trading to the next size is quite clear.

Notice that in this context the phrase "four *goes into* five" has absolutely no meaning. In fact, 5 hundreds are being *put into* four sets. Teachers especially have difficulty abandoning the "goes into" terminology because it has become such a strong tradition while doing long division.

In making up the exercises that students should do with models, you will have to keep different problem types and nuances in mind. Figure 9.19 illustrates some of these. If excessive trades are avoided, most exercises with dividends through three-digits and divisors of one-digit can be easily modeled.

You may wish to stop at this point and try some of these division exercises with your own models. (Use the ones suggested in Figure 9.19.) Pay special attention to the language you are using. Remember you are simply passing pieces out fairly as you would M&M's to friends. Avoid the "goes into" terminology. Also stress the language that explains the trading.

CONNECTING CONCEPTS TO SYMBOLS

The general approach of having students record on paper each manipulative step *as they do it* is the way to connect symbolism to division.

What Should Be Recorded?

Long division is simply a matter of passing out pieces within each place-value column and then trading any that may be left over for the next smaller size. When recording, however, two inbetween steps must also be written down. All four steps are given here. Notice that steps B and C correspond to no action with models.

A. DO: Distribute available pieces in column to the sets. RECORD: The number given to each set.

B. RECORD: The number of pieces given out in all. Find this by multiplying the number given to each times the number of sets.

C. RECORD: The number of that type piece remaining. Find this by subtracting the total given out from the amount you began with.

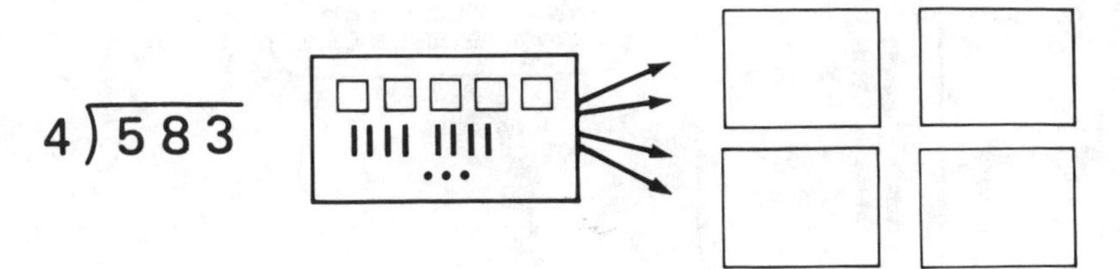

The task is to share these 5 hundreds, 8 tens, and 3 ones among these four sets so that each set gets the same amount.

I'll begin with the hundreds pieces. There are enough so that each set can get 1 hundred. That leaves one hundred left which cannot be shared.

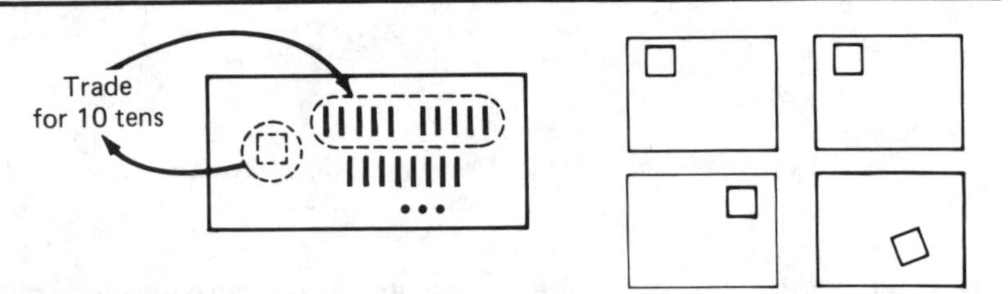

I can trade the hundred for 10 tens. That gives me a total of 18 tens. With 18 I can put 4 tens in each of the four sets and have 2 tens left. Two is not enough to go around to all four sets.

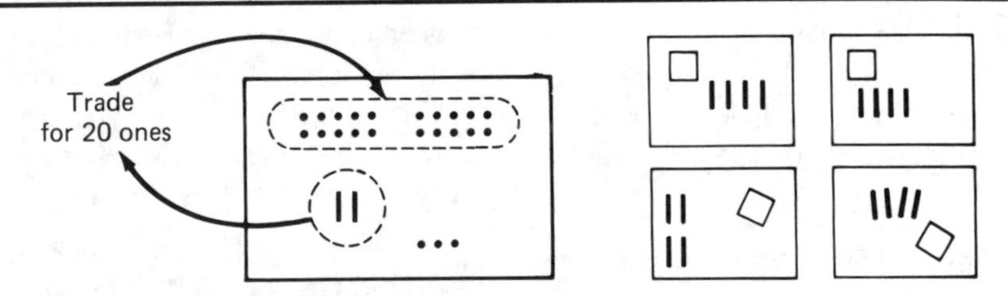

I can trade the 2 tens for 10 ones each or a total of 20 ones. With the 3 ones I already had, that gives me 23 ones. I can put 5 ones in each of the four sets. That leaves me with only 3 ones left over as a remainder.

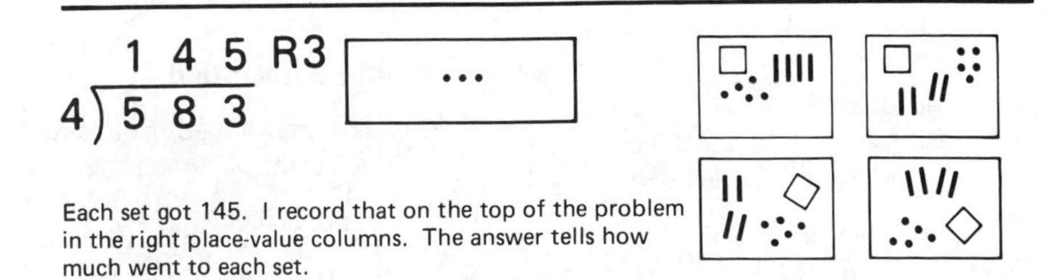

Each set got 145. I record that on the top of the problem in the right place-value columns. The answer tells how much went to each set.

If I added up how much they got altogether, that would be 4 X 145 or 580. The remainder of 3 makes the 583 I started with.

Figure 9.18
Long division at the concept level.

Two-digit dividends.
Not very challenging or interesting.

$4\overline{)67}$

Be careful of excessive trades.

$5\overline{)865}$

3 hundreds for 30 tens (a lot but OK)
1 ten for 10 ones

$5\overline{)745}$

2 hundreds for 20 tens
4 tens for 40 ones (excessively tedious)

No hundreds to distribute.

$4\overline{)372}$

No trades in one or more places.

$4\overline{)852}$ $3\overline{)426}$ $3\overline{)693}$

No tens to distribute. (zero in an answer)

$4\overline{)832}$

Zeros in dividend.

$3\overline{)704}$

Figure 9.19
Consider a variety of special cases to be worked out using models.

D. DO: Trade (if necessary) for equivalent pieces in the next column and combine those with any that may already have been there.
RECORD: The total number of pieces now in the next column.

The point is, that students must write more than what they do or think about when working manipulatively. In the beginning, students frequently omit steps B and C. The value or necessity of this extra writing may seem obscure to some students while they are still using models.

Two Alternatives for Recording

Along with the traditional method of recording this algorithm, an alternative method is suggested here that helps keep the conceptual meaning more clearly tied to the symbolism. Steps A, B, and C are recorded as usual in both methods, as shown in Figure 9.20(a). In the traditional algorithm, the trade is implied when you "bring down" the next digit. Notice that the total is technically in two columns. In the example, the 2 that stood for the remaining hundreds somehow becomes 20 tens when the 6 is brought down [Figure 9.20(b) on the left]. In the alternative version, the trade for tens is made explicit by crossing out the 2, and the total of 26 tens is written completely in the tens column. Next, multiplication indicates that 25 tens were passed out in all. In the traditional scheme the 25 must be recorded part in the hundreds column and part in the tens column. In the alternative method, the 25 tens is written in the tens column, clearly representing what was done [Figure 9.20(c)]. The idea of writing a two-digit number in one column is not entirely new. In subtracting 637 − 281, we write a 13 above the 3 to represent 13 tens.

There are pros and cons to each method. The traditional method is more familiar to adults, although the "bring-down" procedure seems a bit difficult to explain to children. The alternative method is a direct match with the modeling of the procedure. It requires that the digits in the dividend be spaced out more and almost necessitates the use of lines to mark the columns, even as a permanent feature of the algorithm.

Both methods are explored in the section on two-digit divisors.

Avoiding Some Difficulties

Some of the difficulties that young children have with the division algorithm are due to sloppy writing. You can help students by preparing division record sheets. Nine blanks similar to those shown in Figure 9.21 (also Appendix C) can easily fit on one page. This is especially useful if the alternative recording scheme is used. Similar sheets can be prepared with four columns.

Many children forget to record zeros in the answer. An example is shown in Figure 9.22. The practice of drawing place-value columns or using record sheets as just noted is very helpful since the columns mark a space for each place value to be filled in. Perhaps more importantly, children should be encouraged to check that their answer makes sense. In the example, 642 things would have at least 100 in each set. The answer 17 is not even in the ballpark.

The process of doing division problems with models and recording all of the steps is a complex task for young children. Some will get carried away with the materials and forget to record.

Activity 9.12

In groups of 3, one child can be the "doer," one the "recorder," and one the "foreman." The foreman's job is to see that all the steps are being written down and to keep the two processes together. All three children should be talking to each other about the problem as it is being solved. Three

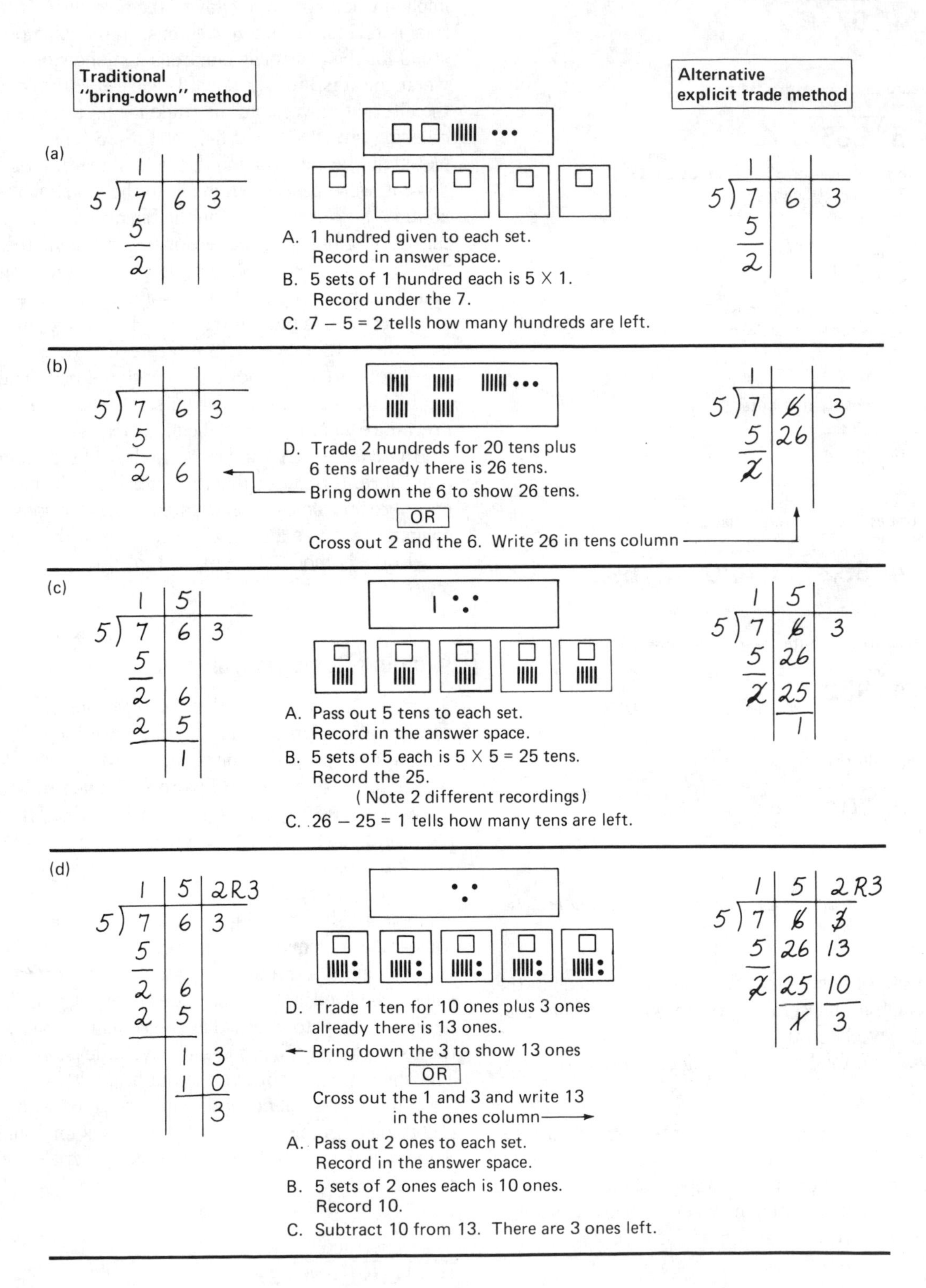

Figure 9.20
Two methods of recording long division.

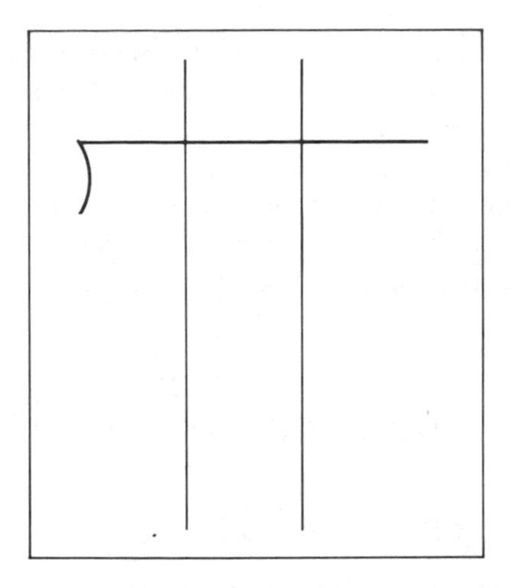

Figure 9.21
Pages with blank charts like this (shown actual size) or with four columns are helpful when children are recording division problems.

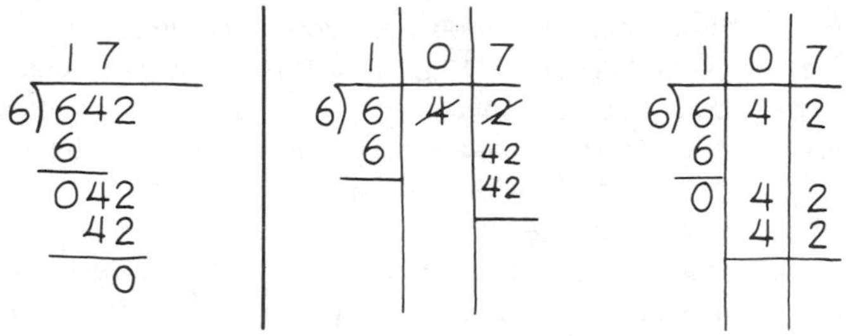

Figure 9.22
Using lines to mark place-value columns can help avoid forgetting to record 0 tens.

problems in a period provide each student the chance to take on each task.

You are ready to move to the symbolic level and leave the models aside when students can take a completed problem and explain the meaning of each number in the problem in terms of base ten pieces (decoding). Children should then be able to complete a problem without models and talk through it conceptually as they go.

Even in the upper grades where long division is a review topic, the complete sequence of modeling and recording exercises with one-digit divisors is well worth two or three days' effort. The seventh-grade teacher cannot assume that students have this understanding. A quick development can prove quite helpful.

TWO-DIGIT DIVISORS

Examining the Difficulties

With a two-digit divisor, some form of rounding or similar trick must be used. Not only is rounding complicated, it does not always work. This causes frustration and lots of erasures. Consider the problem in Figure 9.23. The 63 is usually rounded to 60 or, equivalently, 37 divided by 6 is used. Either approach yields an estimate of 6 tens, which turns out to be too much and must be erased. Sometimes we round up instead of down. In these cases it sometimes happens that the partial quotient is too small and must also be erased. Even when the exercises are contrived to make these devices work, children confuse them and/or have difficulty applying them.

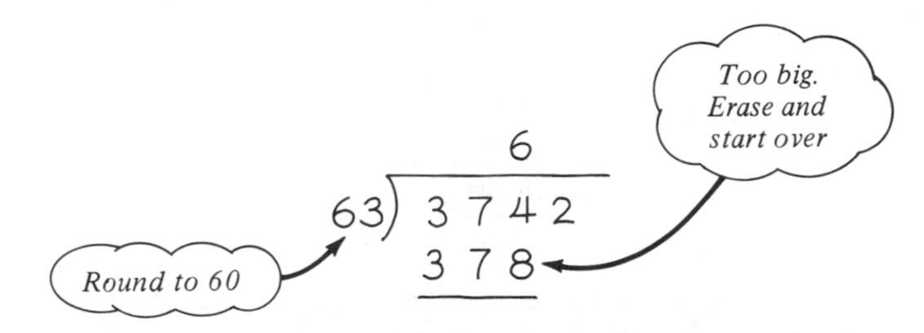

Figure 9.23
Rounding the divisor is not only hard, it does not always work.

An Intuitive Idea

Suppose that you were sharing a large pile of candies with friends. Instead of passing them out one at a time, you conservatively estimate that each could get a least six pieces. You can always pass out more. So each takes six. What if some are left? If there are enough to go around, you simply pass out more. It would be silly to collect those you already have given out and begin all over. Why not apply this commonsense approach to the distribution of base ten pieces, the conceptual basis for long division?

The principal features of the candy example are first to never overestimate the first time and second to distribute more if the first attempt leaves enough to pass around. One way to make an easy but always safe estimate the first time is to pretend there are more sets than there really are. For example, if you have 312 pieces for 43 sets, pretend you have 50 sets instead. It is easy to determine that you can distribute 6 to each of 50 sets because 6 × 50 is an easy product. Therefore, you must be able to distribute *at least* 6 to each of the actual 43 sets. The point is, to avoid overestimates, always consider a larger divisor; *always round up*.

But what if the result shows you could distribute more? Simply do what you would do with the candy situation—distribute some more.

Apply the Ideas to Long Division

The two preceding ideas can be translated to long division as follows: (1) *Always round the divisor up* to the next multiple of 10 to determine the first estimate. (2) If the first estimate is too small, simply distribute some more.

In Figure 9.24, these ideas are both illustrated, using the same problem that presented difficulties earlier. Both the traditional and the explicit trade methods of recording are

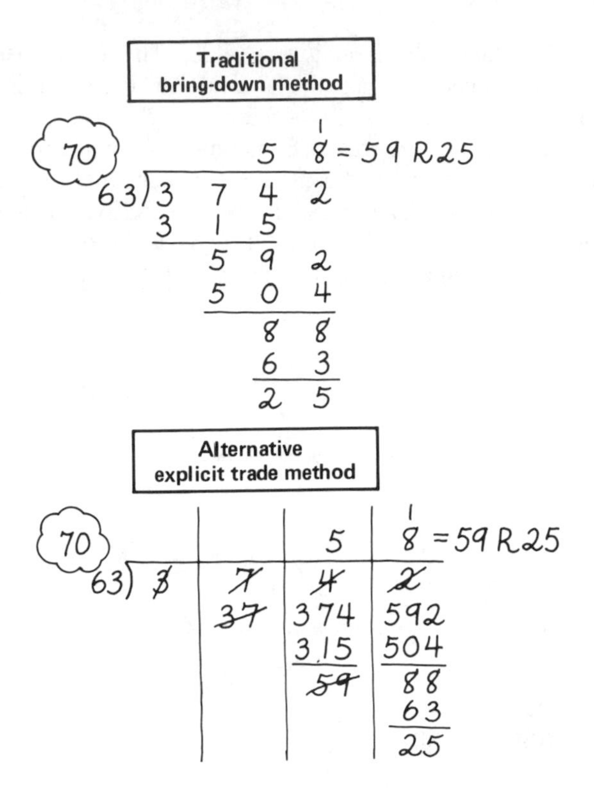

Figure 9.24
Think 70 piles instead of 63. In the ones column, pass around 8 to each pile and then pass around one more.

illustrated. Children should write the 70 in the little "think bubble" above the divisor. It is easy to run through the multiples of 70:

3 × 70 is 210 (too small)

4 × 70 is 280 (too small)

5 × 70 is 350 (close)

6 × 70 is 420 (too big, use 5)

Put 5 tens in each pile. But there are only 63 piles to pass the tens to. Five times 63 tells how many were actually distributed, and subtraction tells how many are left. In this case, 59 tens is not enough to pass out any more, and so a trade is made for ones. Fifty-nine tens is 590 ones, and 2 ones that were already there is 592 ones. Pretending 70 piles instead of 63 will suggest 8 ones in each pile (8 × 70 is 560 and 9 × 70 is 630). Distributing 8 ones to each of only 63 piles leaves 98. Since that is enough to put one more in each pile, do so. Indicate this by putting a 1 above the 8 and repeat the multiply and subtract steps as always. The result of 5 tens and 9 ones (that is 8 + 1) with 25 left can be written to the right.

This suggestion (always round up and distribute more if needed) is not currently taught in most programs, and it may never be. There is tremendous resistance to change of any sort in the traditions of algorithms. At the same time, the approach has proven to be successful with children in the fourth grade learning division for the first time and with children in the sixth and seventh grades in need of remediation. It reduces the mental strain of making choices and essentially eliminates the need to erase. If an estimate is too low, that's OK. And if you always round up, the estimate will never be too high. Nor is there any reason to ever change to the more familiar approach. It is just as good for adults as for children. The same is true of the explicit trade notation. It is certainly an idea to consider.

Incorrect Algorithms or Error Patterns

HOW ERROR PATTERNS DEVELOP

Children's difficulties with algorithms are due to three major factors: carelessness or sloppy recording, incorrect basic facts, and incorrect procedures or algorithms. Of these, the use of a faulty algorithm is by far the most common and the most serious. Through practice, neatness and fact mastery can be developed. But practice will not correct the problem of a faulty algorithm. We should recognize that the algorithms are complex procedures for children. It is relatively easy for these rules to become confused, especially if there has been a premature rush to mastery and not enough time on developing ideas and connections with models.

When children use an incorrect procedure, they tend to continue to use it for other problems as well. The repeated use of an incorrect procedure or algorithm is called an *error pattern*. Some children retain the same error pattern for years, while others sometimes switch from one error pattern to another. Almost all of these incorrect algorithms can be blamed on a faulty connection between procedural and conceptual knowledge. This connection can be incomplete for two different reasons. First, the link between concepts and procedure may not be adequate. Second, the conceptual knowledge, usually of place value or the rationale for the algorithm, may not be there to begin with.

According to what VanLehn (1983) calls *repair theory*, children develop error patterns when they have an incomplete procedure for dealing with a problem. When their procedure brings them to a difficulty, rather than simply quit or ask for help students become inventive; they *repair* their algorithm to deal with the situation. This repaired procedure is frequently used over and over and becomes part of the algorithm. For example, consider a child who has learned to "borrow" and is quite competent with that procedure in problems such as the first one in Figure 9.25. The child's algorithm may be, "Go to the next column, make it 1 less, and make the current column 10 more." When this procedure is applied to the second problem with a 0 in the tens column, the procedure breaks down. The child now "fixes" or repairs the procedure to get on with the problem. For this particular example, different children repair the algorithm differently, thus producing different error patterns as shown.

Many computational error patterns do not produce errors in all problems. The one just described, for example, only

Procedure: When you can't subtract, go to the next column and make it one less.
Make the top number 10 more.

```
  4 12
  952
 -324
  628
```

Procedure works

```
  802     Nothing in the next column.
 -465
```

Stuck! "Repair" the process

These two incorrect "repairs" produce two different error patterns.

(a) Go one more column over. Make it one less. Make the top number 10 more.

```
  7 12
  802
 -465
    7
```

(b) Since next column is zero, it's as small as it can get. Make the top number 10 more.

```
    12
  802
 -465
    7
```

Figure 9.25
An example of how error patterns develop.

occurs when there is a "borrow from zero" situation. Consequently, children complete pages of practice exercises using faulty algorithms, but get only a few wrong. Teachers must be aware of this potential and discern between the child who gets three problems wrong due to carelessness or faulty facts and the child whose algorithm doesn't work in specific cases.

EXAMPLES OF ERROR PATTERNS

Figure 9.26 illustrates only a few of the more common faulty algorithms that children present. For each example work a similar problem to see how the error pattern works. Are there some cases where the error pattern produces correct answers?

REMEDIATION OF ERROR PATTERNS

In order to remediate faulty algorithms, they must first be identified. Look for special cases where errors occur. Have children talk through procedures where they do them correctly and where they do them incorrectly. Try to determine where the specific difficulty is. Correct the cause of the problem, be it incomplete place-value concepts, poor linkage of the procedure with concepts, or a poorly developed algorithm. The most serious error on a teacher's part would be to simply provide more practice for a child who seems to get some problems wrong all of the time. The result may be reinforcement of a faulty procedure. More practice without addressing the linkage with concepts is almost always nonproductive.

PREVENTION OF ERROR PATTERNS

Part of the philosophy of this book is to help children completely develop conceptual knowledge before moving them prematurely to symbolic rules or procedures. In the case of algorithms, this means spending sufficient time with place-value concepts and then devoting careful attention to the concept and connecting levels of the algorithms. Anticipation of faulty algorithms can help to prevent them. Some general suggestions follow:

1. *Introduce trade and nontrade problems at the same time.* This maxim is finally finding its way into the curriculum. It used to be common to have a second-grade chapter on addition without regrouping. Children learned and practiced an incomplete algorithm which then had to be systematically "repaired" in a later chapter. Many children confused the old and new algorithms and could not remember when to "carry" and when not to. Even in the first grade, trade and nontrade problems should be done together, *from the very beginning,* and with models.
2. *Use language that is as conceptual as possible.* Encourage children to talk through their procedures with

Addition Error Patterns

48 + 35 = 713	38 + 7 = 18	52 + 34 = 96 (carry 1)	76 + 8 = 164 (carry 1)	60 + 38 = 90	426 + 298 = 624 (carry 1)
Writes sum of each column	Adds all digits	"Carries" all the time	Re-adds ones digit	Adds to 0 and gets 0	Trade for hundreds not recorded. May depend on rest of first column

Subtraction Error Patterns

63 − 48 = 25	482 − 165 = 327 (2 crossed out, 12 written)	520 − 286 = 240 (5 crossed to 4, 2 crossed to 12)	497 − 135 = 3512 (9 crossed to 8, 7 crossed to 17)	68 − 5 = 13
Smaller from larger	Fails to reduce tens column in trade	0 − n = 0 instead if trade	Trades in ones column when not needed	Subtracts single digit from both columns

Multiplication Error Patterns

38 X 46 = 148 (trade 4)	67 X 4 = 322 (trade 8)	75 X 7 = 505 (trade 1)	49 X 27 = 343, 98, 441 (trade 16)	27 X 9 = 723	67 X 8 = 116 (trade 5)
Only multiplies ones X ones and tens X tens	"Carries" wrong digit	"Carries" a 1 regardless	Records second partial product in wrong columns	Adds trade digit to top number	Adds in tens column when there is a trade

34 X 8 = 242 (trade 3)	65 X 72 = 4210
Fails to use trade number	Only multiples ones X ones and tens X tens with no trades

Division error patterns.
Many division errors are really error patterns in multiplication or subtraction.

8)1624 = 23; 16, 24, 24	7)448 = 46; 42, 28, 28	8)1624 = 230; 16, 24, 24
Fails to record zero tens in quotient	Records answer digits right to left (answers are correct if read backwards).	Fails to record zero in tens digit, puts ones digit in tens column and annexes a zero to fill up space

Figure 9.26
A small collection of frequently seen error patterns.

references to concepts and base ten materials. Include decoding activities to develop and assess connections.

3. *Include all special cases during the concept and connecting levels of your instruction.* Point out special difficulties to children and discuss them in the meaningful context of models rather than deliver rote rules. In subtraction, for example, the difficulties involving zero should be explored before symbolic recording takes place. Similarly, a zero in the quotient of a division problem should occur when models are still used.

For Discussion and Exploration

1. It is becoming more and more popular to introduce computation as a problem-solving endeavor rather than to structure each step of the algorithm from the beginning. That is, let students use models and their own reasoning skills to determine results before showing them how to use the standard procedures. When the concepts involved have been explored, then is the time to introduce the more efficient approaches found in the usual algorithms. Discuss the pros and cons of such an approach. Read Madell's article, "Children's Natural Processes" (1985).
2. The multiplication algorithm is much more difficult to understand and teach conceptually than any of the other algorithms. With the current deemphasis on pencil-and-paper computation, discuss the proposition that the multiplication algorithm may as well be taught as a rote procedure without any attempt to explain it.
3. Talk with teachers in the upper grades (fifth and above) about how much time they spend teaching algorithms. Do they use base ten models? Would a better development prevent some of the need for remediation? What about the idea of not teaching the algorithms at all until about fourth grade, when students can handle them?
4. What is the educational "cost" of teaching students to master pencil-and-paper computational algorithms? Here cost means time and effort required over the entire elementary grade span, in contrast with all other topics taught or which could be taught if more time were available. How much about algorithm skill or knowledge do you think is really essential in an age of readily available calculators?
5. Select one or more of the error patterns in Figure 9.26. What might have been done to prevent it from developing? What is the minimum that must be done to remediate it conceptually? (It is not reasonable to begin at the beginning for every child.) Which error patterns are a result of not understanding trading? For which is a lack of knowledge about the algorithm itself a principal cause? (These questions do not have clear-cut answers.)

Suggested Readings

Ashlock, R. B. (1986). *Error patterns in computation: A semi-programmed approach* (4th ed.). Columbus, OH: Charles E. Merill.

Bidwell, J. K. (1987). Using grid arrays to teach long division. *School Science and Mathematics, 87,* 233–238.

Hazekamp, D. W. (1978). Teaching multiplication and division algorithms. In M. N. Suydam (Ed.), *Developing computational skills, 1978 yearbook.* Reston, VA: National Council of Teachers of Mathematics.

Kamii, C., & Joseph, L. (1988). Teaching place value and double-column addition. *Arithmetic Teacher, 35*(6), 48–52.

Labinowicz, E. (1985). *Learning from children: New beginnings for teaching numerical thinking.* Menlo Park, CA: Addison-Wesley.

Madell, R. (1985). Children's natural processes. *Arithmetic Teacher, 32*(7), 20–22.

Madell, R. (1979). *Picturing multiplication and division.* Palo Alto, CA: Creative Publications.

Madell, R., & Stahl, E. L. (1977). *Picturing addition.* Palo Alto, CA: Creative Publications.

Madell, R., & Stahl, E. L. (1977). *Picturing subtraction.* Palo Alto, CA: Creative Publications.

Merseth, K. K. (1978). Using materials and activities in teaching addition and subtraction algorithms. In M. N. Suydam (Ed.), *Developing computational skills.* Reston, VA: National Council of Teachers of Mathematics.

Robold, A. I. (1983). Grid arrays for multiplication. *Arithmetic Teacher, 30*(5), 14–17.

Stanic, G. M. A., & McKillip, W. D. (1989). Developmental algorithms have a place in elementary school mathematics instruction. *Arithmetic Teacher, 36*(5), 14–16.

Thompson, C., & Van de Walle, J. A. (1980). Transition boards: Moving from materials to symbols in addition. *Arithmetic Teacher, 28*(4), 4–8.

Thompson, C., & Van de Walle, J. A. (1981). Transition boards: Moving from materials to symbols in subtraction. *Arithmetic Teacher, 28*(5), 4–9.

Tucker, B. F. (1989). Seeing addition: A diagnosis-remediation case study. *Arithmetic Teacher, 36*(5). 10–11.

Van de Walle, J. A., & Thompson, C. (1985). Partitioning sets for number concepts, place value, and long division. *Arithmetic Teacher, 32*(5), 6–11.

VanLehn, K. (1986). Arithmetic procedures are induced from examples. In J. Hiebert (Ed.), *Conceptual and procedural knowledge: The case of mathematics.* Hillsdale, NJ: Lawrence Erlbaum.

Weiland, L. (1985). Matching instruction to children's thinking about division. *Arithmetic Teacher, 33*(4), 34–45.

CHAPTER 10

MENTAL COMPUTATION AND ESTIMATION

Alternative Forms of Computation

FOUR CHOICES

There are at least four distinct methods of computation:

1. Use a pencil-and-paper algorithm.
2. Use a calculator or computer.
3. Use mental computation.
4. Use an estimation based on a mental computation.

The first three of these methods provide exact answers or results. The fourth provides only an estimate. To choose a method of computation, then, you must first decide if an exact answer is required or if an approximate answer is sufficient (Figure 10.1). When exact answers are required, the choice of methods is frequently determined by the nature of the numbers involved and the availability of a calculator or computer.

REAL-WORLD CHOOSING

Consider each of the following situations. Decide first if an approximate or exact result is appropriate. If an exact result is called for, what is the most reasonable choice of computation? If the numbers were different, would your choice of methods change?

Melinda walks into a store with $10.00. After picking up 5 packs of paper priced at $1.69 each, she spies a marking pen she has been wanting. It is marked down

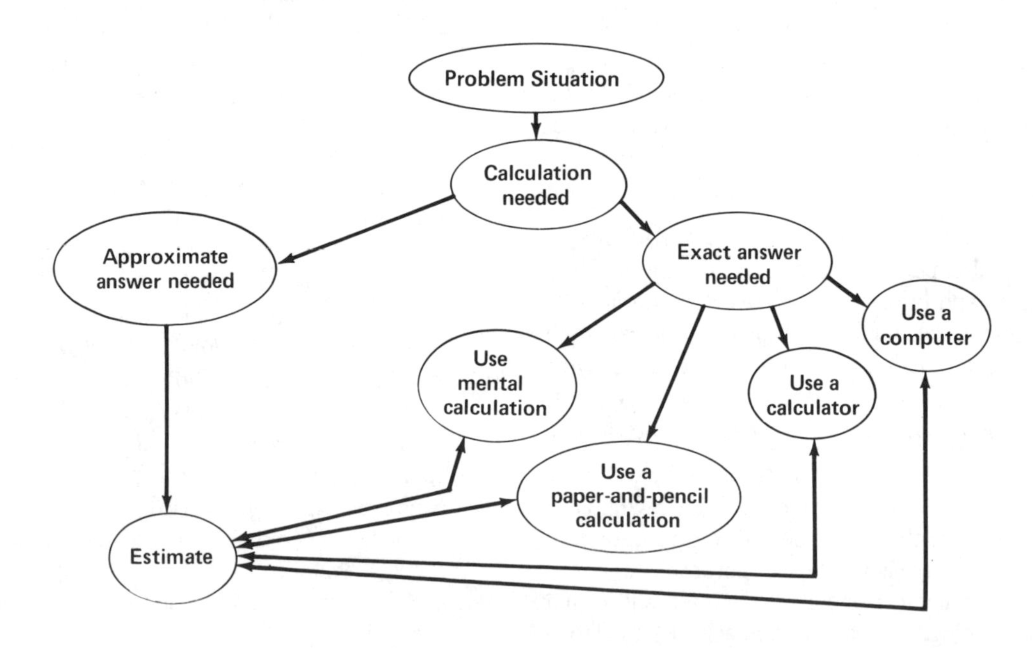

Figure 10.1
Decision about calculation procedures in numerical problems. NCTM Commision on Standards for School Mathematics (1989). Curriculum and Evaluation Standards for School Mathematics. *Reston, VA: National Council of Teachers of Mathematics. Reprinted by permission.*

to $1.35. She needs the 5 packs of paper and does not wish to be embarrassed at the checkout.

Jim works in a restaurant as the cashier. It is his job to figure the total bill when the waiter gives him the prices of the meals purchased. Tax in Jim's city is 4.5%.

In the teacher's lounge, John asks Betty how much she paid for the snacks she bought for her class. Betty had purchased a 19-cent pack of gum for each of her 30 children.

Professor Able is talking with a student in his office about her grade point average. Sanya has 78 hours and 216 quality points. She has with her a student teaching application form that requires the average.

Dana makes $3.80 an hour at the day care center. Last week she worked 20 hours. On her way home she wonders what her paycheck will be (before deductions).

Perhaps what is most interesting when considering real examples of computation is how infrequently pencil and paper is the obvious choice. Mental methods, both for exact and approximate answers, are frequently the most useful of all methods. Would they be the first choice even more often if we had greater mental arithmetic skills? Facility with mental computation is a significant factor in our ability to do computational estimation (Hope, 1986; Leutzinger, Rathmell, & Urbatsch, 1986). Many of us would probably choose to use mental techniques, either exact or approximate, if we only had better skills in mental computation.

MENTAL METHODS IN THE CURRICULUM

Learning and Selecting Methods

The techniques of mental computation and computational estimation are mental algorithms. There are specific methods or procedures which can be taught and practiced the same way that pencil-and-paper algorithms are taught and practiced. You may feel inadequate with mental computation and wonder in awe at people who compute and estimate mentally. Most likely you simply have not been taught. Relatively speaking, mental computation is a very new curriculum area.

In contrast with pencil-and-paper algorithms, there are lots of mental algorithms to learn, not just a few. Furthermore, it is not always clear-cut which methods to use. Given the same situation, different people will utilize different methods. Consider the following:

48 WIDGETS
13 CENTS EACH
HOW MUCH?

How would *you* estimate the total? Ask some friends how they did it. Are there are other ways? Is one method better? What if you wanted an exact answer? Could you do that mentally? How would you approach that task? (Try it before reading on.)

Some possibilities:

Let's see. 48 is about 50. 50 times 10 cents is $5. 50 × 3 cents is $1.50 more. About $6.50.

13 cents is close to 10 cents. 50 at 10 cents is $5.00. Probably more—say $6.00.

Think of 50 times 13. 5 times 13 is 65. Must be $6.50 because $65 or 65 cents don't seem right.

Instead of 48 and 13, use 50 and 15. Now 5 times 15 is 75. But I made both numbers bigger—$7.50 is too big. Probably less than $7.00.

OK, 48 × 10 cents is $4.80. Then I need 3 times 48 more or about $1.50. $4.80 and $1 is $5.80 and 50 cents is $6.30.

In the classroom you can help children learn a new mental algorithm, and you can provide situations in which they practice newly taught procedures. However, you must also provide opportunity for choice of methods and discussion between class members. Children will learn that different methods are acceptable, and that single correct answers to estimations do not exist. When an exact number is required the choice of method depends a lot on the numbers involved and the skills acquired.

Scope and Sequence

At the present time, there is no "standard" curriculum for estimation or mental computation. One of the best ways of working on estimation skills seems to be to integrate them in with other areas of the mathematics curriculum. In this sense, when there is an opportunity to encourage or discuss estimation of a particular computation, we should capitalize on the opportunity. Textbook authors are now making estimation and mental computation regular features of their books at all grade levels. Generally these features tend to be what publishers refer to as "floating strands," with exercises occurring periodically throughout the texts. The use of these features provides the opportunity to relate estimation and mental computation to the other areas. Few books now have chapters or complete units to teach either mental arithmetic or computation.

Hazekamp (1986) argues that mental multiplication should be taught before children master the pencil-and-paper algorithm. He reasons that once the pencil-and-paper method is mastered, children attempt to use a form of mental blackboard when asked to do mental computation. Try to use the pencil-and-paper algorithm mentally to compute 78 × 6. For most it is very difficult to keep track of trades and adding partial products. Many mental algorithms use a "left-hand" approach. For 78 × 6, think 6 × 70 is 420, and 48 more is 468. Hazekamp's argument is equally applicable to the other operations with the possible exception of division.

On the other hand, in the beginning of the last chapter, a

strong argument was made for integrating the mental methods with pencil-and-paper methods. The point made was that the strong emphasis on base ten models and the rationale behind the pencil-and-paper algorithms provides an opportunity to discuss mental strategies at the same time.

The difference in these two arguments is in the word *mastery*. What Hazekamp means is that if we wait to teach mental computational algorithms until after children become proficient, we will lose the flexibility that children have before they become completely attached to one procedure. When the one procedure for computation is a pencil-and-paper algorithm, then there is a problem. Reconsider the product 78 × 6 for a moment. Another approach is to think of 80 times 6 or 480 and then subtract 12 (for 2 × 6). This approach is like building a rectangle that is 2 too wide and then subtracting off two rows of 6 ones. Both methods can be related directly to the area model presented in the last chapter. The very fact that there is more than one way, and that these methods do not interfere but rather complement instruction in the pencil-and-paper algorithm, suggests that at least these methods of computation might be taught together throughout the curriculum.

Long-Term Goals

Mental algorithms develop and improve in both quality and quantity over years of practice. Mental computation and estimation instruction should begin as early as the first grade and continue beyond the eighth grade. As new concepts and skills are developed, so too can mental algorithms. Mental arithmetic is not a three-week unit but a long-term goal. Each teacher at each grade level must help children develop new ideas and provide children with guided practice.

Mental Computation

Each mental algorithm can be presented to the class, discussed, and practiced briefly on different days. As new methods are introduced, some children will select different approaches for the same task. These should certainly be discussed and accepted. Once taught, be sure to encourage continued practice on a daily basis.

MENTAL ADDITION AND SUBTRACTION

Adding and Subtracting Tens and Hundreds

Sums and differences involving multiples of 10 or 100 are easily added mentally, especially if place-value words are used to help keep track of what you are doing.

Example:

300 + 500 + 20

Say: 3 *hundred* and 5 *hundred* is 8 *hundred* and 20 more is 820.

Use base ten models to help children begin to think in terms of tens and hundreds. Early examples should not include any trades. The exercise 420 plus 300 involves no trades while 70 plus 80 may be more difficult.

Activity 10.1

Display some tens and hundreds base ten models on the overhead. Have students say what number is shown. Next, say how much will be added or subtracted as in "minus 300," and have students say the result. Finally move to examples with trades. Show 820. "Minus 50." (Think: "minus 20 is 800 and 30 more is 770.")

Activity 10.2

Play a chain game. Each student has one or more cards with a number written on it. After naming a start number, the teacher calls out for example, "plus 300" or "minus 70." The student holding the result card calls out the result. Then the next change is read by the teacher, and the chain continues. If the cards are prepared so that some numbers occur more than once, then all students must remain alert even after their numbers have been called. Figure 10.2 illustrates the idea with only seven cards.

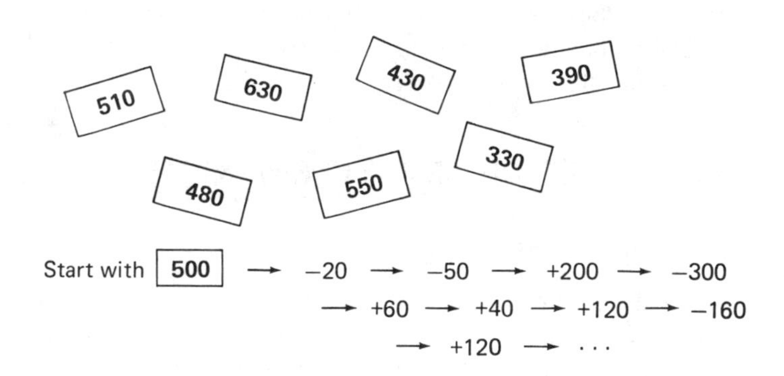

Figure 10.2
A chain game.

When working with thousands and hundreds, it is sometimes useful to think of the thousands as hundreds. That is, 3400 can be thought of as 34 hundred. Then 34 hundred and 8 hundred is 42 hundred.

Adding and Subtracting One-Digit Numbers

Many children who have mastered their addition and subtraction facts continue to use counting techniques for facts such as 56 + 7 or 73 − 4. These are sometimes referred to as *higher decade facts*. Present these facts in conjunction with basic facts as in the following examples.

7 + 8 =	12 − 6 =
47 + 8 =	32 − 6 =
77 + 8 =	82 − 6 =
367 + 8 =	552 − 6 =

Of course, facts that do not bridge a decade are easier such as 48 − 5 or 62 + 7. For those that do cross a decade, the exact strategy may differ with the child or with different numbers. For 44 + 7 you might think of 4 + 7 being 11 and ending with a 1, so the sum must be 51. For 49 + 7 it may be more reasonable to think "one more to 50 and six is 56." Similar differences in thought patterns will appear in subtraction situations. Figure 10.3 shows how a hundreds chart and base ten models can each be used to help students with these ideas.

The use of higher decade facts can be extended to adding or subtracting multiples of 10 or 100. To do this, temporarily ignore the other digits to the right.

Example:

374 + 80 Think: 37 + 8 (or 370 + 80) is
45 → 450 → 454.
The 4 was temporarily ignored.

Front-End Approaches

Mental addition and subtraction most frequently begin with the left-hand digits. When working mentally or orally with the number 472, it is easier to think first of 4 hundred, then 70, and finally 2 more than it is to begin with the 2 and then the 7 tens.

Examples:

46 + 38 = Think: 46 and 30 is 76
76 and 8 is 84

Notice that the first addition converts the problem to a higher decade fact. Here are two more examples of front-end thinking.

382 + 75 = Think: 380 + 70 (or 38 + 7) is
450 → 452, + 5 is 457

342 − 85 = Think: 340 − 80 (or 34 − 8) is
260 → 262, − 5 is 257.

Working with Nice Numbers

Some numbers are "nice" to work with, like 100 or 700 or 50 or even 450. Try adding nice numbers to not so nice numbers. For example, try adding 450 and 27 or add 700 and 248.

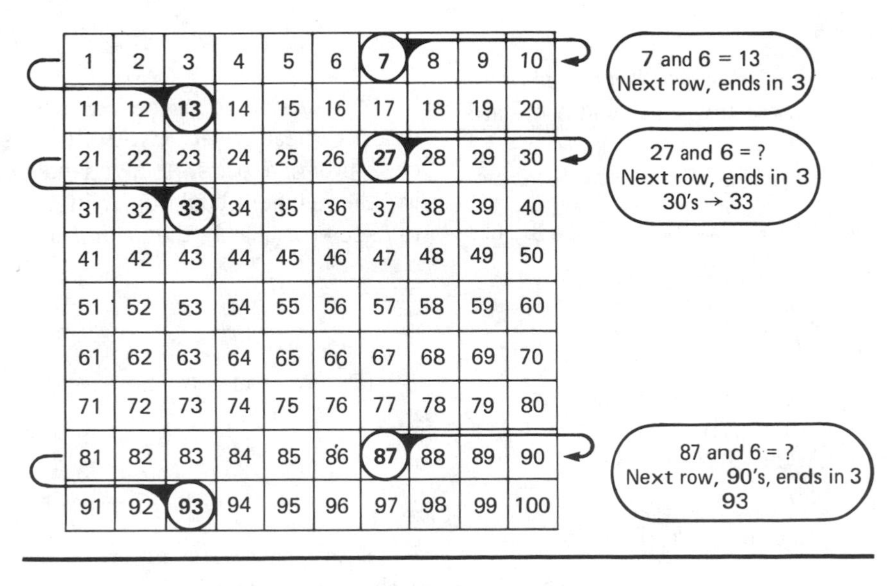

1	2	3	4	5	6	7	8	9	10
11	12	13	14	15	16	17	18	19	20
21	22	23	24	25	26	27	28	29	30
31	32	33	34	35	36	37	38	39	40
41	42	43	44	45	46	47	48	49	50
51	52	53	54	55	56	57	58	59	60
61	62	63	64	65	66	67	68	69	70
71	72	73	74	75	76	77	78	79	80
81	82	83	84	85	86	87	88	89	90
91	92	93	94	95	96	97	98	99	100

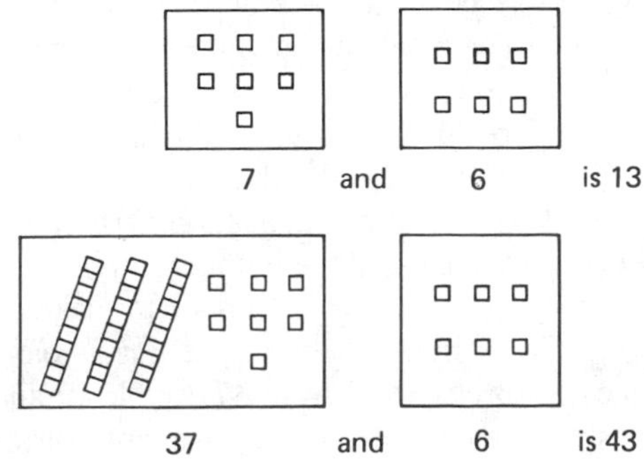

Figure 10.3
Two ways of looking at higher decade addition.

Subtraction with nice numbers is also easy. For 500 − 73, think 73 and what makes 100. For 650 − 85, first take off 50 then 35 more from 600. Notice the two different ways that nice numbers are used in the last two examples. The next two activities help with this type of thinking.

Activity 10.3

Play "The Other Part of 100." For example give students the number 28 and have them determine the other part of 100 or 72. Discuss their thought patterns. (Can you think of two different approaches?) The other part of 50 is just as easy. Later expand the game to get the other part of 600 or 450 or other numbers that end with 00 or 50.

Activity 10.4

Play "50 and Some More." Say a number and have the students respond with "50 and ____." For 63, the response would be "50 and 13." Use other numbers, such as "450 and Some More."

Compensation Strategies

When students begin to get comfortable with "nice" numbers, they can use a *compensation* approach when the numbers are "almost nice."

Examples:

97 + 68 = Think: 100 + 68 is 168. Take off 3. 165.

126 + 298 = Think: Adding 300 is 426. Then 2 less is 424.

473 − 99 = Think: 473 − 100 is 373, so it's 374.

452 − 69 = Think: 450 − 50 is 400 less 19 is (other part of 100) 81 → 381. Add back 2 → 383.

650 − 85 = Think: 650 − 100 is 550, then add back 15 → 565.

The last example was used in the previous section and was done differently. Is one method "better"?

Compatible Numbers

Compatible numbers for addition and subtraction are numbers that go together easily to make nice numbers. Numbers that make tens or hundreds are the most common example. Compatibles also include numbers that end in 5, 25, 50, or 75 since these numbers are also easy to work with. The teaching task is first to get students used to looking for combinations that work together and then to look for these combinations in computational situations.

Activity 10.5

"Searching for Pairs" is an activity that can be used to help children think about "nice combinations" or compatible numbers. In Figure 10.4 several different searches are illustrated.

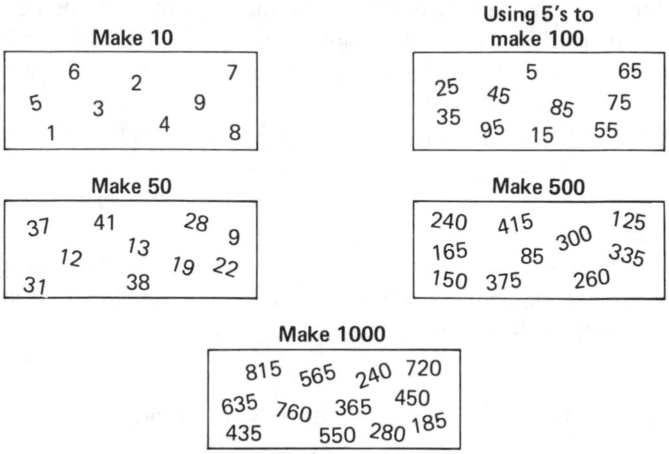

Make 10: 6 2 7 5 3 9 1 4 8

Using 5's to make 100: 25 5 65 45 85 75 35 95 15 55

Make 50: 37 41 28 9 12 13 19 22 31 38

Make 500: 240 415 125 300 165 85 335 150 375 260

Make 1000: 815 565 240 720 635 760 365 450 435 550 280 185

Figure 10.4
Pair searches.

Pair searches could be worksheets or could be presented on an overhead projector with students writing down or calling out appropriate pairs. As a variation for an independent activity, the numbers could be written on small cards. Students can see how quickly they can pair up the cards.

Activity 10.6

Present strings of numbers that use compatibles.

30 + 80 + 40 + 50 + 10
25 + 125 + 75 + 250 + 50
95 + 15 + 35 + 5 + 65

Strings such as these can be approached in two ways, and each should be practiced. One way is to search out compatible combinations such as the 5 and 95 in the last example. The other way is to simply add one addend at a time, saying the result as you go. For the first example, that would be "30, 110, 150, 200, 210."

The compensation strategy can now be combined with the compatible numbers approach. "Break off" a small piece of a not so nice number to make it into a nice compatible. For example, 47 and 35 is not as easy as 45 and 35. Therefore, break off 2 from 47, think 45 and 35 is 80 and 2 is 82. To practice this idea, present pairs with a nice number and a nearly nice number and have students explain their process. This is especially useful with numbers that are close to a 5 (like 137) or close to a 10 (like 62). Alternatively, you can add on a piece. For 37 + 155, add 3 onto 37, 40 and 155 is 195. Now take the 3 off again → 192.

MENTAL MULTIPLICATION

Factors with Zeros

Products such as 400 × 60 that have only a few nonzero digits are prime candidates for mental multiplication. If you write the answers for the exercises below you will notice that counting the zeros and adding them on the end is easy. If done orally, the zero counting is not as easy. Base ten language becomes very important.

Begin by multiplying by 10, 100, and 1000.

Examples:

7 × 100 = 700
6 × 1000 = 6000
23 × 100 = 2300 (23 hundred)
40 × 100 = 4000 (40 hundred—that's 4 thousand)

When students are proficient with this beginning level, begin to use multiples of 10, 100 and 1000. The nonzero part is done first, and then the zeros are added on or the product adjusted orally.

Examples:

8 × 400 = Think: 4 × 8 is 32. So 32 hundred. 3200.
5 × 30 = 150
12 × 3000 = Think: 3 × 12 is 36. So 36 thousand. 36,000.

Next, work with examples where both factors involve zeros as in 40 × 300. In written form this is easy; 4 × 3 is 12 and add 3 zeros. The use of oral base ten products as discussed in Chapter 9 is quite useful.

Examples:

60 × 80 = Think: 6 × 8 is 48. Tens × tens are hundreds. So 48 hundred. 4800.
8000 × 20 = Think: 2 × 8 is 16. Tens × thousands are ten thousands. So 16 ten thousands, that's 160 thousand. 160,000.
or Think: 8 × 20 is 160. That's 160 thousands.
50 × 600 = Think: 5 × 6 is 30 and tens × hundreds are thousands, 30 thousand. 30,000.

Children will need help with dual names for numbers such as 32 hundred for 3 thousand 2 hundred. In the second example, the base ten language produced a nonstandard name, 16 ten thousands. This needs to be discussed with students so that they can use oral alternatives to help their thinking without writing down the numbers. Try writing numbers with zeros and have children think of at least two ways to read each and then produce two factors that would have that product.

Example:

240,000 Read two ways: 240 thousand or 24 ten thousands
Factors: 8 × 30,000 or 80 × 3000 or 800 × 300.

Again, in written form this is easy. Just count zeros. Without seeing the zeros, base ten language is a key factor.

Front-End Multiplying

The product 63 × 8 can easily be thought of as 8 × 60 and 24 more. The idea is to work with the big parts first (front end) and add in the little parts as you work from left to right. The examples below get progressively more difficult.

Examples:

83 × 6 = Think: 80 × 6 is 480 and 18 is 498.
370 × 8 = Think: 8 × 300 is 24 hundred, and 560 (8 × 70) is 29 hundred 60 or 2960.
or Think: 8 × 37. So 8 × 30 is 240 and 56 is 296, times 10 is 2960. Here the zero part is used last.
42 × 300 = Think: 3 × 40 is 120 and 6 is 126. Then 2 zeros (times hundreds) is 126 hundred or 12,600.
or Think: 4 × 300 is 12 hundred, so 40 × 300 is 12 thousand. 6 hundred more (2 × 300) is 12,600.

In working through examples such as these with your class, write down only intermediate results as they are spoken. Figure 10.5 illustrates this idea.

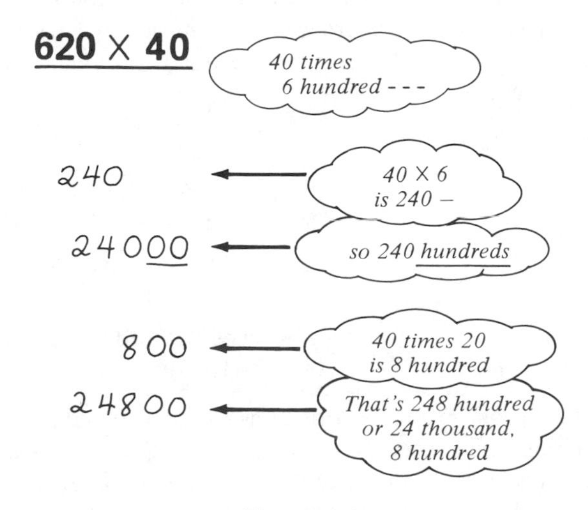

Figure 10.5
Write down intermediate results but not complete calculations.

Sometimes it is useful to use two-digit parts, especially those that involve 25 or other easy products.

Examples:

$3 \times 412 =$ Think: 3×4 hundred is 12 hundred and 36 (3×12) is 12 hundred 36, or 1236.

$725 \times 40 =$ Think: 4×7 hundred is 28 hundred and 4×25 is 1 hundred or 29 hundred. Times 10 is 290 hundred or 29 thousand.

Compensation with Eights and Nines

When one factor ends in an 8 or a 9, it is useful to use the next higher multiple of 10 and then subtract the difference from the product.

Examples:

$7 \times 39 =$ Think: 7×40 is 280. But that is 7 too much. So 273.

$498 \times 6 =$ Think: 6×500 is 30 hundred or 3 thousand. Take back 12, that's 2988.

The use of compensation for eights and nines is more prevalent in money. Prices such as \$5.98 or \$23.99 abound. Even if your discussion of mental computation does not include decimals, money can easily be dealt with. For example, 7 times \$5.98 is 7×6 dollars less 7×2 cents. Alternatively, the dollar and cents parts can be dealt with separately and then added.

Halve and Double

This approach is based on the idea that if one factor is doubled and the other halved, the product is the same. For example, 8×10 is the same as 4×20 or 16×5. (A really nice exercise for group work is to prepare an explanation for why this is so. Would it work for thirds and triples? for $n$ths and n times?) The halve-and-double approach can be applied to any problem with an even factor, but it is most useful with 5, 50, and 500 and also with 25 or 250.

Examples:

$682 \times 5 =$ Think: double 5 is 10 and half of 682 is 341. So 341 times 10 is 3410.

$50 \times 26 =$ Think: double 50 and halve 26 is 100 times 13 or 1300.

$60 \times 25 =$ Think: 4 twenty-fives is 100 and a fourth of 60 is 15, 15 hundred.

In the last example, 4 times is the double-and-halve approach applied twice. An alternative thought process might be that each set of 4 twenty-fives is 100. There are 15 fours in 60. So that's 15 hundred. Similar language can be used with 50 or 500. What if the factor to be halved is odd? For 50×35, try 34×50 and 50 more. Half of 34 is 17 so that's 17 hundred and one more 50 is 1750. What would you do with 25×47 or 500×43?

MENTAL DIVISION

Division facts are closely tied to multiplication. For $42 \div 7$ students are taught to think, "What times 7 is 42?" When long division is introduced this think-multiplication idea is reduced to a digit-by-digit concept. An advantage here not found with the other operations is that both the pencil-and-paper algorithm and the mental process begin on the left. Here is a case where it would obviously pay to work on both mental and paper algorithms at the same time. For example, $936 \div 4$ is done by dividing the 9 hundreds and then trading for tens, and so on. If 900 is completely divided up as four sets of 225 and then the 36 is four sets of 9, then the quotient is 225 plus 9 or 234.

Zeros in Division

When the dividend has trailing zeros and the divisor is a single digit, a good idea is to emphasize base ten language. In written form, the zeros at the end are just added on.

Examples:

$6\overline{)24{,}000}$ Think: 6×4 is 24, so 6×4 thousand is 24 thousand. Must be 4000.

$4500 \div 9 =$ Think: 9×5 hundred is 45 hundred. 500.

When both divisor and dividend have zeros, rely on the base ten product language of multiplication.

Examples:

$40\overline{)3200}$ Think: 8×4 is 32. 40 is 4 tens. The product is hundreds. Tens $\times$ tens are hundreds. 8 tens. 80.

$35{,}000 \div 70 =$ Think: 7×5 is 35. Tens times what are thousands? Hundreds. 5 hundred. 500.

$300\overline{)18{,}000}$ Think: 6×3 is 18. Hundreds times tens are thousands. Must be 60. (Check 60×300).

Some suggest that these examples be done by crossing off as many zeros in the dividend as in the divisor. That is, the last example is equivalent to $3\overline{)180}$. This approach is excellent when the numbers are in written form, but it is difficult when the problem is presented orally.

Working with Part at a Time

By thinking missing factor and using base ten language, many division problems can be dealt with mentally.

Examples:

$7\overline{)637}$	Think: 637 is 630 and 7. 90 × 7 is 630 and 1 more is 91.
189 ÷ 3 =	Think: 189 is 180 and 9. What times 3 is 180? (60) And 3 × 3 is 9, so that's 63.

Notice in the last example, the emphasis is on 180 and not 18, as would be done with pencil and paper. Is that necessary? In the following examples, if the base ten language is not used, either a critical zero will be omitted (first example) or an easy-to-use part will not be considered (second example).

Examples:

$6\overline{)4236}$	Think: 6 × 700 is 4200 and 6 × 6 is 36. That's 7 hundred and 6 more, 706. (Without the base ten language the result may well be 76.)
$6\overline{)198}$	Think: 6 × 30 is 180. That leaves 18. So 6 × 3 is 18 or 33 in all.

Missing-Factor Practice

A significant exercise for both mental division and multiplication is to have students find appropriate factors for given products. In Figure 10.6 two exercises are shown that are of a seek-and-find nature with the possible factors given. It is a good idea to include more factors than are needed. Another approach is to give only the products and have students determine possible factors. This approach will frequently yield more than one pair of factors. For example, 42,000 is 60 × 700, 2100 × 20 or 420 × 100, and many more. Groups might be challenged to find as many factors as possible as long as the products could be computed mentally.

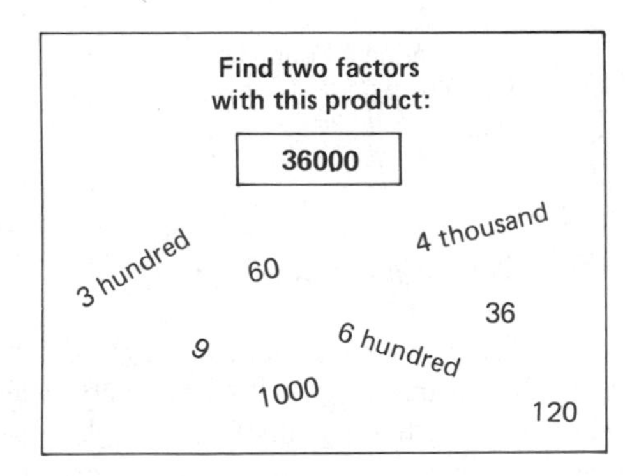

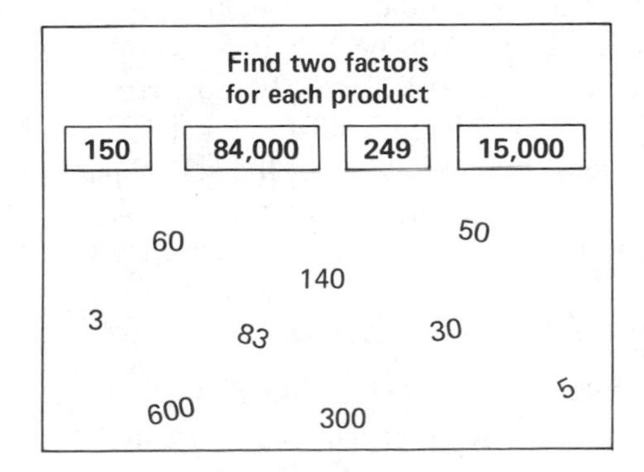

Figure 10.6
Factor searches.

You may have noted that the division examples seem to be contrived. That is, the numbers "work out" nice. In fact, there are relatively few workable divisions that can be done mentally compared with the other three operations. (Why?) That does not mean that division is less important as a mental computational skill. However, mental division is more of a tool for estimation. Many estimations are done by adjusting numbers to make them workable mentally. So "not nice" divisions are sometimes adjusted to those that fall into the categories that have been discussed. The missing-factor exercises are useful in helping students learn to think of appropriate nice numbers.

Computational Estimation

Computational estimation differs from mental computation in that the goal is to quickly determine a result that is adequately accurate for the situation. More often than not, we can be satisfied with an approximate result rather than an exact one. If mental computation was always possible or just as fast as making an estimate, there would be little need for estimation skills. However, many computations do not involve nice numbers, making exact mental computation virtually impossible or difficult. Furthermore, many situations do not require an exact answer, so reaching for a calculator or pencil is not necessary—if one has some good estimation skills. In everyday life, estimation skills prove to be tremendously valuable and time-saving. Teaching these skills to children has become more and more important in recent years.

TEACHING ESTIMATION

Make Estimation a Realistic Topic

Students need to see that estimation is useful in real life. Most exercises should be given in real-life contexts: prices, distances, quantities, and so forth. Have students interview parents and other adults to find out when and where they estimate. Discuss situations where an estimate is satisfactory and contrast those with situations requiring an exact answer. Curriculum materials are becoming available to help you with this task. Figure 10.7 shows two pages from teacher resource books that illustrate these ideas.

Utilize the Language of Estimation

The language of estimation includes words and phrases like the following: about, close, just about, a little more or less than, and between. Estimation exercises should be posed with

TRANSPARENCY 1

Is an Estimate Enough?

1. Is an estimate enough when:
   The waitress figures 5% tax?
   The waitress finds the total?
   The customer figures a 15% tip?
   The customer checks the bill?

2. Is an estimate enough when:
   Shawn decides if he'll make enough money?
   The boss figures out how much to pay Shawn?

3. Is an estimate enough when:
   The accountant figures out how much money was made on ticket sales?
   The newspaper reports how many people attended the game?

GET YOUR MIND IN GEAR

TRANSPARENCY 1

Everyday Estimation

One of my answers was wrong on the quiz. Which answer looks unreasonable?

$47 \times 9 = 423$

$98 \times 16 = 1,516$

$38 \times 12 = 3,516$

$27 \times 32 = 874$

TRY THESE

1. 
   The movie *E.T.* earned $102 million in its first month. About how much money did it make per day?

2. The Thompson's dinner bill totaled $28.75. They plan to leave about a 15% tip. About how much should the tip be?

3. I need 2 pairs of socks and 3 pairs of shoelaces. About how much will they cost?

4. About how many points do I have altogether?

Figure 10.7
R. E. Reys, P. R. Trafton, B. Reys, and J. Zawojeski (1987). Computational Estimation. *Palo Alto, CA: Dale Seymour Publications. Reprinted by permission.*

this type of terminology. Students should understand that they really are trying to get as close as possible but with quick and easy methods. Estimation does not mean, as some students believe, calculating an exact answer and then rounding it off a bit. Demonstrate that in realistic situations that does not make much sense. For example, when deciding if the $10 you have is enough to buy three drugstore items priced $3.95, $2.57, and three for $6.95, the exact sum is not at all important. Finding that first and then rounding off the answer would be foolish.

Build on Related Skills and Concepts

Estimation skills are highly related to mental computation skills, numeration concepts, and real-world number sense. Most estimation strategies are based on the idea of using nice numbers that are close to the numbers in the computation. The nice numbers lend themselves to mental computation. For example, the cost of 30 soft drinks at 69 cents each can be estimated by using 70 instead of 69. Now 30 times 70 is a mental computation skill. If students have not acquired that skill, the use of 70 is not much help. A good estimation program must have a good mental computation component built in.

Conceptual knowledge of number plays a large part in both estimation and mental computation. Specifically, place-value concepts for both whole numbers and decimals, an understanding of fractions, and comfort with percent concepts are critical components of estimation skills.

Finally, what might be called *real-world number sense* is also useful. Is $210, $21, or $2.10 most reasonable for thirty 69-cent soft drinks? Is the cost of a car likely to be $950 or $9500? Could the attendance at the school play be 30 or 300 or 3000? Knowing what is reasonable in familiar situations helps put approximate results in the right ballpark.

Accept a Range of Answers

What estimate would you give for 27 × 325? If you use 20 × 300 you might say 6000. Or you might use 25 for the first factor and divide the second by 4 and get 8100. Or you might use 30 × 300 or 30 × 320 to get 9000 or 9600? Is one of these right and the other wrong?

In the following sections, different algorithms or strategies for estimation will be explored. When different methods are used on the same computation, different estimates result. Making adjustments after the mental computation is another factor. Even two excellent estimators will give different estimates. Experience and conceptual knowledge will eventually help students make better estimates. But it is crucial that children understand that there is no one answer to an estimation.

FRONT-END METHODS

Front-end methods involve the use of the leading or leftmost digits in numbers and ignoring the rest. After an estimate is made based only on these front-end digits, an adjustment can be made by noticing how much has been ignored. In fact, the adjustment may also be a front-end approach.

Front-End Addition and Subtraction

Front end is a very easy estimation strategy for addition or subtraction. This approach is reasonable when all or most of the numbers have the same number of digits. Figure 10.8 illustrates the strategy. Notice that when a number has fewer digits than the rest, that number is ignored completely.

(a) Front end addition—column form

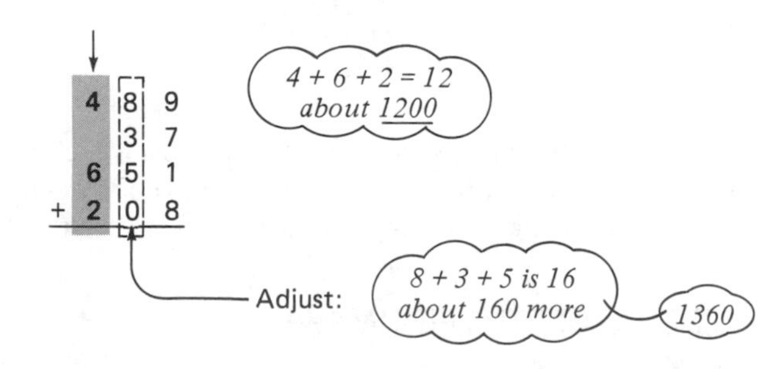

(b) Front end addition—numbers not in columns

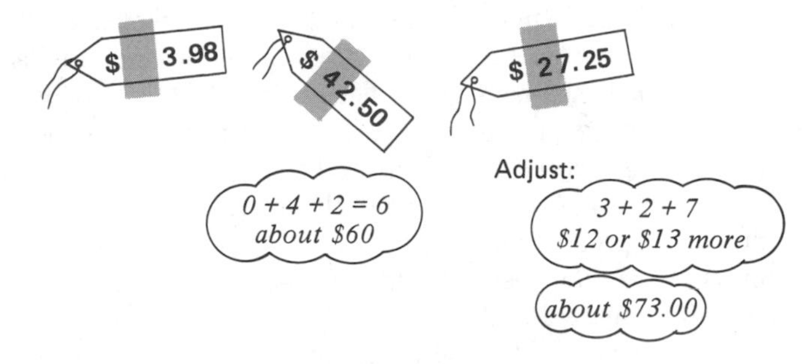

Figure 10.8

After adding or subtracting the front digits, an adjustment is made to correct for the digits or numbers that were ignored. Making an adjustment is actually a separate skill. For very young children, practice first just using the front digits. Pay special attention to numbers of uneven length when not in a column format.

When teaching this strategy, present additions or subtractions in column form and cover all but the leading digits. Discuss the sum or difference estimate using these digits. Is it more or less than the actual amount? Is the estimate off by a little or a lot? Later, show numbers in horizontal form or on price tags that are not lined up. What numbers should be added?

The leading-digit strategy is easy to use and easy to teach because it does not require rounding or changing numbers. The numbers used are there and visible, so children can see what they are working with. It is a good first strategy for children as low as the third grade. It is also useful for older children and adults, especially as they learn to make better adjustments in the front-end sum or difference.

Front-End Multiplication and Division

For multiplication and division, the front-end method uses the first digit in each of the two numbers involved. The computation is then done using zeros in the other positions. For example, a front-end estimation of 48×7 is 40 times 7 or 280. When both numbers have more than one digit, the front ends of both are used. For 452×23 consider 400×20 or 8000.

Division with pencil and paper is almost a front-end strategy already. First determine in which column the first digit of the quotient belongs. For $7\overline{)3482}$, the first digit is a 4 and belongs in the hundreds column over the 4. Therefore, the front-end estimate is 400. This method always produces a low estimate, as students will quickly figure out. In this particular example, the answer is clearly much closer to 500, so 480 or 490 is a good adjustment.

ROUNDING METHODS

Rounding is the most familiar form of estimation. Estimation based on rounding is a way of changing the problem to one that is easier to work with mentally. Good estimators follow their mental computation with an adjustment to compensate for the rounding. To be useful in estimation, rounding should be flexible and well understood conceptually.

Rounding in Addition and Subtraction

When a lot of numbers are to be added, it is usually a good idea to round them to the same place value. Keep a running sum as you round each number. In Figure 10.9, the same total is estimated two ways using rounding. A combination of the two is also possible.

In subtraction situations there are only two numbers to deal with. For subtraction and for additions involving only two addends, it is generally necessary to round only one of the two numbers. For subtraction, round only the subtracted number. In $6724 - 1863$ round the 1863 to 2000. Then it is easy: $6724 - 2000$ is 4624. Now adjust. You took away a bigger number, so the result must be too small. Adjust to

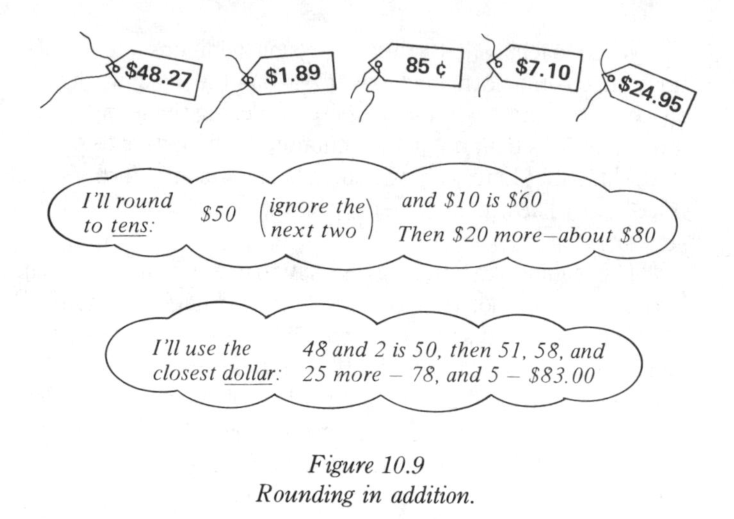

Figure 10.9
Rounding in addition.

about 4800. For 624 + 385, round 385 to 400 or 624 to 600, but there is no need to do both.

Rounding in Multiplication and Division

The rounding strategy for multiplication is no different than for other operations. However, the error involved can be significant, especially when both factors are rounded. In Figure 10.10, several multiplication situations are illustrated, and rounding is used to estimate each.

If one number can be rounded to 10, 100, or 1000, the resulting product is easy to determine without adjusting the other factor.

When one factor is a single digit, examine the other factor. Consider the product 7 × 836. If 836 is rounded to 800, the estimate is relatively easy and is low by 7 × 36. If a more accurate result is required, round 836 to 840 and use a front-end computation. Then the estimate is 5600 plus 280 or 5880 (7 × 800 and 7 × 40). The parts technique relies on the skill of doing the front-end multiplication mentally.

If possible, round only one factor and select the largest one if it is significantly larger. (Why?) For example, in 47 × 7821, 47 × 8000 is 376,000, but 50 × 8000 is 400,000.

Another good rule of thumb with multiplication is to round one factor up and the other down (even if that is not the closest round number). When estimating 86 × 28, the 86 is about in the middle but 28 is very close to 30. Try rounding 86 down to 80 and 28 to 30. The actual product is 2408, only 8 off of the 80 × 30 estimate. If both numbers were rounded to the nearest ten, the estimate would have been based on 90 × 30 with an error of nearly 300.

With one-digit divisors, it is almost always best to search for a compatible dividend rather than to round off. For example, 7)4325 is best estimated by using the close compatible number, 4200, to yield an estimate of 600. Rounding would suggest a dividend of 4000 or 4300, neither of which is very helpful. (Recall the contrived examples in the Mental Division section.)

When the divisor is a two- or three-digit number, rounding

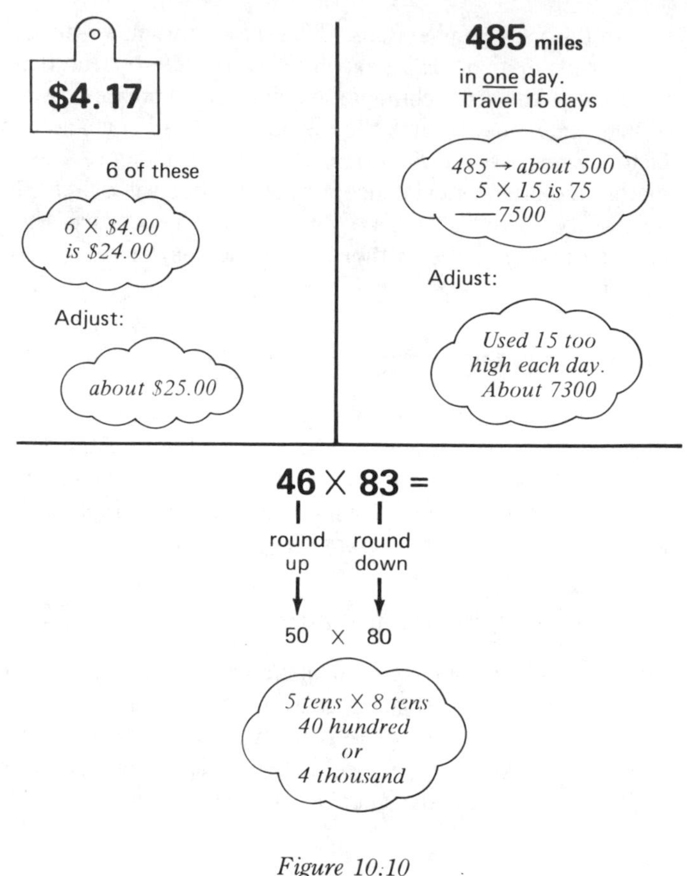

Figure 10.10
Rounding in multiplication.

it to tens or hundreds makes looking for a missing factor much easier. For example, to estimate 425)3890, round the divisor to 400. Then think, 400 times what is close to 3890?

USING COMPATIBLE NUMBERS

Finding Compatibles in Addition and Subtraction

When adding a long list of numbers, it is sometimes useful to look for two or three numbers that can be grouped to make 10 or 100. If numbers in the list can be adjusted slightly to produce these groups, that will make finding an estimate easier. This approach is illustrated in Figure 10.11.

In subtraction, it is frequently easy to adjust only one number to produce an easily observed difference. The thought process may be closer to addition than subtraction, as illustrated in Figure 10.12.

Frequently in the real world, an estimate is needed for a large list of addends that are relatively close. This might happen with a series of prices for similar items, attendance at a series of events in the same arena, cars passing a point on successive days, or other similar data. In these cases, as illustrated in Figure 10.13, a nice number can be selected as representative of each, and multiplication used to determine the total. This is more of an *averaging technique* than a compatible numbers strategy.

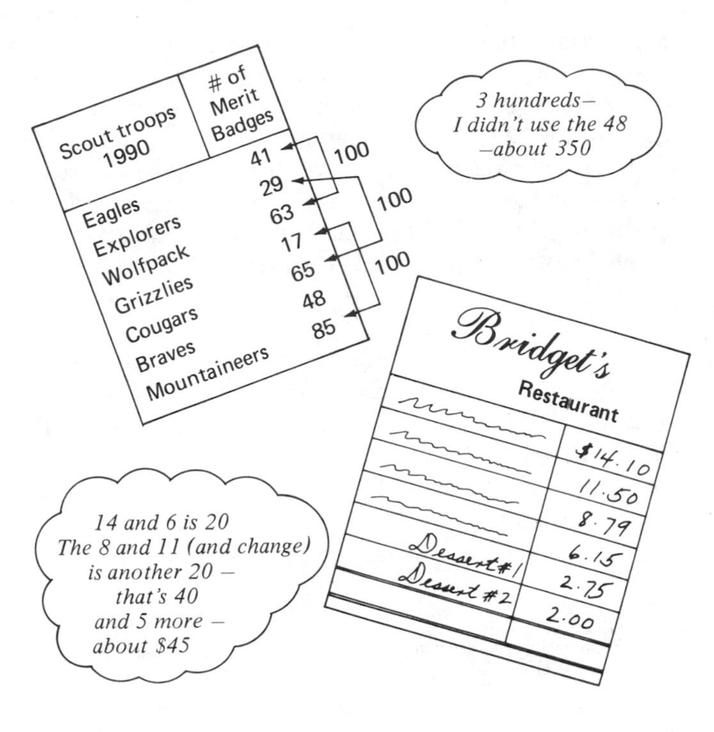

Figure 10.11

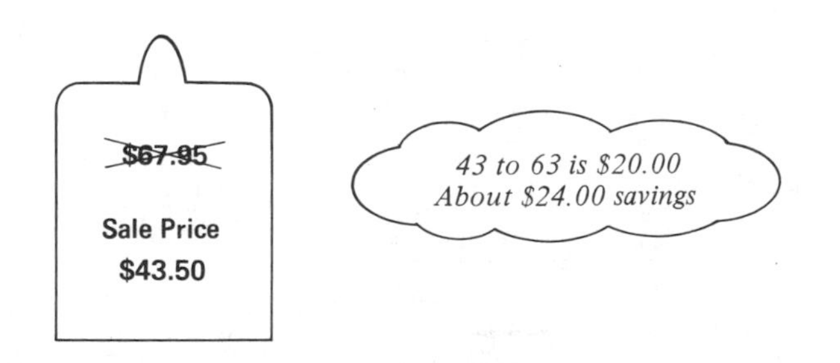

Figure 10.12
Compatibles can mean an adjustment that produces an easy difference.

Cookies sold
Troop 124

1. Marcie ____ 68
2. Sally ____ 42
3. Chris ____ 81
4. Yvonne ____ 35
5. Yolanda ____ 57
6. Andrea ____ 60
7. Meggan ____ 71
8. Jo Ann ____ 63

Looks like about 60 each —
8 × 60 is 480

Figure 10.13
Estimating sums using averaging.

Compatibles in Division

One of the best uses of the compatible number strategy is in division. The two exercises shown in Figure 10.14 illustrate adjusting the divisor and/or dividend to create a division that comes out even and is therefore easy to do mentally. The strategy is based on whole number arithmetic and is not difficult. Many percent, fraction, and rate situations involve division and the compatible number strategy is quite useful, as shown in Figure 10.15.

ESTIMATION EXERCISES

The ideas presented so far illustrate some of the types of estimation and thought patterns you want to foster in your classroom. However, making up examples and putting them into a realistic context is not easy and is very time-consuming. Fortunately, good estimation material is now being included in virtually all textbooks. Text material will most likely provide you with hints and periodic exercises for integrating estimation throughout your program. This exposure, however, will probably not be sufficient to provide an ongoing, intensive program

Figure 10.14
B. J. Reys and R. E. Reys (1983). Guess (Guide to Using Estimations and Strategies) Box II. *Palo Alto, CA: Dale Seymour Publications, Cards 2 and 3. Reprinted by permission.*

73% of our 132 eighth graders watch more than 2 hours of TV per day.

73% → about 75% or $\frac{3}{4}$.
If I use 120, one fourth is 30
and three fourths is 90.
(That's probably a little low.
I used a bigger % but a lot less
students.
About 90 to 95

The chances of getting a winning lottery ticket are about 1 in 8. Zeke bought 60 tickets. About how many "winners" is reasonable.

$\frac{1}{8}$ of 60 ⟶ $\frac{1}{8}$ of 64 is 8
About 7 or 8 winners

A box of 36 Thank You cards is $6.95. How much is that per card?

36 X 2 is 72 → or 36 X 20 is 720
$6.95 is close to $7.20 – these
cards are a little less than 20c each

Figure 10.15
Using compatible numbers in division.

for developing estimation skills with your students. Try adding to the text using one or more of the many good teacher resource materials that are available commercially. Some of these are listed at the end of the chapter.

Teacher's guides, resource books, and professional books and articles on estimation offer good activities or activity models that you can easily adapt to your particular needs without additional resources. The examples presented in this section are not designed to *teach* estimation strategies, but offer useful formats to provide your students with practice using skills as they are being developed.

Calculator Activities

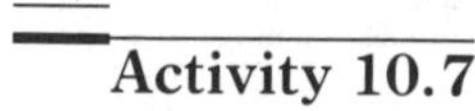

Activity 10.7

"The Range Game" (Wheatley & Hersberger, 1986) is an estimation game for any of the four operations. First pick a start number and an operation. The start number and operation are stored in the calculator. Students then take turns entering a number and pressing [=] to try to make the result land in the target range. The following example for multiplication illustrates the activity: Suppose a start number of 17 and a range of 800 to 830 is chosen. Press 17 [×] 0 [=] to store 17 as a factor. Press a number then [=]. Perhaps you try 25. The result is 425. That is about half the target. Try 50. The result is 850. Maybe 2 or 3 too high. Try 48. Result is 816—in the target! Figure 10.16 gives examples for all four operations. Prepare a list of start numbers and target ranges. Let students play in pairs and see who can hit the most targets on the list.

The "Range Game" can be played with an overhead calculator with the whole class, or by an individual, or two or three children with calculators can race one another. The speed element is important. The width of the range and the type of numbers used can all be adjusted to suit the level of the class.

After entering the start # as shown, players take turns pressing a number, then [=] to try to get a result in the target range.

Addition:

Press: 0 [+] (start #) [=]

START		TARGET
153	⟶	790 → 800
216	⟶	400 → 410
53	⟶	215 → 220

Subtraction:

Press: 0 [−] (Start #) [=]

START		TARGET
18	⟶	25 → 30
41	⟶	630 → 635
129	⟶	475 → 485

Multiplication:

Press: (Start #) [×] 0 [=]

START		TARGET
67	⟶	1100 → 1200
143	⟶	3500 → 3600
39	⟶	1600 → 1700

Division:

Press: 0 [÷] (Start #) [=]

START		TARGET
20	⟶	25 → 30
39	⟶	50 → 60
123	⟶	15 → 20

Figure 10.16
Calculator "Range Game".

Activity 10.8

"Secret Sum" is a calculator activity that utilizes the memory feature. A target number is selected, for example 100. Students take turns entering a number and pressing the M+ key. Each of the numbers is accumulated in the memory, but the sum is never displayed on the screen. If one player thinks that the other player has made the sum go beyond the target, he or she announces "over" and the memory return key is pressed to check. If a player is able to hit the target exactly, bonus points can be awarded. Interesting strategies quickly develop. (Adapted from an example in *Everybody counts,* National Research Council, 1989.)

The "Secret Sum" game can also be played with the M− key. First enter a total amount in the memory. Each player's number is followed by a press of M− and is subtracted from the memory. Here the first to correctly announce that the other player has made the memory go negative is the winner.

Activity 10.9

An excellent calculator activity for estimating quotients is suggested by Coburn (1987). First enter the divisor of a two-digit division problem and press ×. Then enter an estimate of the quotient and press =. The result is compared with the dividend and two subsequent estimates are allowed. There is no need to reenter the divisor.

For example:

Two players can compete by working on the same problem trying to get the smallest difference in 30 seconds. (Can you see how this game is really the same as the "Range Game"?)

Computer Programs

A number of computer programs are available that practice estimation skills. Computer programs can present problems for estimation, control speed and also compare the result to the actual answer. Most allow the teacher or user to adjust the skill level of the exercises. These programs can be effectively used with a full class using a large monitor or projection system. Students can write down estimates on paper within the allotted time frame. In the MECC package *Estimation,* the "Estimate" program has all of the features mentioned previously (Minnesota Educational Computing Corporation, 1984). While not at all fancy or entertaining, it can be quite effective at a wide range of grade and ability levels.

Activities for the Overhead Projector

Activity 10.10

Select any single computational estimation problem and put it on the board or overhead. Allow 10 seconds for each class member to get an estimate. Discuss briefly the various estimation techniques that were used. As a variation, prepare a problem with an estimation illustrated. For example, 139 × 43 might be estimated as 6000. Now ask questions concerning this estimate. "How do you think that estimate was arrived at? Was that a good approach? How should it be adjusted? Why might someone select 150 instead of 140 as a substitute for 139?" Almost every estimate can involve different choices and methods. Alternatives make good discussions, help students see different methods, and learn that there is no single "correct estimate."

Activity 10.11

Make a transparency of a page of drill-and-practice computations. Have students focus on a single row or other collection of five to eight problems. Ask them to find the one with an answer that is closest to some round number that you provide (Figure 10.17). One transparency could provide a week's worth of 5-minute drills.

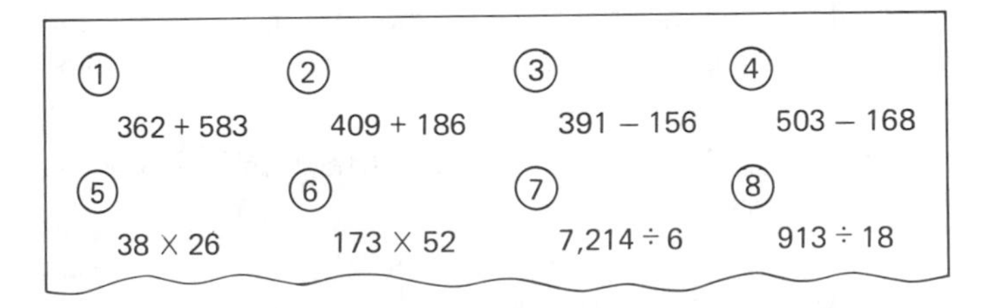

Which of these is CLOSEST TO 600? 1000? 10?

Figure 10.17
Use drill page from your text as an estimation exercise.

Activity 10.12

Also with a page of problems from a workbook, write in answers to six or seven problems. On one or two problems, make a significant error that can be caught by estimation. For example, write 26)5408 = 28 (instead of 208) or 36 × 17 = 342 (instead of 612). Other answers should be correct. Encourage the class to estimate each problem to find those "not even in the ballpark" errors.

Activity 10.13

Make "Is This Enough?" transparencies. Select an amount such as 10 or 50 (or $10 or $50) and write that at the top of a transparency. Below, put a variety of computations

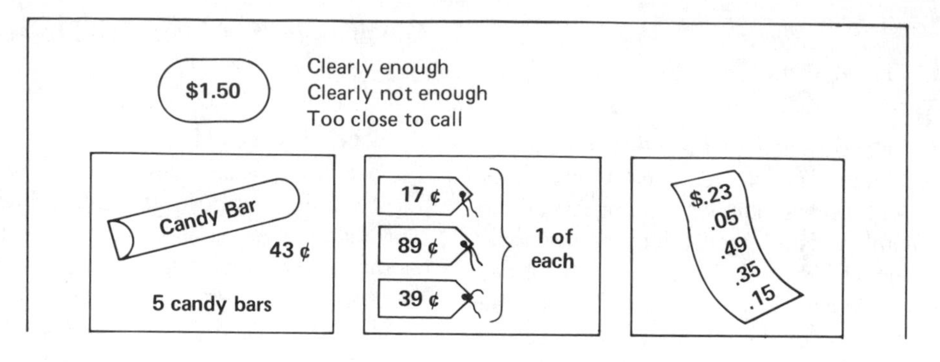

Figure 10.18
"Is This Enough?"

or realistic situations. The task is to decide quickly if the top amount is *clearly enough* (more) or *clearly not enough* (less) or perhaps *too close to call.* An example is shown in Figure 10.18. Show only one example at a time and only for about 10 seconds. This type of estimation is frequently done in real life and does not always call for a very accurate estimate.

ESTIMATING WITH FRACTIONS, DECIMALS, AND PERCENTS

Fractions, decimals, and percents are three different notations for rational numbers. Many real-world situations that call for computational estimation involve the part-to-whole relationships of rational numbers. A few examples are suggested here:

SALE! $51.99. Marked one fourth off. What was the original price?

About 62% of the 834 students bought their lunch last Wednesday. How many bought lunch?

Tickets sold for $1.25. If attendance was 3124, about how much was the total gate?

I drove 337 miles on 12.35 gallons of gas. How many miles per gallon did my car get?

With the exception of a few examples, this chapter has avoided estimations with fractions, decimals, and percents. To estimate with these numbers first involves an ability to estimate with whole numbers. Beyond this, it involves an understanding of fractions and decimals, and what these two types of numbers mean. Calculations with percents are always done as fractions or as decimals. The key is to be able to use an appropriate fraction or decimal equivalent.

In the first of the examples just presented, one approach is based on the realization that if $51.99 (or $52) is the result of 1 fourth off, that means $52 is 3 fourths of the total. So 1 fourth is a third of $52 or a little less than $18. Thus about $52 + $18 = $70 or about $69 seems a fair estimate of the original cost. Notice that the conceptualization of the problem involves an understanding of fractions, but the estimation skill is exactly the same as for whole numbers. This is also the case for almost all problems involving fractions, decimals, or percents.

From a developmental perspective, it is important to see that the skills of estimation are separate from the conceptual knowledge of rational numbers. It would be a mistake to work on the difficult processes of estimation using fractions or percents if concepts of those numbers were poorly developed. In later chapters, where rational number numeration is discussed, it is shown that an ability to estimate can contribute to an increased flexibility or "number sense" with fractions, decimals, and percents.

Evaluating Mental Computation and Estimation

USE FREQUENT PERIODIC TESTING

Just as development of mental computation and estimation skills is an ongoing matter, so also should evaluation be an ongoing part of your program. Tests or quizzes can be given in as little time as 5 minutes and as often as every week.

Early test results may be very discouraging. Do not despair. And do not hold early low scores against your students. Rather, use these first scores as baseline data against which to measure progress. Children do improve at both mental computation and estimation with instruction and practice. The progress will be reflected in the test scores and will reward both you and your students.

CONTROL TIME

Mental computation and estimation is, by definition, to be done mentally, not with pencil and paper. This is extremely difficult for children to accept. In testing situations (and in many practice activities) it is important that you control the amount of time that is allowed for each exercise. This cannot be done when all of the questions are written on test paper and passed out as in a traditional test. Instead, prepare the test items on cards or on transparencies. Show each problem for a 10-second period and then immediately go to the next one. If you allow much longer than 10 seconds, students will attempt to use pencil and paper to compute answers.

Another way to emphasize strictly mental methods is to use prepared numbered answer sheets with only a small blank space next to each number. No other markings are to be written on the paper. Some teachers even cut answer sheets into narrow strips so that there is no room left to write anything but the answer.

CONSIDER ANSWERS AND ANSWER FORMATS

For mental computation, you are looking for exact answers. Tests should almost always be open-ended in format. That is, the student should write the answers rather than select from a multiple-choice list. Some care should be taken to look at incorrect answers and determine what caused the error. On quizzes, you may even ask students in upper grades to select one example where they made an error and explain in writing what they did that caused it.

Estimation tests can involve multiple-choice formats as well as open-ended responses. If you use multiple choices, put the answer choices in numeric order and include foils (wrong answers) that are likely to be made. For example, if a good answer is 2400, then 24 and 24,000 may be good foils. Another way to select foils is to use an inappropriate estimation technique. In estimating 585 + 29 + 714 + 693, it would be inappropriate to round to thousands. For skilled estimators who have learned to adjust estimates, a front-end estimate of 1800 is much too low. For open-response estimation tests, you must decide in advance what will be the range of acceptable answers. There are no single correct responses, even if every student used the same approach, which is unlikely.

USE OCCASIONAL INTERVIEWS

It is not always easy to tell what techniques or thought processes are being used by individual students. For example, some students may get overly attached to a particular approach and not be flexible enough to switch to a more efficient method when the numbers call for it. One way to find this out is to conduct a short test on a one-on-one basis. Give perhaps five items for mental computation or estimation. Immediately afterwards, return to each answer and ask the student to explain how the estimate was made. This form of "listening to your students" will be very valuable in deciding how to pace your program and determining if there are concepts or strategies that require extra emphasis.

For Discussion and Exploration

1. What types of mental computation or computational estimation can be taught at the first and second grades?
2. Choose one of the four operations with whole numbers. For that operation, give one or two examples of how a lesson on the pencil-and-paper algorithm might incorporate a mental computation exercise.
3. When is a pencil-and-paper method of calculation more likely or more reasonable than the use of a calculator? Does your response justify the relative amount of time that is given in school to the mastery of pencil-and-paper computations? Similarly, when is a mental method (exact or approximate) more likely than the use of either a pencil and paper or calculator? How does your response to this question compare with the amount of time spent on each of these methods in the curriculum?
4. Examine a sixth-, seventh-, or eighth-grade textbook for the total program provided for mental computation and computational estimation. You will want to look at the teacher's edition because many of the best suggestions are found in supplementary activities or pages. Find out how estimation and mental computation is integrated into the rest of the content of the text.
5. Examine one of the following teacher resources:
   Computational estimation
   Mental math in the primary grades
   Mental math in the middle grades
   Mental math in the junior high school
   GUESS, boxes I and II (Guide to Using Estimation Skills and Strategies) (See references under Suggested Readings.)
   How would the materials you reviewed be used over a one-year period in your classroom?

Suggested Readings

Cobb, P., & Merkel, G. (1989). Thinking strategies: Teaching arithmetic through problem solving. In P. R. Trafton (Ed.), *New directions for elementary school mathematics.* Reston, VA: National Council of Teachers of Mathematics.

Coburn, T. G. (1987). *How to teach mathematics using a calculator.* Reston, VA: National Council of Teachers of Mathematics.

Coburn, T. G. (1989). The role of computation in the changing mathematics curriculum. In P. R. Trafton (Ed.), *New directions for elementary school mathematics.* Reston, VA: National Council of Teachers of Mathematics.

Driscol, M. J. (1980). Estimation and mental arithmetic. In *Research within reach: Elementary school mathematics.* St. Louis, MO: CEMREL, Inc.

Hazekamp, D. W. (1986). Components of mental multiplying. In H. L. Schoen (Ed.), *Estimation and mental computation.* Reston, VA: National Council of Teachers of Mathematics.

Hope, J. A. (1986). Mental computation: Aliquot parts. *Arithmetic Teacher, 34*(3), 16–17.

Hope, J. A., Leutzinger, L., Reys, B. J., & Reys, R. R.

(1988). *Mental math in the primary grades* Palo Alto, CA: Dale Seymour.

Hope, J. A., Reys, B. J., & Reys, R. (1987). *Mental math in the middle grades.* Palo Alto, CA: Dale Seymour.

Hope, J. A., Reys, B. J., & Reys, R. (1988). *Mental math in the junior high school.* Palo Alto, CA: Dale Seymour.

Hope, J. A., & Sherrill, J. M. (1987). Characteristics of unskilled and skilled mental calculators. *Journal for Research in Mathematics Education, 18,* 98–111.

Leutzinger, L., Rathmell, E. C., & Urbatsch, T. D. (1986). Developing estimation skills in the primary grades. In H. L. Schoen (Ed.), *Estimation and mental computation.* Reston, VA: National Council of Teachers of Mathematics.

McBride, J. W., & Lamb, C. E. (1986). Number sense in the elementary classroom. *School Science and Mathematics, 86,* 100–107.

Reys, B. J. (1988). Estimation. In T. R. Post (Ed.), *Teaching mathematics in grades K–8: Research based methods.* Boston: Allyn and Bacon.

Reys, B. J. (1986). Teaching computational estimation: Concepts and strategies. In H. L. Schoen (Ed.), *Estimation and mental computation.* Reston, VA: National Council of Teachers of Mathematics.

Reys, R. R. (1986). Evaluating computational estimation. In H. L. Schoen (Ed.), *Estimation and mental computation.* Reston, VA: National Council of Teachers of Mathematics.

Reys, R. R. (1984). Mental computation and estimation: Past, present, and future. *The Elementary School Journal, 84,* 547–557.

Reys, R., Trafton, P., Reys, B., & Zawojewski, J. (1987). *Computational estimation: (Grades 6, 7, 8).* Palo Alto, CA: Dale Seymour.

Sowder, J. (1989). Developing understanding of computational estimation. *Arithmetic Teacher, 36*(5), 25–27.

Trafton, P. R. (1986). Teaching computational estimation: Establishing an estimation mind-set. In H. L. Schoen (Ed.), *Estimation and mental computation.* Reston, VA: National Council of Teachers of Mathematics.

CHAPTER 11

THE DEVELOPMENT OF FRACTION CONCEPTS

Children and Fraction Concepts

FRACTIONS ARE DIFFICULT RELATIONSHIPS

Consider the illustrations of two thirds in Figure 11.1. Our adult knowledge confirms that each shows two thirds, yet what is it that these models have in common? They cannot be matched like seven blocks with seven cards or seven fingers. Some are circles, some dots, and some are lines. Some have many elements, and some have only one. If shown a rectangle, you cannot say what fraction it is. Some other shape or rectangle must also be identified as the unit or whole. Even the symbolism is a problem. The relationship represented by $\frac{2}{3}$ is represented just as well by $\frac{6}{9}$.

The point of the previous discussion is simply to heighten your awareness that fractions are not trivial concepts, even for middle school children. A fraction is an expression of a relationship between a part and a whole. Helping children construct that relationship and connect it meaningfully to symbolism is the topic of this chapter.*

FRACTIONS IN THE CURRICULUM

It has been traditional to include some minimal exposure to fractions in grades K to 4, with each successive grade spending just a little more time on fraction concepts than the one before. Usually some limited addition and subtraction of fractions is begun in grade 4 with the development of all of the operations for fractions introduced in grade 5. Continued review and reteaching occurs in grades 6, 7, and 8. This massive explosion of procedural knowledge (symbolic rules) at about the fifth grade is generally not supported by strong conceptual knowledge of fraction meanings, because the curriculum simply has not provided the time for the complex development that fraction concepts require.

As with many other topics in elementary school mathematics, there is a movement to delay the rush toward symbolism and especially computations with fractions. Research efforts are uncovering more of the difficulties children have with fractions and providing suggestions for how to help children construct these ideas (Behr, M. J., Lesh, R., Post, T. R., & Silver, E. A., 1983; Post, T. R., Wachsmuth, I., Lesh, R., & Behr, M. J., 1985; Payne, J. N., 1976; Pothier, Y., & Sawada, D., 1983). This delay in symbolism presents some instructional problems that teachers need to confront:

How do you deal with fractions without symbols?

What models should be used? How do you use them? How can all children have models?

How much can you do with fractions before you add, subtract, multiply, and divide?

How will children learn about fractions without rules?

This entire chapter is about fraction concepts with no mention of adding, subtracting, multiplying, or dividing.

Three Categories of Fraction Models

Models must be used at all grade levels to adequately develop fraction concepts. Further, the variety of representations for fractions suggested by Figure 11.1 indicates that children should have experiences with a wide assortment of models. In this section, three categories of models are presented with numerous examples of each. Most of the activities presented in the chapter can be done with all three model types. A change in the model is usually a significant change in the activity from the viewpoint of the children. As you examine the various ideas in the chapter, consider how the same activity could be done with different models and different types of models.

*To be technically correct, we should say that the relationship between a part and a whole is a *rational number* and that a fraction is one type of symbolism that is used to *represent* a rational number. This number versus symbol distinction is not made in this book in the context of rational numbers. The term *fraction* is used in reference to both the concept of number as well as the symbolism. The context of the discussion will generally make the intent clear. Furthermore, the distinction is not one that is useful for children, especially not before the seventh or eighth grade.

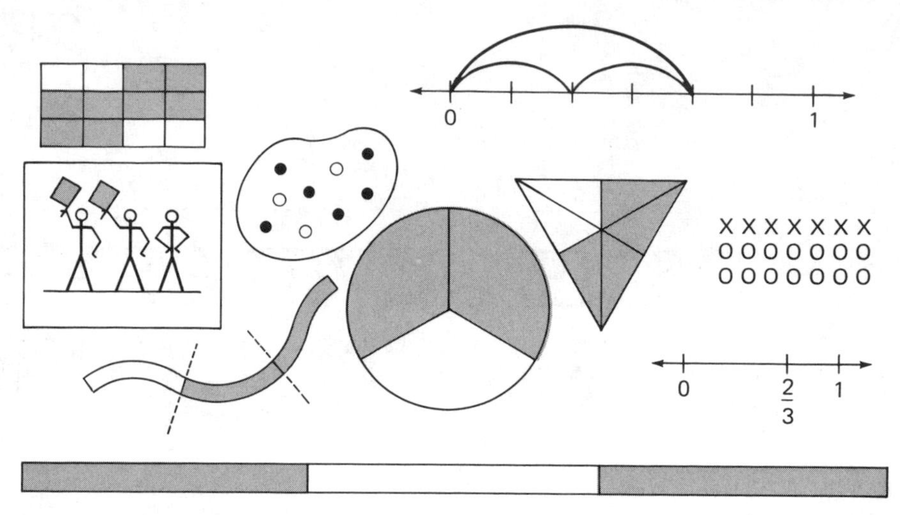

Figure 11.1
Describe how all of these are the same.

REGION OR AREA MODELS

In region models a surface or area is subdivided into smaller parts. Each part can be compared with the whole. Almost any shape can be partitioned into smaller pieces. Figure 11.2 shows a variety of models in this category.

Circular regions and rectangle models can be duplicated on tagboard or construction paper, laminated, and cut into fraction kits kept in plastic bags. (Masters for circular models are included in Appendix C.) Rectangles permit almost any piece to be designated as the whole so that other pieces change in fractional values accordingly. Pattern blocks, geoboards, and grid paper provide the same flexibility. Some

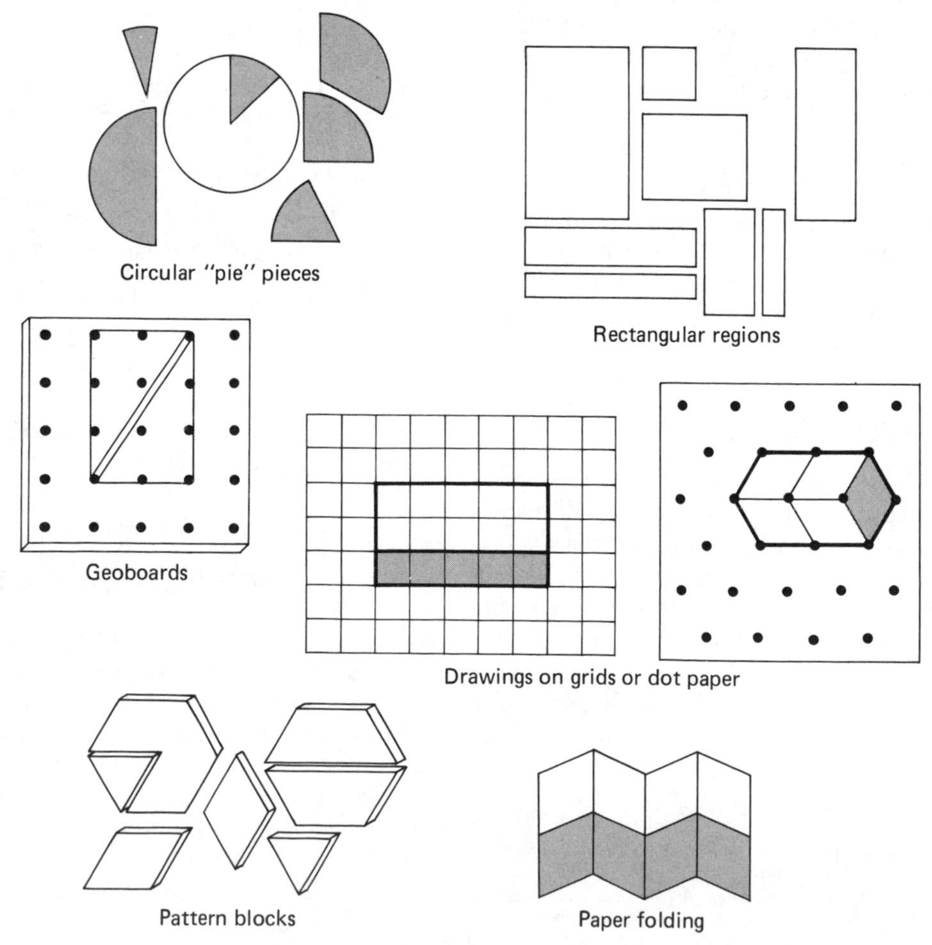

Figure 11.2
Area or region models for fractions.

commercial models for fractions such as the rectangles used in "The Fraction Factory" (Holden, 1986) now come in classroom sets. With grid paper or dot paper, children can easily draw pictures to explore fraction ideas. Paper folding is somewhat limited, but is a model available to everyone.

LENGTH OR MEASUREMENT MODELS

Length or measurement models are similar to area models except that lengths are compared instead of areas. Either lines are drawn and subdivided or physical materials are compared on the basis of length, as shown in Figure 11.3. Manipulative versions provide much more opportunity for trial and error and exploration.

Fraction strips are a teacher-made version of Cuisenaire rods. Both the strips and the rods have pieces that are in lengths of 1 to 10 measured in terms of the smallest strip or rod. Each length is a different color for ease of identification. As an alternative, strips of construction paper or adding machine tape can be folded to produce equal-sized subparts. Older children can simply draw line segments on paper and subdivide them visually.

The rod or strip model provides the most flexibility while still having separate pieces for comparisons and for trial and error. To make fraction strips, cut 11 different colors of poster board into strips 2 cm wide. Cut the smallest strips into 2-cm squares. Other strips are then 4, 6, 8, . . ., 20 cm, producing lengths 1 to 10 in terms of the smallest strip. Cut the last color into strips 24 cm long to produce a 12 strip. If you are using Cuisenaire rods, tape a red 2 rod to an orange 10 rod to make a 12 rod. In the illustrations for this chapter, the colors of the strips will be the same as the corresponding lengths of the Cuisenaire rods as given here:

White	1	Dark Green	6
Red	2	Black	7
Light Green	3	Brown	8
Purple	4	Blue	9
Yellow	5	Orange	10

a pink strip or a "rorange" (red-orange) rod is 12

The number line is a significantly more sophisticated measurement model (Bright, G. W., Behr, M. J., Post, T. R., & Wachsmuth, I., 1988). From a child's vantage, there is a real difference between "putting a number on a number line" and comparing one length to another. Each number on a line denotes the distance of the labeled point from zero. Place the numbers ⅔ and ¾ on a number line and consider how a child would think about these numbers in the context of that model.

SET MODELS

In set models the whole is understood to be a set of objects and subsets of the whole make up fractional parts. For example, 3 objects are one fourth of a set of 12 objects. The set of 12, in this example, represents the whole or one. It is the idea of referring to sets of counters as single entities that contributes to making set models difficult for primary aged children. However, the set model helps establish important connections with many real-world uses of fractions and with ratio concepts. Sets can profitably be explored by grades 3 or 4. Figure 11.4 illustrates several set models for fractions.

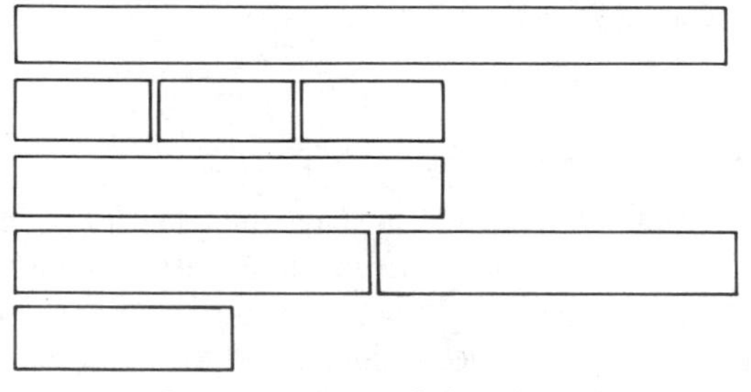

Fraction strips or Cuisenaire rods

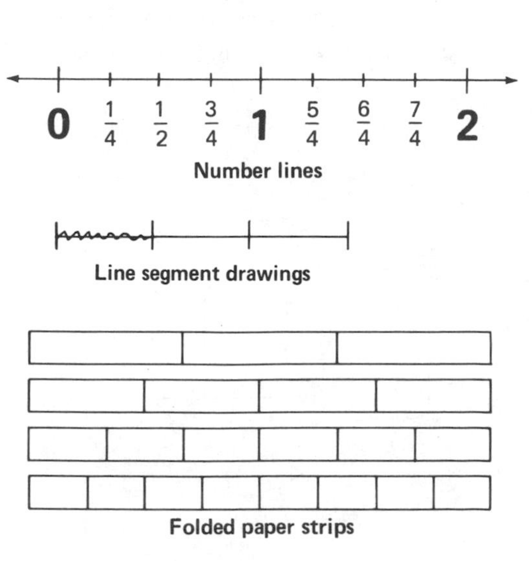

Figure 11.3
Length or measurement models for fractions.

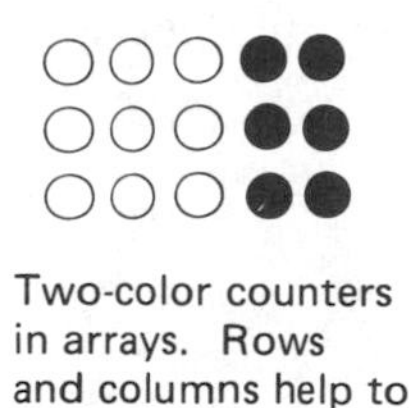

Two-color counters in arrays. Rows and columns help to see parts. Each array makes a whole

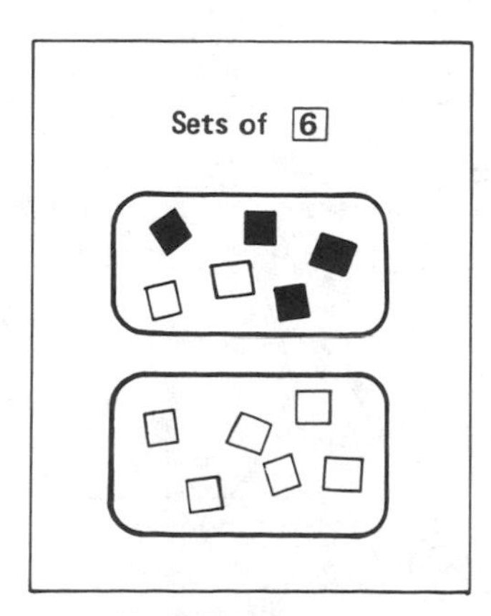

Two-color counters in loops drawn on paper.

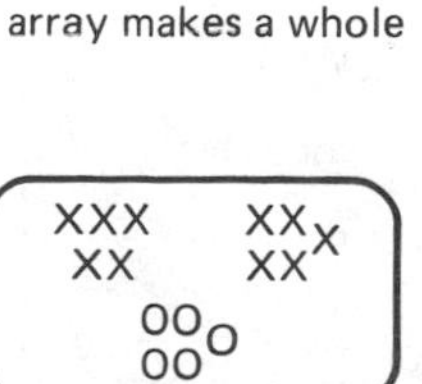

Drawings using X's and O's

Figure 11.4
Set models for fractions.

Any type of counter can be used to model fractional parts of sets, including drawing X's and O's. However, if the counters are colored in two colors on opposite sides, then the counters can easily be flipped to change their color to model various fractional parts of a whole set. Two-sided counters can be purchased, or colored picture mount board can be cut into 2-cm squares. Another simple alternative is to spray-paint lima beans on one side to make two-color beans.

Developing the Concept of Fractional Parts

CONSTRUCTING FRACTIONAL PARTS

Two Fundamental Requirements

In Figure 11.5, some of the regions are divided into fourths and some are not. Those that are not fourths have four parts, but the parts are not the same size, or else the parts are the same size, but there are not four of them. To have fourths, the whole must be divided into four parts *and* the parts must all be the same size.

A first goal in the development of fractions is to help children construct the idea of *fractional parts:* halves, thirds, fourths, fifths, and so on. For any fractional part there are two requirements:

1. There must be the correct number of parts making up the whole.
2. Each of the parts must be the same size (not necessarily the same shape).

While it is correct to simply say "four equal parts" make fourths, the emphasis on *two* requirements may be lost. Also, many children do not have a good concept of the word *equal.*

Children can generalize the notion of fractional parts right from the beginning. A mistake that is frequently made is to assume that halves, thirds, and fourths are somehow easier and prior to sixths, eighths, or even twelfths. By introducing the general notion of fractional parts, children have more to explore and discuss. Once this generalization is made, all fractional parts are available. The traditional artificial restriction to one half, one third, and one fourth in the early grades defeats learning more general relationships.

Activities with Fractional Parts and Fraction Words

Fractional parts should be explored using all available models with the possible exception of sets for the very young. The terminology of *the whole* or *one whole* or simply *one* should also be introduced informally at the same time. In all activities, the *fraction words*—halves, thirds, fourths, fifths, and so on—are used orally and written out. These fractional parts are the nouns or things or objects of fractions. They are the building blocks of virtually all fraction concepts. The early emphasis is on thirds or sixths, not on one third or one sixth.

Activity 11.1

As in Figure 11.5, show examples and nonexamples of specified fractional parts. Have students identify the wholes that are correctly divided into the requested fractional parts and explain why the nonexamples are incorrect. Use length and set models in a similar way.

Activity 11.2

Give students models and have them find fifths, or eighths, and so on, using the model. (Models should never have fractions written on them.) The activity is especially interesting when different wholes can be designated in the same model. That way, a given fractional part does not get identified with a special shape or color but with the relationship of the part to the designated whole. Some ideas are suggested in Figure 11.6.

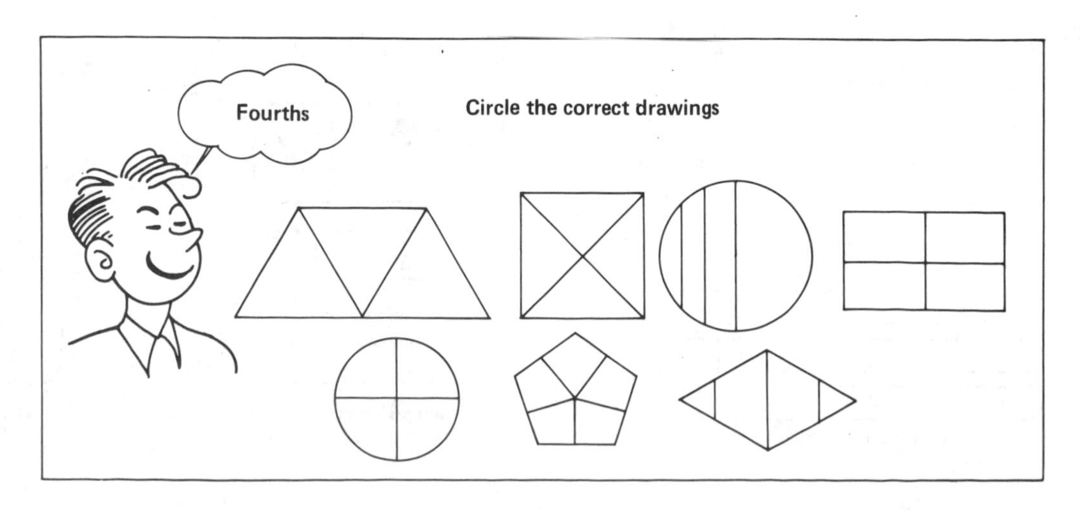

Figure 11.5
Find correct examples of fourths. Why are the nonexamples wrong?

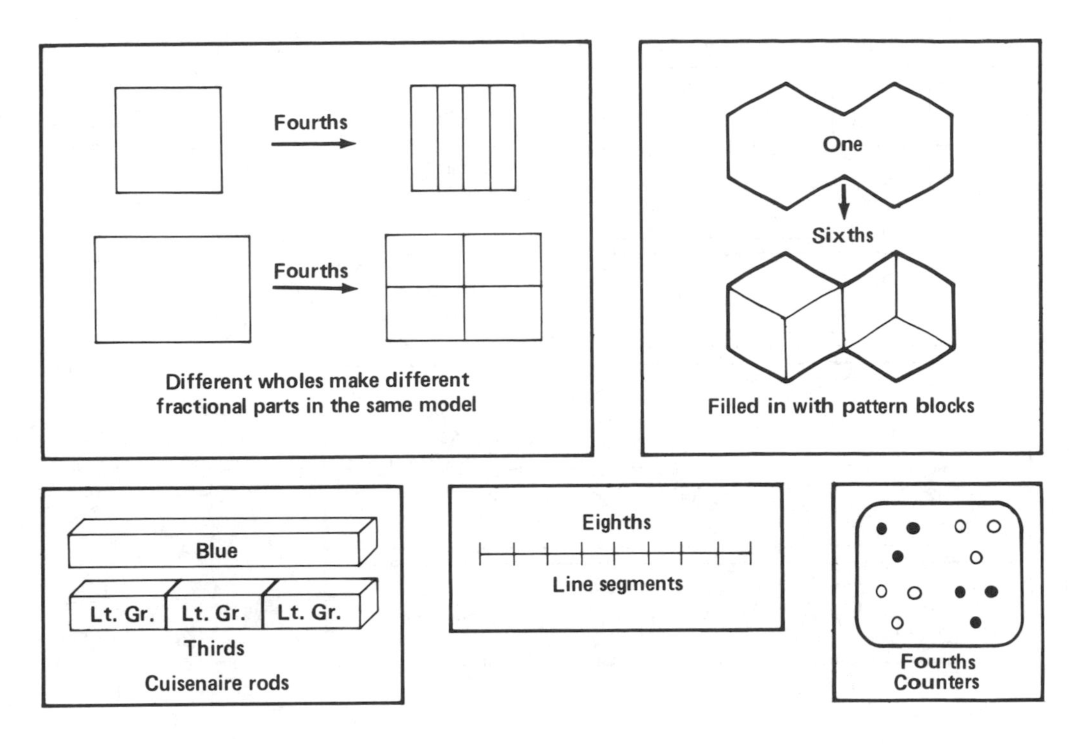

Figure 11.6
Given a whole, find fractional parts.

Activity 11.3

Count fractional parts orally. Once students have identified fourths, for example, count fourths. Show five or six fourths on the overhead. "How many fourths? Let's count: One fourth, two fourths, three fourths, four fourths, five fourths, six fourths." Count other collections of fourths. Ask if a collection of fourths that have been counted is more or less than one whole or more or less than two wholes. As shown in Figure 11.7, make informal comparisons among different counts. "Why did we get almost two wholes with seven fourths, and yet we don't even have one whole with seven twelfths? What is another way we could say seven thirds?" (Two wholes and one third or one whole and four thirds.)

Activity 11.4

Find specified amounts in terms of designated fractional parts. Give students a model and designate the whole if necessary. Then pose questions such as

How many sixths will make this rectangle?

This orange strip is how many thirds? (dark green = whole)

How many eighths will it take to make as much as one and a half?

How many fourths are there if you have 12 counters? (eight counters = whole)

CONNECTING CONCEPTS WITH SYMBOLISM

Fraction symbolism should be delayed as long as possible. The activities in the previous section can all be done orally and with the use of written fraction words such as *7 fourths* or *3 eighths*. Eventually the standard symbolic form must be used.

Always write fractions with a horizontal bar, not a slanted one. The slant used here was dictated by the type style. ($\frac{3}{4}$ *not* 3/4)

Meaning of the Top and Bottom Numbers

Display several collections of fractional parts and have children count each set as discussed earlier and write the count using fraction words as in "3 fourths." Explain that you are going to show how these fraction words can be written much more easily than writing out the words. For each collection write the standard fraction next to the word. Display some other collections and ask students if they can tell you how to write the fraction. Rather than an explanation, use the already developed oral language both with models and when writing the fraction symbols.

Next, have children count by fourths as you write the fractions on the board. Repeat for other parts.

1/4, 2/4, 3/4, 4/4, 5/4, 6/4, 7/4, 8/4, 9/4

1/6, 2/6, 3/6, 4/6, 5/6, 6/6, 7/6, 8/6, 9/6

1/8, 2/8, 3/8, 4/8, 5/8, 6/8, 7/8, 8/8, 9/8

Discuss each row. How are they alike? How different? What part of each row is like counting? Why does the bottom number stay the same as you count fourths, or sixths? Finally, ask students:

What does the top number in a fraction tell you?

What does the bottom number in a fraction tell you?

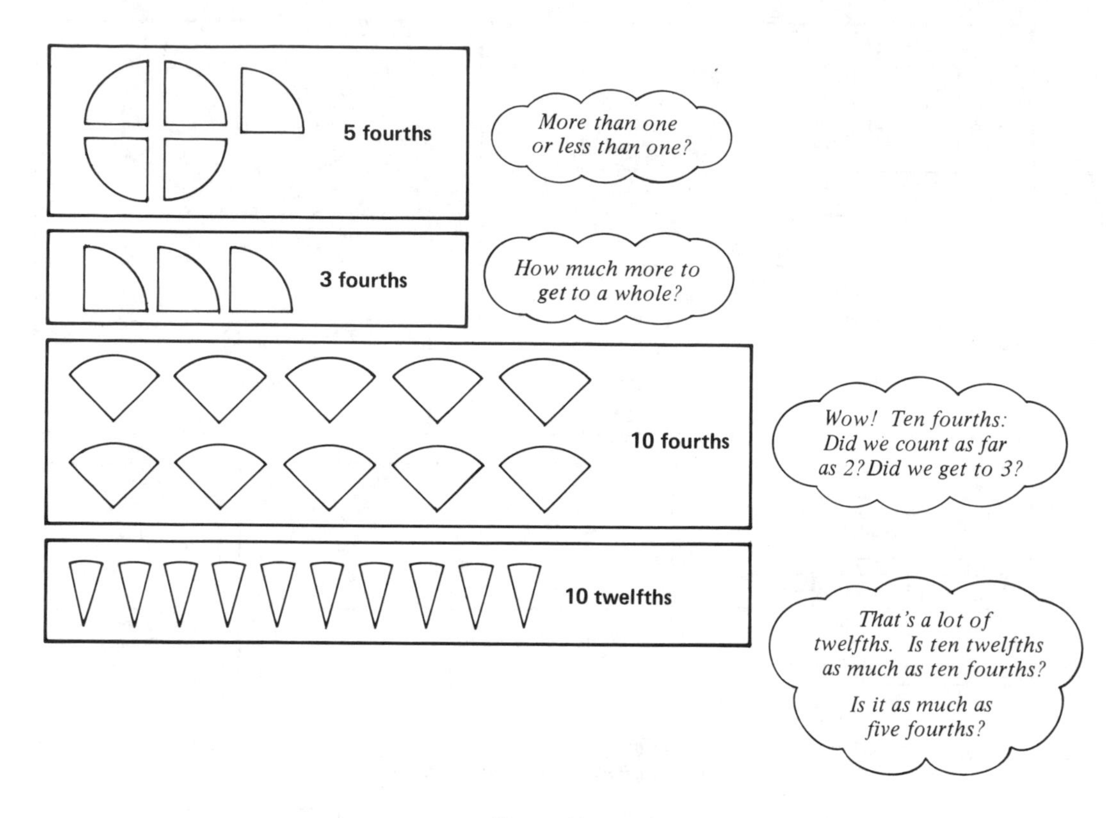

Figure 11.7
Counting fractional parts.

Answer these two questions yourself. Try to think in terms of fractional parts and what has been covered up to this point. Write your explanations for top and bottom number meanings. Try to use children's language. Explore several different ways of saying the meaning. Your meanings should not be tied to a particular model.

Here are some reasonable explanations for the top and bottom numbers.

Top Number: This is the *counting* number. It tells how many things we have. It tells how many have been counted. It tells how many parts are being considered. It counts the parts.

Bottom Number: This tells *what* is being counted. It tells what fractional part is being counted. If it is a 4, it means we counted fourths and if it is a 6, we counted sixths, and so on.

Notice that if the concept of fractional parts is well developed and children can give an explanation for any fraction word, it is not necessary to include that in the meaning of the bottom number. In fact it is clumsy: "It tells how many of the equal parts that are being counted that it takes to make a whole." Not only is that clumsy, it detracts from the two simple but important ideas:

The top number *counts.*

The bottom number tells *what is counted.*

The *what* of fractions are the fractional parts. They can be counted. Fraction symbols are just a shorthand for saying how many and what.

Numerator and Denominator—A Digression

To count a set is to *enumerate* it. Enumeration is the process of counting. The common name for the top number in a fraction is *numerator.*

A $1 bill, a $5 bill, and a $10 bill are said to be bills of different *denominations.* Similarly, the word *denomination* is used to categorize people by religion (Baptists, Presbyterians, Episcopalians, Catholics, etc.). A denomination is the name of a class or type of thing. The common name for the bottom number in a fraction is *denominator.*

Up to this point the terms *numerator* and *denominator* have not been used, as will be the case in most of the rest of the chapter. Why? No child in the third grade would mistake the designations top number and bottom number. The words *numerator* and *denominator* have no common reference for children. Some may feel it is important that children use these words. Whether used or not, it is clear that the words themselves will not assist young children in understanding the meanings.

Mixed Numbers and Improper Fractions

In the fourth National Assessment of Educational Progress, about 80% of seventh graders could change a mixed number to an improper fraction, but fewer than half knew that $5\frac{1}{4}$ was the same as $5 + \frac{1}{4}$ (Kouba et al., 1988a). The result indicates that many children are using a mindless rule that in fact is relatively easy to construct.

Activity 11.5

Use models to display collections such as 13 sixths or 11 thirds. Have children orally count the displays and give at least two names for each. Then discuss how they could write these different names using numbers. They already know how to write $\frac{13}{6}$ or $\frac{11}{3}$. For 2 wholes and 1 sixth a variety of alternatives might be suggested: 2 and $\frac{1}{6}$ or 2 wholes and $\frac{1}{6}$ or $2 + \frac{1}{6}$. All are correct. After doing this with other collections, explain that $2 + \frac{1}{6}$ is usually written as $2\frac{1}{6}$ with the "+" being left out or understood.

Now reverse the process. Write mixed numbers on the board and have students make that amount with models, using only one kind of fractional part. When they have done that, they can write the simple fraction that results. (The term *improper* fraction is an unfortunate yet common term for a simple fraction greater than one.)

Finally, after much back-and-forth between models and symbols using fractions greater than 1, see if students can figure out a simple fraction for a mixed number and a mixed number for an improper fraction. *Do not provide any rules or procedures.* Let students work this out for themselves. A good student explanation for $3\frac{1}{4}$ might involve the idea that there are 4 fourths in one whole, so there are 8 fourths in two and 12 fourths in three wholes. Since there is one more fourth, that is 13 fourths in all, or $\frac{13}{4}$.

For fractions greater than 2 it is a good exercise to have students find other expressions besides the usual. For example, $4\frac{1}{5}$ is not only $\frac{21}{5}$ but also $3\frac{6}{5}$, $2\frac{11}{5}$, and $1\frac{16}{5}$. This idea is extremely useful later when subtracting fractions, as in $5\frac{1}{8} - 2\frac{3}{8}$.

PARTS AND WHOLES EXERCISES

The exercises in this section can help children develop their understanding of fractional parts as well as the meanings of the top and bottom numbers in a fraction. Models are used to represent wholes, and parts of wholes. Written or oral fraction names represent the relationship between the parts and wholes. Given any two of these—whole, part, and fraction—the students can use their models to determine the third.

Any type of model can be used as long as different sizes can represent the whole. For region and area models it is also necessary that single regions or lengths be used to represent nonunit fractions. Traditional pie pieces do not work since the whole is always the circle and all the pieces are unit fractions.

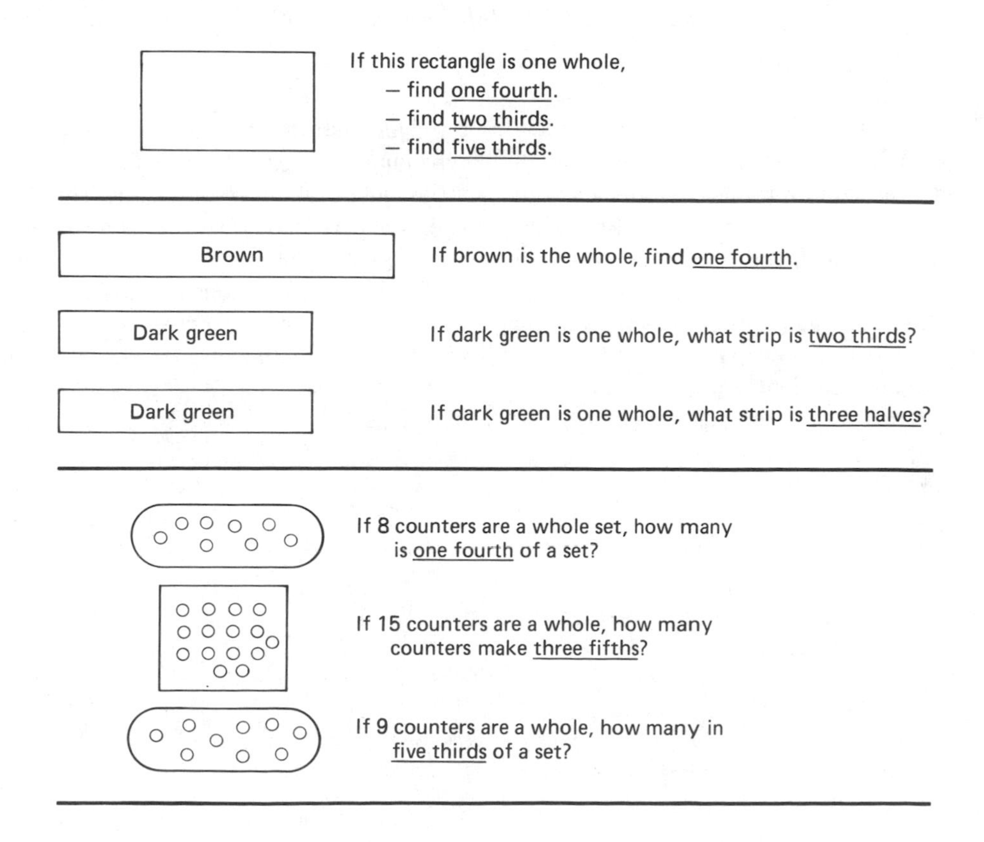

Figure 11.8
Given the WHOLE and FRACTION, find the PART.

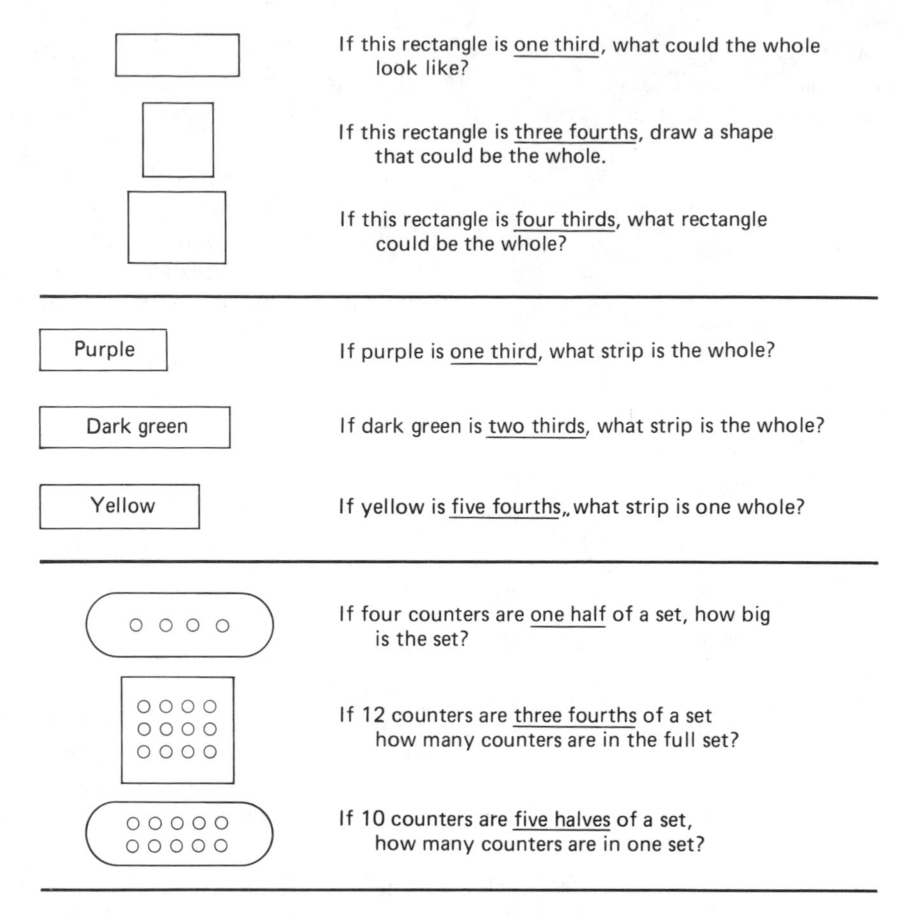

Figure 11.9
Given the PART and the FRACTION, find the WHOLE.

Sample Exercises

In Figures 11.8, 11.9, and 11.10 examples of each type of exercise are provided. Each figure includes examples with a region model (freely drawn rectangles), a length model, (Cuisenaire rods or fraction strips), and set models. It would be a good idea to work through these exercises before reading the next section. For the rectangle models, simply sketch a similar rectangle on paper. For the strip or rod models, use Cuisenaire rods or make some fraction strips. The colors used correspond to the actual rod colors. Lengths are not given in the figures so that you will not be tempted to use an adult-type numeric approach. If you do not have access to rods or strips, just draw lines on paper. The process you use with lines will correspond to what is done with rods.

Answers and explanations are in Figures 11.11, 11.12, and 11.13. The questions that ask for the fraction when given the whole and part require a lot of trial and error and can frustrate young students. Be sure that appropriate fractional parts are available for the region and length versions of these questions.

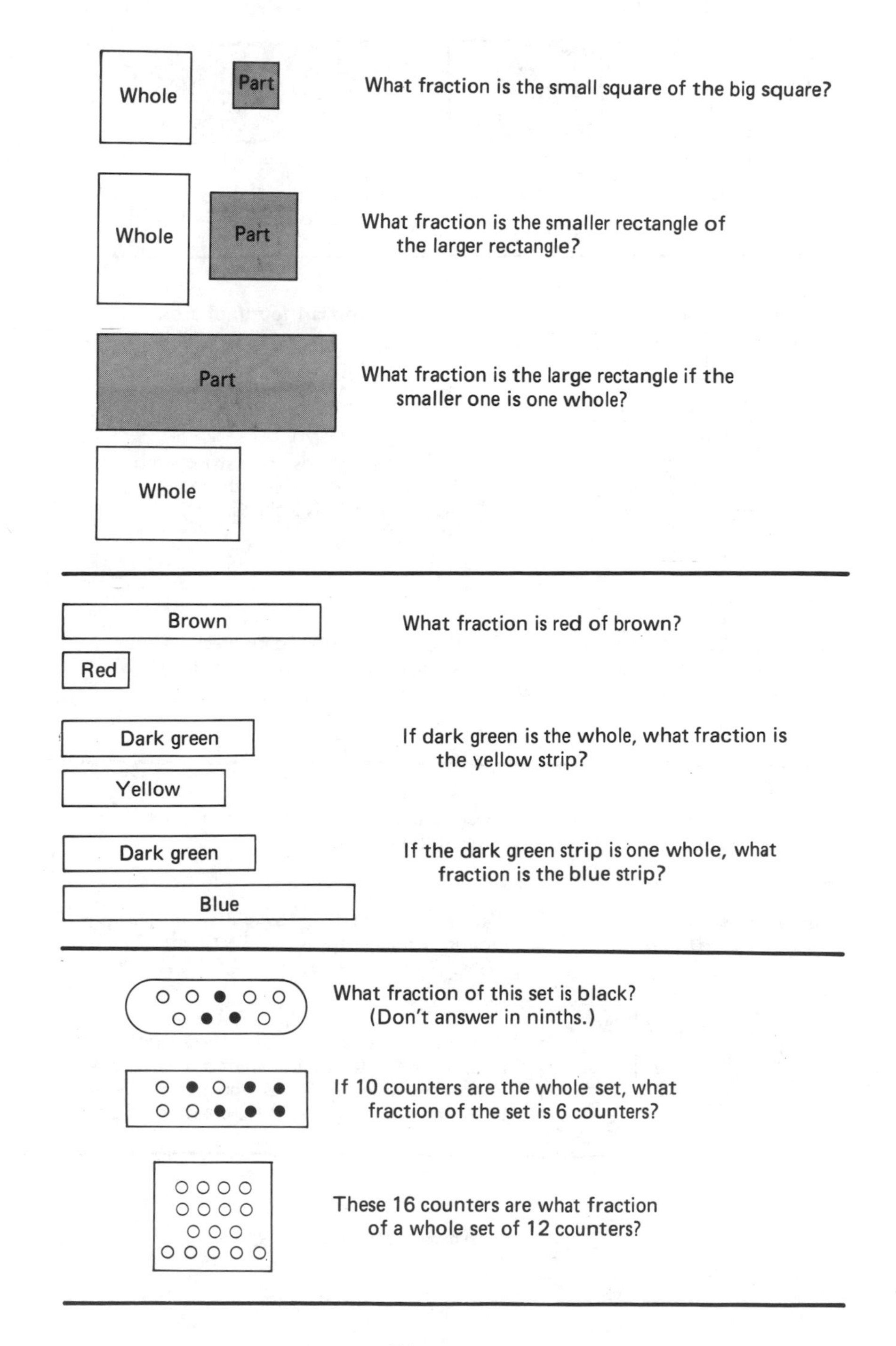

Figure 11.10
Given the WHOLE and the PART, find the FRACTION.

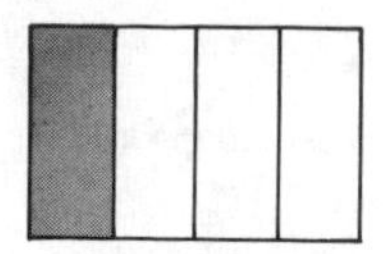

Cut in fourths.
Take 1 fourth.

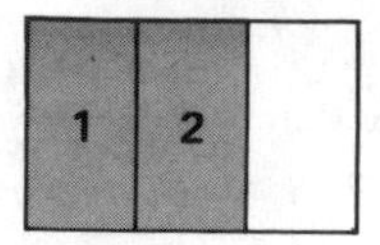

Cut in thirds.
Count 2 thirds.

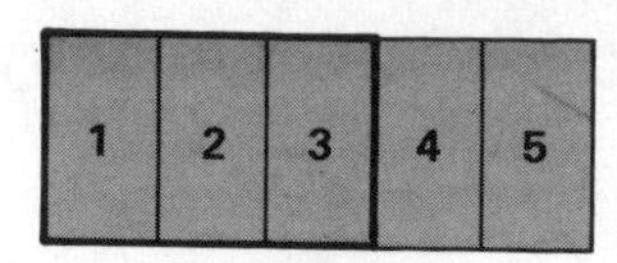

Cut in thirds.
Then count 5 thirds.
Need to add 2 more thirds.

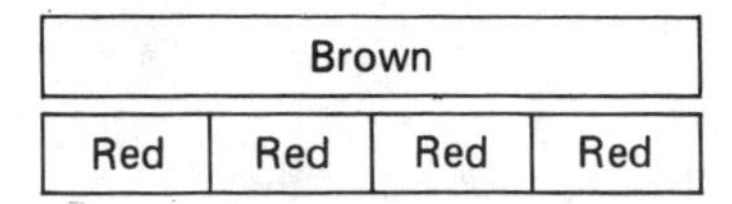

Each red strip is 1 fourth of brown.

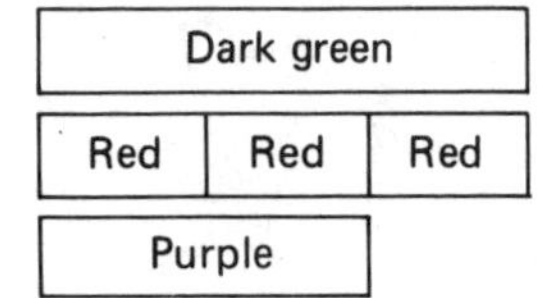

Looking for thirds. Red strips are thirds. Count two reds. That's the same as a purple. Purple is 2 thirds.

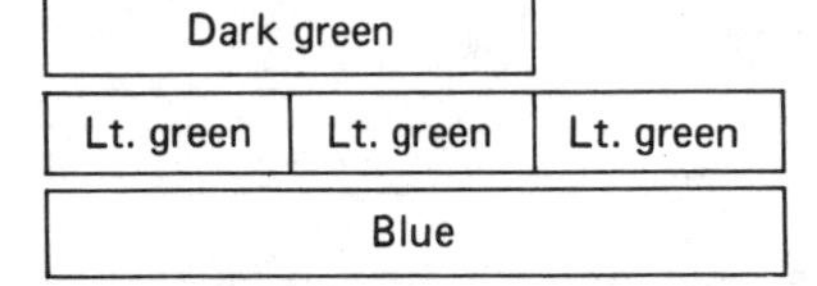

First find halves of dark green. Those are light green. Three halves is 1, 2, 3 light green—or one blue.

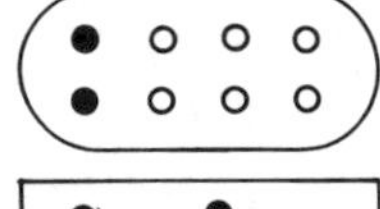

Separate the 8 into fourths or four piles. 1 fourth is two counters.

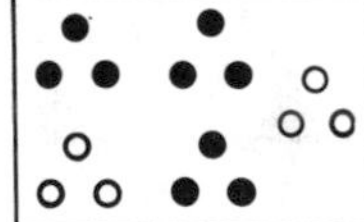

Want fifths, so put the 15 counters in five groups. Each groups is 1 fifth. Count 3 fifths. 3 fifths is 9 counters.

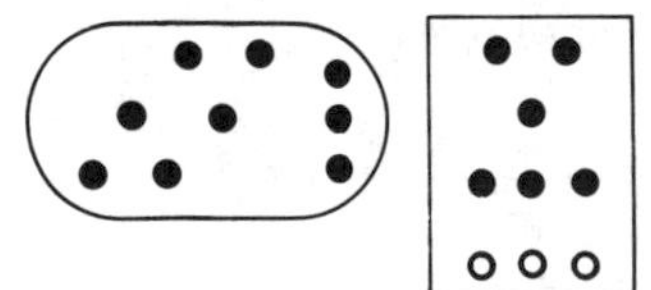

The nine counters is 3 thirds. Make three piles of 3. Each is a third. But want 5 thirds. Make a second whole of nine more counters. Count to 5 thirds, turning over three counters at a time.

Figure 11.11
Answers to Figure 11.8.

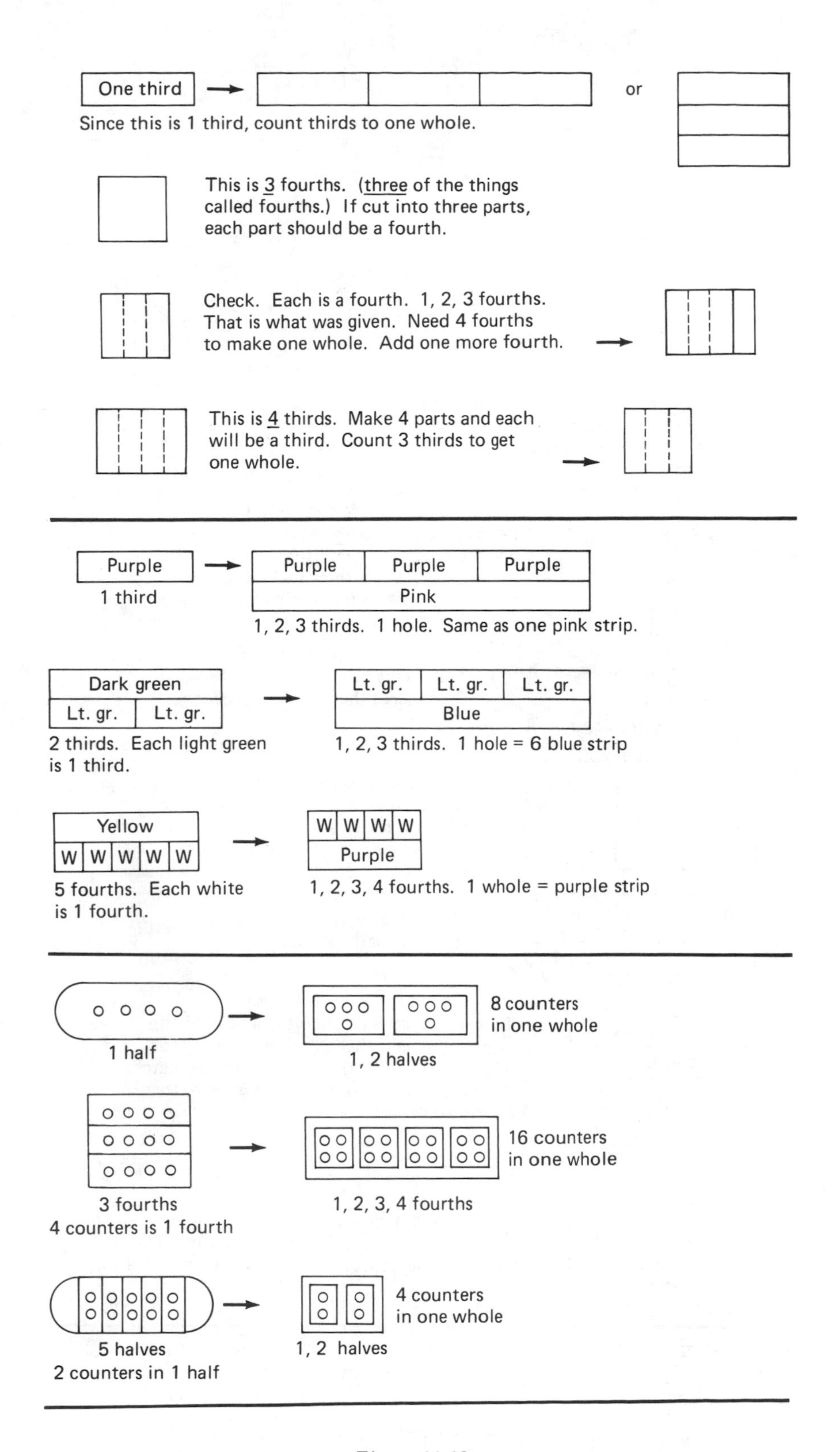

Figure 11.12
Answers to Figure 11.9.

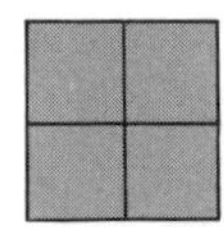

Four small parts fit in the whole. Each part is 1 fourth.

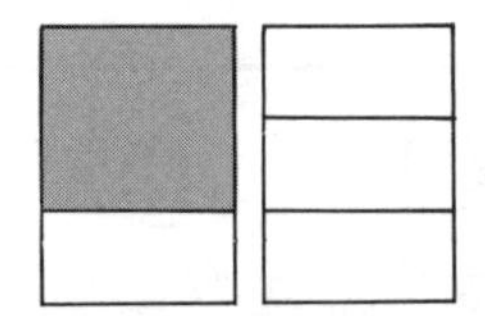

Put the whole on the part. Find pieces that cover both the part and the whole evenly. Three pieces work. Since three make one whole, each is 1 third. The part is 2 thirds.

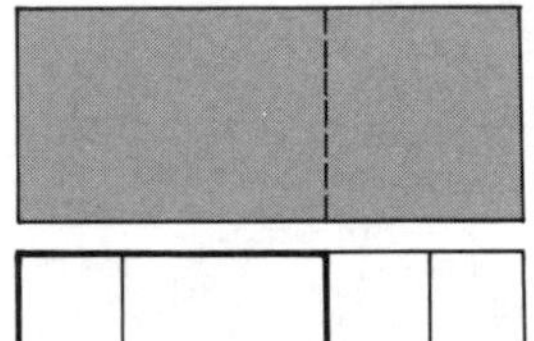

The part is bigger than the whole. Find pieces that cover both evenly.

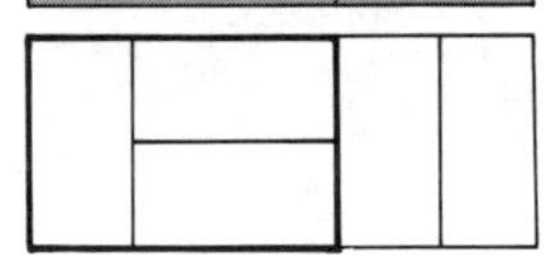

Five rectangles work. Three cover the whole, so each is a third. Five cover the part, so it must be 5 thirds. (smaller parts might have been used resulting in 10 sixths.)

Brown
Red | Red | Red | Red

Four reds (parts) make one brown (whole). Each red is 1 fourth.

Dark green
Lt. green | Lt. green
Red | Red | Red
W | W | W | W | W | W
Yellow

Dark green is the whole. It can be made up of two lt. greens (halves), three reds (thirds) or six whites (sixths). Only the whites match the yellow. Since each white is a sixth, that means yellow is 5 sixths.

Dark green
Lt. green | Lt. green | Lt. green
W | W | W | W | W | W | W | W | W
Blue

If dark green is the whole, then lt. greens are halves and whites are sixths. That means that blue is either 3 halves or 9 sixths.

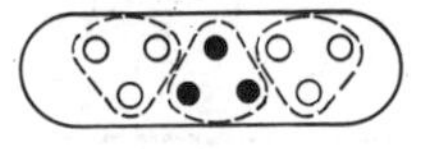

It can be 3 ninths. Or, if the nine are put in groups of 3, then the three black make 1 third.

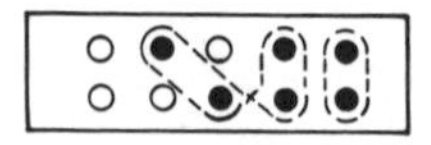

6 tenths is the easy answer. But 10 can be split into 5 sets of 2. Each set would be 1 fifth. Count 1 fifth, 2 fifths, 3 fifths.

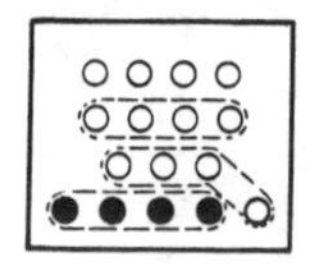

12 counters is one whole. If those 12 are separated into 3 groups of 4, each group is 1 third. Count by thirds. That's 1, 2, 3, 4 thirds in all. (similarly, 8 sixths or 16 twelfths.)

Figure 11.13
Answers to Figure 11.10.

Part and Whole Exercises in the Classroom

The exercises work much better with physical models as opposed to drawings. The models allow students to use a trial-and-error approach and to test their reasoning as they go. Also, younger children simply have limited ability to partition lines or regions into smaller equal parts.

Care must be taken to only ask questions for which there is an answer within the model. For example, if you were using fraction strips, you could ask students "If the blue strip (9) is the whole, what strip is two thirds?" The answer is the strip that is 6 units long, or dark green. You could not, however, ask students to find three fourths of the blue strip, since fourths would each be $2\frac{1}{4}$ units long and no strip has that length. With rectangular pieces of various sizes, you will likewise need to work out your questions in advance and be sure that they are answerable within the set.

Present each exercise to the full class to observe how different children are approaching the task. It may also be good to have students work in groups. After each task, have the students explain or justify their results.

A *unit fraction* is one that designates a single fractional part. In symbolic form, a unit fraction has a 1 in the numerator. Therefore, $\frac{1}{2}$, $\frac{1}{6}$, and $\frac{1}{12}$ are unit fractions and $\frac{3}{4}$, $\frac{7}{3}$, and $2\frac{1}{6}$ are nonunit fractions. Questions involving unit fractions are generally the easiest. The hardest questions usually involve fractions greater than 1. For example, "If fifteen chips are five thirds of one whole set, how many chips are in a whole?" The same question can be asked in terms of a mixed number (15 chips are the same as one and two thirds or $1\frac{2}{3}$). The mixed-number question is more difficult because it must first be translated to an improper fraction.

The part and whole activities can also be presented on worksheets or task cards, or they can be homework exercises. Refer to the specific models by colors and/or drawings and have the students record their answers in a similar manner. It is a good idea to have older students also include an explanation of how they arrived at their results.

Exercise Difficulties

Many students find these parts and wholes activities quite challenging and even confusing. There is absolutely no benefit in providing students with rules for solving them or even telling them what to do next. Rather, try to focus on the concepts of unit fractions and fractional parts. By way of example, the following hypothetical interchange illustrates the type of problem students may have. Sketch a rectangle on a piece of paper and follow along.

The student has a small rectangle with the accompanying question: "If this is four thirds of a box, what might one whole box look like?" The student begins by dividing the box into three parts and then is stumped and does not know what to do.

Teacher: How big is the rectangle?

Student: Four thirds.

Teacher: Does that mean it is four things or three things.

Student: Well, thirds means three. So I divided it into three.

Teacher: So, each of these are thirds?

Student: Yes, thirds.

Teacher: Let's count. (together) One third, two thirds, three thirds. How much is three thirds?

Student: One whole.

Teacher: But the box is not *three* thirds but *four* thirds. How many thirds should be in the box you started with?

Student: Four. It's *four* thirds. I have to make four parts. (Start over and draw the box divided into four parts.)

Teacher: Now count. These are what kind of pieces?

At this point the teacher wants the child to stick with the idea that the parts are thirds and he or she has four of them.

In this and every other example, a unit fractional part comes into play. Once this part is identified in the model, students can count unit parts up to one of the pieces (part or whole) that they had to begin with. This confirms they are correct. If the student counts four thirds, that will agree with what was given, namely that the original box was four thirds. From that point, counting three thirds to find the whole is trivial.

Try to avoid being the answer book for your students. Make students responsible for determining the validity of their own answers. In these exercises, the results can always be confirmed in terms of what is given. Students will learn that they can understand these ideas. There are no obscure rules. It makes sense!

When students do a series of exercises of one type, a pattern will begin to set in because each exercise is similar. When you mix exercises from the first two categories (whole to part and part to whole), students will experience more difficulty. The goal, however, is not to establish routines but to encourage reflective thought.

Comparing Fractions

MISCONCEPTIONS AND DIFFICULTIES

Relative Size Versus Part-to-Whole Relationships

The focus on fractional parts in the previous section is an important beginning since it helps students create and assign a number to the relationship between a part and a whole. However, the activities so far do not require students to reflect on the relative size of fractional parts. Fourths come from dividing the whole into four equal parts and sixths from dividing the whole into six equal parts. It is not until students begin to compare fourths and sixths that they even begin to think about the relative size of sixths and fourths or realize that sixths are smaller than fourths.

Whole Number Mind-set

There is a tremendously strong mind-set that children have about numbers that causes them difficulties with the relative size of fractions. Basically, larger numbers mean more. The tendency is to transfer this whole number concept to fractions incorrectly: Seven is more than four, so sevenths should be bigger than fourths. The inverse relationship between number of parts and size of parts cannot be "told," but must be a creation of each student's own thought process.

THINKING ABOUT WHICH IS MORE

Use Concepts Not Rules

You probably have learned rules or algorithms for comparing two fractions. The usual approaches are finding common denominators and cross multiplication. These rules can be effective in getting correct answers, but require no thought about the size of the fractions. This is especially true of the cross-multiplication procedure. If children are taught these rules before they have had the opportunity to think about the relative size of various fractions, there is little chance that they will develop any familiarity or number sense about fraction size. Comparison activities (which fraction is more?) can play a significant role in helping children develop concepts of relative size of fractions. But we want to keep in mind that it is reflective thought that is the goal and not an algorithmic method of choosing the correct answer.

Before going on to the next section, try the following exercise. Assume for a moment that you know nothing about equivalent fractions or common denominators or cross multiplying. Assume that you are a fourth- or fifth-grade student who was never taught these procedures. Now examine the pairs of fractions in Figure 11.14 and select the larger of each pair. Write down or explain one or more reasons for your choice in each case.

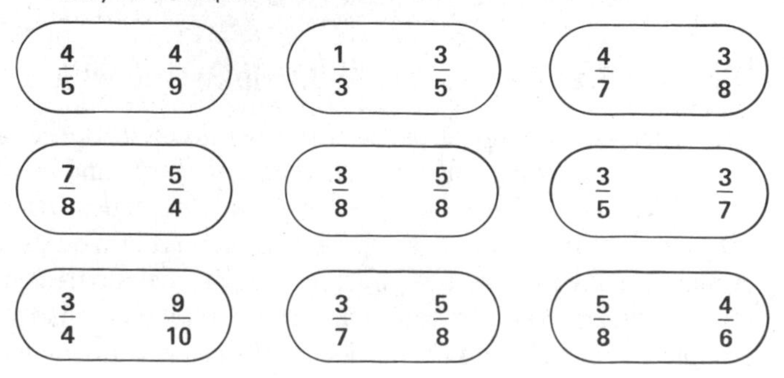

Figure 11.14
Comparing fractions using concepts.

Conceptual Thought Patterns for Comparison

More of the same size parts. When we compare $3/8$ and $5/8$, it is easy to think about having 3 of something and also 5 of the same thing. When the fractions are given orally, this is almost trivial. When given in written form, it is possible for children to choose $5/8$ simply because 5 is more than 3 and the other numbers are the same. Right choice, wrong reason.

Same number of parts but parts are different sizes. This is the case where the numerators are the same, as in $3/4$ and $3/7$. If a whole is divided into seven parts, they will certainly be smaller than if only divided into four parts. As discussed earlier, many children do not understand that more parts in the whole means the parts are smaller. They do understand that they will get less if sharing with more people. Many children will select $3/7$ as larger because 7 is more than 4 and the other numbers are the same. That rule yields correct choices when the parts are the same size, but it causes problems in this case.

More and less than an easy fraction or benchmark. The fraction pairs $3/7$ versus $5/8$ and $5/4$ versus $7/8$ do not lend themselves to either of the previous thought processes. In the first pair, $3/7$ is less than half of the number of sevenths needed to make a whole, and so $3/7$ is less than a half. Similarly, $5/8$ is more than a half. Therefore, $5/8$ is the larger fraction. The second pair is determined by noting that one fraction is less than 1 and the other is greater than 1. The *benchmark numbers* of $1/2$ and 1 are frequently useful for making size judgments with fractions.

Closer to an easy fraction or benchmark. Why is $9/10$ greater than $3/4$? Not because the 9 and 10 are big numbers, although you will find that to be a common student response. Each is one fractional part away from 1 whole, and tenths are smaller than fourths. Similarly, notice that $5/8$ is smaller than $4/6$, since it is only one eighth more than a half, while $4/6$ is a sixth more than a half.

How did your reasons for choosing fractions in Figure 11.14 compare to these ideas?

STUDENT ACTIVITIES

Classroom activities should help children develop informal ideas like those just explained for comparing fractions. However, the ideas should come from student experiences and discussions. To teach "the four ways to compare fractions" would be nearly as defeating as teaching cross multiplication.

Activity 11.6

Count fractional parts with the class. Ask questions that focus on the relative size of the parts. "Why does it take so many more counts to get to one with twelfths than with sixths? If we count by sevenths, when will we get past one half? What if we count by eighths? by fifths? If we count by eighths to one and a half, how many counts will that take? What if we do it by halves? tenths? hundredths?"

Activity 11.7

Have students use models to make comparisons. Designate or fix the model for the whole if necessary. Any two fractions can be compared as long as the pieces are available. If you designate, for example, that the pink strip (12) is the whole, then you can ask for comparisons involving halves, thirds, fourths, sixths, and twelfths. Try to include comparison pairs from each of the categories described before. For older children who have had some experience with fraction models, try pairs in which one or both of the fractions cannot be modeled directly. For example, compare $3/5$ with $6/8$ even though there are no fifths in your pie-pieces kit.

Activity 11.8

Have a "Why We Know It Is More" discussion. Arrange the class in cooperative groups or pairs of students. Provide them with one or more models for fractions. Give the class a pair of fractions to compare. The task is to find as many good explanations for their choice as possible within an allotted time. Explanations can be written down and then discussed as a full class. The same exercise is a very good homework assignment.

Activity 11.9

Play "Catch My Goof." Present a pair of fractions and make a choice of which is more along with a reason for the choice. For example, "I think that three fourths is more than ten twelfths because there is only one more fourth left to get to one." Or, "I think that five eighths is more than three sevenths because the five and the eight are both bigger than the three and the seven." The first example used faulty reasoning and produced a wrong result, while the second was a correct result with incorrect reasons. During the game use correct results and reasons as well as incorrect results and/or reasons. The game is a good way to get students to verbalize the conceptual reasoning that has been described and communicates that it is not rules but reasoning that is important. (What do you think you should do in this activity if no one catches an error you make?)

Activity 11.10

Have students put four or five fractions in order from least to most. In this way, a variety of methods for making comparisons can be included within the same exercise. As with all conceptual fraction activities, limit the denominators to reasonable numbers. There are very few reasons to consider fractions with denominators greater than 12.

A word of caution is implicit in the following situation. Mark is offered the choice of a third of a pizza or a half of a pizza. Since he is hungry and likes pizza, he chooses the half. His friend Jane gets a third of a pizza but ends up with more than Mark. How can that be? Figure 11.15 illustrates how

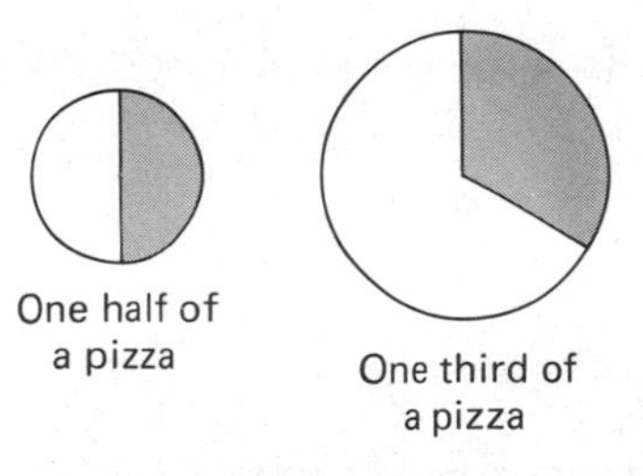

Figure 11.15
The "Pizza Fallacy".

Mark got misdirected in his choice. The point of the "pizza fallacy" is that whenever two or more fractions are discussed in the same context, the correct assumption (the one Mark made in choosing a half of the pizza) is that the fractions are all parts of the same whole.

Comparisons with any model can only be made if both fractions are parts of the same whole. For example, two thirds of a light green strip cannot be compared to two fifths of an orange strip.

OTHER METHODS OF COMPARISON

Certainly not all fractions can be compared by reliance on the conceptual approaches that have been discussed. Most adults would be hard-pressed to compare $2/3$ and $3/5$ without some other methods.

When simple methods fail, a more sophisticated and usually more complex method is required. The most common approach is to convert the fraction to some other form. When the concept and skills related to equivalent fractions are well in place, one or both of the fractions can be rewritten so that both fractions have the same denominator. Another idea is to translate the fraction to a different notation, specifically decimals or percents. Many of these decimal and percent equivalents can and should become second nature and require no computation. In the case of $2/3$ and $3/5$, the equivalents are 66+% and 60%, respectively. More will be said in later chapters about helping students make meaningful connections between decimals, percents, and fractions.

Even when common denominator methods are developed, they should be a last resort. Mental methods are frequently much quicker, and the symbolic methods detract from thinking in a conceptual way. There is little or no justification for asking students to compare $7/17$ with $9/23$.

Equivalent Fraction Concepts

CONCEPTS VERSUS RULES

Question: How do you know that $4/6 = 2/3$?

Why is $\frac{4}{6} = \frac{2}{3}$?

(a)

They are the same because you can reduce $4/6$ and get $2/3$.

(b)

Because if you have a set of six things and you take four of them, that would be 4/6. But you can make the six into groups of two. So then there would be three groups, and the four would be two groups out of the three groups. That means it's $2/3$.

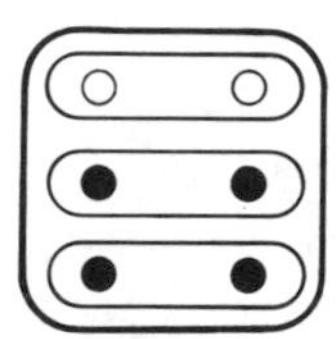

(c)

If you start with $2/3$ you can multiply the top and the bottom numbers by 2, and that will give you $4/6$, so they are equal.

(d)

If you a had a square cut into three parts and you shaded two, that would be $2/3$ shaded. If you cut all three of these parts in half, that would be four parts shaded and six parts in all. That's $2/3$, and it would be the same amount.

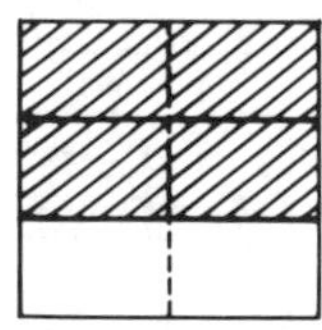

Figure 11.16
Why is $4/6 = 2/3$?

The answers in Figure 11.16, while all correct, provide clear examples of the distinction between conceptual knowledge and procedural knowledge. Responses (b) and (d) are very conceptual, although not efficient. The procedural responses, (a) and (c), are quite efficient, yet no conceptual knowledge is indicated. All students should eventually know how to write an equivalent fraction for a given fraction. At the same time, the rules should never be taught or used until the students understand what the result means.

The concept: Two fractions are *equivalent* if they are representations for the same amount.

The rule: To get an equivalent fraction, multiply (or divide) the top and bottom numbers by the same nonzero number.

The rule or algorithm for equivalent fractions carries no intuitive connection with the concept. As a result, students can easily learn and use the rule in exercises such as "List the first four equivalent fractions for $3/5$," without any idea of how the fractions in the list are related. It becomes an exercise in multiplication. A developmental approach suggests that students have a firm grasp of the concept and be led to see that the algorithm is a meaningful and efficient way to find equivalent fractions.

FINDING DIFFERENT NAMES FOR FRACTIONS

The general approach to a conceptual understanding of equivalent fractions is to have students use models to generate different names for models of fractions.

Area or Region Models

Examples of equivalent fraction representations using area models are illustrated in Figure 11.17.

Activity 11.11

Using the same models that students have, draw the outline of several fractions on paper and duplicate them. For example, if the model is rectangles, you might draw an outline (no subdivisions) of a rectangle for $2/3$, $1/2$, and perhaps $5/4$. Have children try to fill the outlines with unit fraction pieces to determine as many simple fraction names for the regions as possible. In class discussion, it may be appropriate to see if students can go beyond the actual models that they have. For example, if the model has no tenths, it would be interesting to ask what other names could be generated if tenths were available. An easier question involves pieces that can be derived from existing pieces. "You found out that five fourths and ten eighths and fifteen twelfths are all the same. What if we had some sixteenth pieces. Could we cover this same region with those? How many? How can you decide?"

Activity 11.12

Paper folding effectively models the equivalent fraction concept. Have students fold a sheet of paper into halves or thirds. Unfold and color a fraction of the paper. Write the fraction. Now refold and fold one more time. It is fun to discuss, *before opening,* how many sections will be in the whole sheet and how many will be colored. Open and discuss what fraction names can now be given to the shaded region. Is it still the same? Why? Repeat until the paper cannot be folded any longer.

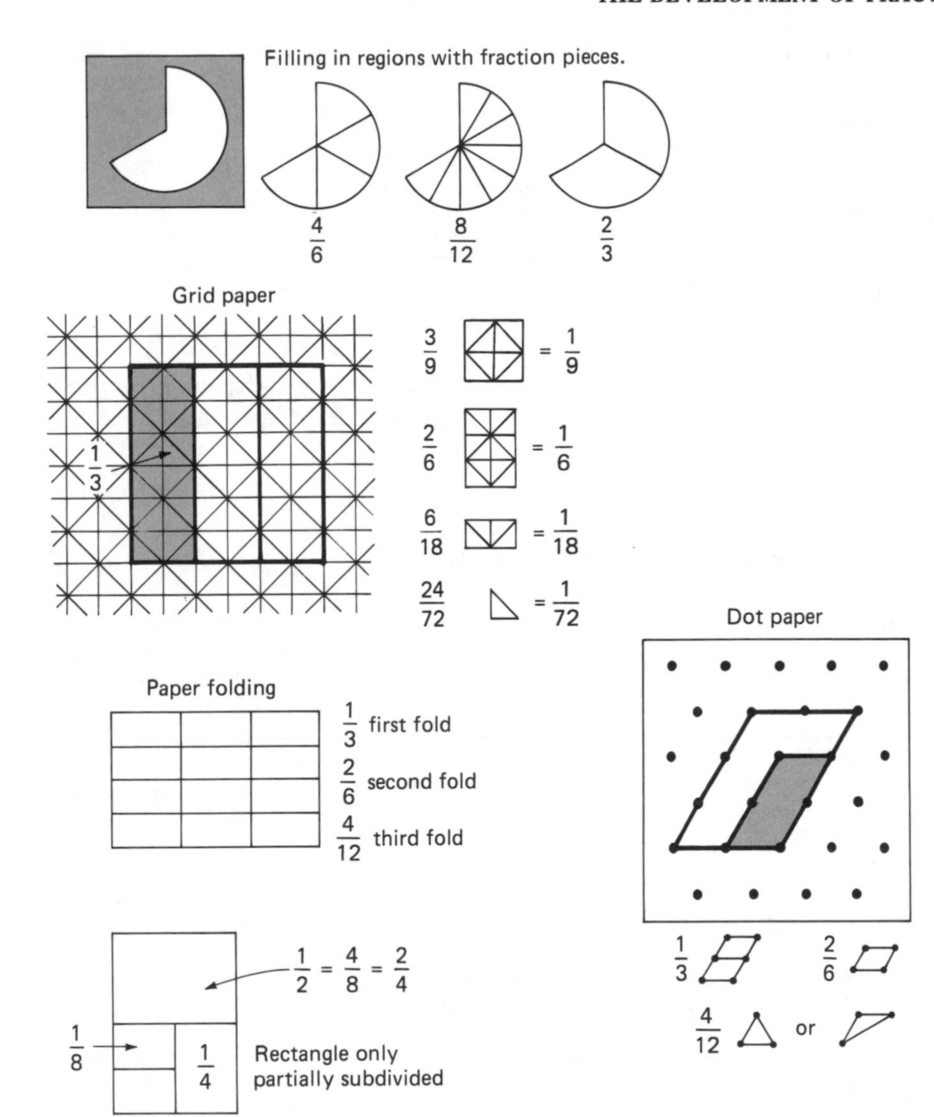

Figure 11.17
Area models for equivalent fractions.

Activity 11.13

Use grid paper or dot paper so that regions can easily be subdivided into many smaller parts. Have students draw a model for a whole and shade in some fraction that can be determined using the lines or dots on the paper. Now see how many different names for the shaded part they can find with the aid of the smaller regions in the drawing. (See Figure 11.17.) Make a transparency of a grid or dot pattern and do this exercise with the full class. Teams or individuals can take turns explaining different names for the shaded part.

While subdivided regions illustrate why there are multiple names for one fractional part, students should learn that existence of the subdivisions is not required. Half of a rectangle is still two fourths even if no subdivisions are present and even if the other half is divided into three parts. Work toward this understanding by drawing models for fractions and then erasing the subdivision lines. "Is this two thirds still four sixths?"

Length Models

Equivalent fractions are modeled with length models in much the same way as area models. One fraction is modeled, and then different lengths are used to determine other fraction names. Some examples are shown in Figure 11.18.

Set Models

The general concept of equivalent fractions is the same with set models as with length and area models, although there are more limitations to how a particular set can be partitioned. For example, if ⅔ is modeled with 8 out of 12 counters, then

Fraction strips

Pink				One whole
Blue				
W W W	W W W	W W W	W W W	
Lt. gr	Lt. gr	Lt. gr	Lt. gr	

Blue = $\frac{9}{12} = \frac{3}{4}$

Folding paper strips

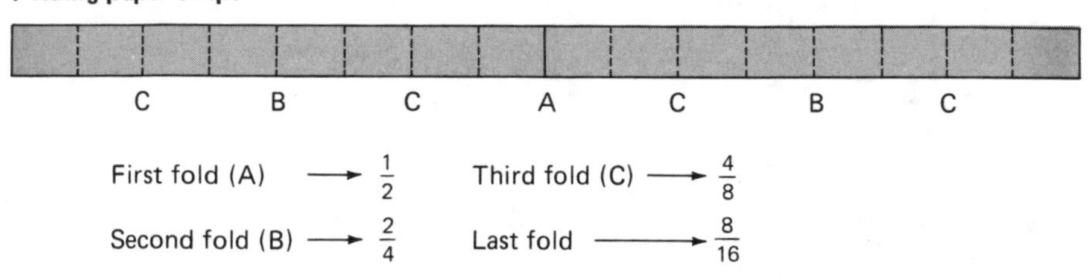

Number lines

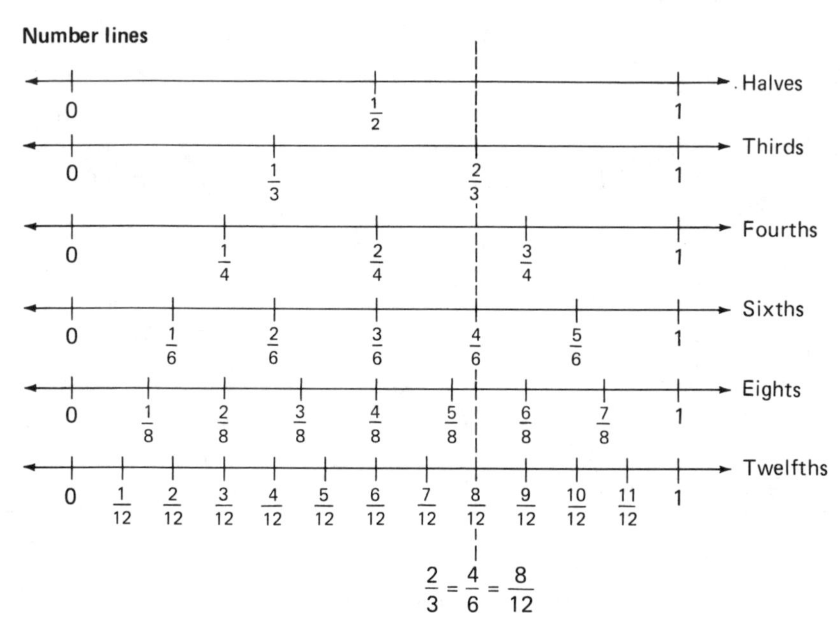

Figure 11.18
Length models for equivalent fractions.

that particular representation shows ⅔ to also be 4⁄6 and 8⁄12. It cannot be seen as 6⁄9 or 10⁄15. As shown in Figure 11.19, a given number of counters in two colors can be arranged in different arrays or subgroups to illustrate equivalent fractions.

Activity 11.14

Have students set out a specific number of counters in two colors. For example, 32 counters with 24 red and the rest yellow. This set will be the whole. Then have them find as many names as they can for each color by arranging the counters into different subgroups. Drawings can be made using X's and O's to produce a written record of the activity.

A Transitional Activity

Each of the following four equations is typical of equivalent fraction exercises found in textbooks. Notice the differences among the examples.

$$\frac{5}{3} = \frac{\square}{6} \qquad \frac{2}{3} = \frac{6}{\square} \qquad \frac{8}{12} = \frac{\square}{3} \qquad \frac{12}{8} = \frac{3}{\square}$$

An excellent exercise is to have students complete equations of this type by using a model to determine the missing number. Some care must be taken to select exercises that can be solved with the model being used. For example, most pie-piece sets do not have ninths. Counters or sets are a model that can always be used. Make students responsible for justify-

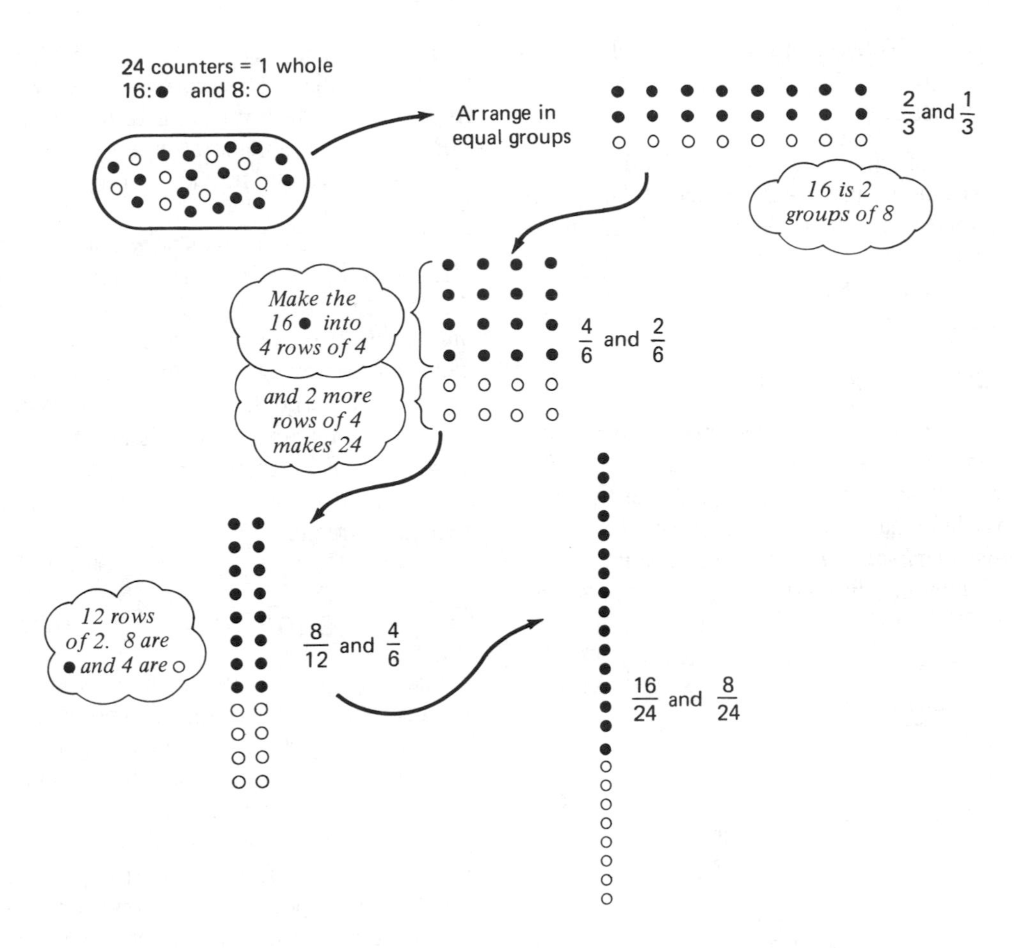

Figure 11.19
Set models for equivalent fractions.

ing their results with the use of models or drawings. Later, when symbolic rules are developed, these same exercises will be completed in a more algorithmic but probably less meaningful manner.

DEVELOPING AN EQUIVALENT FRACTION ALGORITHM

Rectangle Slicing

While there are many possible ways to model the procedure of multiplying top and bottom numbers by the same number, the most commonly used approach is to "slice" a rectangle in two directions.

Activity 11.15

Give students paper with rows of squares about 3 cm on each side. Have them shade the same fraction in several different squares using vertical subdividing lines. Next, slice each rectangle horizontally into different fractional parts as shown in Figure 11.20. Help students focus on the products involved by having them write the top and bottom numbers in

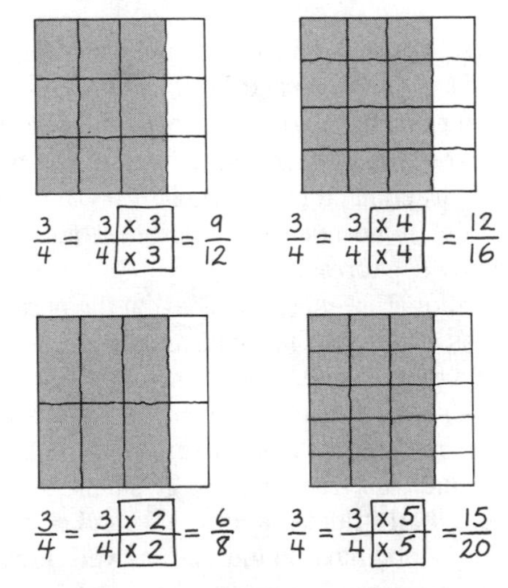

Figure 11.20
A model for the equivalent fraction algorithm.

the fraction as a product. Notice that for each model, the top and bottom numbers will always have a common factor. (Paper folding provides a similar result.)

Examine examples of equivalent fractions that have been generated with other models and see if the rule of multiplying top and bottom numbers by the same number holds there also. If the rule is correct, how can $6/8$ and $9/12$ be equivalent? What about fractions like $2\frac{1}{4}$? How could it be demonstrated that $9/4$ is the same as $27/12$?

Writing Fractions in Simplest Terms

The multiplication rule for equivalent fractions produces fractions with larger denominators. To write a fraction in *simplest terms* means to write it so that numerator and denominator have no common whole number factors. (Some texts use *lowest terms* instead of *simplest terms.)* One meaningful approach to this task of finding simplest terms is to reverse the earlier process, as illustrated in Figure 11.21.

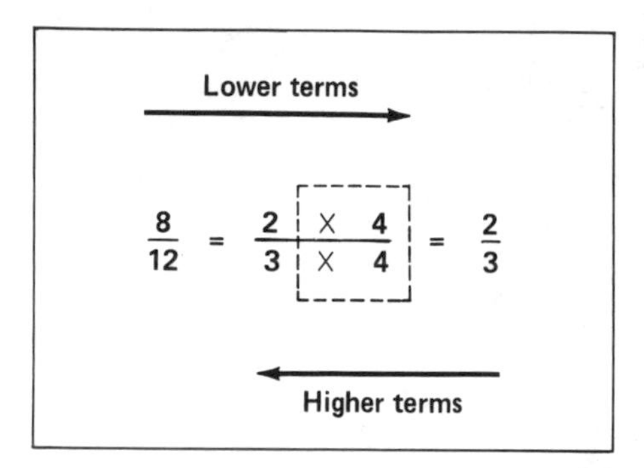

Figure 11.21
Using the equivalent fraction algorithm to write fractions in simplest terms.

Of course, finding and eliminating a common factor is the same as dividing both top and bottom by the same number. The search for a common-factor approach keeps the process of writing an equivalent fraction to one rule: top and bottom numbers of a fraction can be multiplied by the same nonzero number. There is no need for a different rule for rewriting fractions in lowest terms.

Two additional notes: (1) Notice that the phrase "reducing fractions" was not used. This unfortunate terminology implies making a fraction smaller and is rarely used any more in textbooks. (2) Many teachers seem to believe that fraction answers are incorrect if not in simplest or lowest terms. This total assault on fractions not in simplest terms is also unfortunate. When students add $1/6 + 1/2$ and get $4/6$, they have added correctly and have found the answer. Rewriting $4/6$ as $2/3$ is a totally separate issue.

The Multiply-by-1 Method

Many junior high textbooks use a strictly symbolic approach to equivalent fractions. It is based on the multiplicative identity property of rational numbers that says any number multiplied by 1 remains unchanged. The number 1 is the identity element for multiplication. Therefore, $3/4 = 3/4 \times 1 = 3/4 \times 2/2 = 6/8$. Any fraction of the form n/n can be used as the identity element. Furthermore, the numerator and denominator of the identity element can also be fractions. In this way $6/12 = 6/12 \times [1/6 \div 1/6] = 1/2$.

This explanation relies on an understanding of the multiplicative identity property, which most students in grades 4 to 6 do not fully appreciate. It also relies on the procedure for multiplying two fractions. Finally, the argument uses solely deductive reasoning based on an axiom of the rational number system. It does not lend itself to intuitive modeling. A reasonable conclusion is to delay this important explanation until at least seventh or eighth grade in an appropriate prealgebra context and not as a method or a rationale for producing equivalent fractions.

Other Meanings of Fractions

In this chapter the only meaning given to fractions is that of an expression of the relationship between a part and a whole. Unit fractional parts are based on the partitioning of the whole into equal-sized parts (denoted by the denominator), and various fractions are then taken as multiples of these unit parts (denoted by the numerator). Research generally supports the idea that this is the best way to approach fractions from a developmental perspective. However, the fraction notation does have other meanings which are introduced to children at about the sixth grade.

FRACTIONS ARE EXPRESSIONS OF DIVISION

If three people were to share 12 candies, the number that each would get can be expressed by $12 \div 3$. If three people were to share two pizzas, the amount that each would get can be expressed similarly by $2 \div 3$; that is, two things divided three ways. In the pizza example, each person would receive two thirds of a pizza. So, $2/3$ and $2 \div 3$ are both expressions for the same idea: two things divided by three. Similarly, in the candy example, $12/3$ expresses the number each will receive just as well as $12 \div 3$.

These examples are partition situations. In measurement contexts, the fraction also can be used to represent division. If a man walks 3 miles per hour, how many hours will it take to walk 7 miles? He walks $7 \div 3$ or $7/3$ miles.

Put simply, a fraction a/b is another way of writing $a \div b$.

Students understandably find this meaning of fractions unusual. First, it is different from the meaning that has been carefully developed. Second, fractions are commonly thought of as amounts or parts of wholes, not as operations. Similarly, expressions such as $7 \div 3$ are thought of as operations (things to be done), not numbers. In fact, however, 4, $2 + 2$, $12 \div 3$, and $8/2$ are all symbolic expressions for the same number. $12 \div 3$ is not the question and 4 the answer, but both are expressions for 4. Likewise, 2 ÷ 3 and 2/3 are both expressions for the quantity two thirds. Do you find this a little hard to swallow? So do children. Here for the first time in seven or eight years of school you are telling students that a symbol can represent two different ideas. This is a relatively sophisticated idea. Students should be told quite openly that this is a new and different way to think about a fraction. Real-world problems for division should be used. They should be written in both fraction and whole number division notation, and discussed.

FRACTIONS ARE EXPRESSIONS OF RATIOS

Ratios, like part-to-whole fraction concepts, are expressions of a relationship between two quantities. With the part-to-whole concept of fraction, one of those quantities is always fixed and designated as a unit. (Recall the "pizza fallacy.") With ratios there is no fixed unit. If 20 of 30 students in a class are girls, the ratio of boys to girls can be expressed as 10 to 20, as 10:20, or as 10/20. The colon notation is still used, but is much less common than the fraction notation. The fraction notation is also useful in dealing with equivalent ratios or proportions. Two ratios are proportional if the fractions that express them are equivalent. While 6 of 9 counters and 8 of 12 counters each model two thirds, they are not the same amounts but equivalent ratios. (Chapter 14 contains a complete discussion of the difficult concepts of ratio and proportion.)

For Discussion and Exploration

1. A common error that children make is to write 3/5 for the fraction represented in Figure 11.22. Why do you think that they do this? In this chapter, the notation of fractional

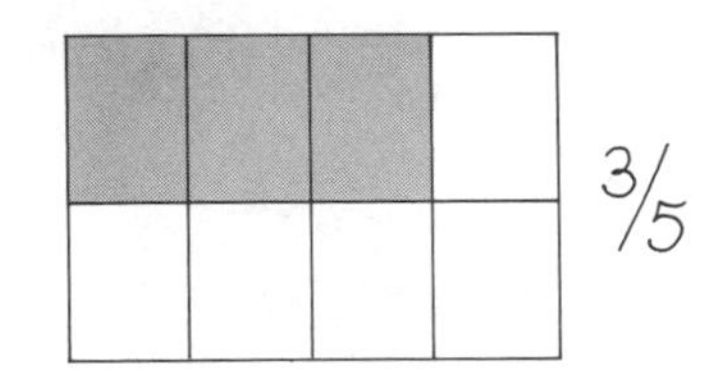

Figure 11.22
A common mistake.

parts and counting by unit fractions was introduced before any symbols. How could this help avoid the type of thinking that is involved in this common error?

2. Use a length model and make up part-and-whole questions for each of the following cases:
   - Given a part and a nonunit fraction less than one, find the whole.
   - Given a part and a nonunit fraction greater than one, find the whole.
   - Given a whole and a nonunit fraction less than one, find the part.
   - Given a whole and a nonunit fraction greater than one, find the part.
   - Given a part and a whole, find the fraction. Make up your example so that the fraction is not a unit fraction.

   Try your questions with a friend. Then change all of the questions so that they are in terms of sets. With sets, be sure that a unit fraction is never a single counter. That is, if the question is about fourths, then use whole sets of size 8 or 12 or more.
3. Make up pairs of fractions that can be compared (largest chosen) without using an algorithm, that is, without cross products or common denominators. Try to find pairs that are close together. Which of the four ideas suggested for comparing fractions did your examples use? Have some friends select the largest from your examples. What strategies did they use?
4. Work with some children. Here are some places to begin to explore their ideas:
   - Use a model they have not seen and try some of the part-and-whole questions. Begin with easy examples involving unit fractions.
   - See if they can use a model to explain why 2 1/3 is 7/3. (Or see if they can write 7/3 as a mixed number and explain their procedure.)
   - Try some comparison questions. Encourage children to produce more than one explanation.
   - Try to find out if children really believe that 2/3 and 8/12 are the same amount. What is their thinking behind this?

   From these initial explorations, you should be able to decide where to proceed for some additional work with your group of children.
5. Read the chapter "Fractions with Cookies" in Marilyn Burns's *A collection of math lessons for grades 3 to 6* (1987). Try these activities with a group of children. While suggested for grade 3, the ideas are easily adapted to grades higher or lower. What is different about this approach to fraction development from the one presented in this chapter?

Suggested Readings

Behr, M. J., Post, T. R., & Wachsmuth, I. (1986). Estimation and children's concept of rational number size. In H. L. Schoen (Ed.), *Estimation and mental computation.* Reston, VA: National Council of Teachers of Mathematics.

Behr, M. J., Wachsmuth, I., & Post, T. R. (1985). Construct a sum: A measure of children's understanding of fraction size. *Journal for Research in Mathematics Education, 16,* 120–131.

Bennett, A. B., Jr., & Davidson, P. S. (1973). *Fraction bars: Teacher's guide.* Fort Collins, CO: Scott Resources, Inc.

Bezuk, N., & Cramer, K. (1989). Teaching about fractions: What, when, and how? In P. R. Trafton (Ed.), *New directions for elementary school mathematics*. Reston, VA: National Council of Teachers of Mathematics.

Delaney, K. (1984). Fraction games. *Mathematics Teaching, 107,* 8–11.

Ellerbruch, W., & Payne, J. N. (1978). Teaching sequence from initial fraction concepts through the addition of unlike fractions. In M. N. Suydam (Ed.), *Developing computational skills.* Reston, VA: National Council of Teachers of Mathematics.

Hiebert, J., & Tonnessen, L. H. (1978). Development of the fraction concept in two physical contexts: An exploratory investigation. *Journal for Research in Mathematics Education, 9,* 374–378.

Holden, L. (1986). *Fraction factory.* Palo Alto, CA: Creative Publications.

Hope, J. A., & Owens, D. T. (1987). An analysis of the difficulty of learning fractions. *Focus on Learning Problems in Mathematics, 9*(4), 25–40.

Kroll, D. L., & Yabe, T. (1987). A Japanese educator's perspective on teaching mathematics in the elementary school. *Arithmetic Teacher, 35*(2), 36–43.

Liebeck, P. (1985). Are fractions numbers? *Mathematics Teaching, 111,* 32–34.

Post, T. R., Behr, M. J., & Lesh, R. (1982). Interpretations of rational number concepts. In L. Silvey (Ed.), *Mathematics for the middle grades (5–9).* Reston, VA: National Council of Teachers of Mathematics.

Post, T., & Cramer, K. (1987). Children's strategies in ordering rational numbers. *Arithmetic Teacher, 35*(2), 33–35.

Pothier, Y., & Sawada, D. (1983). Partitioning: The emergence of rational number ideas in young children. *Journal for Research in Mathematics Education, 14,* 307–317.

Silver, E. A. (1983). Probing young adults' thinking about rational numbers. *Focus on Learning Problems in Mathematics, 5,* 105–117.

Wearne-Hiebert, D., & Hiebert, J. (1983). Junior high school students' understanding of fractions. *School Science and Mathematics, 83,* 96–106.

Williams, D. E. (Ed.) (1984). Rational numbers. Special issue of *Arithmetic Teacher, 31*(6).

Zullie, M. E. (1975). *Fractions with pattern blocks.* Palo Alto, CA: Creative Publications.

CHAPTER 12

COMPUTATION WITH FRACTIONS

Concepts Versus Algorithms

THE DANGER OF RULES

In the short term, the rules for fraction computation can be relatively simple to teach. Students can become quite proficient at finding common denominators during a chapter on adding and subtracting simple fractions. Multiplying fractions is such an easy procedure that many experts suggest it should be taught first. The only requirements are third-grade multiplication skills. Division, following the invert-and-multiply rule, is nearly as easy. Fraction rules can easily become the focus of rote instruction and produce artificial feelings of accomplishment on the Friday quiz.

Focusing attention on fraction rules and answer getting has two significant dangers: First, none of the rules help students think in any way about the meanings of the operations or why they work. Students practicing such rules may very well be doing rote symbol pushing in its purest sense. Second, the mastery observed in the short term is quickly lost. When taken as a group of rules, the procedures governing fraction computation become similar and confusing. "Do I need a common denominator, or can you just add the bottoms?" "Do you invert the second number or the first?"

ALGORITHMS SHOULD BUILD ON CONCEPTS

Without a firm understanding of fraction concepts, the development of computational algorithms for fractions can quickly become superficial, rule-oriented, and confusing. That is why this chapter was written separately from the preceding chapter—to place more emphasis on fraction concepts themselves and to not confuse computation with numeration as an objective.

In the sections which follow, each operation is first explored with the use of models and meanings of the operations derived from whole numbers. This exploration phase is important so that students can connect fraction operations with existing familiar ideas. These beginning explorations are then structured to allow for the development of efficient procedures. You are strongly encouraged to use models and work examples as you read along.

Addition and Subtraction

CONCEPT EXPLORATION

Have students add two fractions using a fraction model. The results should come completely from the use of the model even if some of the students have been exposed to symbolic rules and the idea of a common denominator. For example, suggest that students use their model to find the sum of $\frac{3}{4}$ and $\frac{1}{3}$. Exercises such as this should be explored with area, length, and set models. You must be careful that the problems can actually be worked with the materials the students have available. With circular regions, the assumption is that the circle will always represent the whole. However, as seen in the last chapter, many models allow different representations for the whole. When using these more flexible models (for example, strips or counters) the first thing that must be done is to determine a whole that permits both fractions to be modeled. (Recall the "pizza fallacy" from Figure 11.15.) Figure 12.1 illustrates how beginning addition tasks might be approached using three different models.

Subtraction of two fractions with models is a similar process, as shown in Figure 12.2. Notice that it is sometimes possible to find the sum or difference of two fractions without subdividing either one. Attention is instead focused on the size of a leftover part. When area models are used for addition and subtraction, common denominators are frequently not involved at all. On the other hand, selecting a set size or length that permits modeling of two fractions is mathematically the same as finding a common denominator. (Why?)

When making up addition and subtraction exercises to be done with models, do not be afraid of "difficult" problems. Include problems with unlike denominators, fractions greater than one, and mixed numbers. For subtraction, explore problems such as $3\frac{1}{8} - 1\frac{1}{4}$, where a "trade" of wholes for eighths might occur. At this stage, avoid directing students with a completely formulated method for arriving at a solution. The problem-solving approach along with the use of the models will help students develop relationships. Discuss different solution processes used by different groups. It can be useful to have different groups use different models for their processes.

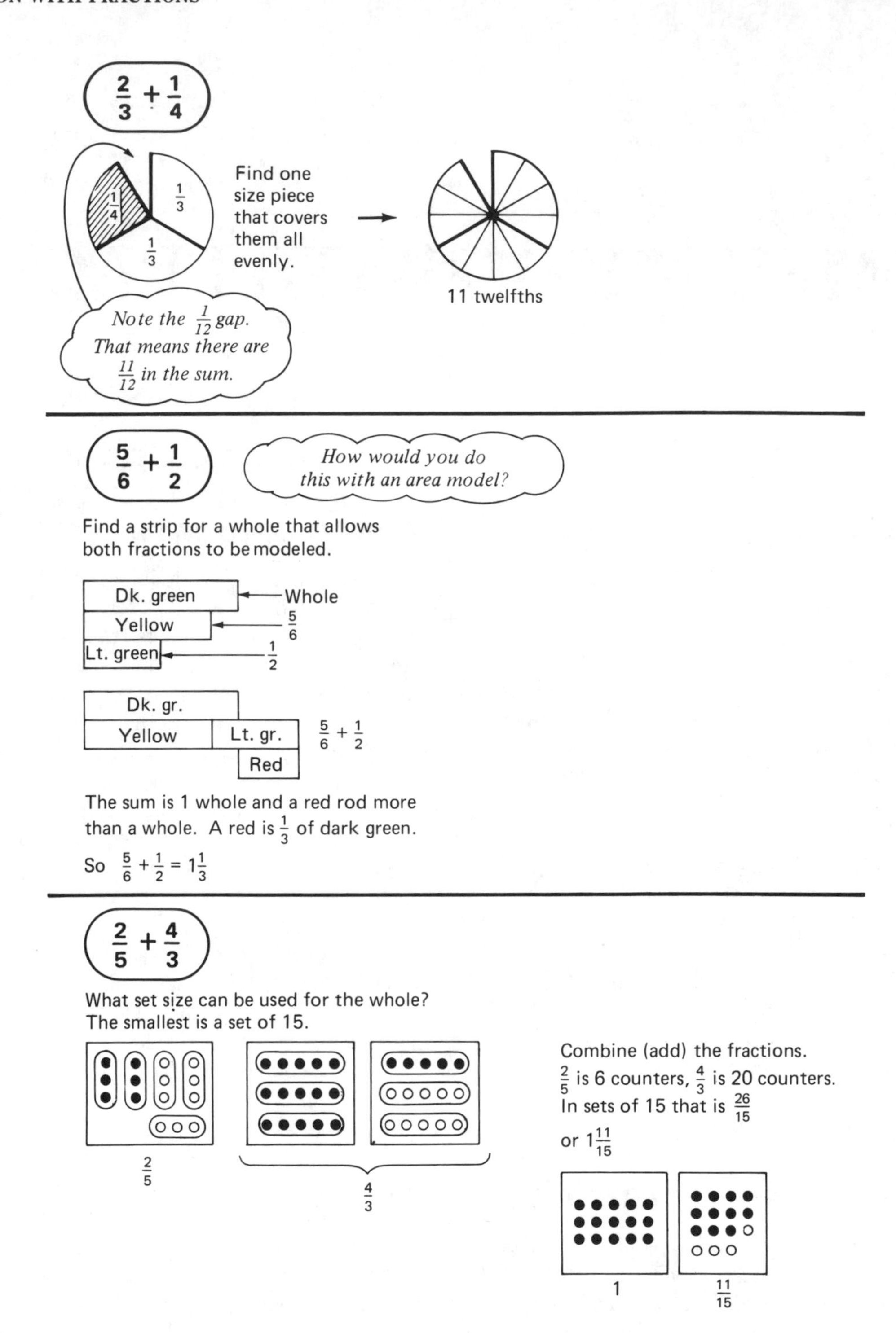

Figure 12.1
Using models to add fractions.

DEVELOPING THE ALGORITHM

Like Denominators

If students have a good grasp of fractional parts and have counted fractional parts with models, it should take no longer than 5 minutes before they can add or subtract two fractions with the same denominator. "Let's add apples. How much is 3 apples and 7 apples?" (10 apples) "If that is true, how much is 3 fifths and 7 fifths?" (10 fifths) "How do you know? What is 3 halves and 7 halves? Now tell me about 3 eighths and 7 eighths? Why is this just like adding 3 apples and 8 apples or 3 cars and 8 cars?" Help students make the connection between 3 *apples* and 3 *eighths* or 3 of whatever fractional part

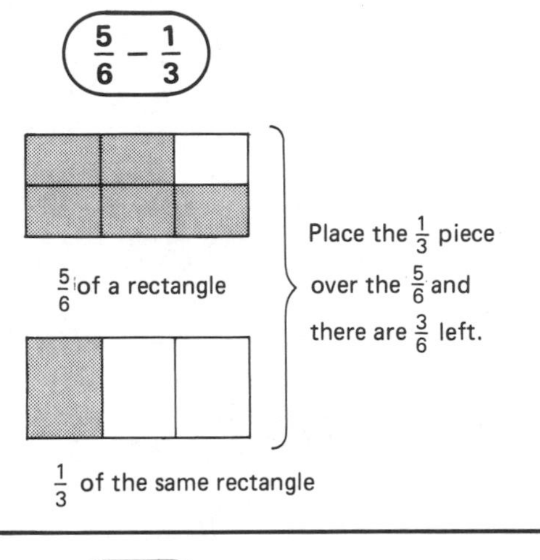

$\frac{7}{8} - \frac{1}{2}$

Find a rod that can be broken into eighths and halves. Brown.

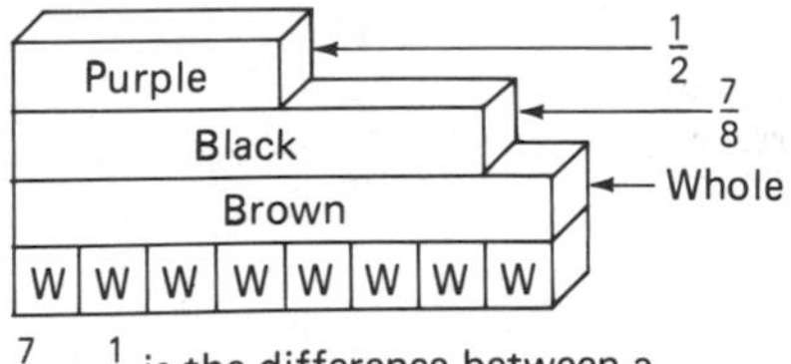

$\frac{7}{8} - \frac{1}{2}$ is the difference between a purple and a black rod. That is three whites or $\frac{3}{8}$. So $\frac{7}{8} - \frac{1}{2} = \frac{3}{8}$.

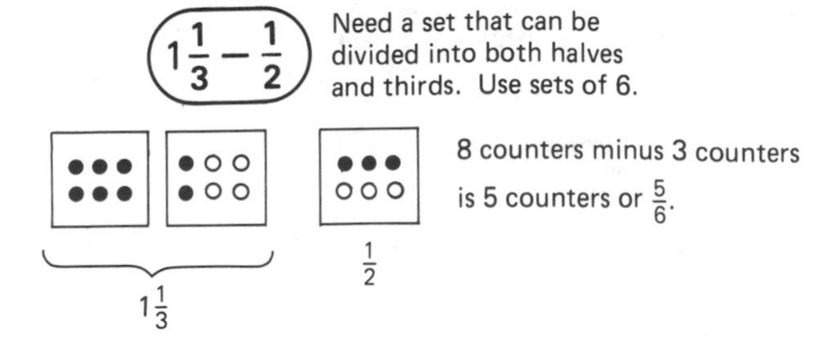

Figure 12.2
Using models to subtract.

you name. Since 3 fourths means three of the things called fourths, if you have 3 of them and 7 of them that will be 10 fourths, just like 3 apples and 7 apples is 10 apples. While this is an apparently trivial exercise, it lays the groundwork for the common denominator algorithm.

Unlike Denominators

After a discussion of adding like fractions, ask how you could, for example, add three eighths and one fourth. The key question at this point is, "How can we change this question into one that is just like the easy ones where the parts are the same?" For the current example, it is relatively easy to see that fourths could be changed into eighths. Have students use models to show the original problem and also the converted problem. The main idea is to see that $3/8 + 1/4$ is exactly the same problem as $3/8 + 2/8$.

Next try some examples where both fractions need to be changed. For example, $2/3 + 1/4$. Be careful that the common denominator can actually be modeled with the materials that the students have. Again, focus attention on *rewriting the problem* in a form that is like adding apples and apples, where the parts of both fractions are the same. Students must fully understand that the new form of the problem is actually the same problem. This can and should be demonstrated with models. However, if your students express any doubt about the equivalence of the two problems ("Is $11/12$ really the answer to $2/3 + 1/4$?"), then that should be a clue that the concept of equivalent fractions is not well understood.

As a result of modeling and rewriting fractions to make the problems easy, students should come to understand that the process of getting a common denominator is really one of looking for a way to change the *statement* of the problem without changing the problem. After getting a common denominator, it should be immediately obvious that adding the numerators produces the correct answer. These ideas are illustrated in Figure 12.3.

Subtraction of two simple fractions follows exactly the same approach. If the denominators are the same, it is like subtracting apples from apples. When the denominators are different, the problem should be rewritten to make it an easy one.

Common Multiple Practice

Many students have trouble with finding common denominators because they are not able to quickly produce common multiples of the denominators. This is a skill that can be practiced. It also depends on having a good command of the basic facts for multiplication. Here are a few activities aimed at the skill of finding least common multiples or common denominators.

Activity 12.1

Practice "Running Through the Multiples." For this oral drill, give students a number between 2 and 16 (likely denominators) and have them list the multiples in order. At first, writing the multiples may be helpful. Work toward the skill of doing this exercise orally and quickly. Students should be able to list the multiples to about 50 with ease.

Activity 12.2

Put two potential denominators on the board. Have students begin to list the multiples of the larger number as quickly as they can as before. This time, however, they should yell "stop" when a number is also a multiple of the smaller number. For example, for 12 and 9, students should say, "twelve, twenty-four, thirty-six, STOP." They should then tell what

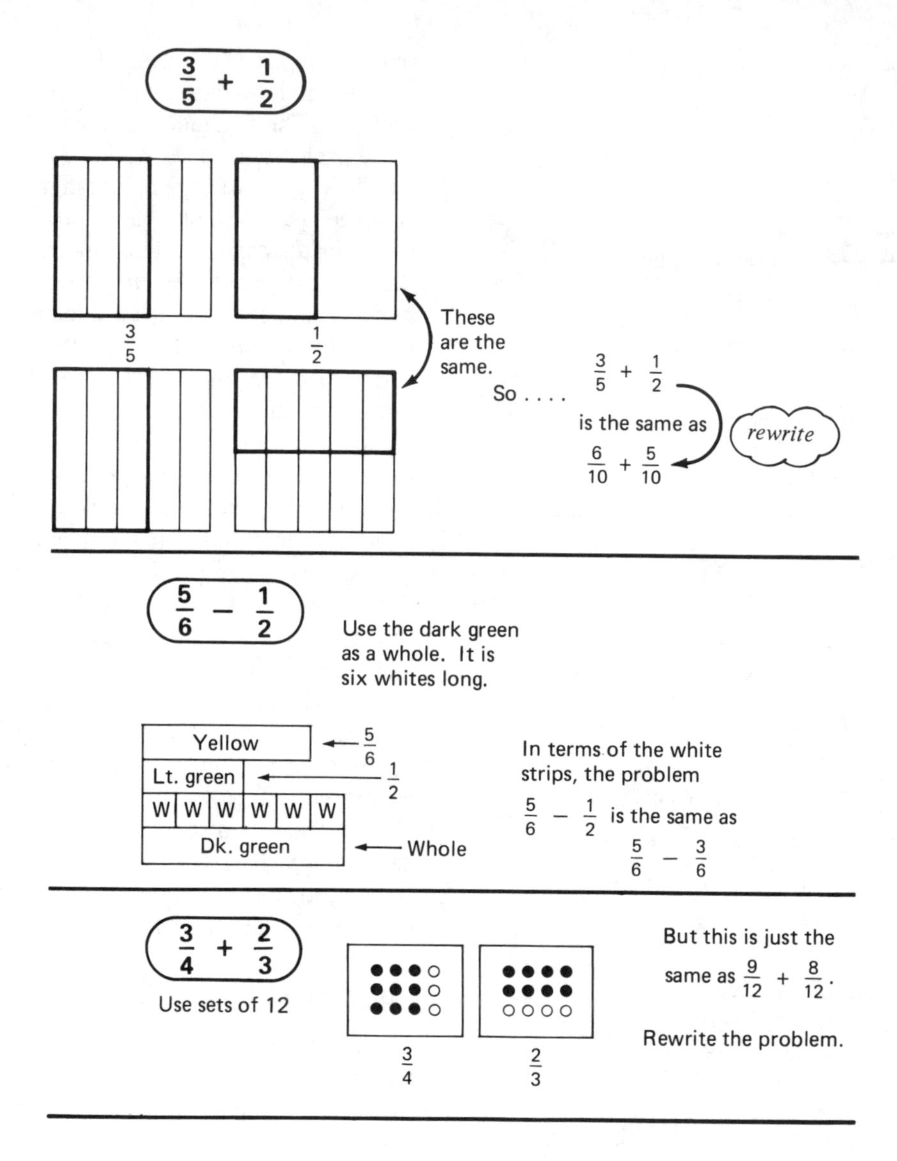

Figure 12.3
Rewriting addition and subtraction problems.

two products will make the common multiple. In this example, 12 × 3 and 9 × 4.

Activity 12.3

Make flash cards with pairs of numbers that are potential denominators. Most should be less than 16 as before. For each card, students try to give the least common multiple (Figure 12.4). Be sure to include pairs that are relatively prime, such as 9 and 5, those where one is a multiple of the other, such as 2 and 8, and those with a common multiple, such as 8 and 12.

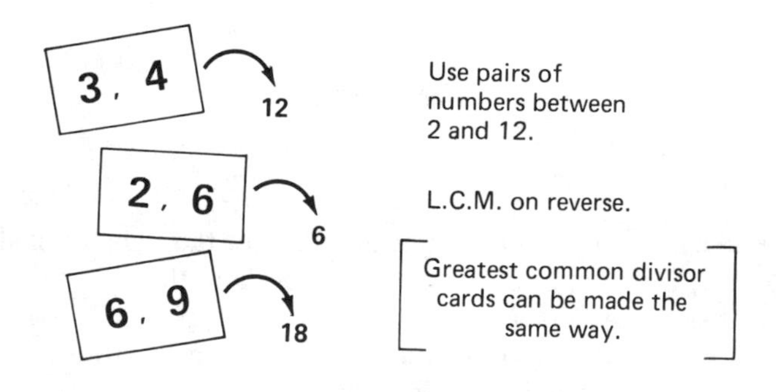

Figure 12.4
Least common-multiple flash cards.

MIXED NUMBERS

When students do addition and subtraction of mixed numbers, they tend to make errors in converting fractions to whole numbers and whole numbers to fractions. This is especially true in subtraction, as illustrated in Figure 12.5. Students making this error are using a base ten place-value idea instead of changing a whole for an equivalent set of fractional parts. The following ideas may help with these difficulties.

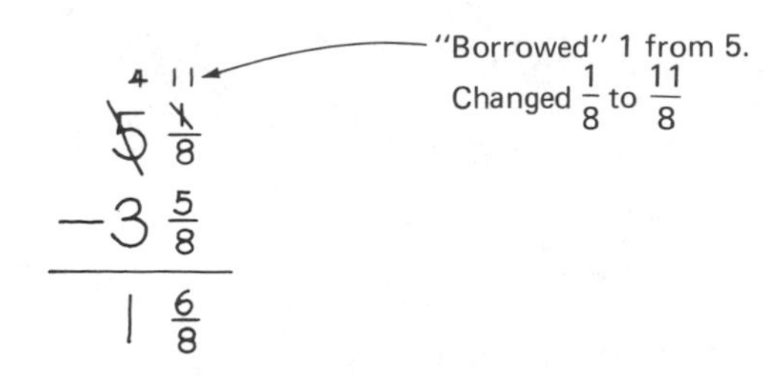

Figure 12.5
A common subtraction error.

Activity 12.4

Do fraction trading activities. Use a two-sided mat and an area model for fractions such as pie pieces. Add fractional parts such as fourths, one at a time, to the right-hand side of the mat. When there are four fourths, discuss the idea of

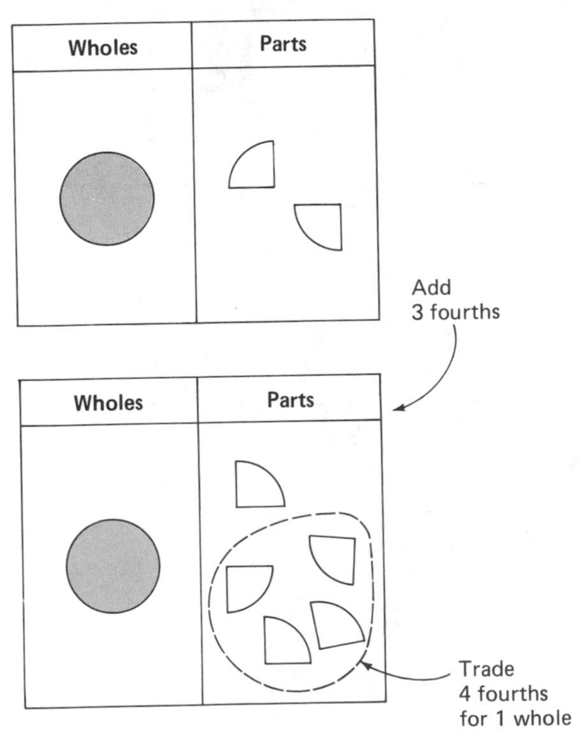

Figure 12.6
A trading activity with fractional parts.

trading for a whole, as shown in Figure 12.6. In a similar manner, start with a model for three wholes on the left side and begin to take off one fourth at a time. Help students see how this is both like and different from trading ones for tens and tens for ones. Remember to use the terminology of "trade" and not the old-fashioned "borrow." Later, allow the numbers and type of part to change at each turn. Roll two dice. Use an ordinary die to tell how many parts to put on or take off of the board. A second die indicates halves, thirds, fourths, sixths, eighths, or twelfths. With the two dice, trading games are played the same as with tens and ones. The first player to accumulate three wholes (forward game) or to clear the board (backward game) is the winner.

Activity 12.5

Have students model addition and subtraction problems involving mixed numbers on the two-sided mat used for fraction trading activities as described earlier. As they do the problems, have them record each step as they go. Put the written fractions on a two-place chart. In essence, make the activity as much like place value as possible, but help students focus on the differences.

Multiplication

CONCEPT EXPLORATION

Recall with students that for whole numbers, 3×5 means 3 sets with 5 in each set, three sets of five each. Have students use fraction models to determine products with the first factor a whole number such as $3 \times 4/5$ or $4 \times 1\,2/3$. The meaning of the two factors remains the same. If you have 3 sets of 4 apples each, that would be 12 apples. Likewise, 3 sets of 4 fifths each is 12 fifths: $3 \times 4/5 = 12/5$.

Next, still using models, try examples with a fraction as the first factor and a whole number as the second factor. The meanings remain the same: $2/3 \times 4$ means $2/3$ of a set of 4. Try modeling a given product in both orders. For example, try $1/6 \times 4$ and $4 \times 1/6$. The answers are the same, but the process is quite different.

Finally, see if students can explain the meaning of a fraction times a fraction. For example what does $2/3 \times 3/4$ mean? Once again the meanings remain the same: $2/3$ of a set of $3/4$. That is, with models, one would first make a set of three fourths and then determine what is two thirds of that set. Notice how easy the last example is compared to $1/4 \times 2/3$. (Why?) When the first factor is greater than one, children may have extra difficulty. For example, try solving $3/2 \times 1/4$ or $4/3 \times 1\,2/3$ using different models and the basic concepts of multiplication. Several examples of how multiplication problems might be modeled are illustrated in Figure 12.7.

Frequently students are perplexed because the answer in multiplication gets smaller instead of larger. When and why a product of two fractions is smaller than either factor is a good question to explore at this preliminary stage.

DEVELOPING THE ALGORITHM

Both Factors Less Than One

While not the only way to generate the multiplication algorithm for fractions, an area model where squares or rectangles are used for the whole is one of the most common. The examples

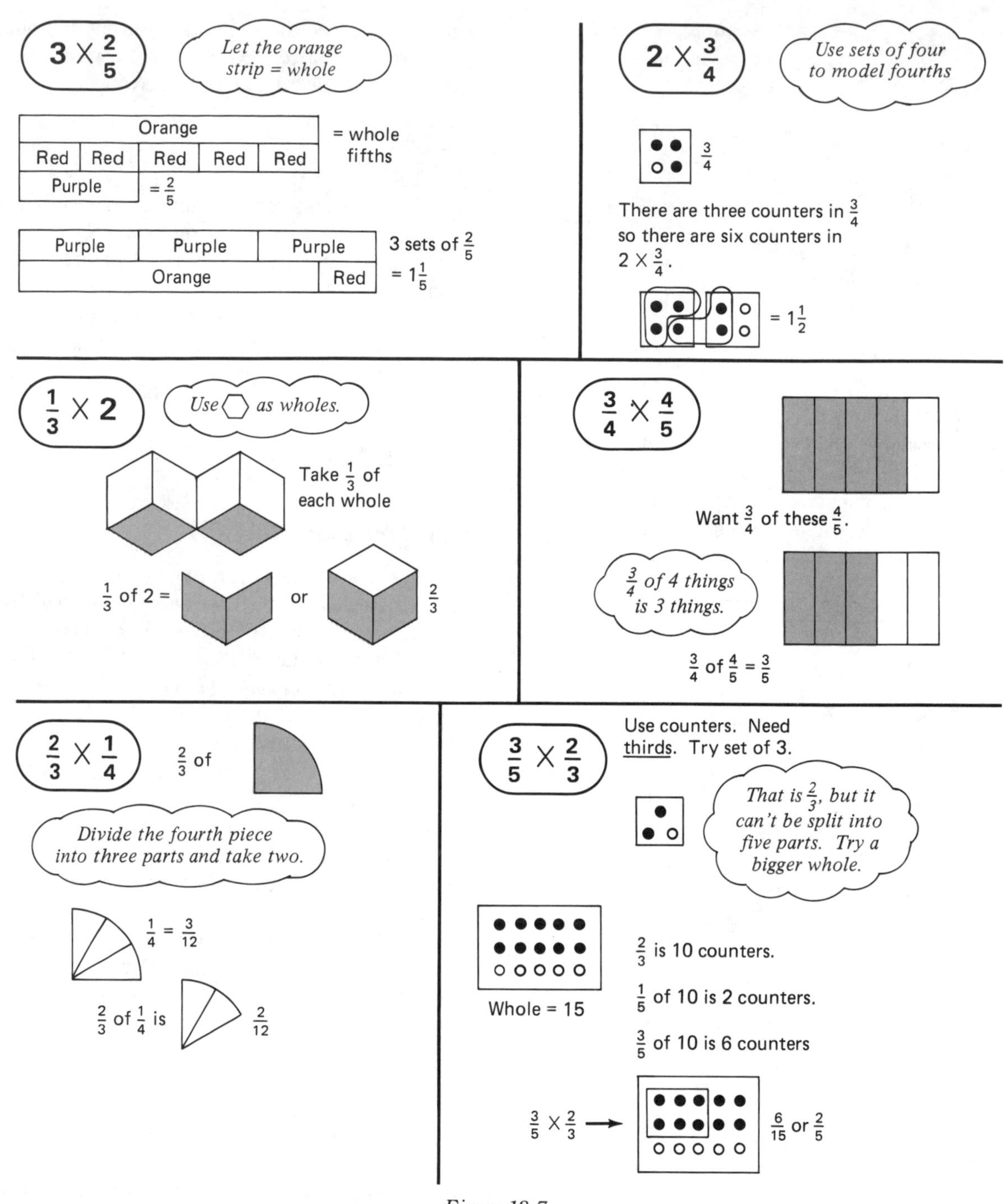

Figure 12.7
Exploring multiplication of fractions.

which are easiest to understand are those where both factors are fractions less than one.

Figure 12.8 shows how the square is first used to model the product and then is used to determine the numerator and denominator of the product in written form. By drawing the lines for each fraction in opposite directions, both the *whole* (the square) and the *product* are arrays made of the same size fractional parts. The numerator of the product is the *number* of parts in the product, the number of rows times the number of columns. The denominator of the product is the *name* of those parts or the number in one whole. Help students see that these two products are also the products of the numerators and the denominators, respectively. After students model a series of fraction products, the rule of multiplying top and bottom numbers will become obvious.

One Factor a Whole Number

For products where the first factor is a whole number, such as $3 \times 4/5$, the intuitive meaning of multiplication is quite easy, as noted earlier. Three sets of 4 fifths each is 12 fifths: $3 \times$

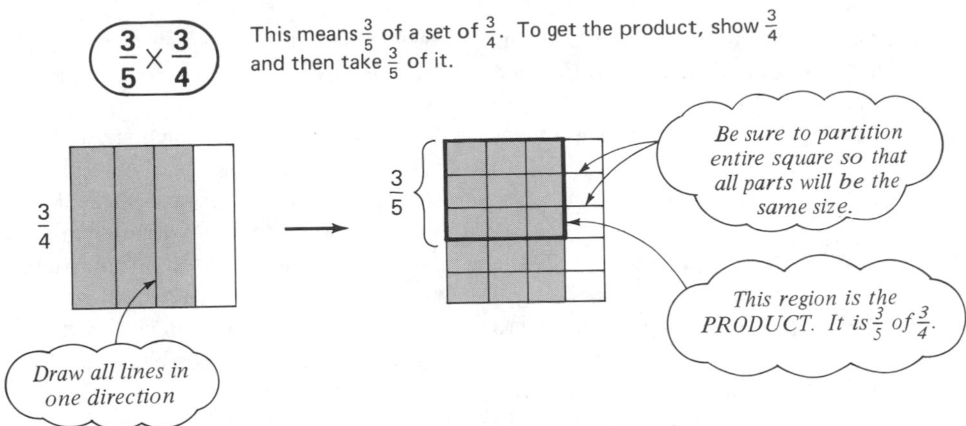

- There are three rows and three columns in the product or 3 × 3 parts.
- The whole is now five rows and four columns so there are 5 × 4 parts in the whole.

$$\text{PRODUCT} = \frac{3}{5} \times \frac{3}{4} = \frac{\boxed{\text{Number}}\text{ of parts in product}}{\boxed{\text{Name}}\text{ of parts}} = \frac{3 \times 3}{5 \times 4} = \frac{9}{20}$$

Figure 12.8
Development of the algorithm for multiplication.

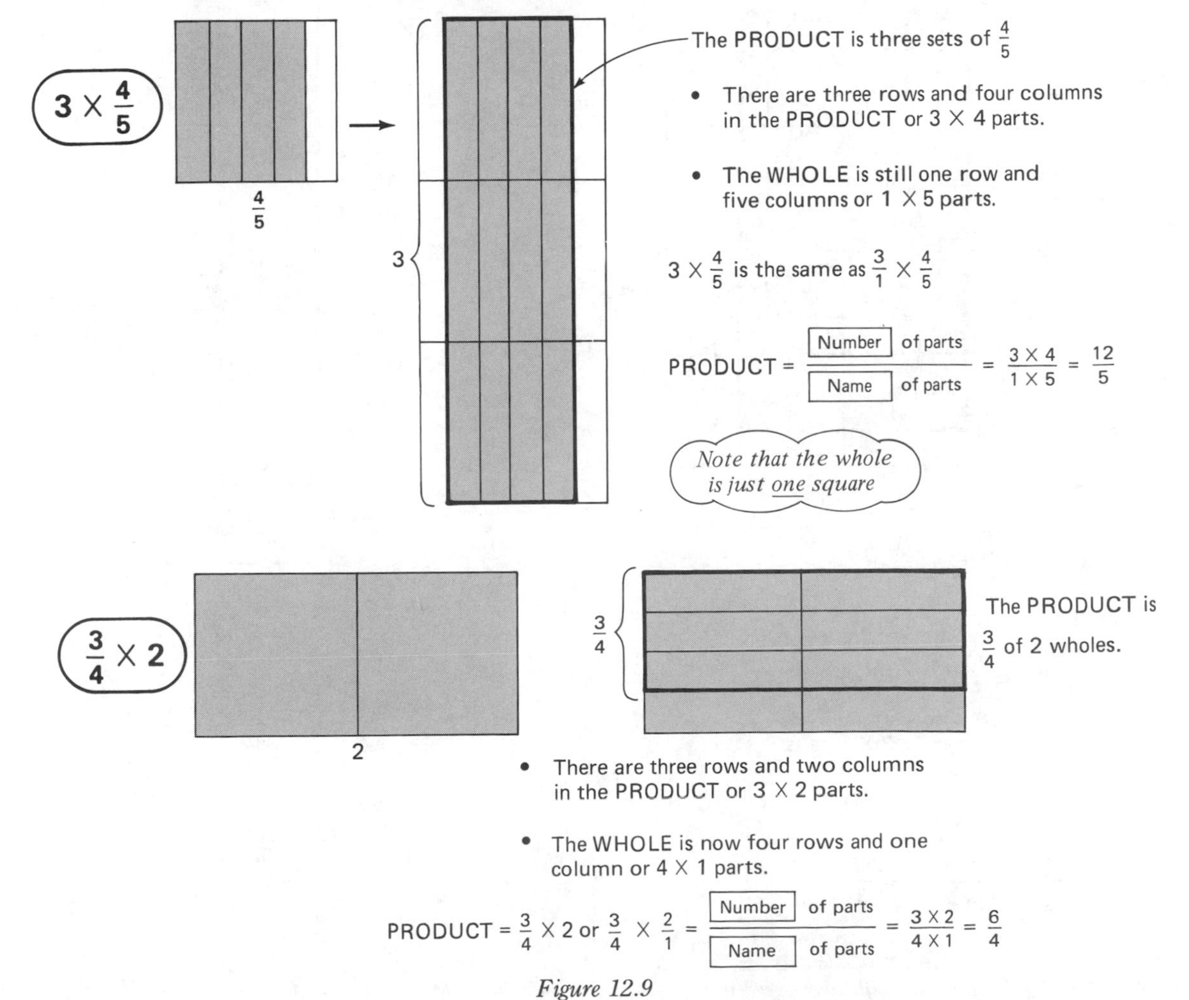

Figure 12.9
The multiplication rule with whole numbers is exactly the same.

$\frac{4}{5} = \frac{12}{5}$. Multiply the whole number by the top number. This rule makes sense even without modeling. Since the 3 can be written as $\frac{3}{1}$, the rule of multiplying tops and bottoms does apply. Figure 12.9 illustrates how this product would be modeled using squares as already discussed.

For products where the second factor is a whole number,

the result is similar, as illustrated in Figure 12.9. It is important that both cases be modeled and discussed. Notice that the name of the parts is always determined by the number of parts in a single whole, not the total number in the illustration.

Mixed-Number Factors

The case where either or both factors is a mixed number is frequently confusing. Consider the example of $3\frac{2}{3} \times 2\frac{1}{4}$. If this product is to be computed by pencil and paper, the factors are first written as improper fractions: $\frac{11}{3} \times \frac{9}{4}$. Figure 12.10 illustrates the same modeling process that has been used with easier examples. The rule of multiplying tops with tops and bottoms with bottoms remains unchanged and makes good sense.

It is interesting to have students explore the same product in terms of mixed numbers. In that form each factor has two parts: a whole part and a fraction part. The result is very similar to the model for multiplying two two-digit numbers. The product region, as shown in Figure 12.11, is partitioned into four partial products which can be determined separately and added. [Compare this with Figure 9.15(c), p. 145.] Helping students see common elements in mathematics is interesting and valuable.

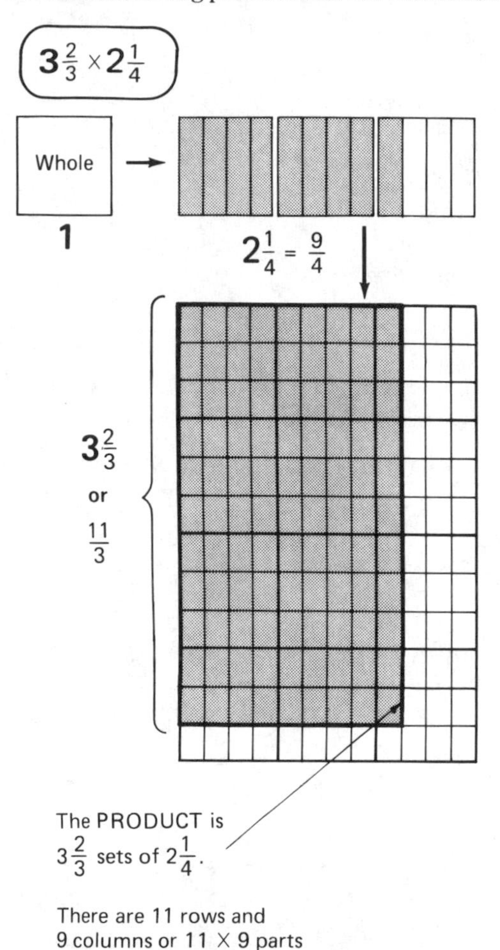

Figure 12.10
The multiplication rule with two mixed numbers.

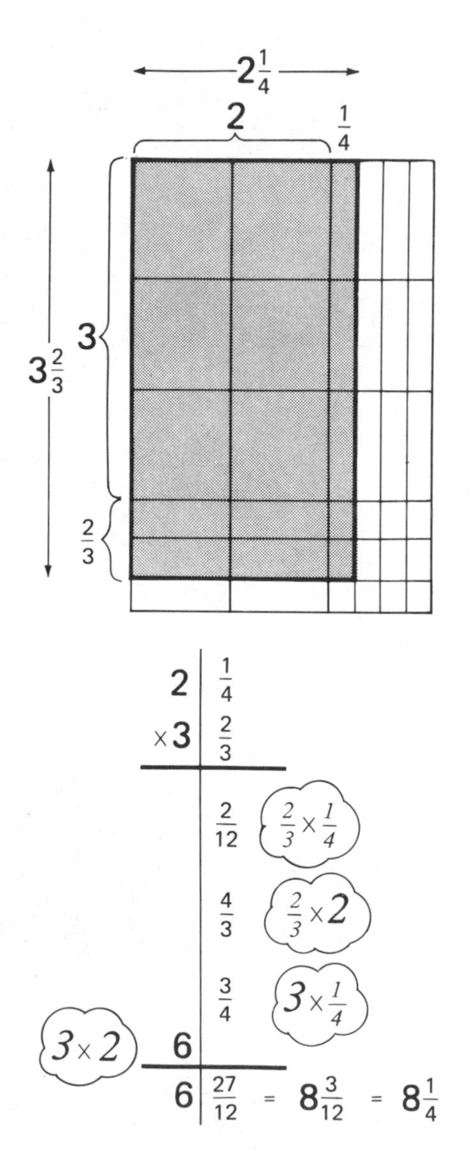

Figure 12.11
The product of mixed numbers as four partial products.

Division

CONCEPT EXPLORATION

Division with fractions can be very confusing. Recall that there are two basic concepts for division: partition and measurement. When modeling fraction division, these two concepts are very different. To help students develop an understanding of division with fractions, it is a good idea to rely heavily on division concepts with whole numbers and to use word problem contexts as well as models to explore meanings.

Partition Concept

In the partition concept for 12 ÷ 3, the task is to separate 12 things into 3 equal parts and determine the amount in each part. Explore this same idea with fractions, first with the divisor a whole number. Consider the following word problem:

> Darlene has 2¼ hours to complete 3 household chores. If she divides her time evenly, how many hours can she give to each?

In this problem, the task is to separate or partition the quantity 2¼ into three equal parts. Since 2¼ is 9 fourths, 3 fourths can be allotted to each task. That is, Darlene can spend ¾ of an hour per chore. These numbers turned out rather nicely. What if Darlene had only 1¼ hours for the three chores? Figure 12.12 shows how this could be modeled with three different models.

Are there partition situations where the divisor is a fraction? Darlene might have finished half of one of her chores. Suppose she has 2¼ hours for 1½ chores? Somewhat harder is 1¼ hours for 1½ chores. Note that 1½ is 3 halves. Use models to figure out how much goes in half of a chore (divide into three parts), and then the hours in two of these half-chores will be the number of hours for one chore. (Go ahead, use your models.)

Partition problems can also be made up where the divisor is less than one. Realistic word problems necessarily must involve contexts where partial sets make sense. Here are a few possibilities:

> Dad paid $2.00 for a ¾ pound box of candy. How much is that per pound? (Divide $2 into 3 parts and then count 4 parts.)

> The runner ran 2½ miles in ¾ of an hour. At that rate, how many miles could he run in 1 hour? That is, what is his speed in miles per hour? (Divide the 2½ miles into 3 parts. That tells how many miles in each quarter hour. Four quarters is a full hour.)

Partition division problems with fraction divisors tend to involve rates of time or price rates. Most of these would probably be done using decimals instead of fractions. The ideas and relationships involved in these explorations may, however, have values in themselves.

Measurement Concept

In the measurement concept for 12 ÷ 3, the task is to decide how many sets of 3 things are in a set of 12 things. What would it mean if the problem was 2¼ ÷ ¾? The meaning is the same. How many sets of ¾ each are in a set of 2¼? This might occur in a problem such as the following:

> Farmer Brown measured his remaining insecticide and found that he had 2¼ gallons. It takes ¾ of a gallon to make a tank of mix. How many tankfuls can he make?

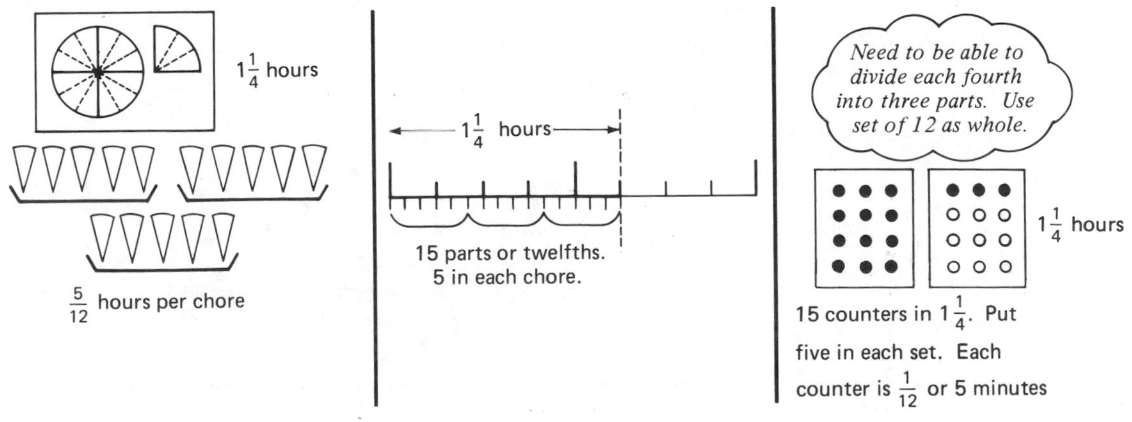

Figure 12.12
Partition division: 3 models.

In this problem both the total amount and the divisor are in the same type of units or parts, namely fourths. The $2\frac{1}{4}$ is the same as 9 fourths. The question then is, how many sets of 3 fourths are in a set of 9 fourth. The result is 3 sets (not 3 fourths). This is relatively easy to model. Try it.

A similar problem is the following:

> Linda has $4\frac{2}{3}$ yards of material. She is making baby clothes for the bazaar. If each pattern requires $1\frac{1}{6}$ yards of material, how many patterns will she be able to make?

To make this easier, convert the $4\frac{2}{3}$ to sixths also. The problem then becomes, how many sets of 7 sixths are in a set of 28 sixths. Once both quantities are in the same units, it really makes no difference what those units are. The problem $5\frac{3}{5} \div 1\frac{2}{5}$ is identical: how many sets of 7 (fifths) are in a set of 28 (fifths).

The answers in the foregoing examples were whole numbers. Explore for a minute what $13 \div 3$ might mean. There are 4 complete sets of 3 in a set of 13 with one left. That one constitutes $\frac{1}{3}$ of a set of 3. What if the 13 and 3 were fractional parts, for example, eighths? There would still be 4 complete sets of 3 eighths and $\frac{1}{3}$ of a set of 3 eighths. The fact that these are eighths is not relevant. Explore this idea using various fraction models and the following exercises:

$$\frac{7}{5} \div \frac{3}{5} = \square \qquad \frac{7}{12} \div \frac{3}{12} = \square \qquad \frac{3}{4} \div \frac{1}{2} = \square \qquad \frac{5}{3} \div \frac{1}{4} = \square$$

As illustrated in Figure 12.13, once both the dividend and

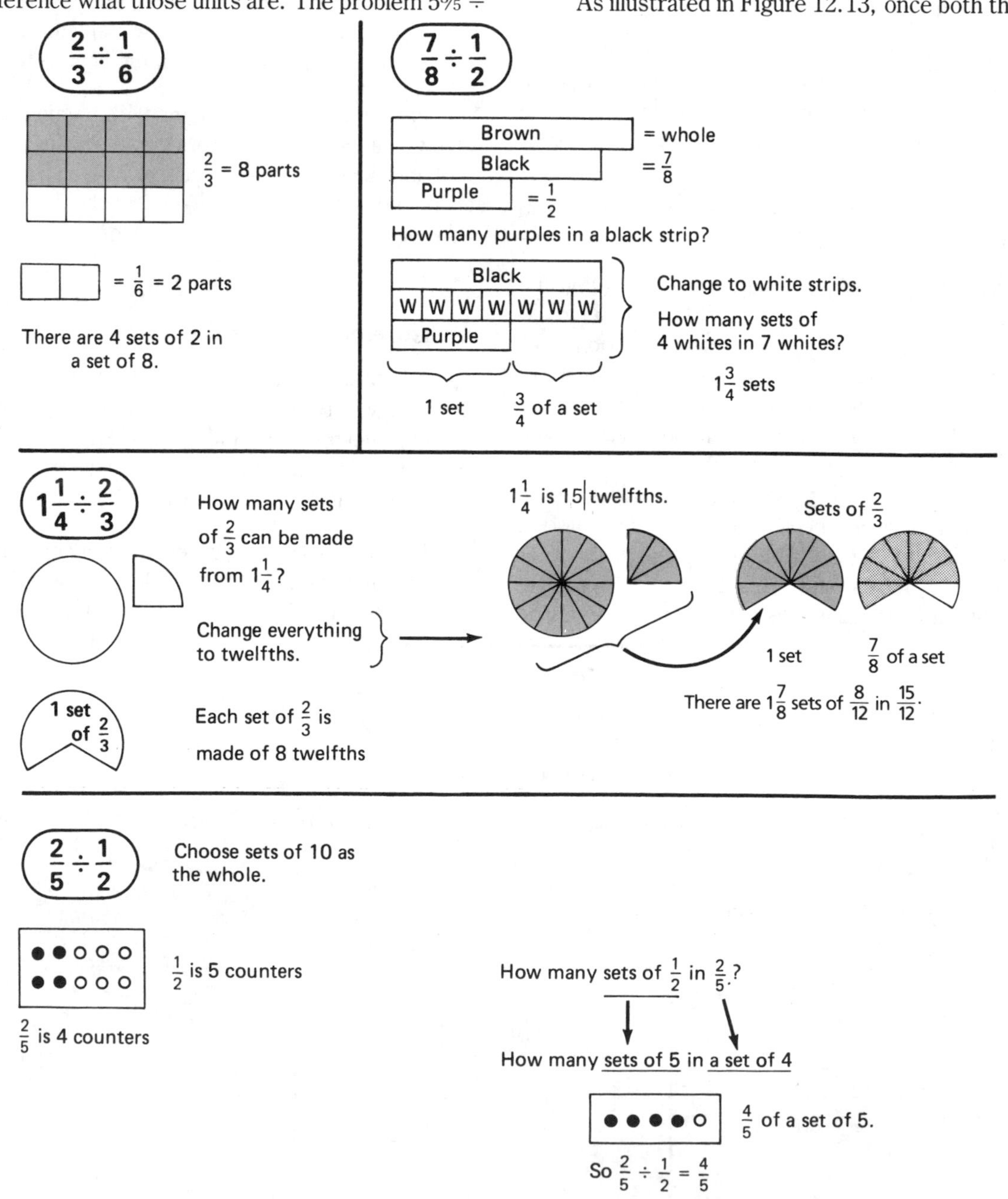

Figure 12.13
Exploring the measurement concept of fraction division.

the divisor are converted to the same amounts, the problem is essentially the same as a whole number division problem. Pay special attention to a problem, such as $\frac{1}{2} \div \frac{3}{4}$ where the answer is less than one.

DEVELOPING THE ALGORITHM

There are two different algorithms for division of fractions. Methods of teaching both algorithms are discussed here.

The Common Denominator Algorithm

The common denominator algorithm relies on the measurement or repeated-subtraction concept of division that was just developed. Consider the problem $\frac{5}{3} \div \frac{1}{2}$. As shown in Figure 12.14, once each number is expressed in terms of the same fractional part, the answer is exactly the same as the whole number problem $10 \div 6$. The name of the fractional part (the denominators) is no longer important, and the problem is one of dividing the numerators. The resulting rule or algorithm, therefore, is as follows: To divide fractions, first get common denominators and then divide the numerators.

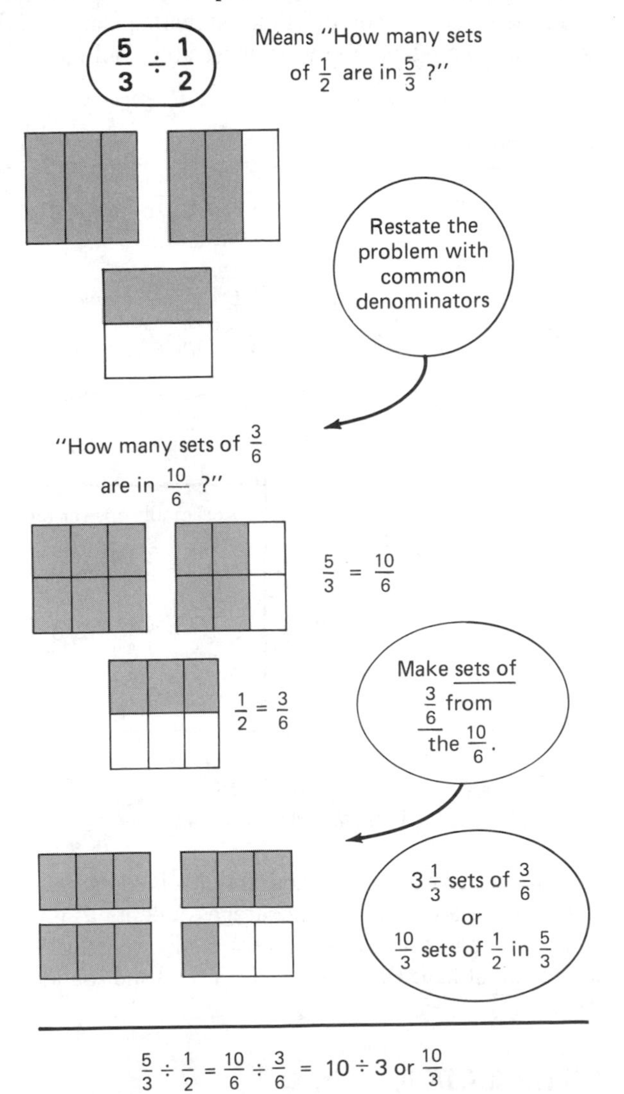

Figure 12.14
Common denominator method for fraction division.

When students learn that a fraction is also a way of indicating division, the rule can be expressed this way: Rewrite the division with common denominators and then $a/c \div b/c = a/b$. Try using pie pieces, fraction strips, and then sets of counters to model $1\frac{2}{3} \div \frac{3}{4}$ and $\frac{5}{8} \div 1\frac{1}{2}$.

The Invert-and-Multiply Algorithm

The more traditional algorithm for division with a fraction is to invert the divisor and multiply. The development of this algorithm relies on a symbolic rationale. One explanation of the invert and multiply approach is outlined in Figure 12.15.

$$\frac{3}{4} \div \frac{5}{6} = \square$$

Write the above equation in an equivalent form as a product with a missing factor.

$$\frac{3}{4} = \square \times \frac{5}{6}$$

Multiply both sides by $\frac{6}{5}$.
$\left(\frac{6}{5} \text{ is the inverse of } \frac{5}{6}\right)$

$$\frac{3}{4} \times \frac{6}{5} = \square \times \left(\frac{5}{6} \times \frac{6}{5}\right)$$

$$\frac{3}{4} \times \frac{6}{5} = \square \times 1$$

$$\frac{3}{4} \times \frac{6}{5} = \square$$

But $\frac{3}{4} \div \frac{5}{6} = \square$ also.

Therefore:

$$\frac{3}{4} \div \frac{5}{6} = \frac{3}{4} \times \frac{6}{5} = \square$$

In general:

$$\frac{a}{b} \div \frac{c}{d} = \frac{a}{b} \times \frac{d}{c}$$

Figure 12.15
To divide, invert the divisor and multiply.

Curricular Decisions

Which of these two division algorithms for fractions is "best"? There are values to each. In favor of the common denominator approach is the fact that it can be explained quite well with models and prior concepts of whole numbers. From a developmental point of view, this is a compelling argument. Very few children below the eighth grade will be able to fully appreciate the explanation for the invert-and-multiply approach. The result may be that they simply learn the rule: turn the second one upside down and multiply. The idea of "turning a fraction upside down" is antithetical to any meaningful approach to mathematics.

For students who are able to understand the symbolic explanation of the invert-and-multiply approach, it is certainly much more quickly explained with no ambiguities or special cases to consider. For some students in the eighth grade, this may be a very appropriate approach.

Another argument turns on the usefulness of the respective algorithms. Invert and multiply is generally the algorithm taught in algebra (although the common denominator algorithm works in algebra almost as well). The common denominator approach and its conceptual rationale may both be more useful for mental estimations of quotients involving fractions.

That brings up an even larger issue. When was the last time you had any need to divide by a fraction where you actually used pencil-and-paper division to obtain an exact result? Perhaps the only real justification for teaching division of fractions at all is as a model for algebra.

Computational Estimation with Fractions

ADDITION AND SUBTRACTION

A frequently quoted result from the Second National Assessment (Post, 1981) concerns the following item:

Estimate the answer to $12/13 + 7/8$. You will not have time to solve the problem using paper and pencil. Response choices and how 13-year-olds answered are listed here.

Responses	*Percent of 13-year-olds*
1	7
2	24
19	28
21	27
I don't know	14

What this result points out all too vividly is that a good concept of fractions is much more significant for estimation purposes than a mastery of the pencil-and-paper procedures. Knowing if a fraction is closer to 0, ½, or 1 proves quite useful. Numbers can either be rounded to the nearest whole or nearest half and then added easily. For example, $2\frac{1}{8} + \frac{4}{9}$ is about the same as $2 + \frac{1}{2}$. A front-end approach is also possible: deal with the whole numbers and then look at the fractions using estimates to the nearest half to make an adjustment. The following activities are useful for developing these ideas.

Activity 12.6

Flash fractions between 0 and 1 on an overhead projector. For each fraction, students should record on their answer sheet 0, ½, or 1, depending on which they think the given fraction is closest to.

Activity 12.7

Flash sums or differences of proper fractions. Response options can vary with the age and experience of the students. More than 1 or less than 1 is one alternative. Closer to 0, 1, or 2 is also an easy option. A more sophisticated option is to give the result to the nearest half: 0, ½, 1, 1½, or 2.

Activity 12.8

Provide short speed drills for estimating sums and differences with mixed numbers as illustrated in Figure 12.16.

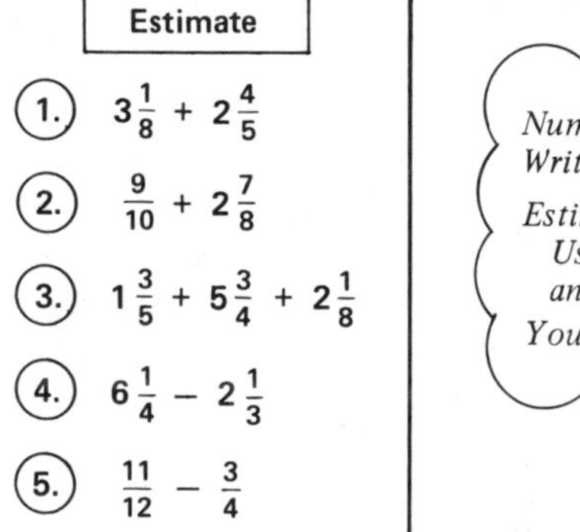

Number your papers 1 to 6.
Write only answers.
Estimate!
Use whole numbers and easy fractions.
You only have 2 minutes.

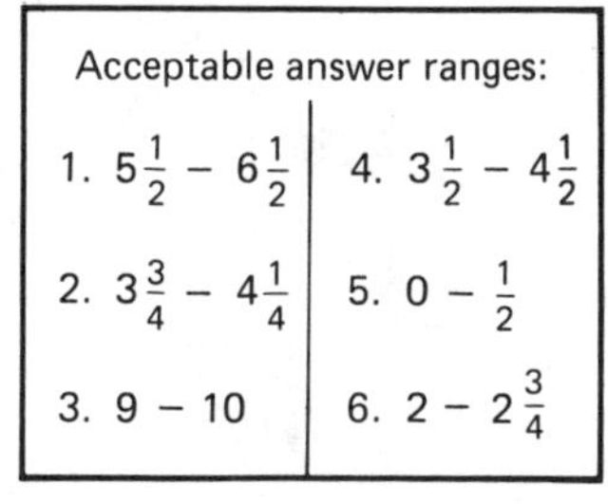

Acceptable answer ranges:

1. $5\frac{1}{2} - 6\frac{1}{2}$	4. $3\frac{1}{2} - 4\frac{1}{2}$
2. $3\frac{3}{4} - 4\frac{1}{4}$	5. $0 - \frac{1}{2}$
3. 9 – 10	6. $2 - 2\frac{3}{4}$

Figure 12.16
A fraction estimation drill.

When more than two addends are involved, an acceptable range may be a bit wider. Encourage students to give whole number estimates at first and later practice refining estimates to the nearest half. Discuss both front-end and rounding techniques.

MULTIPLICATION

Situations that require the estimate of a fraction times a whole number are quite common. For example, sale items are frequently one fourth off, or we read of a one third increase in the number of registered voters. Fractions are excellent substitutes for percents when making an estimate. To estimate 60% of $36.69, it is useful to think of 60% as three fifths or a little less than two thirds.

To mentally multiply $\frac{3}{5} \times 350$, first determine $\frac{1}{5}$ of 350 or 70, then $\frac{3}{5}$ is 3×70 or 210. Although this example has very accommodating numbers, it illustrates a process for multiplying a large number by a fraction. First determine the unit fractional part and then multiply by the number of parts you have. Once again you can see the importance of basic fraction concepts and the meanings of top and bottom numbers.

When the numbers are not so nice as in the preceding example, encourage students to use compatible numbers. To estimate $\frac{3}{5}$ of $36.69, use 35 instead since 35 is compatible with finding fifths: $\frac{1}{5}$ of 35 is 7 and $\frac{3}{5}$ is 21. Adjust by perhaps $0.50 to account for using a smaller number for an estimate of $21.50.

Students can practice estimating fractions times whole numbers or money amounts as soon as they can understand the preceding example. It is quite helpful if they are familiar with such exercises before working with decimals and percents as discussed in the next chapter.

For Discussion and Exploration

1. Give two reasons why the following argument is flawed:

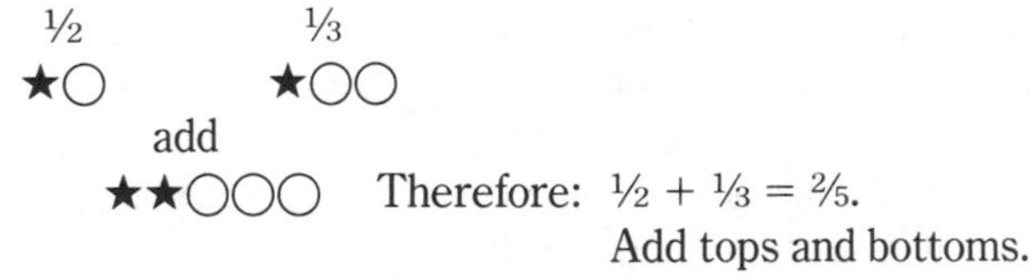

2. Discuss the differences and relative difficulty of the following, especially when they are done with models. Each exercise is different in some way from the others.

   $\frac{3}{4} + \frac{3}{4}$ $\quad \frac{2}{3} + \frac{5}{6}$ $\quad \frac{3}{8} + \frac{1}{2}$ $\quad \frac{2}{3} + \frac{5}{6}$

   $1 - \frac{1}{4}$ $\quad \frac{5}{2} - \frac{1}{4}$ $\quad \frac{3}{4} + \frac{2}{3}$ $\quad 4\frac{3}{4} - 2\frac{5}{8}$

   $3\frac{3}{8} - 1\frac{3}{4}$

3. Explore a textbook series and see where each algorithm is first introduced. Does the preparation for the algorithm seem appropriate? Do you think it is useful to spend time on intuitive approaches using models, or is the time required not worth the effort?

4. Try to develop the multiplication algorithm for fractions using a length model or a set model.

5. Have a debate about the division of fractions. Some issues which relate to the subject are these: Which algorithms should be taught? Why should we teach division of fractions? When should division of fractions be taught?

6. Several calculators are now available that do computations in fractional form as well as in decimal form. Some of these automatically give results in simplest terms. If you have access to such a calculator, discuss how it might be used in teaching fractions and especially fraction computation. If such calculators become commonplace, should we continue to teach fraction computation?

Suggested Readings

Behr, M. J., Wachsmuth, I., & Post, T. R. (1985). Construct a sum: A measure of children's understanding of fraction size. *Journal for Research in Mathematics Education. 16,* 120–131.

Curcio, F. R., Sicklick, F., & Turkel, S. B. (1987). Divide and conquer. *Arithmetic Teacher, 35*(4), 6–12.

Ellerbruch, W., & Payne, J. N. (1978). Teaching sequence from initial fraction concepts through the addition of unlike fractions. In M. N. Suydam (Ed.), *Developing computational skills.* Reston, VA: National Council of Teachers of Mathematics.

Feinberg, M. (1980). Is it necessary to invert? *Arithmetic Teacher, 27*(5), 50–52.

Holden, L. (1986). *Fraction factory.* Palo Alto, CA: Creative Publications.

Post, T. R., Wachsmuth, I., Lesh, R., & Behr, M. J. (1985). Order and equivalence of rational numbers: A cognitive analysis. *Journal for Research in Mathematics Education, 16,* 18–36.

Prevost, F. J. (1984). Teaching rational numbers—junior high school. *Arithmetic Teacher, 31*(6), 43–46.

Sweetland, R. D. (1984). Understanding multiplication of fractions. *Arithmetic Teacher, 32*(1), 48–52.

Thompson, C. S. (1979). Teaching division of fractions with understanding. *Arithmetic Teacher, 25*(5), 24–27.

Trafton, P. R., & Zawojewski, J. S. (1984). Teaching rational number division: A special problem. *Arithmetic Teacher, 31*(6), 20–22.

CHAPTER 13

DECIMAL AND PERCENT CONCEPTS AND DECIMAL COMPUTATION

Connecting Fraction and Decimal Concepts

TWO DIFFERENT REPRESENTATIONAL SYSTEMS

The symbols 3.75 and 3¾ both represent the same relationship or quantity, yet on the surface the two appear quite different. For children especially, the world of fractions and the world of decimals are very distinct. Even for adults, there is a tendency to think of fractions as sets or regions (three fourths *of* something), whereas we think of decimals more like numbers. The reality is, that fractions and decimals are two different systems of representation that have been developed to represent the same ideas. When we tell children that 0.75 is the same as ¾, this can be especially confusing. Even though different ways of writing the numbers have been invented does not make the numbers themselves different.

A significant goal of instruction in decimal and fraction numeration should be to help students see how both systems represent the same concepts. A good connection between fractions and decimals is quite useful. Even though we tend to think of 0.5 as a number, it is handy to know that it is a half. For example, in many contexts it is easier to think about ¾ than seventy-five hundredths or 0.75. On the other hand, the decimal system makes it easy to use numbers that are close to ¾ such as 0.73 or 0.78. Other conceptual and practical advantages exist for each system. For example, an obvious use of the decimal system is in digital equipment such as calculators, computers, and electronic meters.

To help students see the connection between fractions and decimals we can do three things. First, use familiar fraction concepts and models to explore those rational numbers that are easily represented by decimals: tenths, hundredths, and thousandths. Second, extend the base ten decimal system to include numbers less than one as well as large numbers. Third, help children use models to make meaningful translations between fractions and decimals. If two different symbols describe the same model, they must also represent the same idea. These three components are discussed in the following three sections.

BASE TEN FRACTIONS

Fractions that have denominators of 10, 100, 1000, and so on, will be referred to in this chapter as *base ten fractions.* This is simply a convenient label and is not one commonly found in the literature. Fractions such as 7/10 or 63/100 are examples of base ten fractions.

Base Ten Fraction Models

Most of the common models for fractions are somewhat limited for the purpose of depicting base ten fractions. Fraction models for tenths do exist, but generally the familiar fraction models cannot show hundredths or thousandths. It is important to provide models for these fractions using the same conceptual approaches that were used for more familiar fractions such as thirds and fourths.

Two very important area or region models can be used to model base ten fractions. First, to model tenths and hundredths, circular disks such as the one shown in Figure 13.1 can be printed on tagboard. (A full-sized master is in Appendix C). Each disk is marked with 100 equal intervals around the edge and is cut along one radius. Two disks of different colors, slipped together as shown, can be used to model any fractions less than one. Fractions modeled on this hundredths disk can be read as base ten fractions by noting the spaces around the edge, but are still reminiscent of the traditional "pie" sections model.

The most common model for base ten fractions is a 10 × 10 square. These can be run off on paper for students to shade in various fractions (Figure 13.2 and Appendix C). Another important variation is to use base ten place-value strips and squares. As a fraction model, the 10-cm square that was used as the hundred model for whole numbers is taken as the whole or one. Each strip is then 1 tenth and each small square is 1 hundredth. The *Decimal squares* materials (Bennett, 1982) includes squares in which each hundredth is again partitioned into 10 smaller sections. This permits modeling of thousands by shading in portions of the square. Even one more step is provided by the square in Appendix C. It is subdivided into

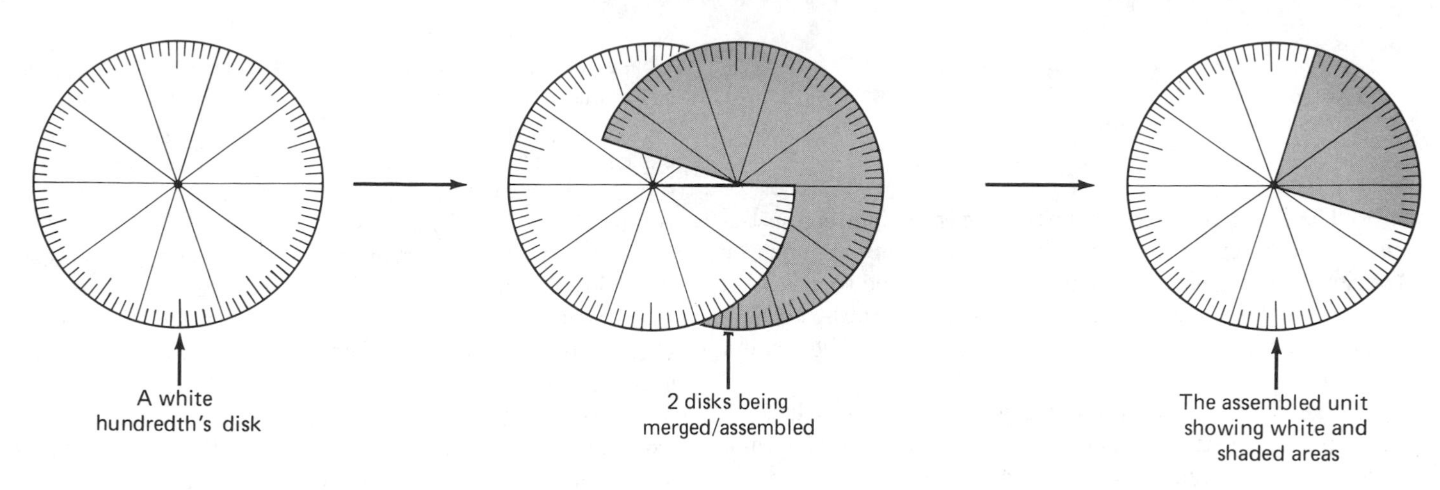

Figure 13.1
A hundredths disk for modeling powers of ten fractions.

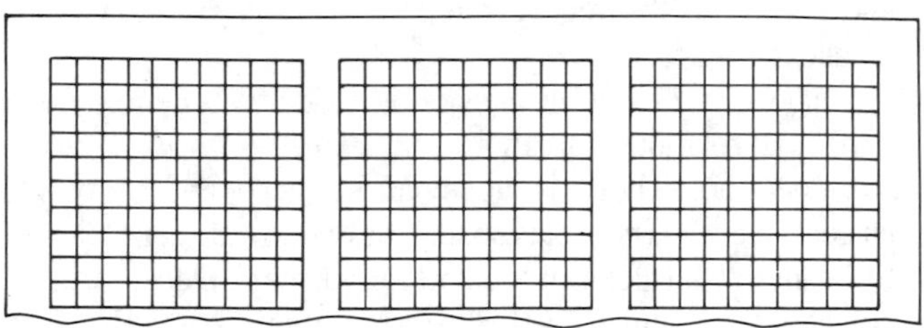

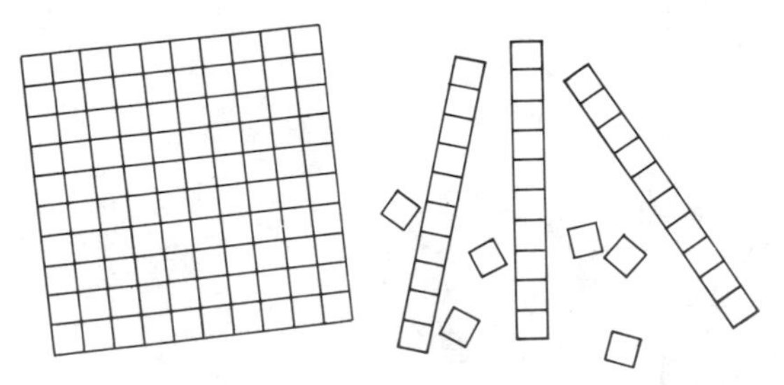

Figure 13.2
10 × 10 squares model base ten fractions.

10,000 tiny squares. While too small for paper reproduction, the master can be used to make a transparency. When shown with an overhead projector, individual squares or ten thousandths can easily be identified and shaded in with a pen on the transparency.

One of the best length models is a meter stick. Each decimeter is 1 tenth of the whole stick, each centimeter 1 hundredth, and each millimeter is 1 thousandth. Any number line model broken into 100 subparts is likewise a useful model for hundredths.

Many teachers use money as a model for decimals and to some extent this is helpful. However, for children, money is almost exclusively a two-place system. That is, numbers like 3.2 or 12.1389 do not relate to money. Children's initial contact with decimals should be more flexible. Money is certainly an important application of decimal numeration and may serve a useful purpose in later activities with decimals.

Multiple Names and Formats

Early work with base ten fractions is primarily designed to acquaint students with the models, to help them to begin to think of quantities in terms of tenths and hundredths, and to learn to read and write base ten fractions in different ways.

Have students show a base ten fraction using any base ten fraction model. Once a fraction, for example $^{65}/_{100}$, is modeled, the following things can be explored:

> Is this fraction more or less than $^{1}/_{2}$? than $^{2}/_{3}$? than $^{3}/_{4}$? That is, some familiarity with these fractions can be developed by comparison with fractions that are easy to think about.
>
> What are some different ways to say this fraction using tenths and hundredths? ("six tenths and five hundredths" or "sixty-five hundredths") Include thousandths when appropriate.
>
> Show two ways to write this fraction. ($^{65}/_{100}$ or $^{6}/_{10} + {}^{5}/_{100}$)

The last two questions are very important. When base ten fractions are later written as decimals, they are usually read as a single fraction. That is, 0.65 is read "sixty-five hundredths." But to understand them in terms of place value, the same number must be thought of as 6 tenths and 5 hundredths. A mixed number such as $5^{13}/_{100}$ is usually read the same way as a decimal: 5.13 is "five and thirteen hundredths." For purposes of place value, it should also be understood as $5 + {}^{1}/_{10} + {}^{3}/_{100}$. Special attention should also be given to numbers such as the following:

$$\frac{30}{1000} = \frac{0}{10} + \frac{3}{100} + \frac{0}{1000}$$

$$\frac{70}{100} = \frac{7}{10} + \frac{0}{100}$$

The expanded forms on the right will be helpful in translating these fractions to decimals. In oral form, fractions or decimals with trailing zeros are sometimes used to indicate a higher level of precision. Seven tenths is numerically equal to 70 hundredths, but the latter conveys precision to the nearest hundredth.

Exercises at this introductory level should include all possible connections between models, various oral forms, and various written forms. Given a model or written or oral fraction, students should be able to give the other two forms of the fraction, including equivalent forms where appropriate.

EXTENDING THE PLACE-VALUE SYSTEM

The 10-to-1 Relationship Works in Two Directions

Before considering decimal numerals with students, it is advisable to review some ideas of whole number place value. One of the most basic of these ideas is the 10-to-1 relationship between the value of any two adjacent positions. In terms of a base ten model such as strips and squares, 10 of any one piece will make one of the next larger, and vice versa. The 10-makes-1 rule continues indefinitely to larger and larger pieces or positional values. This concept is fun to explore in terms of how large the strips and squares will actually be if you move six or eight places out.

If you are using the strip-and-square model, for example, the strip and square shapes alternate in an infinite progression as they get larger and larger. Having established this idea with your students, focus on the idea that each piece to the right in this string gets smaller by one tenth. The critical question becomes: "Is there ever a smallest piece?" In the students' experience, the smallest piece is always the one designated as the ones or unit piece. But what is to say that even that piece could not be divided into 10 small strips? And could not these small strips be divided into 10 very small squares, which in turn could be divided into 10 even smaller strips, and so on and so on. In the mind's eye, there is no smallest strip or smallest square.

The goal of this discussion is to help students see that a 10-to-1 relationship can extend *infinitely in two directions,* not just one. There is no smallest piece and there is no largest piece. The relationship between adjacent pieces is the same regardless of which two adjacent pieces are being considered. Figure 13.3 illustrates this idea with a strip-and-square model.

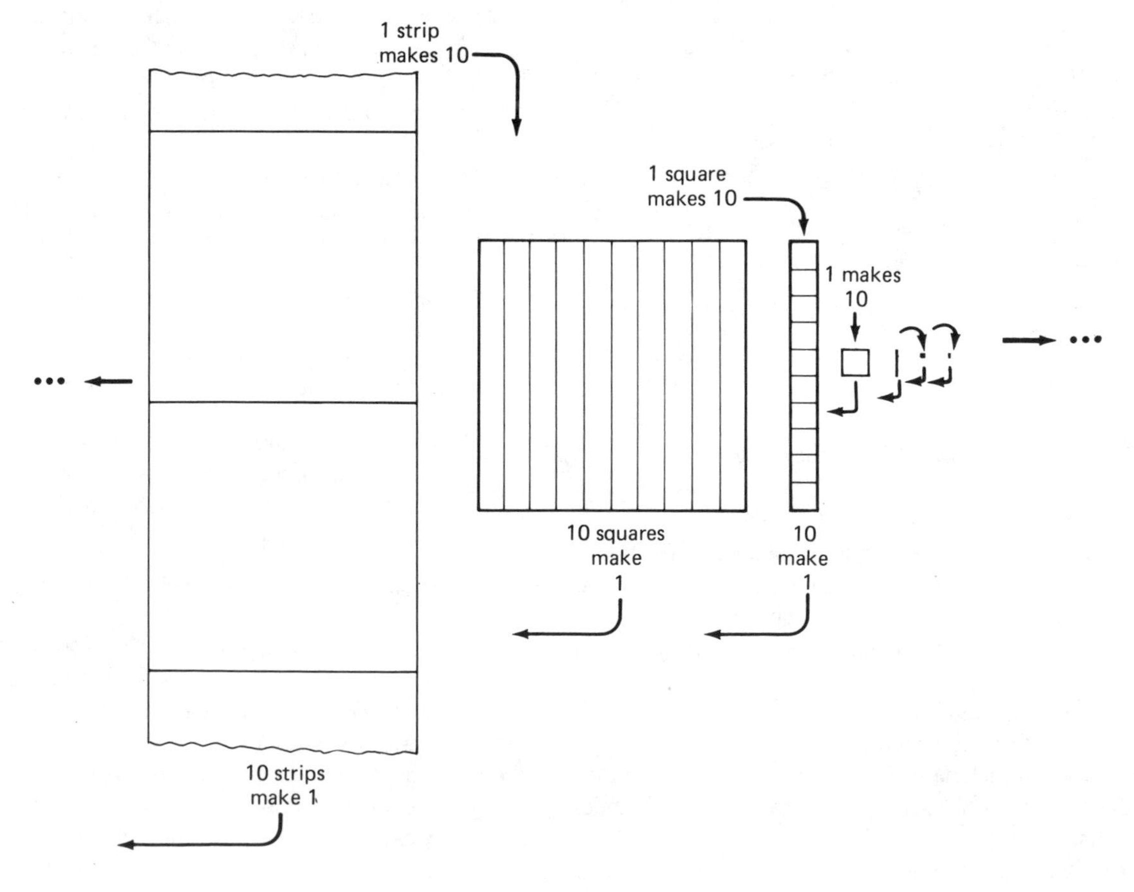

Figure 13.3
The strips and squares extend infinitely in both directions (in the "mind's eye").

Which Is the Unit? The Role of the Decimal Point

An important idea to be realized in this discussion is that there is no built-in reason why any one position should naturally be chosen to be the unit or ones position. In terms of strips and squares, for example, which piece is the ones piece? the small centimeter square? Why? Why not a larger or a smaller square? Why not a strip? *Any piece could effectively be chosen as the ones piece.* The following activity is designed to develop this idea.

Activity 13.1

Connect ten 10-cm squares to make a large strip 1 m long by 10 cm. Call this a "superstrip." Refer to a 10 × 10 square as a "square," a 10 × 1 strip as a "strip," and the small 1 × 1 square as a "tiny." Display on the board a collection of one superstrip, six squares, two strips, and four tinies. Discuss how this amount could be written. The obvious answer is 1624 or one thousand six hundred twenty-four. This is based on the assumption that the tiny is the unit. But suppose that the square was the ones piece and not the tiny. That would make the superstrip a ten and the amount shown would be somewhere between 16 and 17 or 16 + 2/10 + 4/100. Are there other possibilities for what this amount represents? The important conclusion we want to arrive at is that it could represent many different amounts, depending on which piece is designated as the unit. The unit could be a very large square, in which case the amount shown would be very small. Alternatively, the unit might be one of those infinitesimally small pieces, which would mean the amount shown is very large.

The discussion in this activity provides the background for a meaningful role for the decimal point. When a number such as 1624 is written, the assumption is that the 4 is in the unit or ones position. But if a position to the left of the 4 were selected as the ones position, some method of designating that position must be devised. Enter the decimal point. As shown in Figure 13.4, the same amount can be written in different ways, depending on the choice of the unit. The decimal point is placed between two positions with the convention that the position to the left of the decimal is the unit or ones position. Thus, *the role of the decimal point* is to designate the unit position, and it does so by sitting just to the right of that position.

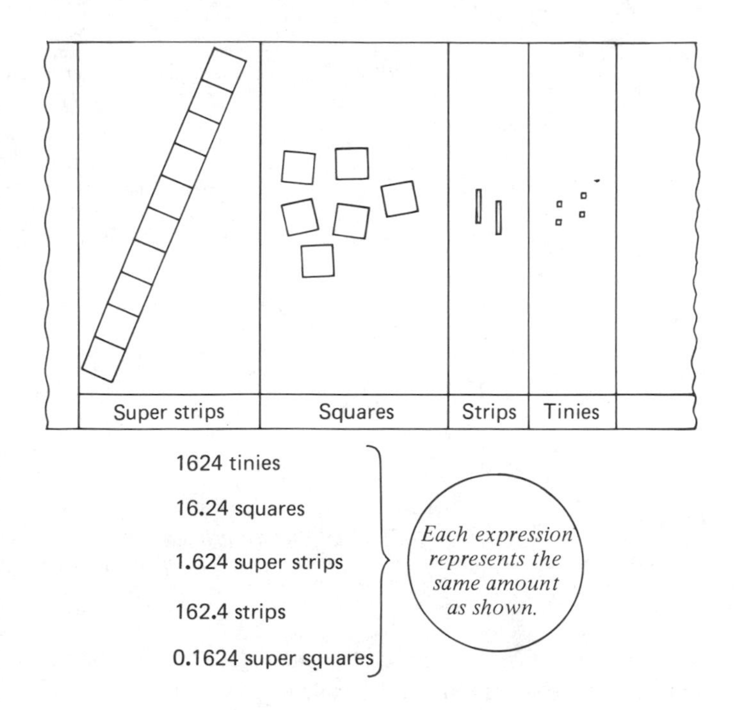

Figure 13.4
The decimal point indicates which position is the units.

A fitting caricature of the decimal is shown in Figure 13.5. The "eyes" of the decimal are always focused up toward the name of the unit or ones. A tagboard disk of this decimal point face can be used between adjacent base ten models or on a place-value chart. If such a decimal point were placed between the squares and strips in Figure 13.4, the squares would then be designated as the units and the written form 16.24 would be the correct written form for the model.

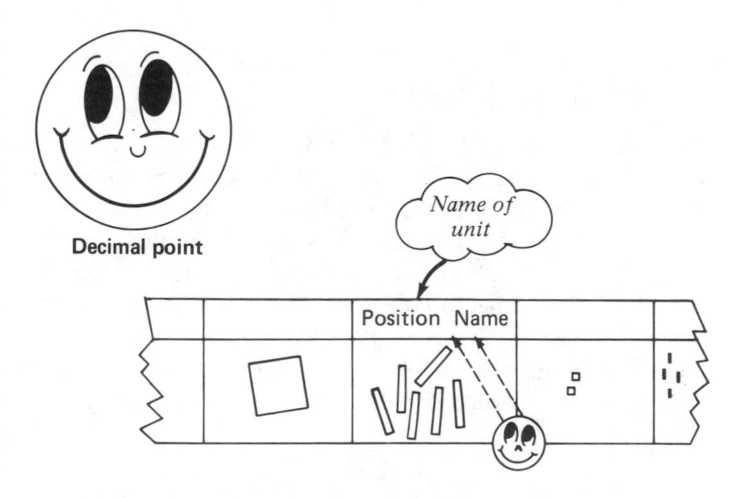

Figure 13.5
The decimal point always "looks up at" the name of the UNIT position.

The notion that the decimal "looks at the units place" is useful in a variety of contexts. For example, in the metric system, seven place values have names. As shown in Figure 13.6, the decimal can be used to designate any of these places as the unit without changing the actual measure. Our monetary system is also a decimal system. In the amount $172.95, the decimal point designates the dollars position as the unit. There are 1 hundreds (of dollars), 7 tens, 2 singles, 9 dimes, and 5 pennies or cents in this amount of money regardless of how it is written. If pennies were the designated unit, the same amount would be written as 17,295 cents or 17,295.0 cents. It could just as correctly be 0.17295 thousands of dollars or 1729.5 dimes. The role of the decimal can be explored in this way using money or metric measures as an example. These

	kilometer	hectometer	dekameter	meter	decimeter	centimeter	millimeter	
			4	3	8	5		

4 dekameters, 3 meters, 8 decimeters, and 5 centimeters =

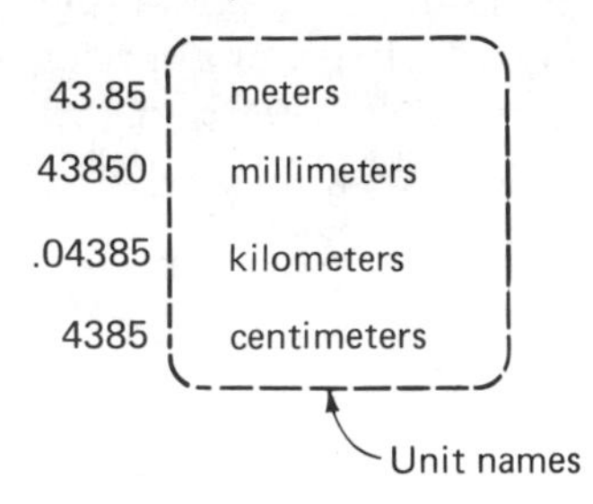

Figure 13.6
In the metric system, each place-value position has a name. The decimal point can be placed to designate which length is the UNIT length.

systems highlight the idea that the choice of a unit is arbitrary but must be designated. In the case of actual measures such as metric lengths or weights, or the U.S. monetary system, the name of the unit is written after the number. You may be 1.62 *meters* tall, but it does not make sense to say you are 1.62 tall.

Reading Decimal Numerals

Consider the example of Figure 13.4. What are appropriate ways to read each of the different designations for these amounts? The correct manner is to read them the same way as one would read a mixed fraction. For example, 16.24 squares is read "sixteen *and* twenty-four hundredths squares." Notice that the word *and* is reserved for the decimal point. It could also be read as "sixteen *and* two tenths and four hundredths squares." The other important feature is that this language is exactly the same as the language that was used for base ten fractions:

$$16.24 \text{ squares} = 16\frac{24}{100} \text{ squares} = 16 + \frac{2}{10} + \frac{4}{100} \text{ squares}$$

Help students see that once a place-value position is selected as the unit, then the next place to the right is 1⁄10 or tenths, and the next is 1⁄100 or hundredths, and so on. As in the previous section on reading fractions, children should become accustomed to reading decimal numerals in two different forms. The oral language then becomes a useful linkage between fraction symbolism and decimal symbols.

Other Decimal Models

Technically, any base ten place-value model can be used as a model for decimals. The strip-and-square model is useful because students can each have a set of these materials at their desks. If the kit includes a tagboard decimal point (Figure 13.5) about 4 cm in diameter, then no place-value mat is required. Students can arrange their base ten pieces in order and put the decimal point to the right of whatever piece is the designated unit. Usually the 10-cm square is selected. This allows students to model decimal fractions to hundredths. Some teachers even cut wooden toothpicks into 1-cm lengths to represent the next smaller strip or thousandths. These show up very nicely on an overhead projector.

The three-dimensional wooden or plastic base ten blocks have four different pieces. If the 10-cm cube or "block" is designated as 1, then the flats, sticks, and small cubes can be used to model decimals to three places.

Any base ten fraction model is also a decimal model. This fact is significant because it points out that decimals and fractions are simply two different symbolisms to represent the same amounts. Three fourths is shown on the hundredths disk as 7 tenths and 5 hundredths. If these pieces could be cut up and put on a place-value chart, they would be shown as 0.75 with the decimal "looking up" at the circles place as in Figure 13.7.

MAKING THE FRACTION-DECIMAL CONNECTION

To connect the two numeration systems, fractions and decimals, students should make concept-oriented translations from one system to another. The purpose of such activities has less to do with the skill of converting a fraction to a decimal than with construction of the concept that both systems can be used to express the same ideas.

Activity 13.2

Start with a base ten fraction such as 35⁄100 or 28⁄10. Have students use at least two different base ten fraction models to show these amounts. Discuss how each model shows the same relationship. Include as one of the models a set of strips and squares with the 10 × 10 square designated as the unit.

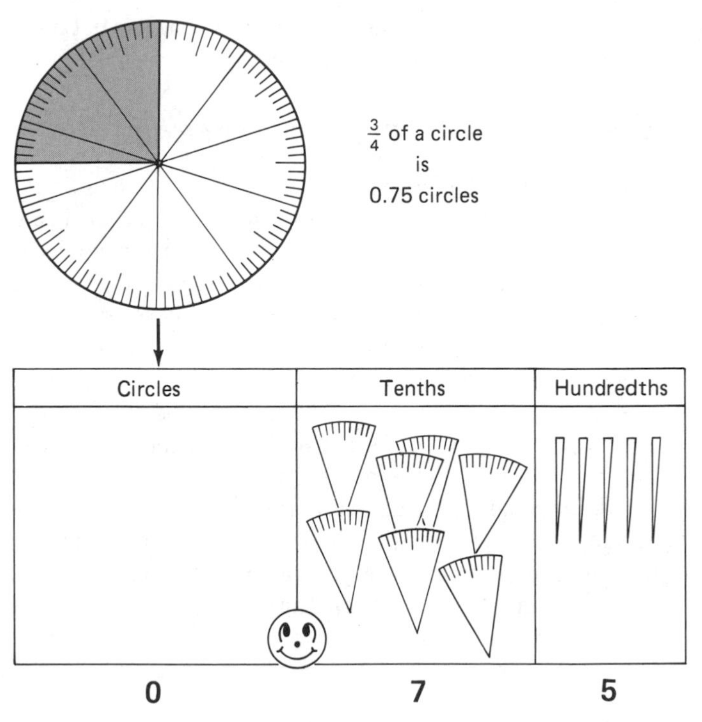

Figure 13.7
Fraction models could be decimal models.

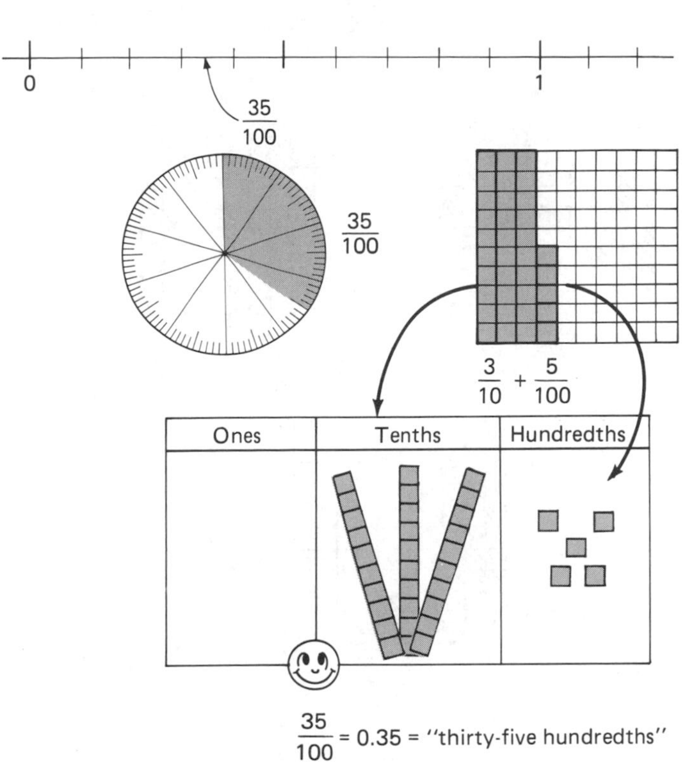

Figure 13.8
Translation of a base ten fraction to a decimal.

The strip-and-square model can then be rearranged in standard place-value format indicating how the fraction can be expressed as a decimal (Figure 13.8).

Once a fraction is modeled and written as both a fraction and a decimal, both symbolisms should be read. The oral language for both fraction and decimals will be the same as has already been discussed.

The previous activity suggests starting with a fraction and ending with a decimal. The reverse activity is equally important. That is, begin with a decimal number and have students use several models to represent it and then read and write the number as a fraction. As long as the fractions are restricted to base ten fractions, the conversions are straightforward.

Developing Decimal Number Sense

FAMILIAR FRACTIONS CONNECTED TO DECIMALS

In Chapter 11 on fraction concepts, an emphasis was placed on helping students develop a conceptual familiarity with simple fractions, especially halves, thirds, fourths, fifths, and eighths. We should try to extend this familiarity to the same concepts expressed as decimals. One way to do this is to have students translate these familiar fractions to decimals by means of a base ten model.

Activity 13.3

Have students shade in familiar fractions on a 10×10 grid. Since the grid effectively translates any fraction to a base ten fraction, these can then easily be written as a decimal. Under each shaded figure, write the fraction in familiar form, then as a base ten fraction, and finally as a decimal as shown in Figure 13.9

For this last activity it is good to begin with halves and fifths since these can be shown with strips of 10 squares. Next explore fourths. Many students will shade one fourth by shading in a 5×5 square of the grid. See if they can then shade the same amount using strips and fewer than 10 small squares. Repeat for other fourths, such as three fourths or seven fourths. Eighths present an interesting challenge. One way to find $1/8$ of a 10×10 square is to take half as much as is in one fourth. Since $1/4$ is $2/10$ and $5/100$, $1/8$ is $1/10 + 2/100 + 5/1000$. The $5/1000$ part is the same as half of $1/100$. (You should try to explain this fact using a 10×10 grid or with the 10,000 grid in Appendix C.)

The exploration of modeling one third as a decimal is a good introduction to the concept of an infinitely repeating decimal. Try to partition the whole square into three parts using strips and squares. Each part receives three strips with one left over. To divide the leftover strip, each part gets three small squares with one left over. To divide the small square, each part gets three tiny strips with one left over. (Recall that

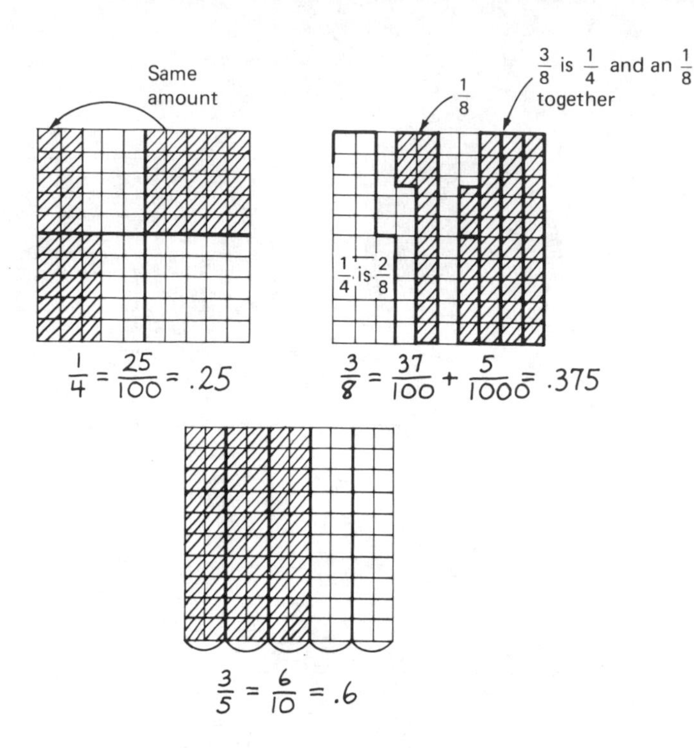

Figure 13.9
Familiar fractions to decimals using a 10 × 10 square.

with base ten pieces, each smaller piece must be a tenth of the preceding size piece.) Each of the three parts will get three tiny strips with one left over. It becomes quite obvious that this process is never ending. As a result, one third is the same as 0.333333 . . . or $0.\overline{3}$. For practical purposes, one third is about 0.333. In a similar manner, two thirds is a repeating string of sixes or about 0.667. Later, students will discover that many other fractions cannot be represented by a finite decimal, and this experience is a good background for that idea.

It is common in many seventh and eighth grades for students to memorize the decimal equivalents for halves, thirds, fourths, fifths, and eighths. In the fourth NAEP (Kouba, et al., 1988a), "60 percent of seventh-grade students could express simple fractions as decimals, but only 40 percent could express an improper fraction as a decimal" (p. 16). An explanation for the discrepancy is that the proper fraction equivalents may have been rotely memorized while to write an improper fraction such as 7/5 as a decimal would require some conceptual understanding. Even the 60% performance speaks to a need for better understanding of these translations. If students know conceptually the decimal translation for the unit fractions 1/3, 1/4, 1/5, and 1/8, and apply to these a "counting by unit fraction" concept, the equivalents for all familiar fractions are nearly immediate. For example, 3/5 is one fifth three times, or 0.2 + 0.2 + 0.2. Even 7/8 can be easily thought of as an eighth more than 3/4 or an eighth less than one. An understanding of mixed fractions and improper fractions can easily be applied to find decimal equivalents for familiar fractions greater than one. For all such exercises, the emphasis should be on conceptual conversions and not on rote memory or other symbolic algorithms.

APPROXIMATION WITH A NICE FRACTION

What do you think of when you see a number such as 7.3962? As is, it is an unruly number to think about, and for many it may even be intimidating. A good approach to these decimal numbers that are "not nice" is to think of a close "nice number." Is 7.3962 closer to 7 or to 8? Is it more or less than 7½? Is 7 or 7½ good enough for your purposes? If more precision is desired, you might look to see if it is close to a nice fraction. In this case, 7.3962 is very close to 7.4, which is 7⅖. The fraction 7⅖ is much easier to think with than a four-digit decimal. In the modern world of digital equipment, decimal numbers are frequently produced with much greater accuracy than is necessary. To have good number sense with these decimals implies that you can quickly think of a simple meaningful fraction as a useful approximation for most any decimal number. This facility is analogous to glancing at your digital watch, seeing 8:48′23″, and announcing that it is about 10 'til nine.

To develop this type of familiarity with decimals, children do not need new concepts or skills. They do, however, need the opportunity to apply and discuss the related concepts of fractions, place value, and decimals in exercises and activities such as the following.

Make a list of decimal numbers that do not have nice fractional equivalents. Have students suggest a decimal number near to the given one that does have a nice equivalent. Try this yourself with this list:

24.8025

6.59

0.9003

124.356

Different students may select different fractions for these numbers. The rationale for their choice presents an excellent opportunity for discussion.

In Figure 13.10, each decimal numeral is to be paired with the fraction expression that is closest to it. The exercise can be made easier by using fractions that are not as close together.

Activity 13.4

Place a "fraction sieve" on the overhead like one of those shown in Figure 13.11. (A master for these sieves is in Appendix C.) About the sieve, write a decimal number between 0 and 1. At each branch students decide if the given decimal is more or less than the fraction at that point. If it is less, they move to the left; if more, to the right.

In these exercises, students should be able to explain and confirm their answers in a meaningful manner. Students who

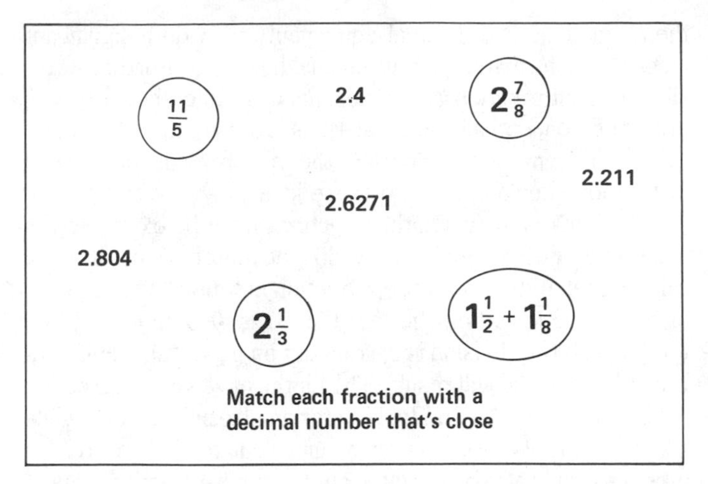

Figure 13.10
Pair fraction expressions.

are not able to deal well with these exercises are probably in need of some conceptual remediation.

ORDERING DECIMAL NUMBERS

Putting a list of decimal numbers in order from least to most is a closely related skill to the one just discussed. In the fourth NAEP (Kouba, et al., 1988a), only about 50% of seventh graders could identify the largest number in the following list: 0.36, 0.058, 0.375, and 0.4. That result is disturbing. It is unfortunately similar to the results of many other studies. (For example, see Hiebert & Wearne, 1986.) The most common error is to select the number with more digits, which is an incorrect application of whole number ideas. Some students later pick up the idea that digits way to the right represent very small numbers. They then incorrectly identify numbers with more digits as smaller. Both errors reflect a lack of conceptual understanding of how decimal numbers are con-

Figure 13.11
Fraction sieves.

structed. The following activities can help promote useful discussion about the relative size of decimal numbers.

Activity 13.5

Present two decimal numbers and have students use models to explain which is larger. A meter stick or a 10 × 10 square are useful for this purpose.

Activity 13.6

Write a four-digit decimal on the board—for example, 3.0917. Start with the whole numbers. Is it closer to 3 or 4? Then go to the tenths. Is it closer to 3.0 or 3.1? Repeat with hundredths and thousandths. At each answer, challenge students to defend their choices with the use of a model or other conceptual explanation.

Activity 13.7

Prepare a list of four or five decimal numbers that students might have difficulty ordering. They should all be between two consecutive whole numbers. Have them first predict the order of the numbers from least to most. Next, have them place each number on a number line with 100 subdivisions, every tenth one marked as in Figure 13.12. As an alternative, have students shade in the fractional part of each number on a separate 10 × 10 grid using estimates for the thousandths and ten thousandths. In either case it quickly becomes obvious which digits contribute the most to the size of a decimal.

OTHER FRACTION/DECIMAL EQUIVALENTS

The only decimal/fraction equivalents that have been discussed so far are those for the nice fractions: halves, thirds, fourths, fifths, and eighths. Also, any base ten fraction is immediately converted to a decimal, and similarly, simple decimals with two or three decimal places are easily converted to fractions. For the purposes of familiarity or number sense, these fractions and decimal equivalents provide a significant amount of information about decimal numbers. Furthermore, all of this information about decimals can and should be approached conceptually without rules, rote memory, or algorithms. It can be argued that the major focus of decimal instruction should center on these ideas.

At times, however, other fractions must be expressed as a decimal. For example, how do you enter $5/6$ or $3/7$ on a calculator that does not accept fraction notation? The answer, of course, is based on the fact that $3/7$ is also an expression for $3 \div 7$. If this division is carried out on paper, an infinite but repeating decimal will result. The ninths have very interesting decimal equivalents, and looking for a pattern is a worthwhile activity. (Try $1/9$, $2/9$, . . . on your calculator.) The division process should also be checked out for familiar fraction equivalents. If students have constructed a good understanding of familiar fractions and decimal equivalents, the concept will confirm the division result. If the division $4 \div 5$ is the only explanation students have for why $4/5 = 0.8$, then a significant lack of a conceptual linkage between fractions and decimals is likely.

Students are frequently shown early in the development of decimal numeration how to use division as a means of converting fractions to decimals. Doing a long division or using a calculator for this purpose does not promote a firsthand familiarity or number sense with fractions and decimals.

In the seventh or eighth grade, students are frequently taught to convert a repeating decimal to a fraction. These conversions serve the purpose of demonstrating that every repeating decimal can also be represented as a fraction and therefore as a rational number. This important theoretical result is primarily useful in making a distinction between rational and irrational numbers. The contribution such tedious activities have for number sense is rather minimal.

Introducing Percents

A THIRD OPERATOR SYSTEM

A major goal that has been stressed so far is to help children understand that decimal numerals and fractions are simply two different symbol systems for the same part-to-whole

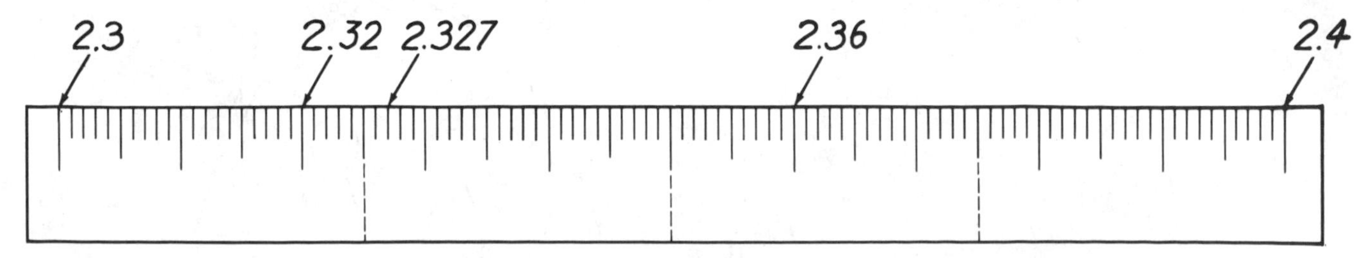

Figure 13.12
A decimal number line.

relationships; that is, they are representations for rational numbers. In this sense, both fractions and decimals are interpreted as real numbers. But when we say "three fifths of a cake" or "three fifths of the students," the number meaning comes from the cake or the set of students. Certainly 3/5 of 5 students is different than 3/5 of 20 students. In fact, it is precisely this operator notion that makes it difficult for children to think of fractions as numbers. They initially learn about them as parts *of* something.

When children have made a strong connection between their concepts of fractions and decimals, the topic of percent can be introduced. Rather than approach percents as a new and strange idea, children should see how percents are nothing more than a different way to write down some of the ideas they have already developed about fractions and decimals. While not a third numeration system, percents are essentially a third symbolism for operators.

Another Name for Hundredths

The term *percent* is simply another name for hundredths. If students can express common fractions and simple decimals as hundredths, then the term *percent* can be substituted for the term *hundredth.* Consider the fraction 3/4. As a fraction expressed in hundredths it is 75/100. When 3/4 is written in decimal form, it is 0.75. Both 0.75 and 75/100 are read in exactly the same way, "seventy-five hundredths." When used as operators, 3/4 of something is the same as 0.75 or 75% of that same thing. Percent is not a new concept, only a new notation and terminology. Connections with fractions and decimal concepts are developmentally appropriate.

Models provide the principal connecting link between fractions, decimals, and percents, as shown in Figure 13.13.

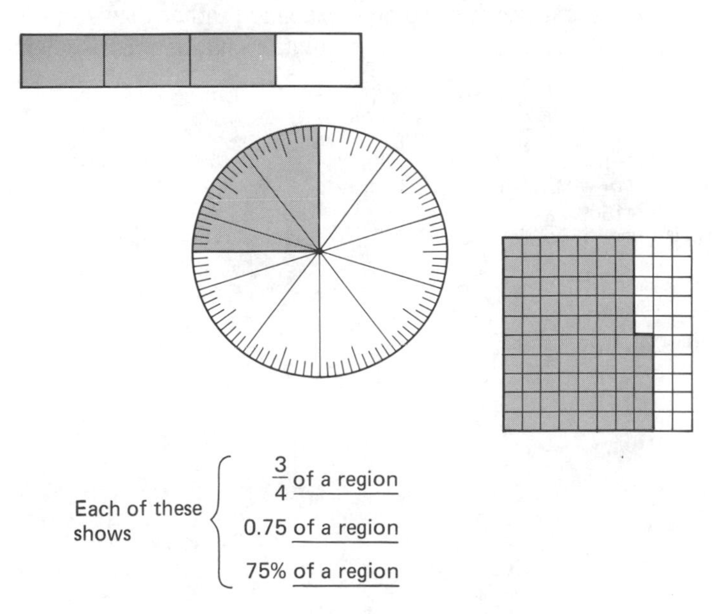

Figure 13.13
Models connect three different notations.

Base ten fraction models are suitable for fractions, decimals, and percents, since they all represent the same idea. (Percent is also a symbol system for ratios, just as fractions are. In this chapter, percents are first connected to fraction and decimal notation and part-to-whole relationships.)

Another helpful approach to the terminology of percent is through the role of the decimal point. Recall that the decimal identifies the units. When the unit is ones, a number such as 0.659 means a little more than 6 tenths of one. The word *ones* is understood. But 0.659 is also 6.59 tenths and 65.9 hundredths and 659 thousandths. In each case the name of the unit must be explicitly identified, or else the unit ones would be assumed. In 6.59 tenths, the interpretation is similar to an operator. It is 6 and 59 hundredths *of the things called tenths.* Since percent is another name for hundredths, when the decimal is "identifying" the hundredths position as the units, the word *percent* can be specified as a synonym for hundredths. Thus, 0.659 (of some whole or one) is 65.9 percent or 65.9% of that same whole. As illustrated in Figure 13.14, the notion of placing the decimal point to *identify the percent position* is conceptually more meaningful than the apparently arbitrary rule: "to change a decimal to a percent move the decimal two places to the right." This rule carries no meaning and is easily confused. "Do I move it right or left?" A better idea is to equate hundredths with percent both orally and in notation.

Use Percent with Familiar Fractions

Students should use base ten models for percents in much the same way as for decimals. The disk with 100 markings around the edge is now a model for percents as well as a fraction model for hundredths. The same is true of a 10 × 10 square. Each tiny square inside is 1% of the square. Each row or strip of 10 squares is not only a tenth but 10% of the square.

Similarly, the familiar fractions (halves, thirds, fourths, fifths, and eighths) should become familiar in terms of percents as well as decimals. Three fifths, for example, is 60% as well as 0.6. One third of an amount is frequently expressed as $33\frac{1}{3}$% instead of 33.3333 . . . percent. Likewise, 1/8 of a quantity is $12\frac{1}{2}$% or 12.5% of the quantity. These ideas should

		Percent	
Ones	Tenths	Hundredths	Thousandths
	3	6	5

.365 (of one) = 36.5 percent (of one)

Figure 13.14
Hundredths are also known as percents.

be explored with base ten models and not as rules about moving decimal points.

REALISTIC PERCENT PROBLEMS

The Three Percent Problems

Junior high school teachers talk about "the three percent problems." The sentence "____ is ____ percent of ____" has three spaces for numbers; for example, *20* is *25* percent of *80*. The three classic percent problems come from this sterile expression; two of the numbers are given, with the students asked to produce the third. Students learn very quickly that you either multiply or divide the numbers given, and sometimes you have to move a decimal point. This approach to percent is doomed from the start. Students have no way of determining when to do what. As a result, performance on percentage problems is very poor. Furthermore, the major reason or perhaps the only reason for learning about percent is that it is commonly used in our society. Sales, taxes, census data, political information, trends in economics, in farming, in production, and so on, are all full of percent terminology. But in almost none of these is the formula version ____ is ____ percent of ____ utilized. So when asked to solve a realistic percent problem, students are frequently at a loss.

In Chapter 11, three types of exercises with fractions were explored. These involved parts, wholes, and fractions. The three types were determined according to which of the three—part, whole, or fraction—was unknown. Students used models and simple fraction relationships in those exercises. Those three types of exercises are precisely the same as the three percent problems. Developmentally, then, it makes good sense to help students make the connection between the exercises done with fractions and those with percents. How can this be done? Use the same types of models and the same terminology of parts, wholes, and fractions. The only thing that is different is that the word *percent* is used instead of fraction. In Figure 13.15, three exercises from Chapter 11 have been changed to a corresponding percent terminology. A good idea for early work with percent would be to review (or explore for the first time) all three types of exercises in terms of percents. The same three types of models can be used. (Refer to Figures 11.8, 11.9, and 11.10.)

Realistic Percent Problems and Nice Numbers

While children must have some experience with the noncontextual, straightforward situations in Figure 13.15, it is important to have them explore these relationships in real contexts. Find or make up percent problems and present them in the same way that they appear in newspapers, television, and other real contexts. In addition to realistic problems and formats, follow these maxims for much of your unit on percents:

Limit the percents (fractions) to familiar fractions (halves, thirds, fourths, fifths, and eighths) or easy percents (1/10, 1/100) and use numbers compatible with these fractions. That is, make the computation very easy. The relationships involved are the focus of these exercises. Complex computational skills are not.

Do not suggest or provide any rules or procedures for different types of problems. Do not categorize or label problem types.

Utilize the terms *part, whole,* and *percent* (or *fraction*). The use of fraction and percent are interchangeable. Help students see these as the same types of exercises they did with simple fractions.

Require students to use models or drawings to explain their solutions. It is better to assign 3 problems requiring a drawing and an explanation rather than 15 problems requiring only computation and answers. Remem-

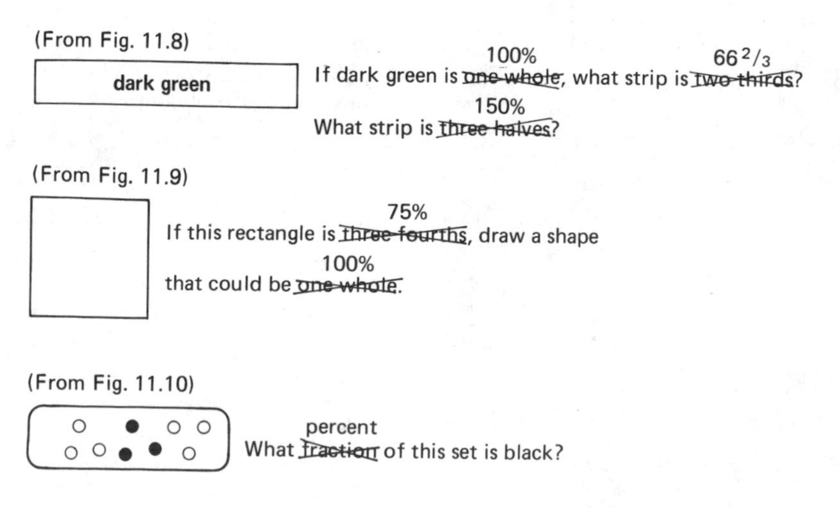

Figure 13.15
Part/whole/fraction exercises can be translated to percent exercises.

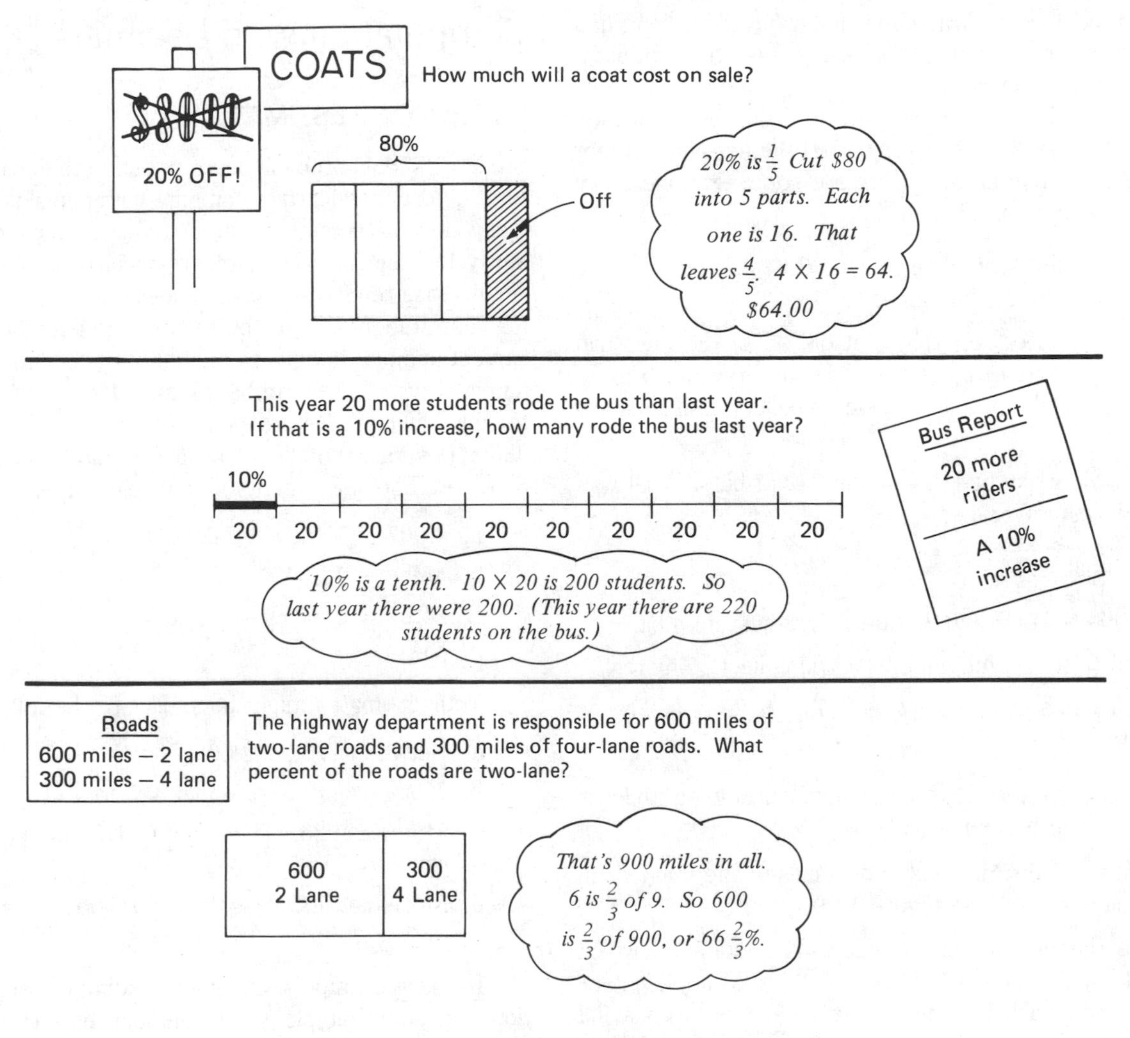

Figure 13.16
"Real" percent problems with "nice" numbers; simple drawings help with reasoning.

ber that the purpose is the exploration of relationships, not computational skill.

Encourage mental arithmetic.

The following example problems meet these criteria for easy fractions and numbers. Try working each problem, identifying each number as a part, a whole, or a fraction. Draw length or area models to explain and/or to work through your thought process. Examples of this informal reasoning are illustrated with additional problems in Figure 13.16.

1. The PTA reported that 75% of the total number of families were represented at the meeting last night. If 320 families are in the school, how many were at the meeting?
2. The baseball team won 80% of the 25 games they played this year. How many games were lost?
3. In Mrs. Carter's class, 20 students or 66⅔% were on the honor roll. How many students are in her class?
4. George bought his new computer at a 12½% discount. He paid $700. How many dollars did he save by buying it at a discount?
5. If Joyce has read 60 of the 180 pages in her library book, what percent of the book has she read so far?
6. The hardware store bought widgets at 80 cents each and sold them for $1 each. What percent did the store mark up the price of each widget?

Estimation in Percent Problems

Of course, not all real percent problems have nice numbers. Frequently in real life, an approximation or estimate in percent situations is all that is required or is enough to help one think through the situation. Even if a calculator will be used to get an exact answer, an estimate based on an understanding of the relationship can confirm that a correct operation was performed or that the decimal was correctly positioned.

To help students with estimation in percent situations, two ideas that have already been discussed can be applied.

First, when the percent (fraction) is not a "nice" percent, find a close percent or fraction that is nice or easy to work with. Second, in doing the calculation, select numbers that are compatible with the fraction involved to make the calculation easy to do mentally. In essence, convert the not-nice percent problem into one that is nice. Here are some examples. Try your hand at estimates of each.

1. The 83,000-seat stadium was 73% full. How many people were at the game?
2. The treasurer reported that 68.3% of the dues had been collected for a total of $385. How much more money could the club expect to collect if all dues are paid?
3. Max McStrike had 217 hits in 842 at-bats. What was his batting average?

Possible estimates:

1. 62,000 (Use ¾ and 80,000 then adjust up a bit.)
2. $190 (Use ⅔ and and $380. Will collect ⅓ more.)
3. A bit more than .250 (4 × 217 > 842. ¼ is 25% or .250)

The following exercises are also useful in helping students with estimation in percent situations.

Compare percents instead of decimals to the fractions in the sieves shown in Figure 13.9.

Choose the closest nice fraction. For example, which fraction is closest to 78%: ½ ⅔, ⅘? Multiple-choice exercises such as this one can be done with a full class or can be made into worksheet exercises. Later, students can give a nice fraction that is close to the percent without a multiple choice. Have students justify their answers.

Work with "easy percents," especially 1% and 10%. Begin with exercises where students give 1% and 10% of numbers. Then show them how to use these easy percents to get other percentages, either exact or approximate. For example, to get 15% of $349, think 10% is about $35, and 5% is half of that or $17.50. So 15% is $35 + $17.50 or $42.50. Similar reasoning can be used to adjust an estimate by 1% or 2%. To find 82% of $400, it is easy to think of ⅘ (80%) as 4 × ⅕ of 400 or 4 × $80, which is $320. Since each 1% is $4, add on $8.

Sometimes an exact result, and therefore some calculation, is required. The emphasis on conceptual thinking, nice fractions, and estimation of results will all pave the way for easy work when an exact result is requested. The use of an estimate will determine if the result is in the correct ballpark or if it makes sense. Frequently, the same estimation process will dictate what computation to do so that the problem can be entered on a calculator.

Computation with Decimals

THE ROLE OF ESTIMATION

Students should probably learn to estimate decimal computations before they learn to compute with pencil and paper. For many decimal computations, rough estimates can be made easily by rounding the numbers to nice whole numbers or simple base ten fractions. In almost all cases a minimum goal for your students should be to have the estimate contain the correct number of digits to the left of the decimal—the whole number part. Select problems for which estimates are not terribly difficult. Before going on, try making easy whole number estimates of the following computations. Do not spend time with fine adjustments in your estimates.

a. 4.907 + 123.01 + 56.1234
b. 459.8 − 12.345
c. 24.67 × 1.84
d. 514.67 ÷ 3.59

Your estimates might be similar to the following:

a. Between 175 and 200
b. More than 400, or about 425 to 450
c. More than 25, closer to 50. (1.84 is more than 1 and close to 2.)
d. More than 125, less than 200 (500 ÷ 4 = 125 and 600 ÷ 3 = 200)

In these examples, an understanding of decimal numeration and some simple whole number estimation skills can produce rough estimates. When estimating, thinking focuses on the meaning of the numbers and the operations and not on how many decimal places are involved. However, students who are taught to focus primarily on the pencil-and-paper algorithms for decimals may find even simple estimations difficult.

Estimation is also a means of locating the decimal point in multiplication or division. For those two operations, one algorithm that is reasonable is the following: *Ignore the decimal points and do the computation as if all numbers were whole numbers. When finished, place the decimal in the result by estimation.* This approach is illustrated in Figure 13.17. In both examples, notice how a shift of the decimal in either direction just one place would give a result that is not even close to the estimate. While some explanation for the estimation method is required, it is quite useful except for very precise computations such as 0.00987 × 0.000103. Those seriously needing a precise answer to such computations almost certainly will have a calculator or computer to help them.

Therefore, a good *place* to begin decimal computation is with estimation. Not only is it a highly practical skill, it helps children look at answers in ballpark terms, can form a check on pencil-and-paper computation, and is one way of placing the decimal in multiplication and division.

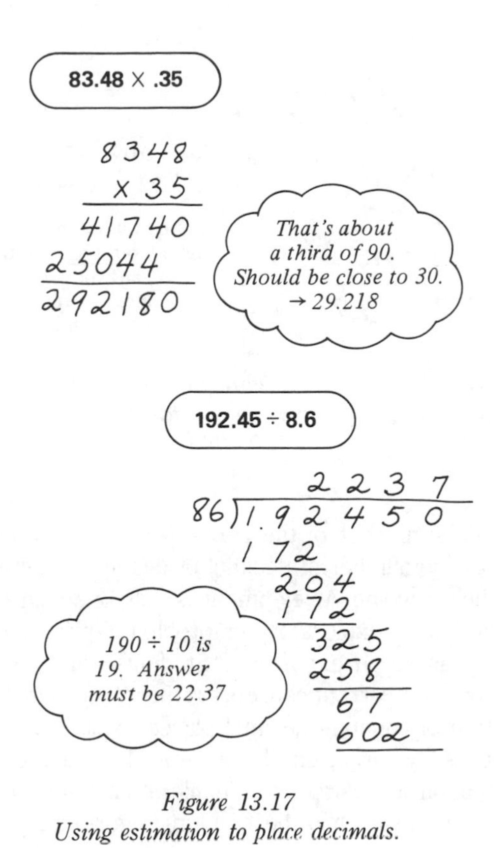

Figure 13.17
Using estimation to place decimals.

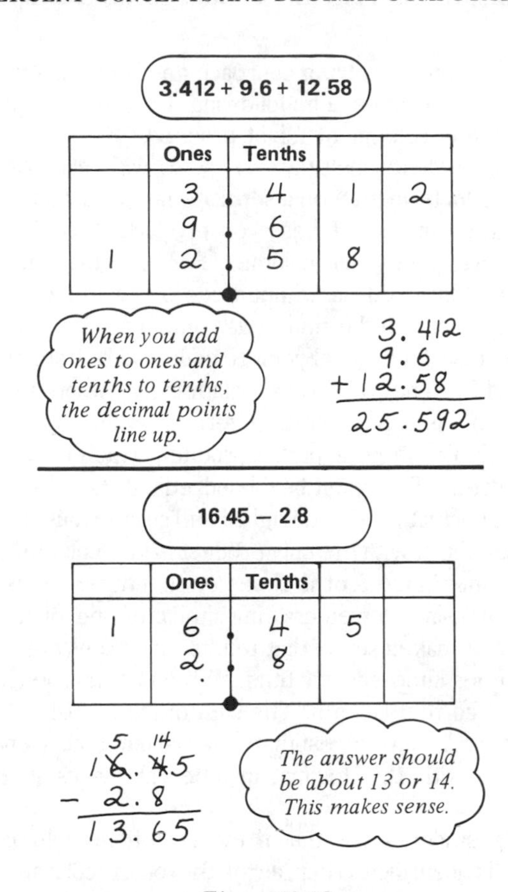

Figure 13.18
Using a place-value chart to develop rules for adding and subtracting decimals.

A good *time* to begin computation with decimals is well after a firm conceptual background in decimal numeration has been developed. Learning the pencil-and-paper algorithms for decimals will do little or nothing to help students understand decimal numeration.

ADDITION AND SUBTRACTION

From a conceptual standpoint, the addition and subtraction algorithms are nearly identical for decimals and whole numbers. The numbers in like place-value columns are added or subtracted with 10-for-1 trades made when necessary. While this is a relatively straightforward application of place value, many students do have trouble. Errors tend to occur when problems are presented in horizontal format or are in word problems with different numbers of digits to the right of the decimal. For example, students might compute the sum of 3.45 + 12.2 + 0.807 by lining up the right-hand digits 5, 2, and 7 as they would with whole numbers.

To help students with addition and subtraction, have them first estimate the answer as discussed earlier. For the example just given, the sum should be about 16. Next, discuss what the numbers mean in terms of a base ten model. What does each digit represent? How would the problem be written if the numbers were written in the appropriate columns of a place-value chart? (See Figure 13.18.) From such a discussion, it should be reasonable that like place values should be added or subtracted. When students follow this procedure, they should compare the results with their estimate. The estimate should be a confirmation of the computation (and not the reverse). Under no circumstances should students use the purely rote rule of "lining up the decimal points" without being able to justify it.

MULTIPLICATION

Before reading further, estimate each of the following products:

347	34.7	3.47
× 2.6	× 2.6	× 2.6

The first is about 900 (2 × 350 is 700 and 6 tenths of 350 is about 200). Similar reasoning would place the second product at about 90 and the third around 9. The whole number product 347 × 26 is 9022, or about 9000. Each of these products uses exactly the same digits even though the decimal point makes the numbers quite different.

A discussion of different decimal products with the same digits, like the products above, can be an interesting way to begin the development of decimal multiplication. It is a bit perplexing that numbers so very different in size are so much alike in terms of the digits in the products. The discussion can lead to a more complete understanding of the algorithm through a problem-solving approach and help students relate old ideas to new ones. The following discussion will illustrate how such a development might proceed.

Have students multiply 347 × 26 with each of the six partial products written on a separate line as in Figure 13.19. Review the concept of base ten products that determines which digits go in which columns. For example, the product 6 ones × 3 hundreds is 18 hundreds or a 1 in the thousands column and an 8 in the hundreds column.

Now insert a decimal point to make the bottom factor 2.6 instead of 26. The place-value columns for the products must be different. Is it possible to determine what any of the columns are? Try the last partial product, which is now 2 *ones* × 3 hundreds. That must be 6 hundreds. If the other columns are in the correct order, the right-hand column must be tenths. Does that check with the only product in that column, 6 tenths times 7 ones? Are tenths times ones tenths? (This can be verified by using an area drawing similar to one for fractions.) Similarly, it makes sense that tenths times tens are ones and tenths times hundreds are tens. (Why?) Each product seems to be in a correct column. The sum of the partial products is 902.2. Recall that the estimate was about 900. Repeat this discussion with the decimal in different places in the two factors.

Help students see that if the right-hand column of the product is identified, then all of the other columns can be labeled in order. If you know the value of the first, or right-hand, column, then you can tell which is the ones column and, consequently, where the decimal belongs. If the very first partial product is written as a base ten fraction, then it is easy to name the right-hand column as illustrated in Figure 13.20.

After discussing several of these products with the same digits, have students write each as the product of two simple fractions rather than in decimal form. For example, 3.47 × 2.6 is the same as $^{347}/_{100} \times {}^{26}/_{10}$. In this form, the product of the numerators is the same for each problem; it is simply the whole number product with no decimals. The denominator product in each instance is a 1 followed by some zeros. The resulting base ten fraction indicates where to place the decimal

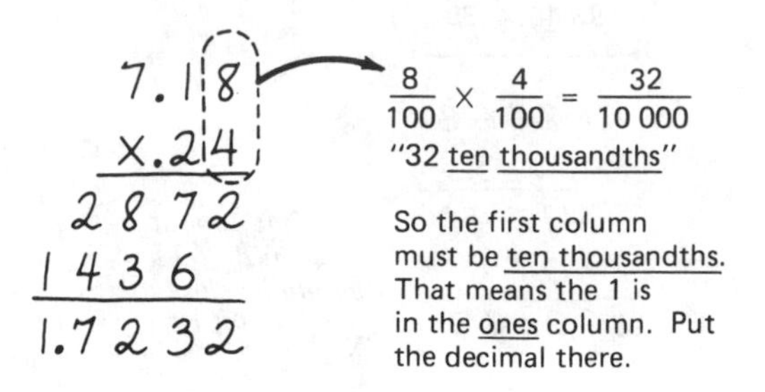

Figure 13.20
By naming the first column you can tell which column is the ones column.

point. (The number of the zeros in the denominators is the same as the number of decimal places in the factors.)

The foregoing development is only one suggestion. It is designed as an exploratory or problem-solving development rather than as an exposition. Students can generate ideas based on their current concepts rather than simply having the algorithm of counting decimal places explained to them and then possibly forgotten. In an era when less emphasis on computation is desired, the exploration itself probably has much more value than the resulting algorithm.

DIVISION

Whole Number Divisors

For whole number divisors, the whole number division algorithm is easily extended to decimal dividends and decimal quotients. In each place-value column, a partition is made and any leftovers are traded to the next column. Since the trade is always 1 for 10 regardless of what two columns are involved, trading in columns to the right of the decimal point is the same as trading to the left of the decimal. Place-value columns in the quotient correspond to those in the dividend. These ideas are illustrated in Figure 13.21.

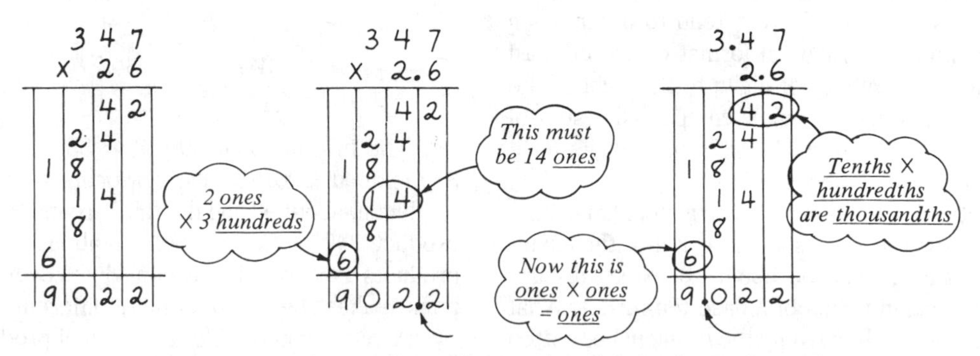

Figure 13.19
Three related products.

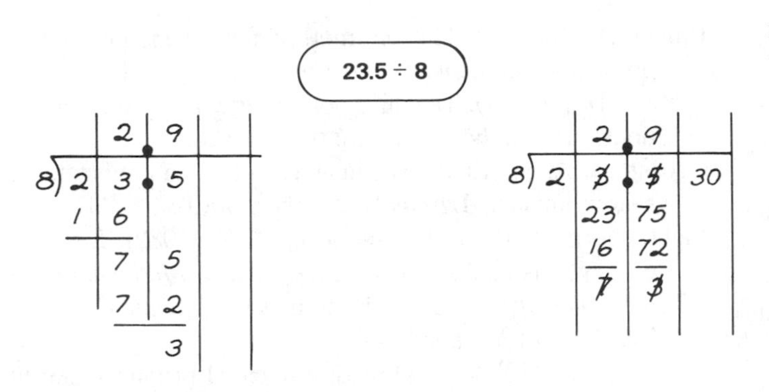

Trade 2 tens for 20 ones, making 23 ones.
Put 2 ones in each group or 16 in all.
That leaves 7 ones.

Trade 7 ones for 70 tenths, making 75 tenths.
Put 9 tenths in each group or 72 in all.

Trade the 3 tenths for 30 hundredths.

(And continue trading for smaller pieces as long as you wish.)

Figure 13.21
Extension of the division algorithm.

Decimal Divisors

It is conceptually difficult to think of partitioning things into fractional sets. When the divisor is a decimal number such as 0.6 or 2.6, the standard algorithm is based on the idea of changing the problem to one that will have the same answer but with a whole number divisor. For example, each of the following problems has the same answer:

5.824)0.26 58.24)2.6 582.4)26

A discussion of these three problems might begin with an estimate of the three answers, as was done for the multiplication algorithm. Others may wish to experiment with a calculator. What if the decimal point were in a different position in the divisor, such as 0.026? Where would the decimal point be in the dividend to produce the same result?

From this initial exploration of related divisions, suggest that fractions could help in understanding where to place the decimal points. Have students express each of the three divisions as a fraction.

$$\frac{5.824}{0.26} \qquad \frac{58.24}{2.6} \qquad \frac{582.4}{26}$$

If each division gives the same answer, then each fraction must stand for the same number. In other words the fractions are equivalent. An examination of these fractions will suggest that if the first is multiplied top and bottom by 100, the result is the third fraction. If the second is multiplied top and bottom by 10, the result is the same. But this, of course, is the way equivalent fractions are generated.

Multiplication of numerator and denominator by the same number is the same as multiplying the dividend and divisor by the same number. Shifting a decimal point to the right is the same as multiplying by a power of 10. (See Figure 13.22.) Once again, the discussion of the rationale may well be more important than mastery of the rule.

$$\frac{.26}{5.824} = \frac{.26 \times 100}{5.824 \times 100} = \frac{26}{582.4}$$

therefore:

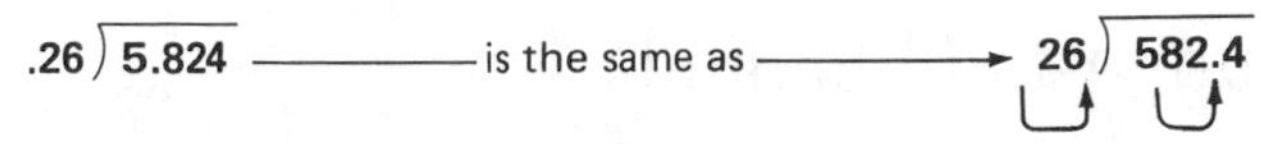

.26) 5.824 —— is the same as ——→ 26) 582.4

Figure 13.22
A rationale for "shifting" the decimal point.

For Discussion and Exploration

1. Examine textbooks for one grade level between fifth and eighth for the development of decimal concepts. How would you modify or add to the textbook presentation of this topic if you were teaching at one of these grade levels? Do not forget to check in the teacher's edition for good teaching suggestions.
2. One way to order a series of decimal numbers is to annex zeros to each number so that all numbers have the same number of decimal places. For example, rewrite

   0.34 as 0.3400

   0.3004 as 0.3004

   0.059 as 0.0590

   Now ignore the decimal points and any leading zeros and order the resulting whole numbers. Discuss the merits of teaching this approach to children.
3. Use models of your choice to show why ⅗ is 0.6 or why ⅔ is about 0.667. Similarly, use models to explain the equivalencies starting with the decimal form. Why is 0.625 = ⅝ or 0.5 = ½?
4. Use a drawing or a model to think through and explain the solutions to the six percent problems on p. 219.
5. Talk one-on-one with some seventh- or eighth-grade students. First find out if they can do simple fraction exercises that ask for the part, the whole, or the fraction given the other two (as in Chapter 11). Encourage them to make drawings and give explanations. Next, ask them to solve simple percent word problems that require the same type of reasoning using simple fractions and compatible num-

bers. Compare students' abilities with these mathematically identical problems.

6. One approach to decimal computation algorithms is to simply provide the standard rules for lining up decimals (addition and subtraction), counting decimal places (multiplication), and moving decimal points (division), and then practice with each rule. The spirit suggested in this chapter is an informal exploration with rules growing out of informal discussion. Between these extremes is a careful exploration of how and why each algorithm works. What are some of the pros and cons of each approach?

Suggested Readings

Allinger, G. D., & Payne, J. N. (1986). Estimation and mental arithmetic with percent. In H. L. Schoen (Ed.), *Estimation and mental computation.* Reston, VA: National Council of Teachers of Mathematics.

Bennett, A. (1982). *Decimal squares: Step by step teacher's guide, readiness to advanced levels in decimals.* Fort Collins, CO: Scott Resources, Inc.

Boling, B. (1985). A different method for solving percentage problems. *Mathematics Teacher, 78,* 523–524.

Glatzer, D. J. (1984). Teaching percentage: Ideas and suggestions. *Arithmetic Teacher, 31*(6), 24–26.

Grossman, A. S. (1983). Decimal notation: An important research finding. *Arithmetic Teacher, 30*(9). 32–33.

Lichtenberg, B. K., & Lichtenberg, D. R. (1982). Decimals deserve distinction. In L. Silvey (Ed.), *Mathematics for the middle grades (5–9).* Reston, VA: National Council of Teachers of Mathematics.

Vance, J. M. (1986a). Estimating decimal products: An instructional sequence. In H. L. Schoen (Ed.), *Estimation and mental computation.* Reston, VA: National Council of Teachers of Mathematics.

Vance, J. M. (1986b). Ordering decimals and fractions: A diagnostic study. *Focus on Learning Problems in Mathematics, 8*(2), 51–59.

Wagner, S. (1979). Fun with repeating decimals. *Mathematics Teacher, 26,* 209–212.

Zawojewski, J. (1983). Initial decimal concepts: Are they really so easy? *Arithmetic Teacher, 30*(7), 52–56.

CHAPTER 14

DEVELOPING THE CONCEPTS OF RATIO AND PROPORTION

Proportional Reasoning

SOME EXAMPLES FOR EXPLORATION AND DISCUSSION

So that the reading of this chapter is based on some common ground, several exercises are provided here for your exploration. Each embodies the type of relationships that make up the concepts of ratio and proportion. Therefore, rather than begin this topic as many basal textbooks do, with a definition of ratio, you can begin with some intuitive notions. It does not matter what your personal mathematical acquaintance with ratio actually is. Try each of the following. You may wish to discuss and explore the exercises in a group or with a friend.

Short & Tall

Mr. Green sells trophies based on their height. A short trophy measures six paper clips tall as shown in Figure 14.1. According to his price chart, this short trophy sells for $4.00. Mr. Green has a taller trophy that is worth $6.00 according to the same scale. How many paper clips tall is the $6.00 trophy?

Laps

Yesterday, Mary counted the number of laps she ran at track practice and recorded the amount of time it took. Today she ran fewer laps in more time than yesterday. Did she run faster, slower, or about the same speed today, or can't you tell? What if she had run more laps in more time?

Similar Shapes

Place a piece of paper over the dot grid in Figure 14.2. Using the dots as a guide, draw a shape that is *like* the one shown but larger. How many different shapes that are like the given shape can you draw and still stay within the grid provided?

Dot Mixtures

On a piece of paper, draw a loop about 2 inches across. Fill the loop with black and white dots in a random mixture that appears to be the same mix of black and white dots as in Figure 14.3. Draw your dots about the same size and with the same spacing as the dots in the figure. Do not count the dots in either drawing until you are done. Try to get both mixes of black and white to look the same. When finished, count and compare

Figure 14.1
The Short and Tall trophy problem.

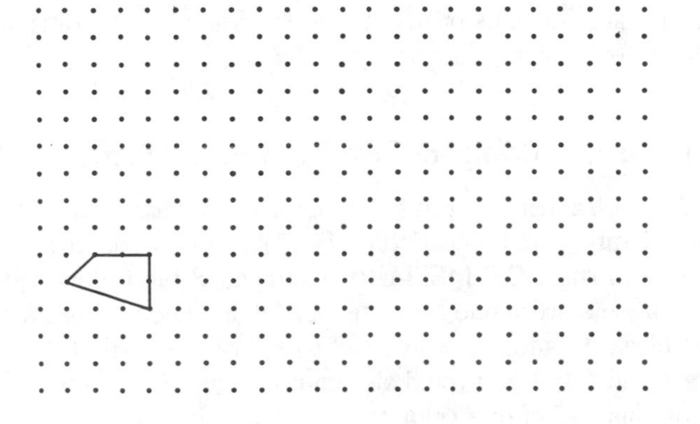

Figure 14.2
Draw some shapes just like this only larger.

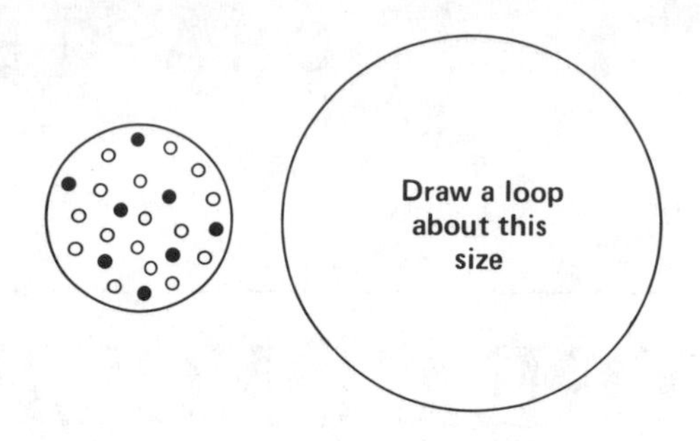

Fill a larger loop with black and white dots in about the same mixture. Do not count the dots until you are finished.

Figure 14.3
Dot mixture.

the number of black and white dots in the book with number of black and white dots in your drawing. How can these counts help you decide if your drawings have the same mixtures? Can you adjust the number of either or both colors so that you could argue that they are in the same mix as those in the book?

EXAMPLES OF RATIOS IN DIFFERENT CONTEXTS

A *ratio* is an ordered pair of numbers or measurements that are used to express a comparison between the numbers or measures. This succinct definition covers a wide range of situations. Mathematically, all ratios are essentially the same. However, from the student viewpoint, ratios in different settings or contexts may present very different ideas.

All Fractions Are Ratios

Both fractions and ratios are comparisons. Fractions are a means of comparing parts to the whole. The fraction 3/4 is a comparison of 3 parts to the 4 parts that make up the whole. Both the whole and the part are "measured" in fourths. Such a comparison of parts to wholes is also a ratio. That is, all fractions are ratios. However, it is not correct to say that all ratios are fractions or that a ratio *is* a fraction. The ratio of 12 to 0, for example, is not a fraction.

Ratios Can Compare Two Parts of the Same Whole

While a fraction is always a comparison of parts to the whole, some ratios are comparisons of one part of a whole to another part. In the DOT MIXTURE example, 8 black dots and 16 white dots are in one whole set of 24 dots. The ratio or fraction of black dots to the whole is 8 to 24 or one third of the dots are black. But you could also compare the black to the white dots instead of one color to the total. A comparison of black to white is not a fraction but a ratio of a part to a part. The black and white dots are in the ratio of 8 to 16 or 1 to 2. Other examples of part-to-part ratios are Democrats to Republicans, boys to girls, sailboats to speedboats, or buses to people.

Ratios Can Be Expressions of Rates

Both part-to-whole and part-to-part ratios are comparisons of two measurements of the same type of thing. The measuring unit is the same for each value. A ratio can also be a rate. A *rate* is a comparison of the measures of two different things or quantities. The measuring unit is different for each value.

In the SHORT & TALL example, the comparison between paper clips and money is an example of ratio that is a rate; in this case, dollars per paper clip. If you figured out the ratio of $2.00 for every three clips, then you were able to determine that the tall trophy should be nine clips tall. All prices are rates and are also ratios; 69 cents each, three for a dollar, or 12 ounces for $1.39. Each is a ratio of money to a measure of quantity.

The LAPS example involves comparisons of time to distance, another example of rate. There are no numbers involved in that example so that the reasoning is qualitative rather than quantitative, but the ratio concept is basically the same. All rates of speed are ratios of time to distance: for example, driving at 55 miles per hour, or jogging at 9 minutes per mile.

Other rates that we may not think of as ratios are miles per gallon, square yards of coverage per gallon of paint, inches of tape per roll, passengers per busload, or roses per bouquet. Changes between two units of measure are also rates or ratios. Examples include inches per foot, milliliters per liter, and centimeters per inch. The distance scales on a map are another example of ratio as rate.

Other Examples of Ratio

In geometry, the ratio of the circumference of a circle to the diameter is designated by the Greek letter π (pi). This is one example of a ratio that is not a rational number. Among any set of similar geometric figures, the ratios of corresponding parts are always the same as in the SIMILAR SHAPES example. The diagonal of a square is always $\sqrt{2}$ times a side. The trigonometric functions can be developed from ratios of sides of right triangles.

The probability of an event occurring is a ratio. The chances of rolling a sum of 7 with two dice are 6 in 36 or 1 to 6. Probabilities are part-to-whole ratios: the number of favorable outcomes to the total number of possible outcomes.

In physics, ratios abound. For example, the volume of a fixed amount of gas is determined by the ratio of the temperature to the pressure (Boyle's law). The laws governing the forces of pulleys, levers, and gears all involve ratios.

In music you find 4 notes to a measure, 8 whole notes or 12 tones in an octave. Pleasant-sounding chords involve tones with wavelengths in special ratios.

In nature the ratio known as the *golden ratio* is found in

many spirals from the nautilus shell to the swirls of a pine cone or a pineapple. Artists and architects have utilized this same ratio in creating shapes that are naturally pleasing to the eye.

PROPORTIONS

As students begin to experience and reflect on a variety of examples of ratios, they will also begin to see different comparisons that are in the *same ratio.* In the SIMILAR SHAPES example, the ratio of any two sides of a small shape is the same as the ratio of the corresponding sides of a larger but similar shape. In DOT MIXTURES, the ratios of black to white dots in two different sets can be seen to be the same even though there are many more dots in one loop than another.

A *proportion* is a statement of equality between two ratios. Different notations for proportions can be used. For example,

$$3{:}9 = 4{:}12 \qquad \text{or} \qquad \frac{3}{9} = \frac{4}{12}$$

These might be read "3 is to 9 as 4 is to 12" or "3 to 9 is in the same ratio as 4 to 12."

A ratio that is a rate usually includes the units of measure when written. For example,

$$\frac{\$12.50}{1 \text{ gallon}} = \frac{\$37.50}{3 \text{ gallons}}$$

12 inches per foot = (is the same as) 36 inches per 3 feet

There is a very real distinction between a proportion and the idea of equivalent fractions. Two equivalent fractions are different symbolisms for the same quantity or amount; they represent the *same rational number* in different forms. If one bag has 3 red and 9 white balls, and another has 4 red and 12 white, the number of red balls in each bag is clearly different but the ratios of red to white and red to total, that is, the comparisons of these quantities, is the same for both bags.

Finding one number in a proportion when given the other three is called *solving a proportion.*

PROPORTIONAL REASONING AND CHILDREN

Proportional reasoning includes, but is much more than, an understanding of ratios. It involves the ability to compare ratios and to predict or produce equivalent ratios. It requires the ability to compare mentally different pieces of information, to make comparisons, not just of the quantities involved but of the relationships between quantities. As you can see from the examples at the outset of this chapter, proportional reasoning involves quantitative thinking as well as qualitative thinking, yet is not dependent on a skill with a mechanical or algorithmic procedure. (Those interested in Piaget's stages of development theory will note that proportional reasoning has generally been regarded as one of the hallmarks of the formal operational thought stage, acquired at about the time of adolescence.)

A Developmental Perspective

Considerable research has been conducted to determine how children reason in various proportionality tasks and to determine if developmental and/or instructional factors are related to proportional reasoning. (For example, see Karplus, Karplus, & Wollman, 1974; Karplus, Pulos, & Stage, 1983; Noelting, 1980; Post, Behr, & Lesh, 1988.) The results of these research efforts suggest that the development of proportional reasoning in children is a complex issue. It is quite clear that proportional reasoning is not an automatic result of natural growth and development. Many adults have not acquired the skills of proportional thinking, and the performance of junior high school students on proportional tasks has been disappointing.

While perhaps not definitive, research does provide some direction for how to help children develop proportional thought processes. Some of these ideas are outlined here.

1. Provide ratio and proportion tasks in a wide range of settings, since how children (and adults) approach tasks is influenced greatly by context. These contexts might include situations involving measurements, prices, geometric and other visual contexts, and rates of all sorts.
2. Encourage reflective thought, discussion, and experimentation in predicting and comparing ratios.
3. Help children relate proportional reasoning to existing processes. Specifically, the concept of unit fractions is very similar to unit rates. Research indicates that the use of a unit rate for comparing ratios and solving proportions is the most common approach used by junior high students even when cross-product methods have been taught. This approach is explained later.
4. Recognize that the use of symbolic or mechanical methods, such as the cross-product algorithm, for solving proportions does not develop proportional reasoning and should not be introduced until students have had considerable experience with more intuitive and conceptual methods.

Need for Experience and Reflective Thought

Of the foregoing ideas, perhaps the most significant are experience and reflective thought—the opportunity to construct the type of comparative thinking that is fundamental to proportional reasoning. Just as you are able to reflect on and talk informally about the experiences suggested at the outset of this chapter, children must have real experiences to think about.

Each of those four beginning activities involved your active participation. The drawings or models were not answer-getting devices but a means of helping you construct relationships. The example of running laps involved comparisons of relative rates in a qualitative sense, since no numbers were involved. The drawings of similar figures on the dot grid and the visual determination of black and white dot ratios were based initially on intuitive ideas that could then be tested numerically through counting. The exercises provided at least some opportunity for you to construct relationships between quantities or measures. You were then able to think about these relationships and apply them to similar but different situations. You had to construct each ratio mentally before it could be applied proportionally to the next measurement. If you worked together with someone, you were also able to put your proportional thinking into words in order to help clarify it or reflect upon it. Children also require these opportunities if they are to develop their abilities to reason proportionally.

A Contrast with a Mechanical Approach

Suppose that you have been given a procedure for "solving proportions" of the following form:

$\frac{a}{b} = \frac{x}{c}$	Step 1. Set up a proportion.	$\frac{a}{b} = \frac{c}{x}$
$a \times c = b \times x$	Step 2. Cross multiply.	$a \times x = b \times c$
$x = \frac{a \times c}{b}$	Step 3. Solve for the unknown.	$x = \frac{b \times c}{a}$

After exercises involving these procedures, you are given problems like SHORT & TALL and guided to set up the proportion

$$\frac{4 \text{ dollars}}{6 \text{ clips}} = \frac{6 \text{ dollars}}{X \text{ clips}}$$

Then ignoring the fact that it does not make a lot of sense to multiply clips times dollars, you cross multiply ($4 \times x = 6 \times 6$) and solve for x by division. How would such an activity help you conceptualize ratio in any general way or even help you solve any additional exercises?

Sixth-, seventh-, and eighth-grade curriculums have traditionally focused on the procedural knowledge associated with ratio and proportion. That includes deciding if two ratios given symbolically are equal or unequal (check the cross products) and solving proportions as in the preceding example. This emphasis has been due partly to standardized testing, which has required such skills, and partly to the difficulty of specifying thought processes, such as proportional reasoning, in lists of behavioral objectives that many school systems have developed. It takes considerable class time to provide students with experiences that permit intuitive thought and reflection. It requires that teachers enter into free and open discussions as students begin to discover relationships and verbalize them in their own way. Situations that are very different on the surface must be explored before commonalities show through. The work of the Rational Number Project (Post, Behr, & Lesh, 1988) has indicated that contextual and numerical differences in problems play a significant role in how children approach proportion tasks and how difficult they are to solve. Prescriptions for how to solve proportion problems focus almost entirely on procedures. Such a focus does not help students think about relationships in any global manner.

Time can be found for these exciting activities over the sixth- to eighth-grade years, and procedural knowledge can easily be connected to the resulting concepts. The ultimate responsibility for providing such experiences rests with teachers.

Exploration of Intuitive Methods for Solving Proportions

Other, more meaningful approaches than cross products need to be explored first, with the symbolic method left until much later. To illustrate some of these other approaches, consider the following examples.

1. Tammy bought 3 widgets for $2.40. At the same price, how much would 10 widgets cost?
2. Tammy bought 4 widgets for $3.75. How much would a dozen widgets cost?

Before reading further, it may be interesting to solve these two problems using an approach for each that seems easiest or most reasonable to you.

In the first situation, it is perhaps easiest to determine the cost of one widget, the unit rate or unit price. This can be found by dividing the price of three widgets by 3. Multiplying this unit rate of $0.80 by 10 will produce the answer. This approach can be referred to as the *unit-rate method* of solving proportions.

In the second problem, a unit-rate approach could be used, but the division does not appear to be easy. Since 12 is a multiple of 4, it is easier to notice that the cost of a dozen is three times the cost of four. Using this *factor-of-change* approach on the first problem is possible but awkward. The factor of change between 3 and 10 is $3\frac{1}{3}$. Multiplying the $2.40 by $3\frac{1}{3}$ would produce the correct answer. While the factor-of-change method is a useful way to think about proportions, it is most frequently used when the numbers are compatible. In other situations, students generally utilize a unit-ratio approach (Post, et al., 1988). Both methods should be explored.

Notice that the unit-rate method is exactly the same as that used in solving many fraction problems. Compare problem 1 with the following fraction exercise: If a given rectangle is $\frac{3}{10}$ of the whole, how big is the whole? Dividing the rectangle into three parts forms a unit fractional part. The unit part can then be "multiplied" to produce any related part, including the whole. (Earlier, we were counting unit fractional parts.) It is developmentally sound to relate new ideas to those that students have already developed.

As illustrated in Figure 14.4, equivalent ratios or rates can be illustrated graphically. Graphs provide another way of thinking about proportions and connect proportional thought to algebraic interpretations. Graphs of equal ratios are a good place for students to see the value of algebraic ideas and to interpret relationships in graphical form. All graphs of equivalent ratios are straight lines through the origin. If the equation of one of these graphs is written in the form $y = mx$, the slope m is always one of the equivalent ratios.

These intuitive approaches to ratio and proportion as well as the procedural algorithm of using cross products should all be explored in the middle or junior high school years. No method should be considered over another, and none should be taught in a rote manner. Rather, activities and guided discussions should help students learn to think proportionally in a variety of ways and contexts in order to make this form of reasoning profitable. Failure to develop an adequate facility with proportional reasoning can preclude study in a variety of disciplines, including algebra, geometry, biology, chemistry, and physics (Hoffer, 1988b).

Informal Activities to Develop Proportional Reasoning

The activities suggested in the following sections each provide a different opportunity for the development of proportional reasoning. While some are easier than others, they should not be interpreted as being in any definitive sequence. Nor are the activities designed to directly teach specific methods or algorithms for solving proportions. Some activities can be modified or repeated using different numbers to produce more or less difficult situations or to suggest different thought processes. Engaging in these and similar activities over the early adolescent years can help children develop proportional thought.

SELECTION OF EQUIVALENT RATIOS

The basic approach in the following activities is to present a ratio in some form and have students select an equivalent ratio from among others presented. The focus should be on an intuitive rationale for why the two ratios selected are the same ratio—why they "look" the same. In some situations, numerical values will play a part and help students begin to develop numeric methods to validate and explain their reasoning. In later activities, students will be asked to construct or provide an equivalent ratio without choices being provided.

Activity 14.1

Choose two fraction strips or Cuisenaire rods to make a pair in some ratio. Present other pairs of strips, some of which are in the same ratio and some not as shown in Figure 14.5. Notice that a unit ratio such as 3 to 1 is easier than a nonunit ratio such as 2 to 5. The same activity can be done with any fraction model that allows for units of varying sizes. Simple lines or rectangles can be drawn on paper instead of using a manipulative model, but the opportunity for exploration is reduced.

Activity 14.2

Have students classify rectangles. Duplicate a collection of 10 or more different rectangles for students to cut from paper. Prepare the rectangles so that each is similar to two or more other rectangles. Make three or four sets of similar rectangles. (Figure 14.6 illustrates an easy way to draw similar rectangles.) Simply have the students put those rectangles together that are "alike." When they have grouped the rectangles, have them make additional observations to justify why

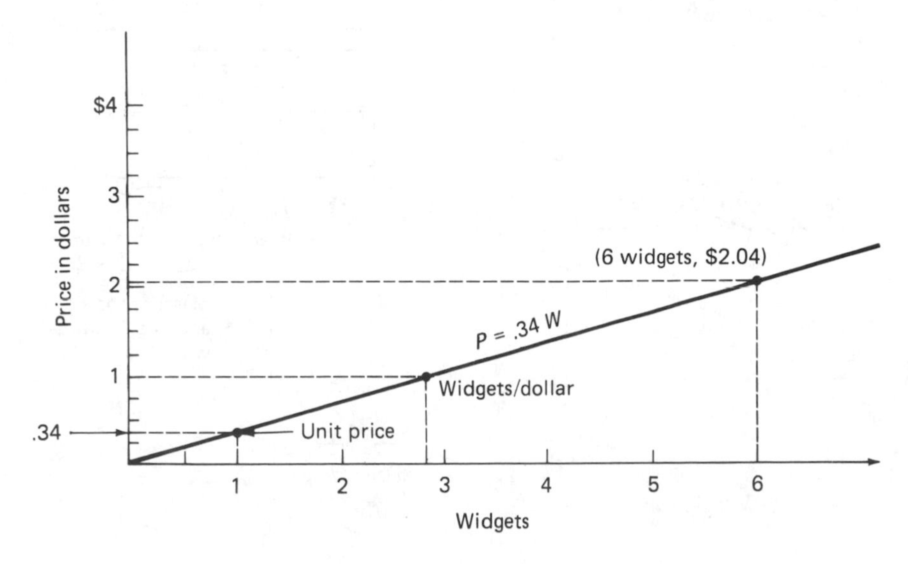

Figure 14.4
Graph of price-to-item ratios.

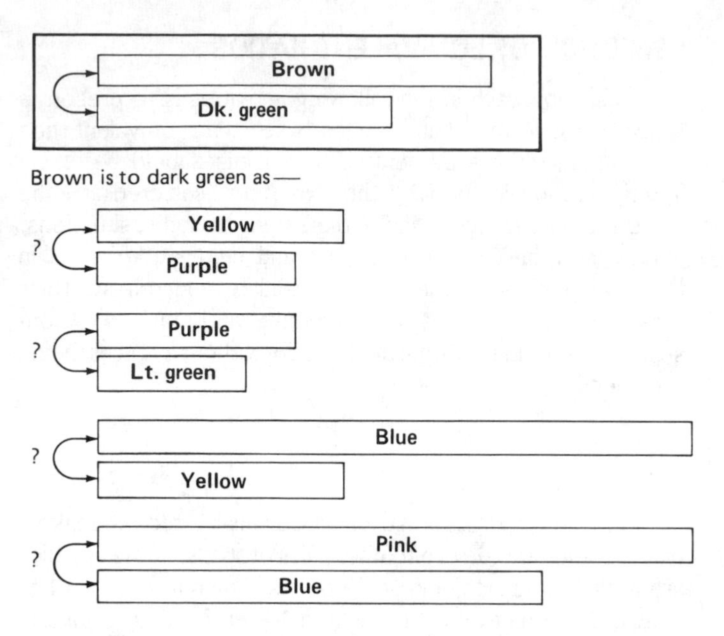

Figure 14.5
Which two *pairs are in the same ratio as brown to dark green?*

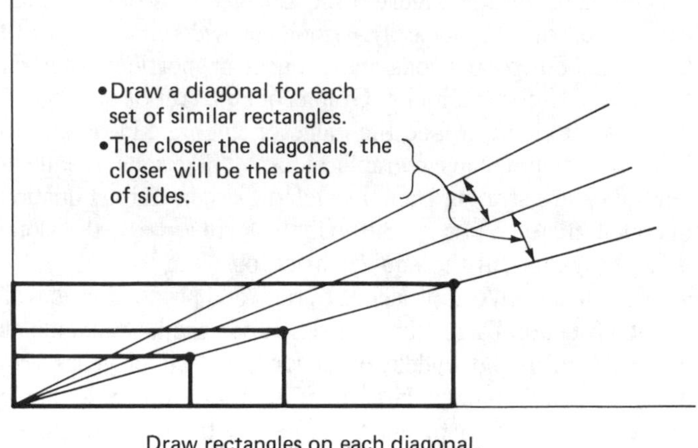

Figure 14.6
Drawing similar rectangles.

they go together: the more obvious the differences in the shapes the easier the task. Encourage as many ways as possible to describe how the rectangles are alike. Some ideas are suggested in Figure 14.7. The same task can be done with right triangles or with any set of triangles as long as there are pairs which are similar.

Activity 14.3

Make cards, transparencies, or worksheets with four or five loops of black and white dots. The dots should be in random arrangements as in the DOT MIXTURES example. Prepare the dot loops so that at least two loops have dots in the same ratio of black to white but with different total amounts. The task is to select loops of dots where the mix-

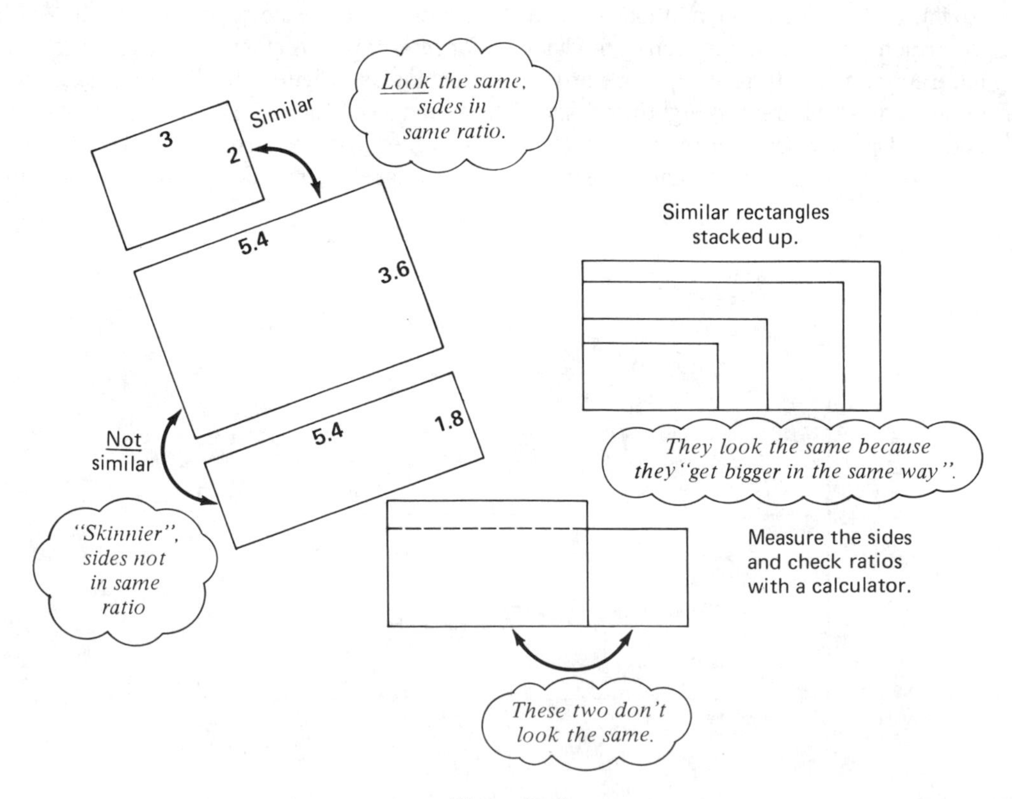

Figure 14.7
Similar rectangles have sides in the same ratio.

tures look to be about the same. Begin with collections where the ratios are very different, such as 1 to 2, 1 to 4, and 1 to 10. In later exercises narrow the differences to ratios such as 1 to 2, 3 to 5, and 3 to 4. Another variation is to adjust the spread of the dots as well as the numbers. (To avoid having to draw new sets for each exercise, prepare lots of different dot sets on separate paper and record the numbers of each color on the back. Assemble different collections on a photocopier to create different exercises.) Students can count dots within sets they have matched visually and compare part-to-part ratios as well as part-to-whole fractions.

Activity 14.4

In an activity similar to the dot mixtures, prepare cards with two distinctly different objects as shown in Figure 14.8. Given one card, students are to select a card where the ratio of the two types of objects is the same. This task moves students to a numeric approach rather than a visual one and is an introduction to the notion of rates as ratios. A unit rate is depicted on a card that shows exactly one of either of the two types of objects. In the example shown, the card with three boxes and one truck provides one unit rate. (A unit rate for the other ratio is not shown. What would it be?) The use of objects paired with coins or bills is a good introduction to price as a ratio.

The last activity provides a good example of how numbers can affect thought processes. Different ratios should be used in various examples to encourage both unit rate and factor-of-change approaches. Each of the following pairs of ratios are equivalent, but the numbers in the ratios vary, depending on whether they are multiples of the other number *within* the ratio or *between* the two ratios.

3 to 3...........5 to 5	Within and 1 to 1
3 to 12.........5 to 20	Within
2 to 6..........8 to 24	Between and within
3 to 7..........6 to 14	Between
4 to 6.........10 to 15	Neither

When there is an even multiple within, the ratio can easily be converted to a unit ratio. For example 3 to 12 is the same as 1 to 4. Pairs that involve between multiples but not within lend themselves to the factor-of-change approach; 6 to 14 is an even multiple of the 3 to 7 ratio.

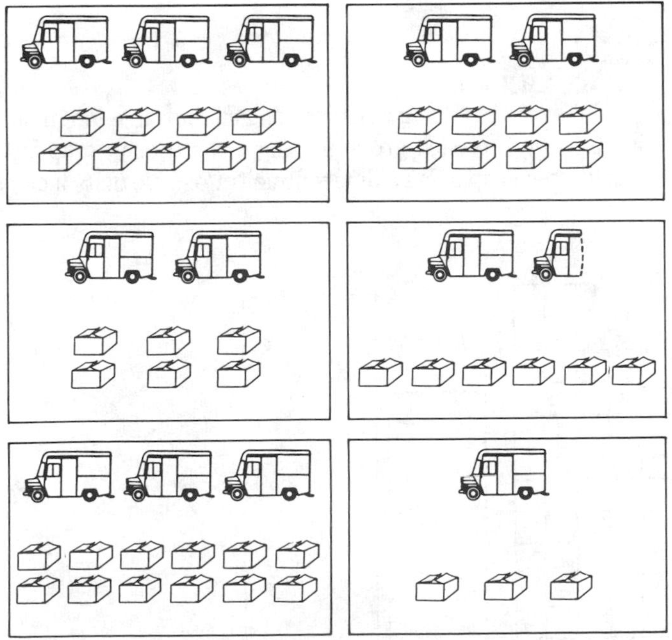

On which cards is the ratio of trucks to boxes the same? Also, compare trucks to trucks and boxes to boxes.

Figure 14.8
Rate cards: Match cards with the same rate of boxes per truck.

SCALING ACTIVITIES

Scaling activities involve filling in charts where paired entries are related in some way. The format of a chart is not important, and different formats can be used. Scaling up is a matter of providing entries with larger numbers, and scaling down is entering smaller numbers. The following are some examples of scaling up:

1 foot→12inches	1 can→\$0.55
2 feet→24 inches	2 cans→\$1.10
3 feet→? inches	3 cans→?

As illustrated, the items in the charts can be related measures such as time (minutes to hours), money (nickels to quarters), weight (pounds to ounces or pounds to kilograms), or common pairings such as crayons to boxes, hands to fingers, wheels to tricycles, and others. They can also be arbitrary ratios, such as a man who eats three bagels for every two bananas.

In a scaling-down activity, one or two later entries in the list or table are given, and students are asked to provide entries earlier or at the beginning of the list as in the following examples:

28 days→4 weeks	12 cups→24 sugars
21 days→3weeks	11 cups→22 sugars
? days→2 week	10 cups→? sugars

Be careful not to make these exercises too long or tedious. Allow students to use repeated addition or subtraction as well as multiplication and division. Also, be sure to permit the use of calculators to make computation easy.

Notice that in the examples so far the unit ratio is easy to determine without use of fractions. As long as unit ratios are easily found, scaling activities can be done at an early grade. Later, the use of fractions and decimals can be used in a scaling-type exercise. Students can scale down with one division to get a unit ratio and multiply to get a particular requested entry. This is illustrated in the following example.

3 boxes for \$2.25→5 boxes for ?
1 box is \$0.75 (unit ratio), so 5 boxes is 5 × \$0.75 = \$3.75

Suppose that the price of nine boxes or some other multiple of 3 was requested in this example. Then scaling can be done using multiples of the given ratio. Scaling can be done up or down without using a unit ratio.

Minutes:	5	?	15	20	25	30	35
Widgets made:	?	14	21	?	?	?	?

The next two activities are related to scaling but are less structured.

Activity 14.5

Play "What's in the Bag," an activity that involves informal probability concepts as well as ratio. Put colored cubes or other counters of two colors in a bag. For example, use four red and eight blue. Explain that there are cubes of different colors in the bag, but do not tell students the number of cubes or number of colors. Shake the bag and have students draw out a cube, record the color, and return the cube to the bag. After 10 or 15 trials, ask students how many of each color they think might be in the bag and record the guesses. After some more trials, you could ask what is the fewest possible number of cubes that they think could be in the bag. Then you can give them one of the following clues: the total counters that are in the bag, or the number of cubes of one of the colors. With this information see if they can predict how many of each color are in the bag. "What if there were more cubes? What are other numbers of each color that might be in the bag?" The discussion is useful even if the students do not guess the correct ratio of colors. You can continue pulling out blocks to see how the ratio stays approximately, but not exactly, the same. Besides changing the ratio of colors on different days, you can also add a third color. After seeing the actual contents of the bag, discuss what other numbers of each color would produce the same result. Groups might explore drawing cubes from bags with equal ratios of colors but different numbers and compare the results.

Activity 14.6

Have students make a graph of the data from a collection of equal ratios that they have scaled or discussed. Figure 14.4 is a graph of price-to-items ratios. The graph in Figure 14.9 is of the ratios of two sides of similar rectangles. If only a few ratios have actually been computed, the graph can be drawn carefully and then used to determine other equivalent ratios. This is especially interesting when there is a physical model to coincide with the ratio. In the rectangle example, students can draw rectangles with sides determined by the graphs and compare them to the original rectangles. A unit ratio can be found by locating the point on the line that is directly above or to the right of the number 1 on the graph. (There are actually two unit ratios for every ratio. Why?) Students can then use the unit ratio to scale up to other values and check to see that they are on the graph as well.

MEASUREMENT ACTIVITIES

In these activities, measurements are made to create physical models of equivalent ratios in order to provide a tangible example of a proportion as well as look at numerical relationships.

Activity 14.7

Cut strips of adding machine tape all the same length and give one strip to each group in your class. Each group is to measure the strip using a different unit. Possible units include

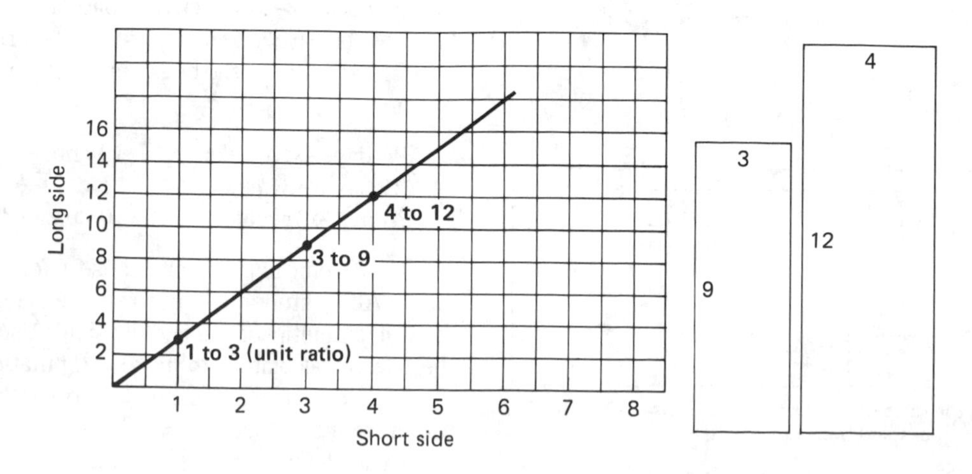

Figure 14.9
Graph shows ratios of sides in similar rectangles.

different Cuisenaire rods or fractions strips, a piece of chalk, a pencil, the edge of a book or index card, or standard units such as inches or centimeters. When everyone has measured the strip, ask for the measure of one of the groups and display the unit of measure. Next, hold up the unit of measure used by another group and have the class compare it with the first unit. See if the class can estimate the measurement that the second group found. The ratio of the measuring units should be the inverse of the measurements made with those units. For example, if two units are in a ratio of 2 to 3, the respective measurements will be in a ratio of 3 to 2. Examine measurements made with other units. Finally, present a unit that no group has used and see if the class can predict the measurement when made with that unit.

Activity 14.8

An expansion of the preceding activity is to provide each group with an identical set of four strips of quite different lengths. Good lengths might be 20 cm, 50 cm, 80 cm, and 120 cm. As before, each group measures the strips using a different unit. Next have each group make a bar graph showing their measurements. Have all groups use the same scale for their graphs so that those with short units will have long bars in their graphs and those with long units will have short bars. Before displaying the graphs, have each group also make a circle graph representing the total length of all four strips. This is easily done by adding the measured lengths of the strips and then using a calculator to divide the length of each strip by the total. The results rounded to two decimal places will be the percent of the total contributed by each strip. By tracing a circle around a hundredths disk (Appendix C) and using the hundredths marks, you can graph the percent of each strip as a part of a circle. Have all groups make their bar and circle graphs with the strips in the same order and color them with the same colors. All of the bar graphs will be different heights due to the use of different units, but the pie graphs will all be nearly identical. Now you can discuss a variety of different ratios and proportions as suggested in Figure 14.10. The fact that certain ratios are the same and that the pie graphs are all the same provides vivid examples of proportionality.

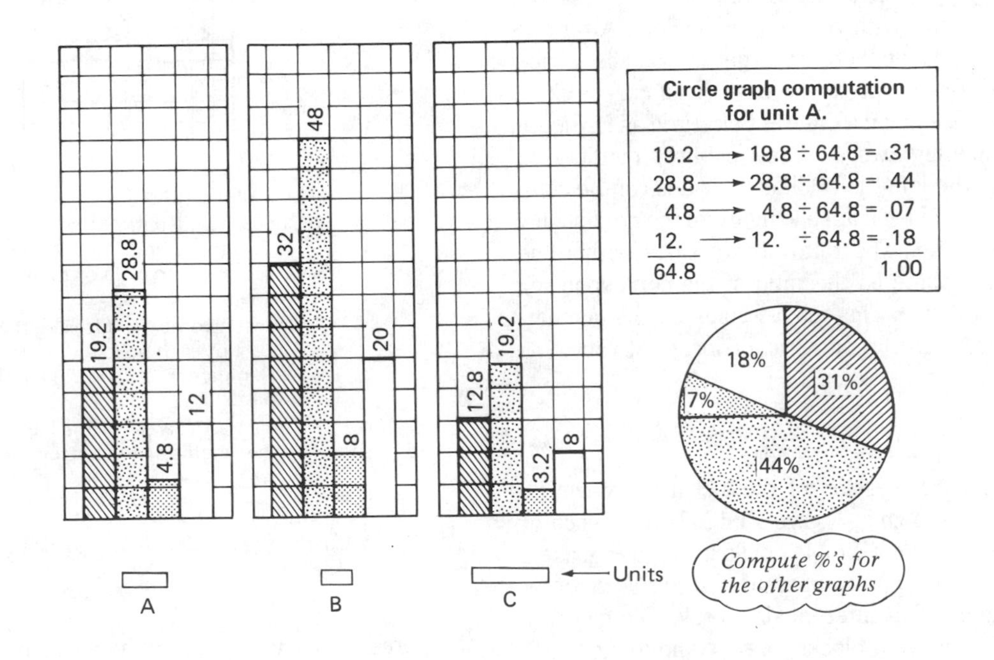

Within-ratios

Example:

$\frac{19.2}{12} = 1.6, \quad \frac{32}{20} = 1.6, \quad \frac{12.8}{8} = 1.6$

Between-ratios

Example:

$\frac{A}{B} = \frac{28.8}{48} = .6 \qquad \frac{A}{B} = \frac{19.2}{32} = .6$

Figure 14.10

Three groups measure 4 strips of 80, 120, 20 and 50 cm. Each group uses a different measuring unit. What like ratios can be observed?

Activity 14.9

An activity involving weight and density can be conducted in a similar manner to the previous one. Instead of four different length strips, provide each group with four small containers of different sizes. The four containers must be the same for each group. Have each group fill their containers with a different "filler." Select fillers that vary greatly in density. For example, use dry oatmeal, rice, sand, and small metal washers. Each group weighs the contents of their containers (not including the containers) to the nearest gram and makes a bar graph of the results. A pie graph is made of the total weight of all containers. The results will be similar to the length experiment. Here, instead of different units of measure, the different densities of the fillers will produce different weights. Since the volumes are the same, the ratios of weights for each group will be the same, and each group should get about the same pie graph.

Activity 14.10

On dot paper (Appendix C), have students draw a simple shape using straight lines with vertices on the dots. After one shape is complete, have them draw a larger or a smaller shape that is "the same" or similar to the first. This can be done either on the same size or on a different size grid as in Figure 14.11. After completing two or three pictures of different sizes, the ratios of the lengths of different sides can be compared. Corresponding sides from one figure to the next should all be in the same ratio. The ratio of two sides within one figure should be the same as the ratio of the corresponding two sides in another figure. This activity connects the geometric idea of similarity with the numerical concept of ratio.*

Activity 14.11

A three-dimensional version of the last activity can be done with blocks, as shown in Figure 14.12. Using 1-inch or 2-cm wooden cubes, make a simple "building." Then make a similar but larger building and compare measures. A different size can also be made using different size blocks. To measure buildings made with different blocks, use a common unit such as centimeters. (Notice that volumes and surface areas do not vary proportionally with the edges of solids. However, there are relationships that are interesting to observe.)

In all of these activities, measurements have been made to observe equal ratios or proportions. In the following activ-

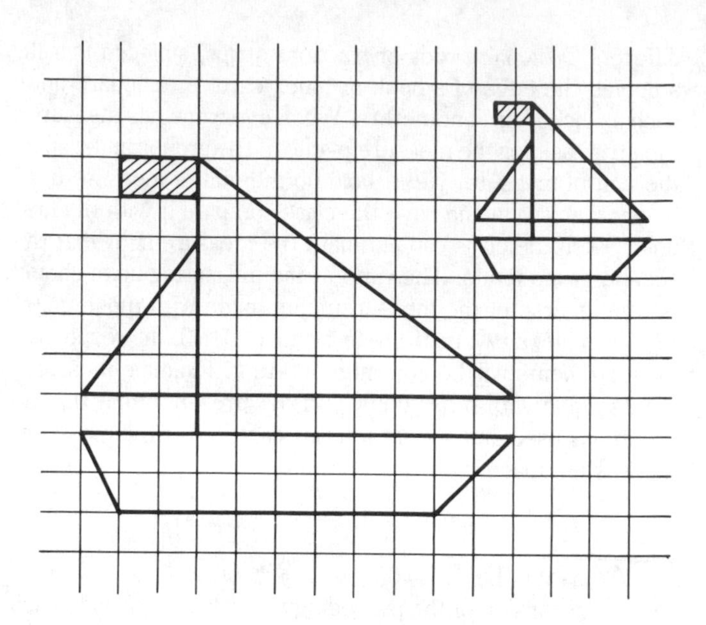

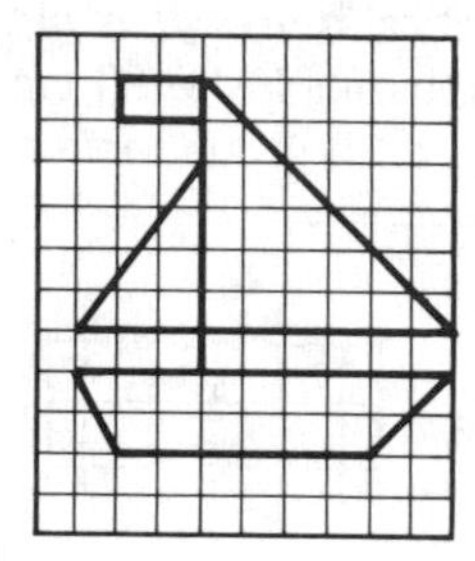

Use a metric ruler

- Choose two lengths on one boat and form a ratio (use a calculator!) Compare to ratio of same parts of the other boats.
- Choose two boats. Measure the same part of each boat and form a ratio. Compare with ratios of another part.
- Compare the areas of the big sails with the lengths of the bottom sides.

Figure 14.11
Comparing similar figures drawn on grids.

ity, perception is used without measurement to create a proportionate length or shape. Measurement follows the perceptual judgment to see how good the estimate is.

Activity 14.12

On the board, draw two lines labeled A and B as in Figure 14.13. Draw a third line, C, that is significantly different from A. Begin drawing a fourth line, D, under C. Have students tell you when to stop drawing so that the ratio of C to D is the same as the ratio of A to B. Measure all four lines and compare the ratios with a calculator. (Notice that here a single decimal number represents a ratio. How do you explain that?)

*If the area of the figures can be easily determined by counting squares or half-squares, then it is interesting to compare the ratios of areas with the ratios of lengths. Areas of plane figures vary with the *square* of the sides and are not, therefore, in the same ratio as the sides. If the ratio of corresponding sides is 1 to 3, the areas will be in the ratio of 1 to 9. (What if the sides are in the ratio of 2 to 3?)

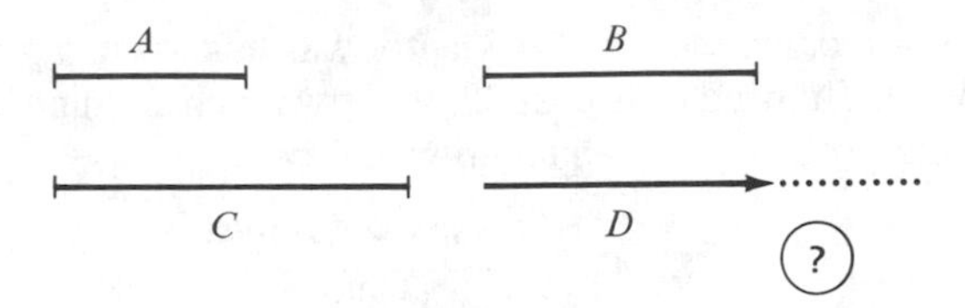

Figure 14.13
Where should line D end so that A:B as C:D?

Similar "buildings" can be made by changing the number of blocks in each dimension (factor of change) or by using different size blocks.

Figure 14.12
Similar constructions.

Other activities that have been explored can also be transformed into estimation activities. For example, prepare a figure on plain paper using graph paper placed underneath as a guide. The figure can be as simple as a right triangle or more complicated such as a drawing of a truck or a house. Make a transparency of the drawing. Students will see a very large figure with no measurements or grid. On paper with a very small grid, perhaps 0.5 cm, students attempt to draw a figure similar to the one on the overhead. While a perfectly similar figure may be difficult, there may be many that are close or even exact.

Solving Proportions

THE CROSS-PRODUCT ALGORITHM

Investigate Known Proportions

Before jumping in with the magic and simplicity of the cross-product algorithm, we can help students examine some situations in which they have already established equal ratios. Data might come from a scaling experience or perhaps from a measuring activity such as one of those discussed in the previous section. For example, suppose that the class had completed the following chart for prices of tennis balls:

Tennis balls:	3	6	9	12	15
Price:	\$2.49	\$4.98	\$7.47	\$9.96	\$12.45

Have students select any two ratios from the chart that they think should be equal and write them in an equation as fractions. One possibility is the following although any pair of ratios from the table could be used.

$$\frac{3 \text{ balls}}{\$2.49} = \frac{9 \text{ balls}}{\$7.47}$$

Now write the cross product, which for this example is

$$3 \text{ balls} \times \$7.47 = 9 \text{ balls} \times \$2.49$$

The numerical part of this equation is correct ($3 \times 7.47 = 9 \times 2.49$). Will the numbers always work out that way in a proportion? The answer is clearly "yes," but students should investigate this with real data and be personally convinced. Students could quickly set up other proportions from the table and check the cross products with a calculator.

But what about the units? Balls times dollars does not make much sense unless you want to invent a unit called "ball-dollars." But the equation in that form is not very useful anyway. A more likely situation occurs when one of the four factors in the equation is unknown. The fact that the cross products are equal can be used to solve for the missing value. Pretend that one of the four factors in the equation is unknown and, with the class, write a corresponding question or problem. For example, if the \$7.47 factor was unknown, a problem might be: "If three balls cost \$2.49, how much are nine balls?" For this problem, the equation would be

$$\$7.47 = \frac{9 \text{ balls} \times \$2.49}{3 \text{ balls}}$$

There are now two meaningful ways to interpret this equation. First, if written in the form

$$\$7.47 = 9 \text{ balls} \times \frac{\$2.49}{3 \text{ balls}} = 9 \text{ balls} \times \frac{\$0.83}{1 \text{ ball}}$$

then the unit rate is apparent. The fraction $\$2.49/3 \text{ balls}$ is equivalent to the fraction $\$0.83/1 \text{ ball}$, which is the cost per ball. It is not really necessary to compute this unit rate before multiplying by 9 to get the cost of nine balls, but students can see that the cross product can make sense in terms of unit rates.

If the equation is grouped in the following way, the factor of change is apparent:

$$\$7.47 = \frac{9 \text{ balls}}{3 \text{ balls}} \times \$2.49 = 3 \times \$2.49$$

The fraction $9 \text{ balls}/3 \text{ balls}$ tells how many sets of 3 balls are in a set of 9 balls. (As noted earlier, ratios of like items are usually written without units.) Three is a factor of change that would probably have been the intuitive choice for solving this particular problem.

Return to the original equation and select any other factor as the "unknown." A similar analysis can be made. After solving for one factor, we can always group the other side of the equation to show a unit rate times a quantity, or we can group it to illustrate a factor of change times a quantity. If both unit rates and factors of change have been used intuitively in previous investigations, then students can begin to make connections between the procedural approach of the cross-product algorithm and their conceptual understanding. (It would be a useful exercise for you to solve this tennis ball/price proportion for each of the other three values. Pretending that one value is an unknown, first make up a word problem and interpret the equation as a unit-ratio approach and then as a factor-of-change approach.)

As an aside, it is easy to see mathematically why the cross product of two equal ratios produces an equality. By definition, two ratios are equal if one is a multiple of the other. Therefore, all equal ratios are of this form:

$$\frac{a}{b} = \frac{ac}{bc} \quad \text{where } c \text{ is a nonzero rational number}$$

The cross product of this expression is clearly an equality.

Find New Values

Students should also investigate the cross-product method within a familiar context to find values that are not part of their original data. Using the same tennis ball example, suppose that balls could be purchased individually. What would the cost of 10 balls be? Help students see that any other known ratio of balls to price can be used along with the incomplete ratio of 10 balls to the unknown price. For example,

$$\frac{6 \text{ balls}}{\$4.98} = \frac{10 \text{ balls}}{x \text{ dollars}}$$

Using cross products and solving for the price of 10 balls, we find

$$x \text{ dollars} = \frac{\$4.98 \times 10 \text{ balls}}{6 \text{ balls}} = \$8.30$$

In this example, the fraction $\$4.98/6 \text{ balls}$ is the unit ratio or unit price, which is multiplied times 10 balls, or the factor of change 10/6 is multiplied times \$4.98 since 10 balls would cost 10/6 times as much as the price of 6 balls. Check to see that the result of \$8.30 is a reasonable "fit" in the original chart. (Compared with the cost of 9 and 12 balls, does this result seem reasonable?) Again, the cross-product approach not only works mathematically, but it can be interpreted in terms of the intuitive experiences of the students. Try solving for other unknown values. Consider the question, "How many balls can be purchased for \$12.00?" The cross-product algorithm produces an answer of 14.4578 balls. Does this answer make sense? Situations like this should be discussed with your students to help them see that any results must be interpreted in real contexts.

WRITING PROPORTIONS FROM A MODEL

Given a ratio word problem, students have great difficulty writing ratios in a correct proportion. The task is to write a proportion of two ratios, one of which includes the missing value. Since many apparently different equations can be written for the same proportion, some students become even more confused.

Draw a Simple Model

Rather than drill and drill in hopes that they will somehow eventually "get it," show students how to sketch a simple model or picture that will help them determine what parts are related. In Figure 14.14, a simple model is drawn for a typical rate or price problem. If a known price-to-weight ratio is used as one of the ratios, then a ratio with the unknown price can easily be determined from the model by choosing the corresponding elements in the same order. These are within ratios. That is, the ratio of two numbers within one situation is equated to the corresponding ratio in the other. (Notice that the reciprocal ratios, weight to price, produce exactly the same cross product.)

It is just as easy and just as reasonable to equate the between ratios. For the same example, a ratio of price to price can be equated to the ratio of weight to weight. That is, if two situations are in proportion, then the ratios of corresponding proportional elements can be equated. Notice that

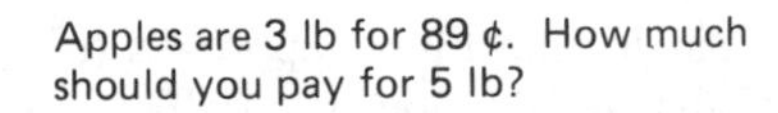

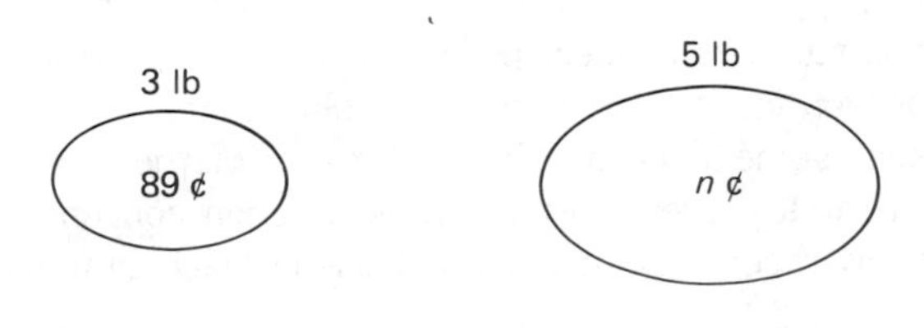

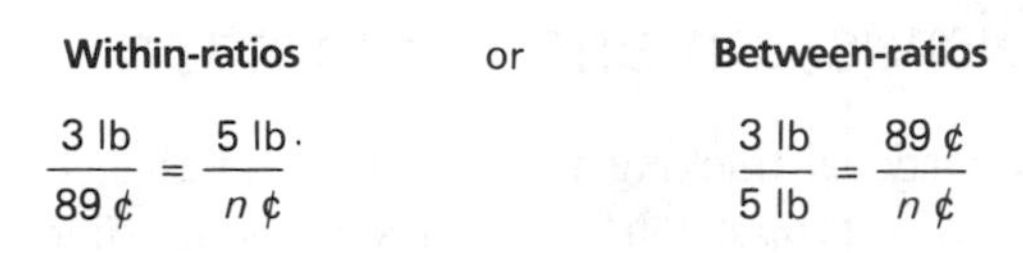

Figure 14.14
A simple drawing helps in a price ratio problem.

Jack can run an 8-km race in 37 minutes. If he runs at the same rate, how long should it take him to run a 5-km race?

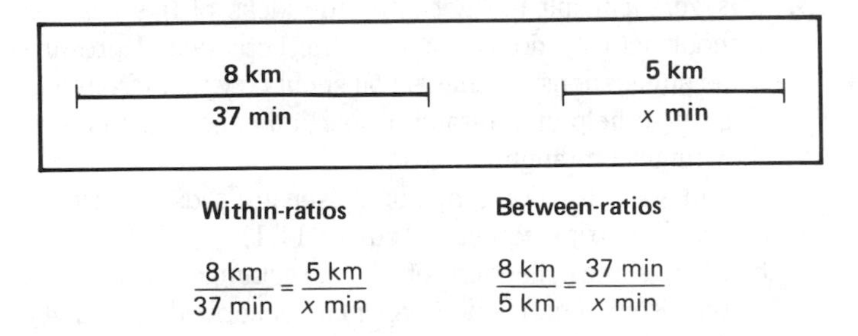

Figure 14.15
Line segments can be used to "model" both time and distance.

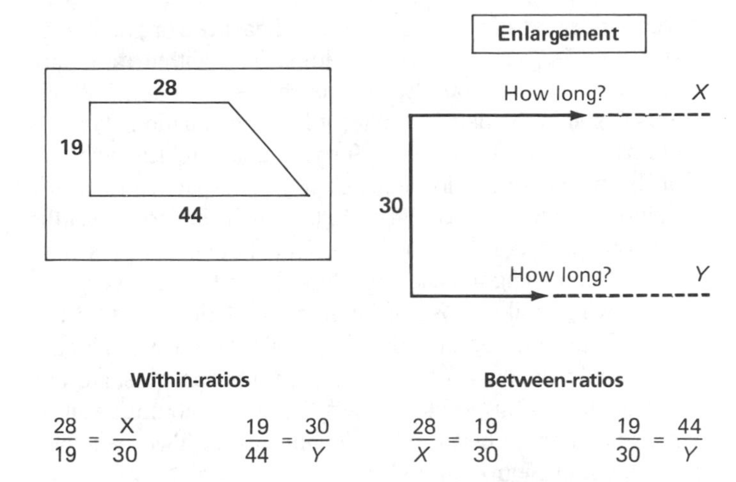

Figure 14.16
Pictures help in establishing equal ratios.

the cross product is the same as for the within ratios and that the reciprocal ratios are also equivalent to the other forms.

In Figure 14.15, a problem involving rates of speed is modeled with a simple line representing the two distances. The distance and the time for each run is modeled with the same line. You cannot see time, but it "fits" into the distance covered. All equal rates of speed problems can be modeled this way. There really is no significant difference from the drawing used for the apples. Again, it is just as acceptable to write between ratios as within ratios, and students need not worry "which one goes on top" as long as the ratios are written in the same order. The model helps with this difficulty.

In Figure 14.16 a scale drawing is being made. As before, with the use of the simple sketch, students can easily find two like ratios. The drawing provides security without being a meaningless trick. It also helps to illustrate that there is not just one correct way to set up the equation.

PERCENT PROBLEMS AS PROPORTIONS

In most sixth- through eighth-grade textbooks, you will find a chapter on ratio, proportion and percent. Percent has traditionally been included as a topic with ratio and proportion because percent is one form of ratio, a part-to-whole ratio. In many older programs, the unit on ratio and proportion focused a lot of time on solving percent problems as a proportion and relatively little time developing proportional thought. In the last chapter, it was shown that the solution to percent problems can be connected to concepts of fractions. In this section, the same part-to-whole fraction concept of percent is extended to ratio and proportion concepts. Ideally, all of these ideas (fractions, decimals, ratio, proportion, and percent) should be conceptually integrated. The better that students connect these ideas, the more flexible and useful their reasoning and problem-solving skill will be.

Equivalent Fractions as Proportions

Before considering percents specifically, consider first how equivalent fractions can be interpreted as a proportion using the same simple models already suggested. In Figure 14.17 a line segment or bar is partitioned in two different ways: in fourths on one side and in twelfths on the other. In the previous examples, proportions were established based on two amounts of apples, two different distances or runs, and two different size drawings. Here there is only one thing mea-

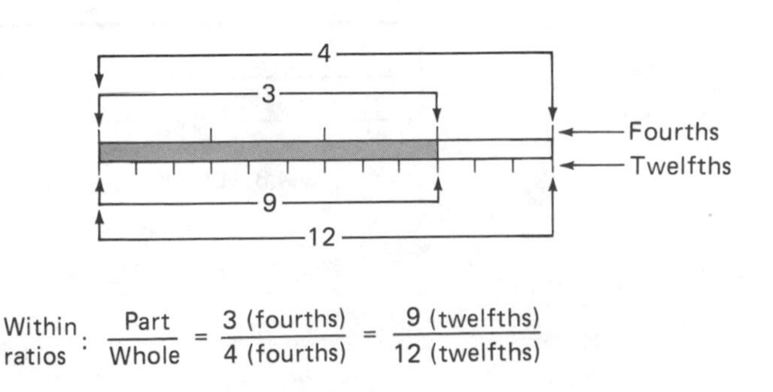

Figure 14.17
Equivalent fractions as proportions.

sured—the part of a whole—but it is measured or partitioned two ways: in fourths and in twelfths. The within ratios are ratios of part to whole "within" each measurement. Within ratios result in the usual equivalent fraction equation, $3/4 = 9/12$ (3 fourths are to 4 fourths as 9 twelfths are to 12 twelfths). The between proportion equates a part-to-part ratio with a whole-to-whole ratio, or $3/9 = 4/12$ (3 fourths are to 9 twelfths as 4 fourths are to 12 twelfths). The between ratios here might be confusing to children. They illustrate, however, how this drawing is like those of Figures 14.14 through 14.16.

A simple line segment drawing similar to the one in Figure 14.17 could be drawn to set up a proportion to solve any equivalent fraction problem, even those that do not result in whole number numerators or denominators. Two examples are shown in Figure 14.18.

Percent Problems

The equivalent fraction examples illustrate how any fraction can be sketched easily on a simple line segment showing the part and the whole measured or partitioned two different ways. All percent problems are exactly the same as this. They involve a part and whole measured in some unit and the same part and whole measured in hundredths—that is, in percents. A simple line segment drawing can be used for each of the three types of percent problems. Using this model as a guide, a proportion can then be written and solved by the cross-product algorithm. Examples of each type of problem are illustrated in Figure 14.19.

It is tempting to teach all percent problems in this one way. That is, whenever there is a percent problem, make a little drawing, set up a proportion, and solve by cross products. Developmentally, such an approach is not recommended. First, and most important, even though the approach is conceptual it does not translate easily to intuitive ideas, mental arithmetic, or estimation as discussed in the last chapter. Second, research does not seem to support the notion of focusing on a single algorithmic approach to solving percent problems (Callahan & Glennon, 1975). The modeling and proportion approach of Figure 14.19 is suggested only as a way to help students connect percent concepts with fraction and proportion ideas and to analyze problems that may verbally present some difficulty. The approach of the last chapter that relates percent to part-whole fraction concepts should probably receive the major emphasis in teaching percent.

For Discussion and Exploration

1. "Proportional thinking is quite different than being able to solve a proportion." Is this statement true? If so, is proportional thinking really important. Why?
2. Examine a teacher's edition of a basal textbook for the sixth, seventh, or eighth grade. How is the topic of ratio developed? What is the emphasis? Select one lesson and write a lesson plan that extends the ideas found on the student pages and actively involves the students.
3. It is very difficult to "get into" the ideas of this chapter without actually doing some of the activities. Here are some suggestions for things you should try to do from this chapter to help understand proportion concepts and how they might be taught.
   a. Get some fraction strips or Cuisenaire rods and try the fraction strips activity (Activity 14.1).
   b. Make some collections of similar rectangles or triangles (cut from construction paper) as described in Activity 14.2. Try sorting them visually. Then make measurements of corresponding sides and areas as suggested. Find both between and within ratios. Can you determine a unit ratio in your shapes?
   c. Translate the dot mixture activity (Activity 14.3) into one using two or three different colored counters. Make a small mixture using 20 to 30 counters and then visually attempt to make a larger mixture in the same

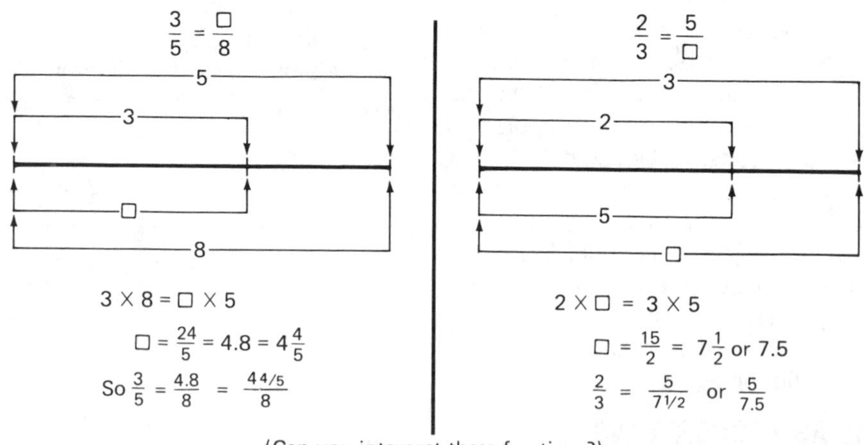

Figure 14.18
Solving equivalent fraction problems by the cross-product algorithm.

In 1960, the U.S. Railroads carried 327 million passengers. Over the next 20 years there was a 14% decrease in passengers. How many passengers rode the railroads in 1980?

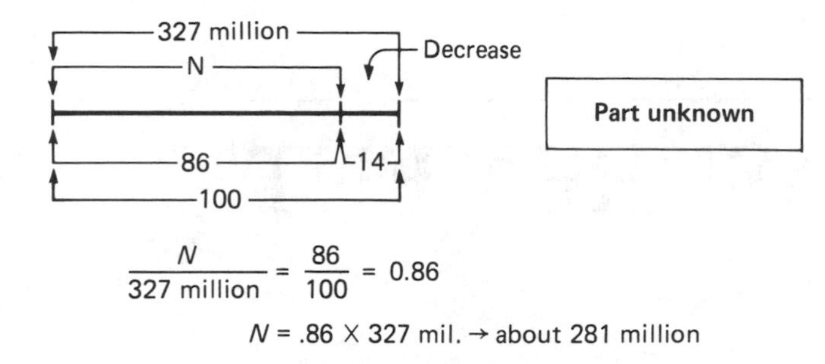

Sylvia's new boat cost $8950. She made a down payment of $2000. What percent of the sales price was Sylvia's down payment?

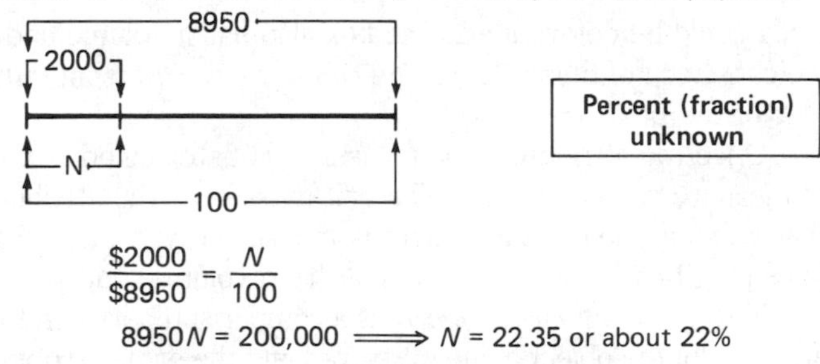

The average dressed weight of a beef steer is 62.5% of its weight before being slaughtered. If a dressed steer weighs 592 pounds, how much did it weigh "on the hoof?"

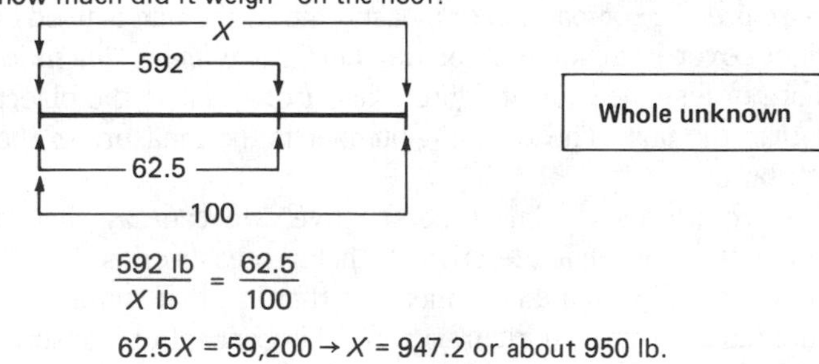

Figure 14.19
The three percent problems solved by setting up a proportion using a simple line segment model.

ratio without counting. Finally, count to test your perception of ratio.

d. Make some cards that could be used in an activity like the one in Figure 14.8 but with different numbers and ratios. (It is not necessary to use fancy drawings, try X's and O's, or black and white dots.) Can you make examples with easy between and within ratios as examples that are neither?

e. Try one of the measurement, bar graph, circle graph experiments (Activities 14.8 and 14.9). Get some friends to each do a measurement with a different unit so that the results will be more interesting. Do not forget to tell all who work with you what scale to use for the bar graphs and in what order to put the different strips or containers.

f. Try making similar drawings on a dot grid. Try it on either an isometric or rectangular grid. Masters are in Appendix C. Can you identify between, within, and unit ratios in this context?

4. Here is a suggestion to check on your understanding of proportion. Identify two equal ratios from any of the experiments suggested in 3 or from a scaling exercise (just make up a ratio and set up a chart). Pretend that one of the measures in the two ratios is unknown and set up a proportion to find it. Use the line segment method to set up the proportion equality as both within ratios and between ratios. Explain how you would solve the proportion in terms of a unit rate. How would you solve it in terms of a factor of change?

5. In the last chapter the three percent problems were developed around the theme of which element was missing—the part, the whole, or the fraction that related the two. In this chapter, percent is related to proportions, an equality of two ratios with one of these ratios a comparison to 100. How are these two approaches alike? Do you have a preference?

6. Get some percent problems from any basal textbook and solve them by using the line segment method of setting up a proportion. In solving percent problems by proportions, are the ratios that are used generally within or between ratios?

Suggested Readings

Dewer, J. M. (1984). Another look at the teaching of percent. *Arithmetic Teacher, 31*(7), 48–49.

Hoffer, A. R. (1988). Ratios and proportional thinking. In T. R. Post (Ed.), *Teaching mathematics in grades K–8: Research based methods.* Boston: Allyn and Bacon.

Lappan, G., Fitzgerald, W., Winter, M. J., & Phillips, E. (1986). *Middle grades mathematics project: Similarity and equivalent fractions.* Menlo Park, CA: Addison-Wesley.

Quintero, A. H. (1987). Helping children understand ratios. *Arithmetic Teacher, 34*(9), 17–21.

Strickland, J. F., & Denitto, J. F. (1989). The power of proportions in problem solving. *Mathematics Teacher, 82,* 11–13.

Vance, J. H. (1982). An opinion poll: A percent activity for all students. In L. Silvey (Ed.), *Mathematics for the middle grades (5–9).* Reston, VA: National Council of Teachers of Mathematics.

Wiebe, J. H. (1986). Manipulating percentages. *Mathematics Teacher, 79,* 21, 23–26.

CHAPTER 15

DEVELOPING MEASUREMENT CONCEPTS

The Meaning and Process of Measuring

CONCEPTUAL AND PROCEDURAL KNOWLEDGE OF MEASUREMENT

Measuring would appear to be a very procedural activity. The tendency is to teach children "how to measure" rather than teaching "what it means to measure." True, it is very difficult to separate the procedural activity of measurement from the concept. The concept of measurement is best embodied by the process. A typical group of first graders measures the length of their classroom by laying strips 1 meter long end to end. But the strips sometimes overlap, and the line weaves in a snakelike fashion around the desks. Which do these children not understand: *how* to measure or the *meaning* of measurement? In the fourth NAEP (Kouba, et al., 1988b), most seventh-grade students could read a ruler to the nearest quarter inch. However, if a line segment was not aligned with the ruler, as in Figure 15.1, fewer than half of the seventh-grade students and very few third-grade students could determine its length. These results point to a similar difference between *using* a measuring device and *understanding* how and why it works.

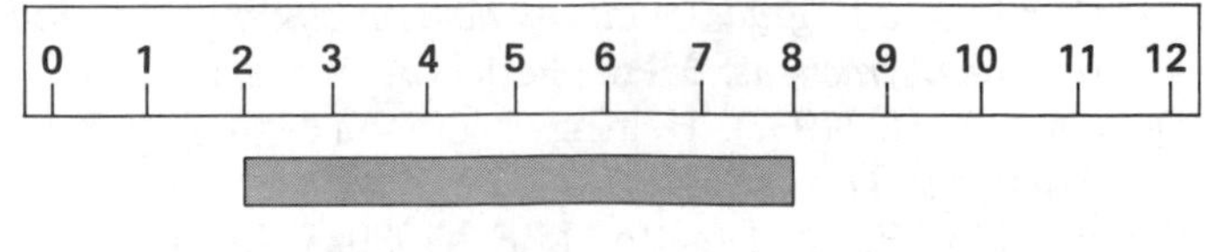

Figure 15.1
"How long is this segment?"

BASIC MEASUREMENT CONCEPTS

Suppose you were asked to measure an empty box. The first thing you would need to know or decide is what about the box is to be measured. Various lengths could be measured such as width, height, or distance around. The area of one or all sides could be determined. The box also has a volume and a weight. Each of these *things that can be measured* is an *attribute* of the box.

Once the attribute to be measured is determined, a unit of measure can be chosen. The unit must have the attribute that is being measured. Length is measured with units that have length, volume with units that have volume, and so on.

Technically, a *measurement* is a comparison between the attribute of the object being measured and the same attribute of a unit of measure. A gross comparison might indicate that the measured attribute was more or less, longer or shorter, heavier or lighter, and so on, than the unit of measure. For a more precise comparison or measurement, the unit is used to fill or cover or in some other way determine *how many times* more or less, heavier or lighter, larger or smaller, the object is than the unit. The resulting number is the measure of the attribute.

From a less technical perspective, we can say that *to measure* means that the attribute being measured is "filled" or "covered" with a unit of measure that has that same attribute (as illustrated in Figure 15.2). This concept of measurement will adequately serve the purposes of this chapter. It is appropriate with this understanding, then, to say that the *measure of an attribute* is a count of how many units are required to fill or cover that attribute of the object being measured.

In summary, to measure something one must

1. Decide on the attribute to be measured.
2. Select a unit that has that attribute.
3. Compare the units by filling, covering, or some other method, with the attribute of the object being measured.

Measuring instruments such as rulers, scales, protractors, or clocks are devices that make the filling or covering process easier. A ruler lines up the units of length and numbers them. The protractor lines up the unit angles and numbers them. A clock can be thought of as lining up units of time and marking them off.

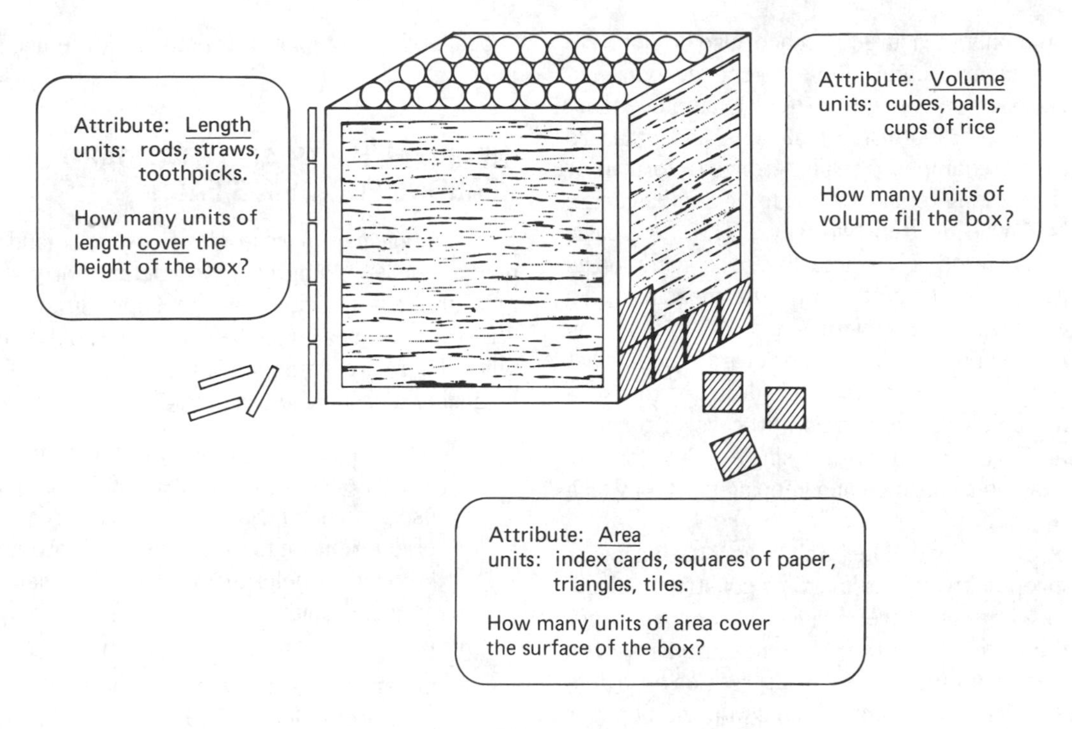

Figure 15.2
Measuring different attributes of a box.

Developing Measurement Concepts and Skills

Return to the children measuring the length of the classroom. Did they understand the concept of length as an attribute of the classroom? Did they understand that each strip of 1 meter had this attribute of length? Did they understand that their task was to fill smaller units of length into the longer one? What they most likely understood was that they were supposed to be making a line of strips stretching from wall to wall (and from their vantage they were doing quite well). They were performing a procedure instrumentally, without a conceptual basis.

A GENERAL PLAN OF INSTRUCTION

A basic understanding of measurement suggests how to help children develop conceptual knowledge of measuring.

Conceptual Knowledge to Develop	Type of Activity to Use
1. Understand the attribute being measured.	1. Make comparisons based on that attribute.
2. Understand how filling or covering an attribute with units produces a measure.	2. Use physical models of measuring units to actually fill or cover the attribute.
3. Understand the way measuring instruments work.	3. Make measuring instruments and use along with unit models.

The different types of instructional activities are discussed briefly in the following sections.

Making Comparisons

When students compare objects on the basis of some measurable attribute, that attribute becomes the focus of the activity. For example, is the capacity of one box more, less, or about the same as the capacity of another? No measurement is required, but some manner of comparing one volume to the other must be devised. The attribute of "capacity" (how much a container can hold) is inescapable.

Many attributes can be compared *directly,* such as placing one length directly in line with another. In the case of volume or capacity, some *indirect* method is probably required, such as filling one box with beans and then pouring the beans into the other box. Using a string to compare the height of a waste can to the distance around is another example of an indirect comparison. The string is the intermediary. It is impossible to compare these two lengths directly.

Constructing or making something that is the same in terms of a measurable attribute is another type of comparison activity. For example, cut the straw to be just as long as this piece of chalk, or draw a rectangle that is about the same size (area) as this triangle.

Using Models of Units

For most attributes which are measured in elementary schools, it is possible to have physical models of the units of measure. Time and temperature are exceptions. (Many other attributes not commonly measured in school also do not have physical units of measure. Light intensity, speed, loudness, viscosity, and radioactivity are just a few examples.) Unit models can be found for both informal units and standard units. For length, for example, drinking straws (informal) or tagboard strips 1 foot long (standard) might be used as units.

Unit models can be used in two ways. The most basic and easily understood method is to actually use as many copies of the unit as are needed to fill the attribute measured. To measure the area of the desktop with an index card unit, you can literally cover the entire desk with index cards. Somewhat more difficult, especially for younger children, is to use a single copy of the unit with an iteration process. The same desktop area can be measured with a single index card by moving it from position to position and keeping track of which area the card has covered.

Another type of activity that helps children focus on measurement in a conceptual way is to make or construct objects with the same measure as a given object. Making (drawing, building, finding) an object with the same measure as a given one is quite different from simply measuring a series of objects and writing the results. For example, to cut a piece of paper that has just as much area as the surface of a can is an effective way to focus on the attribute of area as well as on the meaning of area units of measure. It is also possible when using units to talk about a numeric difference in the attributes of two objects instead of just making more, less, and same comparisons. The capacity of the bucket is 2½ liters larger than the capacity of this box.

It is a good idea to measure the same object with different size units and discuss the results. This will help students understand that the unit used is as important as the attribute being measured. It makes no sense to say "The book weighs 23," unless you tell 23 of what unit. The fact that smaller units produce larger numerical measures, and vice versa, is very hard for young children to understand. This inverse relationship can only be constructed through experience with varying sized units.

Making and Using Measuring Instruments

If students actually make simple measuring instruments using the same unit models that they have already measured with it is more likely that they will understand how an instrument actually measures. A ruler is a good example. If students line up physical units along a strip of tagboard and mark them off, they can see that it is the spaces on rulers and not the marks or numbers that are important. It is quite important that the measurement with actual unit models be compared with the measurement using an instrument. Without this comparison, students may not understand that these are really two means to the same end.

A discussion of student-made measuring instruments for each attribute is provided in the sections which follow. Of course, children should also use standard, ready-made instruments such as rulers and scales. The use of these devices should still be compared directly with the use of the corresponding unit models.

INFORMAL UNITS AND STANDARD UNITS: REASONS FOR USING EACH

It is common in primary grades to use nonstandard or informal units to measure lengths and sometimes area. Unfortunately, measurement activities in the upper grades, where other attributes are measured, frequently do not begin with informal units. There are a number of values in using informal units for beginning measurement activities at all grade levels.

1. Informal units make it easier to focus directly on the attribute being measured. For example, instead of using square inches to measure area, an assortment of different units, some of which are not square, can be used to help understand the essential features of area and units of area.
2. By selecting units carefully, the size of the numbers in early measurements can be kept reasonable. The measures of length for first-grade students can be kept less than 20 even when measuring long distances. An angle unit much larger than a degree can be significantly easier for a sixth grader to use, since a degree is very, very small and individual copies are not reasonable to have.
3. The use of informal units can avoid conflicting objectives in the same beginning lesson. Is your lesson about what it means to measure area or understanding square centimeters? Learning to measure is different than learning about the standard units used to measure.
4. Informal units provide a good rationale for standard units. A discussion of the need for a standard unit can be quite meaningful after each of the groups in your class measured the same objects with their own units.
5. Informal units can be fun.

The use of standard units is also important in your measurement program at any grade level.

1. Students must eventually develop a familiarity with the most common standard units. That is, knowledge of standard units is a valid objective of a measurement program and must be addressed. Students must not only develop a familiarity with standard units, but they must learn appropriate relationships between them.
2. Once a measuring concept is fairly well developed, it is frequently just as easy or even easier to use standard units. If there is no good instructional reason for using informal units, why not use standard units and provide the exposure?

There is no simple rule for when or where to use standard or informal units. Children's initial measurement of any particular attribute should probably begin with informal units and progress over time to the use of standard units and standard measuring tools. How much time should be spent using informal unit models varies with the age of the children and the attributes being measured. For example, first-grade children need a lot of experience with a variety of informal units of length, weight, and capacity. Informal units might be used at this level all year. On the other hand, the benefits of nonstandard measuring units may be diminished in two or three days for measurements of mass or capacity at the middle school level.

EMPHASIZE THE APPROXIMATE NATURE OF MEASUREMENT

In all measuring activities, emphasize the use of approximate language. The desk is *about* 15 orange rods long. The chair is *a little less than* four straws high. Theoretically, *no measurement is absolutely accurate.* The use of approximate language is very useful with younger children using large units because many measurements do not come out "even." Older children will begin to search for smaller units or will use fractional units to try and "measure exactly." Here is an opportunity to develop the idea that *all* measurements include some error. Each smaller unit or subdivision does produce a greater degree of precision. For example, a length measure can never be more than a half unit in error. And yet, since there is mathematically no "smallest unit," there is always some error involved.

The notion of precision related to the size of the unit is an important idea in all measuring tasks. There are times when precision is not required, and a larger unit is much easier to deal with. At other times, the need for precision is significant, and smaller units become important. For example, measuring a pane of glass for a window requires a different precision than measuring a wall to decide how many 4′ × 8′ panels are needed to cover it. An awareness of precision due to unit size and the need for precision in different situations is an important aspect of measurement, especially at the upper grades.

THE ROLE OF ESTIMATION WHILE LEARNING MEASUREMENT

It is frequently useful to have students estimate a measurement before they make it. This is true with both informal and standard units. There are three reasons for including estimation in measurement activities:

1. Estimation helps students focus on the attribute being measured and the measuring process. Think how you would estimate the area of the front of this book with standard playing cards as the unit. To do so, you have to think about what area is and how the units might be fitted into the book cover.
2. Estimation provides intrinsic motivation. It adds fun and interest to measurement activities. It is fun to see how close you can come to estimating a measurement or if your team can make a better estimate than the other teams in the room.
3. When standard units are used, estimation helps to develop a familiarity with the unit. If you estimate the height of the door in meters before measuring, you have to devise some way to think about the size of a meter.

Measuring Activities

For each attribute that we want children to measure, we can identify the three types of activities that have been discussed. Comparison activities should generally precede the use of units and measuring instruments should be dealt with last. Within each of these categories there is also a rough guideline of progression that can be considered as shown in the following chart.

Measurement Activity Sequence

↓		
Comparisons	direct → to → indirect
Use of units	nonstandard → to → standard units
Use of instruments	nonstandard → to → standard units

Almost *all* activities should include an estimation component. Familiarity with standard units is a separate objective related to understanding the measurement process.

Notice especially in the chart that estimation and standard unit familiarity are important considerations. In the sections which follow, the focus is on activities for comparison, use of units, and instruments for each attribute. Separate sections on standard units and estimation follow, pointing to the importance of these objectives as well.

MEASURING LENGTH

Comparison Activities

At the kindergarten level, children should begin with direct comparisons of two or more lengths.

Activity 15.1

Make a sorting-by-length station at which students sort objects as longer, shorter, or about the same as a specified object. It is easy to have several such stations in your room. The reference object can be changed occasionally to produce different sorts. A similar task is to put objects in order from shortest to longest.

Activity 15.2

Go on a length hunt. Give groups of two a strip of tagboard, a stick, a length of rope, or other object with an obvious length dimension. The task on one day might be to find five things in the room that are shorter than (or longer than, or about the same as) their object. They can draw pictures or write the name of the things they find. (Label things in the room to help.)

It is important to compare lengths that are not straight lines. One way to do this is with string or rope. Students can wrap string around objects in a search for things that are, for example, just as long around as the distance from the floor to their belly button. Body measures are always fun. A child enjoys looking for things that are just as long as the distance around her head or waist.

Indirect comparisons are a next step in length comparisons.

Activity 15.3

Give teams of students a pair of things in the room that cannot be directly compared and have them decide which is longer. Their task should include a report to the class about how they made their decisions. Compare things that are at different heights or that are in different orientations so that a simple visual comparison may be difficult. Provide materials such as index cards, straws, Unifix cubes, or lengths of rope. Let students decide on a method of comparison.

Activity 15.4

Make some crooked or curvy paths on the floor with masking tape. Provide students with pieces of rope longer than the total path. Assign teams of two students a starting mark (tape) on the floor and a direction. Have them place a second piece of tape on the floor so that the distance between the marks is just as long as one of the crooked paths. Have students explain to the class how they solved the problem and to demonstrate why they think their straight path is just as long as the crooked one. (This is a good outdoor activity, also.)

The last activity can also be done with small distances at students' desks. A simple worksheet might be prepared like the one in Figure 15.3. Instead of crooked paths, students can make straight paths as long as the distance around simple shapes (perimeter).

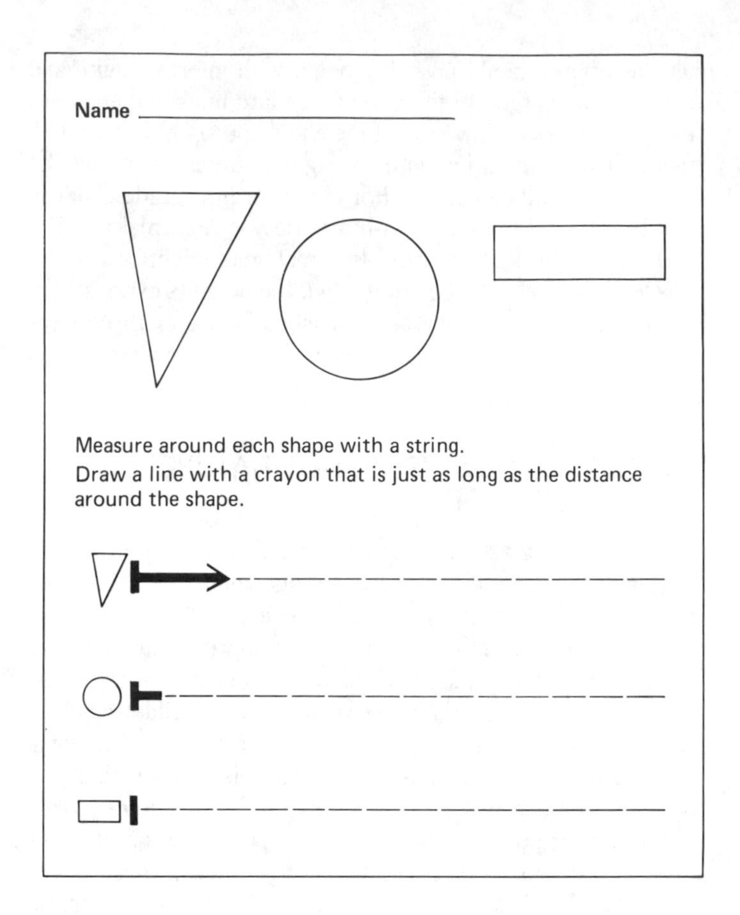

Figure 15.3
Making a straight path just as long as a crooked path.

Using Units

Students can use a wide variety of units to begin measuring length. Some examples of units of different lengths are suggested here:

Giant Footprints: Make about 20 copies of a large footprint about 1½ to 2 feet long cut out of posterboard.

Measuring Strips: Cut strips of poster board about 5 cm wide. Several sets can be made to provide different units. Some can be the long dimension of the poster board, some the short and a third set about 1 foot long. Make each set of a different color.

Measuring Ropes: Cut cotton clothesline into lengths of 1 meter. These are useful for measuring curved lines and around large objects such as the teacher's desk.

Plastic Straws: Drinking straws are inexpensive and provide large quantities of a useful unit. Straws are easily cut into smaller units. A good idea is to string straws together with a long string. The string of straws is an excellent bridge to a ruler or measuring tape.

Shorter Units: Toothpicks, Unifix cubes, strips of tagboard, wooden cubes, and paper clips are all useful as units for measuring shorter lengths. Cuisenaire rods are one of the nicest sets of units since they come in 10 different lengths, are easily placed end to end, and can be related to each other. Paper clips can easily be made into chains of about 20 clips for easier use.

For young children, initial measurements should be along lines or edges. If different teams of students measure the same distances and get different results, discuss why they may have gotten these differences. The discussion can help focus on the reason why units need to be lined up end to end and in a straight line and why units such as ropes must be stretched to their full length.

Activity 15.5

Make lists of things to measure around the room. For younger children, run a piece of masking tape along the dimension of each object to be measured. On the list, designate the units to be used. Do not forget to include curves or other distances that are not straight lines. Distances around desks, doors, or balls are some examples (Figure 15.4). Include estimates before the measures. Young children will not be very good at estimating distances at first.

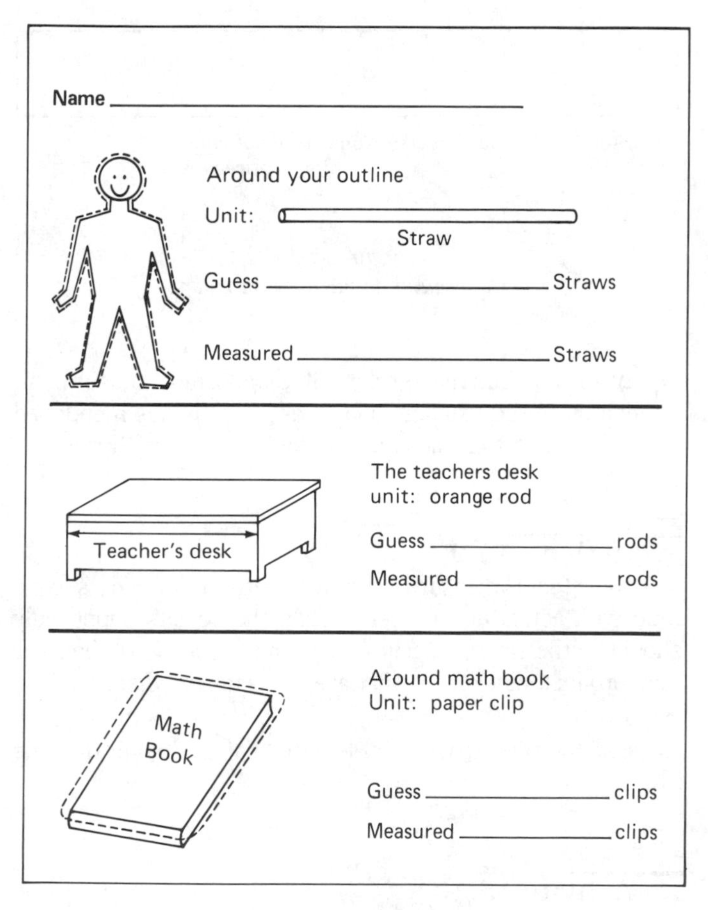

Figure 15.4
Record sheet for measuring with informal length units.

Activity 15.6

Have students measure a distance with one unit and then provide them with a different unit and see if they can predict the measure with the new unit before actually doing the measurement. Students should write down their predictions and an explanation of how they arrived at it. These predictions and explanations when shared and discussed with the class will be the most important part of this activity. The first few times you do this activity, the larger unit should be a simple multiple of the smaller unit. If the two units are not related by a whole number multiple, the task becomes difficult numerically and can be frustrating.

Activity 15.7

Have students measure something such as their desktop and record the results. Later, have them use only the measurements to make a drawing of the object the same size. Let the students select from among several units of different sizes to do the measuring. Compare their drawings to the actual object, and discuss any differences that may appear. Differences may come from a variety of sources. Of special interest in this activity is the idea that the smaller the unit, the more precise will be the measurement.

As children begin to develop a need for more precision, two units can be used at the same time. The second unit should be a smaller subunit of the first. For example, with Unifix cubes, the first unit can be bars of 10 cubes and the second, individual cubes. With measuring strips, make subunits that are one-fourth or one-tenth as long as the longer strip. Cut plastic straws so that an even number of paper clips is equal to one straw. Cuisenaire rods allow for a variety of possibilities. For example, 4 reds make a brown, or 10 whites make an orange. Have children measure with the larger unit until no more of that unit will fit and then add on sufficient smaller units to fill up the distance (Figure 15.5). Report measures in two parts: 8 straws and 3 clips long. For older students, smaller units can simply be fractional parts of longer units.

After a measurement with two related units has been made and recorded, have students figure out how to report the same measurement in terms of either unit. For example, $5\frac{2}{3}$ blue rods or 17 light green rods (3 light green = 1 blue). This provides a readiness exercise for standard units. For example, a measurement of 4 feet 3 inches is sometimes reported as 51 inches or as $4\frac{1}{4}$ feet. The use of two units is also good readiness for subdivision marks on a ruler.

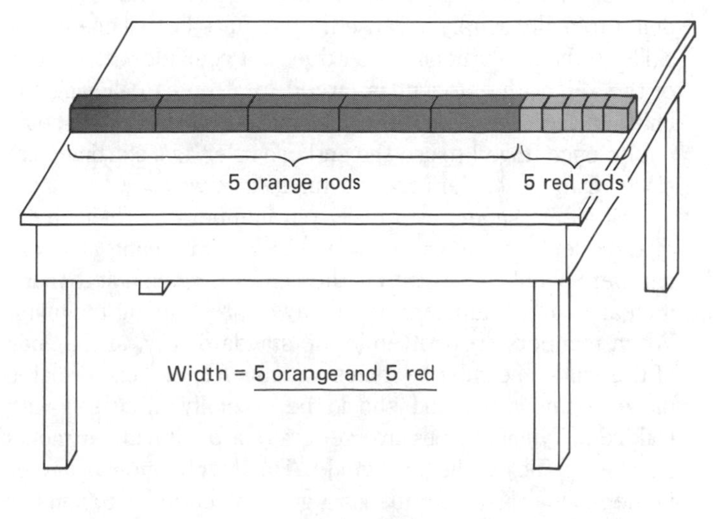

Figure 15.5
Using two units to measure length.

Making and Using Rulers

Rulers or tape measures can be made for almost any unit that students have used. It is important that students have used the actual unit models before making the rulers.

Activity 15.8

After doing some measuring with orange Cuisenaire rods (or any unit not shorter than about 5 cm), give students narrow strips of construction paper in two contrasting colors. Have students use a unit model as a guide and cut the strips into lengths as long as the unit. Discuss how the paper strips could be used instead of the actual units. Have students paste the paper unit strips end to end along the edge of a long tagboard strip about 5 cm wide as shown in Figure 15.6.

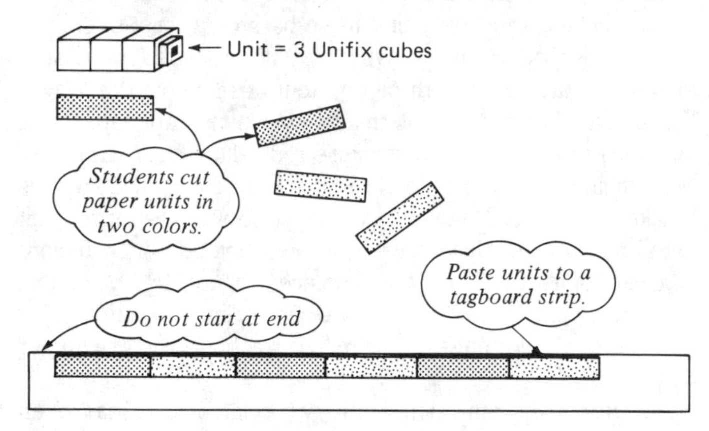

Figure 15.6
Making a simple ruler.

Pasting down copies of the units on a ruler maximizes the connection between the spaces on a ruler and the actual units. Older children can make rulers by using a real unit to make marks along a tagboard strip and then coloring in the spaces. Rulers made with very small units are more difficult for students to make simply because they require better fine-motor skills. If the first unit on a ruler does not coincide with the end of the ruler, the student is forced to attend to aligning the units on the ruler with the object measured. Children should not be encouraged to use the end of a ruler as a starting point since many real rulers are not made that way.

Students should eventually put numbers on their homemade rulers as shown in Figure 15.7. For young children, numbers should be written in the center of each unit to make it clear that the numbers are a way of precounting the units. When numbers are written in the standard way, at the ends of the units, the ruler becomes a number line. This format is more sophisticated and should be carefully discussed with children. (Number lines are generally a poor number model for children below the third grade. The development of a ruler in the manner just described is a good introduction to number lines in general. A number line is, in fact, a *measurement model* for number.)

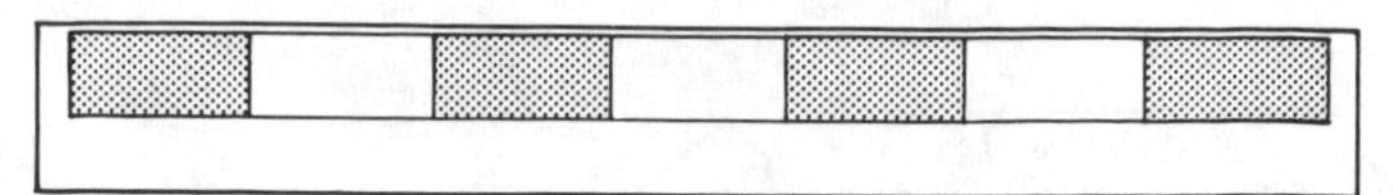

First rulers: Students count units

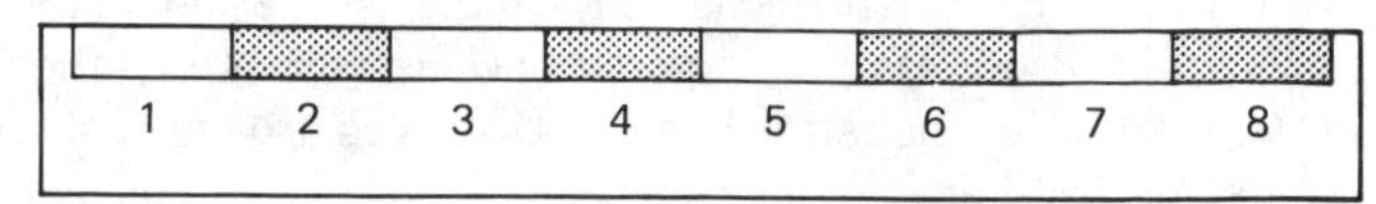

Second rulers: Numbers help to count. Numbers in center of units.

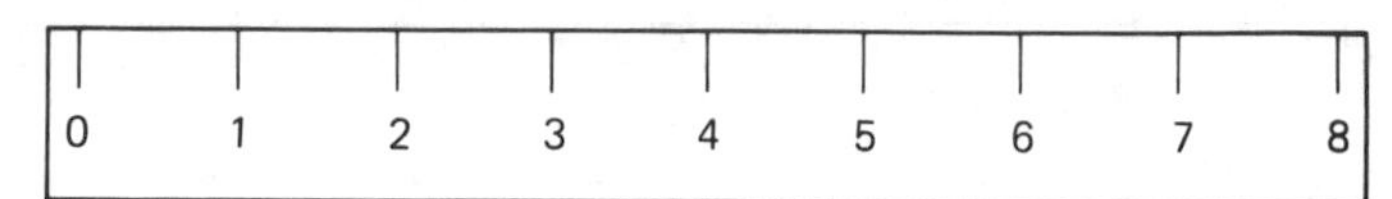

Standard rulers: Numbers are at ends of units. Notice where 0 is.

Figure 15.7
Give meaning to numbers on rulers.

After students make rulers, it is important to use them. In addition to the estimate and measure activities mentioned earlier, some special activities should be done with rulers.

Activity 15.9

Have teams measure items once with a ruler and a second time with actual unit models. While the results supposedly should be the same, inaccuracies or incorrect use of the ruler may produce differences that are important to discuss.

Use the ruler to measure lengths that are longer than the ruler.

Activity 15.10

Challenge students to find different ways to measure the same length with one ruler. (Start from either end, start at a point not at the end, measure different parts of the object, and add the results.)

Tape measures, especially for measuring around objects, can be made in a variety of ways using the same approach as rulers. With long units such as meters, a clothesline can be marked at the end of each meter with a piece of masking tape or a marking pen. Grosgrain ribbon is easily marked with a ballpoint pen. Even adding machine tape can be used to make temporary tape measures.

After working with simple rulers and tapes, have students make rulers with subunits or fractional units. This should follow the use of two units for measuring as described in the previous section.

Much of the value of using student made rulers can be lost

if careful attention is not given to transfer of this knowledge to standard rulers. Give children a standard ruler and discuss how it is alike and how it is different from the ones they have made. What are the units? Could you make a ruler with paper units the same as this? Could you make some cardboard units and measure the same way as with the ruler? What do the numbers mean? What are the other marks for? Where do the units begin?

MEASURING AREA

Comparison Activities

When comparing two areas, there is the added consideration of shape. One of the purposes of early comparison activities with areas is to help students distinguish between size (or area) and shape, lengths, and other dimensions. A long skinny rectangle may have less area than a triangle with shorter sides. This is an especially difficult concept for young children to understand. (Piagetian experiments with conservation of area indicate that many children eight or nine years old do not understand that rearranging areas into different shapes does not affect the amount of area.)

Direct comparison of two areas is nearly always impossible except when the shapes involved have some common dimension or property that makes it possible. For example, two rectangles with the same width can be compared directly as can any two circles. Comparison of these shapes, however, fails to deal with the attribute of area. Instead of comparison activities for area, activities in which one area is "rearranged" are suggested. By cutting a shape in two parts and reassembling it into a different shape, the intent is that students will understand that the before and after shapes have the same area even though they are different shapes. While obvious to adults, the idea is not at all obvious to children in the K to 2 grade range.

Activity 15.11

Give children a rectangle of construction paper and have them fold and cut it on the diagonal, making two identical triangles. Next have them rearrange the triangles into different shapes, including the original rectangle. The rule is that only sides of the same length can be matched up and must be matched exactly. Have each group find all the shapes that can be made this way, pasting the triangles on paper as a record of each shape (Figure 15.8). Discuss the size and shape of the different results. Is one shape bigger than the rest? How is it bigger? Did one take more paper to make, or do they all have the same amount of paper? Help children conclude that while each figure is a different shape, they all have the same *area*. ("Size" in this context is a useful substitute for area with very young children, although it does not mean exactly the same thing.)

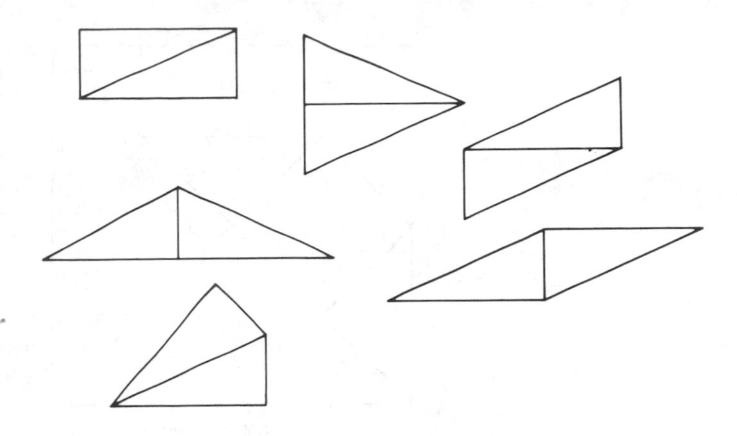

Figure 15.8
Different shapes, *same* size.

The preceding activity can be extended to three or four triangles to produce even more shapes. [If two each of two colors are used, it is also exciting to find all the different color patterns for each shape (Burns & Tank, 1988).] Tangrams, a very old and popular set of puzzle shapes, can be used in a similar way. The standard set of seven tangram pieces is cut from a square, as shown in Figure 15.9. The two small triangles can be used to make the parallelogram, the square, and the medium triangle. Four small triangles will make the large triangle. This permits a similar discussion about the pieces being the same size (area) but different shape (Seymour, 1971). A black-line master for tangram pieces is in Appendix C.

In the following activities, two different methods are used for comparing areas without measuring.

Activity 15.12

Draw the outline of several shapes made with tangram pieces as in Figure 15.10. Let students use tangrams to decide which shapes are the same size, which are larger and which smaller. Shapes can be duplicated on paper and children can work in groups. Let students explain how they came to their conclusions. There are several different approaches to this task, and it is best if students determine their own solutions rather than blindly follow your directions. (You might want to stop here, get a set of tangrams, and make the area comparisons suggested in Figure 15.10.)

Activity 15.13

Duplicate two simple shapes on a piece of paper. Have students cut them out. Suggest that they cut one or both shapes into two pieces so that they can decide if one shape is larger or if they are the same. As shown in Figure 15.11, make the two shapes related in some way so that the comparison requires no measurement.

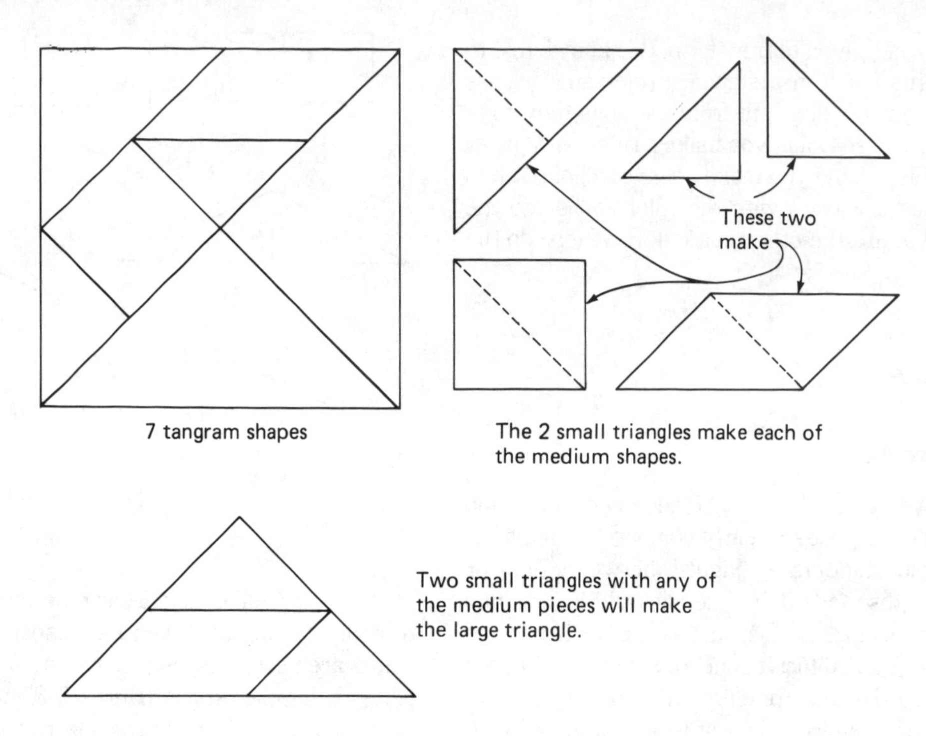

Figure 15.9
Tangrams provide a nice opportunity to investigate size and shape concepts.

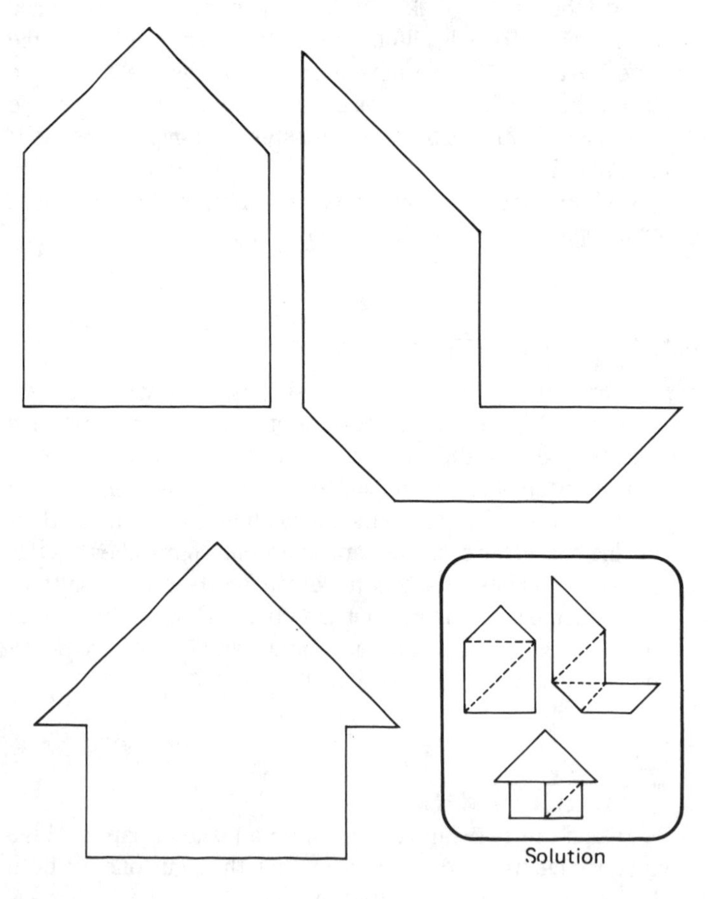

Figure 15.10
Compare shapes made of tangram pieces.

Using Units

Tiles of any type or shape can be used as units of area. Here are some suggestions for area units that are easy to gather or make in the large quantities you will need.

Cut squares or triangles (diagonals of squares or rectangles) from corrugated cardboard. (Use a paper cutter.) Large squares or triangles (about 20 cm on a side) are good for large areas. Smaller units should be about 5 to 10 cm on a side.

Sheets of newspaper make excellent units for very large areas.

Poster board can easily be cut into large quantities of congruent tiles for smaller units. Include 2 by 1 rectangles, equilateral triangles, and right triangles as well as squares. These tiles can be from 2 to 5 cm on a side.

Pattern blocks provide six different units. Five of the six shapes can be related to each other in a manner similar to the tangrams.

Playing cards, index cards, or old business cards make good medium-sized units.

Round plastic chips or pennies can be used. It is not necessary at a beginning stage that the area units fit with no gaps.

Children can use units to measure surfaces in the room such as desktops, bulletin boards, or books. Large regions can be outlined with masking tape on the floor. Use the gym or hallway for very large areas. Small regions can be duplicated on paper so that students can work at their desks. Odd shapes

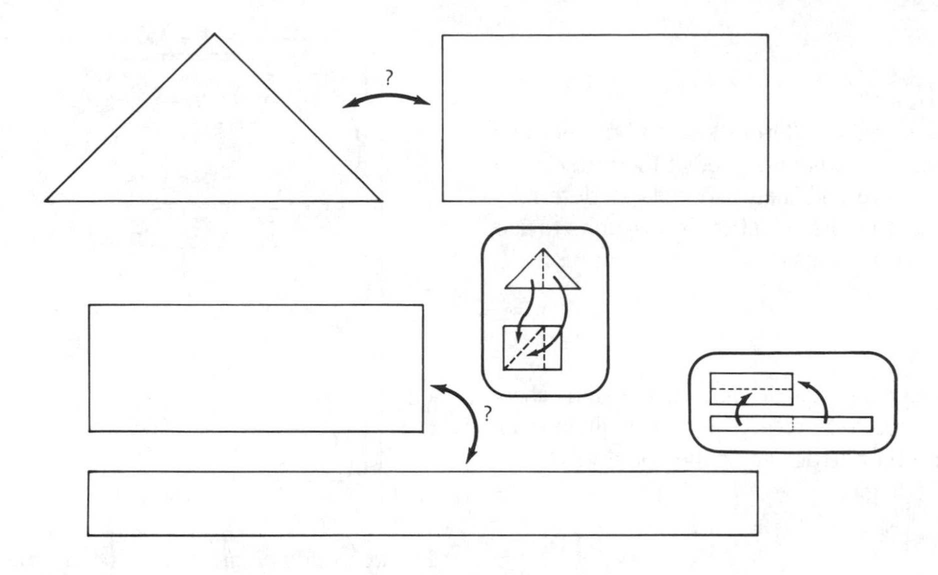

Figure 15.11
Changing a shape to compare area.

and curved surfaces provide more challenge and interest. The surface of a watermelon or the side of the wastebasket are quite difficult, but fun to explore.

For length measurements it is only the "last" unit that may not completely fit, but in area measurements there may be lots of units that only partially fit. You may wish to begin with shapes you have designed so the units "fit." That is, build a shape with units and draw the outline. By third or fourth grade, students should begin to wrestle with counting partial units and mentally "putting together" two or more partial units to count as one (Figure 15.12).

Your objective in the beginning is to develop the idea that area is a *measure of covering.* Do not introduce formulas for areas. Simply have the students fill the shapes and count the units. Be sure to include estimation before measuring (this is significantly more difficult than for length), use approximate language, and relate precision to the size of the units in the same manner as with length. When two or more groups measure the same region, different measures are very likely. Discuss these differences with the children and point to the difficulties involved in making estimates around the edges. Avoid the idea that there is some precise right answer that everyone should have gotten. The following activities are additional suggestions for area measurement.

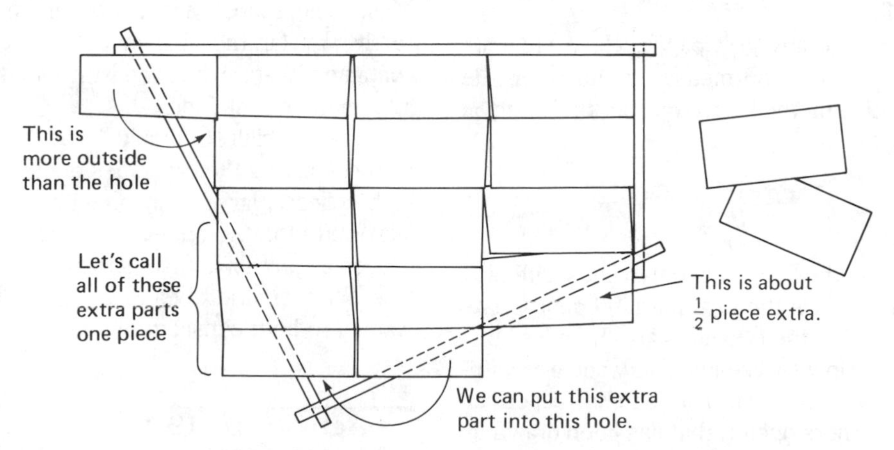

Figure 15.12
Measuring the area of a shape taped on the floor: units are pieces of tagboard all the same shape.

Activity 15.14

Present a series of quite different shapes but with only slightly different areas. Let students predict the order of the shapes from small to large and then use units to determine the correct order. Let students select their own units to use, and discuss results with the class.

Activity 15.15

Have students make a rectangle that has the same shape as another shape that you have given them. The given shape may be a "blob" or other irregular shape, or it may be a triangle or even a rectangle.

Activity 15.16

Measure the same shape with two different but related units. As with length, have students predict the second measure after making the first. Do not forget to have them write an explanation of their prediction.

Using Grids

With the exception of professional drafting equipment and computer methods, there really are no instruments designed for measuring areas. However, grids of various types can be thought of as a form of "area ruler." A grid of squares or triangles does exactly what a ruler does; it lays out the units for you. Square grids, isometric grids (equilateral triangles), and square grids with diagonal lines making smaller triangles, are all available in Appendix C. All can be used for measuring areas with different units.

Activity 15.17

Make transparencies of any grid paper. Have students place the grid over a region to be measured and count the units inside. An alternative method is to trace around a region on a paper grid.

Activity 15.18

For larger regions on the floor, make two lines with tape in an L, as in Figure 15.13. On the tape mark off appropriate units of length. Discuss how the two lines are the edges of a grid and can be used to help visualize imaginary squares without drawing them. Large shapes cut from butcher paper can be positioned into the corners of an L that has been drawn on the floor.

Of course, the most useful grid is one of squares, since most standard units of area are squares derived from standard units of length (square inches, square centimeters, etc.). As children use square grids to determine areas of rectangular

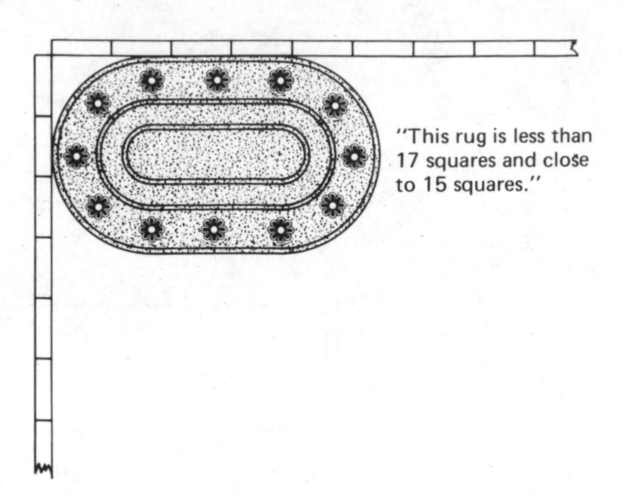

Figure 15.13
Two tape rulers on the floor make an imaginary grid to "measure" areas of big shapes.

shapes, many will begin to see that the product of the numbers of rows times the numbers of columns is an easy way to count the squares. The L approach (which can also be used for small shapes on paper) provides a hint at using lengths of sides to determine areas.

MEASURING VOLUME AND CAPACITY

Volume and *capacity* are both terms for measures of the "size" of three-dimensional regions. Standard units of volume are expressed in terms of length units, such as cubic inches or cubic centimeters. Capacity units are generally applied to liquids or the containers which hold liquids. Standard capacity units include quarts and gallons, liters and milliliters.

Comparison Activities

Most solid shapes and containers must be compared indirectly. By far the easiest comparisons are made between containers that can be filled with something that is then poured into another container.

Young children should have lots of direct experiences with comparing the capacities of different containers. To this end, collect a large assortment of cans, small boxes, plastic jars, and other containers. Try to gather as many different shapes as possible. Also gather some plastic scoops and funnels. Rice or dried beans are good fillers to use. Sand and water are both considerably messier.

Activity 15.19

Sort containers by volume. Select one container as the standard and by pouring fillers from this to other containers, determine which hold less, more, and about the same as the standard. For each container, have students mark a slip of paper MORE, LESS, or ABOUT THE SAME *before* they test it, and then mark the slip again with the actual result.

Activity 15.20

Given a series of five or six containers of different sizes and shapes, have students attempt to order them from least volume to most. This can be quite challenging.

Solids such as a rock, ball, block, an apple, or an eggplant can also be compared in terms of their volumes, but it is a bit more difficult. Some method of displacement must be used. One of several approaches is to use a container that will hold each of the items to be compared. Place an object in the container and fill with rice to a level that will be above all of the objects. Mark this level on the container. Then remove the object and mark the new level. The difference is equal to the volume of the object. Each new object should be "buried" in the container and the rice filled to the same initial level.

Using Units

There are two types of units that can be used to measure volume and capacity: solid units and containers. Solid units are things like wooden cubes or old tennis balls that can actually be used to fill the container being measured. The other type of unit model is a small container that is filled and poured repeatedly into the container being measured. The following are a few examples of units that you might want to collect.

- Thimbles, plastic caps, and liquid medicine cups are all good for very small units.
- Plastic jars and containers of most any size can serve as a unit.
- Wooden cubic blocks or blocks of any shape as long as you have a lot of the same size.
- Styrofoam balls, walnuts, or even marbles can be used, even though they do not pack perfectly. They still produce conceptual measures of volume.
- For large containers, borrow some big cardboard or wooden blocks from the kindergarten room. A large sack of old tennis balls might be collected from parents or from tennis pro shops.

Measuring activities for volume and capacity are similar to those for length and area. Estimation of volumes is a lot more fun as a class activity because it is much more difficult. Finding ways to measure large containers such as a large cardboard carton in terms of a relatively small container-type unit can be an excellent challenge to present to groups of fourth or fifth graders. This can be done long before volume formulas are developed and can even be done with unusual containers such as a wastebasket with slanted sides.

When measuring with solid units such as cubes or balls, it is very difficult to use only one and iterate it, as can be done with length or volume. However, a worthwhile challenge is to determine the volumes of containers given only enough units to cover the bottom.

Making and Using Measuring Cups

Activity 15.21

A measuring "cup" can be made easily from a container-type unit. Select a large transparent container for the cup and a small container for a unit. Fill the unit with beans or rice, empty it into the container, and make a mark indicating the level. Repeat until the cup is nearly full. If the unit is quite small, marks may only be necessary after every 5 or 10 units. Numbers need not be written on the container for every marking. Students frequently have difficulties reading scales in which not every mark is labeled. This is an opportunity to help them understand how to interpret lines on a real measuring cup that are not labeled.

Students should use their measuring cup and compare the measures with those made by directly filling the container from the unit. The cup will very likely produce errors due to inaccurate markings. This is a good opportunity to point out that measuring instruments themselves can be a source of error in measurement. The more accurately made the instrument is, and the finer the calibration, the less error will be attributed to that source.

MEASURING MASS AND WEIGHT

Weight is a measure of the pull or force of gravity on an object. *Mass* is the amount of matter in an object and technically a measure of the force needed to accelerate it. On the moon, where gravity is much less than on earth, an object has a smaller weight but the identical mass as it has on earth. For practical purposes, on the surface of the earth, the measures of mass and weight will be about the same. In this chapter the terms *weight* and *mass* will be used interchangeably.

Making Comparisons

The most conceptual way to compare the weights of two objects is to hold one in each hand, extend your arms, and experience the relative downward pull on each. This is an effective way of communicating to a very young child what "heavier" or "weighs more" means. This personal experience can then be transferred to one of two basic types of scales which can then be used to make other comparisons.

Children should first use their hands to decide which of two objects is heavier. When they then place these in the two pans of a balance, the pan that goes down can be understood to hold the heavier object. Even a relatively simple balance will detect very small differences. If two objects are placed one at a time in a spring scale, the heavier object pulls the pan down further. Both balances and spring scales have real value in the classroom. Figure 15.14 shows a homemade version of each. Simple scales of each type are available

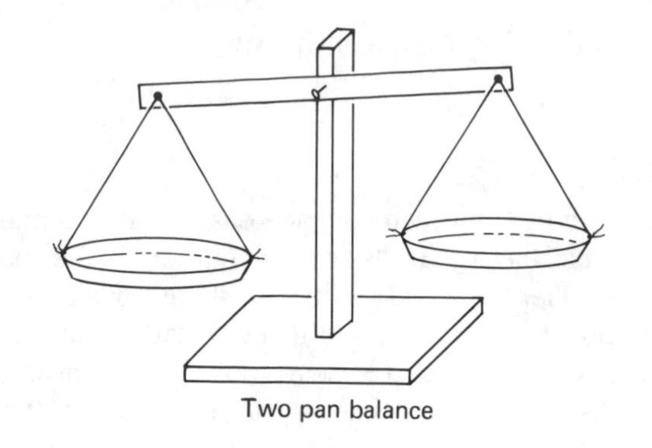

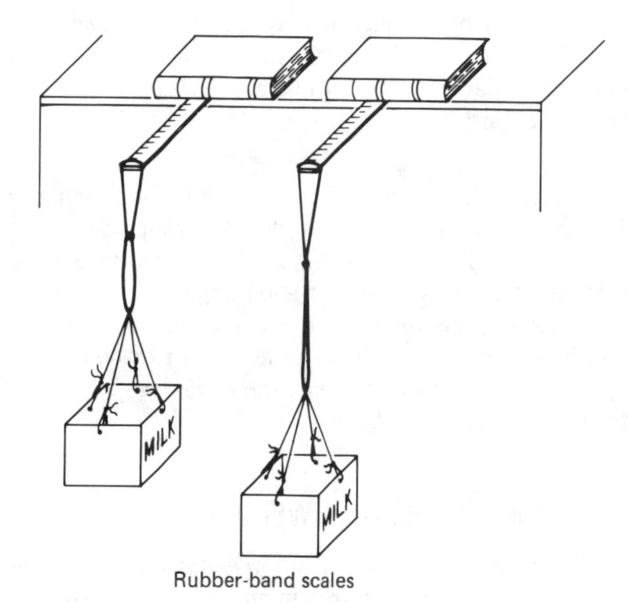

Figure 15.14
Two types of scales.

through school supply catalogs. (Technically, spring scales measure weight, and balance scales measure mass. Why?)

With either scale, sorting and ordering tasks are possible with very young children. For older children, comparison activities for weight are not necessary. (Why?)

Using Units

Any collection of uniform objects with the same mass can serve as weight units. For very light objects, wooden or plastic cubes are quite reasonable. Very large metal washers found in hardware stores are effective for weighing slightly heavier objects. You will need to rely on standard weights to weigh things as heavy as a kilogram or more.

Activity 15.22

In a balance scale the object is placed in one pan and weights in the other until the two pans are balanced. In a spring scale, first place the object in and mark the position of the pan on a piece of paper taped behind the pan. Remove the object and place just enough weights in the pan to pull it down to the same level. Discuss with the children how equal weights will pull the spring or rubber band with the same force.

While the concept of heavier and lighter is learned rather early, the notion of units of weight or mass is a bit more mysterious. At any grade level even a brief experience with informal unit weights is a good readiness for a discussion of standard units and scales.

Making and Using a Scale

Many of the scales that are commonly used are some form of spring scale with a revolving pointer like a clock hand.

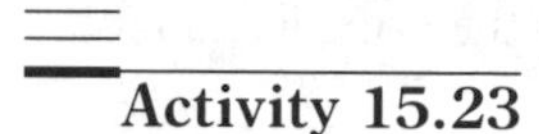

Activity 15.23

Students can use informal weight units and calibrate a simple rubber-band scale like the one in Figure 15.14 to help understand how dial scales work. Mount the scale with a piece of paper behind it and place weights in the pan. After every five weights make a mark on the paper. The resulting marks correspond to the markings around the dial of the standard scale. The pan serves as the pointer. In the dial scale, the downward movement of the pan mechanically causes the dial to turn.

After making the markings on the rubber-band scale, use it to measure objects. Then measure them using the same units, but on a two-pan balance. The pan balance is likely to be more sensitive.

The principal value of this exercise is seeing how scales are made. Even digital readout scales are based on the same basic principle.

MEASURING ANGLES

Comparing Angles

The attribute of an angle that is measured or compared might be called the "spread of the angle's rays," although this is somewhat ambiguous. Angles are composed of two rays that are infinite in length. The only difference in their shape is how widely or narrowly the two rays are "spread apart." (There is no word like *length* or *area* for angular measure.) The relatively simple exercise of making a direct comparison of two angles can help children conceptualize this attribute. Two angles can be compared by tracing one and placing it over the other as in Figure 15.15. Be sure to have students compare angles where the angle sides are of different lengths. A wide angle with short sides may seem smaller than a narrow angle with long sides. This is a common misconception among students.

It is not necessary to spend a lot of time with these activities. As soon as students can tell the difference between

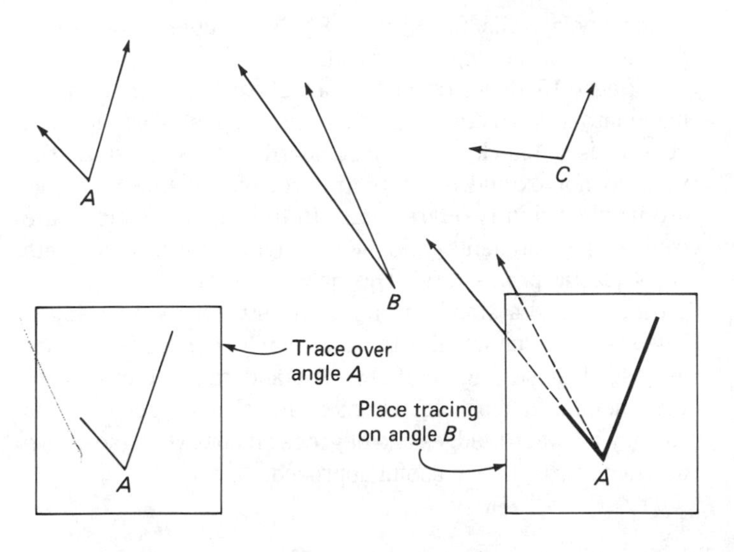

Figure 15.15
Which angle is largest? Smallest? Use tracings to compare.

a large angle and a small one, regardless of the length of the sides, you can move on to measuring angles.

Using Units

A unit for measuring an angle must be an angle. Nothing else has the same attribute of "spread" that we want to measure. (Contrary to popular opinion, it is not necessary to use degrees to measure angles.)

Activity 15.24

Give each student an index card or small piece of tagboard. Have them draw a narrow angle on the tagboard, using a straightedge, and then cut it out. The resulting wedge can then be used as a unit of measure by counting the number that will fit in a given angle. Pass out a worksheet with assorted angles on it and have them use their unit to measure them. Since everyone makes their own unit angles, the results will be different and can be discussed in terms of the size of their units.

The last activity illustrates that measuring an angle is the same as measuring length or area. Unit angles are used to fill or cover the spread of an angle just as unit lengths fill or cover a length. Once this concept is well understood, it is reasonable to move on to measuring instruments.

Making a Protractor

Activity 15.25

Tear off about a foot of ordinary waxed paper for each student. Have them fold the paper in half and crease the fold tightly. Fold in half again so that the folded edges match. Repeat this two more times, each time bringing the folded edges together and creasing tightly. Cut or tear off the resulting wedge shape about 4 or 5 inches from the vertex and unfold. If done correctly, there will be 16 angles surrounding the center as in Figure 15.16. This serves as an excellent

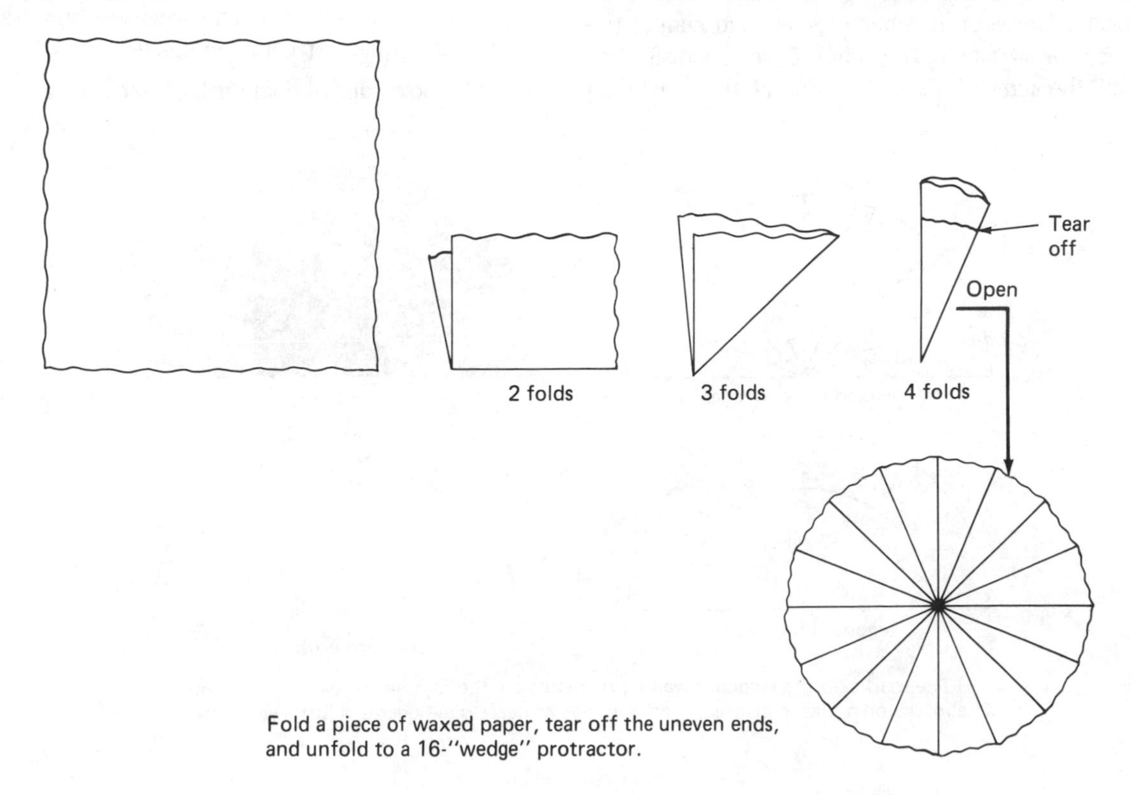

Figure 15.16
Making a waxed-paper protractor.

protractor with a unit angle that is one-eighth of a straight angle. It is sufficiently transparent that it can be placed over an angle on paper, on the blackboard, or on the overhead projector to measure angles, as shown in Figure 15.17.

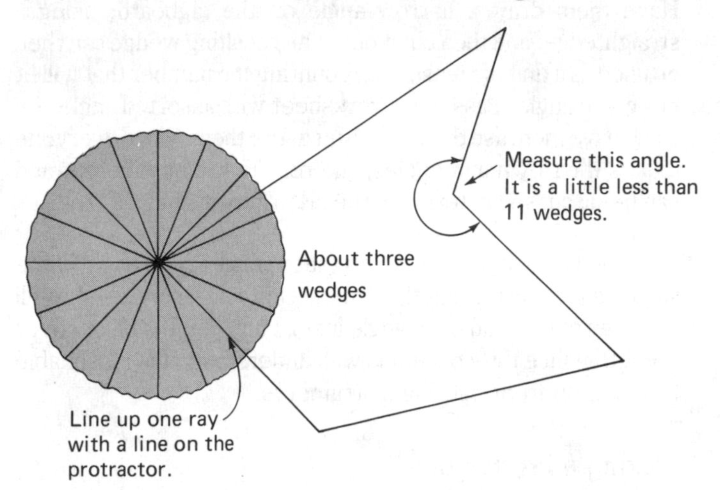

Figure 15.17
Measuring inside angles in a polygon using a waxed-paper protractor.

The waxed-paper protractor makes it quite clear how a protractor fits unit angles into an angle for measurement. When measuring angles, students can easily estimate halves, thirds, or fourths of a "wedge," a possible name for this informal unit angle. This is sufficiently accurate to measure, for example, the interior angles of a polygon and discover the usual relationship between number of sides and sum of the interior angles. For a triangle, the sum is 8 wedges or 8^w. For a quadrilateral, the sum is 16^w. And in general, the sum for an n-sided polygon is $(n - 2) \times 8^w$. The superscript w is a forerunner of the degree symbol.

Figure 15.18 illustrates how a tagboard semicircle can be made into a protractor to measure in wedges. This tagboard version is a bit closer to a standard protractor since the rays do not extend down to the vertex and the markings are numbered in two directions. Both of these features are confusing to students who begin angle measurement with small plastic protractors. The only difference between this protractor and a standard one is the size of the unit angle. The standard unit angle is the *degree*. The degree is simply a very small angle (Figure 15.18). A standard protractor is not very helpful in teaching the meaning of a degree. But an analogy between wedges and degrees and between these two protractors, is a very useful approach.

Introducing Standard Units

INSTRUCTIONAL GOALS

Three broad goals relative to standard units of measure can be identified:

1. Familiarity with the unit. Familiarity means that when a commonly used unit is encountered the student has a basic idea of its size and the attribute it measures.
2. Ability to select an appropriate unit. Knowing what is a reasonable unit of measure in a particular situation is related to a familiarity with the unit, an understanding of measurement and precision, and to some extent, practicality or common sense.
3. Knowledge of relationships between units. It is impor-

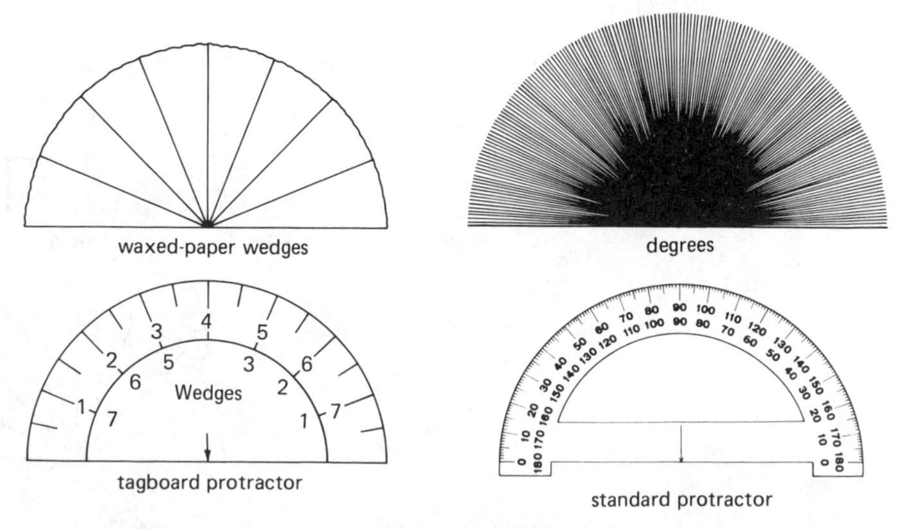

The marks on the tagboard wedge protractor are the rays on the waxed-paper version.
The marks on a plastic protractor are the rays of degrees. A degree is just a very small angle.

Figure 15.18
Comparison of protractors.

tant that students know, for example, how to convert feet to inches and liters to milliliters. The emphasis is on commonly used conversion factors.

Developing Unit Familiarity

The importance of familiarity with standard units of measure is highlighted by the reluctance (even fear) on the part of a majority of U.S. citizens to adopt the metric system. The average adult has developed a "feel" for, a familiarity with, those units he or she has used throughout life: inch, foot, mile, pound, cup, gallon. These *customary* units are certainly not easier to use in any way. They simply are known while *metric* units are not. Helping students develop that same sense of familiarity with the most frequently used units of measure, from *both* the metric and the customary systems, is an important goal of instruction.

Two simple types of activities can be utilized to help develop familiarity with most standard units. First, comparison activities with a focus on a single unit. Second, development of familiar and personal referents or benchmarks for single units or easy multiples of units.

Activity 15.26

Give students a model of a standard unit and have them search for things that measure about the same as that one unit. For example, to develop familiarity with the meter, give students a piece of rope 1 meter long. Have them make lists of things that are about 1 meter. Things that are a little less (or more) or twice as long (or half as long) should be noted in separate lists. Encourage students to find familiar items that are in their daily lives. In the case of lengths, be sure to include circular lengths. Later, students can try to predict if a given object is more, less, or close to 1 meter.

The same activity can be done with other unit lengths. Parents can be enlisted to help students find familiar distances that are about 1 mile or about 1 kilometer. Suggest in a letter that they check the distances around the neighborhood, to the school, shopping center or other frequently traveled paths.

For capacities, students need a container that holds or has a marking for a single unit. They should then find other containers that hold about as much, more, and less as in the length example. Remember that the shapes of containers can be very deceptive when estimating their capacity. Have students look for examples at home as well as at school.

For the standard weights of gram, kilogram, ounce, and pound, students should have ample opportunity to compare objects on a two-pan balance with single copies of these units. It may be more effective to work with 10 grams or 5 ounces. Students can be encouraged to bring in familiar objects from home to compare on the classroom scale.

Standard areas are in terms of lengths such as square inches or square feet so that familiarity with lengths is more important. Familiarity with a single degree is not as important as some idea of 30, 45, 60, and 90 degrees. The turtle graphics of Logo provides an excellent avenue for degree familiarity. (Logo is a very accessible computer language with a simple graphics capability that even first-grade children can control. It is discussed more fully in Chapter 20.)

A second and somewhat different approach to unit familiarity is to begin with very familiar items and use their measures as references or benchmarks. A doorway is a bit more than two meters. A bag of flour is a good reference for 5 pounds. Your bedroom may be about 10 feet long. A paper clip is about a gram and about 1 centimeter wide. A gallon of milk weighs a little less than 4 kilograms.

Activity 15.27

For each unit of measure you wish to focus on, have students make a list of at least five things that they are very familiar with and have them measure them using that unit. For lengths, encourage them to include long and short things; for weight, to find both light and heavy things; and so on. The measures should be rounded off to nice whole numbers. Discuss lists in class so that different ideas are shared.

Of special interest for length are benchmarks found on our own bodies or using our bodies. These become quite familiar over time and can be used as approximate versions of units or rulers in many situations. Perhaps you know some personal reference that you use for an inch, a foot, or a yard. Even though young children grow quite rapidly, it is useful for them to know the approximate lengths that they "carry around" with them.

Activity 15.28

Measure your body. About how long is your foot, your stride, your hand span (stretched and with fingers together), the width of a finger, arm span (finger to finger and finger to nose), height to waist, to shoulder and to head, distance around your wrist and around your waist. Perhaps you cannot remember all of these, but some may prove to be very useful benchmarks and some may be excellent models for single units. (The average fingernail width is about 1 cm and most people can find a 10-cm width in some way with their hands.)

To help remember these references, they must be used in activities where lengths, volumes, and so on, are compared to the benchmarks in order to estimate measurements.

What Unit Is Appropriate

Should the room be measured in feet or inches? Should the concrete blocks be weighed in grams or in kilograms? The answers to questions such as these involve more than simply knowing how big the units are, although that is certainly required. Another consideration involves the need for accuracy. If you were measuring your wall in order to cut a piece of molding or woodwork, you would need to measure it very accurately. The smallest unit would be an inch or a centimeter and you would also use fractional parts. But if you were determining how much paint to buy, the nearest foot or even the nearest meter would probably be sufficient.

Activity 15.29

Find examples of measurements of all types in newspapers, signs, or other everyday occurrences. Present the context and situation to students and see if they can predict what units of measure are used. There may well be disagreement. Have students discuss their choices.

Developing Relationships Between Units

The number of inches in a foot or yard or the number of cups in a quart is the type of information that must eventually be committed to memory. Practice with conversions is something that lends itself well to pencil-and-paper work. This procedural aspect of unit familiarity is another example of objectives that have, in the past, been overworked in the curriculum largely due to the ease of testing rather than the need to know. Conversions between one unit and another should not be overemphasized.

Many students may know that there are 16 ounces in 1 pound, but when they try to determine how many pounds are in 90 ounces they get confused over which operation to use. This is partly a matter of understanding the meanings of operations. However, simple common sense can help a lot also. In this example, since pounds are bigger or heavier than ounces, it is reasonable to end up with fewer pounds than ounces.

The customary system involves an unfortunate variety of conversion factors, and as long as the United States continues to use it teachers will have to deal with helping children commit these factors to memory.

Exact conversions between the metric and the customary system should never be done and are almost nonexistent in textbooks today. From the standpoint of familiarity with these systems, "soft" or "friendly" conversions may even be useful as long as we live in a country that seems bent on having two systems of measurement. For example, a liter is a "gulp more" than a quart, and a meter is a "bit longer" than a yard. The same is true for benchmarks. One hundred meters is about one football field plus one end zone, or about 110 yards.

IMPORTANT STANDARD UNITS

Both the customary and the metric system include many units that are not used for everyday living. The following chart lists those units that are most common.

	Metric System	Customary System
Length	millimeter centimeter meter kilometer	inch foot yard mile
Area	square centimeter square meter	square inch square foot square yard
Volume	cubic centimeter cubic meter	cubic inch cubic foot cubic yard
Capacity	milliliter liter	ounce* cup quart gallon
Weight	gram kilogram metric ton	ounce* pound ton

The Metric System

Unit familiarity with the popularly used units should be the principal focus of almost all instruction with standard units.

Before students have developed a full understanding of decimal notation, there is very little advantage in teaching students all of the very nice relationships in the metric system. While the customary system frequently mixes units (3 pounds, 6 ounces or 6 ft, 2 in.) the standard version of the metric system insists on the use of a single unit for each measure. To adhere to this rubric of the metric system means that a primary-grade child should report a measure in the smallest unit used since decimal notation is not meaningful. For example, a length would be recorded as 235 cm instead of 2.35 m or 2 m and 35 cm. How important this rule is for the second grade may be open to debate.

Perhaps one of the worst errors in metric measurement curriculum is having students "move decimal points" to convert from one metric unit to another prior to a complete development of decimal notation. Faced with confusion, children memorize rules about moving decimals so many places this way or that and the focus becomes rules and right answers.

As children begin to appreciate the structure of decimal notation, the metric system can and should be developed with all seven places; three prefixes for smaller units (deci-, centi-, milli-) and three for larger units (deka-, hecto-, kilo-). With decimal knowledge and familiarity with the basic and popularly used units, the complete decimal system is easy to learn.

*In the U.S. Customary System, the term *ounce* refers to a weight or *avoirdupois* unit, 16 of which make a pound, and also a volume or capacity unit, eight of which make a cup. While the two units have the same name, they are not related.

Making conversions within the metric system can be approached in two related ways. In Figure 13.6 (p. 213), a place-value chart is used to give a metric name to each of seven consecutive places. If the decimal point is understood to always identify the unit position, then given any metric measurement, each digit is in a position with a metric name. The decimal point can be repositioned if the name of the unit is changed accordingly. For example, in the measure 17.238 kg, the decimal indicates that the 7 is in the unit position. The label "kg" indicates that the name of the position is kilograms. Therefore, the 2 is in the hectogram position, the 3 in the dekagram position, the 8 in the gram position. Repositioning the decimal to indicate grams as the unit, makes the same measure read 17,238 g or 17,238.0 g.

An alternate rationale is to think of decimal point shifting as multiplying or dividing by powers of 10. In the example above, since there are 1000 grams in a kilogram, change to grams by multiplying by 1000 or shift the decimal three places to the right.

It should be emphasized once again that unit conversion is perhaps the least important part of learning the metric system or any standard system. It simply is a skill that is not used that frequently.

The Customary System

The "familiar" system of units is technically known as the U.S. Customary System. It is difficult because it lacks any common structure or common conversion ratios. After an attempt by schools during the 1970s to go completely metric, most schools and textbooks have resigned themselves to teaching both metric and customary systems.

Conversions of units within the customary system is difficult for children for two reasons. There are more conversion ratios to memorize, and they are not conveniently related to the decimal system. For example, many children will add 3 feet 8 inches to 5 feet 6 inches and get 9 feet 4 inches instead of 9 feet 2 inches. (Why?) Similar difficulties occur with all conversions in the customary system. Once again, while it is important to know how many inches in a foot and a few other relationships that we use regularly, an overemphasis on the type of conversions indicated by this last example is unwarranted.

Estimating Measures

Measurement estimation is the process of using mental and visual information to measure or make comparisons without the use of measuring instruments. About how long is the fence? Find a one gallon container for the juice. Will this paper cover the box? Each of these involves estimation.

Measurement estimation is a very practical skill. Almost every day we make estimates of measures. Do I have enough sugar to make the cookies? Will the car fit in that space? Can you throw the ball 50 feet? Is this suitcase over the weight limit?

Besides its value outside of the classroom, it was pointed out earlier that estimation in measurement activities helps students focus on the attribute being measured, adds intrinsic motivation, and helps develop familiarity with standard units. Therefore, measurement estimation both helps our measurement instruction and provides students with a valuable life skill.

TECHNIQUES OF MEASUREMENT ESTIMATION

Like computational estimation, specific strategies or approaches exist for making measurement estimations. These strategies can be identified and taught specifically.

1. Develop and use benchmarks or referents for important units. (This strategy was also mentioned as a good way to help students develop a familiarity with units.) Students should have a good referent for single units and also useful multiples of standard units. Referents or benchmarks for 1, 5, 10, and, perhaps, 100 pounds (or something near these amounts) might be useful, but there is little value in a referent for 500 pounds. On the other hand, a referent for 500 milliliters is very useful. These benchmarks can then be compared mentally to objects being estimated. That tree is about as tall as four doorways or between 8 and 9 meters.
2. Utilize a "chunking" procedure when appropriate. Figure 15.19 is a common example. It may be easier to estimate the shorter chunks along the length of the wall than to estimate the whole length as one. The weight of a stack of books is easier if some estimate is given to an "average" book.
3. Use subdivisions. This is a similar strategy to chunking. However here, the chunks are imposed on the object by the estimator. For example, if the wall length to be estimated has no useful chunks, it can be mentally divided in half and then in fourths or even eighths by repeated halving until a more manageable length is arrived at. Length, volume, and area measurements all lend themselves to this technique.
4. Iterate a unit either mentally or physically. For length, area, and volume, it is sometimes easy to visually mark off single units. You might use your hands, make markings, folds, or use other methods of keeping track as you go. For length, it is especially useful to use a body measure as a "unit" and iterate with that. If you know, for example, that your stride is about ¾ meter, you can walk off a length and then multiply to get an estimate. Hand and finger widths are useful for shorter measures.

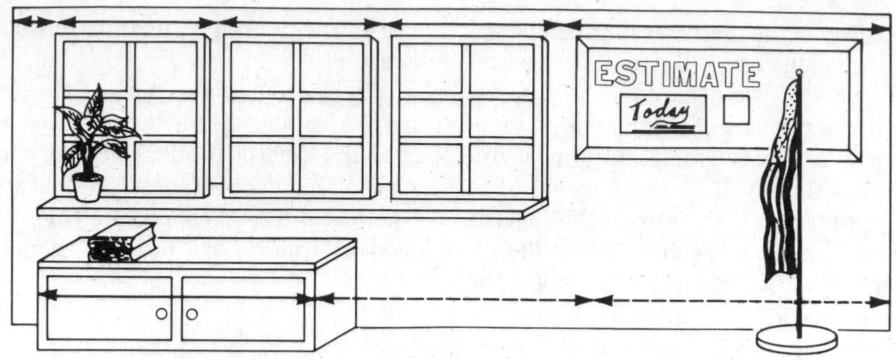

Estimate the room length.
Use: windows, bulletin board, and spaces between as "chunks".
Use: cabinet length – looks like about three cabinets will fit into the room – plus a little bit.

Figure 15.19
Measuring by chunking.

MEASUREMENT ESTIMATION ACTIVITIES

Tips for Teaching Estimation

Each of the strategies listed in the last section should be taught directly and discussed with students. But the best approach to improving their estimation skills is to have them do a lot of estimating. As you conduct estimation activities, keep the following tips in mind:

1. Help students learn strategies by having them use a specified approach. Later activities should permit students to choose whatever techniques they wish.
2. Periodically discuss how different students made their estimates. This will help students understand that there is no single "right" way to estimate and also remind them of different approaches that are useful.
3. Accept a range of estimates. Think in relative terms about what is a good estimate. Within 10% for length is quite good. Even 30% off may be reasonable for weights or volumes.
4. Sometimes have students give a range of measures that they believe includes the actual measure. This is not only a practical approach in real life but helps focus on the approximate nature of estimation.
5. Let students measure to check estimates. However, it is only necessary that one or two students or one team do the measurement if the focus of the activity is on estimation. If all students are required to follow their estimates with a measure, they may correctly wonder why they bothered estimating.
6. Make measurement estimation an ongoing activity during the year. A daily measurement to estimate can be posted on the bulletin board. Students can turn in their estimates on paper and discuss them quickly in a 5-minute period. Older students can even be given the task of making up the things to estimate with a team assigned this task each week.
7. Make an effort to include estimations of all attributes. It is easy to get carried away with length and forget about area, volume, weight, and angles.

Activities

Estimation activities need not be elaborate or involve a lot of effort on your part. Any measurement activity can have an "estimate first" component. For more emphasis on the process of estimation itself, simply think of things that can be estimated and have students estimate. Here are a few suggestions to think about.

Activity 15.30

Select a single object such as a box, a watermelon, a jar, or even the principal. Each day select a different attribute or dimension to estimate. For a watermelon, for example, students can estimate its length, girth, weight, volume, and, perhaps, even its surface area.

Activity 15.31

Conduct measurement scavenger hunts. Give teams a list of measurements and have them find things that are close to having those measurements. Permit no measuring instruments. For example, a list might include

A length of 3.5 meters

Something that weighs more than 1 kg but less than 2 kg

A container that holds about 200 milliliters.

Let students suggest how to judge results in terms of accuracy.

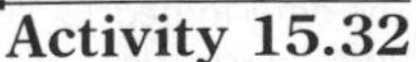

Activity 15.32

Use estimate-measure-estimate sequences (Lindquist, 1987). Select pairs of objects to estimate that are somehow related or close in measure, but not the same. Have students estimate the measure of the first and check by measuring. Then have them estimate the second. Some examples of pairs are

Width of a window, width of wall

Volume of coffee mug, volume of a pitcher

Distance between eyes, width of head

Weight of handful of marbles, weight of bag of marbles

This type of activity is a good way to help students understand how benchmarks are used in estimation.

Developing Formulas

DIFFICULTIES

Perimeter or Area

Figure 15.20 shows results from the fourth NAEP (Kouba, et al., 1988b) concerning area and perimeter. Even the seventh-grade results are disappointing. Those who analyzed the results noted that the most common errors involved computing area for perimeter and vice versa. The discrepancy between the two area items also points to a use of formulas without understanding. Notice that these errors can be attributed to confusion of meanings and a reliance on formulas instead of concepts.

Children get words and meanings confused when they are rushed into formulas they have not conceptualized and understood. To quickly give children a formula and then have them compute areas, perimeters, or volumes without understanding the formulas provides little more than an arithmetic drill. *Perimeter* may be an important word, but "distance around" should certainly be a synonym. The attribute of area should be conceptually understood before formulas are ever considered. (See Figure 15.21.)

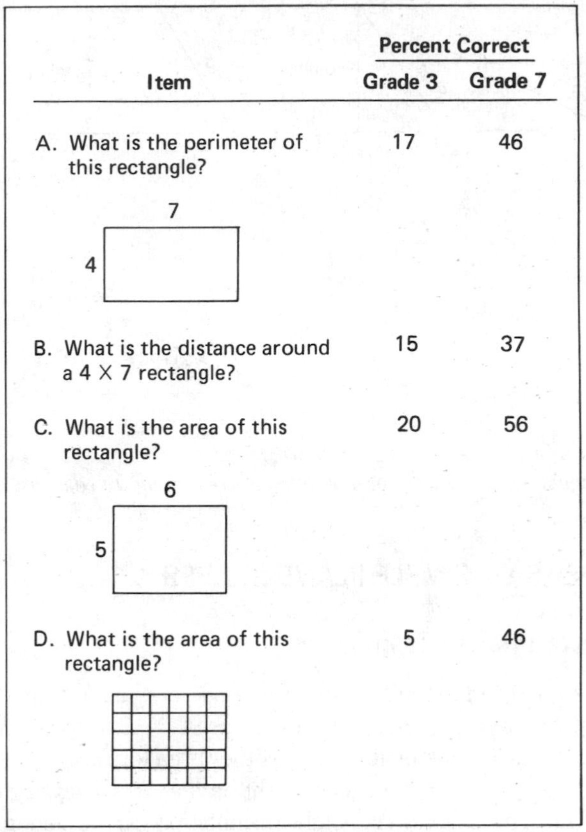

Item	Percent Correct Grade 3	Percent Correct Grade 7
A. What is the perimeter of this rectangle? (7, 4)	17	46
B. What is the distance around a 4 X 7 rectangle?	15	37
C. What is the area of this rectangle? (6, 5)	20	56
D. What is the area of this rectangle?	5	46

Figure 15.20
Data from the Fourth National Assessment of Educational Progress (NAEP), 1988.

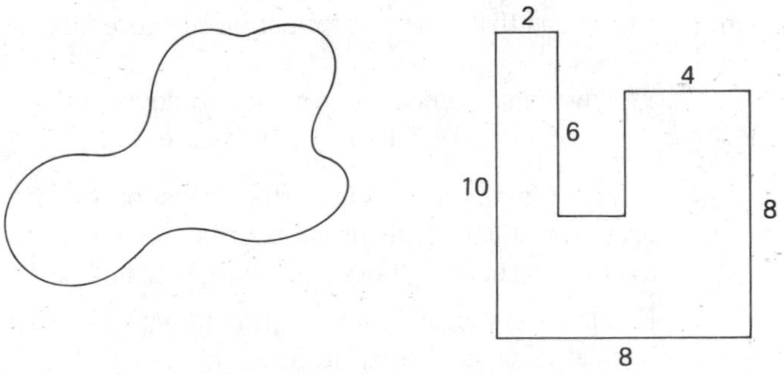

"How would you determine the areas of these shapes?"

Note: Many children believe that such shapes do not have areas or that the areas are impossible to determine because there are no formulas.

Figure 15.21
Understanding the attribute of area.

Height or Side

Another error that is commonly made when students use formulas is due to not conceptualizing the meaning of *height* in geometric figures, both two- and three-dimensional. The shapes in Figure 15.22 each have a slanted side and a height given. Students tend to confuse these two. *Any* side or flat surface of a figure can be called a *base* of the figure. For each base that a figure has, there is a corresponding height. If the figure were to slide into a room on its base, the height would be the height of the shortest door it could pass through without bending over: the perpendicular distance to the base. Children have a lot of early experiences with the $L \times W$ formula for rectangles, in which the height is exactly the same as the length of a side. Perhaps this is the source of the confusion.

Before formulas involving heights are discussed, children should discuss the meaning of height of a geometric figure. They should be able to identify where a height should be measured for any base that a figure has.

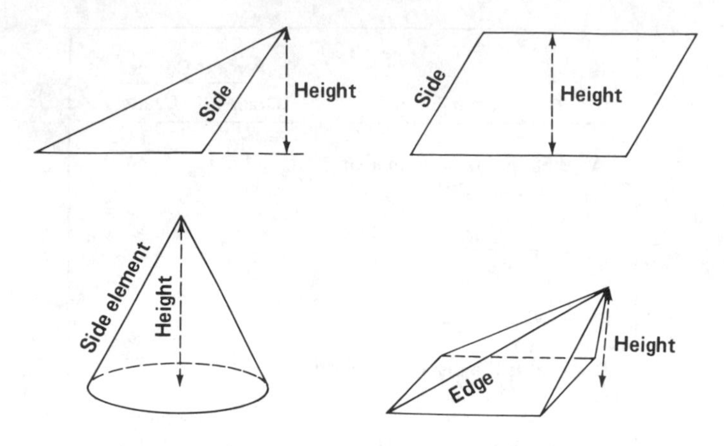

Figure 15.22
Heights of figures are not always measured along an edge or surface.

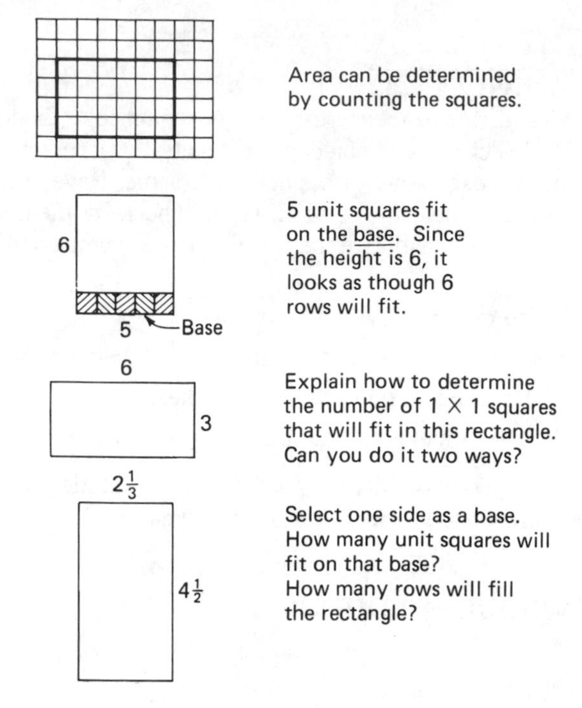

Figure 15.23
Determining the area of a rectangle.

AREAS OF SIMPLE PLANE SHAPES

Rectangles to Trapezoids

The formula for the area of a rectangle is one of the first that is developed and is usually given as $A = L \times W$ or *length times width.* Looking forward to other area formulas, an equivalent but more unifying idea might be $A = b \times h$ or *base times height.* The base times height formulation can be generalized to all parallelograms, and is useful in developing the area formulas for triangles and trapezoids. Furthermore, the same approach can be extended to three dimensions where volumes of cylinders are given in terms of the *area of the base* times the height. Base times height, then, helps to connect a large family of formulas that otherwise must be mastered independently.

The following sequence of exercises to develop the area formula for a rectangle is illustrated in Figure 15.23.

1. Have students determine the areas of rectangles drawn on square grids or geoboards. Some may count every square, and others may multiply to find the total.
2. Examine rectangles not on a grid but with whole number dimensions. Designate one side as the base and line unit squares along this side. How many such rows can fit into the rectangle? On the same rectangle repeat this approach using the other side as the base.
3. Give students rectangles with only the dimensions provided and have them determine the area. Require them to justify their results. Encourage approaches similar to those in step 2.
4. Examine rectangles with dimensions that are not whole numbers. If the base is 4½ units, then 4½ unit squares will fit along the base. If the height is 2⅓ units, then there are 2⅓ rows with 4½ squares in each, or 2⅓ sets of 4½.

As illustrated in Figure 15.24, parallelograms that are not rectangles can be transformed into rectangles having the same

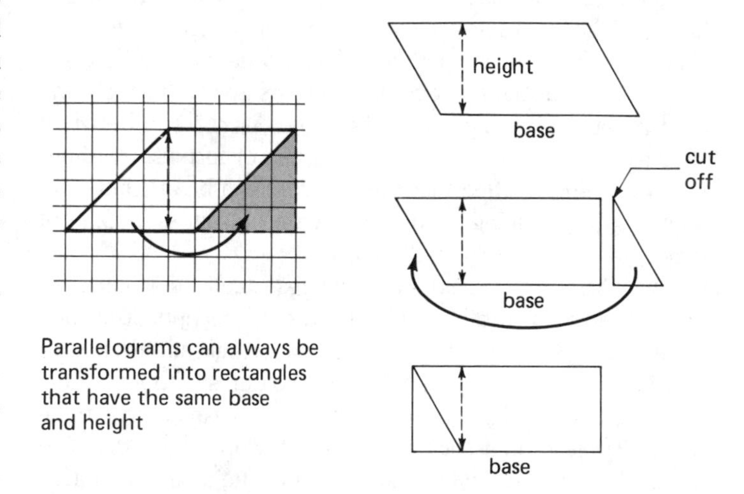

Figure 15.24
Area of a parallelogram.

area. The new rectangle has the same height and two sides the same as in the original parallelogram. Students should explore these relationships on grid paper, on geoboards, or by cutting paper models, and be quite convinced that the areas are the same and that it can always be done. As a result, the area of a parallelogram is base times height, just as for rectangles.

The area of triangles can easily be determined by showing that two identical triangles can always be arranged to form a parallelogram, as shown in Figure 15.25.

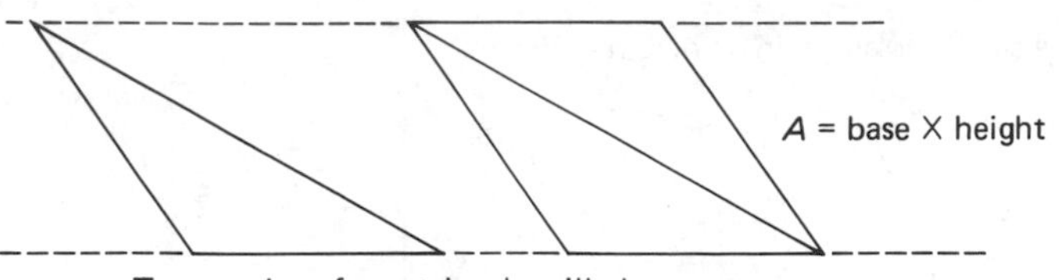

Figure 15.25
Parallelograms of triangles.

Activity 15.33

Have students draw any triangle on a piece of paper and cut out two identical copies. These should then be rearranged into a parallelogram. (If all sides are different, three different parallelograms are possible.) For each parallelogram, have students identify a base and a height. Are these also the base and height of the triangle? The area of the parallelogram is still $b \times h$. What is the area of the triangle?

For trapezoids, explore a similar approach. Have students cut out any two trapezoids and put them together to make a parallelogram and relate the area of that to the area of the trapezoid as in Figure 15.26.

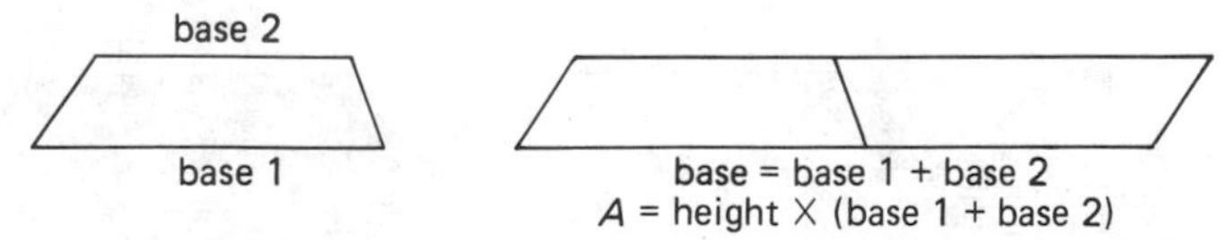

Figure 15.26
Parallelograms of trapezoids.

Students should be able to explain how the areas of triangles, rectangles, parallelograms, and trapezoids are all related. If they cannot, it is a good possibility that they do not understand the development of these formulas. (Do you think that a special formula for squares should be taught? What about formulas for perimeter?)

Circle Formulas

One of the most interesting relationships that children can discover in geometry is that between the *circumference* of a circle (the distance around) and the length of the *diameter* (a line through the center joining two points on the circle). As discussed in the next chapter, students can measure diameters and circumferences of circles of all sizes and use calculators to compare these two measures. The circumference of every circle is about 3.14 times as long as the diameter. The exact ratio is an irrational number *close* to 3.14 and is represented by the Greek letter π. That is, $\pi = C/D$, the circumference divided by the diameter. In a slightly different form, $C = \pi D$ or $2\pi r$, where r is the length of the radius.

In Figure 15.27 an argument is presented for the area formula $A = \pi r^2$. This development is one commonly found in textbooks. Another informal proof is based on the notion that the area of a polygon inscribed in a circle gets closer and closer to the area of the circle as the number of sides increases.*

VOLUMES OF COMMON SOLID SHAPES

Volumes of Cylinders

A *cylinder* is a solid with two congruent parallel bases and sides with parallel elements. There are several special classes of cylinders such as *prisms* (polygons for bases), right prisms, *rectangular prisms,* and *cubes* (See Chapter 16, p. 286). Interestingly, all of these solids have the same volume formula, and that one formula is analogous to the area formula for parallelograms.

Activity 15.34

Provide students with some wooden cubes and square grid paper that matches the face of the cubes. Have students draw a 3×5 rectangle on paper. Have students place 15 cubes on the rectangle. This makes a "box" with a height of one unit. The volume of this box is one times the area of the base. Now place a second layer of cubes on the first. What is the height? The volume? Continue to add layers up to five or six (Figure 15.28). For each new layer notice that the total number of cubes is the number of cubes on the bottom layer times the number of layers. The number on the bottom layer (or any layer) is the area of the base. The number of layers is the height. Therefore, the volume of the solid is $V = A \times h$, the *area of the base times the height.*

Recall how the area formula for rectangles was developed (Figure 15.23) and notice how that development is like the one just presented for volume. The following activity extends the same idea to other cylinders.

Activity 15.35

Make a stack of three or four decks of playing cards (or a stack of any cards or paper). When the cards are stacked straight, they form a rectangular solid. The volume is $V = A \times h$. The area of the base is the area of one playing card. Now if the stack is slanted to one side, a different cylinder is formed as in Figure 15.29. But the new cylinder (or prism) has the

*The relationship between regular polygons and the circle is well known to those who have tried to draw circles with Logo. For them, the latter approach to the area of a circle may be especially appealing.

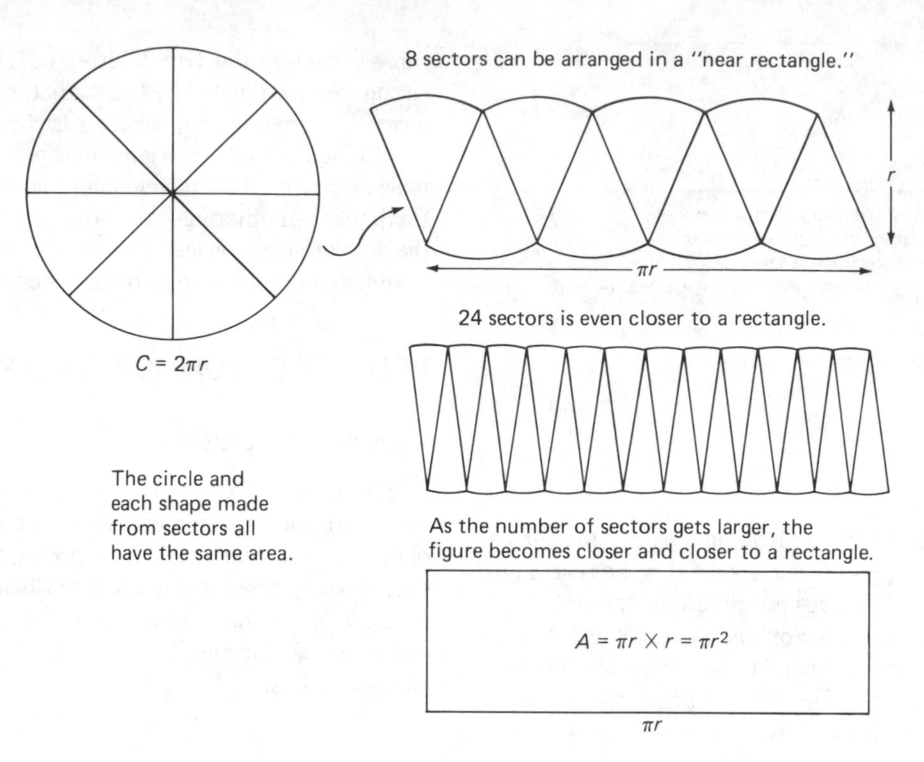

Figure 15.27
Development of circle area formula.

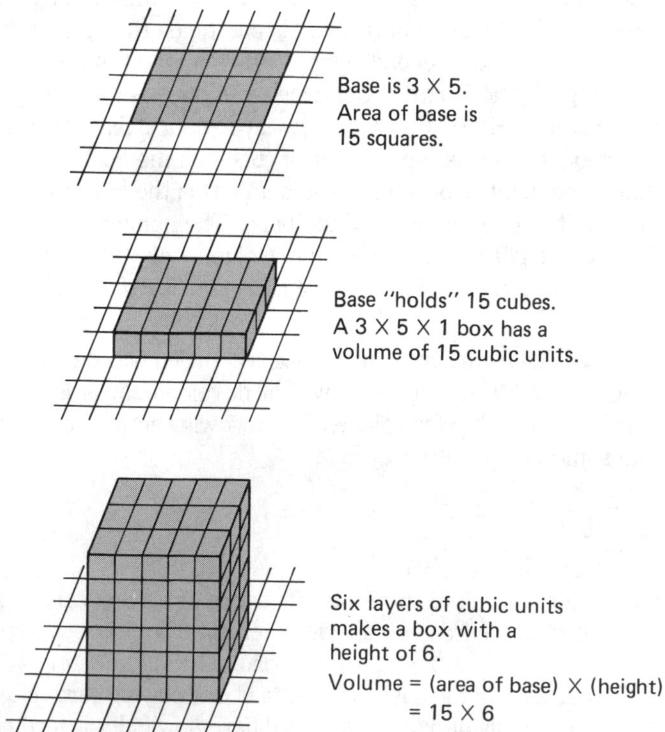

Figure 15.28
Volume of a prism.

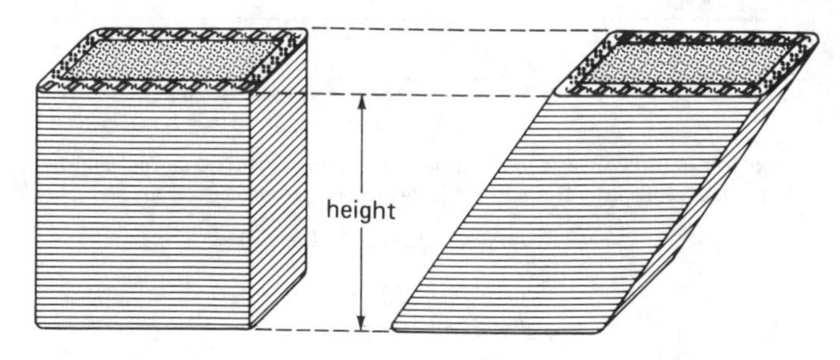

Figure 15.29
Two cylinders with the same base and height have the same volume.

same base, the same height, and obviously the same volume. (Why?) Therefore, the volume formula remains $V = A \times h$.

What if the cards in this activity were some other shape—any shape? If they were circular, the volume would still be the area of the base times the height. If they were triangular, still the same. The conclusion is that the volume of any cylinder is equal to the area of the base times the height.

Volumes of Cones and Pyramids

Just as there is a nice relationship between the areas of parallelograms and triangles, there is a similar relationship between the volumes of cylinders and cones.

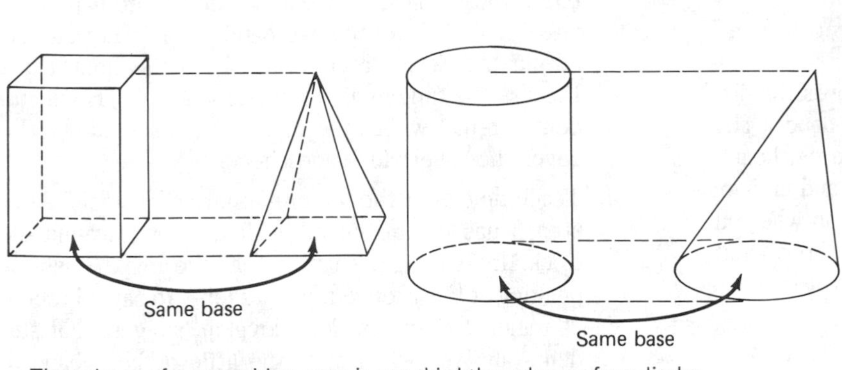

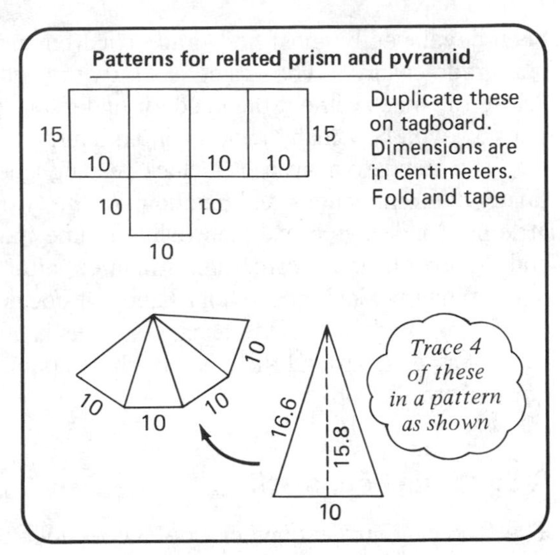

Figure 15.30
Comparing volumes.

Activity 15.36

Help students make a poster board prism and a pyramid with the same height and base. Leave the base of each open. Score fold lines for accurate folds. Dimensions for a square pyramid and prism are given in Figure 15.30. Have students estimate the number of times the pyramid will fit into the prism. Then have them test their prediction by filling the pyramid with beans or rice and emptying it into the prism. They will discover that exactly three pyramids will fill a prism with the same base and height. (An alternative method is to carefully make models from clay and compare the weights.)

The 3-to-1 ratio of volumes is true of all cylinders and cones with the same base and height regardless of the shape of the base or the postion of the vertex. That is, for any cone or pyramid, $V = \frac{1}{3}(A \times h)$.*

Time and Clock Reading

MEASURING TIME

Time is measured in the same way that other attributes are measured: a unit of time is selected and used to "fill" the time to be measured. Time can be thought of as the duration of an event from its beginning to its end. An informal unit of time might be the duration of a pendulum swing, the steady drip of a water faucet, or the movement of the sun's shadow between two fixed points (as on a sundial). To measure time, the units of time are started at the same time as the activity being measured ("timed") and counted until the activity is finished. Thus, the pendulum swings, for example, are "fit into" the duration of time that it takes the child to print his name. Young children enjoy "timing" events with informal units. Older children can appreciate the measurement of time as a process similar to the measurement of other attributes and thus see commonality in all measuring.

CLOCK READING

"Telling time" has very little to do with measurement of time conceptually. The skills of clock reading are related to the skills of reading any meter that uses pointers or hands on a numbered scale. Clock reading is a difficult skill to teach in the first and second grades, and yet nearly everyone learns to tell time by middle school.

Some Difficulties

Young children's problems with clock reading may be due to the curriculum. Children are usually taught first to read clocks to the hour, then the half and quarter hours, and finally to 5- and 1-minute intervals. In the early stages of this sequence, children are shown clocks set exactly to the hour or half hour. Many children who can read a clock at 7:00 or 2:30 have no idea what time it is at 6:58 or 2:33.

Digital clocks permit students to read times easily but do not relate times very well. To know that a digital reading of 7:58 is nearly eight o'clock, the child must know that there are 60 minutes in an hour, that 58 is close to 60, and that 2 minutes is not a very long time. These concepts have not

*The volume of a sphere is a bit more difficult to observe experimentally. The resulting formula, $V = (4/3)\pi r^3$ can be demonstrated using an intuitive argument based on knowledge of the volume of a pyramid. For this development and for truly fascinating explorations of shapes and their measurement, the reader is urged to consider the book *Experiencing geometry* by James Bruni (1977). Bruni's book is aimed at helping the middle school student explore geometric relationships in an intuitive, hands-on manner.

been developed by most first-grade children and not all second-grade children. The analog clock (with hands) can show "close to" times without the need for understanding big numbers or even how many minutes in an hour.

The standard approach to clock reading ignores the distinctly different actions and functions of the two hands. The little hand indicates broad, approximate time (nearest hour), and the long hand indicates time (minutes) after and until an hour. When we look at the hour hand, we focus on where it is pointing. With the minute hand, the focus is on the amount that it has gone around the clock or the amount yet to go to get back to the top.

A Suggested Approach

The following suggestions can be successfully used to help students understand analog clocks and be able to read them.

1. Begin with a one-handed clock. A clock with only an hour hand can be read with reasonable accuracy. Read clocks with only an hour hand and use lots of approximate language. It is about seven o'clock. It is almost four o'clock. It is halfway between two o'clock and three o'clock (Figure 15.31).
2. Discuss what happens to the big hand as the little hand goes from one hour to the next. When the big hand is at 12, the hour hand is pointing exactly to a number. If the hour hand is about halfway between numbers, about where would the minute hand be? If the hour hand is a little past or before an hour (10 to 15 minutes), about where would the minute hand be?

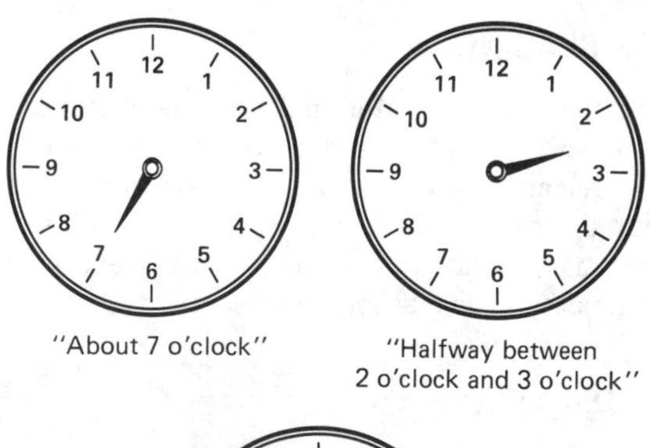

Figure 15.31
Approximate time with "one-handed clocks."

3. Use two real clocks, one with only an hour hand and one with two hands. (Break off the minute hand from an old clock.) Cover the two-handed clock. Periodically during the day, direct attention to the hour hand. Discuss the time in approximate language. Have students predict where the minute hand should be. Uncover the other clock and check.
4. Teach time after the hour in 5-minute intervals. After step 3 has begun, count by fives going around the clock. Instead of predicting that the minute hand is pointing at the 4, encourage students to say it is about 20 minutes after. As skills develop, suggest that students always look first at the little or hour hand to learn *about* what time it is and then focus on the minute hand for precision.
5. Predict the reading on a digital clock when shown an analog clock, and vice versa; set an analog clock when shown a digital clock. This can be done with both one-handed and two-handed clocks.

Related Concepts

Students also need to learn about seconds, minutes, and hours and to develop some concept of how long these units are. If you make a conscious effort to note the duration of short and long events during the day, this will help. Timing small events of 1/2 minute to 2 minutes is fun and useful. TV shows and commercials are a good standard. Have students time familiar events in their daily lives: brushing teeth, eating dinner, riding to school, time in the reading group.

As students learn more about two-digit numbers, the time after the hour can also be related to the time left before the hour. This is not only useful for telling time but helpful for number sense as well. Note that in the sequence suggested, time after the hour is stressed almost exclusively. Time before or 'til the hour can come later.

Problem-solving exercises such as "If it was 7:30 when Bill left home and the trip look 8 hours, what time did he arrive?" are important. Even middle school students have difficulty with these ideas. Adding and subtracting time involves understanding the relationships between minutes and hours and also the two cycles of 12 hours in the day.

For Discussion and Exploration

1. Using the suggestions in this chapter or other ideas you may be able to find, prepare materials for two measuring activities: one for comparison and one involving the use of units. Try your activities yourself first and then with some classmates or even better with some children. Select only one attribute: length, area, capacity, weight, or angles.
2. Measure a length with an informal unit model. Select your unit and object to measure so that the measurement is at least 10 units. Do the measurement in two ways. First use enough copies of the unit that they can be lined up

end to end. Second, measure the same thing using only one copy of the unit. Observe how much more difficult the second method is. Try the same activity with area.

3. Make your own measuring instrument for an informal unit of measure. Select one attribute or make one for each attribute. Use your instrument to measure with and then make the same measurement directly with a unit model. What are some of the values and limitations of each method? Can you see the importance of children doing this both ways?

4. Set up an estimation activity for metric measurement and have the class or your friends try it out. Make it for length, capacity, or weight.

5. Find three good metric benchmarks on your own body. Use them to "measure" some things. Include long distances such as the length of the hall and short distances that you can measure on your desktop. Also use your body benchmarks to estimate measures of distances around objects. Check your estimates made this way with measures made with rulers or tape measures.

6. Get a teacher's edition of a basal textbook for any grade level and look at the chapter(s) on measurement. How would you expand on the activities in the text? What special ideas in the teacher's edition do you like?

7. Read "Chapter 10: Foot Activities" in *A collection of math lessons from grades 3 to 6* (Burns, 1987). Identify two good ideas in the sequence of lessons. Modify the activities to suit your own needs and try them out with a class of children.

8. Make a large poster board clock face and laminate it or cover it with clear contact paper. Attach hands with a brass fastener and use tape to hold them in place when the clock is held vertically. Use your clock to teach a lesson on clock reading in a first or second grade. How well are these children able to estimate time with a one-handed clock (hour hand only) as described in this chapter?

Suggested Readings

Coburn, T. G., & Shulte, A. P. (1986). Estimation in measurement. In H. L. Schoen (Ed.), *Estimation and mental computation.* Reston, VA: National Council of Teachers of Mathematics.

Hart, K. (1984). Which comes first—length, area, or volume? *Arithmetic Teacher, 31*(9), 16–18, 26–27.

Hiebert, J. (1984). Why do some children have trouble learning measurement concepts? *Arithmetic Teacher, 31*(7), 19–24.

Kerr, D. R. Jr., & Lester, F. K. (1976). An error analysis model for measurement. In D. Nelson (Ed.), *Measurement in school mathematics.* Reston, VA: National Council of Teachers of Mathematics.

Lindquist, M. M., & Dana, M. E. (1977). The neglected decimeter. *Arithmetic Teacher, 25*(1), 10–17.

Shaw, J. M. (1983). Student-made measuring tools. *Arithmetic Teacher, 31*(3), 12–15.

Shaw, J. M., & Cliatt, M. J. P. (1989). Developing measurement sense. In P. R. Trafton (Ed.), *New directions for elementary school mathematics.* Reston, VA: National Council of Teachers of Mathematics.

Szetela, W., & Owens, D. T. (1986). Finding the area of a circle: Use a cake pan and leave out the pi. *Arithmetic Teacher, 33*(9), 12–18.

Thompson, C. S., & Van de Walle, J. A. (1985). Learning about rulers and measuring. *Arithmetic Teacher, 32*(8), 8–12.

Thompson, C. S., & Van de Walle, J. A. (1981). A single handed approach to telling time. *Arithmetic Teacher, 28*(8), 4–9.

Van de Walle, J. A., & Thompson, C. S. (1985). Estimate how much. *Arithmetic Teacher, 32*(9), 4–8.

Zweng, M. J. (1986). Introducing angle measurement through estimation. In H. L. Schoen (Ed.), *Estimation and mental computation.* Reston, VA: National Council of Teachers of Mathematics.

CHAPTER 16

GEOMETRIC THINKING AND GEOMETRIC CONCEPTS

An Introduction to Informal Geometry

THREE ACTIVITIES FOR YOU TO EXPERIENCE

It is very likely that most of you have quite different ideas about geometry in the elementary school. In order to provide some common view of the nature of elementary school geometry and how young children approach geometric concepts, three simple activities are offered here for you to do and experience. The activities will provide some idea of the spirit of informal, elementary school geometry as well as background for a discussion of children's geometric thinking. All you will need is a pencil, several pieces of paper, a scissors, and 15 to 20 minutes.

Different Triangles

Draw a series of at least five triangles. After the first triangle, each new one should be different in some way from those already drawn. Write down why you think each is different.

Shapes with Triangles

Place a piece of paper over the dot grid in Figure 16.1 and draw three or four different figures by connecting adjacent dots. Each figure should have an area of 10 small triangles. A few simple rules for drawing your shapes are explained in Figure 16.2. Count to find the distance around each figure or the perimeter and record this next to each drawing. You may want to explore any observations you may have by drawing some more shapes.

A Tiling Pattern

First make at least eight copies of any one of the shapes in Figure 16.3. An easy way to do this is to fold a piece of paper so there are eight thicknesses. Trace the shape on an outside section and cut through all eight thicknesses at once. You may want more copies of these shapes.

Think of the shapes you cut out as tiles. The task is to

Figure 16.1

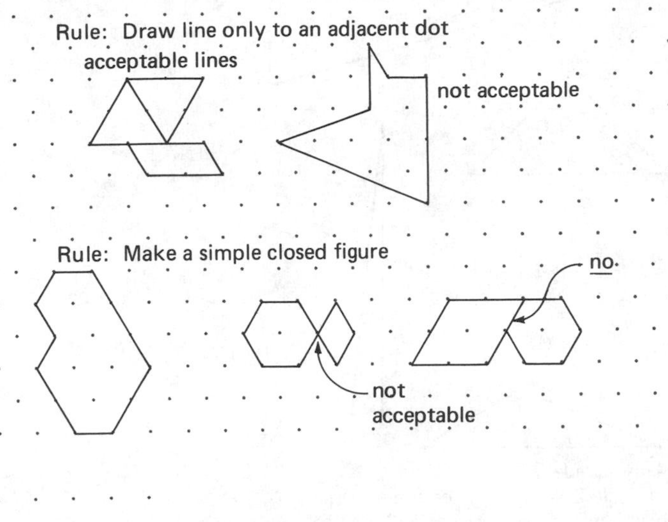

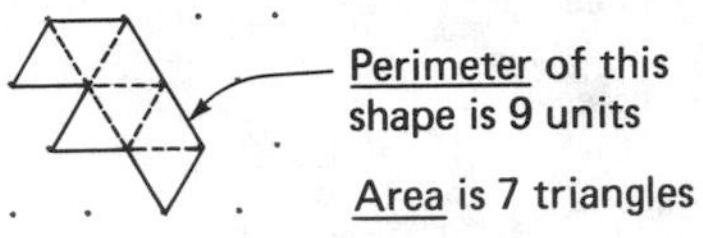

Figure 16.2
Rules for making the shapes in Activity 16.2.

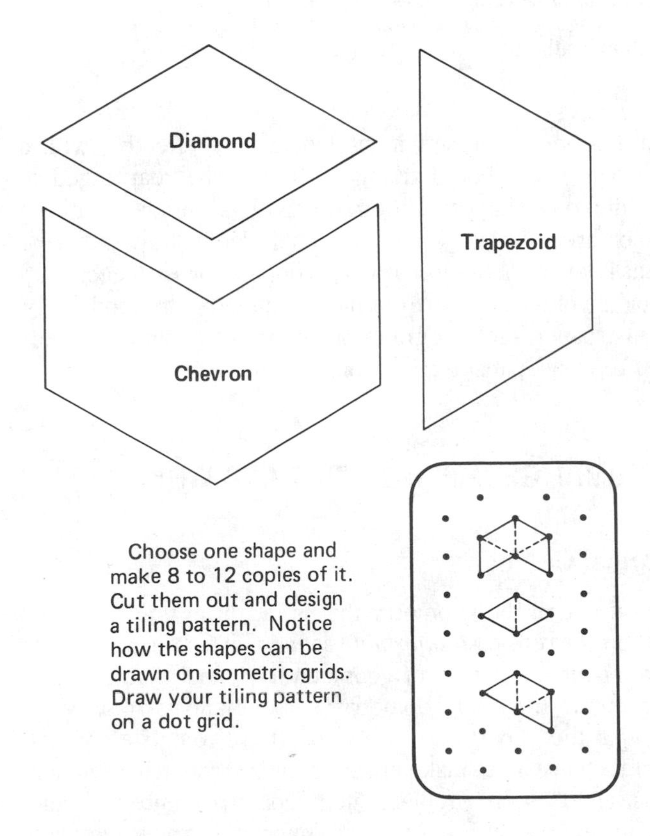

Figure 16.3
Three tile patterns.

use the tiles to make a regular tiling pattern. A tiling pattern made with one shape has two basic properties. First, there are no holes or gaps. The tiles must fit together (without overlapping) without leaving any spaces. Second, the tiles must be arranged in a repeating pattern that could be extended indefinitely. That is, if you were to tile an endless floor with your pattern, the design in one section of the floor would be the same as that found in any other section. Experiment with the paper tiles to decide on a pattern that you like. There are several different tiling patterns possible for each of the three tiles.

Notice that each of the tile shapes is made up of triangles and can be drawn on the dot grid of Figure 16.2. When you have decided on a tiling pattern, place a piece of paper over the dot grid and draw your tiling pattern with small versions of the tile shape using the dots as a guide. Cover most of the grid with your pattern. Finally, suppose your tiles came in two colors and you want to add a color pattern to your tile pattern. With a pen or pencil, shade in some of the tiles to make a regular pattern in two colors.

SOME INITIAL OBSERVATIONS ABOUT THESE ACTIVITIES

Rather than discuss these three activities in detail, the following are some observations that you might reflect on. These comments are about geometry in the elementary school in general as well as reflections or observations about the activities themselves.

Different People Think about Geometric Ideas in Different Ways

Compare your response to the activities on p. 266 with peers. Are there qualitative differences as well as objective differences? How would primary children approach these activities compared to an eighth grader? Figure 16.4 shows how two students, one in the fifth grade and one in the eighth grade, responded to the triangle task. Research indicates that age is not the major criterion for how students think geometrically. The kinds of experiences a child has may be a more significant factor. What were your grade school experiences with geometry?

Explorations Can Help Develop Relationships

The more that you play around with and think about the ideas in these activities, the more there is to think about. A good teacher might be able to extend each of these activities to develop the ideas beyond the obvious. For example:

For *Different Triangles:* How many different ways can two triangles be different? Could you draw five or more *quadrilaterals* that were each different?

For *Shapes with Triangles:* What makes a shape have a smaller perimeter? What if we did this activity with

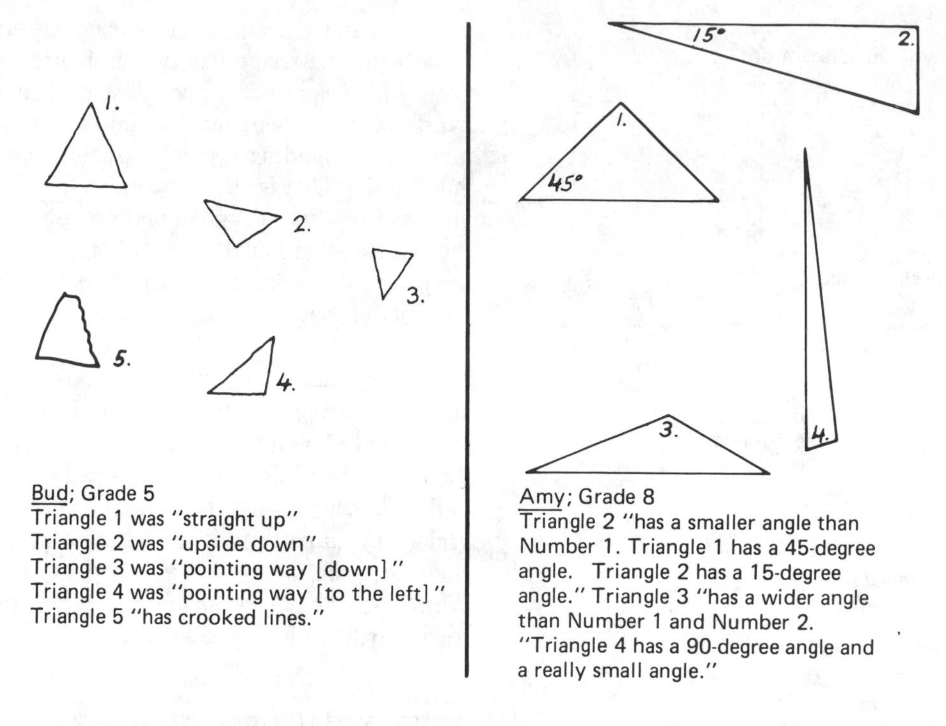

Figure 16.4
Two children show markedly different responses to the task of drawing a series of different triangles. W. F. Burger and J. M. Shaughnessy (1986). Characterizing the van Hiele levels of development in geometry. Journal for Research in Mathematics Education, 17.

rectangles? Do boxes with equal volumes have the same surface area?

For *A Tiling Pattern:* How many different tiling patterns are there for this shape? Can any shape be used to tile with? Can you see any bigger shapes within your pattern?

Geometry Activities Are Also Problem-Solving Activities

Good geometry activities almost always have a spirit of inquiry or problem solving. Many of the goals of problem solving are also the goals of geometry. Reconsider each of the three activities as examples of problem solving. Can you identify at least one problem in each that needs to be solved? In the context of those problems, consider the goals of problem solving, especially perseverance, willingness to take risks, understanding the problem, evaluating the results, and going beyond the initial problem.

Good Geometry Activities Involve Hands-on Materials

Even the simple paper tiles used in *A Tiling Pattern* gave you the opportunity to explore spatial relationships and search for patterns much more easily than without them. Activities on paper such as the dot grid in *Shapes with Triangles,* are a second-best alternative to real physical objects. The same area and perimeter activity is much more effective with a collection of cardboard triangles that can be rearranged to form different shapes. The first activity is the least enticing of the three, but at least you could freely draw pictures. Virtually every activity that is appropriate for K–8 geometry should involve some form of hands-on materials, models, or at least paper such as graph paper or dot paper that lends itself to easy spatial explorations.

INFORMAL GEOMETRY: WHAT AND WHY

Informal Geometry

Most if not all the geometry that is taught in grades K to 8 can be referred to as *informal geometry.* This term suggests more about the nature of geometry in school than its goals or content. Informal geometry is experiential. It provides students the opportunity to explore the geometric environment they live in, to make observations, to construct relationships, and to solve problems in a geometric context. Sometimes it is artistic. Sometimes it involves learning the name of a shape or a property that has been observed. Sometimes it includes building a model or taking one apart. Sometimes it is solving a shape puzzle. Sometimes it is seeing how things are alike or different or if an observation can be generalized. But most of all, informal geometry is a spirit of exploration and observation of the geometric world that surrounds us all.

Why Study Geometry?

Unfortunately, geometry frequently takes a back seat to other topics in school mathematics and so "Why study geometry?" is a fair question. In fact, if you were to decide that you really enjoyed doing informal geometry activities with your students (some of the most fun you can have), you may feel the need to justify your actions. The following list of reasons for why the study of geometry is important is at least a place to begin.

1. Geometry helps people have more complete appreciation of the world in which they live. Geometry can be found in the structure of the solar system, in geological formations, in rocks and crystals, in plants and flowers, and even animals. It is also a major part of our synthetic universe: art, architecture, cars, machines, and virtually everything that humans create has an element of geometric form.
2. Geometric explorations can develop problem-solving skills. Spatial reasoning is an important form of problem solving, and problem solving is one of the major reasons for studying mathematics.
3. Geometry plays a major role in the study of other areas of mathematics. By way of example: Fraction concepts are related to geometric part-to-whole constructs. Ratio and proportion are directly related to the geometric concept of similarity. Measurement and geometry are clearly related topics, each adding to the understanding of the other.
4. Geometry is used by many people daily in their professional as well as their everyday lives. Scientists of all sorts, architects and artists, engineers, and land developers are just a few of the professions that use geometry regularly. In the home, geometry helps build a fence, design a dog house, plan a garden, or even decorate a living room.
5. Geometry is fun and enjoyable. While fun may not be a reason in and of itself, if geometry is a way to entice students into a little more love of mathematics in general, then that makes much of the effort worthwhile.

The Development of Geometric Thinking

THE VAN HIELE LEVELS OF GEOMETRIC THOUGHT

The work of two Dutch educators, Pierre van Hiele and Dina van Hiele-Geldof, is beginning to have an impact on the design of geometry instruction and on geometry curriculum in the United States. Van Hiele's work began in 1959 and was eagerly adopted by the Soviets, but for nearly two decades it received little attention in this country (Hoffer, 1983, 1988a).

The most prominent feature of the van Hiele model is the existence of five different levels of geometric thought within which one learns, talks about, and works with geometric ideas. What follows is a very brief description of the five van Hiele levels.

Level 0: Visualization. Students recognize and name figures in a global or gestalt manner. A square is a square "because it looks like one." The properties of a figure are not understood as defining characteristics. At this level, students are influenced by nonrelevant attributes of a figure such as its orientation. For example, some students would not recognize a triangle that "pointed down." While able to recognize squares and rectangles, they would not understand that a square is also a rectangle.

Level 1: Analysis. Students at this level can begin to analyze various properties of shapes such as the existence of right angles in rectangles or opposite sides parallel in parallelograms. When defining or describing shapes, students are quite likely to give as many properties as they know even though some may be redundant. "A square is a four-sided shape with opposite sides equal and with four right angles and opposite sides are parallel." Students can make generalizations about classes of shapes, such as noting that all rhombuses have four equal sides or that parallelograms have opposite angles equal and have two sets of parallel sides. At the same time, however, the relationship between classes of figures is not made. For example, a student does not understand at this level that a square is a rhombus with a right angle even though the defining characteristics of each may be well known. In short, properties can be observed and analyzed, but deductive arguments that help students see relationships between shapes of different classes are not available at this level.

Level 2: Informal Deduction. Construction of relationships between classes of figures is now possible at this level. Shapes can be classified with the use of minimal defining characteristics. For example, rectangles are parallelograms with a right angle. Since all squares meet these criteria, squares must be rectangles and parallelograms. At this level, informal arguments can be used concerning shapes and relationships between shapes. This does not imply the use of a formal deductive or axiomatic system. Students can follow a proof at this level, but are probably unable to construct one on their own.

Level 3: Deduction. Students at this level can utilize a deductive axiomatic system and construct proofs of their own within that system. They are able to see that two different logical arguments could each be valid for the same theorem. This is the level that is *assumed* by the traditional tenth-grade geometry course.

Level 4: Rigor. At this level, there is an appreciation of the distinctions and relationships between different axiomatic systems. The systems themselves become the objects to work with, not just deductions within a system. This is the level of the mathematician who is studying geometry as a mathematical science.

CHARACTERISTICS

Four related characteristics of the van Hiele levels merit special attention. First, the levels are sequential. That is, to arrive at any level above the first, students must move through all prior levels. To move through a level would mean that the strategies and abilities of that level had been developed.

Second, the levels are not directly age-dependent (as, for example, are the well-known developmental stages of Piaget). A third grader as well as a high school student may both be at level 0 or both at level 2. Many high school students and adults have not reached level 1, and a significant number of adults have not reached level 2. At the same time, age is certainly related to the amount and types of geometric experiences that we have. Therefore it is reasonable to expect children from kindergarten to second grade to be at level 0 and most children in the upper elementary grades to be at level 1.

Third, advancement depends primarily on experience with content and instruction. Activities that permit children to explore, talk about, and interact with content at the next level, while increasing experiences at their current level have the best chance of advancing the level of thought for those children.

Fourth, if there is a mismatch between the textbook or teacher presentation of content and the level of the student, learning may not occur. If instruction requires a child to skip a level or attend above his or her level, learning that does occur may well be rote and temporary. For example, a child can memorize that all squares are rectangles but may not have constructed that relationship (Crowley, 1987). Such learning is instrumental and of little value.

IMPLICATIONS FOR INSTRUCTION

Content Versus Thinking Goals

Even this very brief look into the notion of levels of geometric thinking suggests that teachers essentially have two types of goals in their geometry instruction. The first, more traditional goals, include the content that is prescribed by the curriculum. The second type of goal is much more long term: to advance students' level of geometric thought so that by the end of the eighth grade they will have reached level 2 thinking. On one hand, there are "things to be learned": names of shapes, the concepts of congruence, similarity, symmetry, parallelism, and so on. On the other hand, there is the type of experience that will meet children where they are and help move forward their geometric thought processes.

Teach at the Students' Level of Thought

The level of an activity can be varied even for the same content. It is the type of thinking that children are required to do that makes a difference in learning, not the specific content. Following are some suggestive features of instruction for each of the first three levels.

Level 0 activities should

Use lots of physical models that can be manipulated by the students.

Include many different and varied examples of shapes so that irrelevant features do not become important. (Many students, for example, believe that only equilateral triangles are really triangles or that squares turned 45 degrees are no longer squares.)

Involve lots of sorting, identifying, and describing of various shapes.

Provide opportunities to build, make, draw, put together, and take apart shapes.

Level 1 activities should

Continue to use models as with level 0, but include models that permit the exploration of various properties of figures.

Begin to focus more on properties of figures rather than on simple identification. Define, measure, observe, and change properties with the use of models.

Classify figures based on properties of shapes as well as by names of shapes. For example, find different properties of triangles that make some alike and others different.

Use problem-solving contexts in which properties of shapes are important components.

Level 2 activities should

Continue to utilize models with a focus on defining properties. Make property lists and discuss which properties are necessary and which are sufficient conditions for a specific shape or concept.

Include language of an informal deductive nature: all, some, none, if-then, what if, and so forth. Investigate the converse of certain relationships for validity: For example, the converse of "If it is a square, it must have four right angles," is "If it has four right angles, it must be a square."

Using models and drawings as tools to think with, begin to look for generalizations and counterexamples. Encourage hypothesis making and hypothesis testing.

Most of the content of the elementary school curriculum can be adapted to any of the three levels. An exception may be an inappropriate attention to abstract concepts such as point, line, ray, and plane as basic elements of geometric forms. These abstract ideas are not even appropriate at level 2.

Listen to your children during a geometry activity. Compare the types of comments and observations that they make with the descriptions of the first two van Hiele levels. Be sure that the activities you plan do not require students to reason above their level of thought.

Plan for Growth Opportunities

Most geometry activities in the elementary school will be aimed at the first two van Hiele levels with sixth- through eighth-grade students beginning to explore concepts within level 2 activities. One of your long-term goals should be to move students forward in their level of thinking. To this end many activities should begin to explore ideas one level above where you think most of your students are operating. For example, students at level 0 can be encouraged to observe properties of shapes and describe them informally. For example, lines that "go in the same direction" are parallel, "corners like a square" are right angles, and "is the same if you fold it over" is a rough description of line symmetry. Level 1 children can move toward definitions of shapes by selecting properties that they think might be sufficient to distinguish one shape from another.

Informal Geometry Activities: Level 0

The emphasis at level 0 is on shape and form experiences. While properties of figures are included, they are only explored informally. Remember that a level 0 activity does not necessarily mean a primary-grade activity. Not all students in the upper grades have had sufficient opportunity to experience ideas and develop their thinking beyond a beginning level.

Any categorization of informal geometry activities according to van Hiele levels is likely to be a bit "fuzzy." You may think some activities represent level 1 better and many can be extended easily to include that level.

EXPLORING SHAPES AND PROPERTIES

Sorting and Classifying Shapes

Sorting or classifying shapes using models is a good way to introduce geometric ideas. Names of shapes and properties can be provided as students begin to recognize and discuss them in their own words.

Activity 16.1

Make collections of posterboard shapes similar to those in Figure 16.5. You will want many more shapes than those shown. Have groups of children find sets of shapes that are alike in some way. By preparing the pieces so that the ideas you want children to learn are represented by at least four or five different examples, it is likely that students will notice that concept. When students have sorted out some shapes indicating that they recognize some idea that is common, you have an opportunity to label the concept or provide the proper name of a shape without trying to define it formally. Figure 16.6 suggests some concepts that can be explored in this way. Notice the use of varied examples of each idea.

When students omit shapes from a category they have identified or fail to create a category you hoped they would discover, it is a clue to their perceptual thinking. Interact informally by selecting an appropriate shape and have children discuss why they think it may be different or the same as other shapes. Avoid definitions and right or wrong answers at this level.

Figure 16.5
An assortment of shapes for sorting.

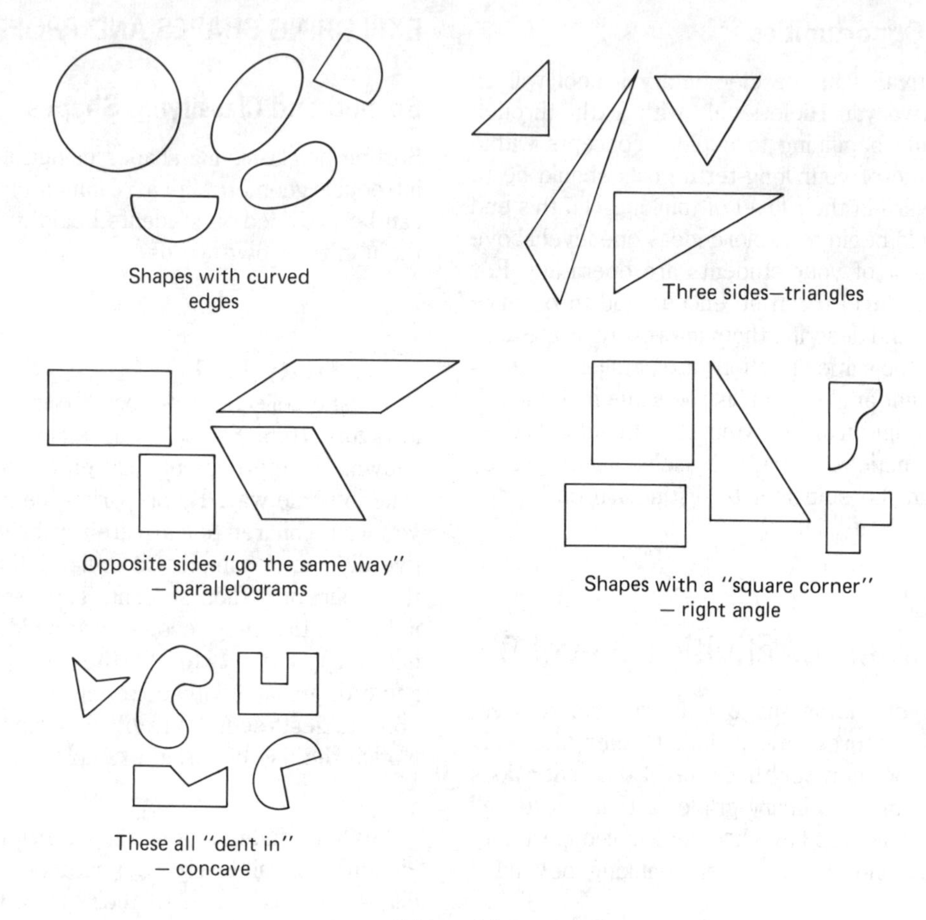

Figure 16.6
By sorting shapes students begin to recognize properties.

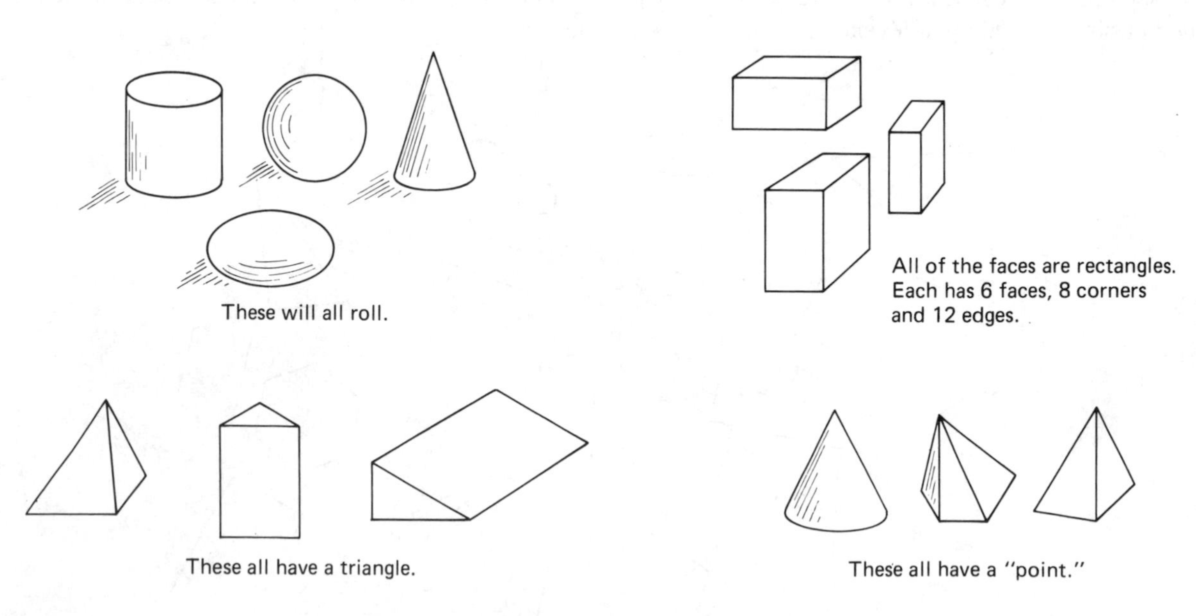

Figure 16.7
Early classifications of three-dimensional shapes.

The same activity can be done with three-dimensional shapes. Wooden or plastic collections are available as one option. Another is to make some solids from tagboard or modeling clay. Real objects such as cans, boxes, balls, or Styrofoam shapes are another source of three-dimensional models. Figure 16.7 illustrates some classifications of solids.

Matching Shapes

Activity 16.2

Prepare two identical collections of shapes. One set is placed in view. Without children seeing, place a shape from the other set in a box or bag. Children reach in the box and feel the shape without looking and attempt to find the matching shape from those that are displayed. The activity can be done with either two-dimensional tagboard shapes or with solids. The shapes in the collection determine what ideas will be focused on and how difficult the activity is.

All shapes have different numbers of sides or faces.

All shapes are different but belong to one category such as quadrilaterals, triangles, pyramids, prisms, curved surfaces, curved edges.

A variation is to have the hidden shapes be small versions of the shapes the students see. Matching a small shape with a larger one provides an informal introduction to the notion of similarity: same shape, different size.

A similar matching activity can be done with the entire class and provides more opportunity for informal discussion of how shapes are alike and what their properties are.

Activity 16.3

Display a collection of shapes for all to see (either two- or three-dimensional). Show the class a shape that has something in common with one or more of the shapes in the collection. Students are to select the shape that is like your shape in some way and explain their choice. The target shape may be another example of a particular shape or it may be entirely different with only some property in common with another shape. There may be excellent choices and reasons that you did not even think about. Figure 16.8 illustrates only a few of the many possibilities for this activity.

Another type of matching activity matches solid shapes with copies of the faces. "Face cards" can be made by tracing around the different faces of a solid shape as in Figure 16.9.

Activity 16.4

Two different matching activities can be done with face cards: given a face card, find the solid; or given a solid, find the face card. If cards are made with only one face per card, students can select from a larger collection, those faces that belong with a particular shape. Another variation is to show only one face at a time as "clues."

Going on a "shape hunt" is a well-worn but still worthwhile activity.

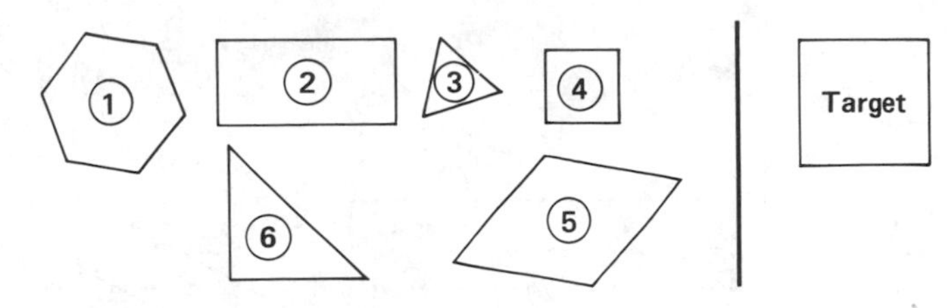

Like (4) – Both are squares.
Like (5) – All four sides same, squeezed in a little.
Like (6) – Has the same kind of corner.
– Looks like half of it.
Like (2) – Same except its longer.

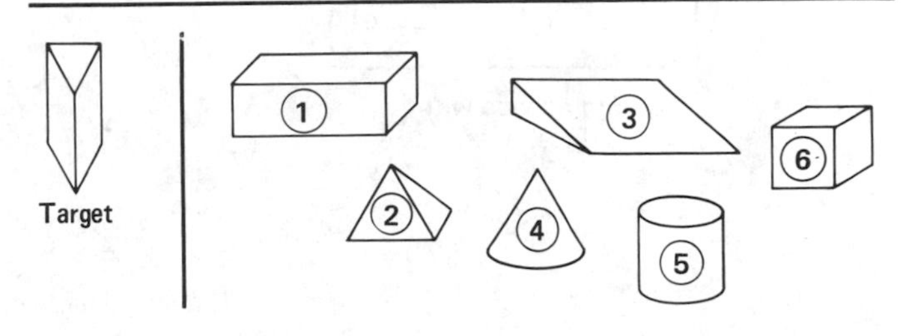

Like (3) – Both are "wedges."
– Both have two triangles.
Like (1), (3) and (6) – All have a rectangle face.
Like (1), (5) and (6) – All have a "flat top"
(i.e., two parallel faces – so does (3)).
Like (2) – Has triangles.

Figure 16.8
Playing "Find a shape that's like the target": There are usually many good solutions.

Activity 16.5

Have students search not just for triangles, circles, squares, and rectangles, but also for properties of shapes. A shape hunt will be much more successful if you let students look for either one thing or a specific list. Different groups can hunt for different things. Some examples of things to search for are

Parallel lines (lines "going in the same direction")

Right angles ("square corners")

Curved surfaces or curved lines

Two or more shapes that make another shape

Circles inside each other (concentric)

Shapes with "dents" (concave) or without "dents" (convex)

Shapes used over and over in a pattern (brick wall, chain link fence)

Solids that are somehow *like* a box, a cylinder (tube), pyramid, a cone

Five shapes that are alike somehow (specify solid or flat)

Shapes that are symmetrical

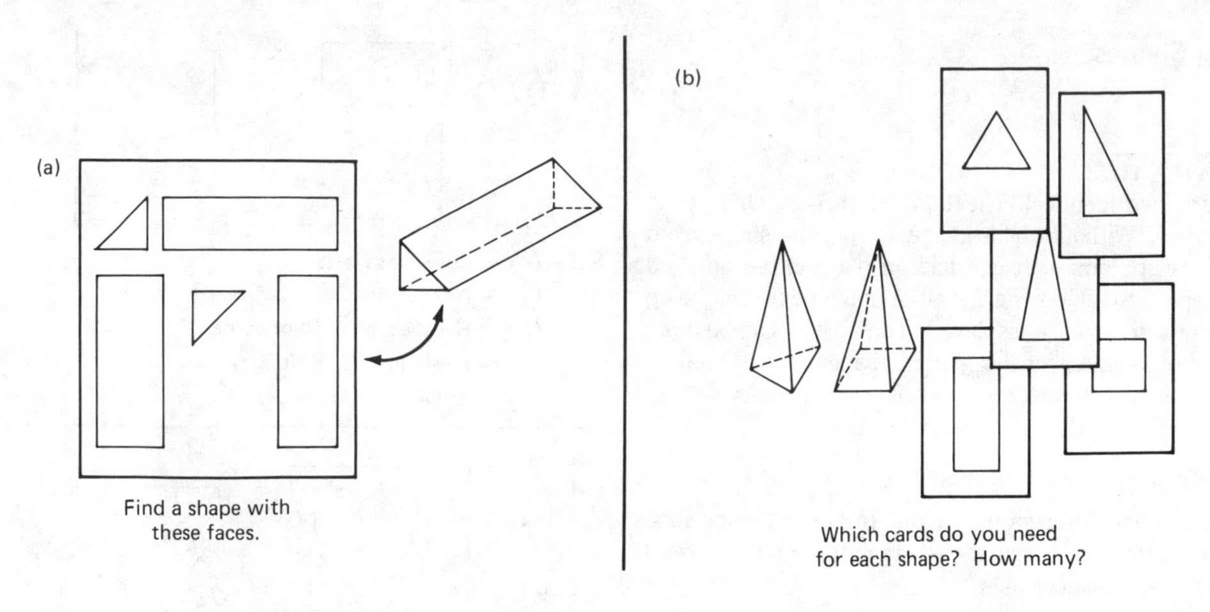

Figure 16.9
Matching "face cards" with solid shapes.

BUILD, DRAW, MAKE

Using Tiles to Make Shapes

A good way to explore shapes at level 0 is to use smaller shapes or "tiles" to create larger shapes. Different criteria or directions can provide the intended focus to the activity. Among the best materials for this purpose are pattern blocks, but many teacher made materials can be used equally well. In Figure 16.10 a variety of different shapes are suggested. Class sets can be cut from poster board and placed in plastic bags for individual or group use. Some of the activities which follow can be repeated using a different set of tiles to provide not only variety but a different perspective.

Activity 16.6

Make a specific type of shape using one or two different types of tiles. Exactly what shapes can be suggested is deter-

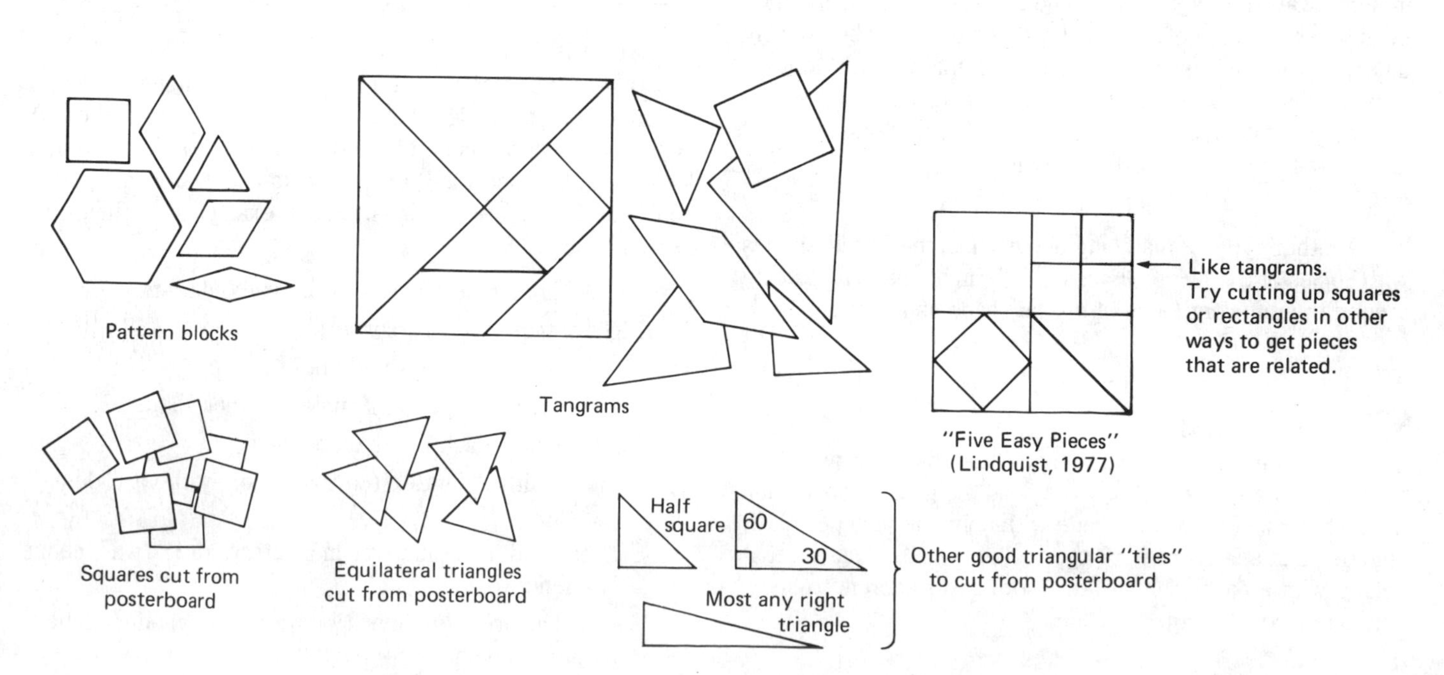

Figure 16.10
Collections of tiles can include an assortment of shapes or can be all the same shape.

mined by the tiles being used. Try some of these ideas with different tiles and make up more of your own.

Make some different triangles. Make big ones and little ones. What do you notice?

Make some rectangles. How many different ones can you make?

Make some shapes with four sides. (Try other numbers.)

Make some parallelograms (or trapezoids, or rhombuses, squares, or hexagons.) How are your shapes alike? How are they different?

Activity 16.7

Make some shapes that ______. Fill this in with different properties. Again, what is possible will vary with the tiles being used.

Make a shape that has parallel sides (or that has no parallel sides, or two, or even three sets of parallel sides) [Figure 16.11(a)].

Make some shapes that have square corners. Make some with two right angles, three, four, . . . [Figure 16.11(b)].

Make a shape that has a line of symmetry. Check it by placing a mirror on the shape [Figure 16.11(c)].

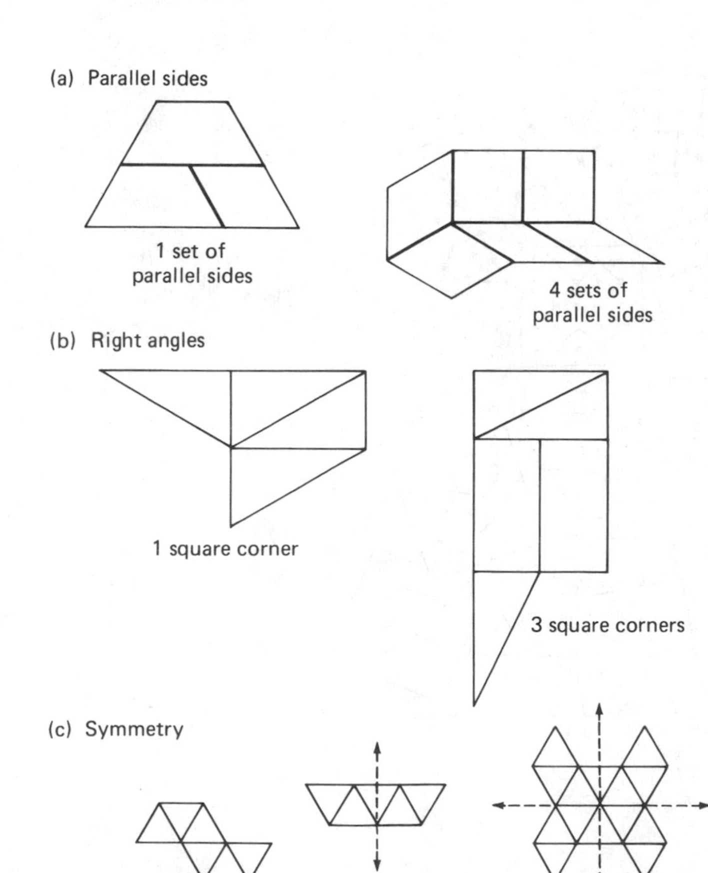

Figure 16.11
Build shapes with special properties using tiles.

Circles and curves are frequently neglected because they are difficult for young children to draw. The three tiles in Figure 16.12 are quarters of two different circles and an oval. Along with square tiles that match the radii of these three tiles, many interesting shapes can be constructed. Children gain informal experiences with the way tangent curves combine to make different complex curves. Just playing with these tiles is a good activity. More structured tasks might include making shapes with symmetry or making two shapes that are somewhat alike but perhaps longer or fatter than the first one that was made. Tracing around the tiles is another way that interesting curved shapes can be made, even by young children. A master for curved tiles, found in Appendix C, can be duplicated onto tagboard as one way to make these tiles easily.

Geoboard Explorations

The geoboard is one of the best devices for "drawing" two-dimensional shapes. Literally hundreds of activities, task cards, and worksheets have been developed for geoboards. Here are just a few activities for a beginning level.

Activity 16.8

Copy shapes, designs, and patterns from prepared cards as in Figure 16.13. Begin with designs shown with dots as on a geoboard and later have students copy designs drawn without dots.

Activity 16.9

Make shapes using smaller shapes (Figure 16.14).

Activity 16.10

Copy a shape from a card and then have students subdivide or cut it into smaller shapes. Specify the number of smaller shapes. Also specify whether they are all to be congruent or simply of the same type as shown in Figure 16.15. Depending on the shapes involved, this activity can be made quite easy or relatively challenging.

Activity 16.11

Challenge students to see how many different shapes they can make of a specific type or with a particular property. (Very young children will feel more comfortable searching for *three* shapes or *four* shapes instead of trying to find many shapes.) Here are some appropriate ideas for level 0.

Make shapes with five sides (or some other number).

Make shapes with all square corners. Can you make one with three sides? four sides? six, seven, eight sides?

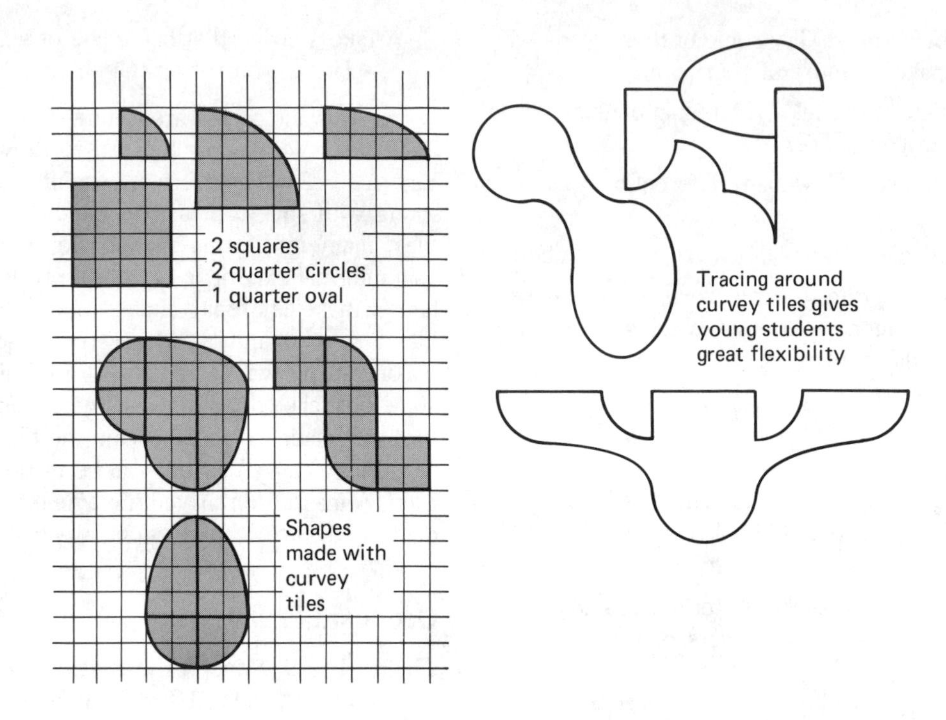

Figure 16.12
Curved tiles can be used to build shapes or to trace the outlines of shapes.

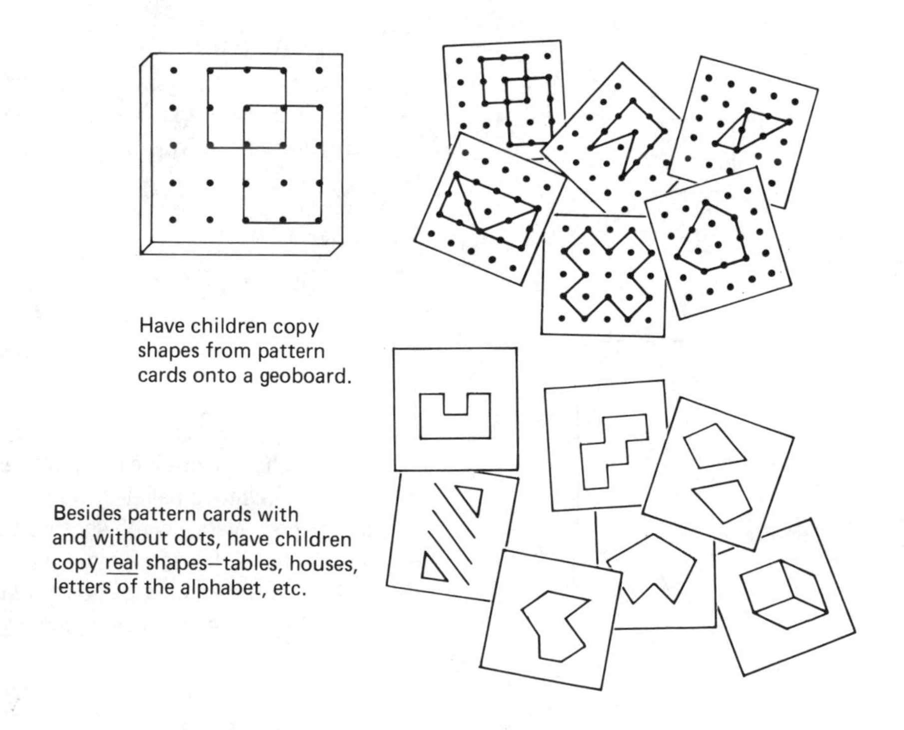

Figure 16.13
Shapes on geoboards.

Make some trapezoids that are all different (or any other shape that students can identify).

Make a shape that has a line of symmetry. Check it with a mirror.

Two practical comments should be made about geoboards. First, have lots of them available in the classroom. It is better for two or three children to have 10 or 12 boards at a station than for each to only have one. This way a variety

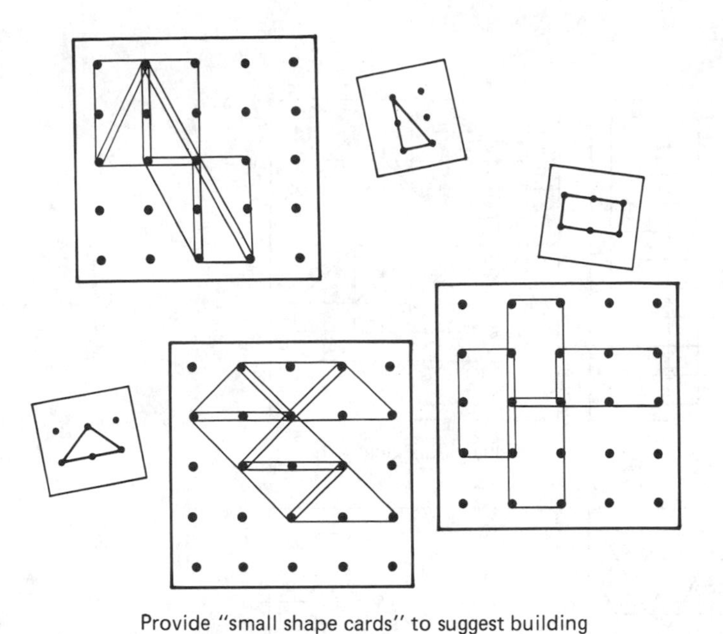

Figure 16.14
Bigger shapes from smaller shapes.

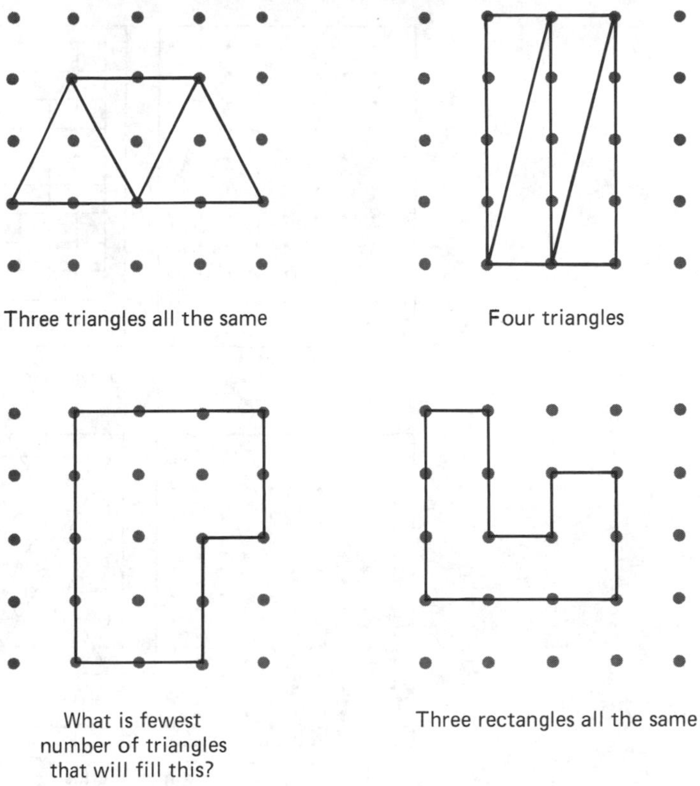

Three triangles all the same

Four triangles

What is fewest number of triangles that will fill this?

Three rectangles all the same

Start with a shape and cut it into smaller shapes. Add special conditions to make the activity challenging.

Figure 16.15
Subdividing shapes.

of shapes can be made and compared before they are changed. That leads to the second point. Teach students from the very beginning to copy their geoboard designs onto paper. Paper copies permit students to create complete sets of drawings that fulfill a particular task. If a student wants to make a series of different six-sided shapes but has only one geoboard, the paper permits her to copy the entire series. Drawings can be placed on the bulletin board for classification and discussion, made into booklets illustrating a new idea that is being discussed, and can be sent home to show mom and dad what is happening in geometry.

Younger students can use paper with a single "geoboard" on each sheet. At the kindergarten or first-grade level, the paper geoboard can be the same size as the real board. Later, a paper board about 10 cm square is adequate. Older children can use centimeter dot paper or a page of small boards, each about 5 cm square.

In the very early grades, children will have some difficulty copying geoboard designs onto paper, especially designs with slanted lines. To help, suggest that they first mark the dots for the corners of their shape. (Example: "second row, end peg.") With the corners identified it is much easier for them to draw lines to make the shape.

Dot and Grid Paper Explorations

Geoboards are excellent due to the ease with which drawings can be made and changed. They do have limitations in terms of size, arrangement, and number of pegs. Assorted dot and grid papers provide an alternative way to make drawings and explore shapes. Virtually all of the activities suggested for tiles and geoboards can also be done on dot or grid paper. Changing the type of grid paper changes the activity and provides new opportunity for insight and discovery. Figure 16.16 shows several possibilities for dot and grid papers that are included in Appendix C.

Besides the geoboard and tiles activities mentioned so far, here are some additional ideas that lend themselves particularly well to dot and grid paper.

Activity 16.12

Isometric dot paper is an effective way to draw solid shapes that are built with cubes. Square dot paper can be used to draw side and top views as shown in Figure 16.17. Building a simple shape with cubes and then drawing plan and perspective views is an excellent activity for middle grade students to help with perspective. (See Winter, Lappan, Phillips & Fitzgerald, 1986, for a series of activities.)

Activity 16.13

Which shapes made of six squares on a square grid will fold up to make a cube? Which shapes with five squares will fold up to make an open box? On an isometric grid, which

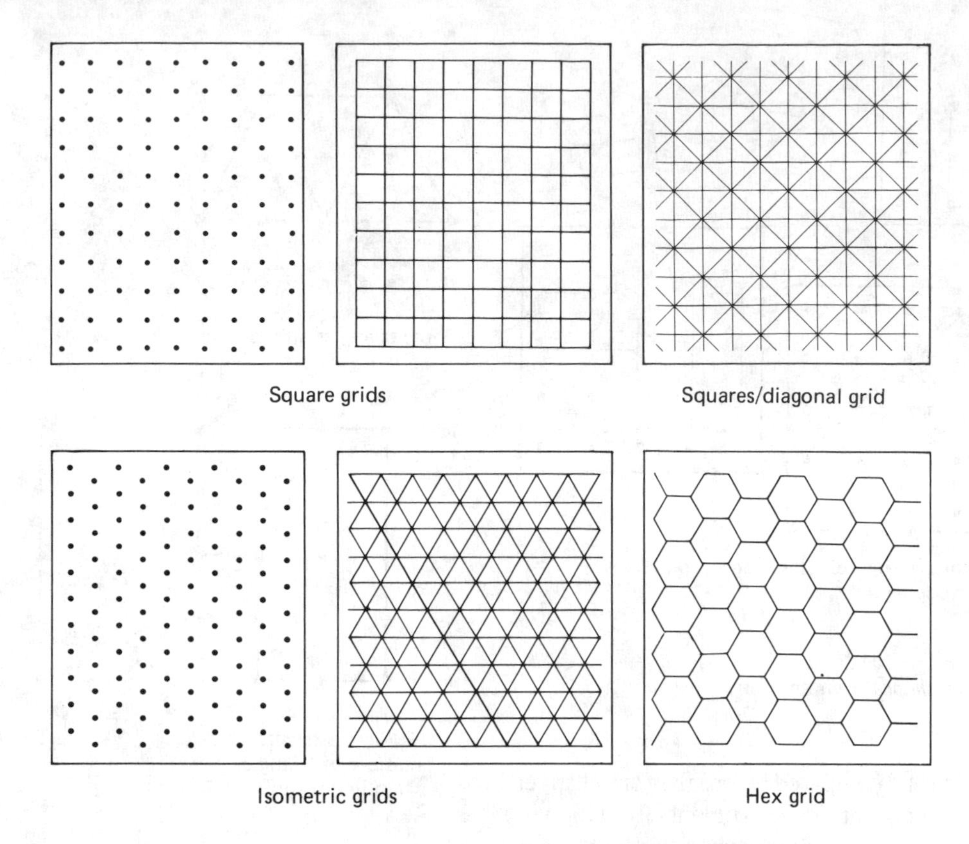

Figure 16.16
Dot and grid paper of various types and sizes can be used for many geometric explorations.

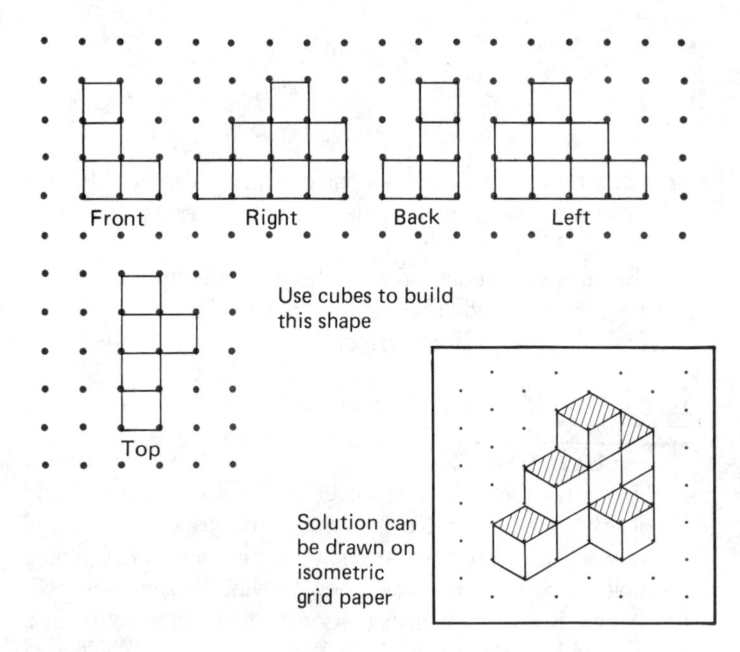

Figure 16.17
Develop perspective and visual perception with cubes and plan views. Draw block "buildings" on isometric dot grids.

shapes with four triangles will fold to make a pyramid: Can you fold up a solid with six triangle faces? with eight faces?

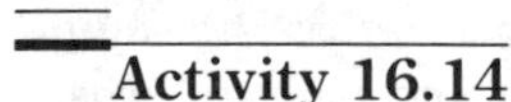

Activity 16.14

Slides, flips, and turns can be investigated on any grid. Start with a simple shape. Draw the same shape "flipped over," or turned, or placed in a different orientation. Trace the original shape and cut it out. This copy can be flipped or reoriented to test the drawings that are made (Figure 16.18).

Building Solids

Building solid or three-dimensional shapes presents a little more difficulty than two-dimensional shapes, but is perhaps even more important. Building a model of a three-dimensional shape is an informal way to get to know and understand the shape intuitively in terms of its component parts.

Activity 16.15

Have children build skeleton models of solid models or common three-dimensional shapes in their environment (Figure 16.19). Pipe cleaners and *D-Stix* are two possible materials for this purpose. Larger skeletal models are surprisingly easy to make by rolling newspaper into long tight tubes. Use about three thicknesses of a large sheet and roll diagonally. A

Figure 16.18
Slides, flips, and turns can be explored on most any type of grid paper.

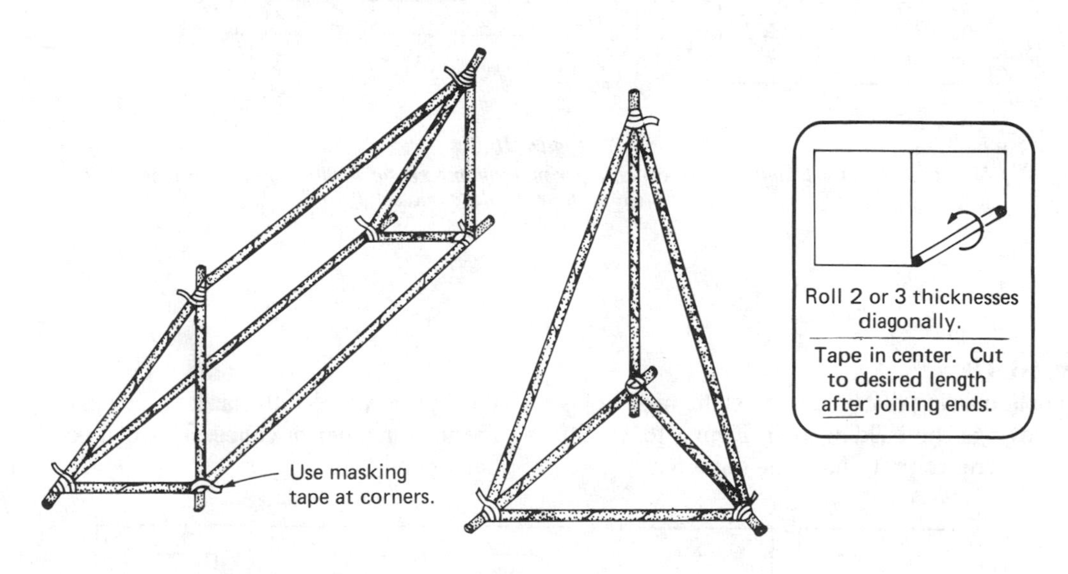

Figure 16.19
Skeletal models of large three-dimensional shapes can be built with rolled up newspaper or pipe cleaners.

piece of tape will hold the roll together. Ends can be bunched together and fastened with tape. Tubes are easily cut to different lengths. (How would you build a skeleton model of a circular cylinder with pipe cleaners?)

Activity 16.16

Have students design and test nets for various solids. (A *net* for a solid is a flat shape that will fold up to make that solid.) If a square centimeter grid is used, parallel lines and angles can be drawn without tedious measuring (Figure 16.20). Paper nets can be traced or pasted to tagboard for a sturdier solid. Circular cones are easily made by cutting a sector from a circle. Experiment with different circle sizes and different sectors. Much of the value in folding up a solid from tagboard is planning what shape the faces will be and where faces can be connected. Encourage groups to solve these problems themselves; provide only as much help as necessary.

Solids can also be built from blocks. Either small wooden cubes (1 inch or 2 cm) or plastic connecting cubes are ideal. The following tasks can make such activities challenging and valuable.

Activity 16.17

In groups, have students see how many different rectangular solids can be made using just 12 cubes in each. (A *rectangular solid* has six faces and each face is a rectangle.) Try other numbers of cubes. When are two rectangular solids congruent (exactly the same)? How would you have to turn one solid to get it in the same orientation as another that is the same shape?

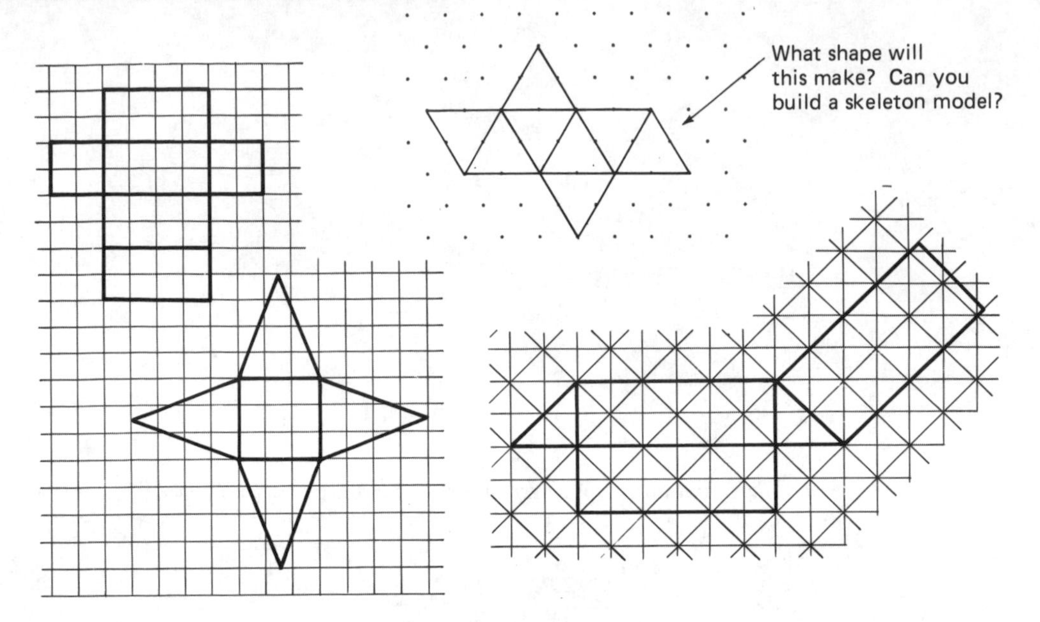

Figure 16.20
Nets are easily drawn on grid paper. After making one net for a solid, try to find others that will fold up to make the same solid.

Activity 16.18

From a simple building that you have built, make a task card with five views of the building as in Figure 16.21. Students use trial and error to try to build the same building from the five views. Slightly harder is to omit the top view but give the total number of cubes. This is a good exercise in spatial perception.

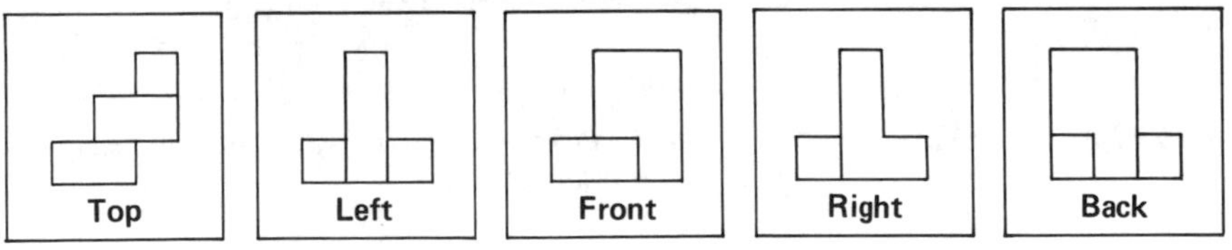

(Lines indicate different faces).

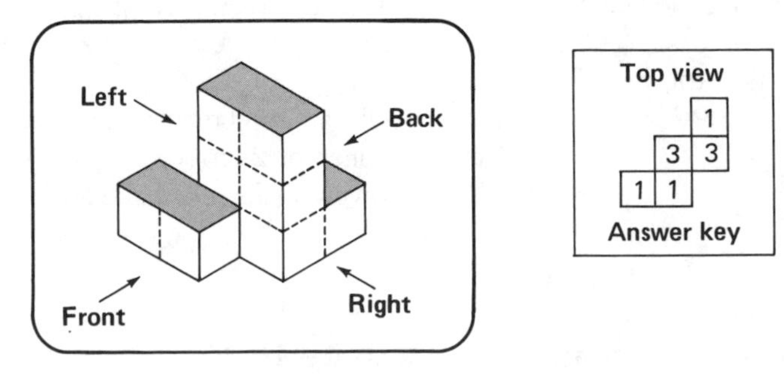

Given the five cards at the top, use wooden cubes to build the building. An answer key can be made by putting the numbers of cubes in each vertical stack on the top view. To make your own task cards, start by building a building and then make the cards.

Figure 16.21
Views of a simple building.

GEOMETRIC PROBLEM-SOLVING ACTIVITIES

Many excellent geometry activities are essentially "spatial process problems." Content of a traditional nature may be minimal while the problem-solving value is significant. These geometric problem-solving activities are just as important as verbal problem solving and also provide opportunity for growth in geometric thinking.

At level 0, geometric problems involve the manipulation, drawing, and creation of shapes of all types. The tasks revolve around global features of shapes rather than any analysis of properties or relationships between classes of figures. Many activities already suggested have a problem-solving orientation. The following are additional suggestions.

Geometric Puzzles

Activity 16.19

Tangrams have been a popular geometric puzzle for years. Figure 16.22 shows tangram puzzles of different difficulties. Easy ones are appropriate even for preschool children. Several good books exist that have a wide assortment of tangram puzzles, for example, Fair (1987) and Seymour (1971).

Activity 16.20

Pentominoes are the set of all shapes that can be made from five squares connected as if cut from a square grid. Each square must have at least one side in common with another square. It is an excellent geometric puzzle to have students try to find the complete set of 12 different pentominoes shown in Figure 16.23. (Do not tell the students how many different shapes there are. Good discussions can come from deciding if some shapes are really different and if all shapes have in fact been found.)

Activity 16.21

Cut the 12 pentomino shapes out of tagboard and use them as puzzle pieces to make shapes in the same way as with tangrams.

Activity 16.22

Play "Pentomino Squeeze" on an 8 × 8 grid of squares. Pieces are dealt out randomly to two players. The player with the cross places it on the board. In turn players place pieces

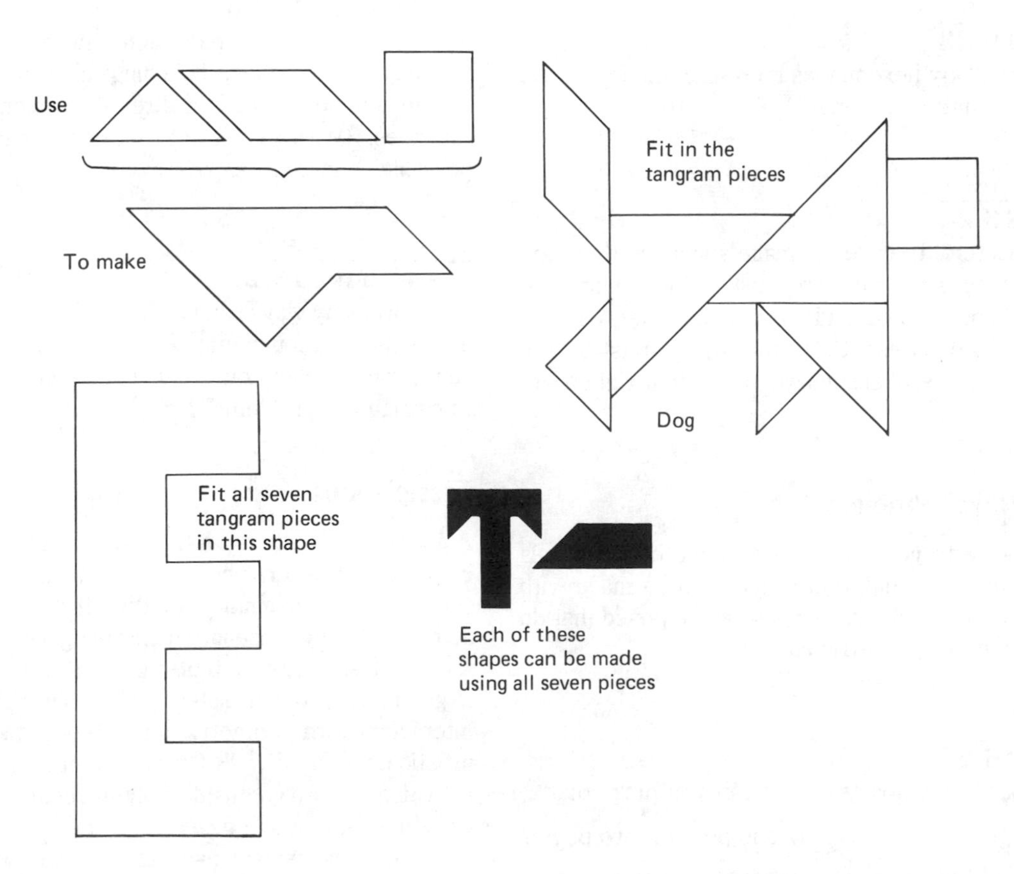

Figure 16.22
Four different types of tangram puzzles illustrate a range of difficulty levels.

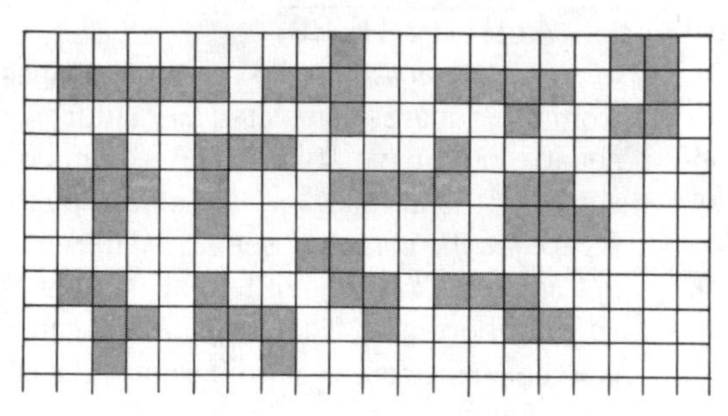

There are 12 pentominoes

Finding all possible shapes made with five squares, or six squares (called "hexominoes"), or six equilateral triangles, etc., is a good exercise in spatial problem solving.

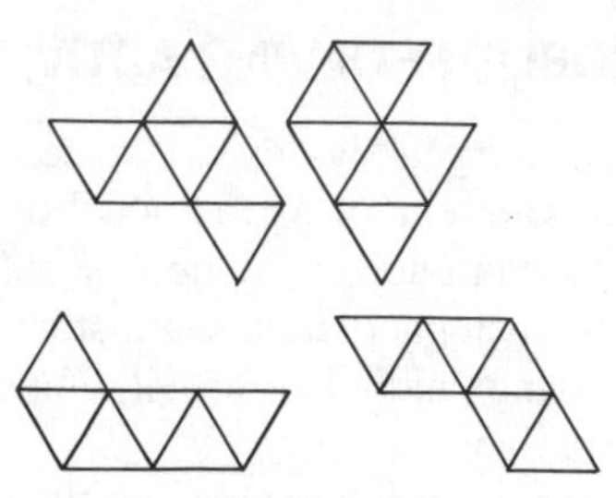

Four of the different shapes that six equilateral triangles make.

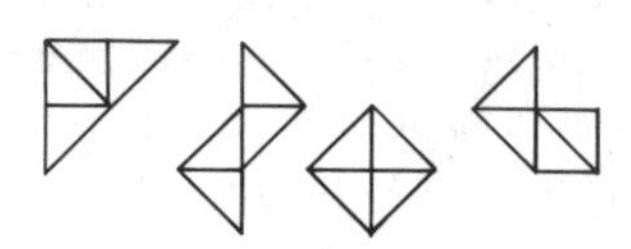

Four of the different shapes that four "half-square" triangles will make.

Figure 16.23
Pentominoes.

on the board without overlapping any piece that is already placed. The last player to be able to play is the winner. Smaller boards can be used to make the game more difficult.

Activity 16.23

Find out how many hexominoes there are (made with *six* squares). This is quite a challenge.

Activity 16.24

Use six equilateral triangles to make shapes in the same way as pentominoes. An even more challenging activity is to see how many shapes can be made from five 45 degree right triangles (halves of squares). Sides that touch must be the same length. (There are 14 shapes when only four right triangles are used.)

"How Many Ways" Problems

Many of the tasks with geoboards, tiles, and other materials involve seeing how many different shapes can be made with a particular property. Similar challenges can be posed that do not involve a standard geometric concept.

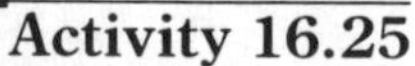

Activity 16.25

On a geoboard, how many shapes can you make that

Have no pegs inside (or exactly one peg, or two pegs)?

Will fit around this shape (any cardboard shape)?

Touch exactly five pegs (or some other number)?

Can be cut into two identical parts with one line on the geoboard?

Activity 16.26

With cubes, how many different buildings can you make with exactly five cubes? Touching cubes must have a complete face in common. This is a three-dimensional version of pentominoes. (With plastic interlocking cubes, the number of possibilities increases. Why?)

Activity 16.27

How many ways can you build a building using eight cubes if they must all touch on a whole face and no more than three cubes can be in any one row? (The numbers can be changed to change the problem.)

Tessellations

A *tessellation* is a tiling of the plane using one or more shapes in a repeated pattern with no holes or gaps. The *Tiling Pattern* activity at the beginning of the chapter was a tessellation activity. Making tessellations, or tiling patterns, is a good way for level 0 students from first grade to eighth grade to engage in geometric problem solving. Not only is there a nonanalytic interaction with geometric form, there is also challenge and artistic interest as well. One-shape or two-shape tessellation activities can vary considerably in relative difficulty and still remain level 0 activities.

Some shapes are easier to tessellate with than others as illustrated in Figure 16.24. When the shapes can be put together in more than one pattern, both the problem-solving

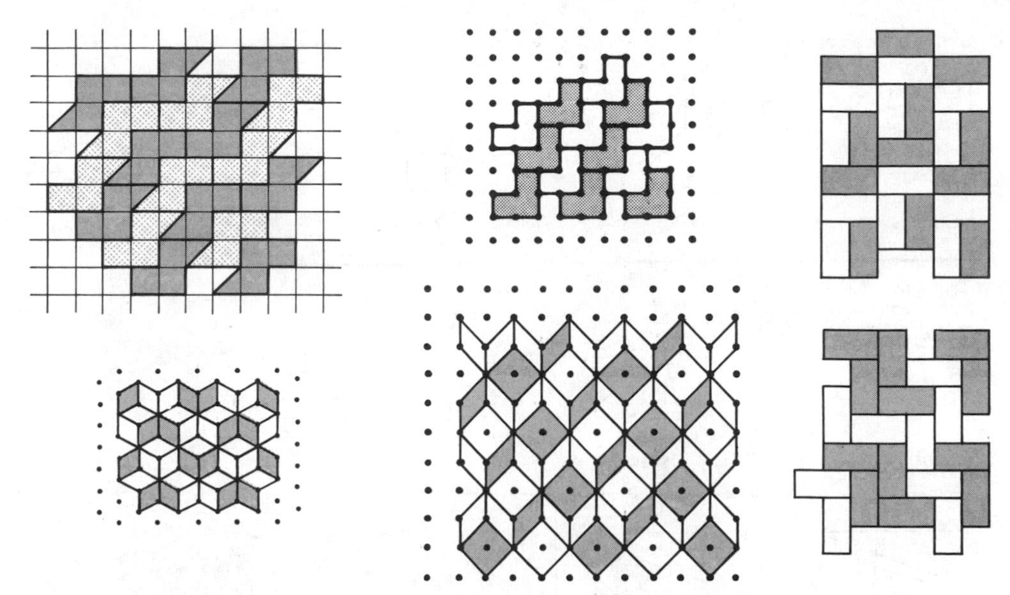

Tessellations can be drawn on grids or made of construction paper "tiles". They provide challenge, art, and an opportunity for spatial reasoning.

Figure 16.24
Tessellations.

level and the creativity increase. Literally hundreds of shapes can be used as tiles for tessellations. Every one of the 12 pentominoes will tessellate. It is fun to "create" shapes on various grid papers and then test them to see if they can be used to tessellate.

Most children will benefit from using actual tiles with which to create patterns. Tiles can be cut from construction paper. Simple tiles can be cut quickly on a paper cutter. Other tiles can be traced onto construction paper and several thicknesses cut at once with scissors. Older children may be able to use various dot or line grids and plan their tessellations with pencil and paper. Spend one period with tiles or grids letting children experiment with various tiling patterns. To plan a tessellation, use only one color so that the focus is on the spatial relationships. To complete an artistic-looking tessellation, add a color design. Use only two colors with younger children and never more than four. Color designs should also be repeated regularly all over the tessellation.

Tessellations can be made by gluing construction paper tiles to large sheets of paper, by drawing them on dot or line grids, or by tracing around a poster board tile. Do not worry about the edges of tessellations. Work from the center out, leaving "ragged" edges to indicate that the pattern should go on and on.

Informal Geometry Activities: Level 1

The activities at level 1 begin to focus more on the properties of shapes including some analysis of those properties. For example, at level 0, triangles might have been sorted by "big" and "little" or "pointy" and "not as pointy" or even those with square corners and those without. At level 1, the same set of triangles can be sorted by the relative sizes of the angles or the relative lengths of the sides. Combinations of these categories produce even more relationships. Most of the materials in the suggested activities are the same as those used at level 0. It is quite reasonable to have several similar investigations proceeding within one classroom with different groups pursuing tasks at different van Hiele levels.

CLASSIFYING SHAPES BY PROPERTIES

Special Categories of Two-Dimensional Shapes

In a level 1 sorting activity, shapes are presented so that specific properties and categorizations are clearly evident. You may have to be more direct in pointing out particular properties or categories that students do not notice. When a classification of shapes has been made, discussed, and is well understood, the appropriate name for the classification can be supplied. Listed below are some important categorizations of two-dimensional shapes. Examples can be found in Figure 16.25.

Simple Closed Curves:

Concave or convex. An intuitive definition of concave might be a "shape with a dent in it." If a simple closed curve is not concave it is convex. A more precise definition of concave may be interesting to explore with older students.

Symmetrical or nonsymmetrical. Shapes can have one or

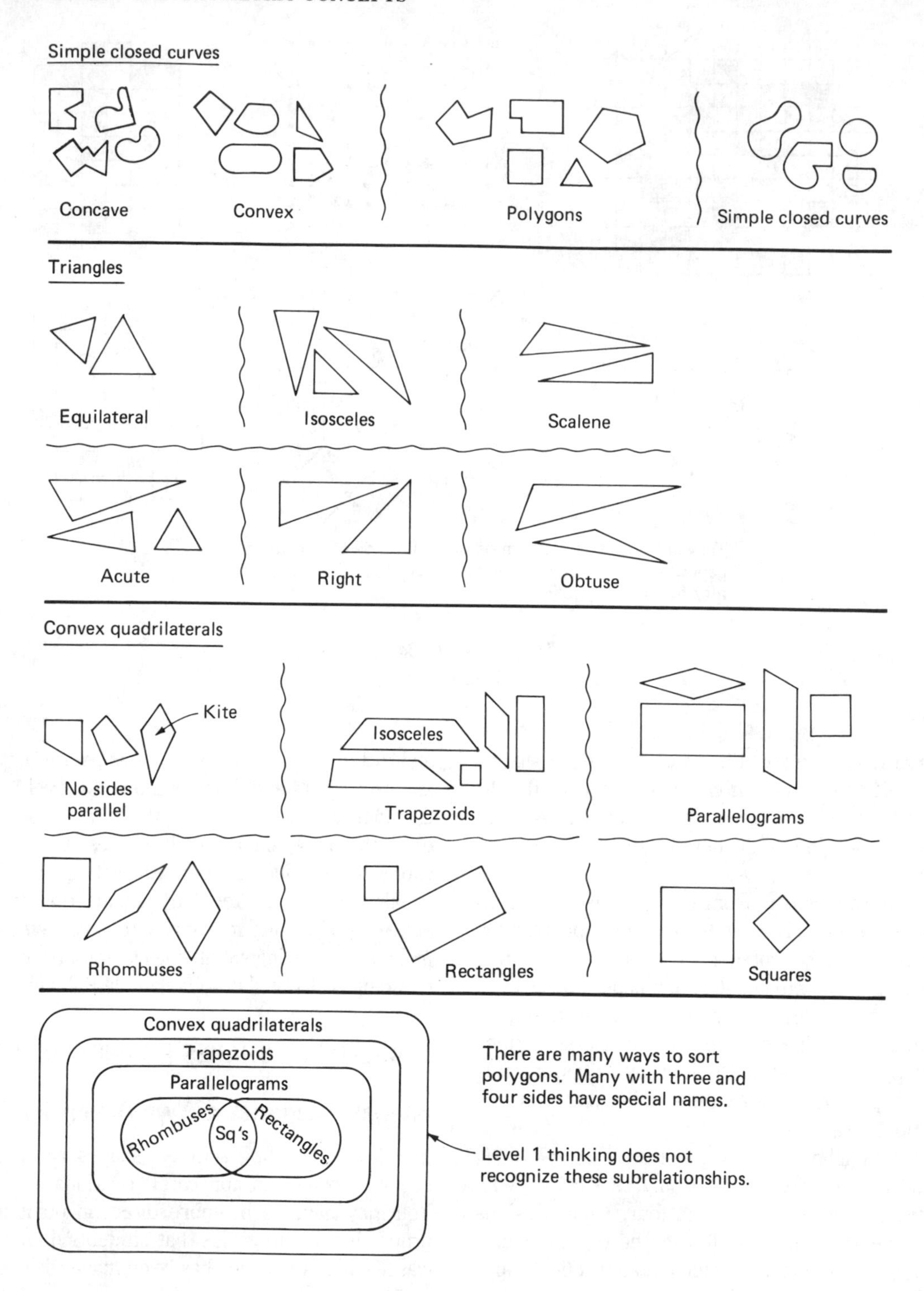

Figure 16.25
Classification of polygons.

more lines of symmetry and may or may not have rotational symmetry. These concepts will require more detailed investigation, as discussed later.

Polygons:

Regular and nonregular. A regular polygon has all sides and all angles congruent.

Triangles:

Classified by sides. Equilateral (all sides congruent), isosceles (only two sides congruent), scalene (no sides congruent).

Classified by angles. Right, acute (all smaller than right angles), obtuse (one angle greater than a right angle).

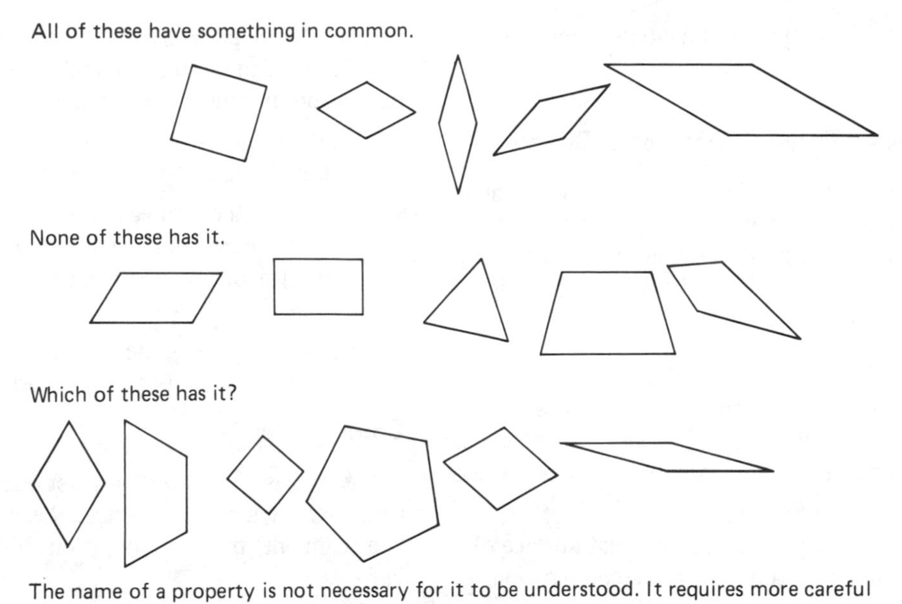

Figure 16.26
All of these/none of these.

Quadrilaterals:

Classified by the number of parallel sides. No sides parallel (no standard name by this scheme), trapezoid (at least one pair of parallel sides, parallelogram (two pairs of parallel sides).

Trapezoids:

Isosceles or not isosceles.

Parallelograms:

Classified by angles. Rectangle (has a right angle).

Classified by sides. Rhombus (has all sides congruent).

Square. Has a right angle and all sides congruent.

In the classification of quadrilaterals and parallelograms, the subsets are not all disjoint. For example, a square is a rectangle that is a rhombus. All parallelograms are trapezoids, but not all trapezoids are parallelograms.* Children at level 1 have difficulty seeing this type of subrelationship. They may quite correctly list all the properties of a square, rhombus and a rectangle and still identify a square as a "nonrhombus" or a "nonrectangle." Is it wrong for students to refer to subgroups as disjoint sets? By fourth or fifth grade, it is only wrong to promote or encourage such thinking. Burger (1985) points out that upper elementary students correctly use such classification schemes in other contexts. For example, individual students in a class can belong to more than one club. A square is an example of a quadrilateral that belongs to two other clubs. Recall that the van Hiele levels are descriptive of a progression in children's thinking, not what they are able to think at different ages. The passage from level to level is more a function of experience than maturation.

Several specific approaches to sorting activities can help students grow in their understanding of how shapes are related to each other.

Activity 16.28

Sort shapes by naming properties and not by names of the shapes. When two or more properties are combined, sort by one property at a time. "Find all of the shapes that have opposite sides parallel." (Find these.) "Now find those that *also* have a right angle." (This group should include squares as well as nonsquare rectangles.) After sorting, discuss what the name of the shapes is. Also try sorting by the same combinations of properties but in a different order.

Activity 16.29

Use an "all of these, none of these" type of activity as in Figure 16.26.

Activity 16.30

Use loops of string to keep track of shapes as you sort them. Have students put all of the shapes with four congruent sides in one string and those with a right angle in the other group. Where do squares go? Let them wrestle with this dilemma until they can see that the two loops must be over-

*Some definitions of trapezoid specify only one pair of parallel sides in which case parallelograms would not be trapezoids.

lapped with the squares belonging to the intersection or overlapping region.

Special Categories of Three-Dimensional Shapes

There is a wide variety of important and interesting shapes and relationships in three dimensions. Some classifications of solids are given below. At least an example or two of each of these can be made with clay or with tagboard. Other suggestions for making solids are given later.

All Solids:

Sorted by edges and vertices (corners). No edges and no vertices, edges but no vertices, edges and vertices. Students can find real-world examples as well as use clay to make unusual examples.

Sorted by faces and surfaces. (A *face* is a flat surface of a solid.) Solids can be sorted by various combinations of faces and curved surfaces. Some have all faces, all curved surfaces, some of each, with and without edges, with and without vertices.

Parallel faces. Find, sort, or make solids with one or more pairs of parallel faces. Since faces are two-dimensional, students can refer to the shapes of the faces. For example, solids with two square faces that are parallel and two pairs of rectangle faces that are parallel.

Cylinders:

Two properties separate cylinders from other solids: *Cylinders* have two congruent faces called *bases* in parallel planes. Second, all lines joining corresponding points on the bases are parallel. These parallel lines are called *elements.* In verbal form this description is quite difficult. Models are very useful for discussion. Examples of cylinders are shown in Figure 16.27.

Right cylinders and oblique cylinders. In a right cylinder the elements are perpendicular to the bases.

Prisms. If the two bases of a right cylinder are polygons, then the cylinder is a prism. If the bases are rectangles, the prism is called a *rectangular prism* or *rectangular solid.*

Cubes. A cube is the only possible solid with all square faces. A cube is a square prism with square sides.

Cones:

A *cone* is a solid with at least one face called the base and a vertex that is not on the face. It is possible to draw a straight line (element) from any point on the edge of the base to the vertex.

Sorted by the shape of the base. If the base is a circle, the cone is a circular cone, the type most people associate with the word *cone.* But the base can be any shape and the figure is still a cone, as shown in Figure 16.28.

Pyramids. A pyramid is a special cone in which the base is a polygon. All of the faces of a pyramid are triangles except, possibly, the base.

It is interesting to note that both cylinders and cones contain straight lines called *elements.* A special type of both cylinders and cones occurs when the base is a polygon: a *prism* is a cylinder with polygon bases; a *pyramid* is a cone with a polygon base.

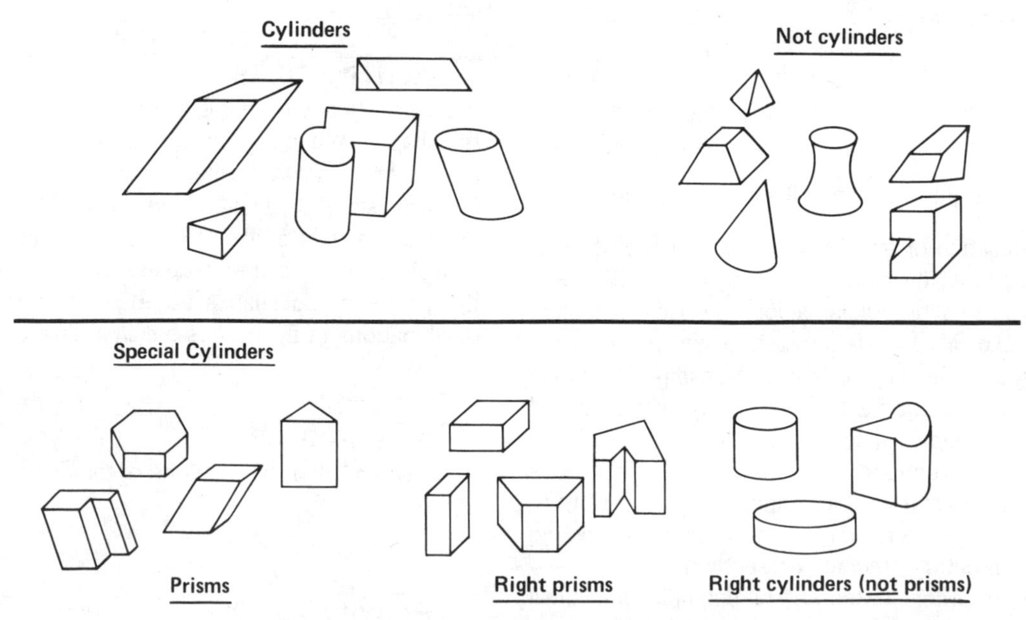

Figure 16.27
Cylinders and prisms.

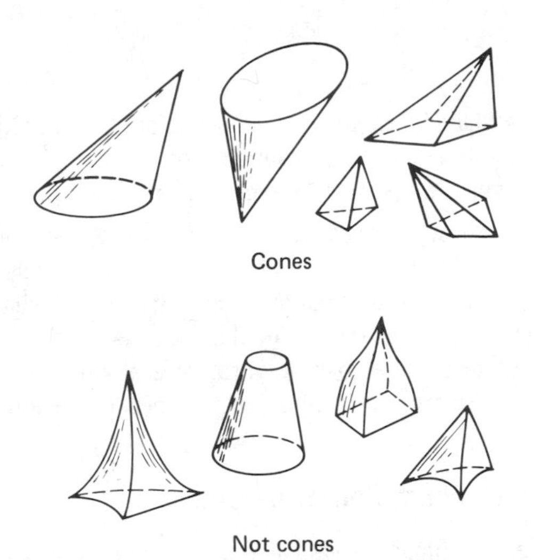

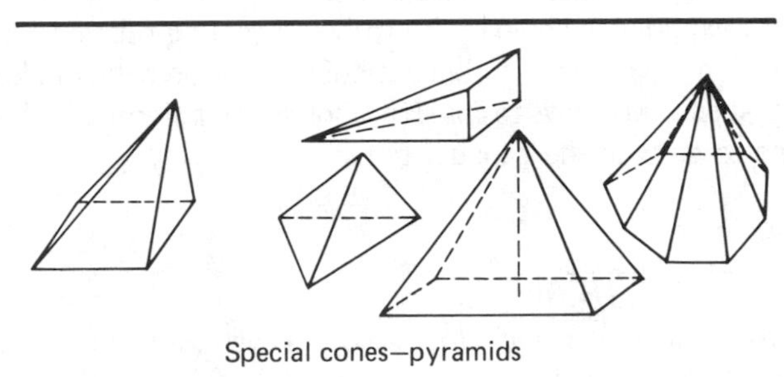

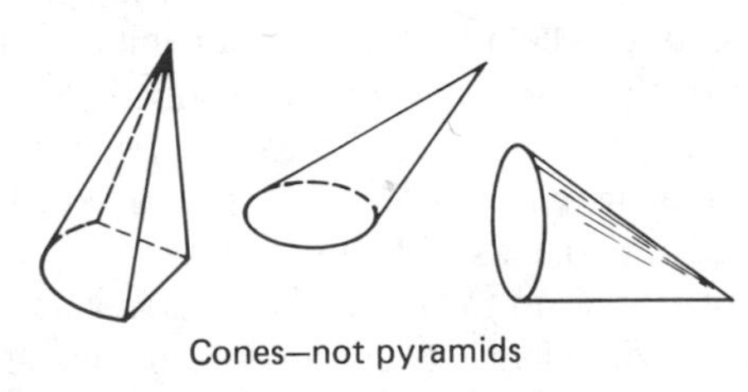

Cones and cones with a polygon base (pyramids) all have straight line elements joining every point of the base with the vertex. (Yes, it's true! A pyramid is just a special type of cone.)

Figure 16.28
Cones and pyramids.

BUILD, DRAW, MAKE, AND MEASURE

Constructing shapes in both two and three dimensions is still one of the most profitable types of activities that can be done in geometry. As children begin to demonstrate level 1 thinking, the tasks for construction activities can be posed in terms of the properties of shapes rather than how they look.

In addition to geometric properties, measurements of shapes can help students develop even more relationships. Specifically, area, perimeter, surface area, volume, angles, radii, and circumferences are examples of things that can be measured on various shapes. For example, students can measure interior angles of various polygons and discover that when the number of sides of two polygons is the same, so is the sum of the measures of the interior angles. Direct comparisons, informal units, and simple student-made measuring devices are sufficient for exploring almost all interesting relationships involving measures. Do not let sophisticated measurements interfere with the activities.

Two-Dimensional Constructions

Tiles, geoboards, and dot and grid paper continue to be the best construction materials for two-dimensional shapes.

Activity 16.31

Give students properties or relationships and have them construct as many shapes as possible that have these properties. Compare shapes made by different groups. For example:

Make a four-sided shape with two opposite sides the same length but not parallel.

Make some six-sided shapes. Make some with one, two, and then three pairs of parallel sides and some others with no parallel sides.

Make shapes with all square corners. Can you make one with three sides? four sides? five sides? six, seven or eight sides?

Make some six-sided shapes with all square corners. Count how many squares are inside each. What is the distance around each?

Make five different triangles. How are they different? (This is also good for four-, five-, or six-sided shapes.)

Make some triangles with two sides equal (congruent). Make some four-sided shapes with three congruent sides. Try five-sided shapes with four congruent sides.

Make some quadrilaterals with all sides equal (or with two pairs of equal sides).

Make a shape with one or more lines of symmetry or that has rotational symmetry. (A longer discussion of symmetry is provided later in the chapter.)

It is quite easy (and fun) to make up tasks for this list. Explore some of these and make up others that suit your needs. One idea is to use combinations of previously explored concepts.

Can you make an isosceles triangle that is obtuse? an equilateral triangle that is a right triangle? (Of the nine possible combinations of triangle types, two are not possible.)

Can you make a triangle with two right angles?

Can you make a rhombus that has a right angle? a parallelogram with only two equal sides?

Combination challenges can also include the notions of perpendicular, angle measurement, area, perimeter, similarity, concave and convex, regularity, and symmetry, just to name a few. Using the materials that your students will use, prepare a series of combination challenges. Also encourage your students to come up with some of their own. Notice that some of the preceding tasks are not possible. Discovering that some combination of relationships is not possible and why, is just as valuable as learning the relationships themselves.

Activity 16.32

Make shapes according to special measurement requirements.

Make at least five different shapes with an area of ______ (appropriate number for your materials). What is the perimeter of each?

Make shapes with a fixed perimeter and examine the areas.

Try to make the shape with the largest area for a given perimeter, or the smallest perimeter for a given area. (For polygons, the largest area for a fixed perimeter is always regular. Try it.)

Angle and side specifications provide lots of opportunities to examine properties. Angles can usually be kept to multiples of 30 and 45 degrees, or informal units can be used. Length measurements can be made using grid dimensions if diagonal segments are avoided.

Make several different triangles that all have one angle the same. Next make some different triangles with two angles the same (for example, 30 and 45 degrees). What did you notice?

Can you make a parallelogram with a 60 degree angle? Make several. Are they all alike? How? How are they different?

Make some parallelograms with a side of 5 and a side of 10. Are they all the same?

Draw some polygons with four, five, six, seven, and eight sides. Divide them all up into triangles, but do not let any lines cross or triangles overlap. What do you discover? Measure the angles inside each polygon.

Draw an assortment of rectangles and draw the diagonals in each. Measure the angles that the diagonals make with each other. Measure each part of each diagonal. What do you observe? Try this exercise with squares, rhombuses, other parallelograms, and kites (quadrilaterals with two pairs of adjacent sides congruent).

Three-Dimensional Constructions

Shapes are so difficult to visualize in drawings that the only way to experience some of the truly interesting relationships in three dimensions is to build models. The activities below suggest different ways to construct solids and potential explorations to go with the constructions.

Activity 16.33

An easily made model for exploring cylinders, cones (and prisms and pyramids) is a string model with poster board bases. Students can help you make your initial collection of models and soon you will have a wide assortment. Directions are in Figure 16.29. When both bases are held together, the vertex can be adjusted up, down, or sideways to produce a family of cones or pyramids with the same base (Figure 16.30). By moving one base up to the knot and adjusting the other base, you can model a family of cylinders or prisms. The bases can also be tilted (not kept parallel) and/or twisted so that

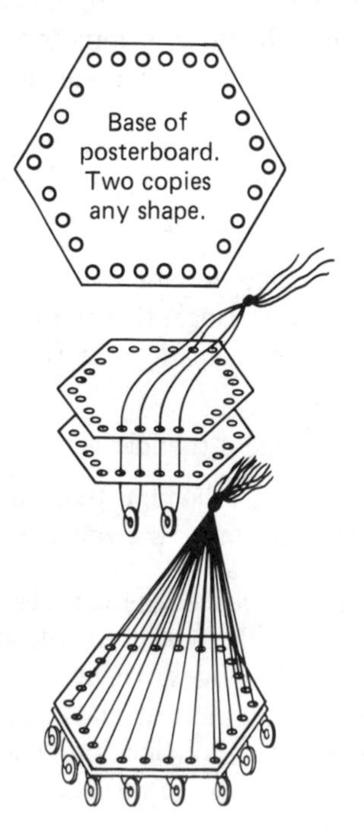

Materials: Soft cotton string or embroidery yarn, metal washers about 3/4 inch in diameter (about 15 per model), poster board, and a hole punch.

Directions:

1. Cut two identical models of the base from posterboard. There are no restrictions on the shape. The size should be roughly 4 to 6 inches across.
2. Place the two bases together and punch an even number of holes around the edges, punching both pieces at the same time so that the holes line up. The holes should be about 1 cm apart.
3. Cut pieces of string about 20 inches long. You will need half as many pieces as holes.
4. Run each piece of string through a washer and thread the two ends up through two adjacent holes of both bases. Pull all the ends together directly above the base and tie in a knot.

Figure 16.29
A model for cylinders and cones.

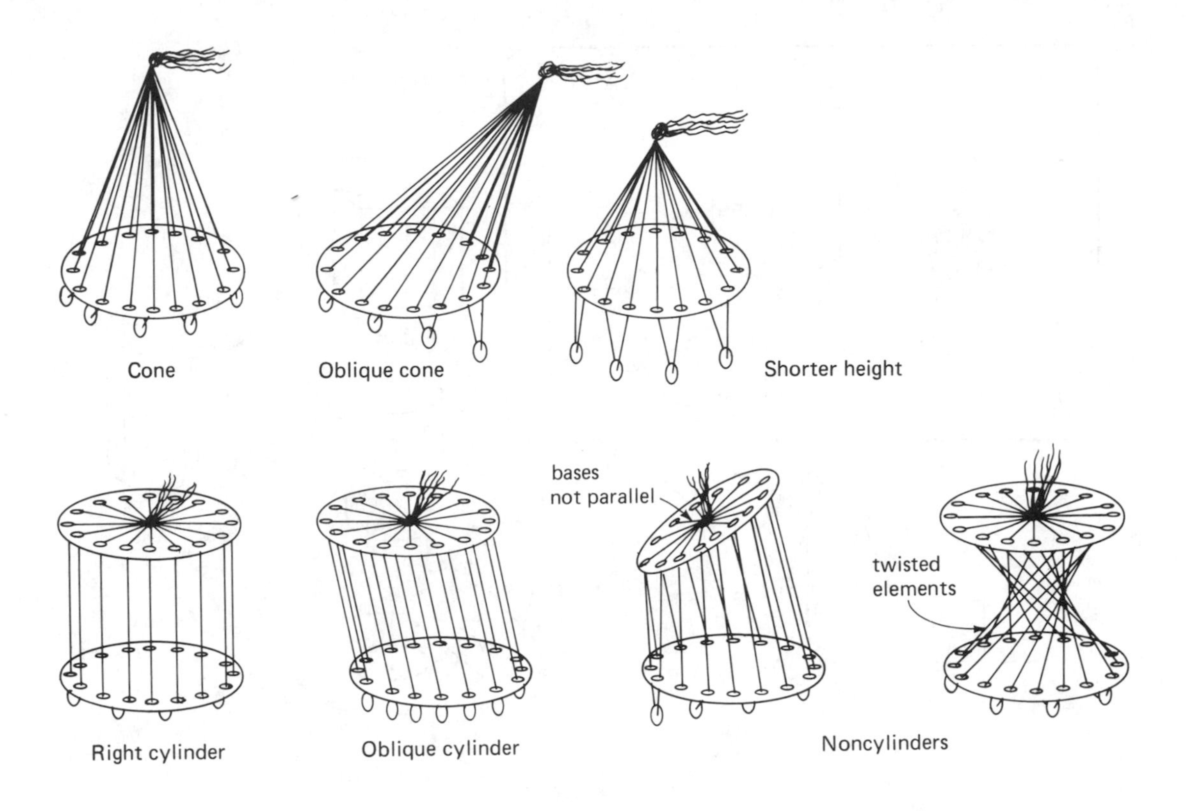

Figure 16.30
The string cone/cylinder model illustrates a variety of relationships and can even show solids that are noncylinders.

noncylindrical shapes are formed. Be sure that students notice that for cylinders and prisms, the elements (strings) remain parallel regardless of the position of the bases. The angles that each makes with the base is the same for all elements. For cones and pyramids, examine how the angles change relative to each other as the vertex is moved. String models like these are well worth the small cost and effort simply for the opportunity to explore how these related solids change as the bases or vertices are moved.

Activity 16.34

Modeling clay is a useful way to make most any shape. Use an oil-based craft clay that will not dry out. Perhaps the best reason to go to this trouble is that clay models can be sliced to investigate the resulting faces on the slices. An inexpensive tool called a "piano wire," a wire with two handles, is used to cut the clay. This works much better than a knife and can be purchased in an art supply store. Have students make cubes, cones, prisms, cylinders, a torus (doughnut shape), and other shapes. Precision is not important as long as the essential features are there. There are two types of challenges. First suggest where to make a slice and see if students can predict what the face of the slice will be. Second, given a solid, ask students to find a way to slice the shape to produce a slice with a particular shape. For example, how could you slice a cube so that the slice is a trapezoid? Will it be isosceles? Always? Figure 16.31 suggests a few ways that different solids can be cut (Carroll, 1988).

Activity 16.35

Students can "generate" imaginary solids by sliding or revolving a plane surface as shown in Figure 16.32. The resulting shapes can be made with clay. Notice that *all* cylinders can be generated by sliding a base. Of all cones, only a right circular cone can be "spun" from a right triangle, but many other shapes can be "made" by spinning a shape about an edge. How would you generate a torus (doughnut shape) or other shapes with holes in them? What shapes around the room could be generated by rotation or sliding a surface? Which ones cannot? How are shapes that can be generated in this manner different from those that cannot? From one triangle or rectangle, how many different shapes can be generated by sliding? how many by rotation? Are the answers the same for all triangles? all rectangles?

Using cubes to build shapes was a suggestion for level 0. At level 1 it is interesting to examine rectangular prisms (box shapes) using cubes because surface area and volume are so

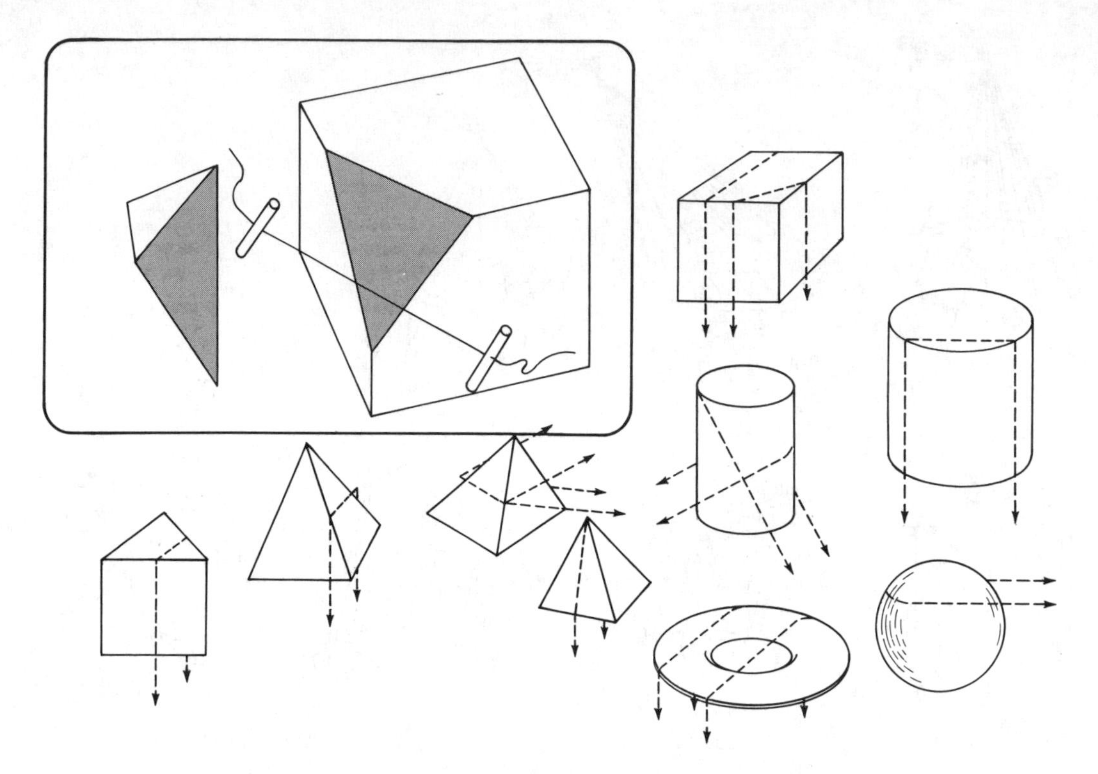

Figure 16.31
Predict the slice face before you cut a clay model with a "piano wire."

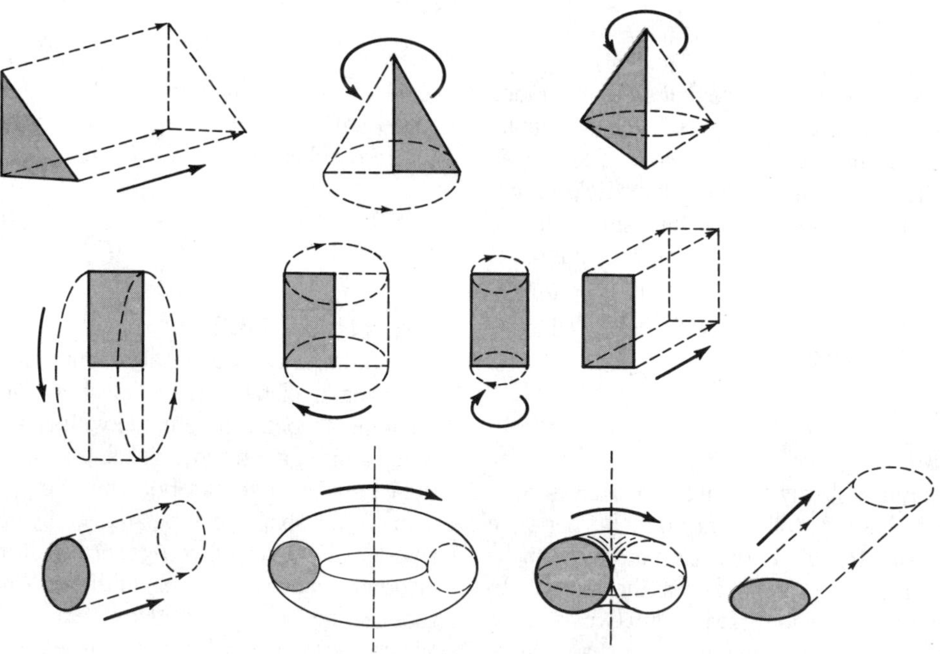

Use a poster-board model of a flat shape and "generate" solids in the air. Slide or rotate the shape. Try to make your generated solid from clay.

Figure 16.32
Generating imaginary solids.

easily determined. Tasks involving surface area and volume are analogous to tasks that involve perimeter and area in two dimensions.

Activity 16.36

How many boxes can be made that have a volume of 36 cubes? What is the surface area of each? Similarly, fix the surface area and find out what volumes are possible. What shapes produce the largest volume with a fixed surface area? Try surface areas of 24, 54, or 96 squares.

EXPLORATION OF SPECIAL PROPERTIES AND RELATIONSHIPS

Line Symmetry

Line symmetry (bilateral symmetry or mirror symmetry) is fun and challenging to explore using a variety of different materials.

Activity 16.37

On a geoboard or on any dot or grid paper, draw a line to be a mirror line. Next draw a shape on one side that touches the line as in Figure 16.33. Try to draw the mirror or symmetric image of the shape on the other side of the line. For two mirror lines, draw the reflections in stages. Mirrors can be used to test the results.

Activity 16.38

On a piece of dot paper use the technique of the previous exercise to create a symmetric drawing. Fold the paper on the mirror line and notice how corresponding points on each side of the line match up. Open the paper and connect several corresponding points with a straight line. Notice that the mirror line is a perpendicular bisector of the lines joining the points. Use this property to create a symmetric drawing on plane paper. Draw the line and half of the figure as before. From several critical points, draw perpendicular lines to the

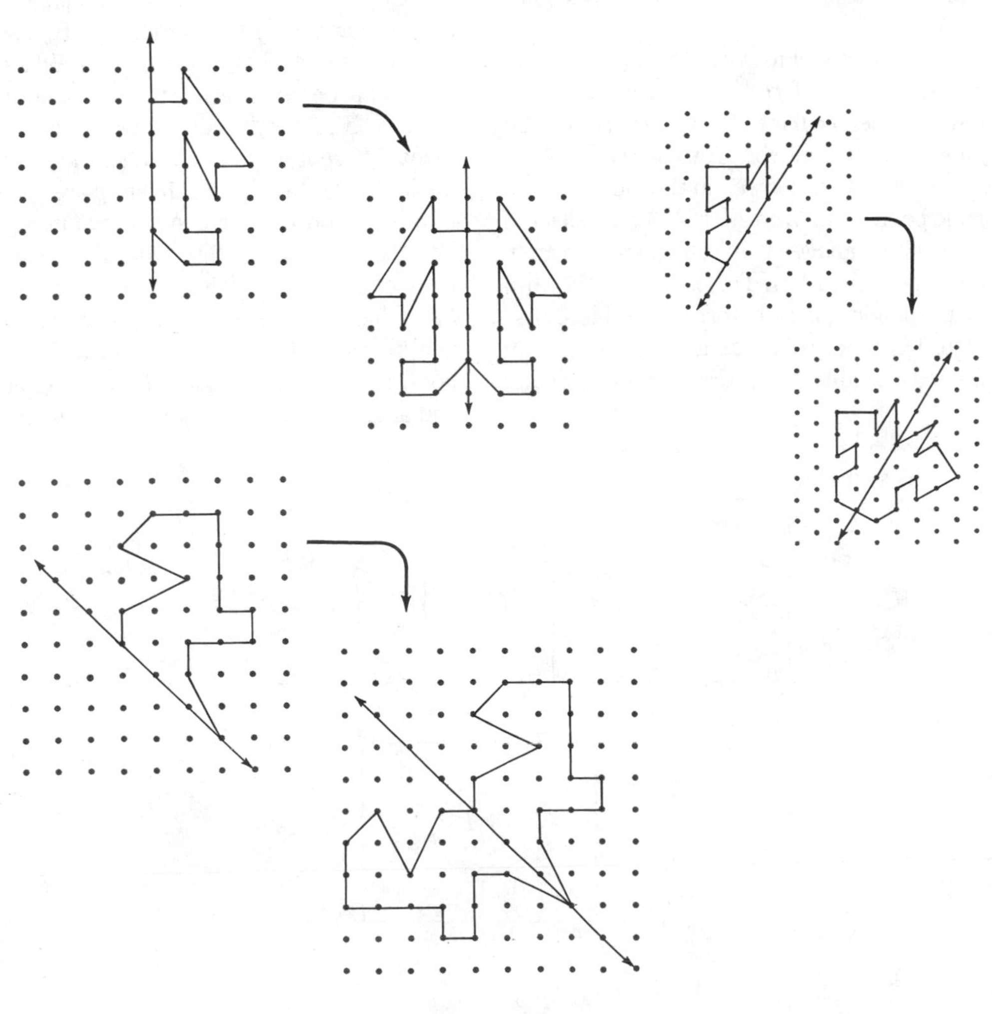

Figure 16.33
Exploring symmetry on dot grids.

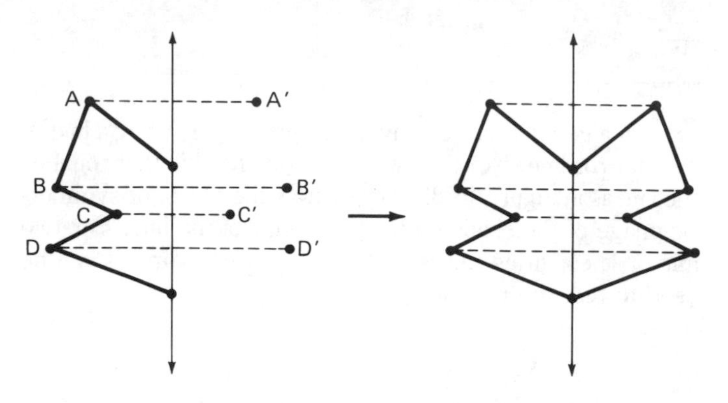

Figure 16.34
From points on one side, draw perpendicular lines to the mirror line and extend them an equal distance beyond.

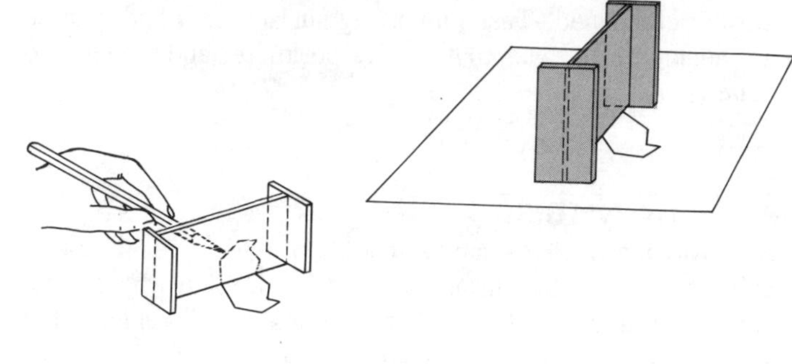

Figure 16.35
The Mira *is a "see-through mirror" that allows you to draw the image that you see in the glass.*

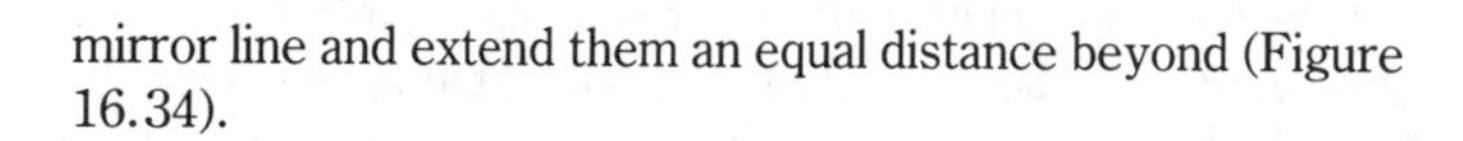

mirror line and extend them an equal distance beyond (Figure 16.34).

A *flip* or *reflection* of a shape through a line is very similar to the concept of symmetry. If you plan to investigate transformations (slides, flips, and turns), the last few activities are good beginnings.

A very useful device for studying symmetry and transformations is the *Mira*©, a piece of red plexiglass. Shown in Figure 16.35, it stands perpendicular to the table surface. The Mira is essentially a see-through mirror. You can reach behind the Mira and draw the image that you see in the mirror. Since you can see the image behind the Mira as well as the reflected image, it is possible to match images with reflections and draw the mirror line at the base of the Mira. This feature of the Mira allows one to draw perpendicular bisectors, angle bisectors, reflect images, and to do a variety of constructions more easily than with a compass and straight edge and in a very conceptual manner.

The following activity offers a slightly different view of symmetry and provides a good preparation for rotational symmetry.

Activity 16.39

Cut out a small rectangle from paper or cardboard. Color one side and label the corners A, B, C, and D so that they have the same label on both sides. Place the rectangle on a sheet of paper and trace around it. Refer to the traced rectangle as a "box" for the cut-out rectangle. The question is, How many different ways can you flip the rectangle over so that it fits in the box? Before each flip, place the rectangle in the box in the initial orientation. As shown in Figure 16.36, each flip into the box is a flip through a line, and these lines are also lines of symmetry. That is, for a plane shape there are as many lines of symmetry as there are ways to flip a figure over and still have it fit into its box. Try with other figures: a square, nonsquare rhombus, a kite, a parallelogram with unequal sides and angles, triangles, regular pentagons, and others.

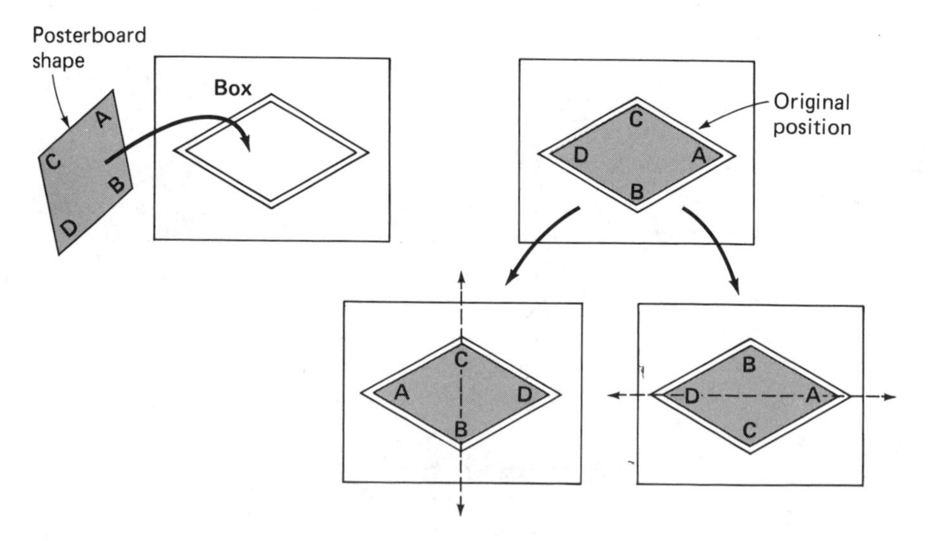

Figure 16.36
There are at least two ways to flip this diamond into its box. Are there more? Cut one out and try it!

Rotational Symmetry

One of the easiest introductions to rotational symmetry is to create a box for a shape by tracing around it as in the last activity. If a shape will fit into its box in more than one way without flipping it over, it has *rotational symmetry*. The *order of rotational symmetry* is the number of different ways it can fit into the box. Thus a square has rotational symmetry of order four as well as four flip lines or lines of symmetry. A parallelogram with unequal sides and angles has rotational symmetry of order two but no lines of symmetry (Figure 16.37).

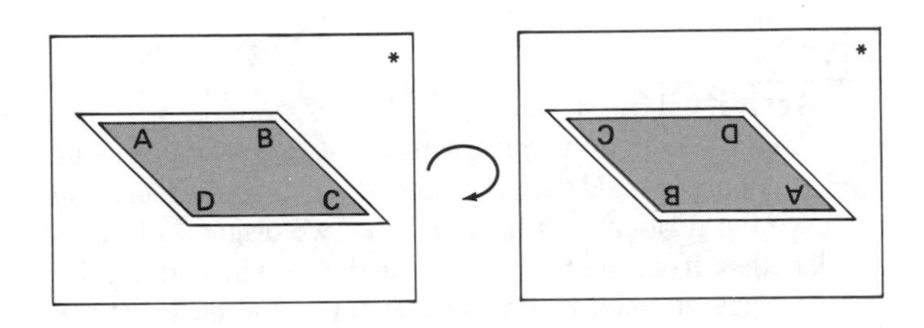

Figure 16.37
This parallelogram fits in its box two ways without flipping it over. Therefore, it has rotational symmetry of order 2.

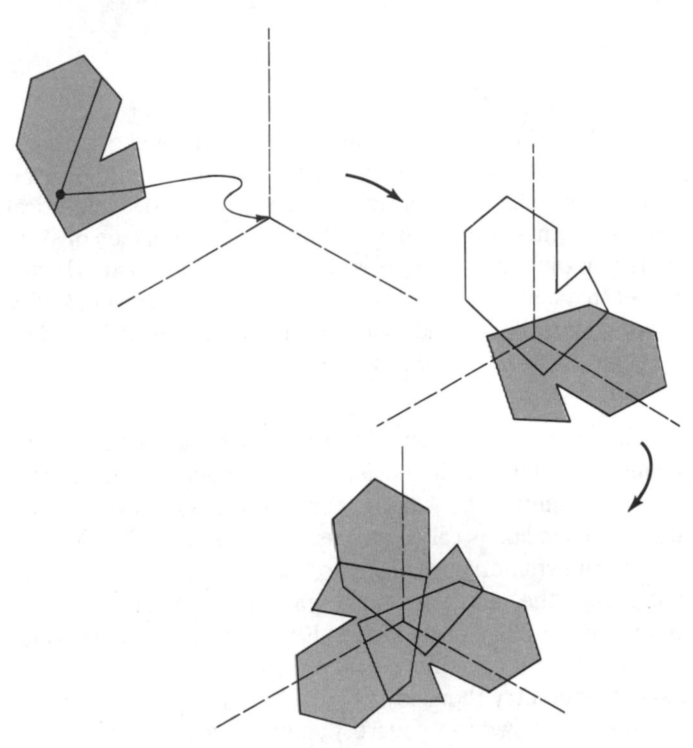

Figure 16.38
How to build a "weird" shape with rotational symmetry of order 3. How would you build one with order 2 or 5?

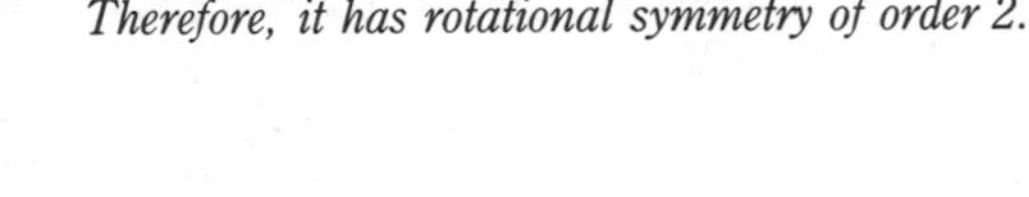

Activity 16.40

Use tiles, geoboards, dot or grid paper to draw a shape that has rotational symmetry of a given order. Except for regular polygons, this can be quite challenging. To test a result, trace around it and cut out a copy of the shape. Try to rotate it on the drawing.

Activity 16.41

On a piece of paper lightly draw three rays from one point separated by 120 degrees. Cut out any shape from a piece of tagboard. Draw a line through it anywhere. Push a pin or compass point through the shape at any point along the line. Place the point at the center of the three rays so that the line on the shape coincides with one of the rays. Trace around the shape. Rotate the shape to each of the other two lines and repeat. By ignoring the interior lines, the three tracings will form a shape with rotational symmetry of order 3. The center point is called the *center of rotation*. Connect any three corresponding points from the three tracings and observe an equilateral triangle (Figure 16.38). The same approach can be used for rotational symmetries of any order.

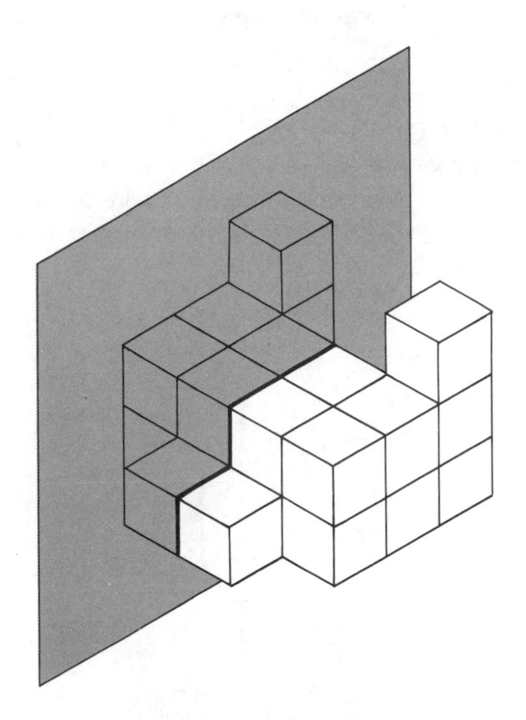

Figure 16.39
A block building with one plane of symmetry.

Symmetries in Three-Dimensional Figures

A plane of symmetry in three dimensions is analogous to a line of symmetry in two dimensions. Each point in a symmetric solid corresponds to a point on the other side. The plane of symmetry is the perpendicular bisector of the line segments joining each pair of points. Figure 16.39 illustrates a shape built with cubes that has a plane of symmetry.

Activity 16.42

Build a "building" with cubes that has a plane of symmetry. If the plane of symmetry goes between cubes, the shape can be "sliced" by separating the building into two symmetrical parts. Try making buildings with two or more planes of symmetry. Examine various prisms that you can build. Do not forget that a plane can slice diagonally through the blocks. Try using clay to build solids with planes of symmetry and slice through these planes with a piano wire.

Rotational symmetry in the plane also has an analogous counterpart in three dimensions. A solid figure can have rotational symmetry about a line through the solid instead of a point. Such a line is called an *axis of symmetry*. As a solid with rotational symmetry revolves around an axis of symmetry, it will occupy the same position in space (its box) but in different orientations. While plane figures have only one center of rotation, a solid can have more than one axis of rotation. For each axis of symmetry there is a corresponding order of rotational symmetry. A regular square pyramid has only one axis of symmetry that runs through the vertex and the center of the square. A cube, on the other hand, has a total of 13 axes of symmetry; three (through opposite faces) of order 4, four (through diagonally opposite vertices) of order 3, and six (through diagonally opposite edges) of order 2.

Activity 16.43

To investigate an axis of symmetry, use a small tagboard model of the solid and insert a long pin or wire through an axis of symmetry. Color or label each face of the solid to help keep track of the different positions. Hold the axis (pin) vertically and rotate the solid slowly, observing how many times it fills the same space as it did in the original position as in Figure 16.40 (Bruni, 1977).

Diagonals of Quadrilaterals

The diagonals of various quadrilaterals provide a wealth of interesting relationships to observe. The two diagonals are either equal or different in length, are perpendicular or are not, and are each bisected or are not. The relationships between the two diagonals are unique for each category of quadrilaterals.

Activity 16.44

Have students draw quadrilaterals of various types using grid or dot paper. Add the diagonals. Have them measure and observe the relationships between the two diagonals (angles and lengths). If different groups within the class investigate different types of quadrilaterals, conjectures and observations about different relationships can be made and tested. Given a particular relationship, two diagonals can be drawn and the end points connected to see what type of quadrilateral results.

Activity 16.45

Connect two strips of tagboard as in Figure 16.41 and explore the different quadrilaterals that can be generated. What happens when the strips are congruent? What if they are joined at the midpoint of one but not the midpoint of the other? If they meet at right angles, do you get a special figure? Suppose one strip is longer than the other? What strips could

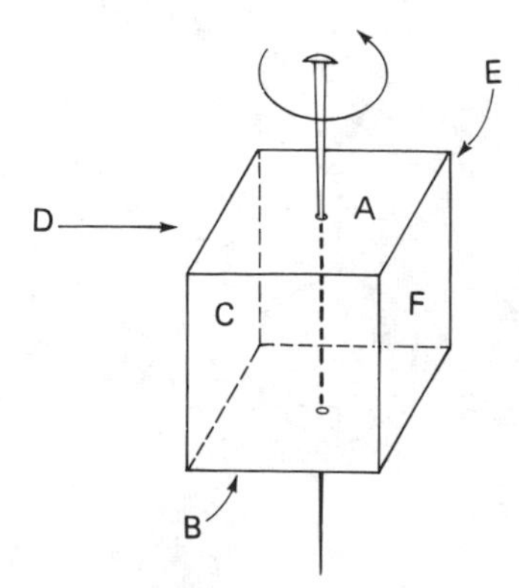

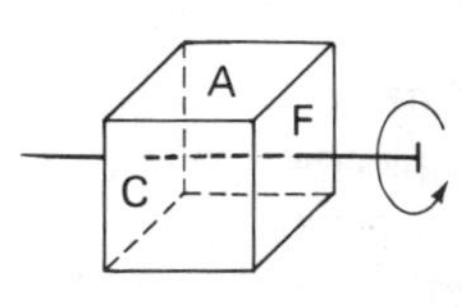

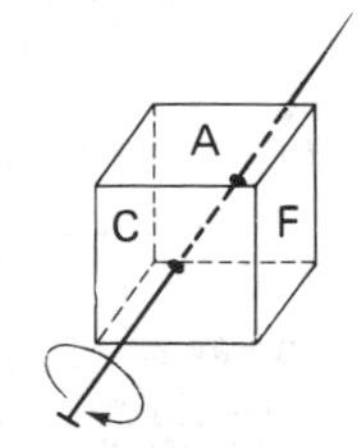

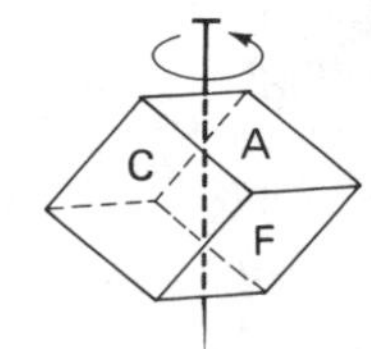

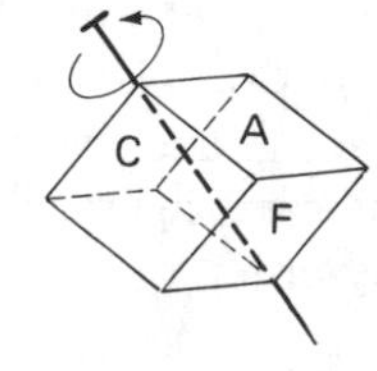

Figure 16.40
Rotations of a cube.

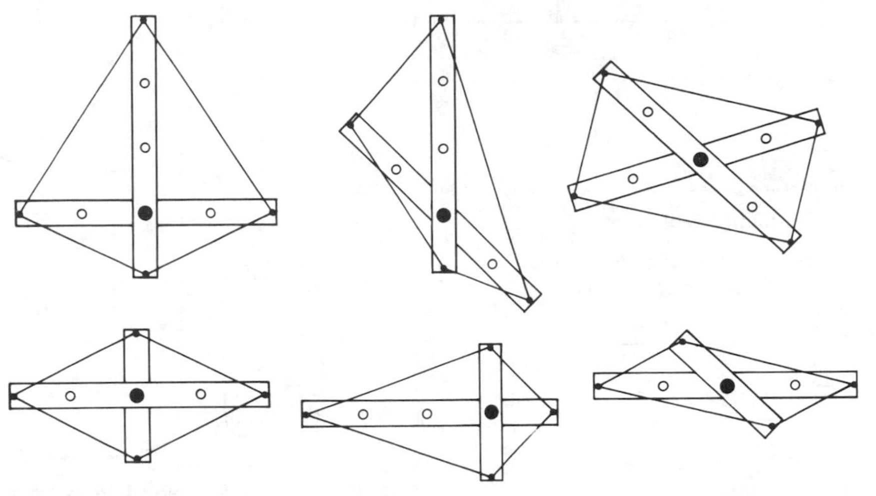

Quadrilaterals can be determined by their diagonals. Consider the length of each, where they cross, and the angles between them. What conditions will produce parallelograms? rectangles? rhombuses?

Figure 16.41
Diagonals of quadrilaterals.

be used and where should they be joined to form a trapezoid? What will make the trapezoid isosceles? Similar explorations are fun to make up and investigate.

Angles, Lines, and Planes

The relationships between angles within a figure can be explored quite well by tracing angles for comparisons, by comparing angles to a square corner, and by using informal units. With only such simple techniques, students can begin to look at relationships such as the following:

The angles made by intersecting lines or by lines crossing two parallel lines. Which angles are equal? Which add up to a straight angle?

The sum of the interior angles of polygons of different types. Is there a relationship in the number of sides and the sum of the angles? What if the shape is concave?

The exterior angles of polygons. Extend each side in the same direction (for example, clockwise) and observe the sum of these angles. How is the exterior angle related to the interior angle?

In three dimensions, students can begin to observe the angles formed by intersecting planes and between lines and planes. How can two lines not intersect but not be parallel? Can a line and a plane be parallel? How could you describe a line that is perpendicular to a plane? Where are there some examples of these relationships in the real world?

Similar Figures

In both two and three dimensions, two figures can be the same shape but be different sizes. At level 0, students can sort out shapes that "look alike." At that level the concept of similar is strictly visual and likely not precise. At level 1, students can begin to measure angles, lengths of sides, areas, and volumes (for solids) of shapes that are similar. By investigation, relationships between similar shapes can be observed. For example, students will find that all corresponding angles must be congruent, but other measures vary proportionately. If one side of a larger yet similar figure is triple that of the smaller figure, so will all linear dimensions be triple those of the smaller figure. If the ratio of corresponding lengths is 1 to n, the ratio of areas will be 1 to n^2, and the ratio of volumes will be 1 to n^3.

Activity 16.46

Have students build or draw a series of similar shapes. Start with one shape and pick one side (or for solids, one edge) to be the control side. Make the corresponding side of the similar figures in easy multiples of the control side. For example, make shapes with a corresponding side two, three, and four times as long as the control (Figure 16.42). Make measures of lengths, areas, and volumes (for solids) and compare ratios of corresponding parts. If the control sides are in the ratio 1 to 3, then all linear dimensions will be in the same ratio, areas in the ratio 1 to 9, and volume in the ratio 1 to 27.

Circles

The circle, an apparently simple shape, is also very important. Consider the world around you. What are some of the ways the circle is used? Why do you think it is so frequently used instead of other shapes? Many interesting relationships can be observed between measures of different parts of the circle (see Bruni, 1977). Among the most astounding and important

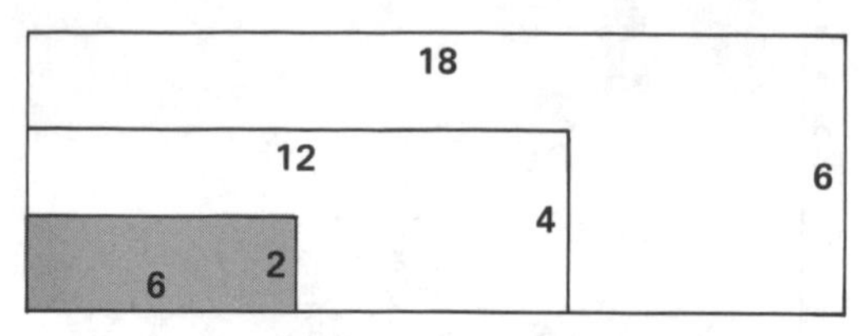

Similar rectangles

Compare the ratios of lengths of sides and ratios of areas.

Example: Small to Large

	Small		Large	
Length	2	to	6	(1 to 3)
Area	12	to	108	(1 to 9)

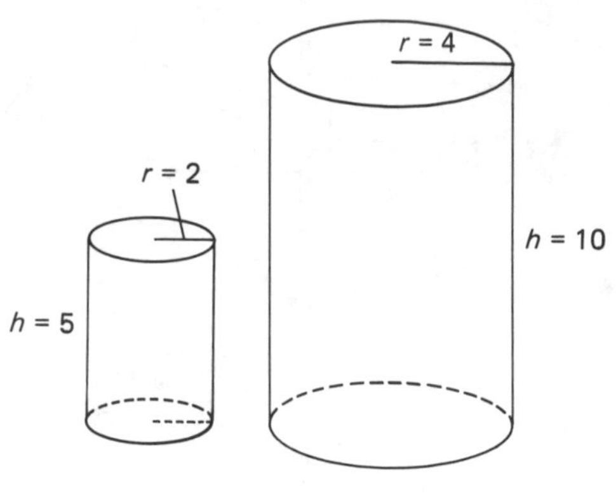

Similar cylinders

Height, radius and circumference are all in ratio 1 to 2.

Surface Areas:
Sides 20π to 80π } 1 to 4
Tops 4π to 16π }
Volumes: 20π to 160π or 1 to 8

Similar shapes have corresponding dimensions in predictable ratios. What measures stay the same? How do lengths, areas, and volumes change? Is that true of all shapes?

Figure 16.42
Similar solids.

is the ratio between measures of the circumference and the diameter.

Activity 16.47

Have groups of students carefully measure the circumference and diameters of many different circles. Each group can be responsible for a different collection of circles.

Carefully wrap string around jar lids, tubes, cans, and similar items, and mark the circumference. Measure the string and the diameters to the nearest millimeter.

Draw larger circles using a string as shown in Figure 16.43. String can be used to measure the circumference.

Measure the large circles found on gym floors and playgrounds. Use a trundle wheel or rope to measure the circumference.

Using a calculator, students should record the ratio of the two measures for each circle. With careful measurement, the results should be close to 3.1 or 3.2. The exact ratio is an irrational number that is represented by the Greek letter π, which is about 3.14159.

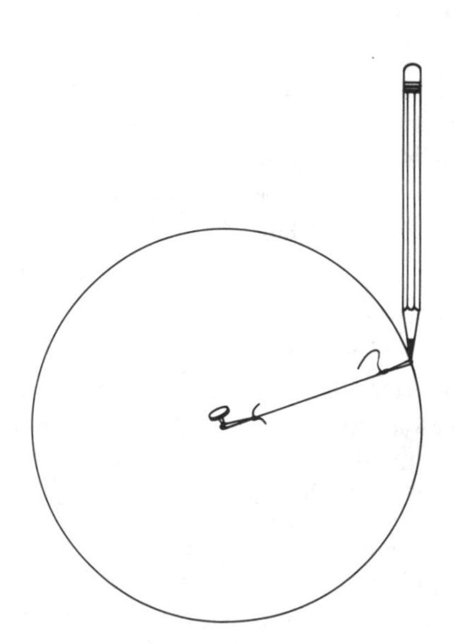

A string with two loops can be used to draw very large circles that are larger than those drawn with a compass. Two students working together will have better success than just one.

Figure 16.43
Drawing a circle.

What is most important in the above activity is that students develop a clear understanding of π as the ratio of circumference to diameter in any circle. The quantity π is not some strange number that appears in "math" formulas.

GEOMETRIC PROBLEM-SOLVING ACTIVITIES

Problem solving continues to be an important feature of geometry activities at level 1. Many activities already suggested are explorations posed as problems. A few additional explorations of a problem-solving nature are suggested in this section.

Tessellations

At level 1, tessellations can be developed with the aid of geometric properties. With the help of grids, students can begin to explore what shapes will fit together in a tessellation and why. Tessellations can be described in terms of the number of lines that come together at each intersection and the properties of the tiles themselves. Through rotations and reflections, simple tiles can be altered to form tessellations similar to those constructed by the Dutch artist M. C. Escher. Figure 16.44 illustrates one example of this approach. Books by Ranucci and Teeters (1977) and Bezuszka, Kenney, and

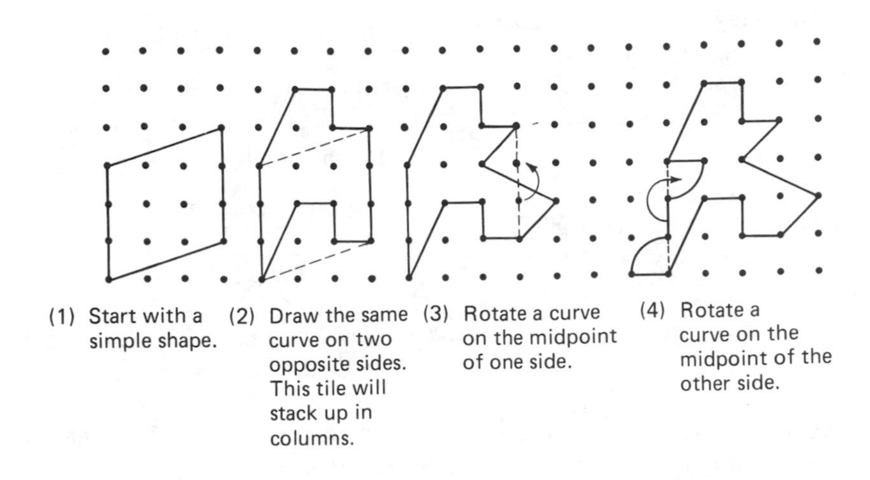

A column of this tile will now match a like column that is rotated one complete turn. Find these rotated columns in the tessellation below.

Figure 16.44
One of many ways to create an "Escher type" tessellation.

Silvey (1977) are among a number of excellent resources that can be used to explore this fascinating topic further with junior high students.

Assorted Problems

Properties of shapes and measurements of shapes provide material for a wealth of problem-solving opportunities. A few such problems are suggested here.

Activity 16.48

Draw any triangle. Choose three measurements of either angles or lengths of sides and use only these to tell a partner how to draw a triangle that is congruent to yours. What combinations of angles and sides will work?

Activity 16.49

Determine the areas of odd shapes such as those in Figure 16.45 for which there are no formulas or for which dimensions are not provided.

Activity 16.50

If a dart lands at all points on a circular target with equal chance, is it more likely to land closer to the center or closer to the edge?

Informal Geometry Activities: Level 2

At the van Hiele level 2 of geometric thinking, students begin to use informal deductive reasoning. That is, they can follow and utilize logical arguments, although they will have a difficult time constructing a proof of their own as in tenth-grade geometry. Physical models and drawings are still important but for different reasons. At level 1 students' explorations lead to inductive conclusions about shapes. For example, the diagonals of parallelograms bisect each other. Students at level 1 are satisfied that such results are so because it seems to happen that way when they try it. At level 2, students can utilize a drawing to help them follow a deductive argument supplied by the teacher. They may also use models to test conjectures or to find counterexamples. Models become more of a tool for thinking and verification than one of exploration.

Many topics in the seventh- and eighth-grade curriculum lend themselves to projects that promote level 2 thinking. It is important that students be challenged to reason through these topics and not just memorize formulas and procedures by rote. Eighth-grade geometry may be one of the last opportunities for good activities that prepare students for the level 3 thinking required in a high school geometry course.

DEFINITIONS AND PROPERTIES

Classification activities at level 2 can begin to focus on the definitions of shapes and how different classes of shapes are related one to another.

Activity 16.51

Consider a particular type of shape, for example a rectangle, and list all of the properties of that shape that students can think of. One possible list is suggested here. What is the shortest list that can be chosen from this one that will still determine a rectangle? Are there different sublists that will work? Which property or properties are necessary if the rectangle is to be nonsquare?

A. Four sides.
B. Four right angles.
C. Opposite sides parallel.
D. Diagonals bisect each other.
E. Adjacent sides are perpendicular.
F. Opposite sides equal in length.
G. Diagonals are congruent.
H. Only two lines of symmetry.
I. Diagonals are not perpendicular.

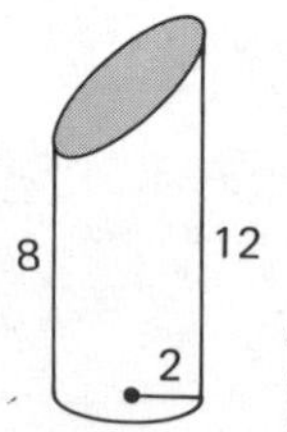

What is the lateral surface area (not including the two faces?)

Hint: Recall how the area of a triangle was derived from the area of a parallelogram.

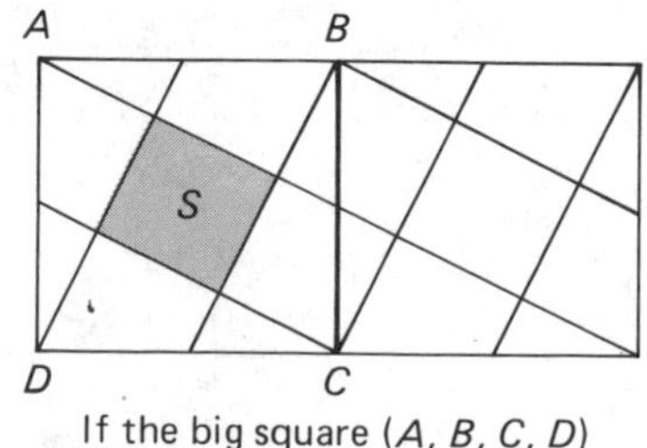

If the big square (*A*, *B*, *C*, *D*) is 1 unit in area, what is the area of *S*?

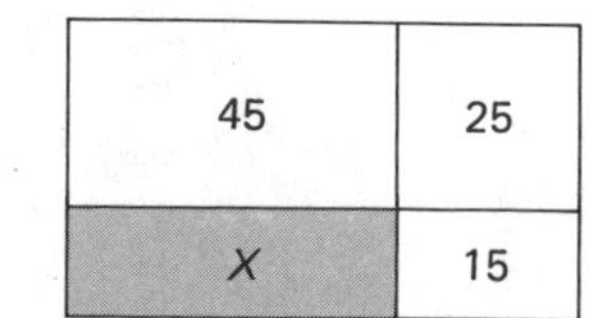

The rectangle is divided into four smaller rectangles with areas 45, 25, 15, and *X*. What is the area of *X*?

(Problems adapted from Milauskas, 1987)

Figure 16.45
Examples of area problems that require some analysis.

Activity 16.52

Explore statements in these forms:

If it is a ________then it is also a ________.

All ________s are ________.

Fill the blanks with names of shapes or statements of properties and relationships. Let students use drawings or models to decide if the statements are true or to find counterexamples. A few examples are suggested here.

If it is a cylinder, then it is a prism.

If it is a prism, then it is a cylinder.

If it is a square, then it is a rhombus.

All parallelograms have congruent diagonals.

All quadrilaterals with congruent diagonals are parallelograms.

If two rectangles have the same area, then they are congruent.

All prisms have a plane of symmetry.

All right prisms have a plane of symmetry.

If a prism has a plane of symmetry, it is a right prism.

Obviously not all of the above statements are true. For those that you think may be false, find a counterexample. You and your students can make up similar puzzles to stump the class.

INFORMAL PROOFS

The Pythagorean Theorem

An area interpretation of the Pythagorean theorem states that if a square is made on each side of a right triangle, the areas of the two smaller squares will together be equal to the area of the square on the longest side or hypotenuse. Figure 16.46 illustrates two proofs that students can follow and begin to appreciate.

Area and Volume Formulas

In Chapter 15, suggestions are given for helping students develop formulas for areas and volumes of common shapes. Formula development is a major connection between measurement and geometry and is also a good bridge between level 1 and level 2 of geometric thinking.

Constructions

Too frequently students are taught to perform constructions with a straightedge and compass but have no idea why the constructions work. In most instances, the constructions represent simple theorems that students can follow and provide reasons for the steps. An example of bisecting an angle is illustrated in Figure 16.47.

The Mira offers a different approach to doing many con-

(a)

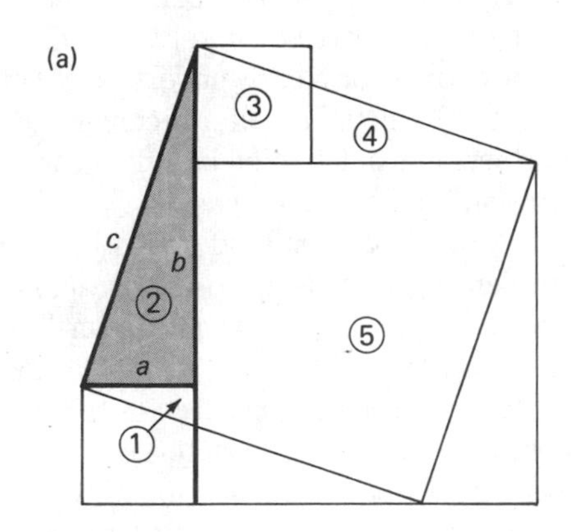

c^2 = region #'s ① + ② + ③ + ④ + ⑤

a^2 = region #'s ① + ③

b^2 = ② + ④ + ⑤

Therefore

$c^2 = a^2 + b^2$

(The second and third equations require some thought. Do you see why they are correct?)

(b)

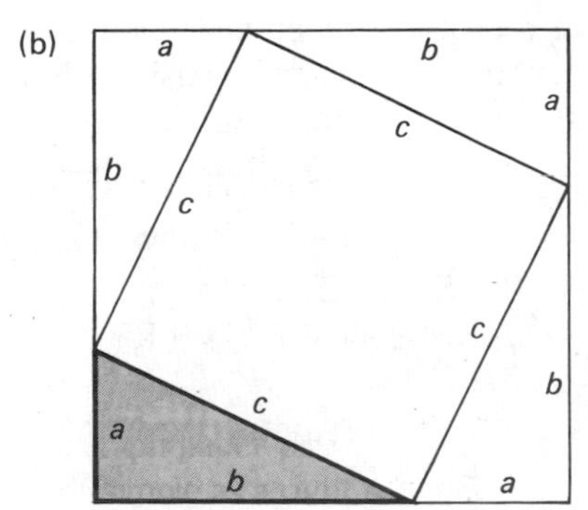

The area of the large square is $(a + b)^2 = a^2 + 2a \cdot b + b^2$

The same area is also c^2 plus 4 times the area of one triangle.

$c^2 + 4\left(\frac{1}{2}a \cdot b\right) = c^2 + 2a \cdot b$

So: $c^2 + \underline{2ab} = a^2 + \underline{2ab} + b^2$

$c^2 = a^2 + b^2$

Figure 16.46
Two proofs of the Pythagorean theorem.

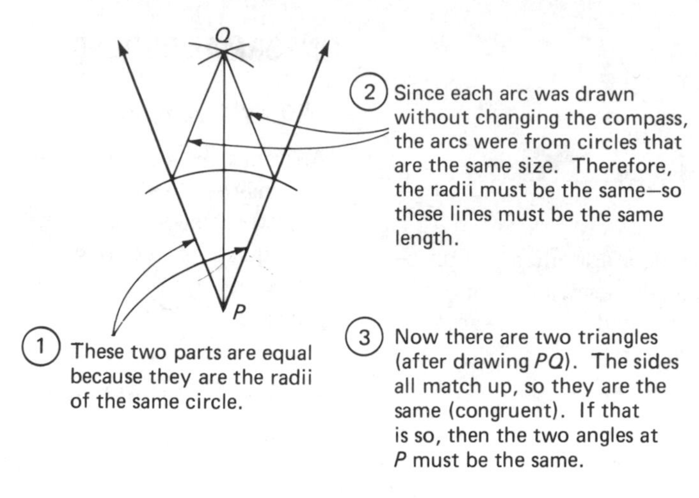

Figure 16.47
A possible informal explanation for bisecting an angle with compass and straightedge.

structions that have traditionally been done with compass and straightedge. With either set of tools, determining on their own how to do these constructions is also a valuable form of deductive reasoning.

Informal Geometry on the Computer

The graphic capabilities of the computer make it an important tool for the study of geometry. The computer language Logo and a family of software programs called the *Geometric Supposers* are such significant applications of the computer to informal geometry that they deserve separate attention.

LOGO AND INFORMAL GEOMETRY

Turtle graphics is only one aspect of Logo, a highly accessible yet powerful computer language that has many uses other than geometry. However, the ease with which Logo can be used to draw pictures on the monitor makes it a natural vehicle for geometric explorations. Since Logo is both very easy to use and very powerful, it can be used by children as young as kindergarten and also be a challenging medium for college students. Very briefly, in Logo graphics, a tiny triangle called a "turtle" can be controlled with simple keyboard commands. It can be made to go forward or backward any distance. A FORWARD 50 command makes the turtle move forward about 5 cm and draw a line in its path. The turtle can also be turned in its place. The command RIGHT 90 turns the turtle 90 degrees to its right. With these and other commands coupled with the ability to put lists of commands into procedures (little programs), children have both power and flexibility with which they can draw on the computer screen.

The purpose of the sections which follow is only to illustrate some of the ways that Logo can be used in an informal geometry program. No attempt is made to explain the details of the Logo programming involved. (For a more detailed explanation of the Logo language, see Chapter 20.)

Shapes in Logo

Children can learn to use regular Logo commands and Logo programming as early as the first grade. Second and third graders can learn to define procedures to create shapes of their own. Within the simplest turtle graphics programming, children develop intuitive understandings of angles and distances.

Consider a child who is trying to write a procedure to draw a rectangle. He has a visual image of rectangle and may even have one drawn on paper. But the turtle must be "taught" to draw each part separately, and the relationship between the parts must be clearly understood. The resulting procedure is a form of definition of a rectangle developed by the child. Further, it is a definition based on component parts and relationships rather than a gestalt visual image.

A more profound difference between visualizing shapes and defining procedures to draw shapes can be observed when watching children who want the turtle to draw a circle. The turtle moves only in straight lines, not curves. A common technique for helping children translate shapes into turtle actions is to have them "play turtle" or walk the shape as if they were the turtle on the screen. How can you walk in a circle if you can only move forward? Children quickly conclude that they should go forward just a little bit and then turn a little bit and repeat this over and over. Logo includes a REPEAT function which makes it easy to do a series of actions over and over. A common "first circle" is produced by the line,

```
REPEAT 360 [FD 1 RT 1]
```

Compare the following Logo procedures:

```
TO SQUARE
  REPEAT 4 [FD 50 RT 90]
  END
TO EIGHTSIDES
  REPEAT 8 [FD 10 RT 45]
  END
TO 36SIDES
  REPEAT 36 [FD 2 RT 10]
  END
```

Each procedure draws a regular polygon. The 36-sided polygon looks exactly like a circle and yet it was conceived as a polygon. "Circles" of different sizes can be made by changing the amount of the turn and/or the forward step. This example illustrates a completely different form of thinking and analysis of shape than is provided in other mediums. How do you know how much to turn? What if you wanted to make a triangle or a hexagon? How is a circle like an octagon or even a square? How do you make different triangles? Can you make a curve that is not a circle?

Activities in the regular Logo environment can be designed for most any level of geometric thought. Most of these will be at level 1 or higher.

"Instant" Logo Activities

Programs in Logo can be written by teachers that permit the very young child to press only a single key, possibly followed by RETURN, to control the turtle. Most of these so-called "instant" Logo routines have a key for each of the following actions: FORWARD 10, BACK 10, RIGHT 30, and LEFT 30.

Activity 16.53

In Figure 16.48, these actions are caused by pressing F, B, R, and L. The keys S, T, and X make the turtle draw a square, triangle, and a rectangle (box). The D key clears the screen. The N and P keys are for PENUP and PENDOWN. Keys can be labeled with stickers to correspond to the turtle action.

With an instant Logo program children can explore drawing shapes and putting simple shapes together on the screen. There is a significant difference between this activity, and using tiles or a geoboard. With tiles and similar materials, children visualize shapes as whole entities. Shapes are easily twisted, turned, and shoved together. When drawing with the Logo turtle, the interaction with shape and position is quite different. Even with an instant program, children must consider how shapes are formed, at what corner of a shape the turtle starts, what direction it turns, and what direction the turtle is facing when it begins. This is not to say that instant Logo is better than tiles or geoboards, only that it provides a significantly different perspective of shape and orientation.

Figure 16.48

Instant programs are easily modified for various purposes. The turns can all be 90 degrees or 45 degrees instead of 30. The shapes can be altered or more shapes added. It is also possible to include a "TEACH" feature that allows the child to save drawings or shapes they have made and redraw them at any time in any orientation (Campbell, 1988). In this way students create their own shapes and then manipulate and combine them.

Like any good geometry activity, instant Logo procedures should be used with some guidance or direction. After initial exposure and free exploration, children should engage in specific tasks. They might be asked to create a particular picture, to reproduce a drawing shown to them on a task card, or perhaps make the turtle draw a picture that they themselves have made off of the computer. If, for example, the six pattern block shapes are programmed, children can make a design with actual blocks and then attempt to draw the design on the computer. The computer task is very different and much more challenging than building with blocks, and yet the two experiences are significantly linked.

All of the major versions of Logo include a simple version of an instant program in the manuals. A complete listing and description of an excellent version is included in Campbell's article, "Microcomputers in the Primary Mathematics Classroom" (1988), or in *Learning with Logo* (Watt, 1983).

Instant Logo activities are appropriate for level 0 thinking and provide a nice bridge to level 1 activities and to regular versions of Logo.

More Ideas with Logo

The few ideas which follow are only intended to give some flavor for what can be done with Logo and geometry.

Activity 16.54

Use variables in procedures to create many shapes with the same characteristics. For example, the following procedure will draw rectangles of any dimensions. Executing RECTANGLE 20 20 will produce a square. How are RECTANGLE 20 60 and RECTANGLE 60 20 alike? How are they different?

```
TO RECTANGLE :L :W
  REPEAT 2 [FD :L RT 90 FD :W]
  END
```

Not only can forward distances be variable but also angles. How would you write a procedure to draw any parallelogram?

Activity 16.55

Symmetry can be explored in Logo to provide a different perspective. A mirror image of a shape (reflection) is achieved by changing all right turns to left, and vice versa. It is also easy to create shapes and images with rotational symmetry.

Activity 16.56

Write a procedure with no variables to create a small shape or drawing. Next write a new procedure from the original in which all distances are multiplied by the same variable. By changing the value of the variable, proportional drawings (similar figures) are produced.

GEOMETRIC SUPPOSERS

A Brief Look at the *Supposers*

The *Geometric Supposers* from Sunburst are software programs designed as tools with which to explore geometric shapes, and to encourage the formulation and testing of conjectures. The first *Supposer* (Yerushalmy and Schwartz, 1985) explores triangles. Later came similar programs for investigating quadrilaterals and circles, and a fourth program called the *preSupposer* that includes most of the drawing capabilities of the other three.

The *Geometric Supposer: Triangles* starts by drawing a triangle. Users can select different types of triangles or can draw their own choosing various combinations of sides and angles. The *Supposer* generates its shapes with random dimensions so that each time an isosceles triangle is selected the program will draw a different isosceles triangle and in a different orientation on the screen. Once a shape is on the screen, the user selects additional constructions or measurements from menus. Angles can be bisected, circles can be inscribed or circumscribed, line segments can be partitioned into equal segments, lines can be extended, reflected, or drawn parallel or perpendicular to other lines, and points can be joined. Lengths, angles, and areas can all be measured, either in terms of a fixed unit or in terms of some other line segment or area. As a result, it is easy to make comparisons or observe ratios of distances or areas. Figure 16.49 shows the *Geometric Supposer: Triangles* after several constructions and measurements have been made.

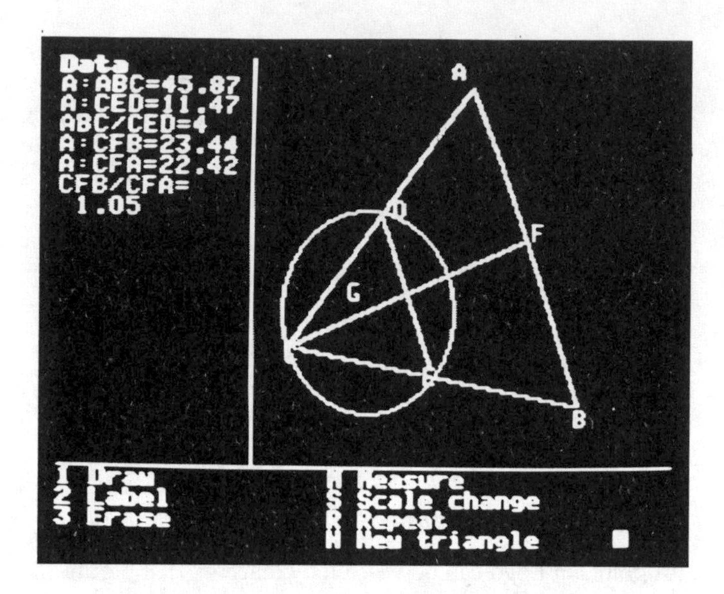

Figure 16.49

One of the most interesting features of the *Supposers* is their ability to repeat a series of constructions on previous figures or on new figures. When students observe a relationship on one figure, they are encouraged to test it immediately on a whole series of figures. With the *Geometric Supposer: Quadrilaterals,* for example, the teacher might suggest joining the midpoints of the sides of quadrilaterals. The first experiment on a random parallelogram produces another parallelogram. Was that just a coincidence? Try it quickly on another parallelogram. What about on trapezoids? How about rectangles, kites, rhombuses, squares, or even a random quadrilateral? Since the *Supposer* draws these shapes quickly and accurately, and because measurements on any shape are easily made by the computer, the student can focus attention on relationships and make observations. Drawing and measuring skills are not required.

In the Classroom

The *Supposers* can be used in a variety of ways. If students have access to computers such as in a computer lab, assignments can be given to individuals or groups of two or three working together. Students can make freehand sketches of the computer drawings noting important results as they work independently. Results and observations can then be discussed later.

With the use of the newer liquid crystal overhead projector display screens, the computer screen can easily be projected by the teacher onto a large screen. In this mode, the *Supposers* provide teachers with an electronic blackboard. Not only can they make drawings quickly and accurately on the screen but they can also make measurements with no effort. As a teaching tool, students can interact with each other and with the teacher concerning conjectures and observations. Suggestions for what type of shape or what construction to look at can come from the students.

The *Supposers* make accessible to students the act of creating geometric theorems in the form of conjectures. Students begin not only to make observations but learn the difference between proof and inference. The role of counterexamples becomes obvious. With some differences in the amount of guidance supplied, *Supposer* activities are appropriate for both level 1 and level 2 geometric thinking. They form an excellent bridge between observations and theorems.

As computer technology improves, there will undoubtedly be new and better forms of geometric tool software in the spirit of the *Supposers.* Programs already exist for some computers that permit students to create drawings of three-dimensional figures, rotate them around different axes, view them from any perspective, slice them, and measure component parts. These computer tools can only enhance geometric thinking and add to the excitement of studying geometry.

For Discussion and Exploration

Note: There are so many activities in this chapter that you are probably overwhelmed. The best way to get excited and comfortable with informal geometry is to start doing activities, the same ones that are designed for students. Do not worry about being "right." The only thing that is right is to *do something,* to think about what you are doing, and have fun. If you get interested, you will begin to search for ideas and relationships and you may even decide you want to do some further reading. Good geometry resource books for teachers are quite plentiful, and articles in the *Arithmetic Teacher* are also quite common. You could not go wrong by simply exploring activities from this chapter, especially those designed for levels 0 and 1. A possible limitation may be materials. An effort has been made to include many activities that use only teacher-made materials. The suggestions which follow are only a few selected ideas for your exploration.

1. Examine the teacher's edition of a basal textbook at any grade level. Select any lesson on geometry. Remember that the authors of the pupil's book are restricted to the printed page by the very nature of books. Teachers are not so restricted. How would you teach this lesson so that it was a *good* informal geometry lesson? Your lesson should include a hands-on activity and have a problem-solving spirit.
2. Many geometry resource books are written around one set of materials rather than for a particular objective or grade level with a variety of materials. One way to choose activities from this chapter is to do some of those that fit one set of materials you may have. Some good choices for materials:
   —Geoboards (Directions for these are in Appendix C.)
   —Wooden cubes.
   —"Tiles": Make a set from poster board or there are three sets in Appendix C that can be easily duplicated on tagboard. Try to get pattern blocks; they are excellent.
   —Paper grids. There are several in Appendix C, and many activities include only drawing on a grid.

   Decide on one set of materials and explore some activities that seem appropriate for a grade level of your interest. After trying the activities yourself, write a lesson plan to teach a lesson with these activities to children. If possible, teach your lesson.
3. Get a partner and make some large skeleton models from newspaper as in Figure 16.19. All you need is a supply of old newspaper and some masking tape.
4. Make some models of geometric solids. Look through the chapter for ways to fold these up from tagboard shapes. Make a string model for cylinders and cones as in Figure 16.29.
5. Get a copy of the book *Tangramath* (Seymour, 1971) or *Tangram treasury,* Book A, B, or C (Fair, 1987), or any other book of tangram activities. Explore one idea that you like and develop a lesson for children around that activity.
6. Make a tessellation. The most exciting and pretty results happen when you cut tiles out of two or three colors of construction paper and paste the tessellation on large white art paper. Make your tiles about 5 cm across. Begin your tessellation in the center of the paper and leave the edges uneven. There are suggestions for tessellations in both level 0 and level 1 activities.
7. Cut out some shapes that have 1, 2, and more than 2 lines of symmetry. Draw their "box" and see how many ways the shapes will fit (Figure 16.36). Use a sheet of grid paper to create shapes with lines of symmetry as described. Try similar activities with rotational symmetry as shown in Figure 16.38. Stretch your knowledge of symmetry into three dimensions by exploring some of the activities suggested with clay or with wooden cubes.
8. Get one of the *Geometric Supposers* and play around with it. There are suggested explorations in the manuals or you might pursue one of the activity booklets that has been designed to accompany the *Supposers* (Houde & Yerushalmy, 1988).
9. Read any chapter of interest from Bruni's *Experiencing geometry* (1977) and try one of the activities that interests you.

Suggested Readings

Barson, A. (1971). *Geoboard activity cards: (Primary, intermediate).* Fort Collins, CO: Scott Resources, Inc.

Bezuszka, S., Kenney, M., & Silvey, L. (1977). *Tessellations: The geometry of patterns.* Palo Alto, CA: Creative Publications.

Bidwell, J. K. (1987). Using reflections to find symmetric and asymmetric patterns. *Arithmetic Teacher, 34*(7), 10–15.

Burger, W. F. (1982). Graph paper geometry. In L. Silvey (Ed.), *Mathematics for the middle grades (5–9).* Reston, VA: National Council of Teachers of Mathematics.

Carroll, W. M. (1988). Cross sections for clay solids. *Arithmetic Teacher, 35*(7), 6–11.

Collier, C. P. (1976). *Geometry for teachers.* Boston: Houghton Mifflin.

Creative Publications. (1986). *Hands on pattern blocks, Books 1, 2, & 3.* Palo Alto, CA: Creative Publications.

Creative Publications. (1986). *Hands on geoboards, Books 1, 2, & 3.* Palo Alto, CA: Creative Publications.

Crowley, M. L. (1987). The van Hiele model of the development of geometric thought. In M. M. Lindquist (Ed.), *Learning and teaching geometry, K–12.* Reston, VA: National Council of Teachers of Mathematics.

Dana, M. E. (1987). Geometry—a square deal for elementary teachers. In M. M. Lindquist (Ed.), *Learning and teaching geometry, K–12.* Reston, VA: National Council of Teachers of Mathematics.

Dana, M. E., & Lindquist, M. M. (1978). Let's try triangles. *Arithmetic Teacher,26*(1), 2–9.

Elementary Science Study. (1974). *Teacher's guide for mirror cards.* New York: McGraw-Hill (available from Delta).

Elementary Science Study. (1974). *Teacher's guide for geoblocks.* New York: McGraw-Hill (available from Delta).

Elementary Science Study. (1970). *Teacher's guide for pattern blocks.* New York: McGraw-Hill (available from Delta).

Hill, J. M. (Ed.). (1987). *Geometry for grades K–6: Readings from the Arithmetic Teacher.* Reston, VA: National Council of Teachers of Mathematics.

Hoffer, A. R. (1988). Geometry and visual thinking. In T. R. Post (Ed.), *Teaching mathematics in grades K–8: Research based methods.* Boston: Allyn and Bacon.

Hoogeboom, S. *Moving on with geoboards.* Palo Alto, CA: Creative Publications.

Kaiser, B. (1988). Explorations with tessellating polygons. *Arithmetic Teacher, 36*(4), 19–24.

Lappan, G., & Even, R. (1988). Scale drawings. *Arithmetic Teacher, 35*(9), 32–35.

Perl, T. (1974). *Relationshapes activity cards.* New Rochelle, NY: Cuisenaire Corporation of America, Inc.

Ranucci, E. R., & Teeters, J. L. (1977). *Creating Escher-type drawings.* Palo Alto, CA: Creative Publications.

Shaw, J. M. (1983). Exploring perimeter and area using centimeter squared paper. *Arithmetic Teacher, 31*(4), 4–11.

Shroyer, J., & Fitzgerald, W. (1986). *Mouse and elephant: Measuring growth.* Menlo Park, CA: Addison-Wesley.

Skinner, J. (1987). Extracts from a teacher's diary. *Mathematics Teaching, 121,* 23–26.

Van de Walle, J. A., & Thompson, C. S. (1981). A triangle treasury. *Arithmetic Tecacher, 28*(6), 6–11.

Van de Walle, J. A., & Thompson, C. S. (1980). Concepts, art, and fun from simple tiling patterns. *Arithmetic Teacher, 28*(3), 4–8.

Van de Walle, J. A., & Thompson, C. S. (1984). Cut and paste for geometric thinking. *Arithmetic Teacher, 32*(1), 8–13.

Walter, M. J. (1970). *Boxes, squares and other things.* Reston, VA: National Council of Teachers of Mathematics.

Willcutt, B. (1987). Triangular tiles for your patio. *Arithmetic Teacher, 34*(9), 43–45.

Winter, M. J., Lappan, G., Phillips, E., & Fitzgerald, W. (1986). *Middle grades mathematics project: Spatial visualization.* Menlo Park, CA: Addison-Wesley.

CHAPTER 17

CLASSIFICATION AND PATTERNING

Introduction

Classifying and patterning (extending, exploring, and creating patterns or sequences) are two basic processes of mathematical thinking. Both involve the creation of relationships. Both can also be viewed as a form of problem solving. Both types of activities provide students with the opportunity to develop logical reasoning abilities. Logical reasoning skills and especially the meaningful use of the language of logic (if-then, and, or, not, all, some) are valuable across all areas of mathematics. From that vantage alone, classification and patterning activities are important.

Classification Activities

ATTRIBUTE MATERIALS

Attribute materials are sets of objects that lend themselves to being sorted and classified in different ways. Natural or *unstructured* attribute materials include such things as sea shells, leaves, the children themselves, or the set of the children's shoes. The *attributes* are the ways that the materials can be sorted. For example, hair color, height, and gender are attributes of children. Each attribute has a number of different *values:* for example, blond, brown, or red (for the attribute of hair color), tall or short (for height), male or female (for gender).

A *structured* set of attribute pieces has exactly one piece for every possible combination of values for each attribute. For example, several commercial sets of plastic attribute materials have four attributes: color (red, yellow, blue), shape (circle, triangle, rectangle, square, hexagon), size (big, little), and thickness (thick, thin). In the set just described there is exactly one large, red, thin, triangle as well as one each of all other combinations. The specific values, number of values, or number of attributes that a set may have is not important.

In Figure 17.1 three teacher-made sets of attribute pieces are illustrated. The attribute shapes are easily made in nice large sizes out of poster board and laminated for durability. A black-line master for the Woozle Cards is in Appendix C. These can be duplicated on tagboard and quickly colored in two colors before laminating. The ice cream cones can be made from construction paper and laminated or could be duplicated on paper and cut out by the students.

The value of using structured attribute materials (instead of unstructured materials) is that the attributes and values are very clearly identified and easily articulated to students. There is no confusion or argument concerning what values a particular piece possesses. In this way we can focus our attention in the activities on the reasoning skills that the materials and activities are meant to serve. Even though a nice set of attribute pieces may contain geometric shapes of different colors and sizes, they are not very good materials for teaching shape, color, or size. A set of attribute shapes does not provide enough variability in any of the shape attributes to help students develop anything but very limited geometric ideas. In fact, simple shapes, primary colors, and two sizes are chosen because they are most easily discriminated and identified by even the youngest of students.

PURPOSES

Attribute activities are designed to help young children develop certain mathematical reasoning and problem-solving processes: logical reasoning ("if-then" type of thinking), classification skills (observing likenesses and differences), and trial-and-error or guess-and-check strategies. Most students find the activities exciting and fun which contributes to their willingness to solve problems. Young children learn to work together, persevere in their pursuit of a solution, make conjectures, and test their own ideas through the use of materials without appealing to a higher authority to determine if they are "right." Interest and problem-solving self-confidence is enhanced.

Notice that all of the objectives of classification or attribute activities are process objectives or affective objectives. None of these should be considered as a skill that your students must master. Process skills develop slowly over an extended time, and not all students will develop them to the same degree. More important is to separate in your mind these problem-solving objectives from the actual activities used to develop that skill. No curriculum should ever list as an objective the ability to complete a particular type of activity with attribute materials. It is the mental process, not the physical activity, that is important.

Classification skills and the development of classification schemes are science process objectives as well as examples

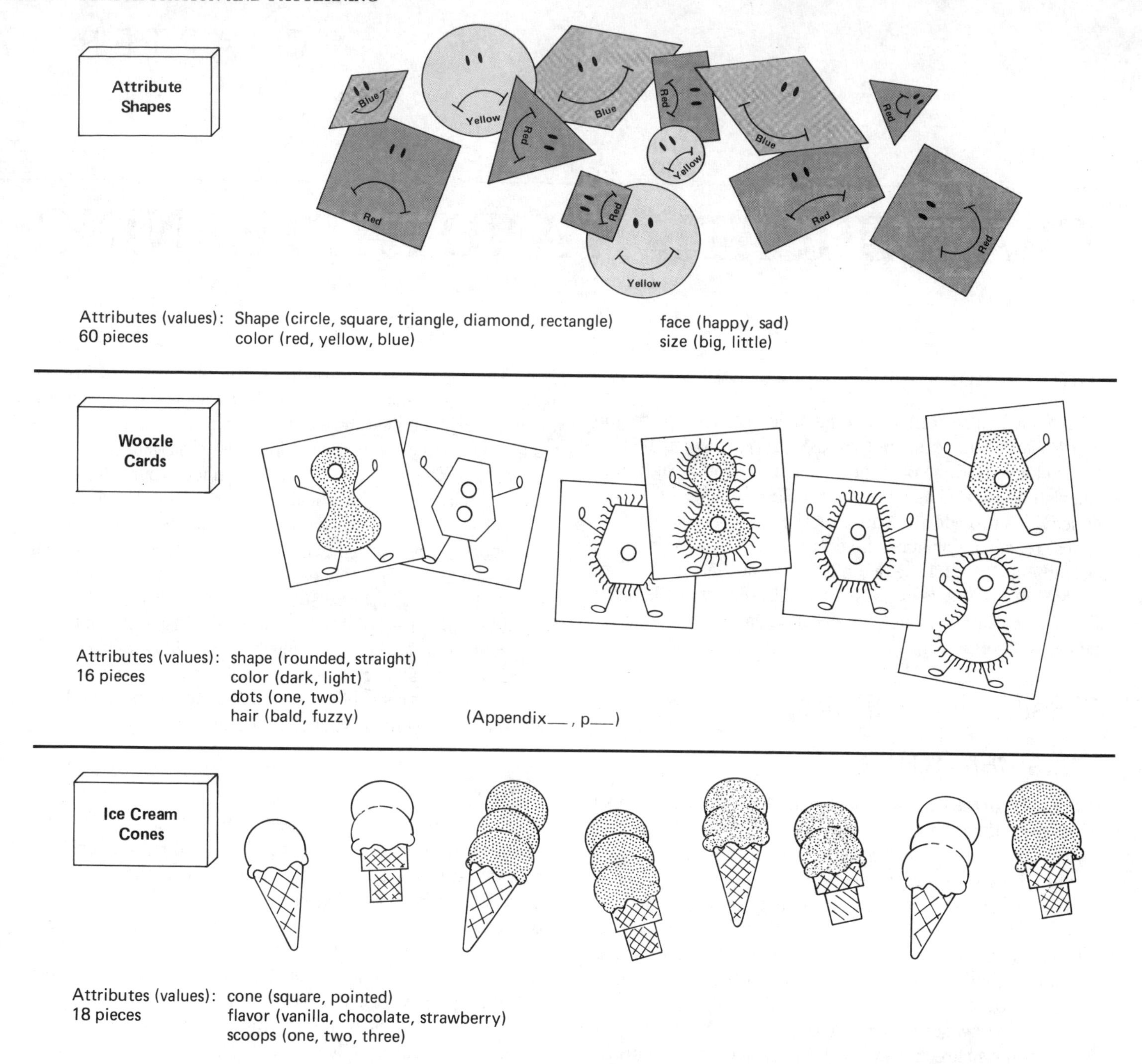

Figure 17.1
Three teacher-made attribute sets.

of good mathematics. Many good science programs, especially those that are process-oriented, involve a heavy emphasis on classification. Classification activities, therefore, provide an excellent opportunity to connect science and mathematics in your classroom.

Classification activities became very popular in the 1960s and 1970s as early childhood activities. This emphasis was largely due to the importance accorded classification by Piaget as a basic cognitive process necessary, in his view, for the development of number concepts. As discussed in Chapter 5, research has demonstrated that the connection between classification and early number understanding is probably not as significant as once was thought. The point is raised here so that you consider the activities in this section as part of the broader K to 8 strand of problem solving and the development of mathematical thinking. As presented here, they are not considered "prenumber" experiences.

Many excellent classification activities can be extended profitably as high as the eighth grade, providing students with significant logical challenges and the opportunity to use logical vocabulary (all, some, and, or, not). The few activities described in the next section are only a sample of the many possibilities for attribute materials. Additional ideas can be found in the resources listed at the end of the chapter.

ACTIVITIES WITH ATTRIBUTE MATERIALS

Most attribute activities are best done in a teacher-directed format. Young children can sit on the floor in a large circle where all can see and have access to the materials. Older children can work in groups of four to six students, each group with its own set of materials. In that format, problems can be addressed to the full class and groups can explore them independently. All activities should be conducted in an easygoing manner that encourages risks, good thinking, attentiveness and discussion of ideas. The atmosphere should be nonthreatening, nonpunitive, and nonevaluative.

Most of the activities in this section will be described using the geometric shapes in Figure 17.1. However, each could be done with any structured set and some could be done with nonstructured materials.

String or Loop Activities

Loops are made of yarn or are drawn on paper to hold a designated class of pieces such as "red" or "not square." When two loops overlap, the section that is inside both loops is for the pieces that have both properties. Children as young as kindergarten can have fun with simple loop activities. With the use of words such as *and, or,* and *not,* the loop activities become challenging for children in the upper grades as well.

Activity 17.1

At the beginning level, give children two large loops of yarn or string. Direct them to put all the red pieces in one string and all triangles in the other. Let the children try to resolve the difficulty of what to do with the red triangles. When the notion of overlapping the strings to create a section common to both loops is clear, more challenging activities can be explored.

Once the idea of how a loop can be used to hold a particular type of piece, "strings" or loops can be drawn on poster board or on large sheets of paper. If you happen to have a magnetic blackboard, try using small magnets on the backs of the pieces and conduct full class activities with the pieces on the board. Students can come to the board to place or arrange pieces in strings drawn on the board with colored chalk.

Activity 17.2

Labeled Loops. Label the overlapping loops with cards indicating values of different attributes. Let children take turns randomly selecting a piece from the pile and deciding in which region it belongs. Pieces belonging in neither loop are placed outside. Let other students decide if the placement is correct and occasionally have someone else explain. Do this even when the choice of regions is correct.

A significant variation of the last activity is to introduce negative attributes such as "not red" or "not small." Other labels that are important include the use of the *and* and *or* connectives as in "red and square," "big or happy." The use of *and, or,* and *not* significantly widens children's classification schemes. It also makes these activities quite difficult for very young children. In Figure 17.2, three loops are used in a string game illustrating some of these ideas.

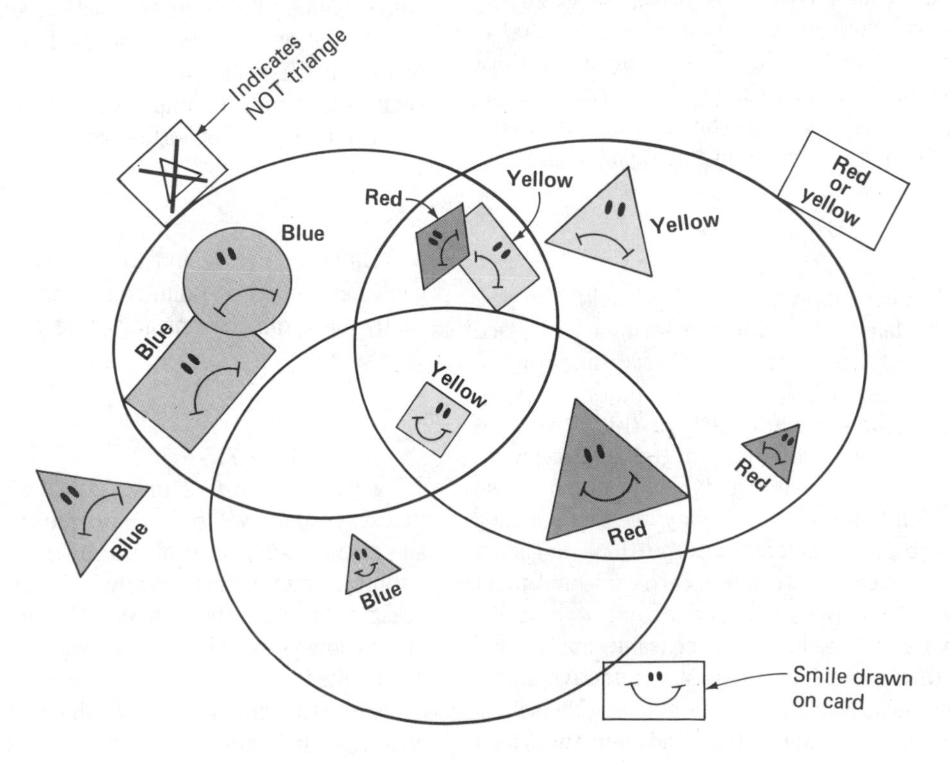

Figure 17.2
A three-loop activity with attribute pieces.

The activities described so far have students attempting to classify materials according to our schemes. That is, the teacher creates a classification, and the children fit pieces into it. While this is important for the purpose of understanding classifications by more than one attribute (the overlapping loops), the activity makes relatively few cognitive demands. All that is required to do these activities is an understanding of the loop method of classification and the ability to discriminate the attributes. When the words *and, or,* and *not* are used, children gain experience with those logical connectives. However, very limited logical reasoning or problem solving is actually going on. A much more significant mental activity is to infer how things have been classified when the scheme is not clearly articulated. The following activities are examples of those that require students to make and test conjectures about how things are being classified. These activities move classification squarely into the domain of problem-solving.

Activity 17.3

Guess My Rule. For this activity try using students as attribute "pieces" instead of shapes. Decide on an attribute of your students such as "wearing blue jeans" or "has stripes on clothing," but do not tell your rule to the class. Silently look at one child at a time and move him or her to the left or right according to this attribute rule. After a number of students have been sorted, have the next child come up and ask students to predict which group he or she belongs in. Before the rule is articulated, continue the activity for a while so that others in the class will have an opportunity to determine the rule. This same activity can be done with virtually any materials that can be sorted. When unstructured materials such as students, students' shoes, shells, or buttons are used, the classifications may be quite obscure, providing an interesting challenge.

The same inference approach can be applied to the string game. Select label cards for the strings and place them face down. Begin to sort pieces according to the turned-down labels. As you sort, have students try to determine what the labels are for each of the loops. Let students who think they have guessed the labels try to place a piece in the string, but avoid having them guess the labels aloud. Point out that one way to test an idea about the labels is to select pieces that you think might go in a particular section. Do not turn the cards up until most students have figured out the rule. Notice that some rules or labels are equivalent; "not large" is the same as "small." With the use of three loops and logical connectives, this activity can become quite challenging, even for middle school students. With simple, one value labels and only two loops it can easily be played in kindergarten.

Be prepared to adjust the difficulty of these and similar activities according to the skill and interest of your children. Remember that children like to be challenged, but an activity that is either too easy or too difficult will likely result in restless children and discipline problems.

Difference Games

As an introduction to the activities in this section, let each child select an attribute piece. Then you hold up one piece for all to see and ask questions such as:

Who has one that is like mine? How is it like mine?

Who has one that is different? Explain how it is different?

Look at your neighbor's piece. Tell how yours is like his or hers. Let your partner explain how his or hers is different from your piece.

Activity 17.4

Difference Trains and Loops. Place an attribute piece in the center of the group. The first student finds a piece that is different from this piece in exactly one way. Let students take turns finding a piece that differs from the *preceding* piece in just one way, creating a "one-difference train." The train can be made as long as the students wish or until no more pieces will fit the one-difference rule.

As a variation, draw a circular track on a piece of paper or poster board with 6 to 10 sections on it. Place the first piece in one of the sections. Subsequent pieces can be placed to the right or the left around the loop, but must differ in one way from the adjacent piece. Placing the last piece may be very difficult or even impossible; it must differ in exactly one way from the piece on either side. A sample is shown in Figure 17.3.

Difference train and loop games can be two-difference games as well as one-difference games. These are not significantly more difficult but add variety.

Activity 17.5

Difference Grids. Draw a 4×5 grid on a poster board so that each space will hold an attribute piece. Select any piece and place it anywhere on the board. In turn students try to put pieces on the board in any space above, below, to the left, or right of a space that is filled. The rule is that adjacent pieces up and down must differ in two ways; to the left and right they must differ in one way. It frequently will be impossible to complete the grid but establishing that fact is a significant challenge. It is important for students to see that not every problem is solvable.

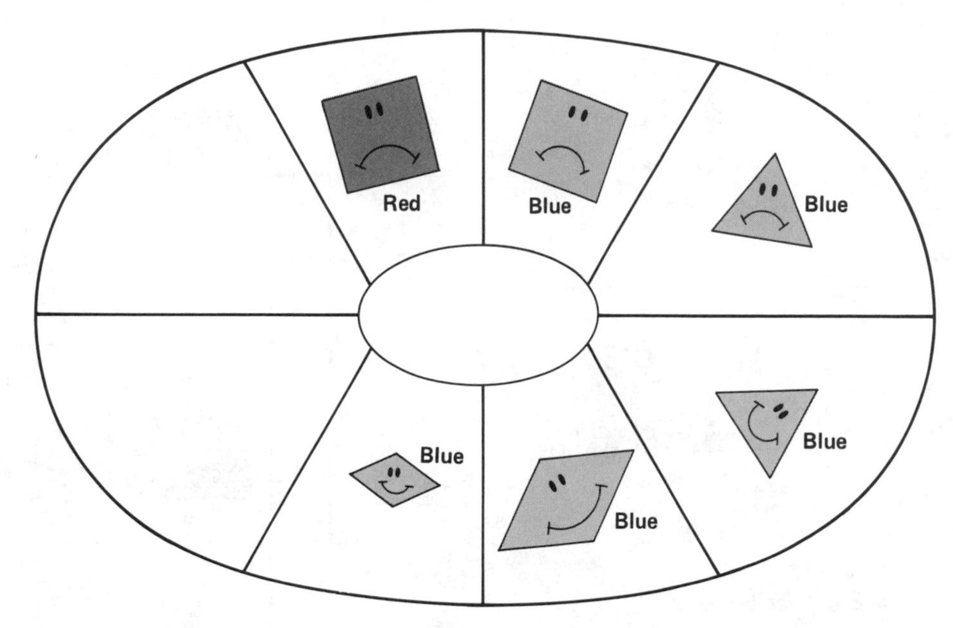

Figure 17.3
Can you finish this one-difference track?

More Classification Challenges

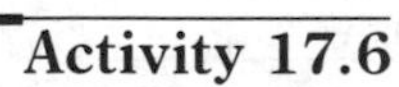

Activity 17.6

Attribute Grids. In Figure 17.4, the cards along the top and left of the grid indicate the attribute values that all pieces in the corresponding row or column must have. Thus, each piece must have two specified attributes. Only one attribute is used for the top cards, and one for the side cards, or else there may be empty cells in the grid (such as a square that is also a circle).

With the cards showing, the students take turns selecting pieces from the pile and finding where they belong in the grid.

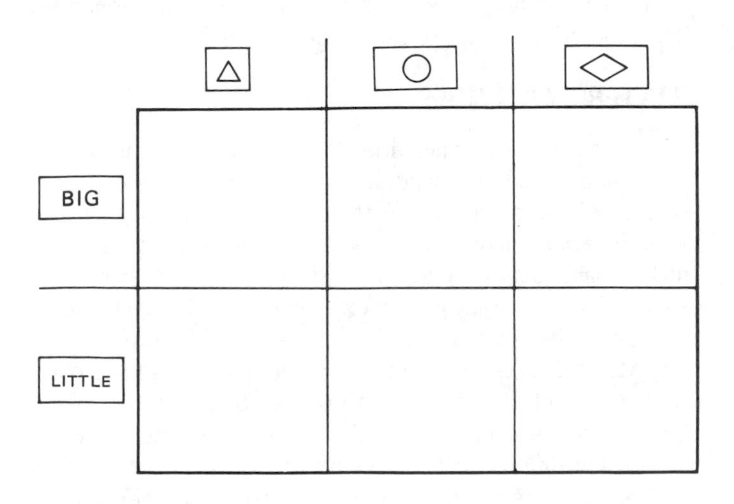

Figure 17.4
A two-way classification grid.

When the grid idea is familiar, the previous activity can be converted to one that requires inference as was done with the loops. Select labels for each row and column, but turn the cards face down. Place pieces on the grid where they belong. The object is to determine what the attribute values (on the cards) are for each row and column. Encourage students to use trial and error to test what they think the cards say.

Activity 17.7

Which Doesn't Belong? The *Sesame Street* game, "One of These Things Is Not Like the Other" is easily conducted with any attribute set. Select four pieces so that three of the pieces have some feature in common that is not a feature of the fourth. The students try to decide which piece is different. Frequently there can be three or even four possible choices, each for a different reason. The students should explain their reasons and classmates should decide if the reason is good. Be sure that you emphasize good reasoning and not "right answers." In Figure 17.5, there are two pieces that are each different from the other three.

Activity 17.8

Sets of Four. Choose one value of any two attributes, such as happy triangles or large red pieces. Select any four pieces that share both values and arrange them in a 2 × 2 array. The challenge is to make more arrays similar to or "like" the original one. To make new arrays, you have to first decide how the pieces in the original area are alike and use those same attributes but with different values to make the other arrays. Look at the example in Figure 17.6. The original

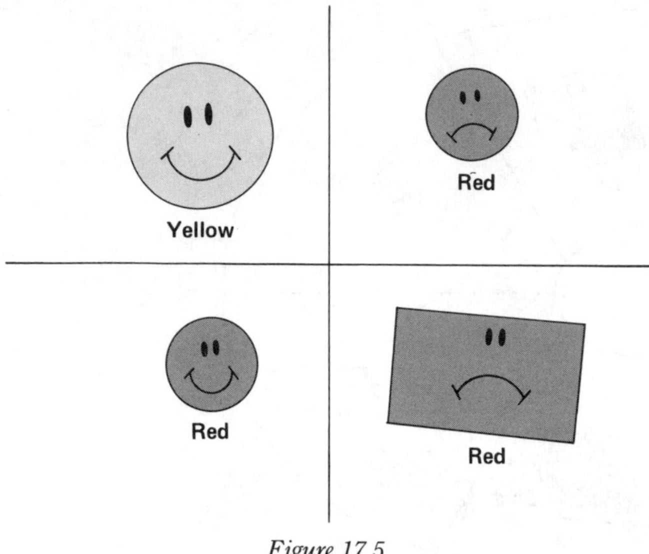

Figure 17.5
"One of these is not like the other."

array has all happy triangles: attributes of face and shape. The values of color and size are mixed. Within each new array, shape and face values are the same. Corresponding pieces each match the original array in color and size. An example is in Figure 17.6. If this activity is done with sets such as the Woozle Cards or the ESS People Pieces (Elementary Science Study, 1966), where each of the attributes has exactly two values, four arrays will use up all of the pieces.

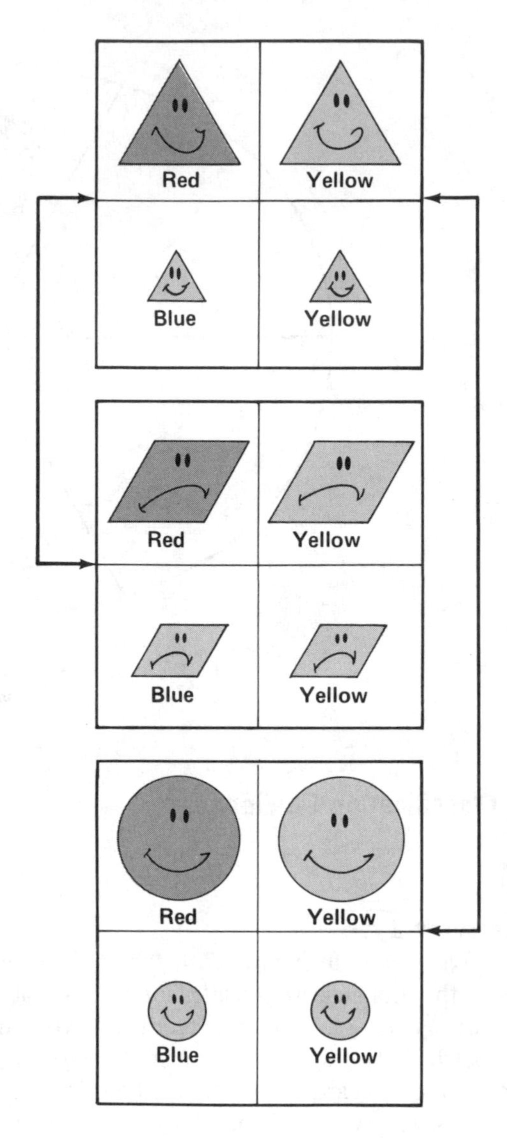

Figure 17.6
Making sets of four.

Patterning

Identifying and extending patterns is an important process in mathematics and mathematical thinking. Simple repetitive patterns can be explored as early as kindergarten using colored blocks, buttons, and a host of other materials. Junior high level students can explore numeric patterns that are progressions or sequences rather than repeating patterns. Here students not only extend patterns but look for generalizations and relationships that allow for determining what is in the pattern at any point along the string.

PURPOSES

Many of the values of patterning are the same as for attribute activities: development of logical reasoning, perseverance in solving problems, willingness to solve problems, and the ability to make conjectures and test ideas. Young children seem to love to extend patterns, such as those made with colored blocks, across an entire room. The internal positive feedback they receive from knowing they are right and successful is significant. Patterning for older children is similarly a matter of testing ideas, extending the patterns to test relationships, and developing general formulas. These are all important aspects of mathematical thinking and problem solving. Numeric sequences are good early examples of the concept of functional relationships. Each term in the sequence or pattern has a unique value, depending on where it is within the sequence or pattern. Thus the value of each term is a function of its position.

PATTERN ACTIVITIES

Almost all patterning activities should involve some form of physical materials to make up the pattern. This is especially true from kindergarten to fourth grade. When patterns are built with materials children can test the extension of a pattern and make changes as they discover errors without fear of being wrong. The materials permit experimentation or trial-and-error approaches that are almost impossible without them.

Many kindergarten and first-grade textbooks have pages where students are given a pattern such as a string of colored beads. The task may be to color the last bead or two in the string. The difference between this and the same activity done with actual materials is twofold. First, by coloring or marking a space on the page the activity takes on a significant aura of "right versus wrong." There is clearly a correct way to finish the pattern. If a mistake is made, correction on the page is difficult and can cause feelings of inadequacy. With materials, a trial-and-error approach can be used. Second, pattern activities on worksheets prevent children from extending patterns beyond the few spaces provided by the page. Most children enjoy using materials such as colored blocks, buttons, Unifix cubes, and so on, to extend their patterns well beyond what could possibly be provided by a printed page. Children are frequently observed continuing a pattern with materials halfway across the classroom floor. In doing so, children receive a great deal of satisfaction and positive feedback from the activity itself. "Hey, I understand this! I can do it really well. I feel good about how I solved my pattern problem."

The same benefit of using materials can be built into patterning activities at the upper grades. There, the satisfaction comes not so much from extending a repeating pattern as it does from seeing how an observed relationship actually exists in a particular design or arrangement of materials. Not only is patterning a form of problem solving and logical reasoning, it can be very satisfying and self-rewarding. It is very important that students connect such positive feelings with mathematical thinking.

Repeating Patterns

The concept of a repeating pattern and how a pattern is extended or continued can be introduced to the full class in several ways. One possibility is to draw simple shape patterns on the board and extend them in a class discussion. Oral patterns can be joined in by all children. For example, "do, mi, mi, do, mi, mi, . . ." is a simple singing pattern. Up, down, and sideways arm positions provide three elements with which to make patterns: up, side, side, down, up, side, side, down, ... Boy, girl patterns or stand, sit, and squat-down patterns are also fun. From these ideas the youngest children learn quickly the concept of patterns. Students can begin to work more profitably in small groups or even independently once a general notion of patterns is developed.

Activity 17.9

Pattern Strips. Students can work independently or in groups of two or three to extend patterns made from simple materials: buttons, colored blocks, Unifix cubes, toothpicks, geometric shapes, and a wide variety of other materials, most of which you can easily gather. For each set of materials, draw two or three complete repetitions of a pattern on strips of tagboard about 5 cm by 30 cm. The students, using actual materials, copy the pattern shown and extend it as far as they wish. Discussions with students can help them verbalize the patterns, uncover errors, and encourage students to make up patterns on their own. Figure 17.7 illustrates some possible patterns for a variety of materials. It is not necessary to have

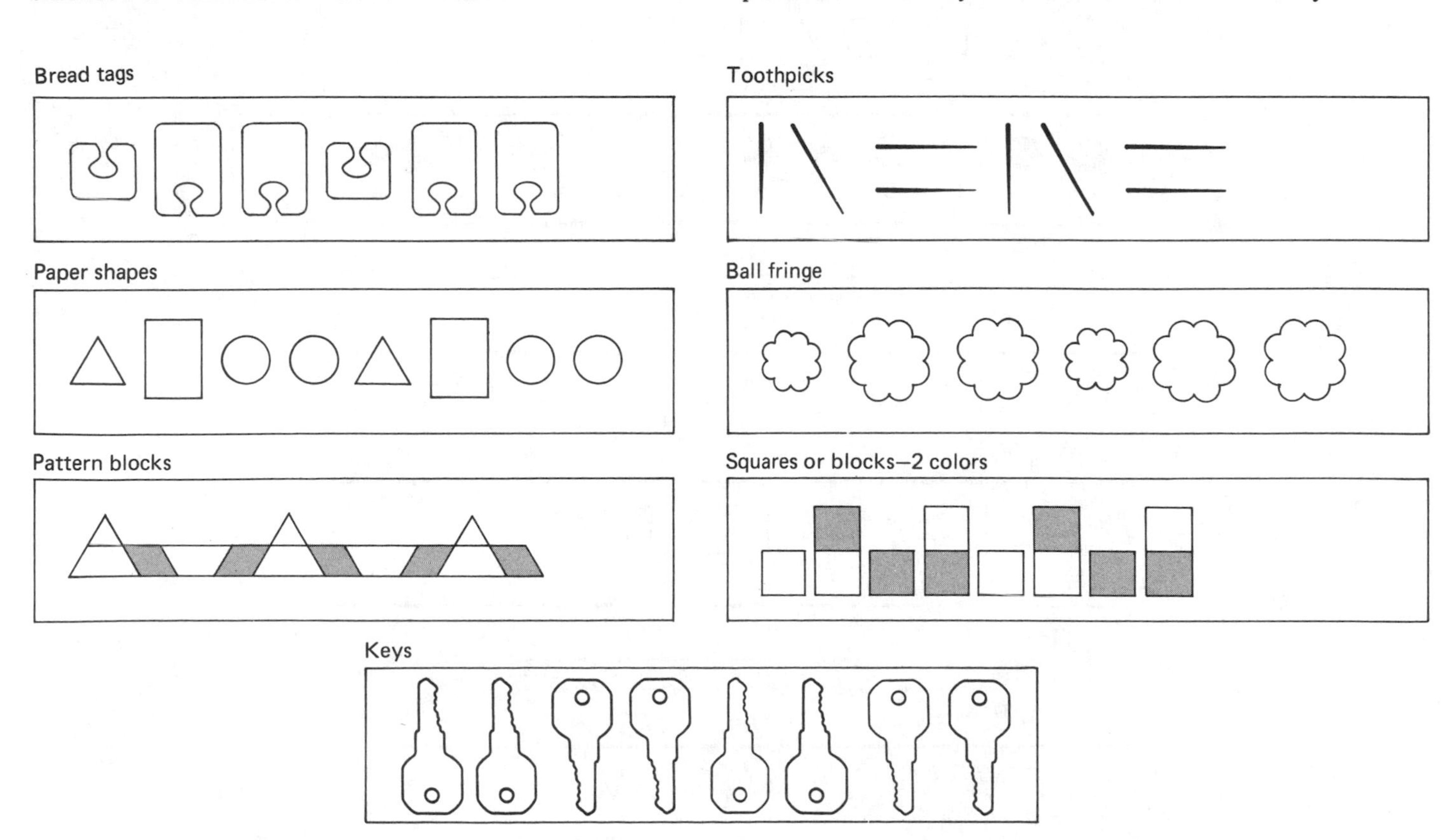

Figure 17.7
Examples of pattern cards drawn on tagboard: each pattern repeats completely and does not split in the middle of a core.

class sized sets of materials. Make 10 or 15 different pattern strips for each set of materials. With six to eight sets, your entire class can work at the same time, with small groups working with different patterns and different materials.

The *core* of a repeating pattern is the shortest string of elements that repeats. Notice in the example pattern cards that the core is always fully repeated and never only partially shown. If the core of a pattern was—oo, then a card might have—oo—oo (two repetitions of the core) but it would be ambiguous if the card showed—oo—oo— or —oo—.

A significant step forward mathematically is to see how two patterns constructed with different materials can actually be the same pattern. For example the first patterns in both Figures 17.7 and 17.8 can be "read" A-B-B-A-B-B- while the second patterns in each figure are A-B-C-C-A-B-C-C- patterns. Challenging students to translate a pattern from one medium to another or to find two patterns that are alike, even though made with different materials, are important activities that help students focus on the relationships that are the essence of repeated patterns.

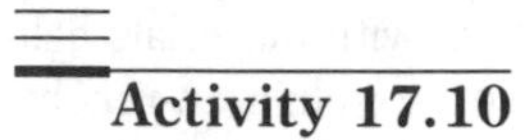

Activity 17.10

On a chalkboard or overhead show four or five patterns with different materials or pictures. Teach students to use an A, B, C method of "reading" a pattern. Half of the class can close their eyes while the other half uses the A, B, C scheme to read a pattern that you point to. After hearing the pattern,

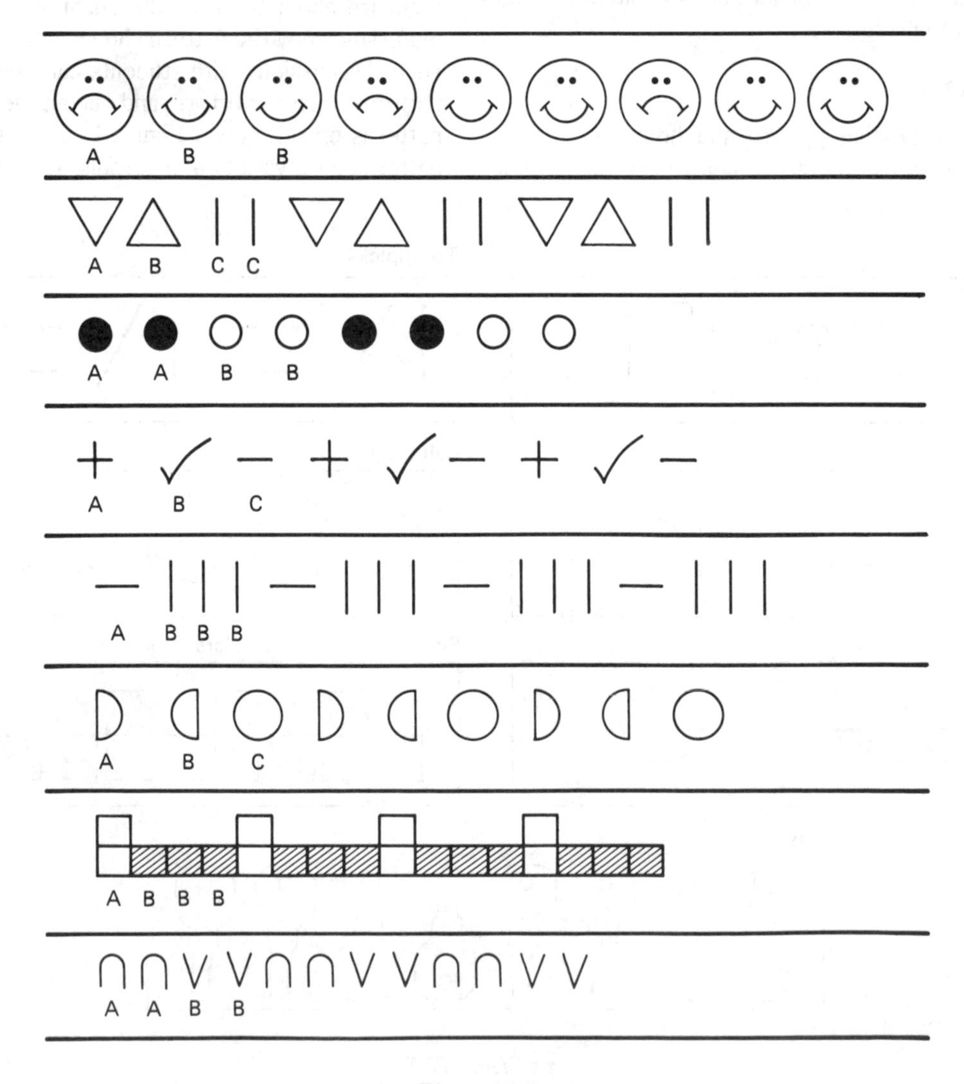

Figure 17.8
More examples of repeating patterns.

the students that had their eyes closed examine the patterns and try to decide which pattern was read. If two of the patterns in the list have the same structure, the discussion can be very interesting.

The following independent activity involves translation of a pattern from one medium to another which is another way of helping students separate the relationship in a pattern from the materials used to build it.

Activity 17.11

Have students make a pattern with one set of materials given a pattern strip showing a different set. This activity can easily be set up by simply switching the pattern strips from one set of materials to another. A similar idea is to mix up the pattern strips for four or five different materials and have students find strips that have the "same" pattern. To test if two patterns are the same children can translate each of the strips into a third set of materials or can write down the A, B, C pattern for each.

Two-Dimensional Patterns

Figure 17.9 illustrates how patterns can be developed on a grid instead of a straight line. Children at the primary level find completion of these patterns quite challenging. Pattern cards can be made by coloring or drawing on a piece of grid paper. If blank grids the same width as the pattern cards are laminated, then students can use colored blocks or colored squares of construction paper on the blank grids. This provides the same trial-and-error potential that was noted with repeating patterns.

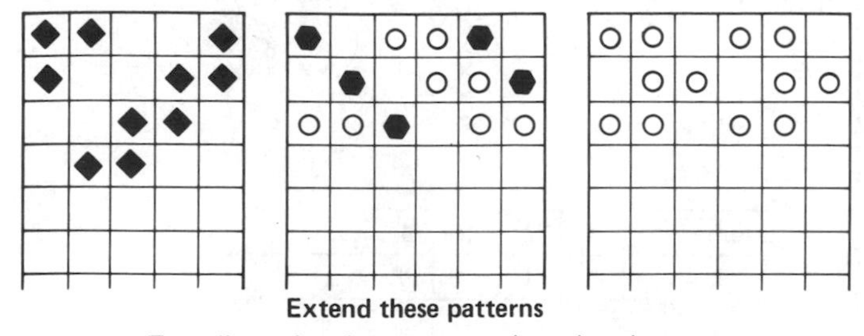

Extend these patterns

Two-dimensional patterns can be colored on 1-inch grid paper. Students can copy and extend them using colored cubes or squares of poster board.

Figure 17.9
Repeating patterns on a grid.

Growing Patterns

A significant step forward in patterning activities involves work with patterns that change from step to step rather than simply repeat over and over. When such "growing" patterns are developed with materials or drawings, each element of the pattern is usually a progression of some sort from the previous element of the pattern. The patterns are best if they can actually be constructed with objects or drawn on grids or dot paper. Several examples of this type of pattern are shown in Figure 17.10. Students should discuss how each "frame" or group in the pattern is different from the preceding group and how it can be formed by building onto the preceding arrangement. (Technically, these "growing" patterns are called *sequences.)*

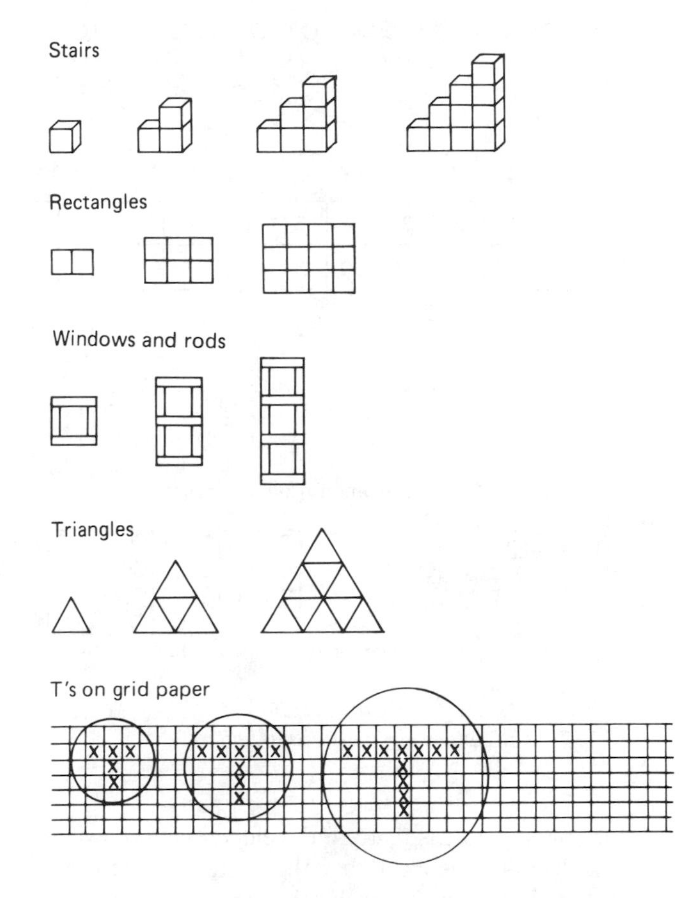

Figure 17.10
"Growing patterns" with materials or drawings.

The number of objects in each frame should be recorded in a chart or table as shown in Figure 17.11. The challenge is to discover not only spatial relationships in the physical patterns as they change or grow from frame to frame but corresponding numeric patterns in the chart. The physical or spatial patterns are described in terms of the objects or drawings. For example, each stairstep in Figure 17.10 can be made by adding a column of blocks onto the preceding stairsteps. There are two different places in the chart where you and your students can look for numeric relationships; first, in the progression from one frame to the next, and second, in the relationship between the frame number and the number of objects in that frame. With the help of the clues provided, see what numeric relationships you can find in Figure 17.11. You should also make charts for the patterns in Figure 17.10 and examine these for relationships.

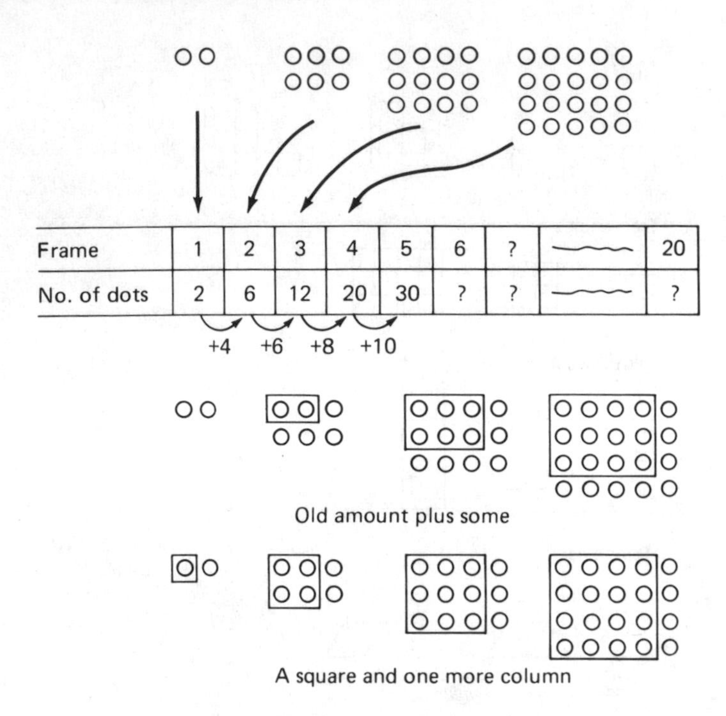

Frame	1	2	3	4	5	6	?	~~~	20
No. of dots	2	6	12	20	30	?	?	~~~	?

Figure 17.11
Relationships in a visual pattern.

With most patterns that are developed with drawings or physical objects, it is clear that the pattern could continue indefinitely. However, it quickly becomes unreasonable to construct all of the frames for lack of space or materials, and the numbers have a tendency to get large quickly. As students first begin to explore or build growing patterns they should be asked to predict what the tenth or perhaps twentieth frame will be like or how many elements it will have. Since the actual construction may not be reasonable, this challenge encourages a search for useful patterns and relationships. A relationship from one frame to the next is often easy to see. For example, with the triangles in Figure 17.10, the value of each successive frame can be found by adding successive odd numbers.

A numeric relationship between the frame number and the number of objects in the frame is the most powerful. If a rule can be discovered that can give the value of any frame in terms of the frame number, the number of objects in any frame can be determined without building or calculating all of the intermediate frames in the pattern. Have students develop numeric expressions for each frame using the frame numbers in the expressions. Work toward finding similar expressions for each frame with the frame number being the only part of the expression that changes. It may take much searching and experimenting for students to come up with an expression that is similar for each frame. Sometimes clues to this relationship can be found in the drawings more easily than in the charts. This search can be an exciting class or group activity. Do not be nervous or upset if students have difficulty. Encourage the search for relationships to continue, even if it takes more than one day. The search for and discovery of relationships is the most significant portion of these activities.

When students have discovered expressions for each frame that are the same except for frame numbers, write them with brackets around the frame numbers as shown in Figure 17.12. Notice that the bracketed numbers "vary" from one expression to the next while the other numbers in the expressions remain the same. Now, the bracketed numbers

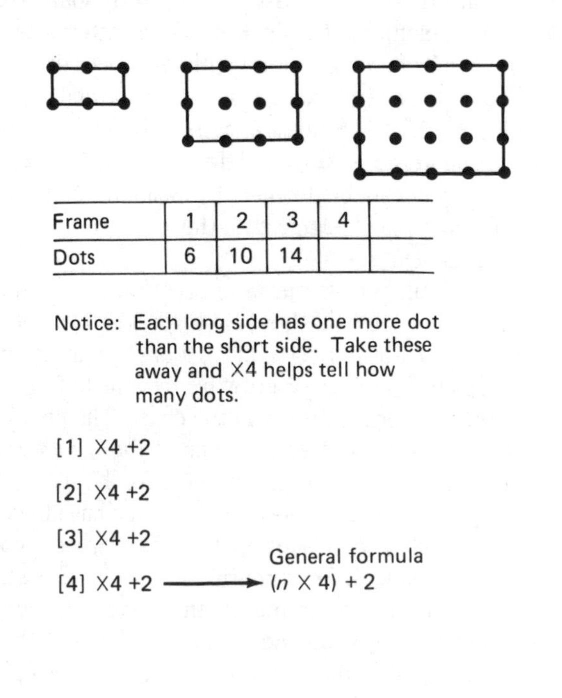

Frame	1	2	3	4	
Dots	6	10	14		

XXX
X

XXXX
XXXX
X
X

XXXXX
XXXXX
XXXXX
X
X
X

XXXXXX
XXXXXX
XXXXXX
XXXXXX
X
X
X
X

Frame	1	2	3	4	
X's	4	10	18	28	

Notice: If the tail part is added to the side of the top part, there is always a square and three more columns.

[1] X [1] + (3 X [1])

[2] X [2] + (3 X [2])

[3] X [3] + (3 X [3])

⋮ General formula

$n \times n + 3n = n^2 + 3n$

Figure 17.12
Generalizing relationships.

can be replaced by a letter or *variable* resulting in a general formula. The formula then defines a *functional relationship* between frame numbers and frame values. Even if these terms are not used at the time, the activity provides an excellent early example of function and variable for students to utilize later when those ideas are encountered.

It would be a good idea for you to explore each of the patterns in Figures 17.10, 17.11, and 17.12 to see if you can find a formula (functional relationship) for the value of each frame in terms of the frame numbers.

The calculator should be utilized at all times to promote discovery of relationships in growing patterns. For example, if the frame-to-frame relationship is observed, the number of items for a large frame number is easily calculated. The result can then be verified by building or drawing that particular frame.

A growing pattern that is just a little bit different is shown in Figure 17.13. It begins with a little square. Each successive frame is formed by building a new larger square onto the previous design. (Can you see how to continue drawing this pattern?) If the side of the first two little squares is 1 each, then the sides of each new square are the numbers of most interest in this pattern. For those squares shown in the figure, the sides are 1, 1, 2, 3, 5, 8, and 13. What would the side of the next square be? This series of numbers, known as the *Fibonacci sequence,* is named for an Italian mathematician, Fibonacci (ca. 1180–1250). The sequence occurs in a variety of living things. For example, if you count the sets of spirals that go in opposite directions on a pineapple or the seeds of a sunflower, the two numbers will be adjacent numbers in the Fibonacci sequence; usually 8 and 13 for a pineapple and 55 and 89 for sunflowers.

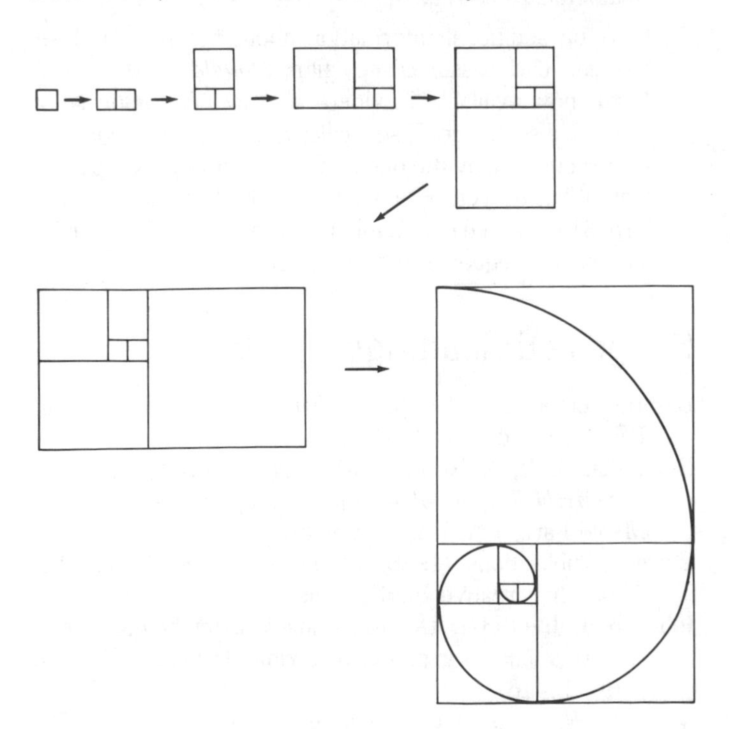

Figure 17.13
A growing pattern of squares: each new rectangle is a little closer to a "golden rectangle."

Another interesting fact about the Fibonacci sequence is that the ratio of adjacent numbers in the sequence gets closer and closer to a single fixed number known as the "golden ratio," a number very close to 1.618. Each larger rectangle in Figure 17.13 has sides in ratio a little closer to the golden ratio. A rectangle in that ratio is called a *golden rectangle,* examples of which can be found in most of the prominent examples of ancient Greek architecture as well as in much art and architecture through the ages. The spiral that is drawn in the last rectangle shown (made from quarter circles drawn in each square) is the same spiral found in the chambered nautilus shell.

Numeric Patterns

Drawings and manipulative objects should be a principal feature of patterning programs since they permit experimentation and trial-and-error approaches without threat. Students can validate their observations and conjectures without recourse to the teacher or answer book. However, many worthwhile patterns can be observed with numbers alone. These can be simple repeating patterns such as 1,2,1,2, ... Even very young children can use numbers in patterns like these. Generally, however, numeric patterns involve some form of progression. The pattern 1,2,1,3,1,4,1,5, . . . is a simple example that even young students can discover. Here is a list of just a few examples of numeric patterns that children can explore.

2, 4, 6, 8, 10, ...(even numbers or "add 2 each time")
1, 4, 7, 10, 13, ...(start with 1, add 3 each time)
1, 4, 9, 16, ...(squares: 1 × 1, 2 × 2, 3 × 3, etc.)
0, 1, 5, 14, 30, ...(add the next square number)
2, 5, 11, 23, ...(double the number and add 1)
2, 6, 12, 20, 30, ...(products of successive pairs of numbers)
3, 3, 6, 9, 15, 24, ...(add the last two numbers—a Fibonacci sequence)

The challenges in these patterns or sequences of numbers are not only to find and extend the pattern but to try to determine a general rule to produce the *n*th number in the sequence. Informal or exploratory approaches are similar to those described for growing patterns in the preceding section.

Patterns with the Calculator

The calculator provides another interesting approach to numeric patterns. Certain sequences of operations will generate patterns that children can explore and discover on their own. Listed here are a few examples, but there are many more.

Activity 17.12

Choose a start number between 0 and 9 and add a constant repeatedly to that number. Remember to use the automatic constant feature. For example, to start with 7 and add 4 repeatedly, you press 7 [+] 4 [=] [=] [=] . . . . What digits appear in the ones place? (1, 5, 9, 3, 7, 1, 5, 9, 3, . . .) How long is the pattern before it repeats? Are all patterns the same length? Are there shorter ones? Can you find one that is length 6? Why not? How does this change when the start number changes? How does it change when the skip number changes?

Activity 17.13

Repeat the preceding exploration, but this time use multiplication instead of addition. What do you notice? For multiplication, the first factor is generally the one stored. Therefore to start with 4, for example, and repeatedly multiply by 7, press 7 [×] 4 [=] [=] [=] . . . . In this exploration the calculator will "overload" relatively quickly. Since you are only interested in the values of the ones digit, how can you continue the pattern?

Activity 17.14

In either of the previous explorations, what happens in the tens digit? Is there a pattern there? Are they always the same? How can you predict what the pattern will be by selecting appropriate numbers? Can you come up with any rules or generalizations? Does the pattern continue into the hundreds?

Activity 17.15

Pick any one-digit number and multiply by 9, then 99, then 999, then 99999. What do you observe? Try other numbers such as 2, then 22, then 222, ...

These calculator patterns tend to be surprising and intriguing. They help students see a beauty and a regularity in mathematics. At the upper grades some exploration into why these patterns occur provides some nice mathematics involving place-value concepts and the use of variables. However, they are fun and valuable in themselves.

For Discussion and Exploration

1. Make or find a set of attribute materials. Explore several of the activities suggested here or found in the readings. For each activity, can you make a modification that will make the activity a little easier? Select an activity and see how challenging you can make it. For example, try the loop activities in a "guess my rule" mode (with the cards turned down) and with one or more of the cards having a logical connective such as *and, or,* or *not.*
2. Try some attribute activities with children. Be sure to be flexible in your lessons and in your planning. Be prepared to modify activities up or down in difficulty to match the abilities of the children.
3. Get a copy of the *Teacher's guide for Attribute games and problems* (Elementary Science Study, 1966). Although quite old, this book still contains some of the best attribute activities. In this book you will find activities for "people pieces," which are a set of 16 tiles after which the Woozle Cards were patterned. Any activity that can be done with people pieces can be done with Woozle Cards.
4. Do some patterning activities with a group of children. Be sure to plan a hands-on approach. For younger children use repeating patterns. For older children, try developing patterns which grow. Be careful not to move too quickly. Children at all ages need ample opportunity to construct ideas about patterns.
5. Invent a growing pattern. Use blocks, or pattern blocks, or grid or dot paper. See how many patterns or relationships you can find in your pattern. Can you generalize any of these numeric patterns? If you are having trouble with ideas, check the *Pattern factory* (Holden & Roper, 1980).
6. In the NCTM *Standards,* the third standard for both the K–4 and 5–8 curriculum is "Mathematics as Reasoning." Read these two standards. Discuss briefly how the activities of this chapter fit those goals. What else is meant by "mathematics as reasoning?"
7. Look up additional information about the Fibonacci sequence. The classic Disney film, *Donald in Mathmagic Land* (now available in video) also includes information about this sequence. Use a calculator and divide numbers in the sequence by the one that comes before. Keep doing that. What do you notice? Why is the last pattern listed on p. 315 referred to as a Fibonacci sequence? Try starting a Fibonacci sequence with other numbers.

Suggested Readings

Baratta-Lorton, M. (1976). *Mathematics their way.* Menlo Park, CA: Addison-Wesley.

Bezuszka, S. J., & Kenney, M. (1982). *Number treasury: A sourcebook of problems for calculators and computers.* Menlo Park, CA: Dale Seymour.

Creative Publications. (1986). *Hands on attribute blocks.* Palo Alto, CA: Creative Publications.

Johnson, J. J. (1987). Do you think you might be wrong? Confirmation bias in problem solving. *Arithmetic Teacher, 34*(9), 13–16.

Masalski, W. J. (1975). *Color cube activities.* Fort Collins, CO: Scott Resources.

Perl, T. (1974). *Relationshapes activity cards.* New Rochelle, NY: Cuisenaire Corporation of America.

Thompson, A. G. (1985). On patterns, conjectures, and proof: Developing students' mathematical thinking. *Arithmetic Teacher, 33*(1), 20–23.

Thompson, C. S., & Van de Walle, J. A. (1985). Patterns and geometry with Logo. *Arithmetic Teacher, 32*(7), 6–13.

Trotter, T., Jr., & Myers, M. D. (1980). Number bracelets: A study in patterns. *Arithmetic Teacher, 27*(9), 14–17.

Van de Walle, J. A. (1988). Hands-on thinking activities for young children. *Arithmetic Teacher, 35*(6), 62–63.

Van de Walle, J. A., & Holbrook, H. (1987). Patterns, thinking, and problem solving. *Arithmetic Teacher, 34*(8), 6–12.

Van de Walle, J. A., & Thompson, C. S. (1985). Promoting mathematical thinking. *Arithmetic Teacher, 32*(6), 7–13.

CHAPTER 18

EXPLORING BEGINNING CONCEPTS OF PROBABILITY AND STATISTICS

Probability and Statistics in Elementary Schools

The related topics of probability and statistics represent two of the most prominent uses of mathematics in our everyday lives. We hear about the possibility of contracting a particular disease, of having twins, winning the lottery, or living to be 100. Simulations of complex situations are frequently based on simple probabilities used in the design of highways, storm sewers, medical treatments, sales promotions, and spacecraft. Graphs and statistics bombard the public in advertising, opinion polls, reliability estimates, population trends, health risks, progress of students in schools and schools in school systems, to name only a few areas.

In order to deal with this information, students should have ample opportunity throughout the school years with informal yet meaningful experiences with the basic concepts involved. The emphasis from the primary level into high school should be placed on activities leading to intuitive understanding and conceptual knowledge rather than computations and formulas.

Organizations such as the Joint Committee on the Curriculum in Statistics and Probability of the American Statistical Association and the NCTM have made a strong effort to promote attention to the topics of probability and statistics in schools. In recent years, a number of factors can be identified that indicate an increase in both the quantity and quality of probability and statistics instruction in the elementary school:

An increased awareness of the importance of probability and statistics concepts and methods

An emphasis on experimental or simulation approaches to probability (instead of rules and formulas)

The use of simplified yet powerful plotting techniques to describe data visually without complicated procedures

The use of computers to (a) conduct thousands of random trials of experiments from flipping coins to simulating baseball batting performances, (b) to do the tedious work of constructing graphs, and (c) to almost instantly perform computations on large sets of numbers

With new approaches and attitudes toward the development of the conceptual knowledge of probability and statistics, it is almost certain that even more emphasis will be placed on these topics in the near future.

Developing Probability Concepts and Strategies

AN INTRODUCTION TO PROBABILITY

Two Experiments

Consider answering the following two questions by actually performing the experiments enough times to make a reasonable guess at the results.

1. *Tossing a cup.* Toss a paper or Styrofoam cup once or twice, letting it land on the floor. Notice that there are three possible ways for the cup to land: upside down, right side up, or on its side. If the cup were tossed this way 100 times, about how many times do you think it will land in each position?
2. *Flipping two coins.* If you were to flip one coin 100 times, you would expect that it would come up heads about as many times as tails. If two coins were tossed 100 times, about how many times do you think that they will both come up heads?

A quick way to conduct these experiments is to work in groups. If ten people each do 10 trials and pool their data, the time needed for 100 trials is not very long. Even if you do not actually do the experiments, jot down your predictions now before reading on.

Theoretical Versus Experimental Probability

In the cup toss, there is no practical way to determine the results before you start. However, once you had results for 100 flips, you would undoubtedly feel more confident in predicting the results of the next 100 flips. If you gathered data on that same cup for 1000 trials, you would feel even more confident. Suppose for example that your cup lands on its side 78 times out of 100. You might choose a "round" figure of 75 or 80 for the next 100 tosses. If, after 200 flips, there were 163 sideways landings you would feel even more confident of the four out of five ratio and predict about 800 sideways landings for 1000 tosses. The more flips, the more confident you become. You have determined an *experimental probability* of ⅘ or 80% for the cup to land on its side. It is experimental because it is based on the results of an experiment rather than a theoretical analysis of the cup.

In a one-coin toss, the best prediction for 100 flips would be 50 heads, although you would not be too surprised if actual results were between 45 and 55 heads. The prediction of 50% heads could confidently be made before you flipped the coin based on your understanding of a fair coin. Your prediction for two heads in the two-coin version may be more difficult. It is quite common for people to observe that there are three types of outcomes: both heads, both tails, and one of each. Based on this analysis they predict that two heads will come up about one-third of the time. (What did you predict?) The prediction is based on their analysis of the experiment, not on experimental results. When they conduct the experiment, however, they are surprised to find that two heads come up only about one fourth of the time. With this experiential base they might return to their original analysis and look for an error in their thinking.

There is only one way for two heads to occur and one way for two tails to occur. However, there are two ways that a head and a tail could result: either the first coin is heads and the second tails, or vice versa. As shown in Figure 18.1, that makes a total of four different outcomes, not three. The assumption that each outcome is equally likely was correct. Therefore the correct probability of two heads is one out of four or ¼, not ⅓. This *theoretical probability* is based on a logical analysis of the experiment, not on experimental results.

When we talk about probabilities we are assigning some measure of chance to an experiment. An *experiment* is any activity that has two or more clearly discernible results or *outcomes*. Both tossing the cup and tossing the two coins were experiments. Observing tomorrow's weather and shooting 10 free throws on the basketball court are also experiments. The collection of all outcomes is generally referred to as the *sample space*. As you have already seen, the toss of two coins has four outcomes in the sample space. Tomorrow's weather can be described in many ways; precipitation or no precipitation, or dry, rain, sleet, or snow. An *event* is any subset of the outcomes or any subset of the sample space. For the two-coin experiment, the event we were concerned with was getting two heads. For the free-throw shooting, we might be interested in the event of getting 5 or more out of the 10.

When all possible outcomes of a simple experiment are equally likely, the *theoretical probability* of an event is

$$\frac{\text{Number of outcomes in the event}}{\text{Number of possible outcomes}}$$

The *experimental probability* of any event is

$$\frac{\text{Number of observed occurrences of the event}}{\text{Total number of trials}}$$

All of the foregoing words and descriptions may be considered as basic or required knowledge of probability. On the other hand, informal activities and investigations over the elementary years can provide students with an experiential background for these basic concepts. This background will help give meaning to these and other definitions and formulas when introduced later at an appropriate time.

First coin	Second coin
Head	Head
Head	Tail
Tail	Head
Tail	Tail

Figure 18.1
Four possible outcomes for two coins.

Implications for Instruction

There are many reasons why an experimental approach to probability, actually conducting experiments and examining outcomes, is important in the classroom. An experimental approach

Is significantly more intuitive and conceptual. Results begin to make sense and certainly do not result from some abstract rule.

Eliminates guessing at probabilities and wondering "did I get it right?" Counting or trying to determine the number of elements in a sample space can be very difficult without some intuitive background information.

Provides an experiential background for examining the theoretical model. When you begin to sense that the probability of two heads is ¼ instead of ⅓, the analysis in Figure 18.1 seems more reasonable.

Helps students see how the ratio of a particular outcome to the total number of trials begins to converge to (get closer and closer to) a fixed number. For an infinite number of trials, the experimental and theoretical probabilities would be the same.

Develops an appreciation for a simulation approach to solving problems. Many real-world problems are actually solved by conducting experiments or simulations.

Is a lot more fun and interesting! It even makes searching for a correct explanation in the theoretical model more interesting.

Whenever possible, then, we should try to use an experimental approach in the classroom. If a theoretical analysis (such as with the two-coin experiment) is possible, then it should also be examined and results compared. Rather than correcting a student error in an initial analysis, we can let experimental results guide and correct student thinking.

Sometimes it is possible to develop theoretical explanations from results of experiments. For example, the results of the cup toss might be compared with the ratio of the height of the cup to the diameter of the opening. The cup can then be cut to different heights and the experiment repeated. A reasonable connection can be made between the ratio of height to top opening and the probability of a side landing. (For a better controlled experiment, try a variety of open-ended cylinders such as paper tubes and tin cans.)

DEVELOPING CONCEPTS OF PROBABILITY

Early Concepts of Chance

Children in kindergarten and primary grades need to develop an intuitive concept of chance. Many young children believe that an event will happen because "it's my favorite color" or "because it's lucky" or "because it did it that way last time." Many games such as *Candy Land* or *Chutes and Ladders* are very exciting for young children who do not comprehend that the outcomes are entirely random chance. When they finally learn that they have no control over the outcome, children begin to look for other games where there is some element of player determination.

Activity 18.1

Ask students to judge events as *certain, impossible,* or *maybe.* For example, consider these:

It will rain tomorrow.

Drop a rock in water and it will sink.

Trees talk to us in the afternoon.

The sun will rise tomorrow morning.

Three students will be absent tomorrow.

George will go to bed before 8:30 tonight.

You will have two birthdays this year.

Activity 18.2

Play simple games where the chance of one side winning can be controlled. Before playing the games, have students predict who will win and why. Afterwards, discuss why they think it happened that way. For example, the hockey game in Figure 18.2 starts with a counter in the center. Two players take turns spinning a spinner. The counter moves one space toward the goal that comes up on the spinner. Play the game with different spinners. As a variation, let students choose a spinner on each turn. Ask them to explain their choices.

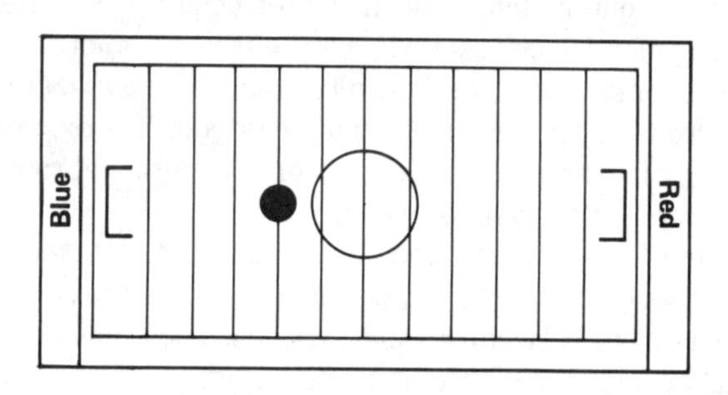

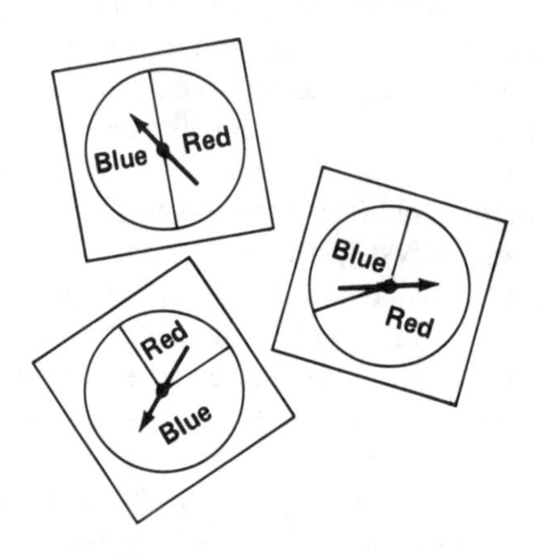

Figure 18.2
A simple game of chance played with different spinners helps young children with basic concepts of chance.

Activity 18.3

Have students make predictions about the outcomes of simple experiments using the terms *more, less, all,* and *none.* For example, show children how many each of red and yellow cubes you have in a bag. You will let children draw cubes one at a time and put them back each time. "If we do this 10 times, will there be more reds, less reds, all reds, or no reds?" Change the number of each color cube and repeat. The same activity can be done with spinners, rolling dice, drawing cards, or any random device that you can adjust. Include situations that are certain, such as using all yellow cubes in the bag.

Determining Probabilities for Simple Events

From a basic understanding that one event can be more or less likely than another, students can begin to predict specific ratios of outcomes of simple events. Before students have worked with part-to-whole ratios, use language such as "65 out of 100" instead of using fractional probabilities. A discussion of reasons for their predictions is always important. The experiment should then be conducted and results compared with expected outcomes.

Figure 18.3 illustrates a number of simple random devices that can be used for experiments. To get large numbers, let groups of students conduct the same experiment and tally their results. Group results can quickly be combined to get larger numbers. For most experiments, 100 or 200 total trials will give reasonable results. As you combine the totals for individual groups, have students notice how the results for smaller numbers of trials are all different and frequently are quite different than what might be expected.

The expression of a probability as a fraction in lowest terms requires some beginning understanding of equivalent ratios. A spinner that is one fourth blue has a probability of ¼ of coming up blue. If spun 100 times, the expectation is that it will come up blue about 25 times. If the same spinner is spun 160 times, the expectation is for about 40 spins of blue. The ratios 1 to 4, 25 to 100, 40 to 160, and 25% are equivalent. Probability provides an excellent opportunity to connect the part-to-whole ratio concept with a meaningful experience for children.

If a coin is tossed repeatedly, the long run will produce a ratio very close to one half heads and one half tails. But what happens in the short run? Do heads and tails alternate? If heads come up six times in a row, what is the probability that tails will come up next? The answers to questions such as these reflect an understanding of "randomness" and are worth discussing.

Occasionally have students list the outcomes of their experiments in a row, to show the order of the outcomes. Interestingly, truly random events do not alternate. They frequently appear in clusters or runs. If eight odd numbers in a row come up on a die roll, the chance remains exactly one half that the next roll will be odd. The die has no memory. The previous roll of a die cannot affect the next roll. It is very unlikely that even numbers will come up seven times in a row (one chance in 128). However, if that does happen, the chance for an even number on the next roll is still one half.

The computer program in Figure 18.4 will quickly generate row after row of X's and dashes and permit the user to adjust the probability of an X appearing. If P is entered as .25,

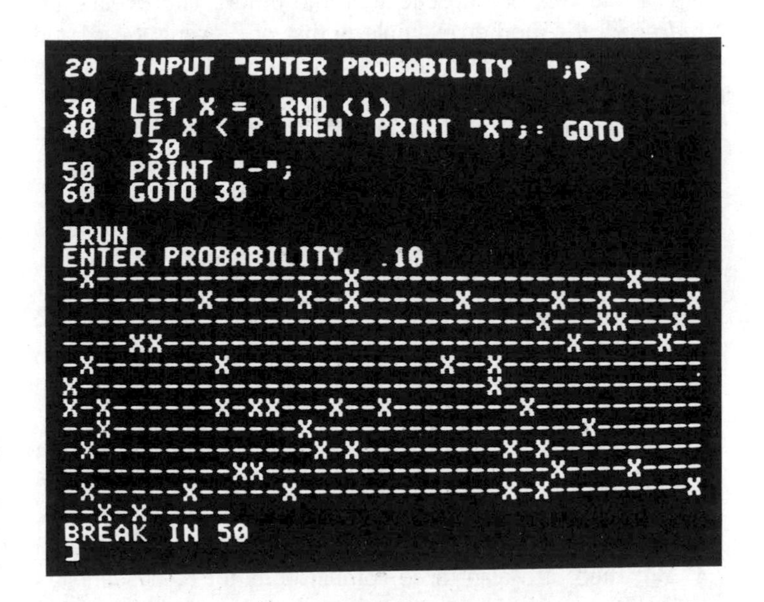

```
20  INPUT "ENTER PROBABILITY  ";P
30  LET X =  RND (1)
40  IF X < P THEN  PRINT "X";: GOTO
      30
50  PRINT "-";
60  GOTO 30

]RUN
ENTER PROBABILITY  .10
-X----------------X-----------------X----
---------X-----X--X------X-----X--X----X
-----------------------------X---XX---X-
----XX-------------------------X-----X--
-X--------X-------------X--X------------
X--------------------------X------------
X-X-------X-XX---X--X--------X----------
--X------------X-----------------X------
-X--------------X-X---------X-X---------
-----------XX------------------X----X---
-X------X-----X-------------X-X--------X
--X-X------
BREAK IN 50
]
```

Figure 18.4
The computer can easily be programmed to produce random outcomes with any probability.

Figure 18.3
There are many simple ways to produce random outcomes.

the X's will appear randomly with a probability of ¼. If the scroll is paused, students can see over 800 outcomes at a time displayed in rows of 40. Similar programs can easily be written to include more than two outcomes and can be designed to count the number of each.

Experiments with Two or More Independent Events

Flipping two coins and observing the result of each is an example of an experiment with two independent events. The flip of one coin has no effect on the other. The events are *independent.* Another example is that of drawing a card and spinning a spinner. Many interesting experiments involve two or more separate, independent events.

Determining the experimental probability of compound events is no different than for simple events. The experiment is performed numerous times, and the number of favorable results is compared to the total trials as before. The challenges come in trying to reconcile experimental results with theoretical ones.

To illustrate, suppose that a class is conducting an experiment to determine the probability of rolling a seven with two dice. They might tally their results in a chart showing each sum from 2 to 12 as a single event as in Figure 18.5(a).

The results of their experiment will show clearly that these events are not equally likely and in fact the sum of 7 has the best chance of occurring. To explain this, they might look for the combinations that make 7: 1 and 6, 2 and 5, and 3 and 4. But there are also three combinations for 8. It seems as though 8 should be just as likely as 7 and yet it is not.

Now suppose that the experiment is repeated. This time, for the sake of clarity, suggest that students roll two different colored dice and that they keep the tallies in a chart like the one in Figure 18.5(b).

The results of a large number of dice rolls indicate what one would expect, namely, that all 36 cells of this chart are equally likely. But there are more cells with a sum of 7 than any other number. Therefore they were really looking for the event which consists of any of the *six* ways, not three, that two dice can make a 7. There are six outcomes in the desired event out of a total of 36 for a probability of 6/36 or ⅙.

When investigating the theoretical probability of a compound event, it is useful to use a chart or diagram that keeps the two independent events separate and illustrates the combinations. The matrix in Figure 18.5(b) is one good suggestion when there are only two events. A tree diagram is another method that can be used with any number of events (Figure 18.6).

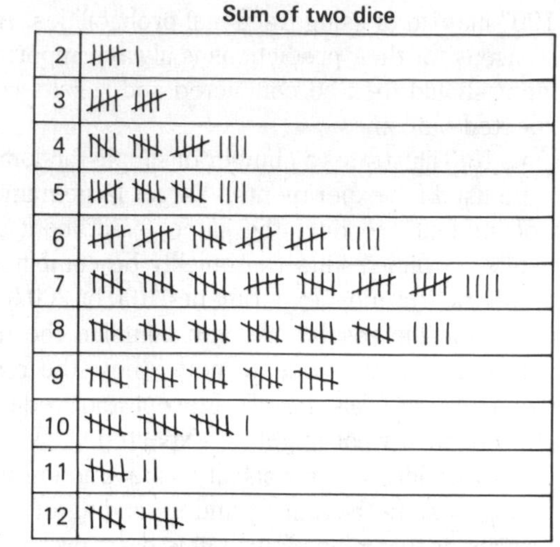

Sum of two dice

Sum	Tallies
2	卌	
3	卌 卌
4	卌 卌 卌 ||||
5	卌 卌 卌 ||||
6	卌 卌 卌 卌 卌 ||||
7	卌 卌 卌 卌 卌 卌 卌 ||||
8	卌 卌 卌 卌 卌 卌 ||||
9	卌 卌 卌 卌 卌
10	卌 卌 卌	
11	卌 ||
12	卌 卌

(a) Tallies that only account for the total.

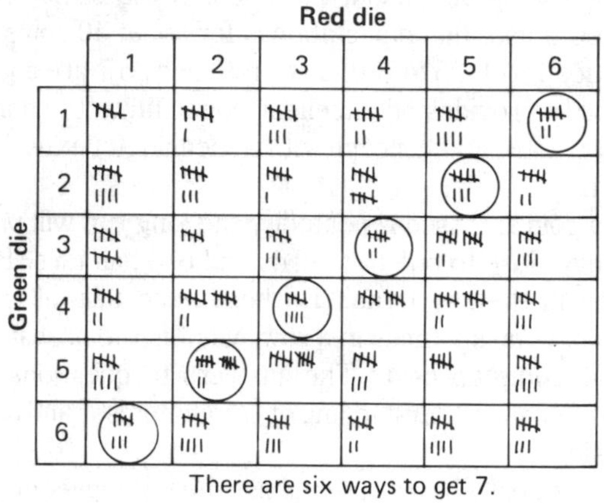

Red die

Green die	1	2	3	4	5	6
1	卌	卌		卌 |||	卌 ||	卌 ||||	卌 ||
2	卌 ||||	卌 |||	卌		卌 卌	卌 |||	卌 ||
3	卌 卌	卌	卌 |||	卌 ||	卌 卌 ||	卌 ||||
4	卌 ||	卌 卌 ||	卌 ||||	卌 卌	卌 卌 ||	卌 |||
5	卌 ||||	卌 卌 ||	卌 卌	卌 |||	卌	卌 ||||
6	卌 |||	卌 ||||	卌 |||	卌 ||	卌 ||||	卌 |||

There are six ways to get 7.

(b) Tallies that keep track of the individual dice.

Figure 18.5
Tally charts.

Activity 18.4

The following are examples of compound experiments with independent events. Determine the probability of

Rolling an even sum with two dice

Spinning blue AND flipping a cup on end

Getting two blues out of three spins (depends on spinner)

Having a tack OR a cup land "up" if each is tossed once

Getting AT LEAST two heads from a toss of four coins.

Words and phrases such as *and, or, at least,* and *no more than* can also cause children some trouble. Of special note is the word *or* since its everyday usage is generally not the same as the strict logical use found in mathematics. In mathematics, *or* includes the case of *both.* So, in the tack-and-cup example, the event includes tack up, cup up, and *both* tack *and* cup up.

Theoretical Probabilities with an Area Model

The method just suggested for determining the theoretical probability of a compound event is to list all possible outcomes and count those that are favorable; that is, those that make up the event. This is very useful and intuitive as a first approach.

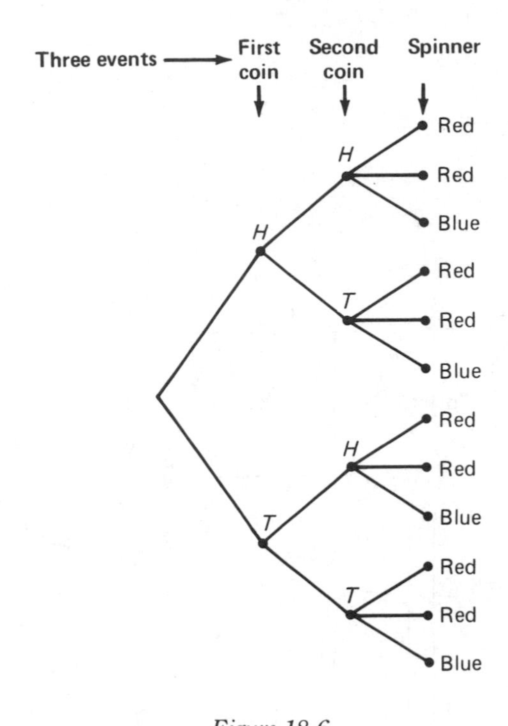

Figure 18.6
A tree diagram showing all possible outcomes for two coins and a spinner that is ⅔ red.

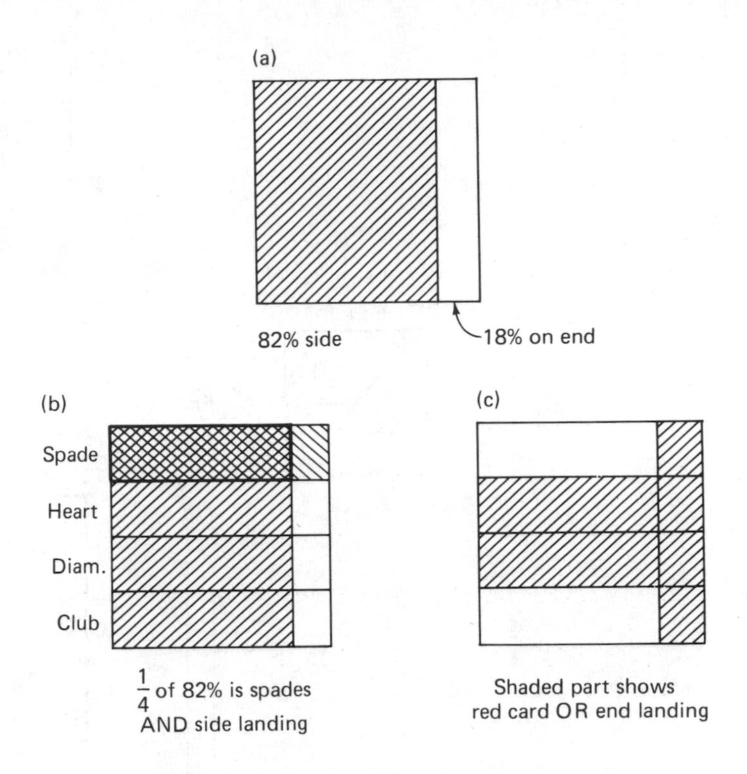

Figure 18.7
An area model for determining probabilities.

However, it has some limitations. First, what if the events are not all equally likely and are not made up of smaller events? An example is the cup toss. Second, it is difficult to move from that approach to even slightly more sophisticated methods. An area model approach has been used successfully with fifth-grade students. It is quite useful for some reasonably difficult problems (Armstrong, 1981). The following example will illustrate how an area model works.

Suppose that after many experiences, you have decided that your cup lands on the side 82% of the time. The experiment is to toss the cup and then draw a card from a deck. What is the probability that the cup will land on the side AND you will draw a spade? Draw a square to represent one whole. First partition the square to represent the cup toss, 82% and 18% as in Figure 18.7(a). Now partition the square in the other direction to represent the four equal card suits and shade ¼ for spades [Figure 18.7(b)]. The overlapping region is the proportion of time that both events, sideways and spades, happen. The area of this region is ¼ of 82% or 20.5%.

You can use the same drawing to determine the probabilities of other events in the same experiment. For example, what is the probability of the cup landing on either end OR drawing a red card? As shown in Figure 18.7(c), half of the area of the square corresponds to drawing a red card. This section includes the case of drawing a red card AND an end landing. The other half of the 18% end landings happen when a red card is not drawn. Half of 18% is 9% of the area. The total area for a red card OR an end landing is 59%.

The area approach is easy for students to use and understand for experiments involving two independent events where the probability of each is known. For more than two independent events, further subdivision of each region is required but is still quite reasonable. The use of AND and OR connectives is easily dealt with. It is quite clear without memorization of formulas, how probabilities should be combined.

Exploring Dependent Events

The next level of difficulty occurs when the probability of one event depends on the result of the first. For example, suppose there are two identical boxes. In one box is a dollar bill and two counterfeit bills. In the other box is one of each. You may choose one box and from that box select one bill without looking. What are your chances of getting a dollar? Here there are two events: selecting a box and selecting a bill. The probability of getting a dollar in the second event depends on which box is chosen in the first event. These events are *dependent,* not independent.

As another example, suppose that you are a prisoner in a faraway land. The king has pity on you and gives you a chance to leave. He shows you the maze in Figure 18.8. At the start and at each fork in the path, you must spin the spinner and follow the path that it points to. You may request the key to freedom be placed in one of the two rooms. In which room should you place the key to have the best chance of freedom?

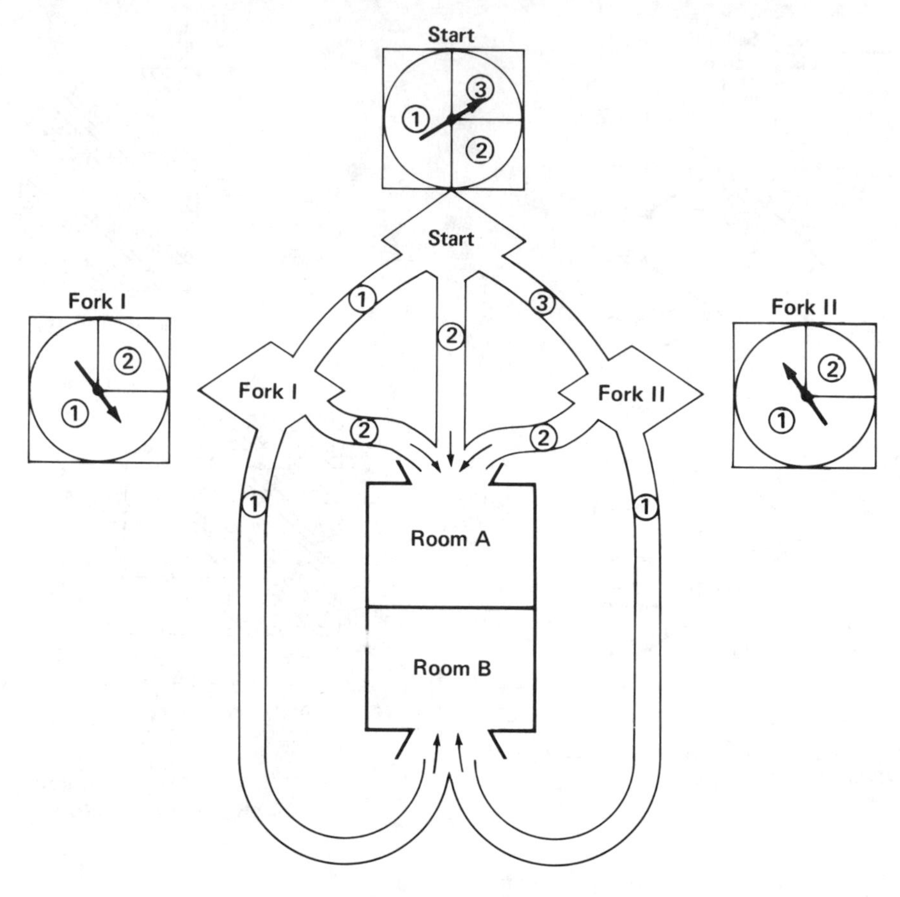

Figure 18.8
Should you place your key to freedom in Room A or Room B? At each fork the spinner determines your direction.

Notice that the probability of ending the maze in any one room is dependent on the result of the first spin.

Either of these two problems could be solved with an experimental approach, a simulation. A second approach to both problems is to use the area model to determine the theoretical probabilities. An area model solution to the prisoner problem is shown in Figure 18.9.

It would be good to stop at this point and try the area approach for the problem of the counterfeit bills. (The chance of getting a dollar is 5⁄12.)

The area model will not solve all probability problems. However, it fits very well into a developmental approach to the subject because it is conceptual, it is based on existing knowledge of fractions, and because more symbolic approaches can be derived from it. Figure 18.10 shows a tree diagram for the same problem with the probabilities of each path of the tree written in. After some experience with probability situations, the tree diagram model is probably easier to use and adapts to a wider range of situations. You should be able to match up each branch of the tree diagram in Figure 18.10 with a section of the square in Figure 18.9. Use the area model to explain why the probability for each complete branch of the tree is determined by multiplying the probabilities along the branch.

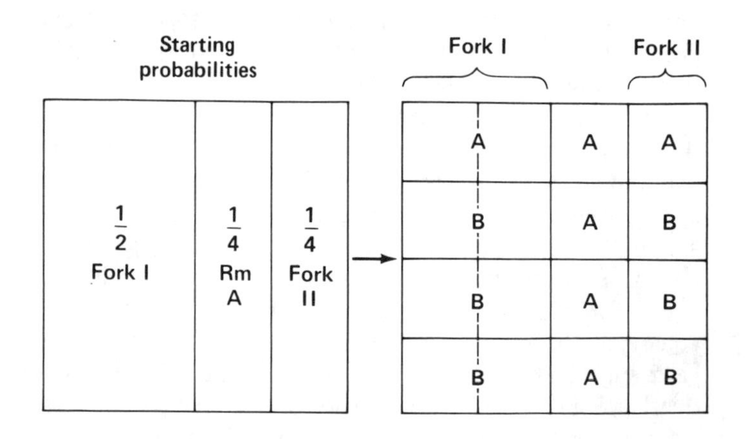

At – Fork I: $\frac{3}{4}$ of the time you will go to room B.
(Note – not $\frac{3}{4}$ of square, but $\frac{3}{4}$ of the times you go to Fork I)
– Fork II: $\frac{3}{4}$ of these times (or $\frac{3}{16}$ of total time) you will go to room B.
You will end up in room A $\frac{7}{16}$ of time, room B $\frac{9}{16}$ of the time.

Figure 18.9
Using the area model to solve the maze problem.

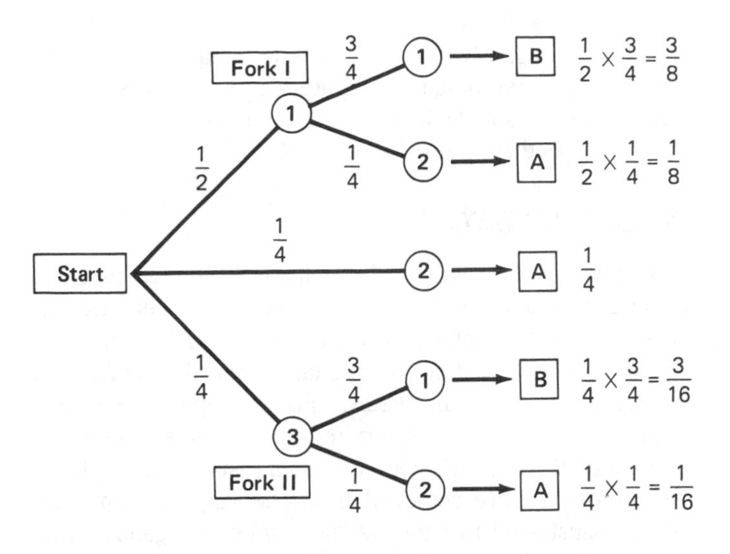

Figure 18.10
A tree diagram is another way to model the outcomes of two or more dependent events.

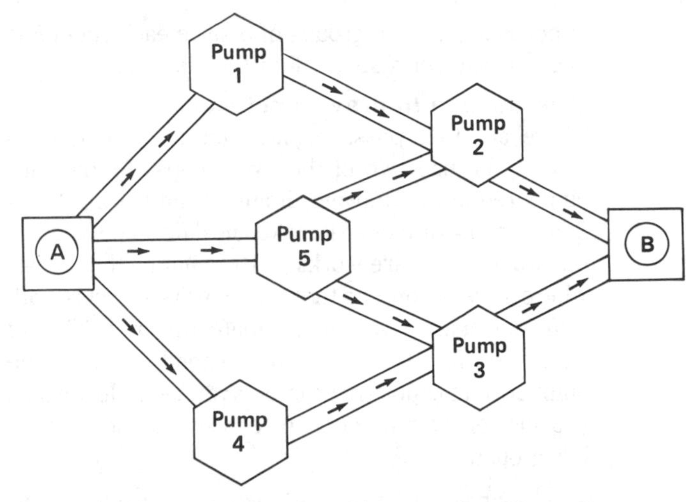

Figure 18.11
Five pumps, each with a 50% chance of failure. What is the probability that some path from A to B is working?

SIMULATIONS

A *simulation* is a technique used for answering questions or making decisions in complex situations where an element of chance is involved. A simulation is very much like solving a probability problem by an experimental approach. The only difference is that one must design a model that has the same probabilities as the real situation. For example, in designing a rocket, a large number of related systems each have some chance of failure. Various combinations of failures might cause serious problems with the rocket. Knowing the probability of serious failures will help decide if redesign or backup systems are required. It is not reasonable to make repeated tests of the actual rocket. Instead, a model that simulates all of the chance situations is designed and run repeatedly, most likely with the help of a computer. The computer model can simulate hundreds or even thousands of flights and an estimate of the chance of failure can be made.

Many real-world situations lend themselves to simulation analysis. In a business venture, the probability of selling a product might change depending on a variety of chance factors, some of which can be controlled or changed and others not. Will advertising help? How much chance is there that a competitor will enter the market? Should high cost materials be used? What location provides the best chance of sales? If a reasonable model can be set up that simulates these factors, then an experiment can be run before actually entering into the venture to determine the best choices.

A Model for Conducting Simulations

The following problem and model are adapted from the excellent materials developed by the Quantitative Literacy Project (Gnanadesikan, Schaeffer, & Swift, 1987). In Figure 18.11, a diagram shows water pipes for a pumping system connecting A to B. The five pumps are aging, and it is estimated that at any given time the probability of pump failure is ½. If a pump fails, water cannot pass that station. For example, if pumps 1, 2, and 5 fail, water can flow through 4 and 3. Consider the following questions that might well be asked about such a system:

What is the probability that water will flow at any time?

On the average, about how many stations need repair at any time?

What is the probability that the 1–2 path is working at any time?

For any simulation, a series of steps or model can serve as a useful guide.

1. **Identify key components and assumptions of the problem.**
   The key component in the water problem is the condition of a pump. Each pump is either working or not. The assumption is that the probability that a pump is working is 1/2.
2. **Select a random device for the key components.**
   Any random device can be selected that has outcomes with the same probability as the key component, in this case the pumps. Here a simple choice might be tossing a coin with heads representing a working pump.
3. **Define a trial.**
   A trial consists of simulating a series of key components until the situation has been completely modeled one time. In this problem, a trial could consist of tossing a coin five times, each toss representing a different pump.
4. **Conduct a large number of trials and record the information.**
   For this problem it would be good to keep the record

of heads and tails in groups of 5 since each set of 5 is one trial and represents all of the pumps.

5. **Use the data to draw conclusions.**
There are four possible paths for the water, each flowing through two of the five pumps. As they are numbered in the drawing, if any of the pairs 1–2, 5–2, 5–3, or 4–3 are open it makes no difference whether the other pumps are working. By counting those trials where at least one of these four pairs of coins both came up heads, we can estimate the probability of water flowing. To answer the second question, the number of tails per trial can be averaged. How would you answer the third question concerning the 1–2 path being open?

Steps 4 and 5 are the same as solving a probability problem by experimental means. The problem solving and interesting aspects of simulation activities are in the first three steps where the real-world situation is translated into a model. Translation of real-world information into models is the essence of applied mathematics.

Here are a few more examples of problems that can be solved by simulation and are easy enough to be tackled by middle school students.

In a true-false test, what is the probability of getting 7 out of 10 questions correct by guessing alone? (Key component: answering a question. Assumption: chance is ½ of getting it correct.) What if the test were multiple choice with four choices?

In a group of five people, what is the chance that two were born in the same month? (Key component: month of birth. Assumption: all 12 months are equally likely.)

Casey's batting average is .350. What is the chance he will go hitless in a complete nine-inning game. (Key component: getting a hit. Assumptions: probability of a hit for each at-bat is .35. Casey will get to bat four times in the average game.)

Krunch-a-Munch cereal packs one of three games in each box. About how many boxes should you expect to buy before you get a complete set? (Key component: getting one game. Assumption: each game has a ⅓ chance. Trial: Use a ⅓ random device repeatedly until all three outcomes appear. The average length of a trial answers the question.) Answer this question: What is the chance of getting a set in six or fewer boxes?

Helping Students Describe and Analyze Data

Statistics includes collecting, organizing, describing, displaying, and interpreting data. It also involves making predictions and judgments based on this information. For children in school, these processes are not only relevant, interesting, and important for daily living, they constitute a real form of problem solving. A variety of simple techniques for graphing and making sense of data are quite simple and accessible to elementary school children. These same techniques can be directly applied to the real world.

COLLECTING DATA

One of the most important rules to follow in conducting graphing and statistics activities is to let students gather their own data. Tables of numbers produced in textbooks tend to be sterile and uninteresting. Real data, gathered by children, is almost always more interesting. For example, one class of students gathered data concerning which cafeteria foods were most often thrown in the garbage. As a result of these efforts, certain items were removed from the regular menu. The activity illustrated to students the power of organized data, and it helped them get food that they liked better.

Sources of Data

There are all sorts of ways for students to gather data that may be of interest to them. Some examples are suggested here but these represent only a few ideas. Use your imagination, the interests of your students, and special events and activities in your class, school, and community.

Activity 18.5

Classroom Surveys. One of the easiest ways to get data in the primary grades is to get one piece of information from each student in your class. The resulting information will have manageable numbers and everyone will be interested. Here are some ideas:

Favorites. TV shows, fruit, season of the year, color, football team, pet, ice cream. When there are lots of possibilities, restrict the number of choices.

Numbers. Bus number, number of pets, sisters, hours of sleep, birthday (month or day in month), time to go to bed.

Measures. Height, weight, and other body measures, long jump distances, length of name, time to button sweater, number of beans in a "handful," seeds in a slice of watermelon, weight of a potato (or measures of any object that students could bring from home).

Activity 18.6

School or Grade-Level Surveys. Help students prepare simple questionnaires that can be used in all classes at your grade level or around the school. By using other classes, data from one class or grade level can be compared to others. Gathering school data can also involve sampling techniques such as randomly selecting 10 students per grade level instead of surveying everyone. Consider school issues such as cafeteria likes and dislikes, preferred lunch order, or use of the playground or gym. Political or social issues that all students may know about are also useful and allow you to integrate other subject areas.

Activity 18.7

School Data. Besides surveys, the school has a wealth of interesting data. Attendance by day of week, by grade level, and by month is one example. What materials are used and how much, how many tests are given in what grades and in what subjects, how many parents attend PTA, are a few additional ideas. Some "people data" can be compared to similar statistics for the population at large. Examples include left-handed people, eye color, race, average family size, or years living at present address.

Activity 18.8

Consumer Data. The ingredients, prices, weights, and volumes of popular grocery items such as cereal, candy bars, paper towels, or laundry soap provide all sorts of interesting data. Catalogs and menus are another method of getting consumer data in the classroom.

Activity 18.9

Other Sources. Count various things in the newspaper (number of letters in headlines, number of vowels in 100 words, number of common words such as "and" used on a page). Almanacs, sports records, and assorted government publications can provide a wide variety of interesting statistics.

PICTURING DATA

Bar Graphs

Bar graphs are one of the first ways to group and present data and are especially useful in grades K to 3. At this early level, bar graphs should be made so that each bar is made of countable parts such as squares, objects, or pictures of objects. No numeric scale is necessary. Graphs should be simple and quickly constructed. As shown in Figure 18.12, simple bar graphs can be real graphs, picture graphs, or symbolic graphs.

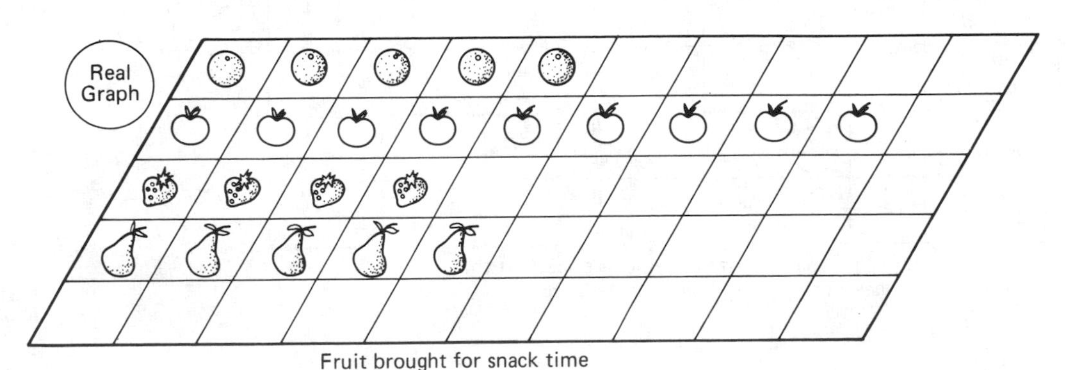

Picture Graph

Favorite farm animals

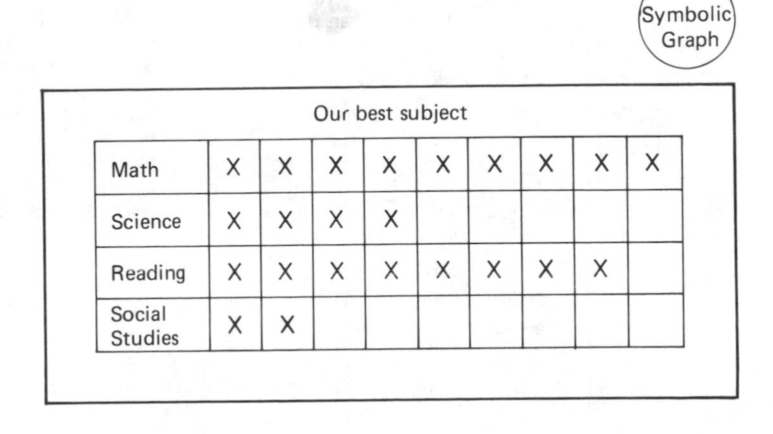

Figure 18.12
Three types of bar graphs: Real, picture, symbolic.

Real graphs use the actual objects being graphed. Examples include types of shoes, sea shells, or books. When making a real graph, each item should be placed in a square so that comparisons and counts are easily made.

Picture graphs use a drawing of some sort that represents what is being graphed. Students can make their own drawings or you can duplicate drawings to be colored or cut out to suit particular needs.

Symbolic graphs use something like squares, blocks, or X's to represent the things being graphed.

It is easier for young students to understand real and picture graphs but exposure to all different types is important. To quickly make a graph of class data, follow these steps:

1. Decide on what groups of data will make the different bars. It is good to have from two to six different bars in a graph.
2. Everyone should decide on or prepare their contribution to the graph before you begin. For real or picture graphs, the object or picture should be ready to be placed on the graph. For symbolic graphs, students should write down or mark their choice.
3. In small groups have students quickly place or mark their entry on the graph. A graph mat can be placed on the floor, or a chart prepared on the wall or blackboard (Figure 18.13). If tape or pins are to be used, have these ready.

By following these steps, a class of 25 to 30 students can make a graph in less than 10 minutes, leaving ample time to use it for questions and observations.

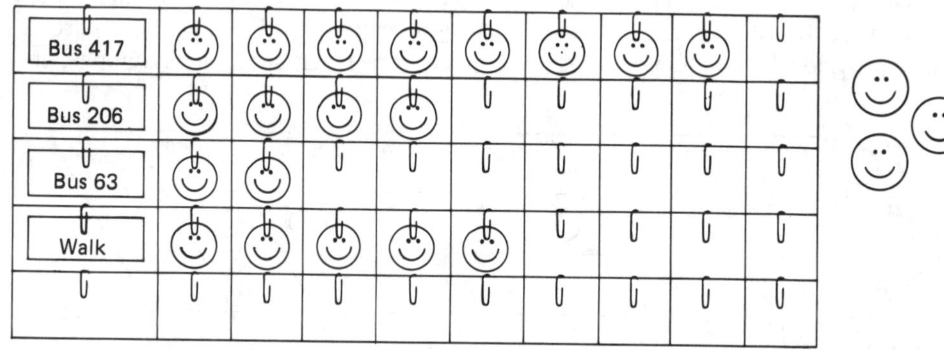

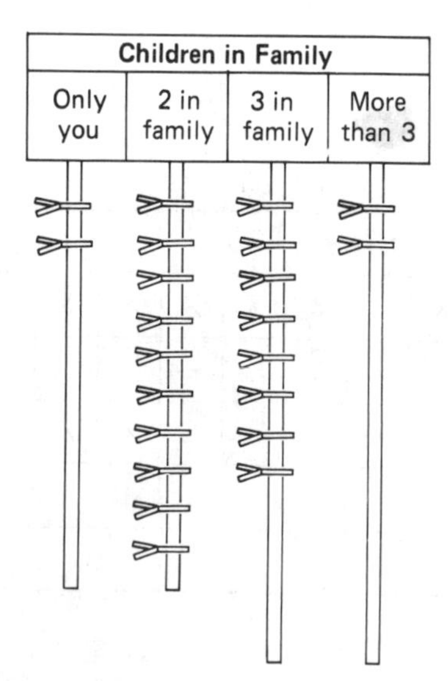

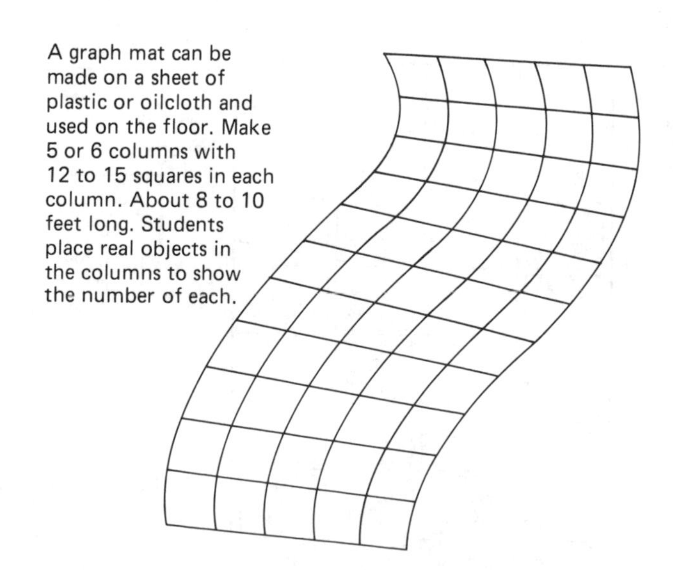

Figure 18.13
Some ideas for quick graphs that can be used over and over.

Questions like the following should be considered for every graph that is made in the early grades:

Which has the most? least?

Are there more _____ or more _____?

Are there less _____ or less _____?

Are there any the same?

How many _____ and _____ are there together?

How many more (or less) _____ are there than _____?

Children in the intermediate grades can also profit from making and using bar graphs. Individual groups can graph data that they gather themselves rather than one element per person. As part of their task, let students decide what the categories should be, what scales should be used, and whether to use pictures, squares, or continuous bars. Single pictures or squares on the graph might represent 5, 10, or 100 things rather than 1.

Line Plots and Line Graphs

When data is grouped along a numeric scale, a line plot can help convey trends or relationships. *Line plots* are useful for *counts* of things along a numeric scale. To make a line plot, a number line is drawn and an X is made above the corresponding number for each value. Numbers can be rounded before they are plotted so that the points will not be too spread out. (Line plots replace the older bar graphs called histograms.)

A *line graph* (very similar to a line plot) is appropriate for indicating trends or relationships when there is a single numeric value associated with equally spaced points along a continuous numeric scale. By *continuous* is meant that values on the graph could exist for any point between numbers on the scale. For example, a line graph might be used to show how the length of the flagpole shadow changed from one hour to the next during the day. Even though a point is plotted only at one-hour intervals, a shadow with some length did exist at every minute or even every second between the hours. An example of a line plot and a line graph are shown in Figure 18.14.

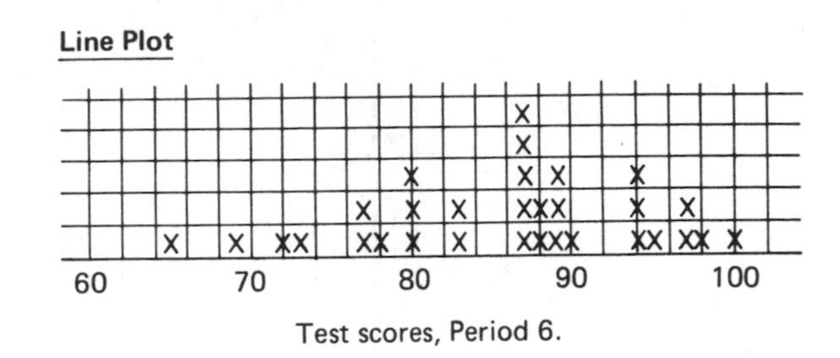

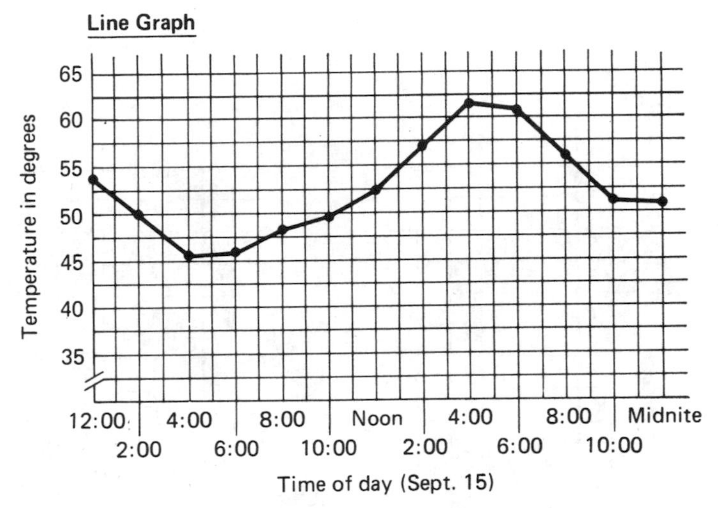

Figure 18.14
Examples of a line plot and a line graph. Notice that the horizontal scale must show some progression and is not just a grouping as in a bar graph.

Stem-and-Leaf Plots

Stem-and-leaf plots are a newer form of bar graph where numeric data is plotted by using the actual numerals in the data to form the graph. By way of example, suppose that the American League baseball teams had posted the following win records over the past season.

Baltimore	45	Milwaukee	91
Boston	94	Minnesota	98
California	85	New York	100
Chicago	72	Oakland	101
Cleveland	91	Seattle	48
Detroit	102	Toronto	64
Kansas City	96	Texas	65

If the data is to be grouped by tens, then list the tens digits in order and draw a line to the right as in Figure 18.15(a). These form the "stem" of the graph. Next go through the list of scores and write the ones digits next to the appropriate tens digit as in Figure 18.15(b). These are the "leaves." There is no need to count or group the data ahead of time. The process of making the graph does it all for you. Furthermore, every piece of data can be retrieved from the graph. (Notice that stem-and-leaf plots are best made on graph paper so that each digit takes up the same amount of space.)

To provide more information, the graph can be quickly rewritten, ordering each leaf from least to most [Figure 18.15(c)]. In this form it may be useful to identify the number that belongs to a particular team, indicating its relative place within the grouped listing.

Stem-and-leaf graphs are not limited to two-digit data. For example, if the data ranged from 600 to 1300, the stem could be the numerals from 6 to 13 and the leaves made of two-digit numbers separated by commas. If the ones digit is not important, round the data to the nearest ten and use only the tens digit in the leaves. Figure 18.16 shows the same data in two different stem-and-leaf plots.

Figure 18.17 illustrates two additional variations. When

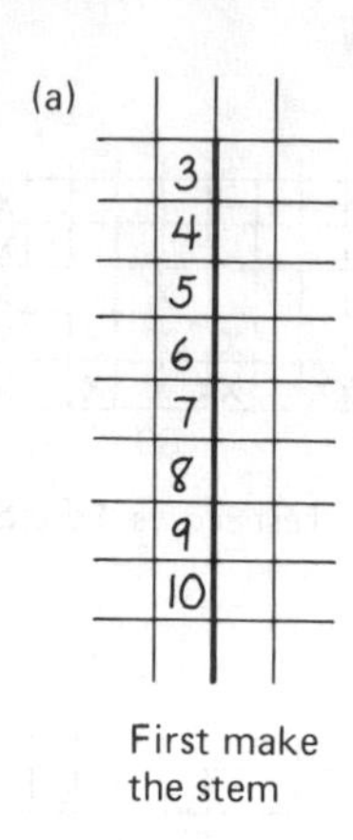

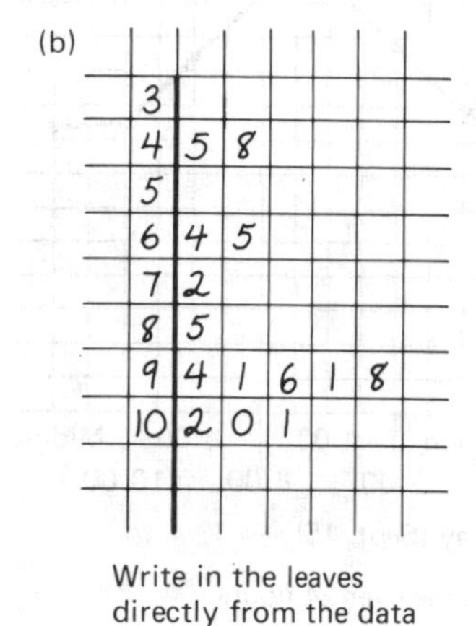

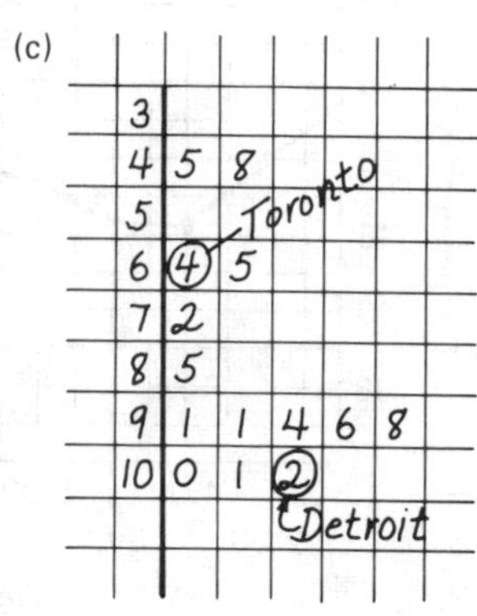

Figure 18.15
Making a stem-and-leaf plot.

(a) **Gallons of water**

6	29, 17, 83, 40, 82
7	68, 08, 95, 47, 12, 03, 18, 50, 54
8	44, 13, 29,
9	14, 83, 91, 27, 56, 30,
10	39, 00, 80, 73, 46, 10, 06
11	93, 11, 09, 25, 36, 66, 68, 57, 32
12	17, 20, 31, 26
13	

(b) **Tens of gallons of water**

6	3, 2, 8, 4, 8
7	7, 1, 5, 1, 0, 2, 5, 5
8	0, 4, 1, 3
9	1, 8, 9, 3, 6, 3
10	4, 0, 8, 7, 5, 1, 1
11	9, 1, 1, 3, 4, 7, 7, 6, 3
12	2, 2, 3, 3
13	

Figure 18.16
In the bottom plot the data from above are rounded to the nearest ten.

Test Scores

Mrs. Day		Mrs. Knight
4	5	
	•	9
2 3	6	
7 7 8	•	5
3 0 4 2 4	7	1 0
7 9 5	•	8 6 9 9
3 4 1	8	4 0 1 3 1 2
5 8 7	•	9 5
	9	3 1 0
9 6	•	7
0 0	10	0

Figure 18.17
Stem-and-leaf plots can be used to compare two sets of data.

two sets of data are to be compared, the leaves can extend in opposite directions from the same stem. In the same example, notice that the data is grouped by fives instead of tens. When plotting 42 the 2 is written next to the 4; for 47, the 7 is written next to the dot below the 4.

Stem-and-leaf plots are significantly easier for students to make than bar graphs, all of the data are maintained, they provide an efficient method of ordering data, and individual elements of data can be identified. For these reasons, stem-and-leaf plots are considered by many to be preferable to bar or line graphs (Landwehr, Swift, & Watkins, 1987).

Circle Graphs

A *circle* or *pie graph* is used when a total amount has been partitioned into parts and interest is in the ratio of each part to the whole and not so much in the particular quantities. In Figure 18.18, two graphs each show the percentages of students with different numbers of siblings. One graph is based on classroom data and the other on school data. Since the pie graphs display ratios not quantities, the small class data can be compared to the large school data. That could not be done with bar graphs.

Circle graphs are almost always made from data converted to percentages of the total. Helping students with the details of constructing pie graphs can easily involve more effort than the resulting graphs merit. After calculating all of the required percentages, each must be multiplied by 360 degrees to determine the size of that sector on the circle. Added difficulties surround the use of a protractor. Even with a calculator (and there is no

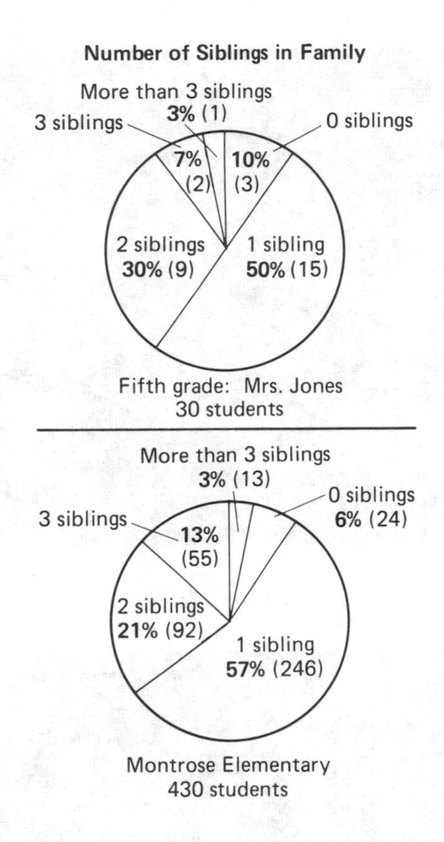

Figure 18.18
Circle graphs show ratios of part to whole and can be used to compare ratios.

excuse for doing this with pencil and paper) most students will feel bogged down with the tedium.

Today there are computer software packages that will make circle graphs directly from raw data. Businesses that require circle graphs to display data most certainly are using computer-generated graphs and the same approach is appropriate for students. A noncomputer solution to making circle graphs is to provide students with tagboard percentage disks duplicated from the master in Appendix C. A circle can be traced around the disk and the percentage marks used as a guide to making the graph without any use of angles.

Computer-Generated Graphs

An assortment of computer software programs are now available that permit students to produce numeric data (a), line (b), bar (c), and circle (d) graphs simply by entering the data and specifying groupings, scale values, and other data. The graphs shown in Figure 18.19 were generated by the *Logo data tool kit* (Fitch, 1987). Not only are these computer graphing packages practical graphing tools, they serve a valuable teaching function. An unfortunate amount of time and effort can go into directing students how to make graphs: how to select appropriate scales, where to put the numbers, how to convert fractional parts to sectors of circles, and so on. What is much more important than this mechanical but tedious and difficult work of making graphs is the before and after thinking that is involved.

Before making a graph, decisions need to be made concerning the data and the purpose of the graph. Is this data appropriate for a line graph, or would a stem-and-leaf plot be better? Will a circle graph convey the information required? How should the data be grouped to best convey the desired message about this data or to best understand it? The computer will not make these decisions but anyone interested in organizing data for personal or public consumption must. Since making a single graph can require a significant amount of student time and effort, teachers tend to make the pregraphing decisions for their students. The computer permits a trial-and-error approach to pregraph decisions. Students can painlessly consider a number of different graphs made with different parameters and discuss the pros and cons of each.

The thinking that goes on after a graph is made is also important. Understanding what a graph tells about a set of data is the skill that is necessary to interpret a significant amount of the data that we see daily. While discussions of graphs can and should happen without a computer, computer graphing programs may encourage more and wider variations in the graphs that students experience while in school.

DESCRIPTIVE STATISTICS

Averages

The term *average* is heard quite frequently in everyday usage. Sometimes it refers to an exact arithmetic average as in "the average daily rainfall." Sometimes it is used quite loosely, as in "she is about average in height." In either situation, an *average* is a single number or measure that is descriptive of a larger collection of numbers. If your test average is 92, it is assumed that somehow all of your test scores are reflected by this number. The mean, median, and mode are each specific types of averages or *measures of central tendency*. While other averages exist, these three are the ones that are generally taught in the elementary school.

The *mode* is that value or values that occur most frequently in the data set. Of these three statistics, the mode is the least useful and could perhaps be ignored completely. Consider the following set of numbers:

1, 1, 3, 5, 6, 7, 8, 9

The mode of this set is 1. In this example, the mode is not a very good description of the set and that is often the case. If the 8 in this string of numbers were a 9, a change of only one, there would be two modes. If one of the 1's were changed to a 2, there would be no mode at all. That is, the mode is a statistic that does not always exist, does not necessarily reflect the center of the data, and can be "unstable" or changeable with very small changes in the data.

The *mean* is computed by adding all of the numbers in the set and dividing the sum by the number of elements added. This is the statistic that is sometimes referred to

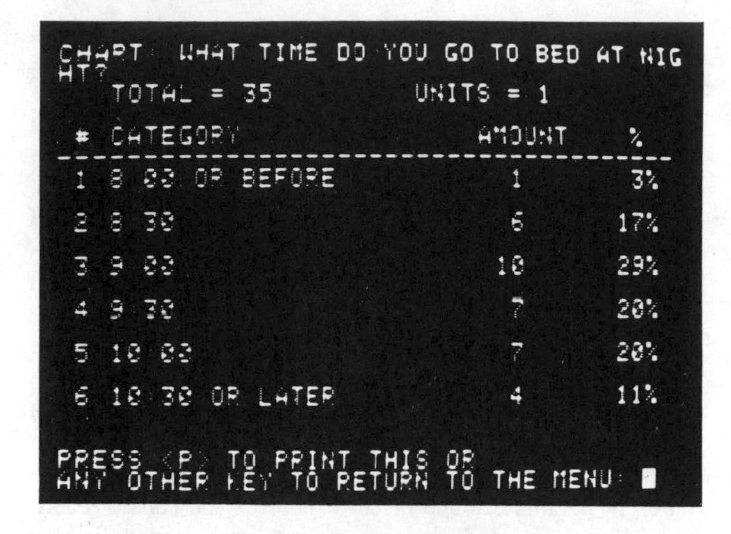

Figure 18.19a

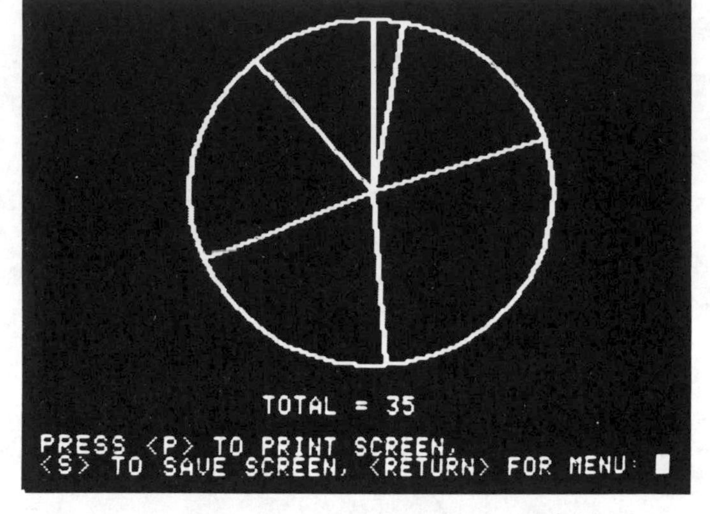

Figure 18.19b

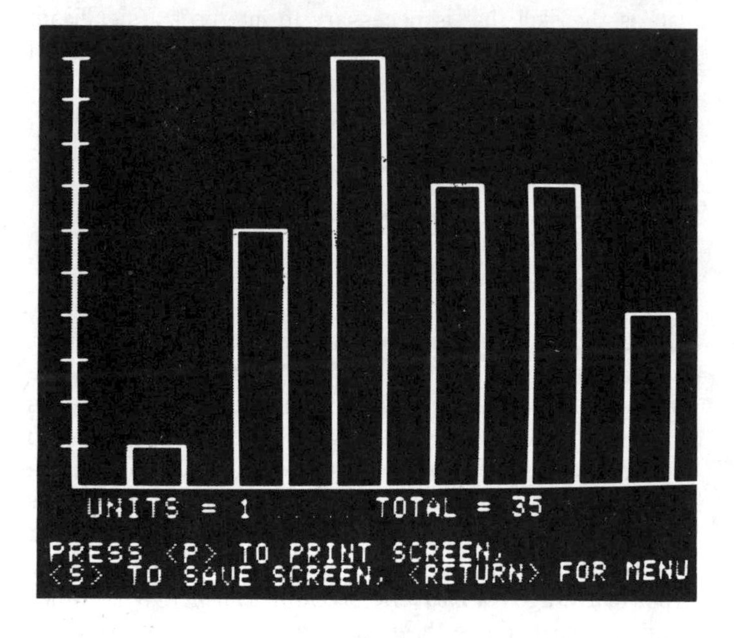

Figure 18.19c

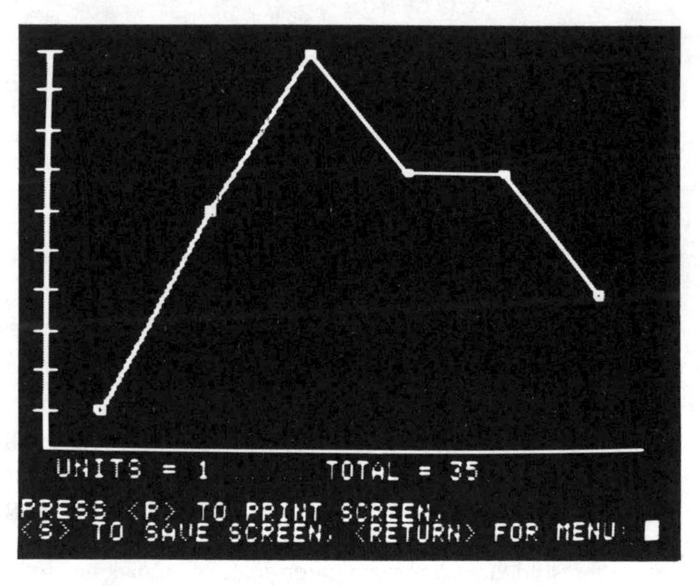

Figure 18.19d

Figure 18.19
Graphs made with the Logo data tool kit.

as the average. The mean of the above set is 5 (40 divided by 8). A bit later in the chapter the mean is discussed in more detail.

The *median* is the middle value in an ordered set of data. Half of all values lie at or above the median and half below. For the eight numbers listed above, the median is between 5 and 6 or 5.5. The Quantitative Literacy Project, in its book *Exploring data* (Landwehr & Watkins, 1987), places a heavy emphasis on the median in preference to the mean or mode. They note that the median is easier to understand, easier to compute, and is not affected by one or two extremely large or extremely small values outside the range of most of the data as is the mean.

Box-and-Whisker Plots

Box-and-whisker plots (or just *box plots)* are an easy method for visually displaying not only the median statistic but also information about the range and distribution of data. In Figure 18.20, the ages in months for 27 sixth-grade students are listed along with stem-and-leaf plots for the full class and the boys and girls separately. Box-and-whisker plots are shown in Figure 18.21.

Each box-and-whisker plot has these three features:

A box which contains the "middle half" of the data, one fourth to the left and right of the median. The ends of the box are at the *lower quartile,* the median of the

The following numbers represent the ages in months of a class of sixth-grade students. The numbers 1 to 14 are the boys, and 15 to 27 the girls.

1.	132	8.	122	15.	140	22.	131
2.	140	9.	130	16.	129	23.	128
3.	133	10.	134	17.	141	24.	131
4.	142	11.	125	18.	134	25.	132
5.	134	12.	147	19.	124	26.	130
6.	(137) Joe B.	13.	131	20.	129	27.	127
7.	139	14.	129	21.	(125) Whitney		

All students

12 | 2, 4
• | 5, 5, 7, 8, (9), 9, 9
13 | 0, 0, 1, 1, (1), 2, 2, 3, 4, 4, 4
• | (7), 9
14 | 0, 0, 1, 2
• | 7

Boys

12 | 2,
• | 5, 9
13 | (0), 1, 2, 3, | 4, 4
• | 7, (9)
14 | 0, 2
• | 7

Girls

12 | 4
• | 5, 7, | 8, 9, 9
13 | (0), 1, 1, 2, | 4
• |
14 | 0, 1
• |

Figure 18.20
Ordered stem-and-leaf plots grouped by 5. Medians, upper, and lower quartiles are found on the stem-and-leaf plots. Medians and quartiles are circled if data elements represented by a bar (|), fall between two elements.

lower half of data, and the *upper quartile,* the median of the upper half of the data.

A line inside the box at the median of the data.

A line extending from the end of each box to the *lower extreme* and *upper extreme* of the data. Therefore each line covers the upper and lower fourths of the data.

Look at the information these box plots provide at a glance! The box in comparison to the lengths of the lines provides a quick indication of how the data is spread out or bunched together. Since the median is shown, this spreading or bunching can be determined for each quarter of the data. The entire class in this example is much more spread out in the upper half than in the lower half. The girls are much more closely grouped in age than are either the boys or the class as a whole. It is immediately obvious that at least three fourths of the girls are younger than the median age of the boys. The *range* of the data (difference between upper and lower extremes) is represented by the length of the plot and the extreme values can be read directly. It is easy to mark and label entries of particular interest. For example, Joe B. and Whitney might be the class officers.

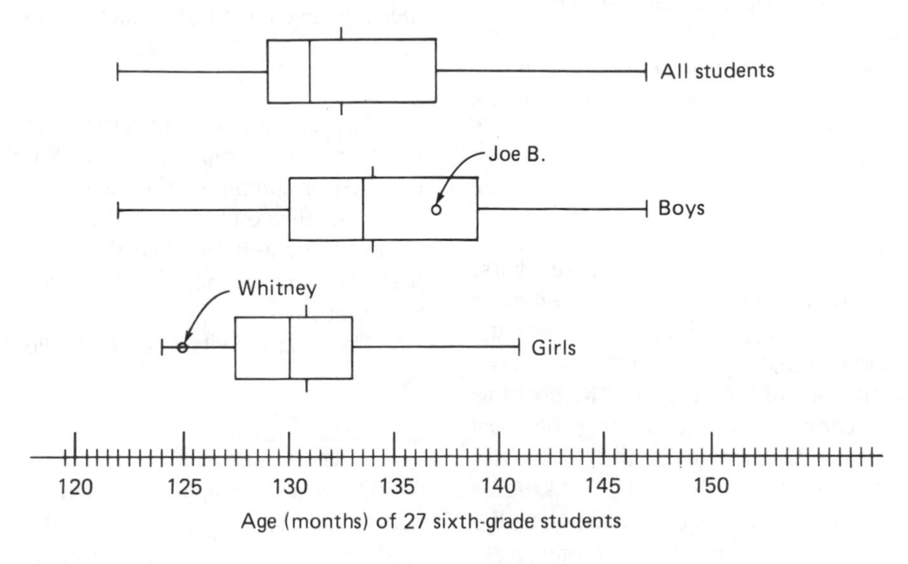

Figure 18.21
Box-and-whisker plots show a lot of information.

Activity 18.10

Making box-and-whisker plots is quite simple. First, put the data in order. An easy and valuable method is to make a stem-and-leaf plot and order the leaves, providing another visual image as well. Next, find the median. If the data is listed in a single row or column, young students can put their fingers on the ends and move each toward the middle one step at a time. Older children will simply count the number of values and determine the middle one. This can be done directly on the stem-and-leaf plots as was done in Figure 18.20. To find the two quartiles, ignore the median itself and find the medians of the upper and lower halves of the data. Again, this can be done on the stem-and-leaf plots. Mark the two extremes, the two quartiles and the median above an appropriate number line. Draw the box and the lines. Box plots can also be drawn vertically.

Note that the means for the data in the above example are each just slightly higher than the medians (Class = 132.4, Boys = 133.9, Girls = 130.8). Even with a calculator, means are somewhat tedious to compute and errors are easy to make. For this example, the means themselves do not provide nearly as much information and are approximately the same as the medians.

Understanding the Mean

Due to ease of computation and stability, the median when compared to the mean has some advantages as a practical average. However, the mean will continue to be used in popular media and in books, frequently along with the median. For smaller sets of data such as your test scores, the mean is perhaps a more meaningful statistic. Finally, the mean is used in the computation of other statistics such as the standard deviation. Therefore, it remains important that students have a good concept of what the mean tells them about a set of numbers. How do you describe the mean other than how to compute it? The activities in this section will help students construct intuitive ideas about the mean.

Activity 18.11

Make a bar graph using wooden cubes for the bars. Choose a graph with five or six bars with lengths of no more than 10 or 12. (The graph in Figure 18.22 represents the "prices" of six toys chosen so that the total of the prices is a multiple of 6.) Sketch a picture of the graph on the board as a record. Have students compute the mean using the usual numeric methods. Discuss with them what they think this number called the mean tells them about the data. After listening to their ideas, compare the process of adding up the numbers to the process of piling all the cubes into one stack. Dividing the sum by the number of values is the same as separating the one big stack of block into equal stacks. Re-

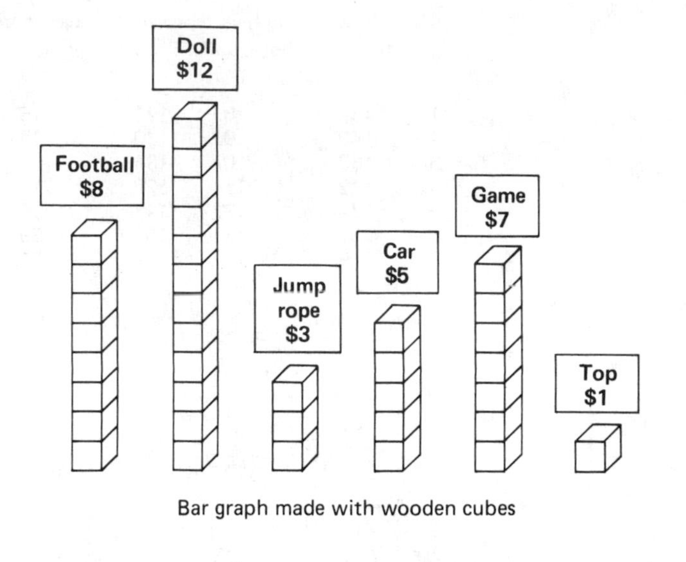

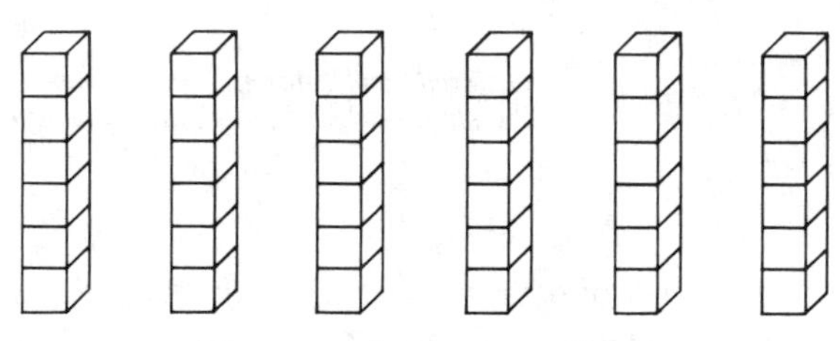

Figure 18.22
Understanding the mean.

arrange the cubes into bars of equal length. The mean, then, is the number you get if all of the values are "leveled off." Make different graphs with cubes and let students find the mean by leveling out the stacks rather than using any computation.

This bar graph/mean activity can also be done by making the original bar graph on paper. Numbers of any size can be used. After computing the mean, cut out all the bars and tape them together end to end. The length of the total strip is the sum of the numbers. Fold the long strip into as many equal parts as there are bars in the graph. The length of each part is the mean.

The next activity is a good follow up to the last one.

Activity 18.12

Duplicate a simple bar graph on graph paper and pass it out to the class. *Without doing any computation,* ask each student to draw a line across the graph where they estimate the mean to be. Next, using the scale on the graph, have them add up the pieces of the bars that are above their estimated mean line and record this number. Similarly, for bars that are below their estimated mean line, add up the spaces between the tops of the bars and the mean line (Figure 18.23). Compare the two sums. Should they be the same, different, or is it impossible to know? Now have them compute the actual mean with a calculator, draw a new line, and check the totals of the bars above and below as before. The two sums should be exactly the same when the mean line is drawn accurately. Notice that the pieces that stick up above the mean could be cut off and fit into the spaces below the line to make the bars all equal (as with the blocks.) Visually estimating the mean of a bar graph is a fun challenge.

The computer provides teachers with an effective method of examining the mean with their classes. Many computer spreadsheets are unbelievably easy to use for things such as adding lists of numbers, ordering them, and computing the mean. In the example in Figure 18.24, data values can be changed as quickly as they can be typed and all the sums and means will be recomputed immediately. With this tool it is easy to add one "far out" number to the data, delete numbers, or change the data in any way at all to observe the effect on the mean. Since spreadsheets will also put data in order, the median and extremes can be found easily. Many spreadsheet programs are coupled with graphing capabilities, so graphs can be examined as the data is changed.

```
File: Mean exp.              REVIEW/ADD/CHANGE                  Escape: Main Menu
========A========B========C========D========E========F========G========H====
10|
11|                       500                             34
12|                        54                             89
13|                        90                             67
14|                       129                            120
15|                       301                             23
16|                       346                             44
17|                       665
18|                        44
19|                       765
20|                       973
21|                      1000
22|
23|         sum:          4867         Sum:              377
24|         Mean:     442.4545         Mean:        62.83333
25|
26|         Combined Mean of Both:          308.4705
27|
--------------------------------------------------------------------------------
F24: (Value) @AVG(F11...F21)
```

Figure 18.24
Screen display from AppleWorks Spreadsheet. In this example, column sums and column means are computed instantly for any data in those two columns.

For Discussion and Exploration

1. Set up a simulation experiment for each of the four examples on p. 327. If you do not use a computer to generate random numbers, you will need to decide on a random device such as coins or spinners or drawing cards from a

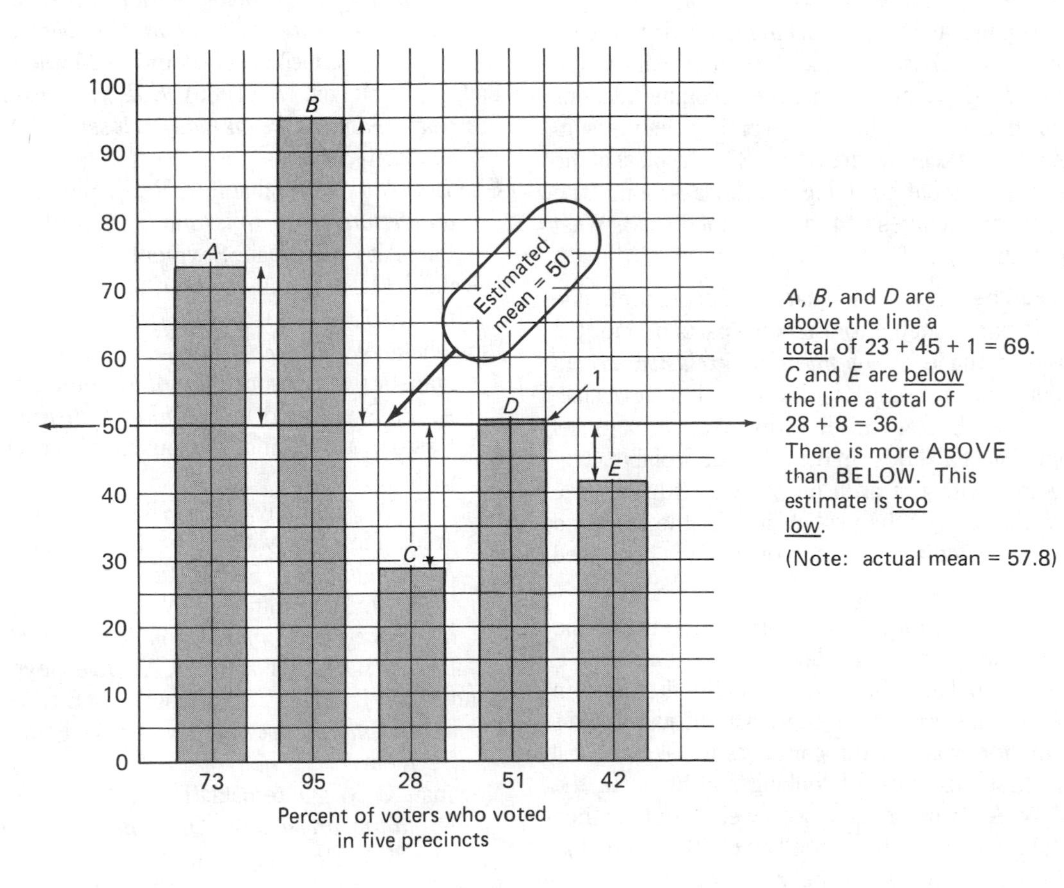

Figure 18.23
Estimating a mean on a bar graph.

hat. A computer is not necessary for any of these exercises. Run at least 10 trials of your experiment. Was the outcome about what you expected?

2. Write a computer program in BASIC or Logo that will print out rows of X's and dashes (–) using a random number generator. Design the program so that the probability of getting an X instead of a dash is controlled by the user (see Figure 18.4). What problems of interest to school aged children could your program be used to simulate? For example, if the home team is winning only 40% of their games, how often might you expect to see a five-game winning streak?
3. Design a two- or three-lesson sequence on experimental probability that is appropriate for a grade range you are working on. Try your plans out with children. How good is their intuitive understanding of probability?
4. Make up two lists of 30 numbers each ranging from 50 to 100. Make one list with most of the numbers clustering in the 60's and a median of about 65. Make the second list with a median of 75 and the numbers somewhat evenly distributed across the range. Make box-and-whisker plots of both sets of numbers. Explain how the box plots describe your two sets of scores. How would you go about adding 40 more scores to one of the sets of data without having to change the box plot?
5. Read the two-page article, "Teaching Statistics: Mean, Median, and Mode" (Zawojewski, 1988) and react to the statements made there concerning the recommendations for statistics instruction. One of several problems suggested by Zawojewski in the article (p. 26) is typical of the spirit of what we should be doing in the classroom: The mean of five brothers' ages is 4, and the mode is 3. What are some possible ages for the five brothers?
6. The area model begins with a square that represents one whole and it is partitioned into fractional parts of one. If a sample space is made up of *mutually exclusive* events (having no outcomes in common), the sum of the probabilities should be one. Explain this in terms of the area model for determining probabilities. In the prisoner problem (Figure 18.8), show with the area model and with the tree diagram how the sum of the probabilities at the start and at each fork is one and also the sum of all of the combined probabilities is one.
7. Use an area model and a tree diagram to determine the theoretical probability for the following experiment:

   Dad puts a $5 bill and three $1 bills in the first box. In a second box he puts another $5 bill but just one $1 bill. Junior, for washing the car, gets to take one bill from the first box without looking and put it in the second box. After these have been well mixed, he then gets to take one from the second box without looking. What is the probability that he will get $5?
8. Design a simulation for the problem in number 7 and try it out. Does it agree with your theoretical probability?

Suggested Readings

Armstrong, R. D. (1981). An area model for solving probability problems. In A. P. Shulte (Ed.), *Teaching statistics and probability*. Reston, VA: National Council of Teachers of Mathematics.

Baratta-Lorton, M. (1976). *Mathematics their way*. Menlo Park, CA: Addison-Wesley.

Berman, B., & Friederwitzer, F. (1989). Algebra can be elementary . . . when it's concrete. *Arithmetic Teacher, 36*(8), 21–24.

Bohan, H., & Moreland, E. J. (1981). Developing some statistical concepts in the elementary school. In A. P. Shulte (Ed.), *Teaching statistics and probability*. Reston, VA: National Council of Teachers of Mathematics.

Bright, G. W., Harvey, J. G., & Wheeler, M. M. (1981). Fair games, unfair games. In A. P. Shulte (Ed.), *Teaching statistics and probability*. Reston, VA: National Council of Teachers of Mathematics.

Bruni, J. V., & Silverman, H. J. (1986). Developing concepts in probability and statistics—and much more. *Arithmetic Teacher, 33*(6), 34–37.

Bryan, E. H. (1988). Exploring data with box plots. *Mathematics Teacher, 81*, 658–663.

Choate, L. D., & Okey, J. K. (1981). Graphically speaking: Primary-level graphing experiences. In A. P. Shulte (Ed.), *Teaching statistics and probability*. Reston, VA: National Council of Teachers of Mathematics.

Fair, J., & Melvin, M. (1986). *Kids are consumers, too! Real-world mathematics for today's classroom*. Menlo Park, CA: Addison-Wesley.

Gnanadesikan, M., Schaeffer, R., & Swift, J. (1987). *The art and techniques of simulation: Quantitative literacy series*. Palo Alto, CA: Dale Seymour.

Horak, V. M., & Horak, W. J. (1983). Take a chance. *Arithmetic Teacher, 30*(9), 8–15.

Juraschek, W. A., & Angle, N. S. (1981). Experiential statistics and probability for elementary teachers. In A. P. Shulte (Ed.), *Teaching statistics and probability*. Reston, VA: National Council of Teachers of Mathematics.

Kelly, I. W., & Bany, B. (1984). Probability: Developing student intuition with a 20 × 20 array. *School Science and Mathematics, 84*, 598–604.

Landwehr, J. M., Swift, J., & Watkins, A. E. (1987). *Exploring surveys and information from samples: Quantitative literacy series*. Palo Alto, CA: Dale Seymour.

Landwehr, J. M., & Watkins, A. E. (1987). *Exploring data: Quantitative literacy series*. Palo Alto, CA: Dale Seymour.

Newman, C. M., Obremski, T. E., & Schaeffer, R. L. (1987). *Exploring probability: Quantitative literacy series*. Palo Alto, CA: Dale Seymour

Russell, S. J., & Friel, S. N. (1989). Collecting and analyzing real data in the elementary classroom. In P. R. Trafton (Ed.), *New directions for elementary school mathematics*.

Reston, VA: National Council of Teachers of Mathematics.

Schultz, H. S., & Leonard, B. (1989). Probability and intuition. *Mathematics Teacher, 82,* 52–53.

Shaw, J. M. (1984). Making graphs. *Arithmetic Teacher, 31*(5), 7–11.

Shulte, A. P., & Choate, S. A. (1977). *What are my chances (Books A & B).* Palo Alto, CA: Creative Publications.

Watkins, A. E. (1981). Monte Carlo simulation: Probability the easy way. In A. P. Shulte (Ed.), *Teaching statistics and probability.* Reston, VA: National Council of Teachers of Mathematics.

CHAPTER 19

PREPARING FOR ALGEBRA

What is Prealgebra?

For many years, the content of middle-grade mathematics courses has been largely review. Flanders (1987) examined three popular textbook series and concluded that only about half of the material in the sixth grade could be considered "new" and even less was "new" in the eighth grade. In the report of the Second International Mathematics Study, *The Underachieving Curriculum* (McKnight, et al., 1987), the U.S. curriculum is criticized for spending too much time on previously taught, lower-level skills with little intense effort on new topics, especially those related to algebra. For example, U.S. eighth-grade teachers estimate spending 18% of their time on fractions and 14% of the time on ratio, proportion, and percent—nearly one third of the year. Japanese teachers report spending only 6% of the time on these two topics together. In contrast, U.S. teachers reported teaching algebraic topics 20% of the time, compared with 37% in Japan. A more recent trend in middle school mathematics curriculum is to increase the emphasis on problem solving, geometry, probability, statistics, and measurement as well as to develop the important foundational concepts of algebra. (Refer to the summary of emphasis changes suggested by the NCTM *Standards* for grades 5–8 in Appendix A.)

All of the topics in this chapter—primes and factors, integers, variables, and functions—will be found in most middle-grade curriculums. Some of the number theory concepts such as odd and even numbers and primes are commonly explored in lower grades as well. The middle school curriculum should begin to prepare students for the more abstract curriculum of the high school. This chapter focuses on those topics that are most important for the study of algebra. The objective is to show how students can develop and explore the basic building blocks of algebra using intuitive approaches connected to their understanding of arithmetic so that algebra can be a natural progression rather than an abrupt change.

Exploring Topics in Number Theory

Number theory is the study of relationships found among the natural numbers. At the elementary level, number theory includes the concepts of prime number, odd and even numbers, and the related notions of factor, multiple, and divisibility. Prime factorization is frequently connected with finding common denominators and reducing fractions. More importantly the concepts of prime, factor, and multiple are also used in algebraic expressions. Students should develop an intuitive understanding of these topics with numbers so that the algebraic generalizations can be built upon them. Number theory topics also provide an opportunity for problem solving and for student discovery of many fascinating relationships.

PRIMES AND FACTORIZATION

Discovering Primes and Other Factors

Prime numbers can be viewed as fundamental building blocks of the other natural numbers. Simply defining a prime number and searching for primes can be a rather dull experience. The activities described here are examples of interesting things we can explore with children and still have fun while developing basic concepts of prime, factor, and multiple.

Activity 19.1

Build or draw rectangles. Have students work in groups using square tiles or cubes or just square grid paper. Begin with the number 12 or 16. Have students find as many different rectangles as possible made up of that many squares. Share results. Students should agree that 1×12 and 1×16 rectangles should be included and that a square (4×4) is also a special kind of rectangle. When the process is understood completely, have them try to build as many rectangles as possible for each number up to 100. Give different groups different numbers to work on. Draw pictures on graph paper for each number and make a display. Students will discover that some numbers (primes) have only one rectangle. The dimensions of each rectangle are two factors of that number. Use this idea to develop a definition of factor and of prime number. (A number other than 1 is a *prime* number if its only factors are itself and 1.) Help students use the idea of rectangular dimensions to develop definitions of prime numbers and factors. Rectangles are a way to think of numbers as a product of two factors. Rectangular solids are a model for a number expressed as a product of three factors.

Activity 19.2

With cubes, find ways to build boxes for a number. For example, 12 cubes can be arranged into a $2 \times 2 \times 3$ box (Figure 19.1). What can you say about a number for which that many cubes can be used to

Build only one box?

Build only two different boxes?

Build only one box that has no edges of length 1?

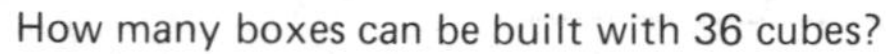

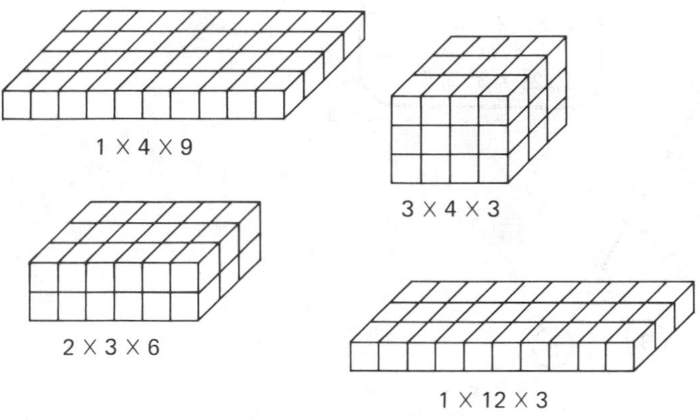

Can you find four more?

Figure 19.1
Building boxes.

Once students have an idea of what a prime number and a factor are, they can explore a variety of other relationships involving primes. Calculators should be available for all explorations.

Given a number, is it a prime or not a prime? (Nonprime numbers other than 1 are called *composite numbers.)* What are some ways to test if a number is a prime? This is a good computer programming exercise in either BASIC or Logo.

Explore Goldbach's conjecture: Any even number greater than 2 can be written as the sum of two prime numbers. For example: $38 = 31 + 7$. Goldbach's conjecture has never been proved or disproved. (Goldbach lived between 1690 and 1764.)

Have students make up and explore different conjectures of their own. They might consider the difference of primes, or the sum of three primes. Can they find a formula that will always produce primes? For example, $2^n - 1$ works for quite a few values of n. The formula $n^2 - n + 41$ yields prime numbers for any value of n between 1 and 40 inclusive. Let students try to find other "prime generators."

Have students write down all of the prime factors of a number, including repeats. Write the number as a product of these prime factors placing them in order from least to most. For example, $360 = 2 \times 2 \times 2 \times 3 \times 3 \times 5$. Before comparing results from different groups, discuss the various strategies or approaches that were used. If a different approach is used, will it still result in the same factorization? (The answer, of course, is "yes." This result is known as the Fundamental Theorem of Arithmetic.) Some different ways to find the prime factorization of a number are shown in Figure 19.2.

The terms *divisor* and *factor* are synonyms. For a nonprime number, find all of the *proper divisors* (divisors other than the number itself) and add them up. (See which groups can come up with the most clever ways of finding all of the factors of a number.) If the sum is less than the number, it is called *deficient,* if more than the number, it is called *abundant,* if it is equal to the sum of its proper divisors, it is called *perfect.* For example, 6 is perfect, 8 is deficient, and 12 is abundant. Those are the easy ones. What are the next three of each type? Students might also explore looking for twin primes, those that are two apart as are 11 and 13.

Activity 19.3

Give students two numbers that are already factored into primes. Suggest that these primes can be helpful in determining if a number is either a factor or a multiple of each number or a common factor or common multiple of them both. For example:

$$A = 2 \times 2 \times 3 \times 7 \qquad B = 2 \times 3 \times 5$$

Test these (fill in the blanks):

	Factor			*Multiple*		
Number	*of A*	*of B*	*of Both*	*of A*	*of B*	*of Both*
2×3	yes	yes	yes	—	—	–
2×5	no	—	no	no	—	no
$2 \times 3 \times 5 \times 7$	—	—	—	—	yes	no
—	no	no	no	yes	no	no
—	no	no	no	no	no	yes

Let students discover and articulate their own ideas about how the prime factors help find these numbers or fill in the chart. From this exercise, let them find ways to get the *least common multiple* (LCM) or *greatest common divisor* (GCD) using the primes. A helpful suggestion is to write each prime factor on a small piece of paper as in Figure 19.3. Using paper like this may not seem very practical. How can the same approach be used by just listing each of the primes?

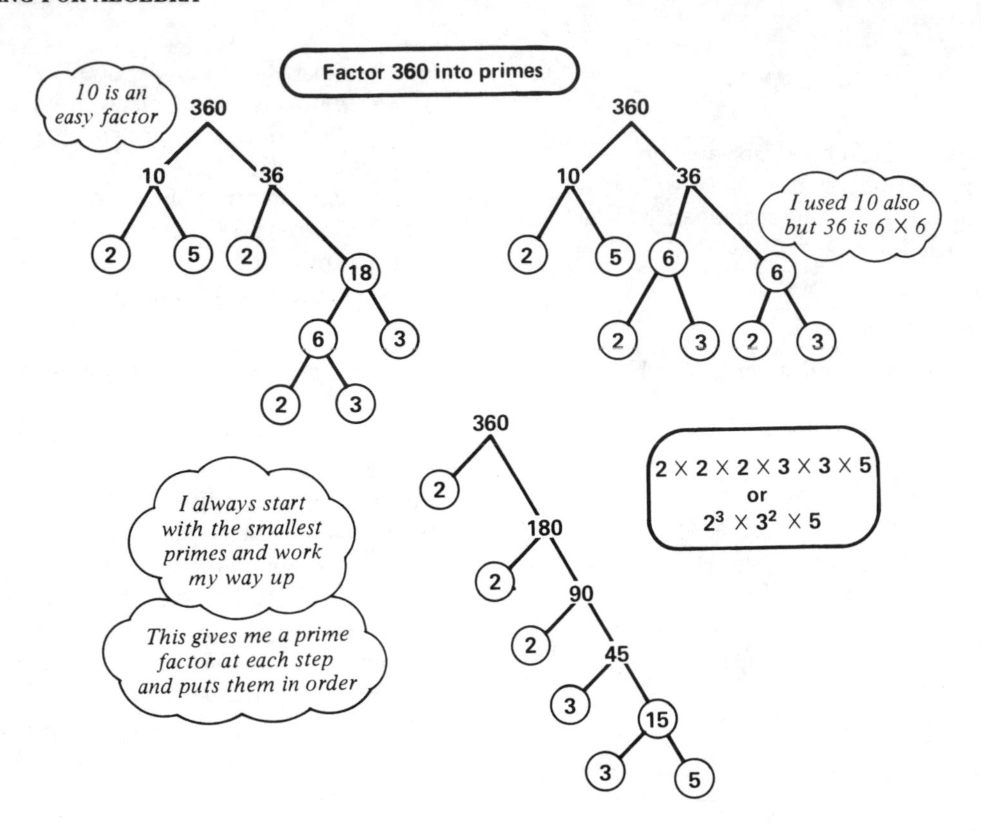

Figure 19.2
Several different routes to the same factors.

Fine the G.C.D. and L.M.C. of 90 and 84.

Write each prime factor on a separate slip of paper. Mark the slips in order to keep track of which factors belong to which numbers.

90 2 3 3 5

84 (2) (2) (3) (7)

A common multiple must have all of the factors of both numbers—but there is no need to have duplicates.

Pair up the primes from each number that match. Put those that don't match at the end.

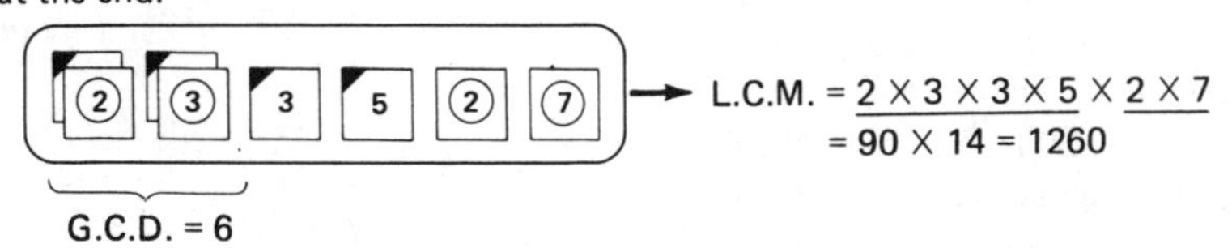

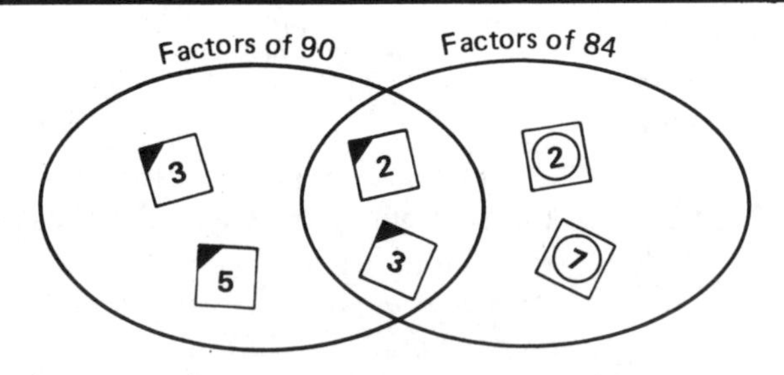

If you put the slips in a diagram like this, the (2) and (3) are not necessary (or they could be paired with 2 and 3).

Where are the G.C.D. and L.C.M.?

Figure 19.3
Using prime factors to think about G.C.D. and L.C.M.

PATTERNS ON A HUNDREDS CHART

One of the truly fascinating aspects of mathematics is the way that numbers tend to appear in patterns. In Chapter 17, the calculator was suggested as a good way to explore some interesting numeric patterns. The Fibonacci sequence was also discussed as an intriguing source of patterns and relationships. In the following activities students can use a hundreds chart to discover other interesting phenomena of numbers. As you explore these ideas, consider how prime numbers and factors play a role in the results. How are these patterns like the calculator patterns in Chapter 17? Encourage students to explore some of these ideas and make their own observations. A self-discovered pattern is much more exciting than one that we point out to students.

Activity 19.4

Duplicate lots of small copies of a hundreds chart (or 0 to 99 chart). Start with a number less than 10 and have students color in that number and all multiples of that number. For every number, some pattern will emerge (Figure 19.4). Which numbers make diagonal patterns and which only produce columns or parts of columns? Notice that the pattern for 6's is made up of numbers that are also in the 3's pattern. How can you tell if one pattern will be a part of another? Do they ever overlap or is one always completely inside of another? What happens if you start a jump by 3's pattern at 2 instead of at 0?

1	2	3	4	5	6	7	8	9	10
11	12	13	14	15	16	17	18	19	20
21	22	23	24	25	26	27	28	29	30
31	32	33	34	35	36	37	38	39	40
41	42	43	44	45	46	47	48	49	50
51	52	53	54	55	56	57	58	59	60
61	62	63	64	65	66	67	68	69	70
71	72	73	74	75	76	77	78	79	80
81	82	83	84	85	86	87	88	89	90
91	92	93	94	95	96	97	98	99	100

Circle the three's.
Put X's on the six's.
What columns are these in?

Figure 19.4
Patterns of multiples.

Activity 19.5

Change the dimensions of the chart in the last activity by changing the lengths of the rows (Figure 19.5). If the rows have six squares, which numbers will make patterns in rows and which in columns? If there are seven squares in the row, only one number will make a column pattern, the multiples of 7. Why? Try other row lengths including rows with more than 10 squares. All patterns seem to be either columns or diagonals. What does the row length have to do with the patterns involved? Consider looking at primes and factors.

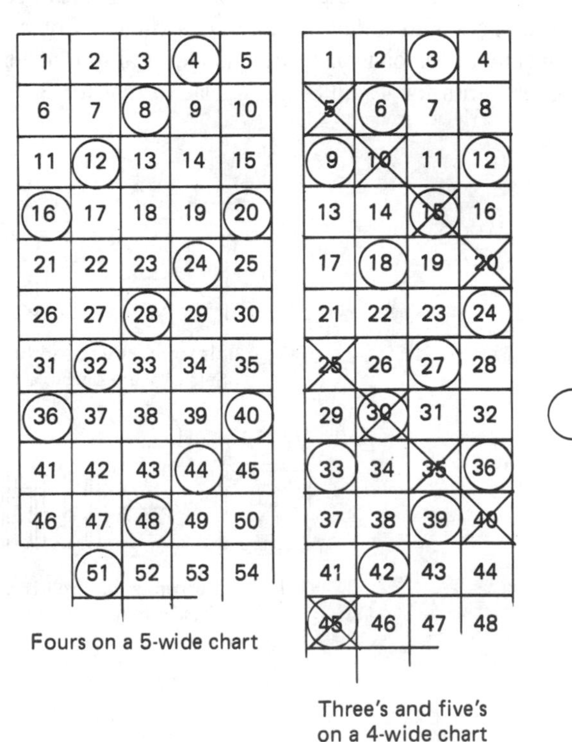

Figure 19.5
Looking for patterns on different hundreds charts.

TESTS FOR DIVISIBILITY

It is easy to tell whether or not a number is divisible by 10. For example, 198,456 is not but 650,270 is. The test for divisibility by 10 is to look at the last digit, the digit in the ones position. You probably know this already. Experience indicates that products with a factor of 10 end in zero. Divisibility by 5 is similar: if the last digit is either 5 or 0, then the number is divisible by 5.

There are similar tests for divisibility by other numbers.

A number is divisible by

3 if the sum of the digits is divisible by 3
9 if the sum of the digits is divisible by 9
2 if the last digit is even (or divisible by 2)
4 if the last two digits are divisible by 4

8 if the last three digits are divisible by 8

6 if it is divisible by 3 and 2

Many students find these intriguing. The limited practical value of such tests indicates that they should be investigated as a problem-solving task rather than a topic of mastery. The more intriguing and more valuable question is, why do these tests work? Since the answer lies in the base ten place-value representation of the numbers, divisibility rules also provide an opportunity to review place-value concepts (Figure 19.6). One approach is to give students the divisibility rule and suggest that they use base ten materials or drawings to figure out why it works. Having discovered one rule, the students can then be asked to find a rule that is similar. For example, the argument for divisibility by 3 is identical to that for 9 and the arguments for 4 and 8 are similar to that for 2.

EXPONENTS

Concepts and Conventions

In algebra classes students get very confused trying to remember the rules of exponents. For example, when you raise numbers to powers, do you add or multiply the exponents? Here is an excellent example of procedural knowledge that is frequently learned without supporting conceptual knowledge. Before algebra, students should have ample opportunity to explore working with exponents on whole numbers rather than with letters or variables. By doing so, they are able to deal directly with the concept and actually generate the rules themselves.

A *whole number exponent* is simply a shorthand symbolism for repeated multiplication of a number times itself; for example, $3^4 = 3 \times 3 \times 3 \times 3$. That is the only conceptual understanding that is required.

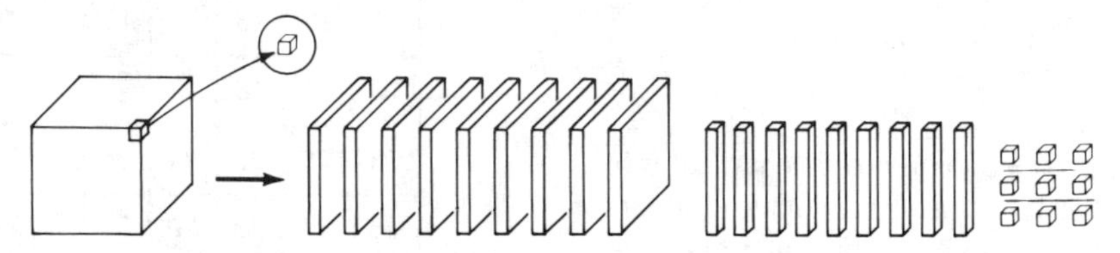

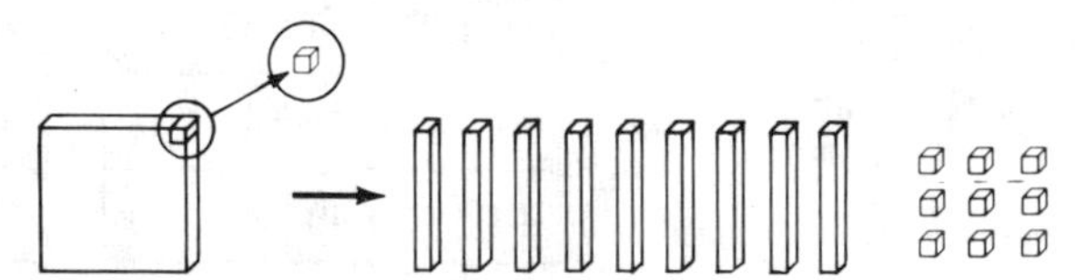

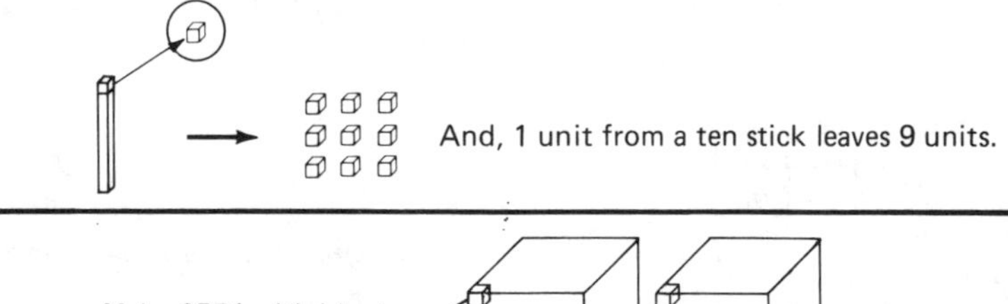

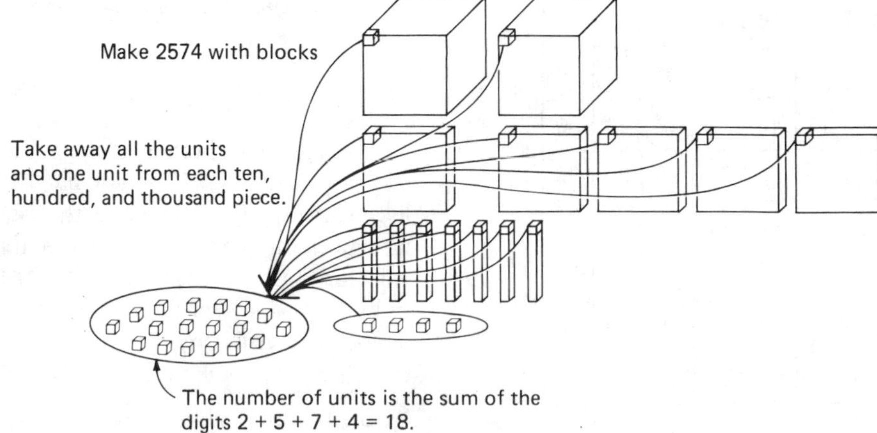

Figure 19.6

Divisibility by 9 and 3 shown with base ten blocks. Can you show that every block larger than a ten can be divided by 4? Use this fact to establish divisibility by 4.

There are also conventions of symbolism that must be learned. As conventions, they are arbitrary rules with no conceptual basis. The first is that an exponent applies to its immediate base. For example, in the expression $2 + 5^3$, the exponent 3 only applies to the 5, so the expression is equal to $2 + (5 \times 5 \times 5)$. However, in the expression $(2 + 5)^3$, the 3 is an exponent of the quantity $2 + 5$ and is evaluated as $(2 + 5) \times (2 + 5) \times (2 + 5)$ or $7 \times 7 \times 7$.

The other convention involves the order of operations; multiplication and division are always done before addition and subtraction. Notice that since exponentiation is repeated multiplication, it also is done before addition and subtraction. In the expression $5 + 4 \times 2 - 6 \div 3$, 4×2 and $6 \div 3$ are done first. Therefore the expression is evaluated as $5 + 8 - 2 = 13 - 2 = 11$. If done in left-to-right order the result would be 4. Parentheses are used to group operations that are to be done first. Therefore, in $(5 + 4) \times 2 - 6 \div 3$, the addition can be done inside the parentheses first or the distributive property can be used and the final result is 16. The phrase "*p*lease *e*xcuse *m*y *d*ear *A*unt *S*ally" is sometimes used to help students recall that operations inside *p*arentheses are done first, then *e*xponentiation and then *m*ultiplication and *d*ivision are done before *a*ddition and *s*ubtraction.

Using Calculators

Most scientific calculators, such as those now popularly used in high school, employ "algebraic logic" that will evaluate expressions correctly and also allow grouping with parentheses. However, the simple four-function calculators generally used in elementary school process symbols as they are keyed in. Thus, if the expression $5 + 3 \times 2$ is keyed in as written, the result will be 16 instead of 11. To evaluate $5 + 7 \times 2 - 28 \div 7$, intermediate results must be written down or stored temporarily in the calculator memory.

Even with this limitation, the simple four-function calculator is a powerful tool for exploring exponents, finding roots, and even helping students deal with the difficulties of order of operations. To evaluate 3^8 on a calculator, press 3 [×] [=] [=] [=] [=] [=] [=] [=]. (The first press of [=] will result in 9, or 3×3.) With the calculator students can observe how quickly numbers grow. It is fascinating to enter a number, press [×], and then continually press the [=] button.

Explorations

Students should have ample opportunity to explore expressions involving mixed operations and exponents with only the conventions and the meaning of exponents as their guides. No rules for exponents should be promoted. When experience has provided students with a firm background, the rules of exponents will make good sense and should not require rote memorization.

Activity 19.6

Provide students with numeric expressions to evaluate with their calculators. Some examples of the type of expressions that can be valuable are

$3 + 4 \times 8$ $4 \times 8 + 3$	$3^6 + 2^6$ $(3 + 2)^6$	$3^4 \times 7 - 5^2$ $(3 \times 7)^4 - 5 \times 2$	$3^4 \times 5^2$ $(3 \times 5)^6$

$\dfrac{5^3 \times 5^2}{5^6}$	$4 \times 3 - 2^3 \times 5 + 23 \times 9$	$\dfrac{4 + 3^5}{2} \quad 4 + \dfrac{3^5}{2}$

When experiencing difficulty, students should write equivalent expressions without exponents and/or include parentheses to indicate explicit groupings. For example,

$$\begin{aligned}(7 \times 2^3 - 5)^3 &= (7 \times (2 \times 2 \times 2) - 5) \\ &\quad \times (7 \times (2 \times 2 \times 2) - 5) \\ &\quad \times (7 \times (2 \times 2 \times 2) - 5) \\ &= ((7 \times 8) - 5) \times ((7 \times 8) - 5) \\ &\quad \times ((7 \times 8) - 5) \\ &= (56 - 5) \times (56 - 5) \times (56 - 5) \\ &= 51 \times 51 \times 51\end{aligned}$$

When discussing results, place all of the emphasis on the procedures rather than the answer. The fact that two groups got the same result does not help a group that got a different result. For many expressions there is more than one way to proceed, and one may be easier to do or to understand than another.

The following activity involves an understanding of exponents, estimation, and is also a good problem-solving task.

Activity 19.7

Before class write down any simple expression involving exponents and evaluate it. Give the expression to the class with at least two of the numbers in the expression replaced by a star. Give students the value of the expression and see which group, using their calculators, can find correct values for the stars first. For example:

$$(* + 7^*)^3 = 41{,}063{,}625$$

The size of the result and the fact that it ends in 25 are clues that may help. There is also more than one solution to this one, so do not be too quick to tell students their answers are not correct.

Roots and Irrationals

INTRODUCING THE CONCEPT OF ROOTS

How could you use a calculator to estimate the side of the square or the edge of the cube in Figure 19.7? (Assume there is no square root key on the calculator.) These are excellent challenges for students and provide a good introduction to the process of finding the root of a number. The solutions will satisfy these equations:

$$\square \times \square = 45 \text{ or } \square^2 = 45 \quad \text{and}$$
$$\square \times \square \times \square = 30 \text{ or } \square^3 = 30$$

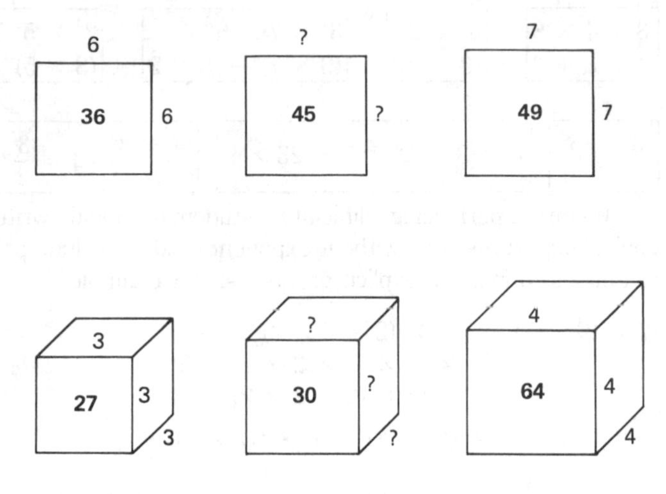

Figure 19.7
A geometric interpretation of square roots and cube roots.

The calculator permits students to guess at a solution and quickly test to see if it is too big or too small. For example, to solve the cube problem, you might start with 3.5 and find that 3.5^3 is 42.875, much too large. Quickly you will find that the solution is between 3.1 and 3.2. But where? Again, try halfway: 3.15^3 is 31.255+. The next try should be lower. By continued trial and error a simple calculator can get the result correct to six decimal places.

From a simple introduction such as this, students can be challenged to find solutions to equations such as $\square^6 = 8$. These students are now prepared to understand the general definition of the *nth root* of a number *N* as that number which when raised to the *n*th power equals *N*. The square and cube roots are simply other names for the second and third roots. With this approach, the concept of the *n*th root is developed in general. The notation involving the radical sign can come last in this development. It should then be clear that $\sqrt{6}$ is a number and is not an exercise to be "done." The cube root of 8 *is the same as* $\sqrt[3]{8}$, which *is the same as* 2.

REAL NUMBERS

While it may be open to some argument, the eighth-grade student probably does not need to have a very sophisticated knowledge of the real number system. There are a few powerful ideas, however, that can be explored or discussed informally.

Some Numbers Are Not Rational

There are two characterizations of a rational number. One is that it is a number that can be written as a fraction of two whole numbers. Alternatively, it is true that a number is rational if and only if, when written as a decimal, the decimal part is either finite or repeats infinitely. Thus 3.45 and 87.19363636. . . are rational numbers and can each be converted to their fractional forms. But what about a decimal number that just goes on and on and on with no repetition? Or, what about the number 3.101001000100001000001. . . ? These never repeat and are not finite and therefore are not rational. A number that is not rational is called *irrational.*

The numbers π and $\sqrt{2}$ are both irrational numbers. The number π is a ratio of two measures in one of the most common shapes we know, the circle. While it is not possible to prove the irrationality of π at this level, the fact that it is irrational implies that it is impossible to have a circle with the lengths of both the circumference and the diameter rational. (Why?) A proof that $\sqrt{2}$ is irrational is generally explored at the junior high level. The usual argument assumes that $\sqrt{2}$ is rational, which then leads to a contradiction.

Rational Numbers Are Dense

The property of being *dense* means that between any two numbers in the system, there is another one. It is impossible to find two rational numbers that are "next" to each other because there are always some more in between. The denseness of the rationals is fun and easy to demonstrate and is a powerful concept for students to experience. In fraction form, consider $7/15$ and $8/15$. They seem relatively "close." But there are clearly at least nine fractions between $70/150$ and $80/150$, the same two numbers. And for any two of those, say $73/150$ and $74/150$, the same process can be used again. It never ends. There are always more in between.

While the density of the rationals is a fairly impressive idea in itself, what is even more astounding is that the irrationals are also dense. And the irrationals and the rationals are all mixed up together. The density of the irrationals is not as easy to demonstrate and is not within the scope of the elementary school.

Only Some Roots of Numbers Are Irrational

One reason it is difficult to comprehend irrational numbers is that we have very little firsthand experience with them. The irrational numbers we are most familiar with are roots of numbers. For example, it has already been noted that students are frequently shown a proof for the irrationality of $\sqrt{2}$. An intuitive notion of that fact is difficult to come by. Most calculators will only show eight digits, requiring at least some leap of faith to accept that the decimal representation is infinite and nonrepeating.

What can unfortunately happen is that whenever students see the radical sign, they think the number is irrational. A possible approach is to consider the concept of roots from the opposite direction. Rather than ask what is the square root of 64 or the cube root of 27, we might suggest that every number is the square root, the cube root, the fourth root, . . . , of some number. (For example, 3 is the square root of 9, the third root of 27, etc.) From this vantage point, students can see that "square root" is just a way of indicating a relationship between two numbers. That the cube root of 27 is 3 indicates a special relationship between 3 and 27.

Integer Concepts and Operations

MEANING FOR THE NEGATIVE INTEGERS

Intuitive Examples of Signed Quantities

As with any new types of numbers that students encounter, some form of real model or examples must be used. Negative numbers or situations that model negative numbers do exist that students can relate to. It is a good idea to discuss some of these with your class before jumping directly into computation with signed numbers.

Bills and Cash *or* Debits and Credits. Suppose you are the bookkeeper for a small business. At any time your records show how many dollars the company has in its account. There are always so many dollars in cash (credits or receipts) and so many dollars in bills (debits). The difference between the debit and credit totals can tell the value of the account. If there is more cash than bills, the account is positive or "in the black." If there are more bills, the account is in debt or shows a negative cash value. Suppose further that all transactions are handled by mail. The mailman can bring mail, a positive action, or take mail, a negative action.

With this scenario, it is easy to discuss addition and subtraction of signed quantities. An example is illustrated in Figure 19.8.

Football. The standard football field has a 50-yard line in the center extending to 0 at each goal. A "mathematical" football field might have the center be the 0-yard line with one goal being the +50 goal and the other being the −50 goal. Any position on the field is determined by a signed number between +50 and −50. Gains or losses are like positive and negative quantities. A positive team moves toward the positive goal, and a negative team toward the negative goal. If the negative team starts on the −15-yard line and has a loss of 20 yards, they will be on the +5-yard line. (You should sketch a picture of this example and convince yourself that it is numerically the same example as the one in Figure 19.8.)

Credits		Debits		Balance
Cash in	Cash out	Bills in	Bills out	Begin 0
50				+50
		30		+20
	10			+10
		50		−40
25				−15
			20	+5

Figure 19.8
A ledger sheet "model" for integers.

Contrived situations such as mailing debits and credits and the football field are suggested as introductory discussion models. They can help students to think intuitively about what happens to quantities when an action causes them to be less than zero. They also provide an example of a joining or positive action and a removal or negative action of both positive and negative quantities. With these models there are two specific types of questions we can pose for students.

1. Give students a beginning and an end value and have them describe different ways that the change might have occurred. For example, how could the ball get from the +40-yard line to the −10-yard line if the negative team has the ball?
2. Give students either a beginning or an ending value and one or more actions and have them determine the value not given. For example if the company received \$20 in cash and \$35 in bills, resulting in a balance of negative \$5, what did they have to begin with?

The calculator is another model that might be explored early in the discussion of signed numbers. It gives correct and immediate results that students seem to believe in. The major drawback is that no rationale for the result is provided.

Activity 19.8

Have students enter a small number such as 10 in the calculator and press [−] 1 [=] [=] [=] . . . until the display shows zero. Discuss what will happen on the next press of [=]. After pressing the equal button several more times, discuss what they think the result means. (Notice that the negative sign appears in different places on different calculators.) Similarly, have students explore subtraction problems such as 5 − 8 = ? and discuss the results.

A negative number is entered on a calculator by first pressing the number and then pressing the change-of-sign key (usually [+/−]). Thus to add −3 and −8, press 3 [+/−] [+] 8 [+/−] [=] Students can benefit by using the calculator along with the intuitive examples mentioned earlier, or they can use the calculator to answer questions of both types previously mentioned. For example, how can you get from −5 to −17 by addition? or 13 minus what is 15?

Mathematics Definition of Negative Numbers

The mathematician defines a negative number in terms of whole numbers. Therefore, the definition of negative 3 is the solution to the equation 3 + □ = 0. In general, the *opposite of n* is the solution to *n* + □ = 0. If *n* is a positive number, then the opposite of *n* is a *negative number*. The set of *integers*, therefore, consists of the positive whole numbers, the opposites of the whole numbers, or negative numbers, and zero, which is neither positive nor negative. This is the definition that you may have learned in a college mathematics class. It is also found in student textbooks. Like many things in mathematics, abstract or symbolic definitions are best when there is some intuitive or conceptual framework with which to link the idea. For one who has never thought about the concept of a negative quantity, the symbolic definition of "the opposite of a number" is relatively vague if not totally strange. Virtually all new ideas in mathematics develop originally from "playing around" with familiar ideas and relationships, frequently aided by real models. It is a major error to let students believe otherwise.

OPERATIONS WITH THE INTEGERS

Symbolism

Up until students encounter the integers, the plus and minus signs are used only for the operations of addition and subtraction. Notation for signed numbers represents a real problem for many students. For example, the sum of 3 and negative 7 can be written as $3 + (-7)$ or as $3 + {}^{-}7$. The latter form might be clear in a printed book, but may be obscure in handwritten form. The use of parentheses is awkward, especially in expressions already involving parentheses.*

Models for the Operations

Two models are popular for helping students understand how the four operations (+, −, ×, and ÷) work with the integers. One model consists of counters in two different colors, one for positive counts and one for negative counts. Two counters, one of each type, cancel each other out. Thus if reds are positive and blues are negative, five reds and seven blues is the same as two blues, each representing negative two (Figure 19.9). It is important with this model that students understand that it is always possible to add to or remove from a pile a pair consisting of one positive and one negative counter without changing the value of the pile. (Intuitively, this is like adding equal quantities of debits and credits.) The actions of addition and subtraction are the same as for whole numbers; addition is joining or adding counters and subtraction is removing or taking away counters.

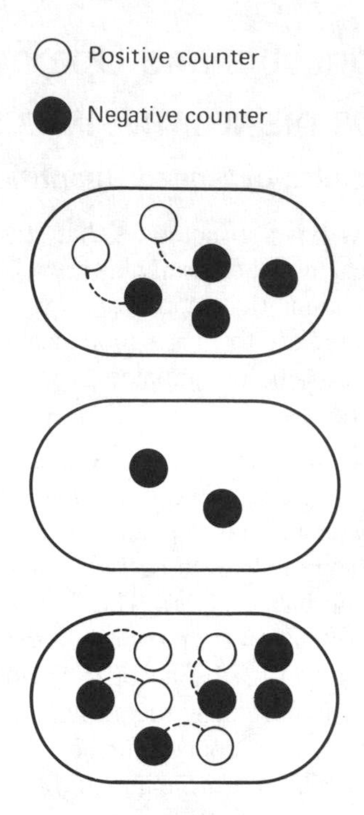

Figure 19.9.
Each collection is a model of negative 2.

The other commonly used model is the number line. It is a bit more traditional and "mathematical," and yet many find it somewhat confusing. The football field model provides an intuitive background. Positive and negative numbers are measured distances to the right and left of zero. It is important to remember that signed values are *directed distances* and not points on a line. The points on the number line are not models of integers; the directed distances are. To emphasize this for students, represent all integers with arrows and avoid referring to the coordinates on the number line as numbers. Poster board arrows of different whole number lengths can be made in two colors, red pointing to the right for positive quantities and blue to the left for negative quantities (Figure 19.10). The physical arrows help students think of integer quantities as directed distances. A

*The Comprehensive School Mathematics Program (CSMP) uses a chevron over a numeral to represent a negative number. Thus negative four is written $\hat{4}$, and $3 - (-7)$ becomes $3 - \hat{7}$. This notation avoids much of the confusion that students have, especially with the notation for subtraction. While you could easily adopt this notation, it is unlikely that it will ever commonly be used.

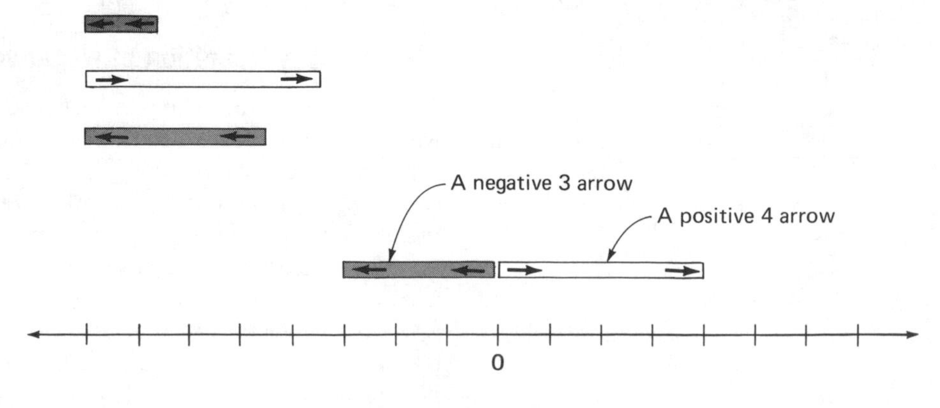

Figure 19.10
Number line model.

positive arrow never points left, and a negative arrow is never pointing to the right. Furthermore, each arrow is a quantity with both length (magnitude or absolute value) and direction (sign). These properties remain for each arrow regardless of its position on the number line.

Before discussing how these models are used to explain the operations, it is worth pointing out that the rules (procedural knowledge) for operations on the integers are generally easier to simply "give" to students than are explanations with a model. The conceptual explanations do not make the rules easier to use, and it is never intended that students continue to think in terms of these models as they practice integer arithmetic. Rather, it is important that students do not view the procedural rules for manipulating integers as being arbitrary and mysterious. Here then is a case where we must remember to make students responsible for the conceptual knowledge. If we emphasize only the procedural rules, there is little reason for students to attend to the conceptual justifications. In your discussions with students, do not be content to get "right answers" but demand explanations. You might even try *giving* students the answers and having them explain with a model why they must be so.

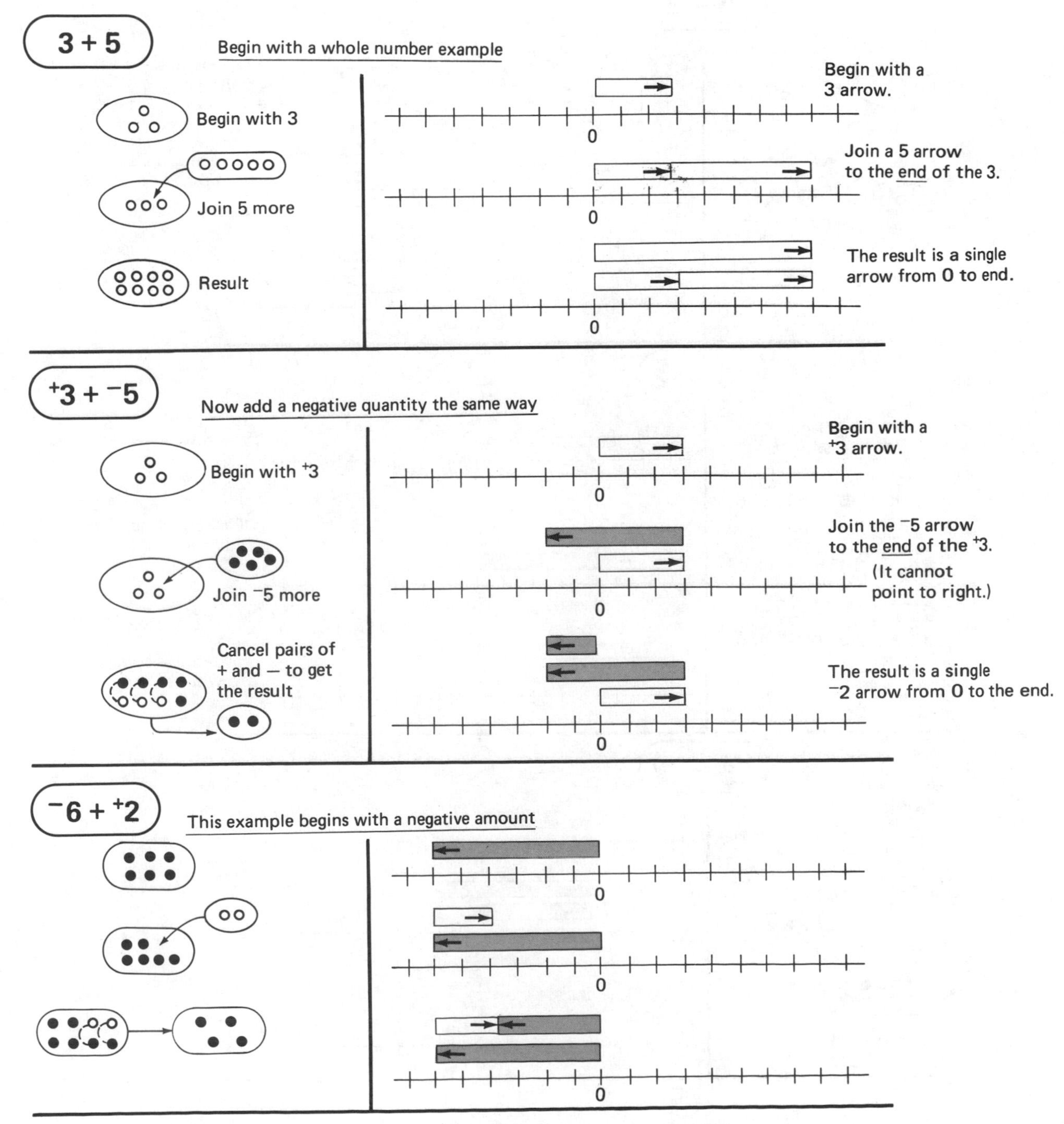

Figure 19.11
Relate integer addition to whole number addition.

Addition and Subtraction

Adding or subtracting integers with the models is straightforward and analogous to the corresponding debit/credit model or the football model. Since middle school or junior high students will not have used counters or number lines for some time, it would be good to begin work with either of these models using positive whole numbers. After a few examples to help them get familiar with the model for addition or subtraction with whole numbers, have them work through an example with integers using exactly the same reasoning. Remember, the emphasis should be on the rationale and not how quickly students can get correct answers.

In Figure 19.11 several examples of addition are modeled, each in two ways: with "positive" and "negative" counters and with the number line and arrow model. First examine the counter model. After the two quantities are joined, any pairs of positive and negative counters "cancel" each other out, and students can remove these making it easier to see the result.

To add using the arrow model, note that each added arrow begins at the point end of the previous arrow. If you help students with the analogy between these arrows and the

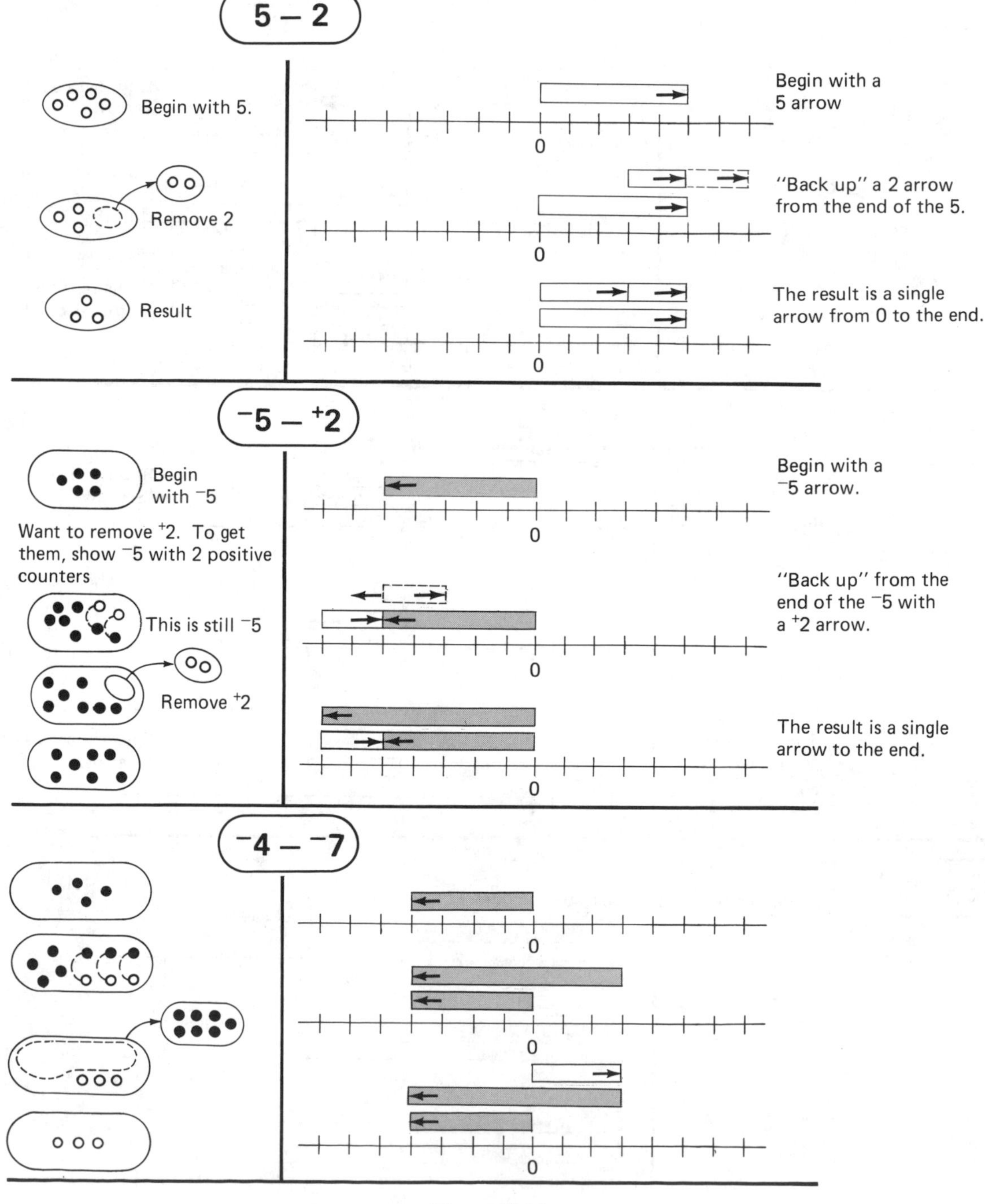

Figure 19.12
Integer subtraction is also related to whole numbers.

football situation, when the arrows change direction, that is like the ball changing teams. Addition is the advance of a team from the previous position. In the $^{+}3 + {}^{-}5$ example, the positive arrow (+ team) starts at zero and ends at positive 3. From that point, the negative arrow begins (the − team takes over and *advances* toward the negative direction). The result, then, is an arrow beginning at zero and ending where the second arrow ended. The same change of direction (change of ball possession) takes place in the $^{-}6 + {}^{+}2$ example. If a negative were added to a negative, the arrows would each go toward the left or negative (no change of possession) just as there was no change of direction for the 3 + 5 example.

Subtraction is interpreted as "remove" in terms of the counter model and "back up" in terms of the arrow model. In Figure 19.12, for $^{-}5 - {}^{+}2$ both models begin with a representation of $^{-}5$. In order to remove two positive counters from a set that has none, a different representation of $^{-}5$ must first be made. Since any number of neutral pairs (one positive, one negative) can be added without changing the value of the set, two pairs are added so that two positive counters can be removed. The net effect is to have more negative counters. This is like removing credits from your ledger if you are already in debt. The result is to leave you further in debt. A similar change in the representation of the beginning amount is always necessary when you need to subtract a quantity of a different sign.

With the number line and arrow model, subtraction means to "back up" or to "move in the opposite direction." Using the football field analogy, either team moves backwards when they are penalized or lose yardage. They move in the opposite direction from their own goal. In the example of $^{-}5 - {}^{+}2$, the first arrow ends at $^{-}5$. If the $^{+}2$ were to be added, it would be in the dotted position, ending at $^{-}3$. But it backs up instead. The result of the operation is an arrow from zero to the back end of the $^{+}2$ arrow. In the football analogy, the ball is first at the $^{-}5$-yard line, changes hands to the positive team that proceeds to lose two yards, leaving the ball at the $^{-}7$-yard line. In the second example ($^{-}4 - {}^{-}7$) the ball does not change hands, but the negative team loses yardage.

You want your students to draw pictures to accompany computations that they do with integers. Set pictures are easy enough. They may consist of X's and O's, for example. For the arrow model, there is no need for anything elaborate either. Figure 19.13 illustrates how a student might draw arrows for simple addition and subtraction exercises without even sketching the number line. Directions are shown by the arrows and magnitudes are written on the arrows. For your initial modeling, however, the poster board arrows in two colors will help them see that negative arrows always point left, and that addition is a forward movement for either type of arrow while subtraction is a backward movement.

An effort is usually made to see that $^{+}3 + {}^{-}5$ is the same as $3 - {}^{+}5$ and that $^{+}2 - {}^{-}6$ is the same as $^{+}2 + {}^{+}6$. With the method of modeling addition and subtraction described here, these expressions are quite discernible and yet have the same result as they should have.

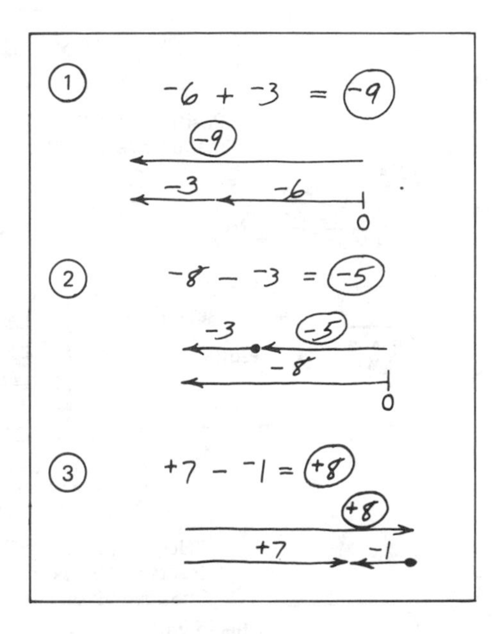

Figure 19.13
Students do not need elaborate drawings to think through the number line model.

Multiplication and Division

Multiplication of integers should be a direct extension of multiplication for whole numbers just as addition and subtraction were connected to whole number concepts. Whole number multiplication we frequently refer to as repeated addition. The first factor tells how many sets there are or how many are "added" in all, beginning with zero. This translates to integer multiplication quite readily when the first factor is positive, regardless of the sign of the second factor. The first example in Figure 19.14 illustrates a positive first factor and a negative second factor.

What could the meaning be when the first factor is negative, as in $^{-}2 \times {}^{-}3$? If the first factor positive means repeated addition (how many times added to zero) the first factor negative should mean repeated subtraction (how many times subtracted from zero). The second example in Figure 19.14 illustrates how multiplication with the first factor negative can be modeled. The success of your students in understanding integer multiplication depends on how well they understood integer addition and subtraction. Notice that there really are no new ideas, only an application of addition and subtraction to multiplication. (The same should be true for the introduction of multiplication in the third grade.)

The deceptively simple rules of like signs yield positive products and unlike signs negative products, are quickly established. However, one more time, it is not as important that your students be able to produce answers correctly and skillfully but that they be able to explain a rationale for those answers.

With division of integers it is again a good idea to explore the whole number case first. Recall that 8 ÷ 4 with whole numbers has two possible meanings corresponding to two

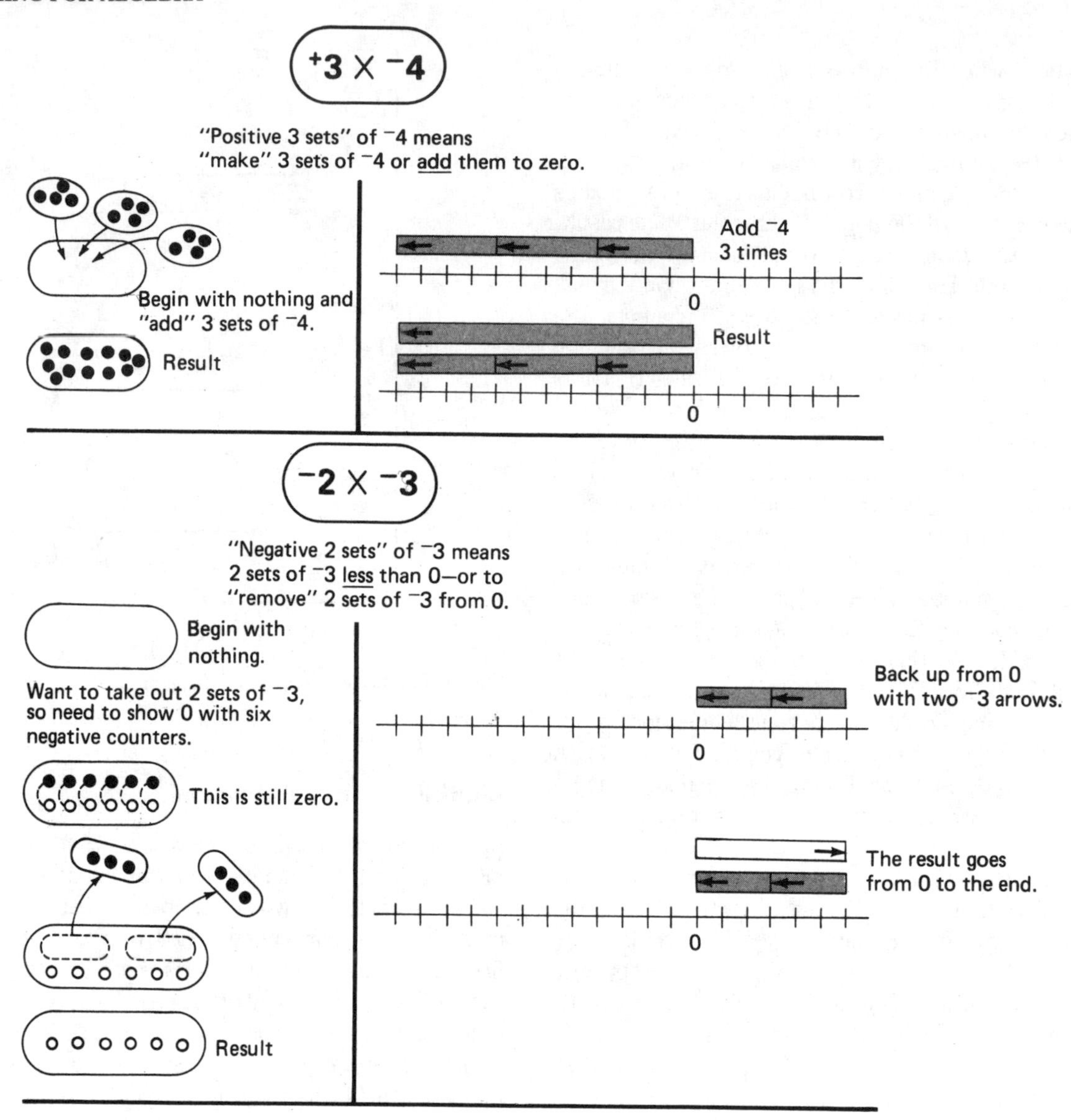

Figure 19.14
Multiplication by a positive first factor is repeated addition. Multiplication by a negative first factor is repeated subtraction.

missing-factor expressions: $4 \times \square = 8$ asks "Four sets of $\boxed{?}$ make eight?" while $\square \times 4 = 8$ asks "How many fours make eight?" Generally, the measurement approach ($\square \times 4$) is the one used with integers, although both concepts can be exhibited with either model. It is helpful to think of "building" the dividend with the divisor from zero in the same way that we fill units into an amount when we measure it. The first example in Figure 19.15 illustrates how the two models work for whole numbers. Following that is an example where the divisor is positive but the dividend is negative. How many sets of $^{+}2$ will make $^{-}8$ or $\boxed{?} \times {}^{+}2 = {}^{-}8$. With the set model, if we try to add positive counters, the result will be positive, not negative. The only way to use sets of $^{+}2$ to make $^{-}8$ is to remove them from zero. This means we must first change the representation of zero as illustrated. For the arrow model, consider how $^{+}2$ arrows can be placed end to end to result in a distance 8 to the left of zero. Starting at zero, the arrows must be "backed up." In both models, the arrows are being repeatedly subtracted or "added a negative number of times." Try now to model $^{+}9 \div {}^{-}3$ using both models. The approach is very similar to this example. Then try a negative divided by a negative. That case is much easier to understand. However, the entire understanding of integer division rests on a good concept of a negative first factor for multiplication and a knowledge of the relationship between multiplication and division.

There is no need to rush your students on to some mastery of use of the models. It is much better that they first think about how to model the whole number situation and then figure out with some guidance from you, how to deal with integers.

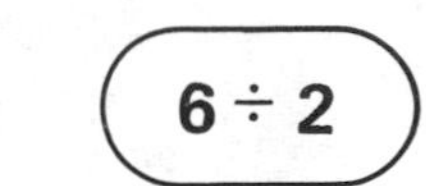

How many sets of 2 will make a set of 6?

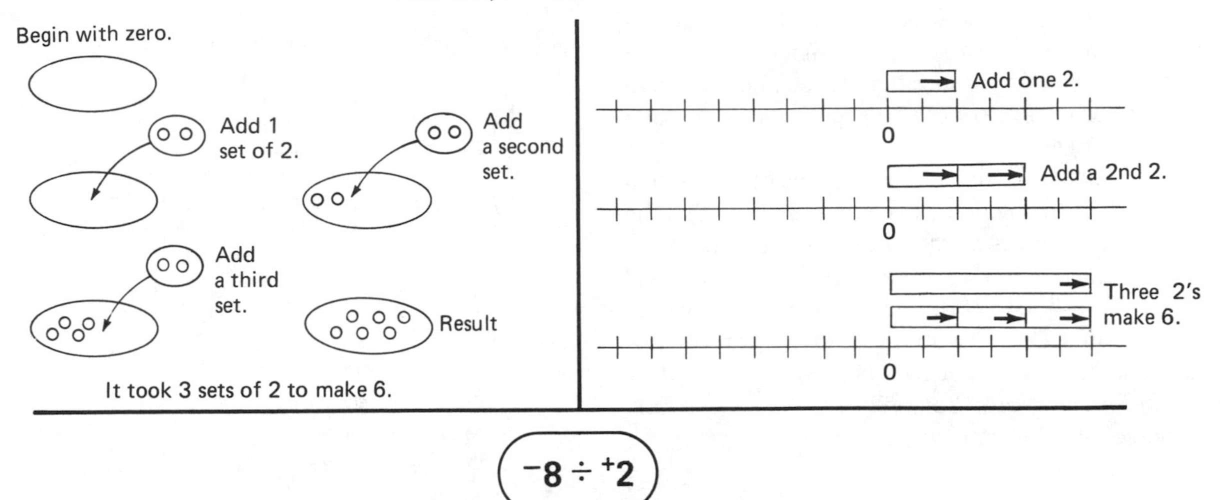

How many sets of $^{+}2$ will make $^{-}8$?
Adding $^{+}2$ to zero a positive number of times will result in a positive amount.
If $^{+}2$ is added a negative number of times (repeatedly subtracted), the result will be negative.

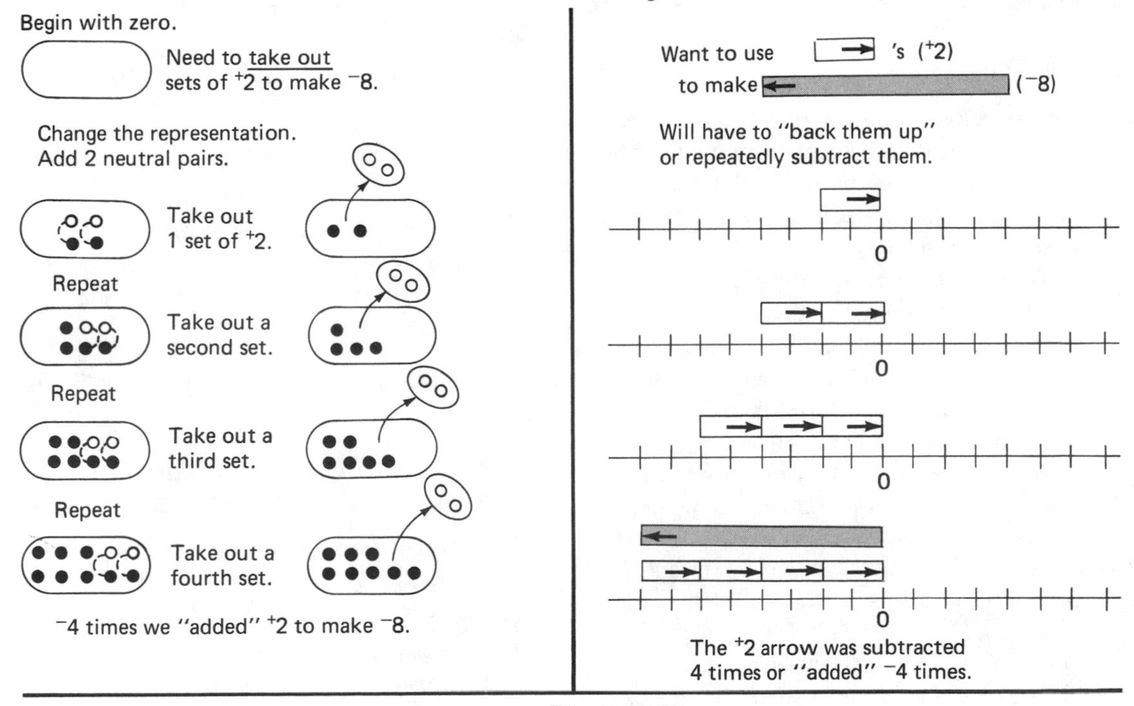

Figure 19.15
Division of integers following a measurement approach.

NEGATIVE RATIONALS AND FRACTIONS

An opposite exists for every positive real number, rational or irrational. When expressed as decimals or as variables, there is little difficulty in extending the ideas developed for whole numbers. Operations and symbolism for the negative real numbers are the same as for negative integers. Students do have notational difficulties with negative rational numbers expressed as fractions. As fractions, rational numbers involve three separate quantities, the numerator, denominator, and the fraction or rational number itself. Each can potentially be either positive or negative. Not just one but three distinct signs must be dealt with.

The concept of a negative rational number presents no

real difficulty. It is just as easy to think about a bill or debit of $3.75 or a $^-3.75$ quantity as it is to think about a debit of $4 or a $^-4$ quantity. All of the models suggested earlier will support rational numbers as well as integers. Therefore, it is reasonable that fractions exist where the sign of the fraction itself can be either positive or negative. In fact, both positive and negative rational numbers can be expressed as the ratio of two positive integers.

Recall that fractions also represent an indicated division (a/b means $a \div b$). In that context it is also reasonable that the top and/or bottom numbers might be negative. This is where the rule about a fraction having three signs comes in to play. That rule says that any two of these signs may be changed at any time. This can seem very strange and arbitrary to students. The case where one of the three signs is negative is explored in Figure 19.16. You and your students should convince yourself of the case where two of the signs are negative.

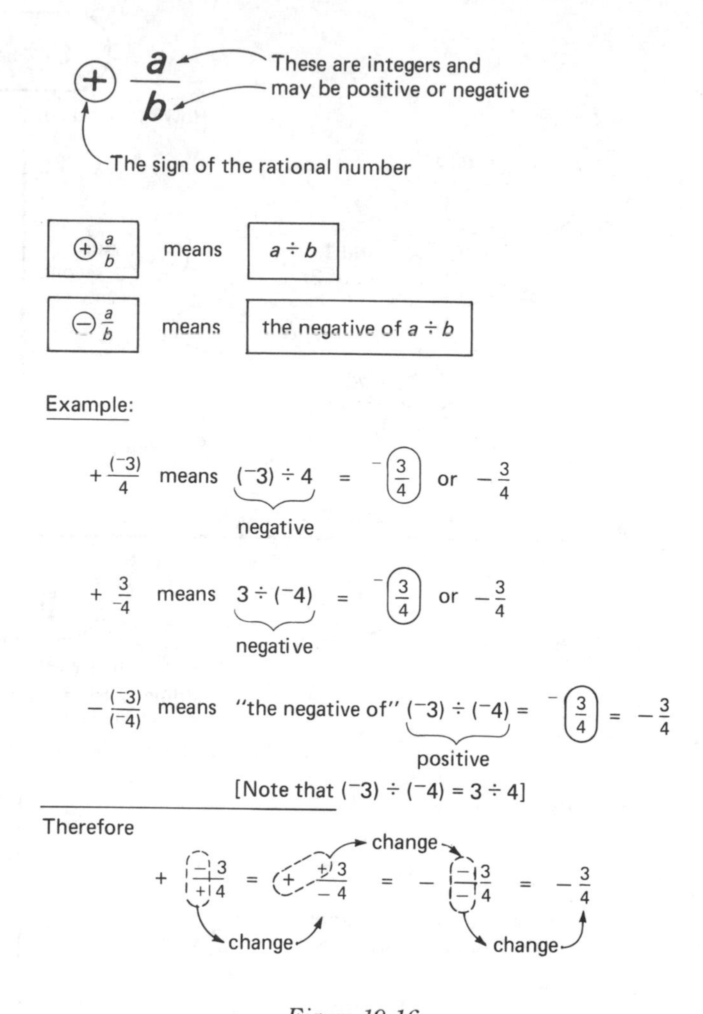

Figure 19.16
All four expressions are the same and each differs from any other by a change of two signs.

Developing Concepts of Variable

DIFFERENT MEANINGS OF VARIABLE

Misunderstandings

Even though students are exposed throughout the elementary years to boxes and letters in arithmetic expressions, studies indicate that most children have a very narrow understanding of variable (for example, see Booth, 1988; Chalouh & Herscovics, 1988). The most common interpretation is that a letter stands for a particular number. Wagner (1981) found many 12- to 17-year-old students believe that $7w + 22 = 109$ and $7n + 22 = 109$ will have different solutions. Some students might accept L for length but would have difficulty using the letter r instead. Some assume that the letter represents an object rather than a numeric value. For example, one child explained that $8y$ could possibly mean "8 yachts or 8 yams" (Booth, 1988). Notation used with variables only compounds the difficulties. For example, in algebra, $3n$ means "three times the value of n" but in arithmetic 37 means 3 tens and 7 ones. So in the expression, $5n$ if n is 2, many students will interpret the result as 52.

That these and other difficulties with variables exist suggests that we should take extra care to help students develop appropriate meanings for variable as a readiness for algebra and the many places in mathematics where variables are commonly used.

Meanings and Uses of Variables

Some of the difficulties with the notion of variable stems from the fact that variables are used in different ways. Meanings of variables vary with the way they are used. Usiskin (1988) identified four uses and corresponding meanings for variable. Of these, three meanings commonly occur in middle school mathematics.

1. *A specific yet unknown quantity.*
   Example: If $3x + 2 = 4x - 1$, solve for x.
2. *A representative from a range of values on which other values depend.*
   Example: $y = mx + b$ or $p = 1/t$.
3. *As a pattern generalizer.*
   Example: For all real numbers, $a \times b = b \times a$ or $N + 0 = N$.

The first meaning is closest to the one students tend to develop during the early elementary school years. In expressions such as $3 + \square = 8$ the $\square$ has a single "correct value." When moving to the other two meanings, the variables are not "unknowns" but may take on or represent any value from among a specified domain of values. In the second meaning, the variables are used to express relationships. In the equation $A = L \times W$, there are no numbers at all, but we can talk about what happens to the area for example, as the length doubles, then triples, and so on. Here, as Usiskin puts it, the "variables vary." In the third meaning, the letters again represent any value in a domain, but the purpose is to illustrate a property or pattern. Things that are known from arithmetic are generalized.

Students need to have experiences with these different meanings and uses of variables and become comfortable with them in different contexts. It is neither necessary nor appropriate to expect students to articulate different meanings or definitions of variable. At the same time, it is also not appropriate to offer definitions that are inaccurate or incomplete. For example, it is overly narrow to say "a variable is an unknown quantity" or "a variable stands for a number." These "definitions" relate almost exclusively to the first use of a variable as representing an unknown quantity, but hinder the understanding of the variable as used in the second meaning. In the equation $y = f(x) = 3x + 2$, the y and the x represent numbers, but not specific ones and there is no unknown to be found.

Related Knowledge from Arithmetic

Consider the following expressions involving variables:

$$3B + 7 \quad \text{and} \quad 3B + 7 = B - C$$

The first is an expression of a quantity. It may be a specific quantity or it may be a whole range of quantities related to a range of values of B. In the second expression, the same quantity is set equal to another quantity involving variables. The equal sign means that the quantity on the left is the same as the quantity on the right even though they are expressed with different letters and different operations. In order to interpret these expressions in this way, students must also interpret similar arithmetic expressions with only numbers in the same way. Expressions such as $3 + 5$ or 4×87 must be understood as single quantities and the equal sign (=) must be understood as a means of showing that two quantities represented are the same.

Unfortunately, while we use both of these ideas and assume all through elementary school that students understand them, a large majority of students do not completely comprehend them (Herscovics & Kieran, 1980; Kieran, 1988; Behr, Erlwanger, & Nichols, 1976). Students tend to look on expressions such as $3 + 5$ and 4×87 as commands or things to do. The + tells you to add and students think of adding as a verb. The idea that $5 + 2$ is another way of writing 7 is not considered. The equal sign is commonly thought of as an operator button: something like pressing ⊟ on a calculator. As students read left to right in an equation, the = tells them, "Now give the answer." In a similar sense, students think of = as a symbol that is used to separate the question or problem from the answer.

The following simple activities are suggested as ways to help students with these basic concepts. They each initially use only numbers but can easily be extended to include variables.

Activity 19.9

Names for Numbers. Challenge students to find different ways to express a particular number, for example 10. At first suggest that they use only + or only − expressions such as $5 + 5$ or $12 - 2$. Later, devise more challenging tasks such as "How many names for eight can you find using only numbers less than 10 and using at least three different operations." One solution is $(5 \times 6 + 2) \div 4$. Make it a point to emphasize that each expression is a way of representing a number. Restrictions on which operations and how many and which numbers can be used, can produce real challenges. Groups can compete with each other for the most expressions in a specified time.

Activity 19.10

Tilt or Balance. On the board draw a simple two-pan balance. In each pan write a numeric expression and ask which pan will go down or will it balance [Figure 19.17(a)]. Challenge students to write expressions for each side of the balance that will make it balance. For each, write a corresponding equation or inequality. Soon the scale feature can be abandoned, and students can simply write equations according to the directions given. For example, use at least three operations on each side and do not use any number twice.

These two activities are clearly related. To understand the balance or equation idea requires an understanding that the expressions in each pan or side of the equation represent quantities.

After a short time with these two activities, variables can be added to each. Instead of names for numbers, have students write expressions for quantities such as the following:

- Two more than three times a number ($2 + 3\square$ or $2 + 3 \times N$)
- Any odd number ($2N + 1$)
- The average age in this class; use S for the sum of our ages ($S/23$)
- The average age of students in any class (S/n)
- A number cubed less another number squared ($A^3 - B^2$)

Notice in these examples, that variables are used as both specific unknowns and as unspecified quantities.

Activity 19.11

In the two-pan balance, have students write expressions in either or both pans that include variables. Use directions similar to the ones that were just listed. Students may then use a trial-and-error approach to find numbers that can be substituted for the variables to make the equation balance (true equation) or tilt (a false equation). Use calculators for tedious calculations [Figure 19.17(b)].

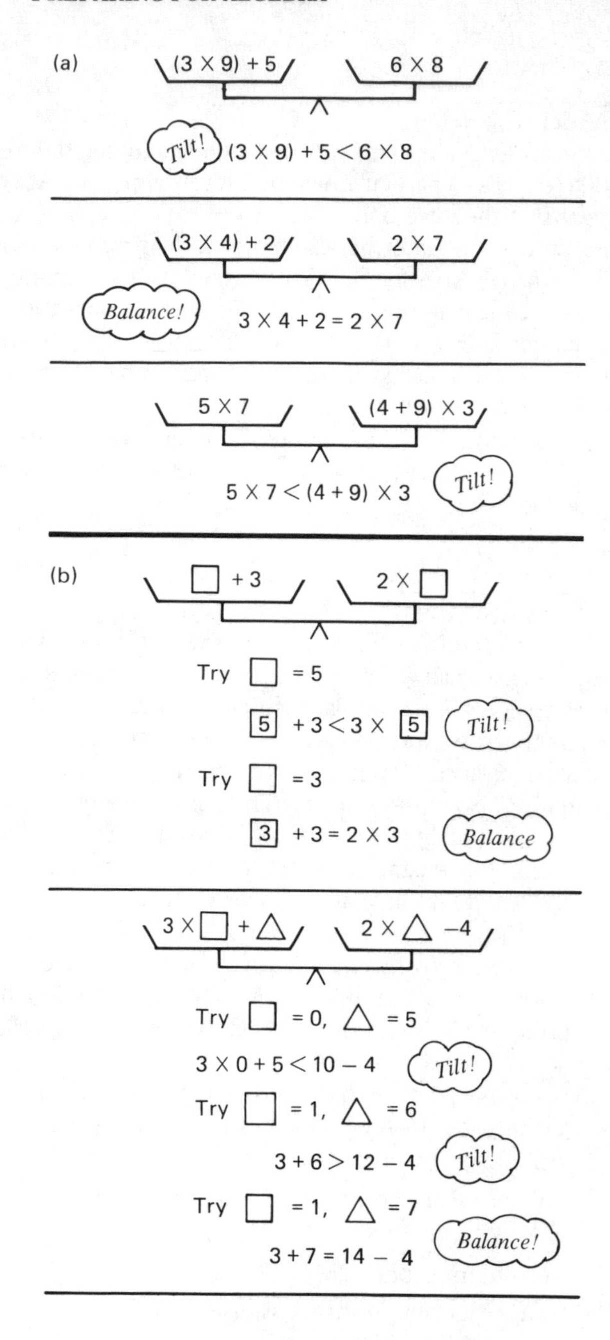

Figure 19.17
Equations, expressions, and variables.

n	$(n+2)$	$3(n+2)$	$3n+6$
0			
2			
−1			
−2			

b	$2b$	$6b$	$6b \div 2b$
0			
1			
2			
3			

x	y	$x+y$	$3x$	$(x+y)/3x$	$1+y$	$(1+y)/3$
1	0					
−1	1					
2	2					
−2	−1					

Figure 19.18
Complete these charts. What ideas can you discover?

MORE EARLY EXPERIENCES WITH VARIABLES

Evaluating Expressions and Formulas

In Figure 19.18, examples of charts or tables are shown. Each chart has a separate column for the variable and the variable expressions across the top of the chart are all related. Students can evaluate the expressions and fill in the charts. In this type of activity, several important ideas about notation and properties of the number systems can be explored informally. For example, $4n$ and $4 \times n$ can be placed in separate columns to help students see this convention for expressing multiplication. Other ideas include

$8x - 3x$ with $5x$ (many students think $8x - 5x$ is 5)
$b(3+b)$ with $3b$, b^2, and $3b + b^2$
$(3 + 2x) - x$ with $3 + x$
$y/5$ with $y \div 5$ (students may not realize these mean the same thing)
$-x$ and $(-1)x$

$$3k+\frac{1}{k} \text{ with } (3k+1) \text{ and } \frac{3k+1}{k}$$

The entries in the charts can also be expressions that appear in geometric formulas or formulas from other areas such as science. As an example, while reviewing volume and surface area formulas, you might have students help you build a chart for a right circular cylinder with the following expressions:

$$r \quad h \quad \pi \quad 2\pi r^2 \quad \pi r^2 h \quad 2h\pi r$$

With two variables, you may want to have students hold one variable fixed for several values of the other. In the preceding example, observe the values of the various columns as r changes from 1 to 5 and h remains fixed at 10. Then change h to 20 and repeat.

Another variation with these tables is to enter values in the columns and have students determine the value of the variable. Notice with this variation two different uses of variable are now included in the same example: letting the value of the variable vary and also solving for a fixed value.

Solving Equations

Students can learn appropriate techniques for solving simple linear equations before they reach a formal algebra class. A good technique is to capitalize on the two-pan balance approach that was suggested earlier. With the balance, the idea of doing to one side what you do to the other can be developed meaningfully.

In Figure 19.19, a balance scale is set up with two numeric expressions. A box is drawn around both of the fives. The task is to manipulate the numbers so that there is only one boxed 5 on the left. You can operate on numbers or on boxed 5's but you may not combine the 5 in the box with any other

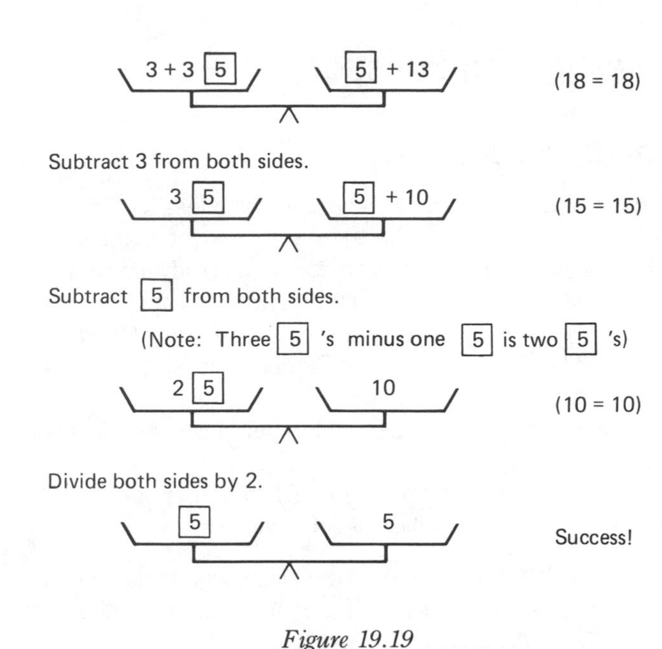

Figure 19.19
Learing how to solve an equation.

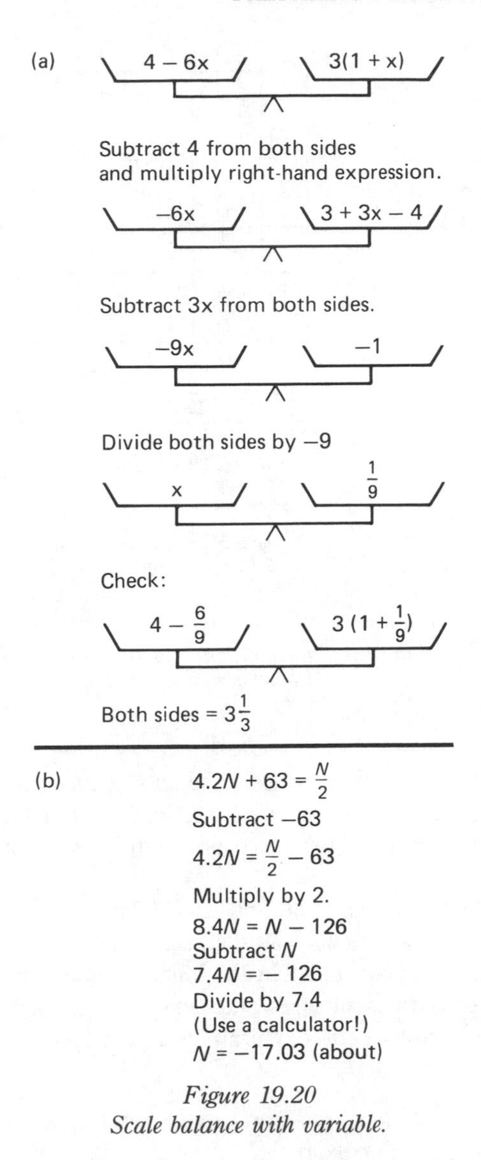

Figure 19.20
Scale balance with variable.

numbers. Obviously, in order to keep the scale balanced, you must do the same things to both sides of the balance. At each step along the way, the balance can be confirmed.

After one or two similar examples, put variable expressions in the balance pans and have students decide how to isolate the variable on one side. The other side should all be numeric.

Even after you have stopped drawing the balance, it is a good idea to refer to the scale or pan-balance concept of equation and the idea of keeping the two sides balanced. Operations on both sides of the equation should be done as illustrated in Figure 19.20, underneath the last equation. Division of both sides by a constant is best done with fraction notation.

A goal in these activities is to promote the use of inverse operations to solve an equation. Teachers are frequently frustrated when students prefer to solve simple equations such as $3N + 2 = 11$ by inspection or trial and error. One effective tactic is to use large numbers and avoid "nice" whole number answers. For all computations, have students use a calculator to avoid tedium.

Graphing Equations with Two Variables

When an equation has two variables, there are usually many solutions. Exploring the solutions to equations such as $3x + 5 = 2y - x$ can have several values for your students. First, it provides a good introduction to graphing equations. Second, it provides experience with variables that can take on a range of values in contrast to the use of variable as an unknown fixed quantity. Third, graphing equations can be fun, interesting, and demonstrate the form and beauty of mathematics.

Begin with simple linear equations and approach the task as one of finding solutions to the equation. Compare this process with solving equations with only one variable. Now instead of one solution, there are infinite solutions. In the beginning, let students find their own methods for finding solutions. That can be an interesting problem-solving task if they are not familiar with a procedure. Eventually, show them how to select a value for one variable and then solve for the other. Demonstrate how a chart can be used to keep track of the solutions.

When a number of solution pairs have been found, show students how to plot each pair on a graph. Emphasize that the graph is simply a visual record of the solutions. When enough solutions are plotted, it will become obvious that there is a pattern to the points: a straight line for linear equations and curved lines for others. An important investigation is to see if the graph of some points can help find other solutions to the equation (Figure 19.21). Pick points that fit the pattern of the graph but have not been plotted and check to see if those coordinates are solutions to the equations. With the use of calculators to handle decimal computation, students can easily check points that are not integers. While it obviously is impos-

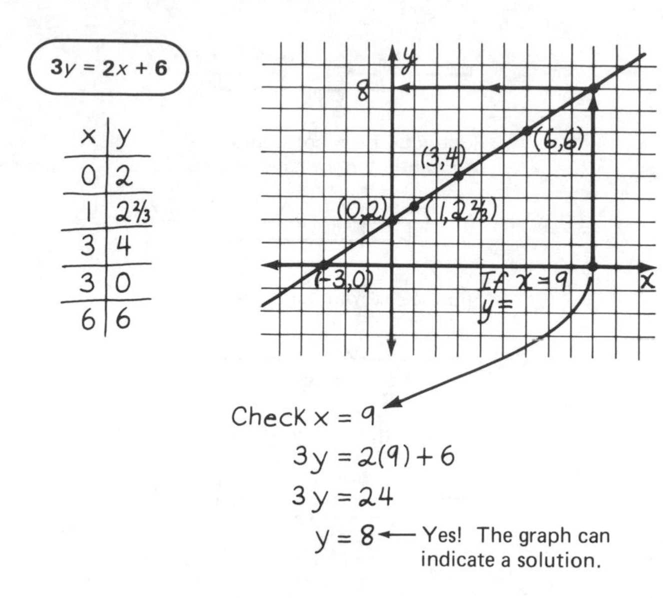

Figure 19.21
Plotting points that satisfy an equation. The line can then be used to find other solutions.

sible to plot an infinite number of points, you want students to discover that if a continuous line joined all of the points, every point on the line corresponds to a solution to the equation.

It is not necessary at this prealgebra level to analyze graphs and properties such as slopes and intercepts. The idea is to explore the concept that the infinite solution pairs to an equation tend to form lines and curves when plotted on a coordinate graph. Your explorations might include some of the following:

Plotting solutions to several equivalent forms of the same equation. For example, $3y + 6x = 12$, $y = -2x + 4$, and $2x + y - 4 = 0$. Have different groups work on different forms and compare and discuss results.

Plotting solutions to two or three equations on the same graph. What is the significance of the points where the graphs cross?

Changing one number in an equation and observing the effects on the graph. For example, in the equation $3x - y = 4$, what if the 4 is changed to a 0 or 2? How do the graphs compare? What if the minus sign is changed to a plus? Try similar changes on other graphs. Keep the discussions and explorations informal.

Plotting solutions to equations that are not linear. There is no reason why students cannot use these simple direct techniques to explore other types of equations. It is helpful to present these equations in an easy-to-use form, usually with one variable already isolated on the left of the equal sign and to use calculators. For example, explore equations similar to these:

$y = 5/x \qquad y = 3x^2 + 1 \qquad y = 2^x$

Investigating Functions

GOALS AND OBJECTIVES

The concept of function is one of the "big ideas" or "common threads" that pervade almost every area of mathematics. While the function concept is critical to the study of mathematics, it is not an easy concept for students to completely understand, even in the ninth-grade algebra course (Markovits, Eylon, & Bruckheimer, 1988). Before a formal and symbolic understanding of function can be appreciated, students need to have a background of informal experiences with functions in a variety of forms.

With respect to the concept of function, students at the middle or junior high school level should

1. Explore meaningful relationships as functions, including the informal use of the language of functions
2. Have experiences with a variety of ways to express function relationships

IMPORTANT FUNCTION CONCEPTS

What Is a Function?

A *function* is a rule of correspondence between the elements of one set, called the *domain* of the function, and another set, known as the *range,* such that each element of the domain corresponds to one and only one element in the range.

Rules of correspondence (properly called *relations*) do exist that are not functions because they violate the uniqueness property that is required for a relation to be a function. An objective of most beginning algebra courses is for students to be able to distinguish between relationships that are functions and those that are not. Students also are required to identify the domain and range sets for a given function. At the elementary level, it is more important to work with and experience functions in a variety of contexts and to develop familiarity with different ways to represent functions.

Different Representations of Functions

At least four different representations of functions are commonly used: "function machines" or flowcharts, arrow diagrams, algebraic equations, and graphs. Examples are illustrated in Figure 19.22. Each of these models or representations of functions provides a slightly different way of thinking about a function and consequently all forms are useful and important.

Students can make up their own functions using any of these formats. With the function machine, they can put simple operations in a flowchart or make up a verbal rule.

There are two ways to show functions with arrows. With two loops, the domain (set where the arrows begin) and range are shown separately. In the other form, the domain and range

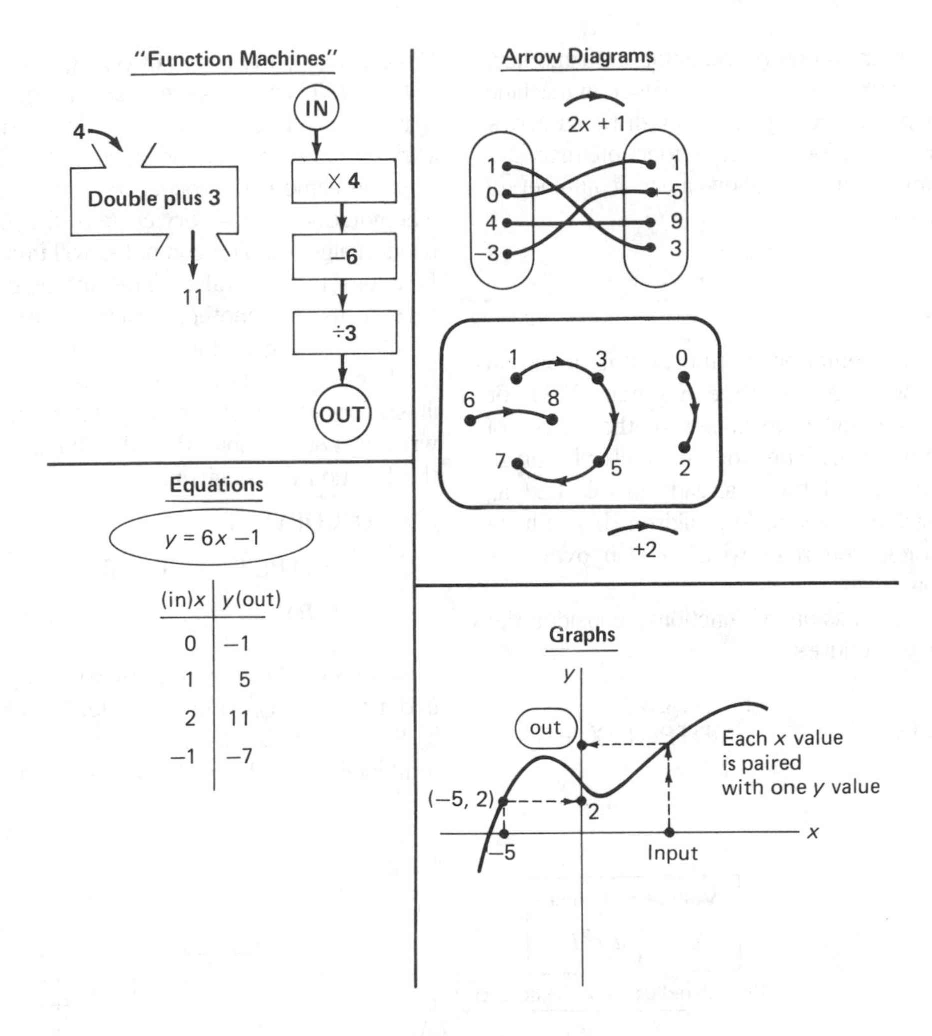

Figure 19.22
Four different representations of functions.

are mixed. Any arrow drawing will define a function as long as there is only one arrow going *from* each dot or number. A rule such as "double the number and subtract five," can be used to determine where the arrows should be drawn. Another idea is to use an existing function machine rule. An input number is the beginning of an arrow and an output number is the end of the arrow.

An equation with two variables can determine a function because it has solutions which are pairs of numbers. The chart of solution pairs shows the correspondence between particular elements. Generally, when an equation is used to determine a function, it is solved for one variable in terms of the other, as in $y = 3x + 2$. Values of the isolated variable are the range elements (the outputs of the machine, the end points of the arrows). Later, expressions like "$f(x)$" will replace the isolated variable or second element. This symbolism is confusing in the beginning, however. The point to be made here is that the equation can be a "rule of correspondence."

Since every point on a graph has two coordinates, it is a way of pairing up two sets of numbers. It is not even necessary for the graph to have an equation. Students could draw any line or curve on a coordinate graph and the line is a picture of a relationship between two sets of numbers: the first and second coordinates of the points on the graph. In order for the correspondence rule shown by a graph to also be a function, the uniqueness property of functions must be met. Graphs will determine functions whenever no two points anywhere on the graph are directly above each other. (Why?)

There are a variety of ways that each of these function representations can be explored informally.

Activity 19.12

Give students any single representation of a function. You can put it on the chalkboard, the overhead, or on a handout. With this representation you can have students

Find the output of a function for particular inputs.

Find the value or values that will produce a particular output.

State a rule that goes with the function you have given.

Translate a function from one representation to another. For example, write an equation for a function machine rule and then graph the equation, or draw an arrow diagram for an equation or for a function machine. (Arrow diagrams can only show a small number of correspondences.)

Functions in Logo

Another powerful representation of function is found in the Logo programming language. (See Kovatch, 1984, or Goldenberg, 1986, for detailed examples of the power of Logo for exploring functions.) The Logo computer language has been referred to several times already in this text as a powerful yet accessible language for children. If you have no knowledge of Logo, you may wish to skip over this short section for now.

Returning to the discussion of functions, consider the following two simple procedures:

```
TO F :X
  OUTPUT 3 * :X + 5
  END
```

```
TO S :X
  OUTPUT :X - 5
  END
```

Once defined, the result of typing F 4 is 17 and the result of S 4 is −1. There are several advantages to using Logo to look at functions. They are easy to define. Students who have used variables in Logo will understand the difference between the function name (the procedure name) and the variable name. The notation used is very close to the $f(x)$ notation commonly used in algebra. The computer will immediately evaluate the functions for any value. The output can be printed on the screen, used in another formula or function, or plotted on a graph (with a few additional procedures to do the graphing.) A different procedure can be used to define the same function, illustrating that it is the rule that is important, not how it is written. For example, the following procedure is the same as the function F just given:

```
TO COPYF :T
  OUTPUT 3 * :T + 5
  END
```

Functions can be composed (like putting one function machine under the other). For example, F (S 8) takes the result of S 8, which is 3, and uses it as input for F, with the combined result being 14. [What is the result of S (F 2)?]

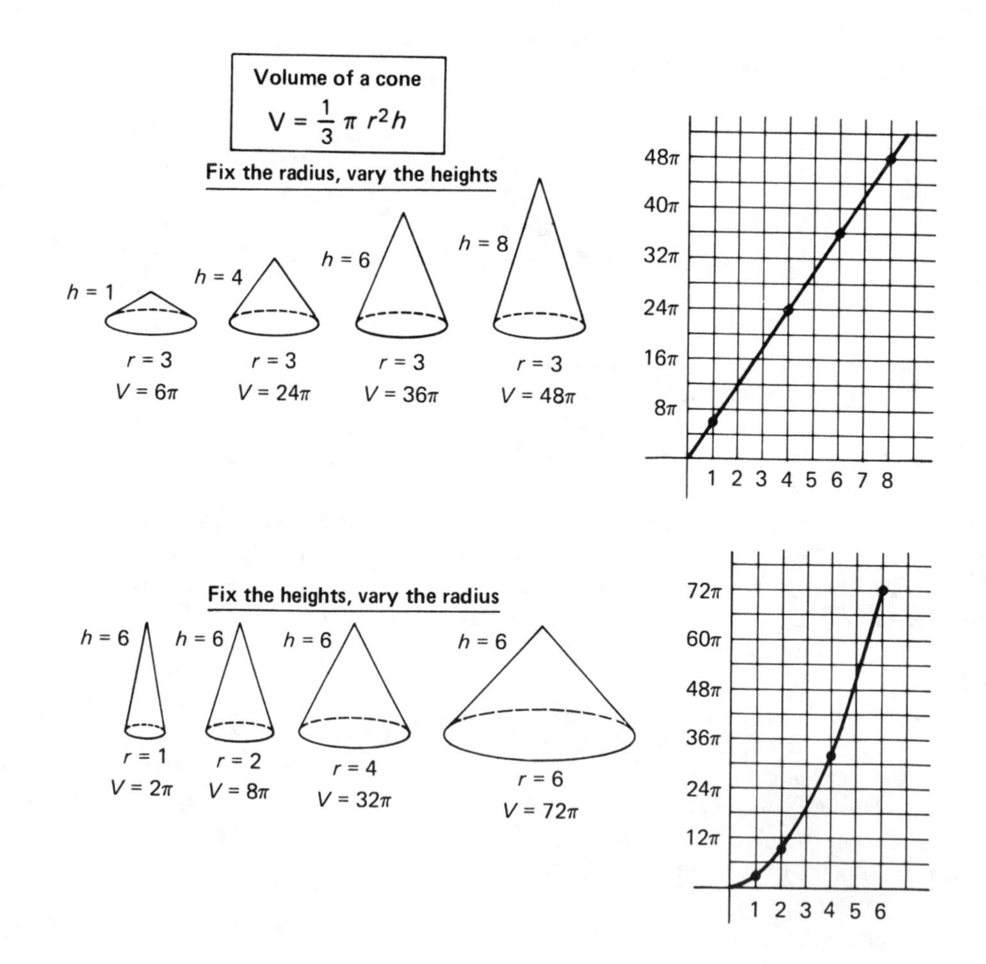

Figure 19.23
Volume as a function of height or radius. If the radius is fixed the changes in heights produce a straight line graph but if the height is fixed the radius changes produce a curved line.

EXPLORING FUNCTIONAL RELATIONSHIPS

Students should see that functional relationships are not simply symbolic formulas. Functions exist in a variety of real situations. A valuable type of activity is to have students define functions that reflect some real experience, measurement, or other observation.

Functions in Geometry and Measurement

In most geometric figures, in either two or three dimensions, the change in one dimension affects the measurement of another. This provides a good example of a functional relationship.

Consider any formula for measuring a geometric shape with which students are familiar. For example, $V = (1/3)\ \pi r^2 h$ is the formula for the volume of a circular cone. Figure 19.23 shows the graphs of volume as a function of height for a fixed value of the radius and as a function of the radius for a fixed value of height.

Activity 19.13

Show the class a particular shape either flat or solid. Let different groups select any two measures that are not directly related by a familiar formula. By systematically letting one of these measures change by making different drawings and measuring with rulers and protractors, they can observe the relationship between the two measures. Suggest that they make large drawings. The results can be graphed even if a formula cannot be found (Figure 19.24).

Draw a whole family of isosceles triangles on centimeter graph paper. Keep the height fixed. Let the base (between the equal sides) vary from one to 20. How will the angle at the top change?

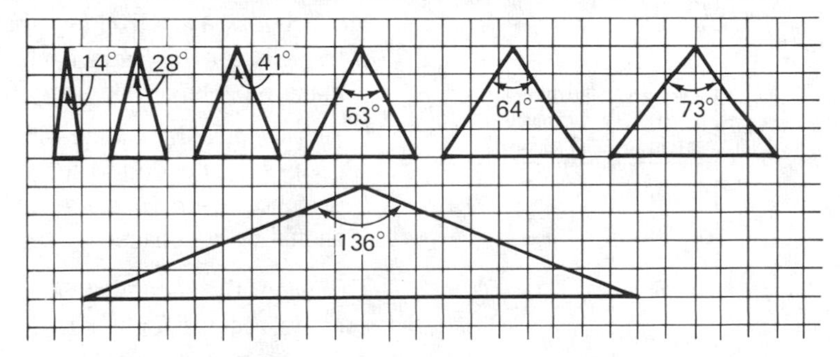

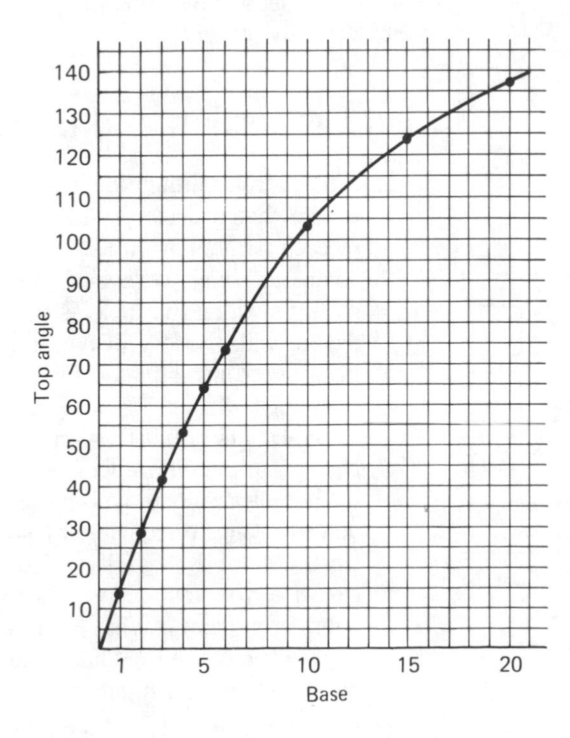

If the heights of these triangles are at least 10 cm, the angles can be measured surprisingly accurately. No formulas are required, and the graph paper makes the triangles easy to draw.

Figure 19.24
One angle of an isosceles triangle as a function of the base.

Functions from Real Data

Students can take measurements from an experiment and plot the data for two related measures on a graph. From this plot of points, they can draw a smooth line that approximates the points. This line can then be used to predict the results for other trials. It is not necessary that an equation for the curve be determined.

Activity 19.14

One example of an experiment that involves functional relationships is determining the number of swings of a pendulum in 10 seconds compared with the length of the string (Figure 19.25). Variations would be

The height of students in the school compared with the height of their head (or arm span). Get data from students in different grades to find different lengths. When two or more students have the same height, use their average head height.

The time it takes a toy car to roll down a long inclined plane as compared to the angle of inclination or the height of the end of the board.

The length of a shadow cast by a pole as compared to different times of day.

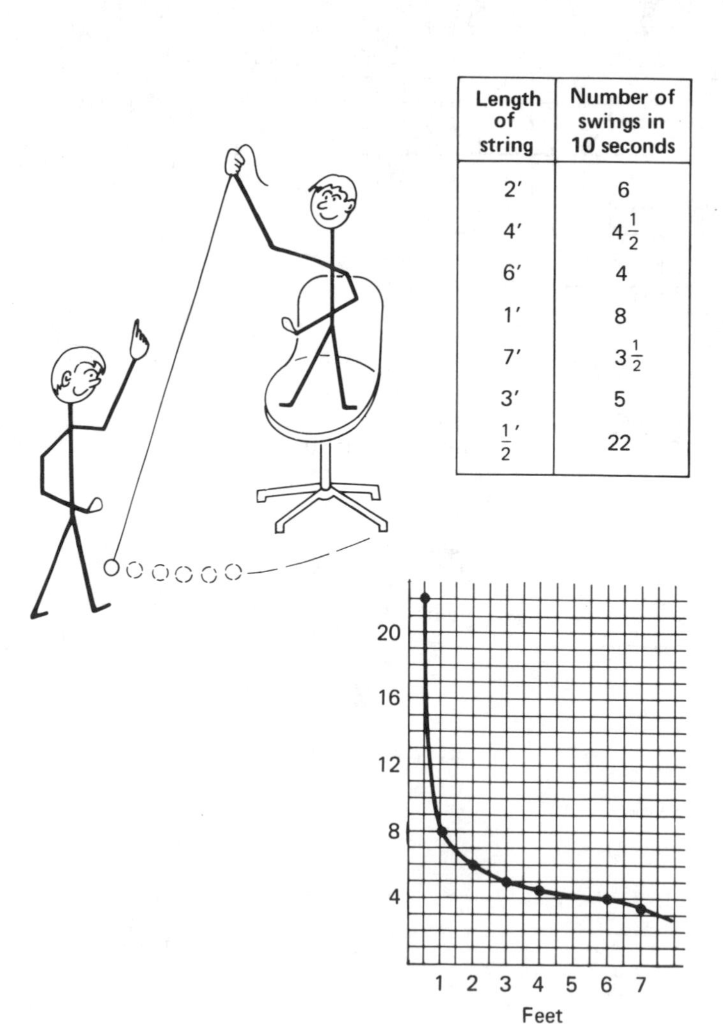

Length of string	Number of swings in 10 seconds
2′	6
4′	$4\frac{1}{2}$
6′	4
1′	8
7′	$3\frac{1}{2}$
3′	5
$\frac{1}{2}$′	22

Figure 19.25
Swings of a pendulum as a function of the length of the string.

Functions as General Solutions to Word Problems

Consider the following word problem:

Two out of every 3 students who eat in the cafeteria drink a pint of white milk. If 450 students eat in the cafeteria, how many gallons of milk are consumed?

In this problem there is a fixed number of students and a single answer to the problem. A more practical problem is to write an equation that will give the number of gallons of milk "as a function of" the number of students. For this problem the equation $G = (\frac{2}{3}S) \div 8$ is a solution. (There are 8 pints in a gallon and two thirds of the students drink milk.)

The following problems lend themselves to similar solutions.

Mr. Schultz pays $4.00 for a box of 12 candy bars, which he sells for 45 cents each. How much profit will he make on n boxes of candy bars?

If each recipe of lemonade will serve 20 people, how many recipes are needed to serve n people? If it takes 3 cans of concentrate to make one recipe, how many cans should be purchased to serve n people?

To answer the second question, write two functions, one that outputs recipes for n people and another that outputs cans for recipes. Now substitute one function in place of the variable in the other.

For Discussion and Exploration

1. A teacher in a fifth grade in Michigan asked her students to use their calculators to make a table of the square of each number from 1 to 100. When the tables were complete, they were challenged to look for as many patterns within the table as they could find. Over three days a wide variety of patterns were discovered, and many were investigated further. Most focused on the ones digit of the numbers or patterns within the last several digits rather than on the magnitudes of the numbers or on the differences between the numbers. Even further discussion, hypothesizing, and exploration involved trying to explain why some patterns happened. (Institute for Research on Teaching, 1988). Try this exercise yourself or preferably with several of your peers or students. What patterns can you find? Can you explain one of them? What value do you think there may be in such an investigation?
2. Read the article, "Why Elementary Algebra Can, Should, and Must Be an Eighth-Grade Course for Average Stu-

dents" (Usiskin, 1987). In this article Usiskin argues for moving the curriculum forward to algebra in the eighth grade for all students. His position is somewhat different from that of the NCTM *Standards*. What is your view?

3. Find the LCM and GCD of 7020 and 44,550 using the methods described in this chapter. (First factor into primes, write the prime factors on slips of paper, and then use one of the techniques shown in Figure 19.3.) What is the value or purpose of such an activity? Examine a junior high textbook for similar activities listed under number theory.

4. Provide an informal argument for the divisibility-by-8 rule.

5. a. Use a calculator to estimate the cube root of 10 to five decimal places.
b. What do you think will happen if you enter 1000 in your calculator and then press [÷] 2 [=] [=] [=] . . . ? Try it.
c. What do you think will happen if you enter 1000 in your calculator and then continually press the square root key? Before you try it, try to explain why you think it will happen.
d. With the help of your calculator see if you can find a prime number that is between 500 and 1000 and prove that it is prime.

6. Use both the arrow model and the counter model to demonstrate the following:
$^{-}10 + {}^{+}13 = {}^{+}3$ $\quad ^{-}4 - {}^{-}9 = {}^{+}5$ $\quad ^{+}6 - {}^{-}7 = {}^{+}13$.
$^{-}4 \times {}^{-}3 = {}^{+}12$ $\quad ^{+}15 \div {}^{-}5 = {}^{-}3$ $\quad ^{-}12 \div {}^{-}3 = {}^{+}4$.

7. Complete the tables in Figure 19.18. Explain what students are supposed to be observing when doing this exercise.

8. Make up a simple flowchart similar to the one in Figure 19.22. Determine the outputs of your function machine for every integer between $^{-}5$ and 5. Draw an arrow diagram for these 11 pairs in your function and plot the corresponding 11 points on a graph. Explain how your graph can operate like a function machine. Can you use your graph to predict the output if the number 6 or the number $^{-}6$ is put in your flowchart? What about the number 3.5?

9. Think up an experiment that produces a set of data like the drawings of triangles or the pendulum swing experiment. Can you find a functional relationship in your experiment? If so, make a graph of your function. What is the domain variable? What is the range variable?

10. Examine the contents of a seventh- and eighth-grade textbook for the topics of integers, functions, and variables. How much emphasis are these topics given? Describe the best idea you can find for each of these topics. Teacher's editions will be best.

11. Examine a textbook labeled "prealgebra" for the same three topics: integers, functions, and variables. How much emphasis are these topics given? If you did 10, make a comparison between a regular eighth-grade book and a prealgebra book. Describe one good idea you found for teaching each of these topics. Use a teacher's edition.

Suggested Readings

Battista, M. T. (1983). A complete model for operations on integers. *Arithmetic Teacher, 30*(9), 26–31.

Bennett, A., Jr. (1988). Visual thinking and number relationships. *Mathematics Teacher, 81,* 267–272.

Blubaugh, W. L. (1988). Why cancel? *Mathematics Teacher, 81,* 300–302.

Coxford, A. F. (Ed.) (1988). *The ideas of algebra.* Reston, VA: National Council of Teachers of Mathematics.

Dirks, M. K. (1984). The integer abacus. *Arithmetic Teacher, 31*(7), 50–54.

Driscol, M. (1982). The learning of algebra. In *Research within reach: Secondary school mathematics.* Reston, VA: National Council of Teachers of Mathematics.

Edwards, F. M. (1987). Geometric figures make the LCM obvious. *Arithmetic Teacher, 34*(7), 17–18.

Fitzgerald, W., Winter, M. J., Lappan, G., & Phillips, E. (1986). *Middle grades mathematics project: Factors and multiples.* Menlo Park, CA: Addison-Wesley.

Herscovics, N., & Kieran, C. (1980). Constructing meaning for the concept of equation. *Mathematics Teacher, 73,* 572–580.

Lenchner, G. (1983). *Creative problem solving in school mathematics.* Boston: Houghton Mifflin.

Martinez, J. G. R. (1988). Helping students understand factors and terms. *Mathematics Teacher, 81,* 747–751.

Osborne, A., & Wilson, P. S. (1988). Moving to algebraic thought. In T. R. Post (Ed.), *Teaching mathematics in grades K–8: Research based methods.* Boston: Allyn and Bacon.

Schoenfeld, A. H., & Arcavi, A. (1988). On the meaning of variable. *Mathematics Teacher, 81,* 420–427.

Sobel, M. A., & Maletsky, E. M. (1988). *Teaching mathematics: A sourcebook of aids, activities, and strategies* (2nd ed.). Englewood Cliffs, NJ: Prentice Hall.

Whitman, B. S. (1982). Intuitive equation-solving skills. In L. Silvey (Ed.), *Mathematics for the middle grades (5–9).* Reston, VA: National Council of Teachers of Mathematics.

CHAPTER 20

TECHNOLOGY AND ELEMENTARY SCHOOL MATHEMATICS

Calculators in the Classroom

In 1976, NCTM published a special issue of the *Arithmetic Teacher* on the use of calculators. Since that time numerous professional groups, including NCTM, have published recommendations encouraging a full integration of calculators into the mathematics classroom at all levels. See, for example, the official NCTM position statement on calculators. In 1987 a second focus issue of the *Arithmetic Teacher* was dedicated to use of calculators. The quality of calculators has increased, prices have gone down, batteries have been made unnecessary, and virtually every business, industry, and home now relies heavily on their use.

During the same time period, the elementary school mathematics curriculum, what is actually taught in schools, has seen the most modest of changes with respect to calculators. Only scattered and limited use is made of the calculator as a teaching tool. Only miniscule changes in content have occurred to reflect society's clear and pervasive use of the calculator as the favored method of exact computation. In contrast to the stampede to have computers in schools, a much more costly and perhaps debatable move, the failure to maximize the potential of the calculator is perplexing. While many calculator activities have been provided throughout this book, suggesting a variety of appropriate uses, the following section sadly remains a required feature of a text of this nature.

REASONS FOR USING CALCULATORS

Common Usage in Society

The fact is, almost everyone uses calculators in almost every facet of life that involves any sort of exact computation—except school children. The traditional reasons for teaching

Calculators in the Mathematics Classroom

The following statement is an official NCTM position. It was developed jointly by the Technology Advisory Committee, the Instructional Issues Advisory Committee, and the Committee to Implement the Recommendations of An Agenda for Action *and adopted by the Board of Directors.*

The National Council of Teachers of Mathematics recommends the integration of the calculator into the school mathematics program at all grade levels in classwork, homework, and evaluation. Although extensively used in society, calculators are used far less in schools, where they could free large amounts of the time that students currently spend practicing computation. The time gained should be spent helping students to understand mathematics, to develop reasoning and problem-solving strategies, and, in general, to use and apply mathematics.

At each grade level every student should be taught how and when to use the calculator. To use calculators effectively, students must be able to estimate and to judge the reasonableness of results. Consequently, an understanding of operations and a knowledge of basic facts are as important as ever. The evaluation of student understanding of mathematical concepts and their application, including standardized tests, should be designed to allow the use of the calculator.

The National Council of Teachers of Mathematics recommends that all students use calculators to—

- concentrate on the problem-solving process rather than on the calculations associated with problems;
- gain access to mathematics beyond the students' level of computational skills;
- explore, develop, and reinforce concepts including estimation, computation, approximation, and properties;
- experiment with mathematical ideas and discover patterns;
- perform those tedious computations that arise when working with real data in problem-solving situations.

The National Council of Teachers of Mathematics recommends that publishers, authors, and test writers integrate the use of the calculator into their mathematics materials at all grade levels. (April 1986)

Data supporting this position statement can be found in the article "Effects of Hand-Held Calculators in Precollege Mathematics Education: A Meta-Analysis," by Ray Hambree and Donald J. Dessart (*Journal for Research in Mathematics Education,* March 1986).

pencil-and-paper computation, especially with numbers involving more than two or three digits, have all but evaporated. It is more than a little hypocritical to forbid the use of calculators.

It also makes good sense that students should know how to use this popular tool effectively. Many adults have not learned how to use the memory keys, how to do a chain of mixed operations, how to utilize the automatic constant feature, or how to quickly judge if a gross error has been made. These are important practical skills that can easily be learned over the school years if the calculator is simply there for open, everyday use.

Affective and Indirect Benefits

The overwhelming conclusions of numerous research studies have found that students' attitudes towards mathematics are better in classrooms where calculators are used than where they are not (Hambree & Dessart, 1986; Reys & Reys, 1987). Students using calculators tend to be enthusiastic and more confident and persistent in solving problems.

In addition to positive affective results, students using calculators discover a wide variety of interesting ideas that might otherwise remain unnoticed. For example, decimal numbers and negative numbers are almost inescapable, and children learn to explore these ideas at an early age. The number of digits that result in some computations is very different than in others. For example, $2 \div 7$ fills up the display but $1245 \div 5$ has only three digits. Calculators act strangely sometimes. On many calculators $5 \div 7 \times 7$ results in 4.9999999 instead of 5. These are just two examples of simple results that students tend to explore and learn from. Students quickly find that it is easy to make errors on a calculator and develop a real appreciation for estimation. Students also learn that it is frequently easier to do a mental computation than to search for a calculator and press buttons. All of these outcomes are positive benefits that can result without any direct instruction. They can happen just by having calculators on every desk, every day, all of the time, at every grade.

Instruction Enhanced

Not only does the calculator make suspect the value of spending significant portions of the school curriculum teaching pencil-and-paper computation, the calculator is as much a teaching tool as the chalkboard or overhead projector. That is, the calculator not only replaces much of the tedious computation required in school, it can be used to help develop concepts and skills in other areas of mathematics. While activities for teaching with the calculator have been suggested throughout the book, a few additional examples are presented here as examples of teaching concepts:

Activity 20.1

Have students find a method for determining the whole number remainder in a division problem using only the calculator. Suggest that they might begin with divisors that end in 2 or 5 so that the complete result will be displayed.

Activity 20.2

Give students a strip of paper exactly 22 cm long and a centimeter ruler. Have them measure and record the length. Next, carefully fold the paper in half and measure the result as accurately as possible. On the calculator, divide 22 by 2 and compare this result with the measurement. Fold the paper in half one more time and again measure carefully. Divide the last result on the calculator by 2 and compare. Do this fold, measure, and divide sequence at least one more time. How do the results of the measurement compare with the results of dividing by 2 on the calculator? Try this with other lengths. (This activity, suggested by a second-grade teacher, is reported by Shumway, 1988.)

In the last activity, there is a vivid contrast between symbolic manipulation and physical reality. What other ideas might develop at least informally from that exploration? What happens if after the first division by 2, you continue to press [=] over and over?

Almost any problem-solving activity that involves computation is enhanced by letting students use a calculator. Besides encouraging trial and error or exploration of different approaches to a solution, students also learn the value of recording intermediate results as they go along. Many interesting process problems arise from the use of the calculator itself. The following is an example of such a problem.

Activity 20.3

Find a way to use the calculator to compute the product of two numbers such as 3456 and 88888. (The standard method of entering this product on the calculator causes an overload.)

Activity 20.4

A two-digit "keypad pair" is one that can be entered by pressing two adjacent keys on the calculator—vertically, horizontally or diagonally in any direction. The numbers 48, 63, 12, and 21 are keypad pairs while 73 and 28 are not. Find two different pairs of keypad pairs that have the same sum. (Or, find two different pairs of keypad pairs that have the same product.)

The following exploration is an appropriate beginning point for both third and eighth graders, although it may end up at different places depending on how it is pursued.

Activity 20.5

Enter [+] 9 in the calculator and begin pressing [=]. What things can be noticed? Find at least one interesting pattern and explore it. See if your pattern continues into three and four digits. Can you explain why the pattern works? What else do you notice? What would happen with numbers other than 9?

Many books and resources exist to help you find other ways to exploit the calculator to teach mathematics. It would be a shame to pass up this valuable resource.

Not Using Calculators Wastes Time

Pencil-and-paper computation is time-consuming, especially for young students who have not developed a high degree of mastery. Why should time be wasted having students add numbers to find the perimeter of a polygon when the lesson is on geometry? Why compute averages, find percents, convert fractions to decimals, or solve problems of any sort with pencil-and-paper methods when pencil-and-paper skills are not the objective of the lesson? Defending laborious and time-consuming computations in noncomputational lessons is indeed difficult.

WHEN AND WHERE TO USE CALCULATORS

Always Have Calculators Available

To have calculators available to students at all grades can easily mean different things to different people. The position taken in this book is that calculators are as available to students as are pencils, paper, and books. Calculators should be in or on students' desks at *all times* at *all grade levels.* This position may seem quite radical to some. Here are a few arguments to support immediate availability at all times:

The need to make a special effort to use calculators for any activity is diminished. Throughout this text, activities have been interspersed that utilize calculators. If we have to stop the flow of our lessons to pass out calculators, we are likely not to take the time. Instead we will save calculators for special "calculator lessons," or we may not even take the time at all. Many excellent calculator explorations will happen spontaneously and/or will take up only a few minutes of class time. These activities simply will not get done if we must stop to distribute and collect calculators.

Ready availability of calculators allows students to choose on their own when it seems appropriate to use calculators. There are many times when it is much easier or quicker to use a mental computation or estimation or even to use the pencil that happens to be in our hands rather than to reach for the calculator. How can students ever learn to make these choices if we decide for them when to use calculators by keeping them out of sight unless otherwise directed?

It simply does no harm to have calculators available! This is very difficult for many teachers, prospective teachers, and parents to accept. But the fact of the matter is that there is virtually no research to suggest that students fail to develop basic skills when taught in the presence of a calculator. Even basic facts and pencil-and-paper skills will still be learned with the calculator on the desk.

Calculators have been promoted by mathematics educators for over 20 years. They are inexpensive. There are excellent resource materials directing their use. There is no evidence that they do any harm. There is lots of evidence that they can enhance learning in many ways. They are still not accepted. Perhaps the reason is that not enough teachers have simply put them on the students' desks.

One Place Not to Use Calculators

There is one place where calculators should not be used: in correcting or checking computational exercises done with pencil and paper. It would be a serious error to have students do a lengthy list of computations by hand and then ask that they check the results with the calculator. This use of the calculator would send very mixed messages to students about why they did the computation by hand in the first place. How much pencil-and-paper computation should be taught in school and what degree of skill should be required remains open to debate. For each operation there is a different continuum of difficulty levels. Societal uses are also different for different operations. But if you have undertaken to teach an algorithm, while you are doing so, let students know that pencil-and-paper methods are an *alternative* to the calculator.

Notice that this suggested restriction is to the *checking* of computation, not necessarily teaching pencil-and-paper computation. An experienced teacher may well be able to integrate some aspects of all four forms of computation into the same lessons. In Chapter 10, the calculator is shown to be used in both estimation and mental computation activities, and some suggestions are made for including mental computational methods along with pencil-and-paper algorithm instruction. While discussing the multiplication algorithm, for example, the calculator might well be used to help children investigate *base ten facts.* It could also be used to show how front-end methods produce reasonable estimates of computation.

PRACTICAL CONSIDERATIONS

What Kind of Calculator

Simple four-function calculators are all that are necessary for all of the activities that have been suggested in this book. The one important feature that is not on all simple calculators is the automatic constant for addition and subtraction. (That is the one that allows you to enter [+] 1 [=] [=] [=] . . . and have the calculator count.) Almost all calculators have this feature for multiplication and division. It is especially important to note that automatic constant features do not operate exactly the same on all calculators.

Other features that might be of interest, especially if you are going to have students bring in their own calculators, include the following:

The size of the keys, especially for young children.

Light-powered (solar) calculators are less likely to need batteries, but even that is a rare event with new, long-life batteries.

The position of the negative sign in the display varies among different calculators.

The number of keys for memory.

Separate *clear all* and *clear entry* buttons exist on some calculators.

Calculators that are completely transparent are now available from several suppliers. These can be placed on the overhead projector so that students can see not only the display but all of the "buttons" that are being pressed. One of these can prove quite valuable as a teaching tool, especially with younger children.

While the only type of calculator referenced in this book is the simple four-function calculator, more powerful calculators are also becoming quite inexpensive. Especially around the fifth-grade level and up, many of the features provided by a fancier calculator offer excellent opportunities for important mathematics investigations. Those generally referred to as "scientific" calculators include features such as parentheses, reciprocal, and exponential functions that allow for interesting explorations of powers, roots, fractions, and functional relationships and can contribute to students' general "number sense" with more sophisticated expressions. The *TI Explorer* calculator permits entry and computation of common fractions, simplifies fractions, switches between fractions and decimals, and has other interesting features useful at the middle school level.

Demana and Osborne (1988) make a strong case for the use of calculators that include algebraic logic. In essence, that means that expressions are evaluated according to the correct order of operations rather than as they are entered. If [5] [+] [3] [×] [4] [=] is keyed in on the standard four-function calculator, the result will be 32 while on a calculator with algebraic logic the result will be 17, which is correct. At the very least we should be aware of these differences.

Another level of calculator sophistication is also becoming quite affordable, namely calculators with graphics capabilities. These calculators are capable of plotting the graphs of all sorts of equations and examining them in a variety of ways. These same capabilities are included in many computer software packages, but it is intriguing to imagine students in a junior high classroom with the power to instantly get a picture of a complex graph right at their seats on a pocket-size calculator. It is a safe prediction that this will happen in future classrooms where "future" is not as distant as you may suspect.

Availability

Calculators should be available for use at *all* times to *all* students in the classroom at *every* grade level. The price of a calculator is now so low that it is entirely reasonable to require that each of your students provide his or her own calculator at the start of the year the same way that pencils, notebooks, and other supplies are required. Be sure to provide parents with a short list of particular calculators that you recommend. Include both brand and model number when appropriate. In this way you can be assured that every student has the features you desire and that all of the calculators in the room operate the same way. Nothing is more frustrating than having to stop in the middle of an activity to provide special directions to one or two students with a nonconforming calculator. Another good idea is to have a few extra calculators available to fill in when one breaks or is left at home.

A possible exception to the suggestion for uniformity in the classroom is in order. By the time students reach the fourth or fifth grade, they should be aware that all calculators are not alike and do not even use the same logic. Having some calculators around that perform differently may provide some interesting opportunities for learning. For example, if some calculators employed algebraic logic as discussed earlier, the results on that calculator could be contrasted with the results on a simple four-function calculator. The need for a rule concerning order of operations will become apparent in this context.

In some school districts, classroom sets of calculators are being purchased by the schools. Some book publishers have also provided calculators at reduced prices as part of their sales promotions. Major manufacturers have designed calculators specifically for this market and sell them at reasonable prices in large quantities. There is no need, however, to wait for someone else to supply your room with calculators. (There is no excuse, either!)

Teaching Mathematics with Computers

PERSPECTIVE

During the early 1980s, the microcomputer hit schools and teachers with a sense of awe, amazement, and hope for moving education firmly into the technological age. "Computer literacy" was all the rage as students were told about such things as the history of computing machines and how the binary number system worked inside a computer. In order to help teachers use these new machines, and of course to sell them, companies began to produce a vast collection of software so that children could sit in front of a computer and do something flashy and exciting that looked like school material. The large majority of this software was directed at the lowest levels of learning. It provided plenty of drill and practice and was initially attractive. Parents wanted their children to have a chance to "be on a computer" even though they had not the slightest idea what that meant or what they would be learning. School systems proclaimed to the public the number of computers available in their schools.

Now that some of the mystery and excitement that initially surrounded computers has subsided, it appears that the microcomputer will continue to be a factor in education for the foreseeable future. Several things have contributed to this widely accepted prediction. The excitement of simply using a computer has diminished so that educators tend to be more objective in deciding how to use them. While software continues to proliferate, it also continues to improve. The pace at which computer technology improves has shown no signs of slowing down. Improved technology means more capability for programmers to create quality software. A significant example of hardware improvement outside of the computer is the liquid crystal display screen that sits on an ordinary overhead projector. With this device it is relatively easy for the teacher to use a single computer in the classroom with the entire class viewing a large screen projected by the overhead. This or some similar equipment may well change the way teachers view computers as a teaching tool, especially if new software is devised that capitalizes on that mode.*

In 1980, the NCTM recommended in its *Agenda for action* that "mathematics programs take full advantage of the power of calculators and computers at all grade levels" (p. 1). The more recent NCTM position on computers is reflected in the *Standards:* ". . . a computer should be available in every classroom for demonstration purposes; every student should have access to a computer for individual and group work; and students should learn to use the computer as a tool for processing information and performing calculations to investigate and solve problems" (p. 8). This position highlights the appropriate use of the computer as a tool for learning rather than an object to learn "about." It is worth noting that there is no standard in the document that speaks to learning about computers or learning how to program. Computer "literacy" is not a topic of mathematics education.

In the remainder of this chapter, the different ways that computers can be used in learning mathematics will be discussed. The purpose is to help you understand the different uses of this instructional tool so that you can begin to make reasoned rather than emotional decisions about how to include computers in your instructional program.

If you have never used a computer for one of the purposes described, the best recommendation is to find some software in that category and try it out. Be careful not to make quick judgments based on limited experiences. Talk to friends, teachers, and other educational specialists who have experience with computers. Ask for their opinions and ideas about particular software programs to try. Find out how they are best used with children: what works well and what does not. If possible, try different software and different computer uses with children. At the end of each section a few popular software packages are listed. These listings are only to provide examples. They are by no means recommendations of the "best there is." Be aware that new software is constantly being generated.

*There are quite a few of these display screens on the market and not all are of the same quality. Heat from the projector can cause the image to fade, so some screens include small fans. With some screens it is possible to project color images in this manner as well.

INSTRUCTIONAL SOFTWARE

Several terms with similar meanings are commonly used when talking about computers and instruction. *Computer-assisted instruction* (CAI) refers to a method of teaching which uses the computer to present instructional material. Students using a CAI program interact with the information on the screen. The learning situation might be drill and practice, tutorial (where concepts or skills are developed), a simulation, or an instructional game or problem-solving situation. The terms *computer-assisted learning, computer-based instruction,* or other variants are frequently used instead of CAI.

The programs that enable the computer to present instructional information are examples of *software* or *courseware.* Generally, software refers to commercially produced programs although that is not a requirement. *Software* is a broad term that encompasses not only CAI programs but also "tool programs" such as spreadsheets, graphing programs, and programming languages such as Logo or Pascal.

Drill-and-Practice Software

Drill-and-practice programs do exactly that; they provide students practice with skills or concepts that are assumed to be learned elsewhere. More drill-and-practice programs exist than any other type of software for elementary mathematics. In general, a drill-and-practice program poses questions which the user answers either by directly entering the response or selecting from options. The program evaluates the response and reacts accordingly. Most packages (several different but related programs sold as a unit on one or more separate disks) include a series of programs on one or more related topics. You can expect that the package includes several levels of difficulty or content. Frequently, different levels of difficulty can be selected within a single program.

Computers have proven to be an effective but not necessarily superior method of providing drill and practice (see, for example, Suydam, 1984; Carrier, Post, & Heck, 1985; Fuson & Brinko, 1985). Some advantage may exist in the provision of motivation and change of pace provided by attractive graphics and action, the ability of the computer to provide immediate feedback, and the possibility of selecting drills specifically to

The Use of Computers in the Learning and Teaching of Mathematics

The following statement is an official NCTM position. It was developed by the Instructional Issues Advisory Committee and the Technology Advisory Committee and adopted by the Board of Directors.

Computer technology is changing the ways we use mathematics; consequently, the content of mathematics programs and the methods by which mathematics is taught are changing. Students must continue to study appropriate mathematics content, and they also must be able to recognize when and how to use computers effectively when doing mathematics. Teachers must know how and when to use the tools of computer technology to develop and expand their students' understandings of mathematics.

It is the position of the National Council of Teachers of Mathematics that the computer is an appropriate tool that can be used in a variety of ways for the enhancement of mathematics learning, teaching, and evaluation. Changes are therefore needed in mathematics curricula, instructional methods, access to computer hardware and software, and teacher education:

- The content of school mathematics courses must be modified to reflect the changes brought about by computer technology. Curricular revisions should provide for the deletion of topics that are no longer useful, the addition of topics that have acquired new importance, and the retention of topics that remain important. In implementing revised curricula, educators must ensure that the time and emphasis allocated to topics are consistent with their importance in an age of increased access to technology. Instructional materials that capitalize on the power of computers must continue to be developed for students at all levels.
- Teachers should use computers as tools to assist students with the exploration and discovery of concepts, with the transition from concrete experiences to abstract mathematical ideas, with the practice of skills, and with the process of problem solving. In mathematics education computers must be instructional aids, not the object of instruction. Similarly, computer programming activities in mathematics classes should be used to support mathematics instruction; they should not be the focus of instruction. The amount of classroom time spent by mathematics students in learning a programming language must be consistent with the expected gains in mathematical understanding.
- Schools should be equipped with computers, peripherals, and courseware in sufficient quantity and quality for them to be used consistently in the teaching and learning of mathematics. Every classroom in which mathematics is taught should be equipped with computing hardware that includes a large-screen display device. Computer laboratories should be available to all students on a regular basis for the extended exploration of mathematical topics by individuals or groups. School systems must budget for the ongoing acquisition, maintenance, and upgrading of hardware and courseware for use in classrooms and computer laboratories at all grade levels.
- All preservice and in-service teachers of mathematics should be educated on the use of computers in the teaching of mathematics and in examining curricula for technology-related modifications. Teachers should be prepared to design computer-integrated classroom and laboratory lessons that promote interaction among the students, the computers, and the teacher. Mathematics teachers should be able to select and use electronic courseware for a variety of activities such as simulation, generation and analysis of data, problem solving, graphical analysis, and practice.

Mathematics teachers should be able to appropriately use a variety of computer tools such as programming languages and spreadsheets in the mathematics classroom. For example, teachers should be able to identify topics for which expressing an algorithm as a computer program will deepen student insight, and they should be able to develop or modify programs to fit the needs of classes or individuals. Keeping pace with advances in technology will enable mathematics teachers to use the most efficient and effective tools available.

Changes in curricula and in the availability of hardware and courseware are not sufficient to guarantee that teachers use computers appropriately. Ongoing in-service programs must be readily available to help teachers take full advantage of the unique power of the computer as a tool for teaching and learning mathematics.

(September, 1987)

meet individual needs. Another positive feature offered by many programs is the ability to keep records of individual student's progress. At the present time, the significant disadvantages of cost and limited access to computers in the classroom must be kept in mind when considering these values.

Programs can be found for practicing all sorts of mathematics skills. For very young children, there are programs that provide counting practice, matching numerals with sets, and identifying simple shapes. Basic facts have been imbedded into all sorts of exciting games and activities, many resembling the popular video arcade games. Ironically, one of the most popular forms of mathematics software provides drill in pencil-and-paper algorithm skills. The irony is in the use of space-age technology to develop skill with the oldest and slowest form of computation known.

Besides basic fact practice, one of the best uses of the computer for drill and practice is for computational estimation and mental arithmetic (see Chapter 10, pp. 168–169). Here the computer can control response time, immediately evaluate responses in terms of percent of error, and students can respond without a pencil, thus eliminating the temptation to do the computations by hand instead of mentally.

How the program responds or gives feedback is one of the most important features to consider when selecting a drill program. At one extreme, the program repeats the question until a correct response is provided or simply gives the correct response after one or two incorrect trials. Other programs will provide helpful or conceptual information in response to a wrong answer. For example, a 4×8 array of squares may be provided for help with the 4×8 fact. A wrong answer to a computation problem may cause the program to proceed through the process step by step, evaluating each portion of the response. The most sophisticated drill programs are coordinated with conceptual tutorials. If a student misses a certain percentage of questions, the program suggests stopping the drill and moving into a different program that provides conceptual or developmental assistance. These latter programs are few in number at the present time. The desired type of feedback will vary with different types of drills.

Examples of software in this category:

Addition Logician, MECC, grades 2–4, 1984

Teasers by Tobbs, Sunburst, grades 3–6, 1985

Math Blaster, Davidson, grades 1–7, 1983

Math Strategies—Estimation, SRA, grades 5–7, 1985

Tutorial Software

Tutorial software is designed to teach rather than to simply drill. Teaching or conceptual development of almost any topic is significantly more complex than providing drill. To design good tutorial programs requires not only good pedagogical understanding and anticipation of wide varieties of student responses but also a high level of programming skill. The enormous amounts of time required to design and program good tutorial programs has left quality software of this type quite scarce. Richard Shumway, a firm believer in the use of computers at every grade level, notes that "the use of a computer to simulate teachers is not yet realized and very difficult" (1988, p. 339). Since conceptual development in mathematics is highly dependent on models, tutorial software has its greatest potential in providing graphics to stand in place of physical models. Many programs provide static pictures of counters, fraction pieces, base ten place-value models, or geometric figures. Some better programs show movement, such as the formation of a ten-stick from 10 single squares. The best of these programs allow students to freely manipulate the computer model with keystrokes or a joystick, thus simulating actual manipulation of physical materials. For example, students may be able to place as many ones, tens, and hundreds pieces on the screen as they wish. Once there, the pieces might be moved, grouped, or separated. The student can then utilize this manipulation to respond to a question posed by the program. The value of many programs can be improved considerably if used in conjunction with a good conceptual development utilizing similar materials.

As with drill and practice, the quality of feedback to the student is of primary importance in evaluating tutorial software. Good programs should anticipate more than one student response and branch accordingly. Some responses may suggest that a student should cycle back to a complete development of the topic while others might require only a hint or partial explanation. Some programs allow the student the option of asking for help rather than simply responding automatically.

Some programs require far too much reading. A program that is simply an electronic page turner is not much better than a good book. When reading is required, it should be minimal and at a level in keeping with the grade level of the material being presented. Unfortunately, many programs err in this area.

Some of the most promising programs to date are designed to run only on larger, more sophisticated computers with faster processors, more memory, and better graphics capabilities. Certainly at some point in the not too distant future, these computers will be quite commonplace in schools and the quality of tutorial software will improve accordingly.

Examples of software in this category:

Balance!, HRM, grades 8–10, 1985

Decimal Squares Computer Programs, Scott Resources, grades 5–8, 1987

Learning Place Value, Mindscape, grades 1–5, 1987

IBM Math Concepts, IBM, grades K–8, 1987

Problem-Solving Software

Programs in which games or activities require the use of logical reasoning and spatial or geometric reasoning have been among some of the most innovative software packages yet developed. These differ from tutorial software in that they present visual, numeric, or logical challenges or problems rather than drill or instruction. Activities requiring identifying and extending patterns and assorted logic activities with attribute pieces can be found in several programs. Other programs involve students with more sophisticated if-then reasoning and the use of *and* and *or* connectives. Numeric challenges involving number theory, and rules of inference are also part of this genre of software.

The argument can be made that most of these activities can be conducted without a computer. Especially for primary children, patterning and attribute activities should certainly be done with actual physical materials before moving students to the computer, and some teachers have found ways to combine physical activities directly with the computer programs.

On the other hand, the computer provides an element of motivation, especially with interesting graphics displays. It also allows students to work independently or in groups of two or three on one computer without constant teacher intervention. This sort of independent involvement is difficult to provide in most noncomputer logic activities.

A different sort of problem-solving software is designed to teach specific problem-solving heuristics such as guess-and-check or make an organized list. Other programs are designed to help children solve translation problems. These higher-order skills are difficult to develop outside of the rich environment of a class discussion led by a skilled teacher. Software of this type may best be used with an entire class rather than by individual students.

Examples of software in this category:

Problem Solving Strategies, MECC, grades 4–9, 1983

The Factory, Sunburst, grades K–12, 1985

Heath Math Worlds: Strategies, Data Analysis, grades 4–6, 6–9, 1986

King's Rule, Sunburst, grades 4–10, 1985

Pond, Sunburst, grades 2–6, 1985

Gertrude's Secrets, The Learning Company, grades K–4, 1985

USEFUL TOOL SOFTWARE

Another important category of computer software is designed not to teach but to perform tedious tasks quickly and efficiently. Some of the most common types of these computer tools or utilities are described next.

Spreadsheets

Spreadsheets are programs that can manipulate rows and columns of numeric data. Formulas can be entered easily so that, for example, the average, highest, and lowest values from each of three different columns of numbers is automatically entered into another position. Every time a value is changed, all other values which depend on that one are recalculated. Words can also be written into a spreadsheet, so that for example, a class of students names could be entered and a column for each of their test grades listed as in a grade book. Besides determining both row and column averages or other statistics that may be of interest, the spreadsheet could be used to order any row or column either numerically or alphabetically. *Spreadsheet* is a generic term, with many different companies producing spreadsheet software. Many exist that are simple to use in the classroom.

In addition to their obvious use as an aid in statistics investigations, spreadsheets are a powerful way to investigate functions numerically. Figure 20.1 was generated by the *Appleworks* spreadsheet and shows comparative values of several simple and not so simple mathematics functions. The left-hand column contains a list of values for which each function was evaluated.

```
File: CH20 FIG 1              REVIEW/ADD/CHANGE             Escape: Main Menu
========B========C========D========E========F========G========H========I====
 4|
 5|      X      X^2      X^3  X^3-X^2         (X^3+2X)/(3+X)
 6|==================================================================
 7|     -5       25     -125     -150               67.50
 8|     -4       16      -64      -80               72.00
 9|     -3        9      -27      -36               ERROR
10|     -2        4       -8      -12              -12.00
11|     -1        1       -1       -2               -1.50
12|      0        0        0        0                0.00
13|      1        1        1        0                 .75
14|      2        4        8        4                2.40
15|      3        9       27       18                5.50
16|      4       16       64       48               10.29
17|      5       25      125      100               16.88
18|      6       36      216      180               25.33
19|      7       49      343      294               35.70
20|      8       64      512      448               48.00
21|
-----------------------------------------------------------------------------
G20: (Value) ((D20)+(2*(B20)))/(3+(B20))
```

Figure 20.1
Spreadsheet

Database Systems

A database system is like an electronic storage file, management, and retrieval system. It permits both numeric and textual data to be filed away for large numbers of individual "cards" or files. This generic computer tool is not as useful in mathematics as it is in social studies and science applications. However, it is an excellent device to help students gather and store information from a survey, a series of measurements, or a series of experiments. The data base system can then be used to sort, order, and print out information. School packages such as the *Logo Data Toolkit* (Terrapin, 1987) also include graphing tools to visually display data.

Graphing Packages

Software that quickly converts numeric data into graphs is one of the most powerful tool uses of the computer in the mathematics classroom. Statistical graphing packages typically produce pie charts, bar graphs, and line graphs from any sets of data that are entered. *Graphing* (MECC) and *Exploring Tables and Graphs* (Weekly Reader Family Software) are two popular examples. Algebraic graphing packages graph the curves of functions and equations that are entered by the user. *Green Globs* (Sunburst) is one of the best-known algebraic graphing packages.

Both types of graphing programs are powerful teaching tools because they demonstrate changes in the graphs very quickly with different changes in the data or functions to be plotted. Students' attention can be focused on understanding and interpretation of graphs and relationships without the tedium of plotting points, determining percentages, measuring bars, and so on. Most programs allow several line graphs and curves to be plotted on the same axis in order to make contrasts and comparisons. Figure 20.2 was generated by *Green Globs.*

Geometry Utilities

The family of four programs called the *Geometric Supposers* (Sunburst, 1985) were discussed briefly in Chapter 17. These programs permit a shape to be quickly produced on the screen and any series of constructions or measurements performed on that shape. For example, the side of a triangle can be subdivided into as many as nine equal parts and lines drawn from the opposite vertex to each subdividing point. Angles, lengths, and areas can then be measured and compared. Furthermore, the same series of constructions can then be repeated on a different triangle for the purpose of making comparisons and conjectures. The programs encourage students to explore their own ideas about geometric relationships and actually "invent" theorems of their own. This type of activity is much closer to the spirit of real mathematics. Students create ideas rather than demonstrate proofs of those ideas that were constructed by someone else.

A simpler program, *Elastic Lines* (Sunburst, 1988), is

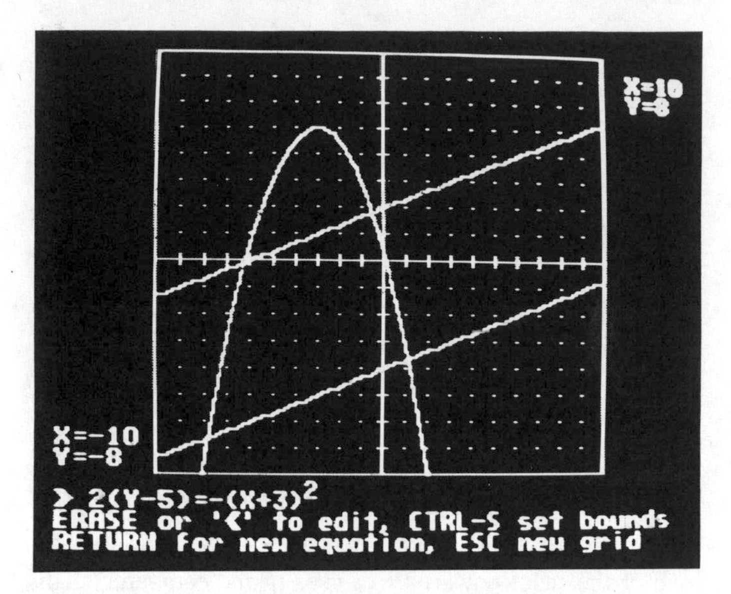

Figure 20.2

essentially an electronic geoboard that allows shapes to be drawn on a 10×10 dot grid, produces flips and rotations, and makes some measurements.

The number of programs that permit shapes to be easily drawn, manipulated, and measured will certainly increase in the near future. This is especially true for computers at the level of the MacIntosh, the IBM, or IBM-compatible. The greater speed and superior graphics of more powerful machines will give teachers "blackboards" to draw and move shapes in ways they never would even have dreamed of.

Modes of Computer Use

Someday, perhaps not nearly as far in the future as you may think, computer technology will have reached a state of sophistication and reduced price that computers for every desk and every child will be commonplace. Until that time, most elementary schools will have a limited number of computers. Most schools currently deal with the limited numbers of machines in two basic modes: either one to four computers are available to teachers in their classrooms, perhaps on a permanent basis, or there is a computer lab where teachers can take an entire class.

COMPUTERS IN THE REGULAR CLASSROOM

At first glance it seems impossible that a single computer could be of much value to a teacher. This is certainly not the case. Even one computer can be a valuable instructional tool.

Full Class Usage

The microcomputer was originally conceived of and is most often used as a tool for individual use. However, one of its most powerful uses is as an electronic blackboard—a teaching tool to be used by the teacher with the entire class in the same way as chalk or the overhead projector. The difference is that the computer can not only put numbers and words and drawings in front of the class, but it can be used interactively as you, the teacher, control what is happening on the screen.

To use the computer as a very fancy blackboard, you must first find a way for students to see what is on the screen. At least three possibilities exist. For smaller classes, simply put the monitor up a bit higher than the desks and have students crowd around in their chairs. While not the best approach, this method can accommodate a class of about 20 students. A slightly better method is to use one or two large-screen TV monitors. This is quite effective for any class, although sometimes a bit clumsy. The newest entry into projection equipment is a simple glass screen that lays on an ordinary overhead projector and connects directly to the computer. The liquid crystal display provides a large-screen view, although usually not in color, that everyone can easily see. While the cost at present is a bit high (in the \$600 to \$1200 range), this new equipment is quite manageable and can easily be shared among several teachers.

With a mind-set that the computer screen can be seen by a class rather than just one or two students, you can begin to think about your new and existing software being used under your control with the full class. Drills can be presented that may be much better than flash cards and can go well beyond basic fact practice. Tutorial software may be a better way to put meaningful models in front of the class than you can do in any other way. The tool or utility software described in the previous section represents some of the best ways to use the computer with the class. The computer in this mode is not a slide show or something to sit back and watch. It provides a vehicle with which you can generate lively and valuable class discussions. Even with software that was designed originally for the individual, responses from the class can be entered by you or a student aid at the keyboard.

Individual Student Use

With just a few computers in the room, some planning must take place for individual student use. If there is a program or package that you want everyone to experience, set up a schedule that cycles students to the computer either alone or in pairs until everyone has used the program. This works best in self-contained classrooms where time is more flexible. Many programs lend themselves to work in pairs or even in groups of three. Frequently the time spent at the computer is even more valuable with two students able to interact with each other about the activity.

Many of the instructional software packages can be used as remedial instruction and can provide extra help when you do not have time to give it. Other programs can provide enrichment or extensions of ideas that not everyone is ready for. Software also exists that provides diagnostic testing and evaluation. For any of these reasons, students can be assigned to go to the computer instead of doing work that may be less beneficial to them.

For independent student use of the computer in your room, a few practical guidelines should be considered.

1. Be sure students know how to use the program and what their responsibilities are before assigning them to the computer. Introduce programs that all are to use in a full class mode.
2. Arrange the computers so that they will not be a distraction to other students in the classroom. Turn the monitor screen away from the rest of the students.
3. If a program has an audio feature, be sure that it has been turned off. Many programs make this "switch" available only to the teacher. Noisy software without such a switch should be avoided.
4. Arrange for some form of recording to be done. This might be data stored on a record disk by the program, a worksheet to be filled out in conjunction with the computer activity, or simply a "log entry" made by the student indicating the program used and progress made.
5. Establish appropriate guidelines for students working in pairs or triples. You should not have to monitor students' behavior at the computers.

COMPUTER LABS

If your school has a room with computers, then plan ahead to make maximum use of the time that your class is there. Most importantly, have a reason for your class to use the computers. Do not just go to the computer room because it is your turn or to provide a break. Let that time in the lab serve a useful purpose.

In addition to the guidelines for using computers in the classroom, the following are suggested when taking students to a lab.

1. Be sure students know what they are to do and how to use the software before you get to the lab. Once there, the distraction of the computers will be overwhelming.
2. If the lab does not have enough computers to accommodate your entire class, have an activity ready for part of the class while they wait their turn. Let students know when they can expect to work at the computers.
3. For activities that require preplanning, especially programming, try to allow time for that planning before going to the lab. It is likely that your time in the lab will be limited. Make the best use of it.

Learning Through Programming

When computers first became popular in schools, there was a heavy emphasis on teaching programming. Most often this was (and is) done under the title of "computer literacy." Computer literacy refers to practical knowledge about computers that every person in our society should know. As such, it is not a topic of mathematics as much as it is social studies. Today, the emphasis has fortunately shifted from teaching students to use computers to using computers for learning. From a practical standpoint very few people in society need to know how to program a computer. Is there any value, then, in teaching grade school children to program? How much programming knowledge do children need to have? What can be learned by learning to program other than a social utility knowledge of how computers work? These and similar questions are still being researched and debated.

LEARNING MATHEMATICS THROUGH PROGRAMMING

Many mathematics educators believe that students can learn mathematics through the process of programming (Shumway, 1987, 1988; Smith, 1984). This might happen in two different but related ways. First, if students, even very young students, are given a computer program to explore, some good mathematics can be learned in the process. Second, if students are asked to program a computer to perform some task, they (not the computer) must understand any mathematics that might be involved.

Exploring a Simple Program

By way of example, consider the following two programs written in BASIC.

```
10 FOR N = 1 to 12
20 PRINT N, N+N, N+N+N+N
30 NEXT N
```

```
10 GR
20 COLOR = 7
30 FOR N = 0 to 39
40 PLOT N, N
50 NEXT N
```

The first program (on the left) will print 3 columns of 12 numbers very quickly. The second (written for an Apple computer) will graph 40 large "points" in a diagonal line down the screen from the upper left corner. The screens produced by these two programs are shown in Figures 20.3 and 20.4.

The first point to make is that even first graders can develop some understanding of how these programs work, can run them, examine their output, make changes, and compare results. Any programming knowledge that is required can be taught as new examples are presented. It is not necessary for children to first "learn to program."

Can you think of any mathematics that might be learned from either of these programs? Here are some possibilities:

For the left program, students might

Notice that all of the numbers in the second and third columns are even. What other formulas could be put in line 20 to produce even numbers? What about odd numbers?

Notice that some of the numbers in the second column

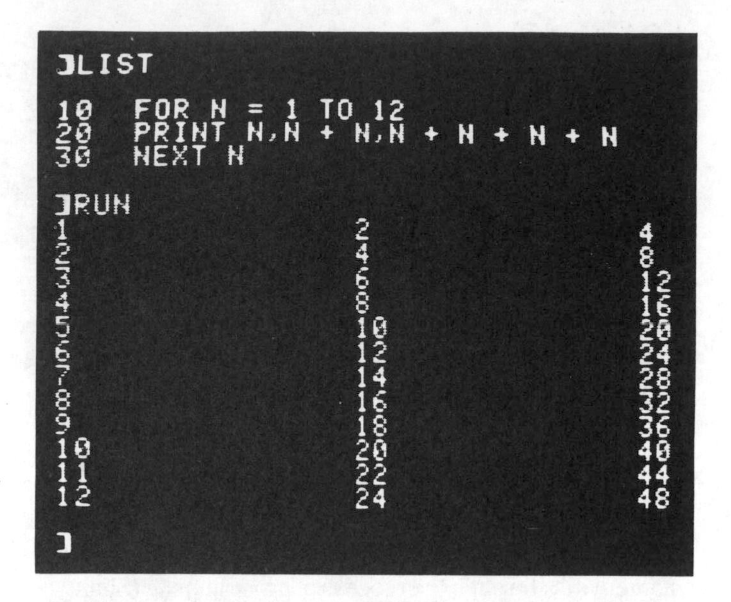

Figure 20.3

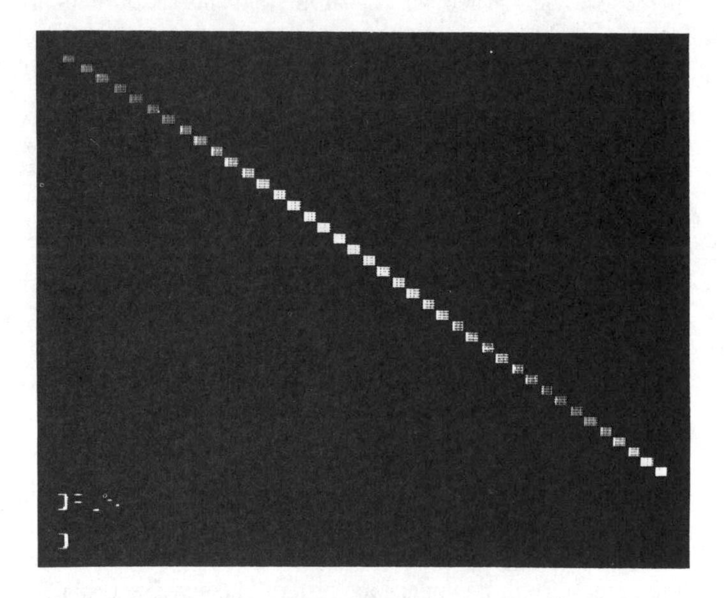

Figure 20.4

are also in the third but much higher up. What is the pattern? How often does that happen? What if line 20 were changed?

For line 20, try

```
20 PRINT N, 2 * N, N+N,
```

or

```
20 PRINT N, 4 * N, N+N+N+N
```

Second graders could begin to explore multiplication this way.

Try

```
20 PRINT N, 1/N, 1/(N * N)
```

and begin to look at the limiting process in the seventh or eighth grades. Change line 10 to allow for longer lists and/or different steps. For example, try

```
10 FOR N = 0 TO 20 STEP .1
```

For the right-hand program, students might

Wonder what caused the program to print the dots where it did. To check, they might change line 30, or line 40 or both to see if they can figure out how this works. That is a mathematical investigation.

Change line 40 to

```
40 PLOT N, N/2,
```

or

```
40 PLOT N, N/3.
```

This provides an informal exploration into the notion of slope of a line. Why do the little rectangles get longer? Does "PLOT" round off the numbers or throw away the decimal parts?

Add a line 45.

```
45 PLOT N, N+3
```

Compare the results. Parallel lines have the same slopes. What could be done to make them intersect?

More significantly, in both programs students are using variables in a meaningful manner. In the graphics program the relationships between variables, real numbers, and coordinates on the screen are powerful ideas. The computer provides a medium within which these ideas are explored at an informal, experimental level.

Teaching the Computer

As students develop some minimal skills with a computer language, they can be given programming tasks that will require that they wrestle with and develop a firm understanding of a particular mathematics concept. Here are a few examples:

Activity 20.6

Write a program that will list all of the prime numbers up to 100. (The computer must decide if the number is a prime.)

Activity 20.7

Write a program that can be used to round any number to the nearest ten, hundred, and thousand.

Activity 20.8

Write a program that will find the least common multiple of two numbers.

Activity 20.9

Write a program that will draw a right triangle given the lengths of the two legs.

These program tasks can be explored using either BASIC or Logo. In working on tasks such as these, preferably in pairs or small groups, students not only learn about place value, primes, factors, the Pythagorean theorem, and angles, but they are also engaging in significant problem solving.

At the present time most schools and teachers have not made any significant use of computer programming in the sense described in this section. It remains to be seen if this approach to developing ideas in mathematics will be accepted as a viable or important way for students to learn.

PROGRAMMING TO DEVELOP PROBLEM-SOLVING SKILLS

Programming Viewed as Problem Solving

Consider the things you might have to do and think about how to program a computer to perform one of the following tasks:

Print in order all of the even numbers from 1 to 100. To make it a bit harder, also include all of the numbers ending in 5.

Play "Guess My Number." The computer selects a number between 1 and 1000, and the player tries to guess it. Each guess should get an appropriate and helpful response from the computer.

Draw a picture of a house or a boat.

Now compare the different stages and thought processes you might have to go through with the following list of generic problem-solving skills:

Understanding or analyzing the problem

Breaking the problem down into smaller parts

Selecting and planning a solution strategy

Monitoring process toward the solution

Evaluating solutions

Extending solutions

When students are taught how to approach a programming task in a thoughtful and well-planned manner, each of the foregoing skills can generally be identified as part of that process. Many educators believe that if students are engaged in programming experiences that these higher-order process skills will transfer from the specific domain of computer programming to the broader area of solving nonroutine problems (for example, Blume & Schoen, 1988; Dalbey & Linn, 1985; Wells, 1981). The computer, in essence, provides a whole class of extremely rich and interesting process problems within which to develop the skills of problem solving.

Research evidence supporting transfer of computer problem-solving skills is at this time inconclusive and sketchy. The methods of instruction used in teaching programming, the amount of time required to develop transferable skills (months or years?), and the ability to detect effects in controlled research settings are all factors that make it difficult to be definitive about the possible effects of programming.

Problem Solving and the Logo Language

The most vocal proponents of the programming-for-problem solving position are the advocates of Logo (for example, Au, Horton, & Ryba, 1987; Clements, 1985a,b,c; 1986; Papert, 1980). Unlike BASIC, Logo permits students to easily design relatively powerful programs using user-defined words. The turtle graphics feature, only one aspect of the Logo language, can be used to create complex designs, draw pictures, and produce geometric shapes. The power of Logo is accessible to very young children, is captivating in its graphics capabilities, and yet is powerful enough to be used in high school and college.

Originally developed at M.I.T. by Seymour Papert, Logo has been the object of numerous research efforts. Much of this interest is due to the claims of Papert (1980) and his followers for the cognitive gains and problem-solving capabilities of children who have had experience with Logo programming. Papert's basic thesis is that the computer is a tool to think with. By exploring or analyzing an idea on the computer, a learner has an object that can be manipulated, viewed, modified, combined, expanded, and, in a sense, "played" with. In fact, Papert's original explorations during the development of Logo were with a robot "turtle" that moved around on the floor and could draw pictures with a marker pointing down from the bottom of the turtle. A variety of robot floor turtles are now available that are controlled with Logo. The screen version of Logo uses a small triangular shape about 1 cm tall that is called a *turtle*. It is controlled the same way that the floor turtle would be. The computer thus adds a personal reality to a child's experimentation and should, according to Papert, enhance his logical reasoning abilities.

Papert's arguments are so inviting and the Logo language so captivating and powerful that troops of teachers and mathematics educators have been on a headlong and enthusiastic pursuit of Logo. What have been the results? In a review of Logo-related research, Clements (1985c) included these observations:

Programming appears to facilitate the development of specific problem-solving behaviors.

Younger students may benefit more than older students.

Logo may enhance social interaction, positive self-images, positive attitudes toward learning, and independent work habits.

Researchers are very cautious about making definitive claims for Logo. Exactly what the teacher and students are doing in a Logo experience is not clear in many of the studies that have been conducted. The transfer of thinking processes to situations that do not involve Logo is not always well supported. After nearly two decades of Logo in the classroom, much of the interest in Logo emanates from Logo enthusiasts who are deeply convinced of the values of Logo more by their intuition than by careful research. For example, Clements says, "I believe that, over several years, working with Logo has changed the way I think. Years, not weeks. I have every respect for the students who made the significant leap from their Logo work to the non-Logo problem-solving tasks they were given. I have every respect for their teachers who, unlike too many researchers, are trying to find better ways to work with Logo. I have fewer illusions, but . . . I have every hope for the future" (1986, p. 25).

The most significant drawback to programming (in any language) as a vehicle for problem-solving development is lack of time in the school day. Students need time to be at the computer. Most schools have not yet determined an efficient way to expose all children systematically to Logo (or any other language) so that by say fourth or fifth grade a certain level of programming knowledge can be assumed. An hour or two per week with Logo programming and exploration on the computer would be more than adequate if done continuously over all grade levels and with trained teachers. Where should this time come from and do the benefits merit the trade-offs? The jury is still out.

A BRIEF LOOK AT LOGO

It is difficult to appreciate the enthusiasm that Logo tends to generate without some firsthand experience. If you have never "played around" with Logo, you are *strongly encouraged to get some firsthand experience.* Even if you have had a negative experience with programming in another language such as BASIC, you owe it to yourself to at least try Logo. It is a truly different way to work with computers, and almost everyone has fun.

One minimal approach is to work through the examples and explorations provided in the following section while actually sitting at a computer. A better idea is to get one of the many books about Logo and teaching Logo and work through some of it on your own or with a friend. A few such books are suggested at the end of the chapter.

Some Logo Turtle Graphics Experiences

This section is strictly for those who have never experienced Logo.

To begin you will need a Logo language disk. This section will not provide you with all of the details of getting the language booted up, correcting typos, or other nuances. A good idea is to find someone who knows a little about Logo to help you get started. If you are using MIT Logo (or Terrapin Logo), you will be able to follow the directions given here exactly. If you are using Apple Logo (or, equivalently, IBM Logo) or some other "dialect" of Logo, there are a few minor differences. These will be pointed out along the way.

Simple Turtle Commands. To get Logo to show a screen where the turtle draws pictures, type DRAW and press RETURN. (In Apple Logo type ST for SHOWTURTLE.) There are few commands the turtle (the little triangle on the screen) "understands" without being taught. Here are some of them.

FD Short for FORWARD. You must also say how far forward you want the turtle to go. For example, FD 100 or FD 30 will make the turtle go forward. Notice the space between FD and 100. Spaces are important in Logo.

BK Short for BACK. BK works like FD and needs an "input" or a number that tells how far, like BK 140.

RT Short for RIGHT. The turtle can turn, but you have to tell it how much, like RT 30 or RT 90 or RT 200. You will soon discover that those turn numbers are degrees.

LT LEFT. Works the same as RT.

Try these commands and make the turtle scribble around. You have to press the RETURN key (⟨R⟩) before it will do anything. When you want to start all over with a clean screen, type DRAW. (In Apple Logo, type CS for ClearScreen.) The turtle has a "pen" that is usually down causing the line to be drawn when it moves. If you want the turtle to move without drawing a line, type PU for PENUP (and press ⟨R⟩). To draw again, type PD for PENDOWN.

Before going any further, play with these commands. Try to draw something: your initials, a rectangle or triangle. What happens if you send the turtle really far, like FD 2000? Try that when the turtle is turned just a little from "north." Can you figure out what happens when the turtle goes off the side of the screen?

There are four lines of text at the bottom of the screen. Sometimes the turtle gets "under" these lines and cannot be seen. Don't worry. It is still there.

Before you go any further, try to draw a simple house like the one in Figure 20.5. All of the distances should be the same. Plan on paper the list of turtle commands you want to use. Test out your list and revise it if necessary.

Teaching the Turtle. One of the nice things about Logo is that there are very few words to remember. You can define the language as you go along, using your own words for whatever you want them to mean. This is called *defining procedures.* Young children might call it "teaching the turtle."

By way of illustration, teach the turtle how to do some simple little squiggle: for example, FD 60 LT 45 BK 20.

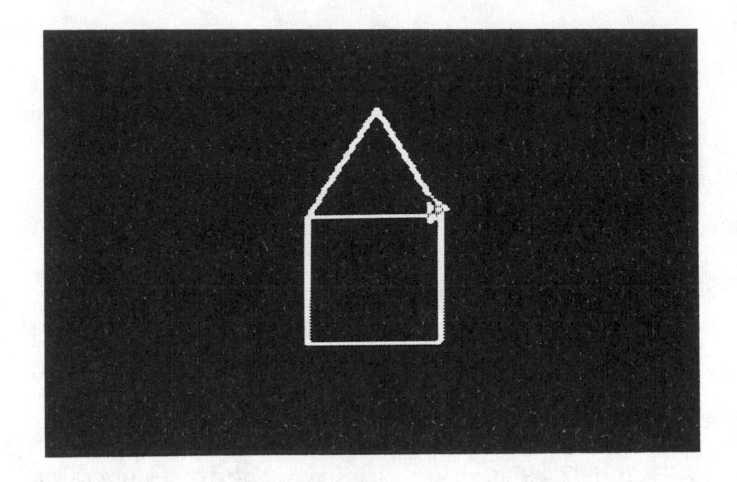

Figure 20.5
A typical "house" drawn with Logo.

Since this is just two lines at an angle, you can call it anything you want, like GEORGE, CHECK, SQUIGGLE, or even YZR or S3.

Suppose you decide on SQUIGGLE. Type TO SQUIGGLE and press ⟨R⟩. (In Apple Logo type EDIT "SQUIGGLE. Be sure to put the quotes in front but *not* at the end of SQUIGGLE.) The word TO (or EDIT) tells Logo you want to define the word that follows it. The screen now looks different. This is called the *edit mode* where procedures are defined. Now type each of the three commands FD 60, LT 45, and BK 20. It is a good idea to press ⟨R⟩ after each. Then type END for the end of the procedure. If you made any mistakes, you can use the arrow keys and the delete key to make changes. When you are done, press Ctrl-C. (Hold the control key down and press C.) Logo responds, SQUIGGLE DEFINED. Now the turtle knows the word SQUIGGLE.

Now try out the new word. Type SQUIGGLE. Try it again. Use SQUIGGLE along with the other commands. Figure 20.6 shows some experiments with SQUIGGLE.

SQUIGGLE is an example of a procedure that you defined. It is actually a little Logo program. Define some other words. Define your initial or a dashed line or some simple shape. If you try to define a word that Logo already knows, Logo will tell you. Just change the word a bit. If you want to change SQUIGGLE, just type TO SQUIGGLE (EDIT "SQUIGGLE in Apple Logo) and then use the arrow keys to add to it or make changes. Remember Ctrl-C to finish or "define the procedure."

Before you go any further, try to define two more procedures. Call them BOX and TRI. Define BOX to be a square and TRI to be a triangle so that all of the sides of BOX and TRI are the same length.

Putting Procedures Together. Once you have defined a procedure, that word can be used just like any other Logo word. That means it can be used inside of another procedure. For example, here are four procedures. Each of the first two

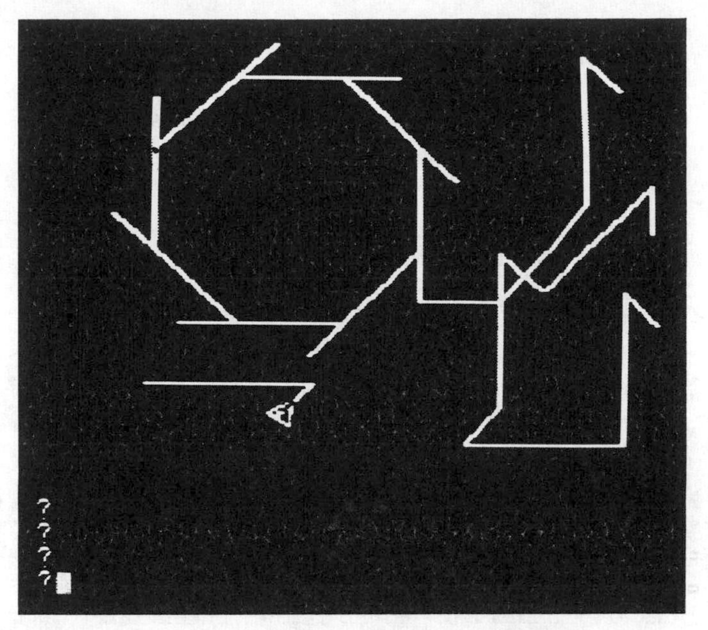

Figure 20.6
Playing with Logo and the SQUIGGLE procedure.

are used in the third, and the third is in the fourth. Use a pencil and paper, pretend you are the turtle, and see if you can predict what each will do. Then define these procedures on the computer and try them out.

```
TO CANE
   FD 60
   RT 90
   FD 20
   END

TO CORNER
   FD 30
   RT 90
   FD 30
   END

TO RECTS
   CANE
   LT 90
   CORNER
   RT 90
   CORNER
   LT 90
   CANE
   END

TO DOWNTOWN
   RECTS
   BK 50
   RT 90
   RECTS
   BK 50
   RT 90
   CANE
   RT 90
   CANE
   END
```

If you were successful, your DOWNTOWN procedure should draw what looks a little like a building. Try putting the building in different places on the screen. Use CANE, CORNER, and RECTS to make other procedures. Add some ideas of your own.

Finally, before you continue, can you use your BOX and TRI procedures to make HOUSE as in Figure 20.5? Try to make a procedure that draws three houses.

This was a very limited exposure to Logo. You should not generalize or make major conclusions about Logo from this experience. If you are at all intrigued by this little exercise, or if you just had fun, there is much, much more that Logo has to offer you. Your next explorations should involve the REPEAT command and also variables. Get a book and a friend. Take a course or go to a workshop.

Logo Versus BASIC

There are strong arguments for Logo as a language of preference over BASIC. Logo has the following features that distinguish it from BASIC:

Logo is procedural.

Logo is interactive.

Logo is recursive.

Logo is designed as a "thinking tool."

The *procedural structure* of Logo refers to the way small programs or "procedures" can be defined independently of each other. This allows a task to be broken down into smaller subtasks and each one dealt with individually. These procedures can then be put together, even one inside of the other. The language can "grow" through user definition of new capabilities. As a result, even very young programmers can construct powerful programs with very simple building blocks that they themselves design. BASIC, on the other hand, consists of only a few primitive words that must be chained together. While many interesting explorations can arise from very simple, short programs, even moderately complicated projects tend to result in long confusing lists. Pascal and many other modern languages are also procedural.

Logo is *interactive* because each command or procedure can be executed immediately and independently of other parts of the program. Most BASIC commands are also executed immediately. To run only one part of a lengthy BASIC program, however, is sometimes a bit tricky.

An explanation of *recursion* is more involved than can be dealt with here. This feature of the Logo language, also found in languages such as Pascal and Lisp, permits a procedure to "call itself." The problem-solving skill of identifying within the original problem a smaller but similar problem is the key to using the recursive power of Logo effectively.

The existence of turtle graphics in Logo provides a context in which logical thinking can take place without the numeric contexts that are common in BASIC programming exercises. Turtle graphics provides a powerful and quite different approach to geometric concepts as well. It should be clear that the purpose of having students use turtle graphics is not the pictures that they draw but the logical processes and the geometric ideas with which they are experimenting. The graphics are a vehicle for learning, not the object of instruction.

Even the error messages are helpful in Logo. For example, if you defined HOUSE and then typed HOSE, Logo responds, THERE IS NO PROCEDURE NAMED HOSE. If your misspelled HOSE was inside of HOUSES, Logo would tell you that also.

A popular saying is, "Logo has no threshold and no ceiling." That is, Logo provides opportunities for useful explorations for the kindergarten child but can be just as useful to the research physicist or mathematician. Logo tends to encourage active participation. In Papert's view, the child or the adult learns by actively constructing ideas. (Papert studied with Piaget and bases much of his philosophy of Logo on a constructivist view of learning.) Logo is designed, then, to be an object to "play" with, to give the mind something to work with. Logo is simultaneously concrete and powerful.

For Discussion and Exploration

1. The February 1987 issue of the *Arithmetic Teacher, 34*(6), is all about teaching mathematics using calculators. Find at least one article of interest to you in that issue and read it.
2. Volumes 35 and 36 of the *Arithmetic Teacher* (1987–88 and 1987–89) have in each issue a two-page feature called "Teaching Mathematics with Technology." Look at some of these features and find an idea of particular interest to you. Prepare a lesson or sequence of lessons using the idea(s) you find.
3. Examine one or more of the resource books for calculator activities in the following list. Try the ideas you find with a calculator. Frequently when we read a calculator activity its value escapes us until we actually try it. Choose several of your favorite ideas and plan a lesson around calculators.
   The *Keystroke* series (Reys, et al., 1980).
   Calculator activities for the classroom, Books 1 and 2 (Immerzeel and Ockenga, 1977).
   How to develop problem solving using a calculator (Morris, 1981).
   How to teach mathematics using a calculator (Coburn, 1987)
4. Ask one or more classroom teachers to share their viewpoint concerning calculators in their classrooms. How do they use them? (Or why do they not use them?)
5. Go to a local school and find out how computers are used throughout the school. If there is a computer room or computer lab, what types of computer activities for mathematics go on in the lab? If teachers have computers in their classrooms, ask them to explain how they use computers to teach mathematics. What mathematics software is available in the school?
6. Richard Shumway's approach to computers with children in the classroom is based on the idea that a lot of mathematics learning can take place by examining some very simple computer programs, even with very young children. To investigate some of his ideas, examine the book, *101 ways to learn mathematics using BASIC (K–8)* (1987). Shumway and others shared similar ideas in the *Arithmetic Teacher* between September 1983 and May 1987, in a feature called the "Computer Corner." Some of these features also include ideas for using Logo.
7. a. If you have never ever tried to play around within the Logo language, get a Logo language disk and at the very least work through the introductory exercises beginning on p. 376.

b. If you do know something about Logo, think of some concept in mathematics that possibly could be explored through a Logo activity. Sketch out some of your ideas and share them with someone else who knows about Logo.

8. Find out what software you have available in your school or media lab. Try at least one piece of software in one of the following categories (you will probably want your instructor to help you identify software that is available in these categories):

a. Drill and practice. Is any tutorial help provided? Would you want to use this in your classroom?

b. Problem solving or logic. How involved will students be in this software program? What benefits can you see in having students use this? What are the major strengths and weaknesses? How could you add to the value of the program with related noncomputer activities.

c. Tutorial software. How effective is the teaching strategy that is used? Do students interact in a passive and receptive mode or are they in more of an active and control mode? What physical models would be good to use at the same time with this software?

Suggested Readings

Aieta, J. F. (1985). Microworlds: Options for learning and teaching geometry. *Mathematics Teacher, 78,* 473–480.

Barnes, B. J., & Hill, S. (1983). Should young children work with microcomputers—Logo before Lego? *The Computing Teacher, 10*(9), 11–14.

Battista, M. T., & Clements, D. H. (1988). A case for a Logo-based elementary school geometry curriculum. *Arithmetic Teacher, 36*(3), 11–17.

Bezuszka, S. J., & Kenney, M. (1982). *Number treasury: A sourcebook of problems for calculators and computers.* Menlo Park, CA: Dale Seymour Publications.

Cathcart, D. C. (1987). Exemplary software for the mathematics classroom. *School Science and Mathematics, 87,* 501–508.

Clements, D. H. (1989). *Computers in elementary mathematics education.* Englewood Cliffs, NJ: Prentice-Hall.

Coburn, T. (1987). *How to teach mathematics using a calculator.* Reston, VA: National Council of Teachers of Mathematics.

Corbitt, M. K. (1985). The impact of computing technology on school mathematics: Report of an NCTM Conference. *Arithmetic Teacher, 32*(8), 14–18, 60.

Heck, W. P., Johnson, J., & Kansky, R. J. (1981). *Guidelines for evaluating computerized instructional materials.* Reston, VA: National Council of Teachers of Mathematics.

Hambree, R. (1986). Research gives calculators a green light. *Arithmetic Teacher, 34*(1), 18–21.

Immerzeel, G., & Ockenga, E. (1977). *Calculator activities for the classroom.* Palo Alto, CA: Creative Publications.

Lilly, M. W. (Ed.) (1987). Calculators. Special issue of *Arithmetic Teacher, 34*(6).

Lindahl, R. (1983, October). Electronic spreadsheets: What they are; what they can do. *Electronic Learning,* 44–50.

Maddux, C. D. (Ed.) (1985). Logo in the schools. Special issue of *Computers in the Schools, 2*(2 & 3).

McDonald, J. L. (1988). Integrating spreadsheets into the mathematics classroom. *Mathematics Teacher, 81,* 615–622.

Morris, J. (1981). *How to develop problem solving using a calculator.* Reston, VA: National Council of Teachers of Mathematics.

Moursund, D. (1981). *Calculators in the classroom with applications for elementary & middle school teachers.* New York: Wiley.

Papert, S. (1980). *Mindstorms: Children, computers, and powerful ideas.* New York: Basic Books.

Reys, B. (1989). The calculator as a tool for instruction and learning. In P. R. Trafton (Ed.), *New directions for elementary school mathematics.* Reston, VA: National Council of Teachers of Mathematics.

Reys, R. R., et al. (1980). *Calculator activities for young students.* Palo Alto, CA: Creative Publications.

Riedesel, C. A., & Clements, D. H. (1985). *Coping with computers in the elementary and middle schools.* Englewood Cliffs, NJ: Prentice-Hall.

Shumway, R. J. (1988). Calculators and computers. In T. R. Post (Ed.), *Teaching mathematics in grades K–8: Research based methods.* Boston, MA: Allyn and Bacon.

Shumway, R. J. (1984). Young children, programming, and mathematical thinking. In V. P. Hansen (Ed.), *Computers in mathematics education.* Reston, VA: National Council of Teachers of Mathematics.

CHAPTER 21

PLANNING FOR DEVELOPMENTAL INSTRUCTION

Teaching is both a science and an art. The science portion comes from knowledge of subject matter, theories of learning and development, and knowledge of instructional materials and activities that research and practice have found effective. The artistic portion comes from adding to this scientific base the human qualities of both the teacher and the students in the classroom and the design of specific lessons and interactions that will promote effective and pleasurable learning.

Recognition of the artistic nature of teaching suggests that there is no single formula or plan that can be set down and followed mechanically. Different content requires different strategies. Young children require different approaches from those for older children. Review is distinct from initial development. And each teacher and class is a unique combination of human factors. The suggestions in this chapter are not rigid lesson plan formulas. Rather, they point toward broad guidelines and practical ideas that you might consider while maintaining an individual style.

Conceptual and Procedural Knowledge: Balance and Pace

Relational understanding is the result of developing conceptual knowledge of mathematics and clearly connecting that knowledge to procedural, symbolic knowledge where appropriate. Tradition, pressures of testing, and value systems that prize computational skills are influences causing many classroom teachers to emphasize the procedural aspects of mathematics at the expense of conceptual knowledge and relational understanding. A better balance is required.

PLAN WITH LONG-RANGE GOALS

To develop relational understanding children need time for reflective thought. They need to construct new ideas and relationships, test them, integrate them with existing knowledge, verbalize them, and make them their own. They need to make mistakes and have the opportunity to learn from them. While an individual lesson may have a specific goal, it is important that each lesson be seen as contributing to the larger, more complex set of concepts and skills. Understanding is a unit objective, not a lesson objective (Burns, 1987; Trafton, 1984).

BALANCE

In Chapter 2 it was suggested that at least 50% of our time should be devoted to conceptual development and/or connection of concepts to procedural knowledge. This maxim of 50% applies to the unit goal, not to each lesson. That means that roughly half of the lessons in a unit might well be conceptually oriented (not half of each lesson). To significantly alter this ratio in favor of procedural knowledge is to risk teaching rules and procedures without concepts.

The 50% rule is a departure from what is found in most traditional classrooms. It is not an easy rule to follow, and it is almost completely violated by following page by page through a basal textbook. Many teachers fear that time spent on conceptual development is simply not available. There is, however, evidence to support the notion that if "up front" time is given to development of concepts and explicit efforts are made to make connections with procedural skills, that the time required at the procedural level will be significantly reduced (Suydam & Higgins, 1977; Madell, 1985; Baroody, 1987; Hiebert & Lefevre, 1986; Good & Grouws, 1979). Meaningful skills simply take less time to master than those not conceptually supported. Even more time is made up by reducing the need for remediation. The investment in conceptual development will, in the long term, more than make up for the time spent.

PACE

A common lament of student teachers is, "I had to cut the lesson short because the children were getting restless" (or they "were tired," or they "were misbehaving," or they "were distracted"). There are times when children really are tired or distracted, and we have to adjust for that. But it is interesting that children do not seem to get tired or distracted as often for the good, experienced teachers. Children and adults enjoy learning. Children and adults also get uncomfortable with boring activity and with activities they cannot or do not understand.

Knowing when to move forward requires real skill at listening to children. The interest and enthusiasm of students is one indicator that can be generalized across content. Children who are asked to do activities requiring no new conceptual growth soon get bored and restless. Children who are asked to do activities that are too difficult or not well connected to conceptual knowledge will likewise soon become fidgety and disinterested.

New relationships can only be constructed in terms of existing ones. If concepts are moved too quickly, required conceptual anchors to current knowledge will not be available. The artistic "trick" is to keep children moving forward without getting beyond them, but not so slow as to bore them.

PACING OF PROCESS OBJECTIVES

Many unit objectives culminate in an observable skill; the ability to do something. Changes in emphasis within the curriculum have created a focus on problem solving, estimation, number "sense," and mental computation. These topics cannot and should not be dealt with entirely in units or chapters. We never really master problem solving. We can only improve at it over a lifetime. The real objectives of problem solving and other process-oriented skills, are mental processes or complex interactions of mental processes and concepts. Attitudes and beliefs are also objectives.

These process objectives require that we attend to them on a continual basis virtually the entire school year and at every grade level. It is best that some specific activities related to problem solving and estimation/mental computation are included every week if not every day. Process skills develop slowly and continuously over time. Current abilities and progress can be assessed periodically even though a "mastery level" approach is inappropriate.

Guidelines for Developmental Lessons

A GENERAL MODEL TO WORK FROM

A constructivist view of learning suggests that students should be actively engaged in the learning process. This implies more than simply "doing work." Mechanistic or routine activity, even with physical models and diligently performed, does not promote reflective thought or growth in conceptual knowledge. Effective teachers tend to utilize a significant portion of their class time in some way that encourages this interactive involvement and development of ideas and relationships (Good, Grouws, & Ebmeier, 1983).

A Daily and Weekly Model

In their model for fourth-grade teachers (Figure 21.1), Good and his colleagues suggest that at least 20 minutes of a lesson be devoted to "development" and 15 minutes to "seatwork."

<table>
<tr><th>Monday</th><th>Tuesday</th><th>Wednesday</th><th>Thursday</th><th>Friday</th></tr>
<tr><td rowspan="2">Weekly Review
(20 Min.)</td><td>Homework,
Review,
Mental Computation (8 Min.)</td><td>Homework,
Review,
Mental Computation (8 Min.)</td><td>Homework,
Review,
Mental Computation (8 Min.)</td><td>Homework,
Review,
Mental Computation (8 Min.)</td></tr>
<tr><td rowspan="2">Developmental
(20 min.)</td><td rowspan="2">Developmental
(20 min.)</td><td rowspan="2">Developmental
(20 min.)</td><td rowspan="2">Developmental
(20 min.)</td></tr>
<tr><td rowspan="2">Development
(10 Min.)</td></tr>
<tr><td rowspan="2">Seat Work
(15 min.)</td><td rowspan="2">Seat Work
(15 min.)</td><td rowspan="2">Seat Work
(15 min.)</td><td rowspan="2">Seat Work
(15 min.)</td></tr>
<tr><td>Seat Work
(10 Min.)</td></tr>
<tr><td>Lesson Conclusion &
Homework Assign.
(2 Min. Max)</td><td>Lesson Conclusion &
Homework Assign. (2 Min. Max)</td><td>Lesson Conclusion &
Homework Assign. (2 Min. Max)</td><td>Lesson Conclusion &
Homework Assign. (2 Min. Max)</td><td>Lesson Conclusion
(2 Min. Max.)</td></tr>
</table>

Figure 21.1

A general model for weekly lesson guide. Fourth grade. T. L. Good, D. A. Grouws and H. Ebmeier (1985), Active Mathcmatics Teaching. *New York: Longman. Reprinted by permission.*

The first 8 minutes of their model includes a brief review, checking homework, and a mental computation activity. A maximum of 2 minutes is allotted for concluding the lesson and assigning homework.

We should not take the weekly plan or the lesson outline as rigid models. Many other models for instructional planning have merit. Good, Grouws, and Ebmeier designed theirs for fourth- through eighth-grade classrooms. However, the success of their model and its emphasis on active student involvement suggests that basic features of it be given serious attention, especially the use of a development and consolidation (seatwork) portion.

Development and Consolidation

The major distinction between development and seatwork is the degree of teacher involvement with the full class. The *development portion* of a lesson is interactive, involving students and teacher. During the development portion the teacher also prepares the way for the seatwork portion. The seatwork portion may be labeled more broadly as a *consolidation period.* During this consolidation time students independently explore the concepts and relationships or practice the procedures that have just been taught. The teacher is active, working with small groups or individuals, diagnosing, receiving feedback, and determining the direction of subsequent lessons.

The next two sections offer guidelines for development and consolidation activities.

(a)

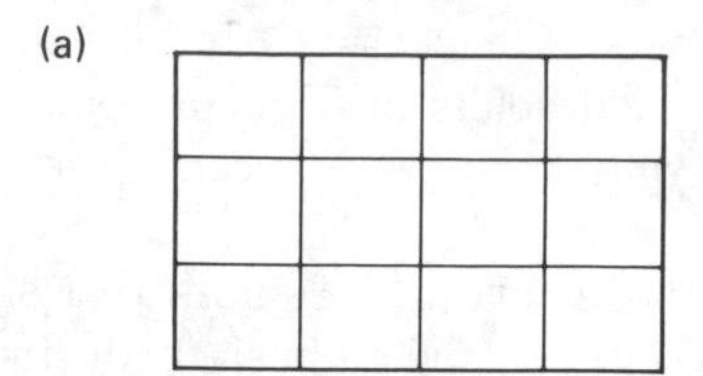

How many different ways can you slice this cake in half?
What if you don't stay on the lines?

(b)

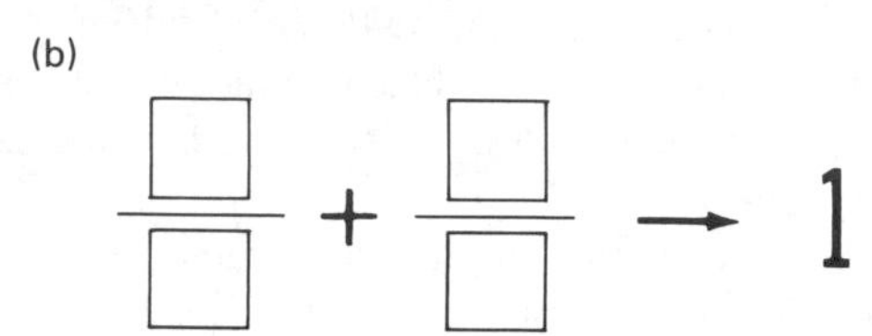

Use four of these numbers in the boxes.
How close can you get the sum to 1?

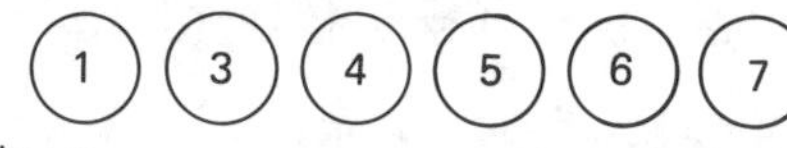

Perhaps:
(Also try it using (11) but no (1).)

Figure 21.2
Two possible "lesson starters."

DEVELOPMENT PORTION OF A LESSON

Purpose

The teaching objective for the development portion of the lesson is always the exploration of a new or developing idea or skill. The content may include conceptual and/or procedural knowledge. A problem-solving flavor or approach is appropriate most of the time.

Getting Started

"Today we are going to talk about fractions. Who can tell me something about fractions?" This opening may appear to be reasonable, but it is certainly not one that would cause students to get excited, care, or begin to really think about the topic of the day. Many will yawn and leave the response to the classroom whiz kid. Good lessons need good beginnings; something that causes students to sit up and take notice and entices them to become involved. A number of possibilities exist.

Start with an interesting question or problem. Figure 21.2 shows two examples of questions that could be posed as a lesson beginning. In the primary grades the cake problem could lead to a discussion of what "half" means, or it might begin an area lesson in measurement. The fraction challenge could be used as an introduction to estimation with fractions or common denominators and computation.

Begin with new or exciting materials. As an example of this beginning, a new set of attribute pieces with different properties might be kept in a bag and pulled out one at a time. Likenesses and differences could be discussed as you proceed. Consideration of what the rest of the bag might hold is both profitable as well as captivating at the K–2 level. Distribution of base ten pieces, geoboards, or fraction strips, along with a quick and easy challenge is frequently a good mind capture. Challenges with most any of the materials in this book can provide interesting lesson starters.

Start with a model or picture on the board or overhead projector. If you are not ready for your students to actually have materials in their hands, they can still be presented to the class. Many materials show up nicely when placed on the overhead projector. Fraction strips, counters, base ten pieces, and geometric shapes or drawings are just a few examples. Computer software is another alternative that can be used with an overhead display screen. As an example of starting with materials, the dot grid and drawing in Figure 21.3 could be prepared on a transparency. Students could have geoboards at their seats to explore the question posed. Many teachers make poster board models of their materials and attach small pieces of magnetic tape to the backs. These are easily presented and manipulated on a magnetic blackboard.

Begin with a quickly played game. The game should relate to the lesson to follow. Max/Min is a good example of

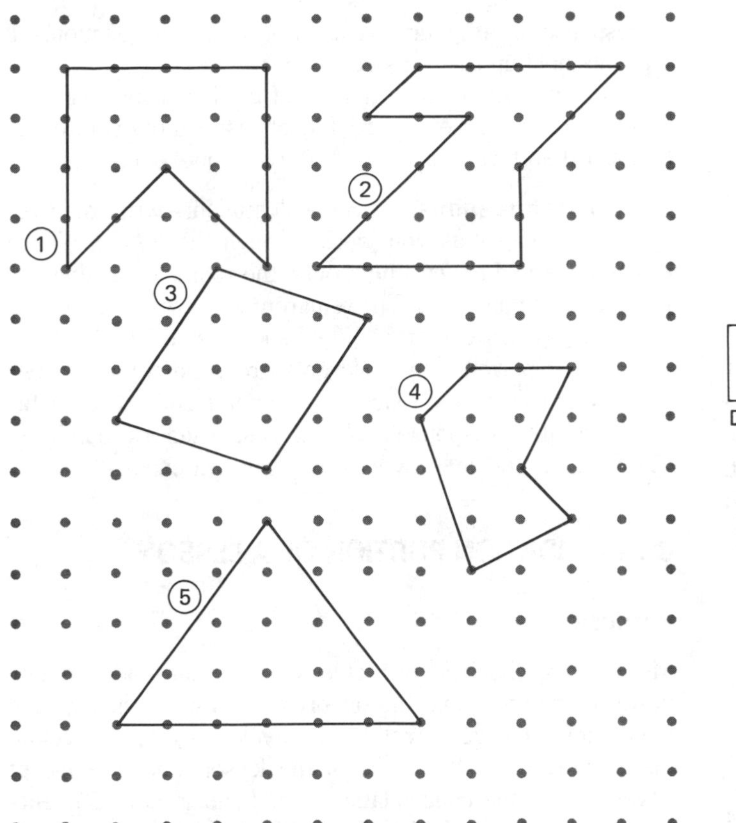

"What easy shapes would help us find the areas of some of these shapes?" "Which shape is largest? Next largest?"

Figure 21.3
Geoboard shapes.

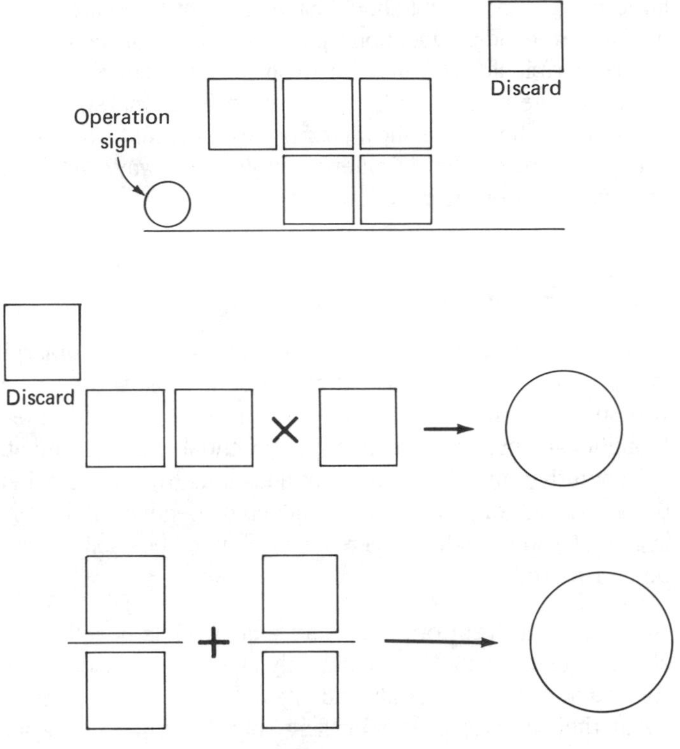

"Max", "min," and "target" games are good lesson starters. Numeral cards 0 to 9 are drawn. Students immediately choose a box to write the number in to work toward the greatest/smallest result or get close to the target.

Figure 21.4
Max/Min boards and target games.

a game with a lot of potential and is easily played with several variations. All students have a drawing of one of the boards in Figure 21.4. Number cards with the digits 0 through 9 are drawn at random. As each card is drawn students must immediately write the number in one of the blanks. Which blank they select is their choice but once written, no changes can be made. Suppose that the game was "Max Sum." When seven numbers had been drawn (one for the discard), each student would have two three-digit numbers on his or her paper. These are added and the student with the greatest sum is the winner. The game can be played to try to get the smallest (Min) result or can be played as "Target" where each student tries to make their result closest to some designated number. These games might be used to begin lessons on computation, estimation, probability (What if the numbers are not replaced or there are twice as many small numbers?), or numeration. There are other examples of games throughout the book that provide good lesson starters.

Maintaining Involvement

Once you have students involved with a good lesson beginning, the next trick is to keep that level of interest and enthusiasm throughout the lesson. In planning a lesson, the key is to consider the students' perspective. While planning your lesson the question, "What will the students be doing?" is much more important than "What will I do next?" If your plans revolve around phrases such as "I will say . . ." or "I will do . . . ," then the students in your lesson will very likely be passive watchers or listeners. Most children do not do either very well and certainly not for very long. The key is to get the students *doing* something as quickly as possible.

Whenever possible, think about how to get manipulative materials into students' hands early in the lesson. If distributed to individuals or groups of students at their desks, a directed activity can and should begin immediately. Students should be given a problem to solve or a concept to demonstrate with the materials as soon as they get them. If the materials are on the desks, that is also where the students' attention will be. Do not pass out materials until you are ready to use them. Activities involving such things as counters, base ten materials, number lines, fraction models, calculators, measurement devices, or geometric materials can be conducted by modeling step-by-step procedures as the students work along with you. Each new example is then planned with increased student independence.

Another approach with younger children is to use a single set of materials that all can see when seated on the floor in a large circle. Each child should have some of the materials in his or her hands. Questions posed or activities conducted should permit all children to be involved. For example, if each child has one piece of a large set of attribute materials, the question can be: "Think how your piece is different from mine. Who has a piece that is different in just one way? Who has one that is different in two ways?"

Student Responses

How you get responses from individuals in a full class or large group is very important. Children frequently follow the leader without thinking. Others are shy or insecure with their thoughts and prefer no response to potential embarrassment. A lesson that involves a series of questions from the teacher to the full class or group can be conducted effectively if thought is given to how children will respond. Several possibilities can be considered.

Accept many responses. After a question is posed to the class, students should respond only after being called upon. Allow sufficient time for all students to have a chance to think about their answers. Avoid calling only the first hand raised or only the children you are confident will have the correct answer. Develop the habit of responding to the first one or two answers in a neutral or noncommittal manner. "That's a good idea. Who else has an idea?" Accept student answers that have already been offered. These repetitious answers may well be personal, genuinely independent thoughts, and not just the result of "following the leader." Slower students frequently are thinking very hard while the quicker children are answering. They deserve to have their ideas heard as well.

Observe responses with physical materials. If the children can answer your questions by selecting an object from those on their desk or by doing something with objects on their desks, you will be able to observe most of the students on each question simply by moving about the class. Using base ten materials to show the number you say is a good example.

Use response cards. Whenever you plan a series of questions with simple quick responses, some method in which every child can respond by holding up the answer works well. If the answers to a series of questions are to be numbers, students can have numeral cards arranged on their desks to be held up simultaneously on your signal. This way, each child must respond and you can check on the response of all members of the class. Two sets of cards from 0 to 9 will allow responses up to 99. Yes/no, true/false, greater/less than, and other paired responses can all be coded by using two colors of construction paper, each color assigned to one of the responses. For example, give each child a square of red and a square of blue construction paper. On the board make it clear that red stands for "yes," and blue for "no" (or whatever pair of responses). At your signal, after asking a question, all students hold up their answer card.

An advantage of this method of getting responses over oral answers is that every child must respond to every question and therefore must attend to every question.

Use written responses. Have all students write their responses on paper as you work through the lesson. These can be observed by walking around the room. Collecting and grading is not important. The requirement to write a response suggests to each child that he or she must think about the question. It permits shy children to respond without public scrutiny and allows the teacher to reinforce correct, thoughtful, or creative responses. Not all responses are numbers. They can be drawings, words, or even sentences.

CONSOLIDATION PORTION OF A LESSON

Purpose

Most lessons should involve at least a short consolidation time in which students have the opportunity to work through and think about the ideas that were developed in the developmental portion. If the focus of the lesson was on concept development, the consolidation period should have students continuing to do activities that promote and develop those concepts. Even if this second phase of a concept lesson begins to connect procedural knowledge to the concepts, the models and/or verbal aspect of the lesson should not be abandoned. If the first phase of the lesson was directed at connecting conceptual and procedural knowledge, then the consolidation should also give students the opportunity to make these connections. Consolidation should only be completely drill or practice when drill, practice, or review was the intent of the first half of the lesson.

The activity in this part of your lesson should be a familiar one to the students or a direct follow-up to the development portion. If pie pieces were the model for the fraction activity conducted with the full class, then pie pieces should probably be used in the consolidation part of the lesson. If you choose to change models, say to fractions strips, that model should at least be familiar. Different groups or individual students may work with different but previously introduced models, games, or activities. The content of the lesson remains the same.

Too frequently a lesson begins with an excellent discussion of a concept or relationship and then shifts dramatically to "here's how you do these problems." Students develop the idea that their real task in mathematics class is to learn how to do the exercises, how to "do the page." Older students even tell teachers, "Don't bother with all that explanation. Just show us how to do the problems." If, however, the consolidation portion of each lesson is at the same level as the development, then the goal of the lesson will be clear in the development. The practice time is a time to work on what was introduced. Students can and should be made responsible for demonstration of conceptual knowledge as well as procedural knowledge.

Specific strategies for consolidation activities are suggested in the following sections.

Worksheets with Manipulatives

A common misconception concerning use of manipulative materials is that any activity done with them must be a group or class activity directed by the teacher. This simply is not the case. One of the best ways to approach the consolidation portion of a concept or connecting level lesson is to design a worksheet that requires the use of physical models. Such worksheets guide the specifics of a manipulative activity to be done at a desk or table. The students either draw simple pictures to show what they have done and/or record numeric results of a manipulative activity. A few examples are provided here to suggest the idea.

Figure 21.5 is a worksheet for number combinations that might be used in first grade. Counters are placed in the area at the top of the page and separated into two parts, and dots are drawn in the small versions at the bottom. Textbooks are beginning to incorporate physical models into first- and second-grade lessons in this way.

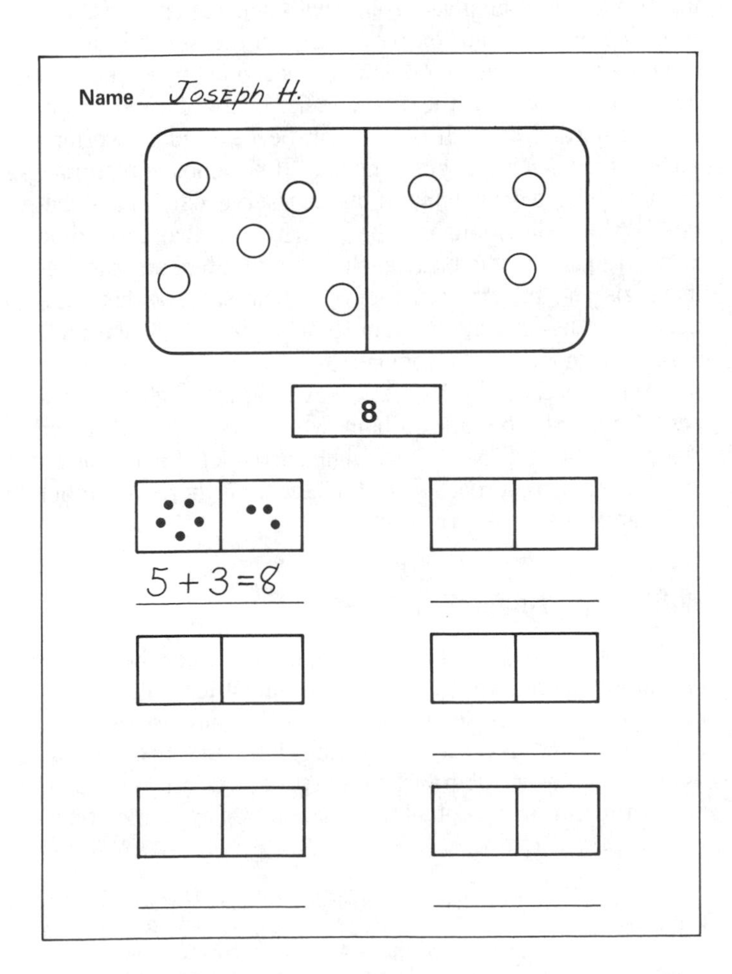

Figure 21.5
A worksheet with counters.

Many activities on worksheets can involve base ten materials with the recording being done as a drawing or with numerals. The worksheet in Figure 21.6 is designed to follow a full class activity involving trading with base ten models. Students draw small dots, sticks, and squares to represent the base ten materials that they worked with to do the exercise.

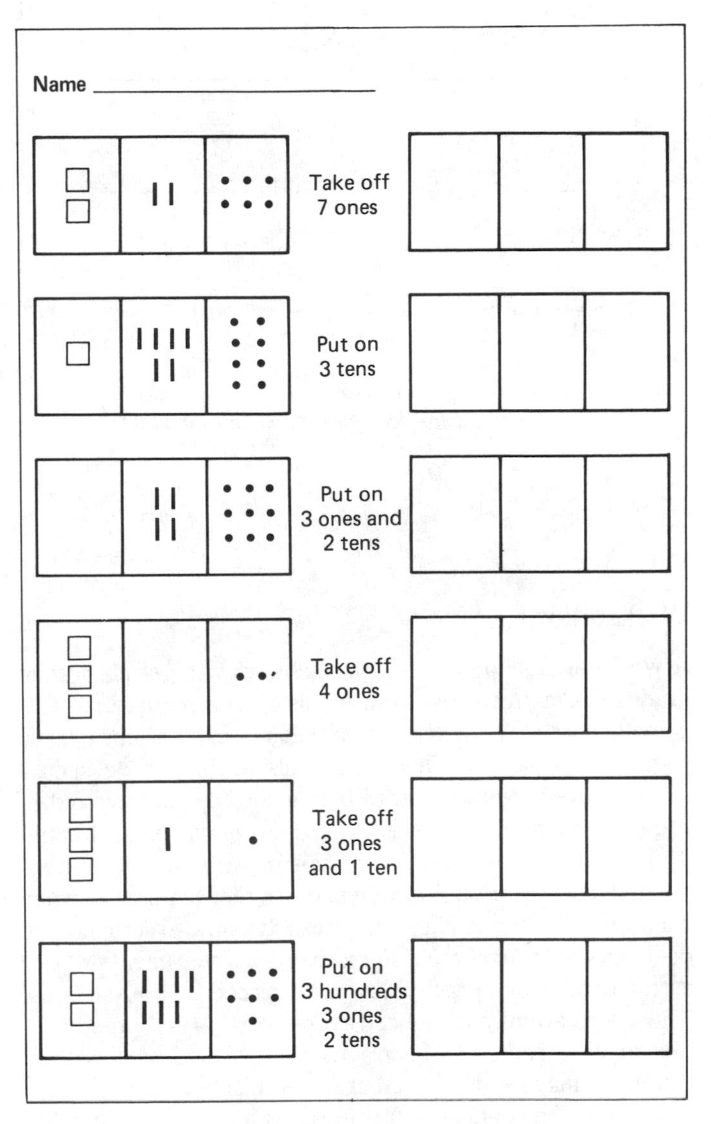

Figure 21.6
A worksheet with base ten materials.

A seventh-grade example in Figure 21.7 involves percentages approximated by familiar fractions. The students could use pie pieces and the hundredths disk (see Figure 13.1, p. 210) to determine the approximate equivalences without doing any computation. Several examples of the worksheet activity could be explored as part of the developmental portion of the lesson.

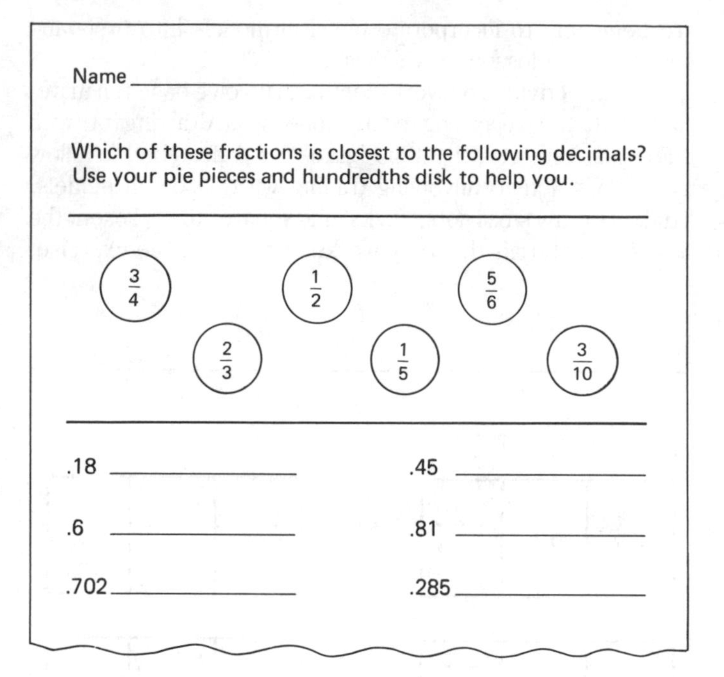

Name ____________

Which of these fractions is closest to the following decimals?
Use your pie pieces and hundredths disk to help you.

$\frac{3}{4}$ $\frac{2}{3}$ $\frac{1}{2}$ $\frac{1}{5}$ $\frac{5}{6}$ $\frac{3}{10}$

.18 ________ .45 ________

.6 ________ .81 ________

.702 ________ .285 ________

Figure 21.7
A worksheet for decimal to fraction estimation.

Workstations

A workstation approach is a profitable way to get children in grades K–2 actively involved. In this approach, materials that go with a particular activity are placed in separate tubs (plastic dishpans) or boxes. These materials might include special manipulatives, work cards to help guide the activity, and, if required, such things as scissors, paper, or paste. Each activity is taught and practiced in groups or with the full class for several days before being assigned to a tub. In this way when children select an activity, they already know exactly how to do it and what is expected. For a given topic you might prepare as many as 8 or 10 activities to be placed in workstations. These are carefully introduced over several days. Some activities may be games to be played by two or three children. Activities may be duplicated at more than one station.

The same containers for the activities are used throughout the year. Each tub is labeled and assigned to an area of the room: a space on a carpet, a table, or collection of desks shoved together. A sign at each area indicates which tub belongs there. Helpers can set out the tubs at the appropriate places in the room. Students can then freely select a station or they can be assigned. This procedure can remain the same throughout the year with the activities within the tubs changing as they are taught to the class.

Several additional comments about the workstation approach are appropriate. It is important that all students go to the workstations at once or that any group not going to a workstation have a similar activity to do at their seats. In this way you are available to interact with and monitor the activity at the stations.

Early in the year, station activities should be so easy that all children will have no difficulty. During this early period you can focus attention on the form of acceptable behavior at the stations, how to get them out, and how to put them away.

Some form of recording of the activity should be included when possible. This may involve drawing pictures, pasting down paper counters, or recording symbolically what was done manipulatively. The recording serves several purposes. Most significantly it represents accountability and thereby lends an air of responsibility and importance to the activity. Second, recording is a way to connect symbolic activity to conceptual activity. Finally, the records are lasting. They can be displayed on bulletin boards, sent home for parents, or kept as records of progress in your files.*

Workstations in the Upper Grades

Variations on the workstation approach are useful even in middle school. The idea is simply to get all students involved independently or in small groups with familiar activities. For example, during a unit on percentage in the seventh grade, a number of different models and conceptual activities may be introduced during the first week. In the second week, a short portion of each period can be devoted to working on activities involving these models. If not enough materials are available, different materials can serve different activities. Written directions can be prepared so that class time is not required to explain such activity. Students can get materials and return with them to their seats without the need for tubs or stations. The teacher is free to interact, assist, diagnose, and remediate.

Many teachers in the sixth, seventh, and eighth grades permit students to work on homework during the last 15 or 20 minutes of class. More challenging activities can be utilized in a workstation approach since the teacher is there to interact and materials can be provided.

Writing an Explanation

At the upper grades, students are able to express their ideas in writing. Rather than finish a conceptual lesson with the usual symbolic exercises in the text, select only two or three to be done. However, require that they be done with manipulatives or with the aid of drawings. On paper, the students' task is to write out a conceptual explanation of why the exercise

*The workstation/tubbing approach was adapted from the *Mathematics Their Way Newsletter V* (Center for Innovation in Education, 1977). This method has been used effectively around the country in kindergartens and primary classrooms as a way to get students actively involved with materials. It provides choice and variety with many materials without the need for classroom sets of each type.

was done the way it was or why the answer makes sense. Encourage the use of drawings in the explanation. To emphasize to the students that the responsibility is for the conceptual explanation and not for the answer, sometimes provide answers along with the exercises.

Practical Considerations

DEVELOPING STUDENT RESPONSIBILITY FOR CONCEPTS

Communication of Values

How do students know what is important and what is not? The answer lies in those things for which we hold them accountable. In school, accountability is determined by quizzes, worksheets, tests, and other forms of evaluation. It is the results of these things that go home to mom and dad.

Most children are not able to see beyond the short-term goal of surviving Friday's test. They do not appreciate the value of understanding when all we evaluate is "how to do it." If the development of conceptual knowledge is as important as we think it is, we had better let students know it, and we had better back that up with what we evaluate. Students know that procedural knowledge is important because we constantly test and grade it. By not doing the same with conceptual knowledge, we deliver a clear, albeit unintended, message that it is not important. By fourth or fifth grade, students actually resist conceptual activities because they are not seen as important.

Suggestions

The following are ways we can tell our students, "It is important to understand this!"

Be interested and enthusiastic about conceptual activities. Spend entire class periods on developmental activities, not just the first 10 minutes. Interact with student discussions and discoveries. Praise students verbally and often for "making a connection" or discovering a relationship. Let students know how hard you have worked to make materials.

Keep checklists of student involvement. During developmental activities, keep a grade book handy and give students credit for their efforts (not just right answers). Let students know that a part of their grade is determined by their cooperation with activities and their participation in class discussions.

Collect products from conceptual activities. Many developmental activities have products: something made, a recording worksheet, a drawing, a record of a game, and so on. If you have students do these things then let them know they are important. Give credit for completing activities, even when they involve errors. Reward efforts. Make helpful comments on these products before returning them instead of grading them down for being incorrect.

Assign conceptual activities for homework. Almost any type of conceptual activity that can be done independently at a student's desk can be done at home. Instead of assigning exercises "1 to 39 odd" for homework, select two or three to explain with models. (Homework is discussed in greater detail later in the chapter.)

Include conceptual activities on tests. Older children can be required to write explanations and draw pictures of what they have thought through with models. The models can and should be available during the test or quiz. Younger students can also be taught simple ways to make responses that demonstrate understanding. If tests or quizzes are administered orally, children can use materials on their desks in response. With groups of 8 to 10, you can see what everyone is doing. Such quizzes need not be lengthy, and you can quickly evaluate the entire class.

Share conceptual activities in your class with parents. Let your room visually reflect a commitment to concepts. Models should be visible, and student work that shows conceptual knowledge should be prominently displayed. Keep examples of students' concept activities and share it with parents during conferences.

Make your grading scheme reflect conceptual understanding. Let students and parents know how the various evaluations suggested here will be used to determine grades. We are fortunately in a time when almost all parents will be highly supportive of this approach.

COOPERATIVE LEARNING GROUPS

For both developmental and consolidation activities, organizing the class into small cooperative learning groups is an excellent idea for maximizing student involvement and interaction. Groups have been suggested for many activities throughout the book. In this section some suggestions for setting up and working with groups are offered.

Size and Composition

There is no magic number for the size of a group. Most experts (for example, Burns, 1987; Worth, 1987) seem to think that four is a good number to aim for, but groups of two, three, or five are occasionally used when the situation calls for it. Students within a group obviously need to work together so some accommodation in the physical arrangement of the room is usually required. Desks can be shoved together or tables and chairs arranged to allow groups of students to work together.

A good case can be made for forming groups heterogeneously; that is, with a range of ability levels within the groups. This permits less able students the chance to hear ideas expressed by others and thus learn from them. At the same

time, more capable students can benefit from explaining their ideas to the group. Ability grouping will quickly create a feeling of inferiority for those in the slower groups. For the purposes of interactive group work there is no benefit to homogeneous grouping.

Random assignment of students to groups on a daily or weekly basis is one method of easily obtaining heterogeneous groups. If group composition changes daily or at least weekly, quarrels over who gets to be with whom can be minimized. Tomorrow it will be different. When groups are fixed for a longer period of time, an unhealthy competitiveness can develop. Some individuals may not be able to adjust to their group members. There is little advantage to permanent groups. Burns suggests dealing out playing cards, using four each of as many numbers ace, two, three, and so on, as are required to give one card to each class member. The ace group, the two group, and so on, are thereby quickly and randomly established. This or similar methods will be seen as fair and are quickly managed.

Responsibilities of Group Members

Working cooperatively and productively in groups is not something that children do automatically or easily. Students need direction and practice with working together. Three simple but important rules for groups have been suggested in a number of places and are offered here, since they have proven to be quite effective:

1. Each student in the group is responsible for his or her own work and his or her own behavior.
2. Each student must be willing to help any group member who asks for help.
3. The only time during group work that a student may ask the teacher for help is when every student in the group has the same question.

Each of these rules is designed to help both students and teacher. Each requires discussion and periodic review with the class. The first is a basic rule of individual responsibility. It prevents students from complaining that they are being penalized for behavior of others and therefore encourages students to monitor both their own work and the group work. The rule does *not* mean work independently inside the group. Rather, it means each child has a responsibility to see that the group works effectively. Answer questions when asked; ask questions when you need help; listen to the ideas of others during discussions. If there is individual work to be done, that work is *yours*. It may not be borrowed. But group projects are group responsibilities and each group member is responsible for seeing that the group does its best.

The second rule not only cuts down on bothersome procedural questions, but lets each student know that there are helpers near by. It also is a way of telling each student to help others. Helping is part of everyone's responsibilities.

The last rule is the most significant. It not only cuts down on procedural questions but promotes interaction within the group. This rule will be the most difficult of the three for *you* to keep. As teachers we all want to be responsive and helpful. If a child asks you a question during group work, you must remind them to check with their group. In the long run you are helping children to become more independent and responsible. The fact that the group members are there to help gives children a means to get assistance without needing the guidance of an adult.

When groups are given a task, the group is responsible for the task. One very effective way to ensure that the group works together is to allow only one worksheet or one record per group. The group can decide on a recorder to write down results. A presenter can be appointed to share results of the group work with the class, but each student in the group should be prepared to contribute. Where appropriate, other tasks such as gathering data or taking measurements can be assigned within each group. When every group member has a specific job or responsibility, the group has a better chance of working together. Allow the groups to make these decisions after offering suggestions. Group work does not mean four students working independently in a circle.

These general rules are a principal basis from which to develop effective cooperative group work. The more you regiment groups and make decisions for them, the more time you will spend on management. Invest efforts in helping children understand the rules of working in groups and the advantages that groups have for them. Expect cooperation from them and help them work together without your constant intervention.

During a lesson you can shift from full class discussion to group work several times. Just because the groups have been formed at the beginning of a lesson does not mean that all work must be done by groups. Individual tasks can still be assigned and questions can be posed to the whole class. At the same time, when an activity would benefit from reflection or interaction, the group structure is there to support it.

When group activities become familiar and commonplace, they will add a new dimension to your instruction.

HOMEWORK

Purpose

We have traditionally used homework as a means of providing extra drill and practice on the procedures that have been taught that day. Drill-oriented homework is an obvious extension of lessons that end with a procedural "how to do it" recipe. In a developmental approach to instruction, many lessons do not have this mechanical goal. Rather, the instruction requires reflection and students validating their results through the use of models or other methods. Drill-oriented homework is not appropriate to follow these lessons.

If symbolic drill and practice is not appropriate, does it follow that no homework can be assigned? Certainly not! Homework can serve the same purpose as the consolidation portion of a lesson. It can be a time when students are required to reflect upon the ideas that have been explored during the lesson.

Homework is also an effective way to communicate to both students and to parents the importance of conceptual understanding. If only drill lessons are followed by homework, students may well infer that drill is more important than understanding. When students take work home that involves models, requires writing or developing oral explanations, or gathering data, parents can see that their children are being required to do more than drill. Homework is a parent's window to your classroom. Students may discuss homework with their parents, and most parents will help see that it is done. If developmental lessons are not followed by developmental homework assignments, parents will not see this portion of your instruction.

Homework also builds self-reliance as students are forced to grapple with ideas apart from the teacher's guidance. In this sense, homework is a way of communicating to students, "I know you are able to do this on your own."

Finally, practice *is* an important part of learning and requires time. When you have activities that require practice, from basic fact drill to solving verbal percent problems, homework is an effective way to provide that practice.

Suggestions for Alternative Homework Assignments

The design of good homework follows much the same approach as the consolidation portion of the lesson described earlier. Students can use manipulatives, draw pictures, write explanations, find and correct errors, or prepare explanations for class. Return to the ideas presented for consolidation activities and modify them for homework.

The manipulative worksheet approach is the easiest to adapt for home use. Many times the only materials that are required are simple counters. Children can use lima beans, buttons, or pennies as counters. If materials such as base ten pieces or fraction pieces are required, paper versions that students can cut out and take home are usually adequate. For younger children, a periodic letter to parents may be required to explain how to help with cutting out models or providing manipulatives. Take the opportunity to tell the parent what your students are learning and enlist their support.

Drawing pictures and/or writing out explanations for three or four problems is a much better homework assignment for a connecting-level lesson than having students do 10 or 20 symbolic exercises. There is no law that says that all of the exercise problems in the text must be worked. Select a few and require students not only to produce the answers but also (and more importantly) the explanations. The responsibility to understand is a significant feature of such an assignment.

Of course there are times when the lesson is at the symbolic level and a strictly procedural homework assignment is appropriate. For such assignments it is important to keep the purpose of the assignment in perspective. Provide sufficient practice to meet the needs of the students, but avoid creating a tedious burden just to show that you are giving homework.

Using Homework in the Classroom

How you attend or do not attend to what was done for homework sends a clear value message about that homework to your students. However, homework checking can easily become a major time-consuming burden to your plans. It is important to find ways of dealing with homework that are quick yet effective.

Be sure that it was done. This does not mean it is necessary to take all homework up and grade or evaluate it. Simply checking that the homework assigned was completed is a minimal activity. This can usually be accomplished by simply requiring that it be placed out on the desks for you to see. Make it clear that you do keep a daily record of homework done and not done.

If an assignment consists of practice exercises, it is rarely necessary to go over each individual item. Try randomly selecting one or two exercises for quick review. For those assignments that required explanations or other conceptual work, a few students can be selected to share their ideas with the class. Not every student needs to share every day as long as you acknowledge that their work was done. Models, drawings, and written explanations should be out on desks while others are making presentations. Instead of having all students explain or present all of their work, ask students to compare what they did with what was shown first. "Martha, I see a different drawing on your paper. Can you tell us about that part?" Or, "Pete, it looks like you may have done about the same thing. Is that right, or did you do something differently?"

Sample solutions (or, in the case of drill, an answer list) can be prepared ahead of time for display on the overhead projector. Students can check their work against the ideas or answers presented.

When homework is conceptual, it is not necessary to provide answer keys or solutions to every assignment. When students begin to appreciate that they are responsible for their understanding and that it is important both to you and to their grades, they will soon learn to check their own ideas against those shared in class. Tests and quizzes *must* follow up on this responsibility by requiring similar explanations and demonstration of concepts. If students are not held accountable for conceptual activity and are only graded on instrumental, procedural exercises, it is nearly impossible to get them to take conceptual knowledge as serious and important subject matter. But homework is their chance to practice and consolidate ideas—to learn. Tests and quizzes evaluate progress.

The Basal Textbook and Developmental Teaching

The basal textbook has been by far the most significant factor influencing instruction in the elementary classroom. In order to make decisions about the use of the textbook, it is good to have an objective view of textbooks and the role they can serve in instruction.

AN OVERVIEW OF THE BASAL MATHEMATICS TEXTBOOK

How Are Textbooks Developed?

It is worthwhile to remember that publishing textbooks is a business. If the very best ideas from mathematics educators were incorporated into a textbook that did not sell, those excellent books sitting in warehouses would be of no value to students and would cost the publisher millions of dollars.

Most publishers enlist as authors mathematics educators and teachers who are quite knowledgeable about teaching mathematics. They also do extensive market research to determine what will sell and what teachers think they want in a book. There is frequently a significant gap between what the authors think would be good and what the publisher thinks will sell. Compromise between author and market becomes the rule. Rather than be forward with new ideas for content or pedagogy, publishers make a serious attempt to offend no one. As a consequence, there is frequently a significant time lapse between the state of the art in mathematics education and what appears in textbooks.

Experience has demonstrated to publishers that if the text requires the teacher to provide additional materials or models, then there is an increased likelihood that the text will not sell. As a result, authors are somewhat limited in writing textbooks that incorporate all the principles of learning mathematics they know to be effective. As we move into the 1990s, there is finally enough demand for models to be used that publishers are trying to respond. Punch-out cardboard models come with many books. Kits of materials, a very high cost item for publishers, are made available as a purchase option.

Textbook authors are also limited by other factors. The textbook is a printed medium. Pictures can be put on pages, but manipulative models cannot. To convincingly illustrate movement and relationships such as grouping 10 ones to make a single ten requires a lot of page space. Young children have a great deal of difficulty following time-lapse-sequence drawings that show materials in several stages of an activity. To use more pages simply creates a book that is too expensive and too heavy. The result is frequently a simple illustration followed by symbolic exercises which require less space.

The Teacher's Editions

The teacher's editions provide considerably more author freedom. Some publishers have provided one or two extra pages for each lesson as well as additional pages with ideas, explanations and so on, in the front and back of each chapter. Frequently, the teacher's editions suggest excellent ways to teach the content of the lesson completely apart from doing the activity presented on the student pages. Teachers should take advantage of this information, especially the suggestions for activities. The package of off-page activities and student page exercises is meant by the authors as the suggested manner of teaching the concepts or skills of the lesson or unit. Too many teachers interpret the textbook curriculum as "getting students through the student pages" when, in fact, the real objectives require the much broader scope presented in the teacher editions. Pupil pages are just one tool for instruction. Pupil pages are not the objective nor are they the curriculum.

The Two-Page Lesson Format

The usual textbook lesson in the pupil book is presented on two pages. An observable pattern to these lessons can be seen in almost all popular textbook series. A portion of the first page consists of pictures and illustrations that depict the concepts for that lesson. The teacher is to use this section of the page to discuss the concepts with the students. Next are well-explained examples for the students to follow or an exercise guided by the text. Finally, the lesson has a series of exercises or practice activities. Thus, many lessons move from conceptual development to symbolic or procedural activities rather than a unit or chapter moving gradually over a period of days from concepts to procedures.

This three-part characterization of a textbook lesson is unfairly oversimplified. At the K–2 level where the children write directly in consumable workbooks, what the student writes for all or most of the two pages may be closely tied to meaningful pictures or even simple hands-on models. The clear adherence to the two-page lesson is not always evident in seventh and eighth-grade texts. As noted earlier, there is a definite movement in the textbook industry to develop pupil pages that require some form of hands-on models. When this is done, authors can write activities that are much like the consolidation portion of a lesson, activities that really are connected to conceptual development.

The traditional two-page format sends a clear message to students that the pictures, concepts, and discussion part of a lesson can be ignored. They begin to "tune out" until we begin to explain "how to do the exercises." Following page by page and assigning procedural exercises from every lesson can easily negate all other efforts to communicate the importance of conceptual understanding.

SUGGESTIONS FOR TEXTBOOK USE IN A DEVELOPMENTAL APPROACH

Avoid the Two-Page Syndrome

Perhaps the best maxim for avoiding the textbook focus on procedures is *teach the content, not the page.* Our task as teachers is to help children construct relationships and ideas, not to get them to do pages. We should look on the textbook simply as one of a variety of teaching tools available in the classroom, not the object of instruction.

Suggestions

If one considers the limitations of the printed medium and understands that the authors and publishers had to make compromises, then the textbook can be a source of ideas for designing lessons rather than dictate what the lessons will be. Here are some suggestions:

Consider chapter objectives rather than lesson activities. The chapter or unit viewpoint will help focus on the learning objectives rather than the activity required to complete a page.

Let the pace of your lessons through a unit be determined by student performance and understanding rather than the artificial norm of two pages per day.

Use the ideas in the teacher's edition.

Consider the conceptual portions of lessons as ideas or inspirations for planning more manipulative, interactive, and reflective activities. Think about how these practice exercises could be modified if students were to use models, write explanations, or discuss outcomes or approaches. The exercises are there for our *use.* We should use them wisely, not let them rule how we teach.

Remember that there is no law saying every page must be done or every exercise completed. Select activities that suit your instructional goals rather than designing instruction to suit the text.

Choose a few selected exercises to be done at a concept or connecting level with manipulatives, drawings, discussions, or written explanations, rather than requiring students to complete all exercises.

Feel free to omit pages and activities you believe to be inappropriate for the needs of your students and your instructional goals, assuming of course, the change is consistent with your district objectives.

If the general approach in the text for a particular unit is not the same as the approach you prefer, omit its use for that unit altogether or select only exercises that provide appropriate practice after you have developed the concepts with your method.

Other Uses of the Textbook

The text is usually a good guide for scope and sequence, especially for computational objectives. There is no reason that you as a teacher should be required to be a curriculum designer. If exercises are selected from pages covering the objective we are teaching we can be reasonably sure that they have been designed to work well for that objective. It is not easy to make up good exercises for all activities. Take advantage of the text.

When drill and practice at the symbolic level are desired, look to the textbook as one source of such activity.

Problem solving, estimation, and mental arithmetic are areas that are becoming more and more visible in textbook series. These are difficult topics to teach, and we need all of the help we can get. There is no need to do these activities only when they appear in the book. Good problems and activities for these areas are difficult to come by. Pull them from all over your book as suits the needs of your class.

Textbooks and other supplementary materials (ancillaries) supplied by publishers usually include evaluation instruments that may be of use for diagnosis, for guiding the pace of your instruction, or for evaluation. At the present time such tests are more likely to assess computational skills rather than conceptual understanding, problem solving, or other process skills. However, there is a definite effort to develop better testing methods for these higher-order skills. The availability of good tests provided by a publisher would be of significant value to the classroom teacher.

Get Involved in Text Adoption

In 1987, the state of California chose not to adopt any textbook series that was offered. They informed every textbook publishing company (with one exception) that their books did not meet the standards set down in their guidelines for instruction (California State Department of Education, 1985). As a result a number of publishers spent huge sums of money to produce revisions or in some cases, special California editions of their series.

The point of this story is that it is the market that drives what is in textbooks much more so than does the mathematics education profession. Mervin Keedy, founder and president of the Textbook Authors Association, and former textbook author himself, notes that many publishers are owned by foreign interests who are only concerned with profit and care little about the quality of U.S. education. The challenge is for teachers to act professionally and demand (as California did) real quality in textbooks. Keedy believes that if we demand that books actually be written by professional mathematics educators with teaching experience (many are) and that the subject matter meet the approval of professional organizations such as NCTM, then the publishing industry will respond to that marketplace (Keedy, 1989).

In the same year that California was demanding such things as the use of manipulatives, open-ended questioning, group work activities, and "situational lessons," many independent districts around the nation were adopting books without even looking at the teacher's editions and/or were comparing content lists with outdated objectives in their local curriculum. Many series that same year were not adopted because they did not have enough drill or because they were "too difficult for the teachers to teach." The publishers have a fiscal obligation to try to sell their books to these groups as well as others. The "common denominator" among textbook adoption committees needs to rise to a higher standard.

Become an active professional teacher in your district. Get involved in the textbook adoption process. Demand that books be adopted that really will be tools to help you teach

developmentally. Don't adopt books that are simply easy to teach.

For Discussion and Exploration

1. Examine a textbook for any grade level. Look at a topic for a whole chapter and decide on one or two major objectives that the chapter addresses. Then look at each individual lesson. How do lesson objectives relate to the unit objectives? How are students guided to work on conceptual development and connection of concepts to skills? Are any lessons predominantly conceptual? Are any clearly focused at connections?
2. Select a lesson from a textbook and try to structure a developmental activity for the lesson based on ideas found either within the lesson itself or on the pages of the teacher's edition. What materials would you need? How could you get the lesson started with a problem-solving emphasis? See how you can use the exercises in the textbook lesson to your advantage. Remember it is not necessary to assign every exercise.
3. For the same lesson chosen in 2 or for another lesson, design either a consolidation activity or a homework assignment that is not purely symbolic or procedural.
4. Look through some back issues of the *Arithmetic Teacher* and try to find activities that would be good lesson starters. A file of these activities is a good resource to start and build on throughout your career. When planning a lesson it is not always easy to remember all of those neat games, problems, and other ideas you have seen in the past.
5. In a classroom where you are teaching, establish the cooperative group approach and teach some lessons which capitalize on its use. Compare the amount of student involvement in these lessons with lessons directed at the class as a whole.
6. Read Keedy's one-page article in the March 1989 *Arithmetic Teacher,* "Textbooks and Curriculum—Whose Dilemma?"
7. Compare at least two current textbook series for one grade level on the basis of conceptual development rather than on a comparison of content coverage. Make your comparison first by examining only the pupil books. Then examine the teacher's editions looking for added suggestions and comparing the quality of those suggestions.

Suggested Readings

Behounek, K. J., Rosenbaum, L. J., Brown, L., & Burcalow, J. V. (1988). Our class has twenty-five teachers. *Arithmetic Teacher, 36*(4), 10–13.

Bell, A. (1986). Diagnostic teaching: 2 developing conflict-discussion lessons. *Mathematics Teaching, 116,* 26–29

Driscol, M. (1982). Effective mathematics teaching. In *Research within reach: Secondary school mathematics.* Reston, VA: National Council of Teachers of Mathematics.

Ellis, A. K. (1988). Planning for mathematics instruction. In T. R. Post (Ed.), *Teaching mathematics in grades K–8: Research based methods.* Boston, MA: Allyn and Bacon.

Johnson, D. W., & Johnson, R. T. (1989). Cooperative learning in mathematics education. In P. R. Trafton (Ed.), *New directions for elementary school mathematics.* Reston, VA: National Council of Teachers of Mathematics.

Keedy, M. L. (1989). Textbooks and curriculum—Whose dilemma? *Arithmetic Teacher, 36*(7), 6.

Leinhardt, G. (1989). Math lessons: A contrast of novice and expert competence. *Journal for Research in Mathematics Education, 20,* 52–75.

Lindquist, M. M. (1984). The elementary school mathematics curriculum: Issues for today. *The Elementary School Journal, 84,* 595–608.

Slavin, R. E. (1987). Cooperative learning and individualized instruction. *Arithmetic Teacher, 35*(3), 14–16.

Trafton, P. R. (1984). Toward more effective, efficient instruction in mathematics. *The Elementary School Journal, 84,* 514–528.

Walberb, H. J. (1985). Homework's powerful effects on learning. *Educational Leadership, 42*(7), 76–79.

Weber, L. J., & Todd, R. M. (1984). On homework. *Arithmetic Teacher, 31*(5), 40–41.

CHAPTER 22

ASSESSMENT IN THE CLASSROOM

The Need for Assessment

There are specific needs within the instructional process that dictate the assessment of students. Most of these needs are summarized in the following list:

1. Teachers need to monitor daily and weekly the effectiveness of specific instructional practices. There are dangers in moving forward before concepts or skills have been well constructed or in staying with a topic too long causing disinterest. New methods of instruction should also be evaluated.
2. Teachers must convey to parents and others the progress of individual students. Regardless of philosophy, grades should be based on a careful system of evaluation.
3. Teachers need to diagnose difficulties that individual students or groups may be demonstrating with a particular topic in order to design an appropriate method of remediation.
4. Schools and school districts require periodic data to determine how well objectives are being met.

The broader needs of schools or districts to evaluate the curriculum will generally be met by the use of some form of standardized tests. Each of the other needs dictates that teachers make an individual effort to assess students on a regular basis.

Types of Tests and Their Purposes

STANDARDIZED TESTS

A *standardized test* is one that includes carefully prescribed methods of administration and scoring to ensure that "standard" procedures are used. Statistical data is generally provided to the users of standardized tests to indicate how different groups of students might be expected to perform on the test *(norms),* as well as statistics indicating how well the test actually measures the objectives it is supposed to measure *(validity).* Standardized tests usually come from a commercial testing company, but school districts may design their own tests and standardize at least the administration and scoring procedures. The term "standardized," then, refers to procedures of administration and scoring rather than purpose.

ACHIEVEMENT TESTS

The large majority of tests used by teachers are *achievement tests,* tests that attempt to measure what a student knows or has learned. There are several types of achievement tests, each with specific purposes, strengths, and weaknesses.

Norm-Referenced Tests

A *norm-referenced* test measures achievement in relationship to others who have taken the same test. The results are usually reported in terms of percentiles or grade equivalents. For example, if a student receives a score indicating the 70th percentile among fourth-grade students based on national norms, that score can be assumed to be higher than 70% of fourth-grade students across the nation if they were to take the same test. The same raw score might fall in the 40th percentile of eighth-grade students.

For the classroom teacher, normative data provides only limited information about the general performance level of the class or students within the class. Generally, the curriculum being taught includes much more than what is tested on a norm-referenced test.

Criterion-Referenced Tests

A *criterion-referenced* test is designed to measure achievement of a specific list of objectives. Different test items or subtests each measure a particular skill or concept. Criterion-referenced tests are valuable to teachers in helping them determine specific strengths and weaknesses of their class or individual students within the class. Many standardized tests are criterion-referenced. However, with standardized tests designed for broad usage, specific concepts, and skills of special interest to you are likely to be omitted. Teachers need to be aware that standardized tests only sample standard curriculum areas.

Tests found at the end of the chapters in most basal texts are good examples of criterion-referenced tests, with specific items to test each objective listed for that chapter. School

districts at both the state and local level frequently design criterion-referenced tests for the objectives in their particular curriculum.

Diagnostic Tests

Diagnostic tests are a form of achievement test designed to help the teacher determine where strengths and weaknesses may occur. They are usually designed for individual administration. A good diagnostic test should provide information on a comprehensive array of related concepts and skills relative to a particular area of mathematics.

Among commercial diagnostic tests which are most popular are the *Key Math—Revised* (Connolly, 1988), the *Woodcock-Johnson Tests of Achievement* (Woodcock & Johnson, 1989), *Stanford,* and the *Diagnostic Mathematics Test* (Beaty, Madden, Gardner, & Karlsen, 1984). The *Test of Early Mathematics Ability* (TEMA) (Ginsburg & Baroody, 1983) is one of several designed strictly for early childhood use (ages 4 to 8). The *Mathematics Assessment/Prescriptive Edu-Disks* (Reader's Digest Service, 1983) is an example of computer software diagnostic tests. Most commercial diagnostic tests provide norms for a variety of subtests. Below grade-level performance on a particular subtest is an indication of a weakness in that area. Specific responses can then be examined more carefully to pinpoint weaknesses.

In the *Key Math—Revised* and the TEMA, items within a particular subtest are arranged in increasing order of difficulty. Administered individually, the teacher begins at a level where the student will likely succeed and proceeds until a certain number of items are missed in succession. *Key Math—Revised* (1988) is a significant improvement over the earlier *Key Math Diagnostic Arithmetic Test* (Connolly, Nachtman, & Pritchett, 1976). Among commercial diagnostic tests, it is one of the most useful to the classroom teacher or specialist needing an initial diagnosis of an individual.

Scores from tests such as the *Stanford* or the *Woodcock-Johnson—Revised* will only be able to determine areas of strength and weakness rather than a complete diagnosis. A careful examination of missed items may provide additional information. Within a particular area a more detailed analysis beyond the test will usually be required in order to plan a course of remediation.

Daily Assessment in a Developmental Approach

WAYS OF LISTENING TO CHILDREN

Teaching developmentally, as described in Chapter 2 and illustrated throughout this text, is a child-oriented approach. To know what knowledge is being or has been constructed by children we must continually "listen" to our students. To really listen to children means that we must build into our instructional plans a variety of listening strategies.

Questioning and Responding

We have a tendency in mathematics classes to ask questions that have very specific right/wrong-type answers. A further tendency is to explore or follow up only on incorrect responses. Students whose answers are challenged are more likely to change them than to defend them. A questioning technique which looks only for correct answers will fail to help you see how students are constructing ideas. It will intimidate less able students and will be dominated by the top students in your class.

In order to get more information during classroom interchanges we need to (1) develop an open-ended questioning technique, and (2) utilize a more inquiring form of response, encouraging students to defend or explain both correct and incorrect answers.

Here are some examples of closed and open questioning for the same situations:

Second-grade subtraction algorithm: $\begin{array}{r} 72 \\ -45 \\ \hline \end{array}$

Closed: You can't take 5 from 2, so what should we do with the 7?

Open: Is there anything special about this problem? If we are going to begin in the ones column, what do you think we should do first?

Fourth-grade one-step translation problem (or simple word problem).

Closed: What kind of a problem is this: add, subtract, multiply, or divide?

Open: What could we do to help us understand this problem? Could you draw a picture, perhaps using a number line? Explain how your picture is the same as the problem. What operation goes with your picture?

Sixth-grade measurement. Selecting an appropriate unit.

Closed: What unit should be used to measure the length of the room?

Open: How would you measure the length of this room? What choices of units do you have? Why would some units seem more appropriate than others?

An open question will encourage students to think about several related ideas. It will implicitly or explicitly indicate that there may be more than one way to respond. It may indicate that the answer can be figured out from what is known and does not require rote memorization.

The second part of good questioning is to respond to students in a manner that helps them think and lets you see what they are thinking. You may like some of the following techniques, or develop some similar ones of your own:

Waiting or Nonjudgment. An immediate judgment of a response stops any further pondering or reflection on the part of the students. If, after a student answers, you simply nod thoughtfully, or say, "Uhuh," or "Hmm," the student is encouraged to elaborate on the response. Other students can join in, offer alternative ideas, or continue to think. A similar response is, "That's a good idea. Who else has an idea about this?" When used consistently, this response tells students that the time for thinking is not yet over.

Request for Rationale. Most students have learned to expect judgmental responses from teachers and to have teachers tell them if their answer is correct. These students are initially shocked when teachers begin to respond to their answers with phrases such as the following:

"That's a good idea. Why do you think that might be right?"

"Are you sure? How do you know that?"

"I see. Can you explain your answer to the rest of us?"

"Show us how you figured that out. Maybe you could use these base ten blocks to help you."

These responses should be used frequently and with both correct and incorrect answers. Soon the request for self-validation will become the expected norm and students even in the lower grades will begin to offer unsolicited explanations along with their answers.

Search for Alternative Ideas. A similar approach is to accept the first response without judgment and then ask if other students may have a different idea. "I see. That's one possible idea. Who else has an idea about this?" Alternative responses by different classmates can then be evaluated in a discussion that looks at the rationale for each. Allow plenty of wait time for the second and third responses, because it is during this time that slower or even average students are given an opportunity to think through the idea. Accept repeat answers, since they are quite likely original thoughts and not copies. Students frequently pay little attention to the response of others because they are busy working through the idea on their own.

Use of Models

One of the most startling effects of using models in a classroom is the amount of information it provides the teacher almost at a glance. When students are engaged in a task at their desks with a model, casually walking about the classroom permits the teacher to see how each student is dealing with the concept at hand. When students are comfortable with an idea, the use of the model will very likely be self-assured and direct. When, on the other hand, they are experiencing some difficulty, it will become very evident. It is hard to cover up or disguise lack of understanding when a model is being used.

Written Work

Written work provides a ready source of information about your students. Most of the time this information is much more than just the number of items correct.

In Figure 22.1(a), two exercises from the same worksheet are incorrect. However, the first is a simple counting error, while the second indicates some misconnection between base ten concepts and the written numerals. Figure 22.1(b) shows several errors that might be made on a worksheet involving integer computation. Some of these may be due to carelessness, to basic fact difficulties, incorrect application of rules for integers, or a variety of conceptual problems. A look at similar exercises would help decide what sort of difficulty the student may be having. In either example, to simply count each exercise as wrong and to ignore the distinction between the errors is not taking advantage of feedback that should be used to guide instruction or even to assign grades.

WHAT TO LISTEN FOR

Relational Understanding

Knowing how to listen is only half of the task. It is equally important to know what to listen for. Most tests are designed to assess what children can *do*. It is important to go beyond simple performance and be sure that we know what children

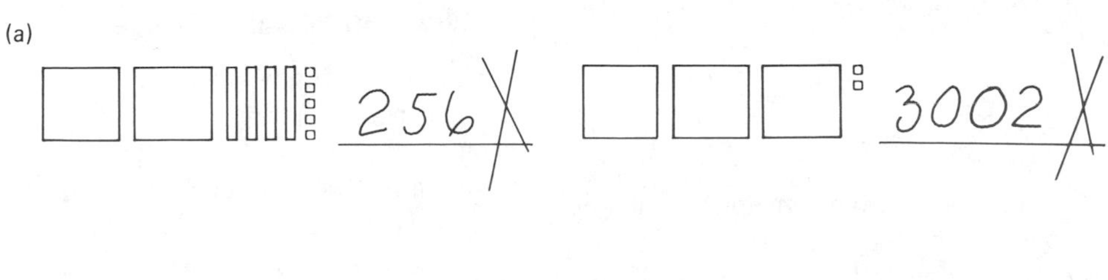

Figure 22.1
Worksheet errors.

understand. In Chapter 2, relational understanding was presented as a major goal of teaching mathematics developmentally. Relational understanding was described as consisting of (1) conceptual knowledge, (2) procedural knowledge, and (3) a connection between these two domains (Figure 22.2). These three components of relational understanding can help guide assessment in the classroom.

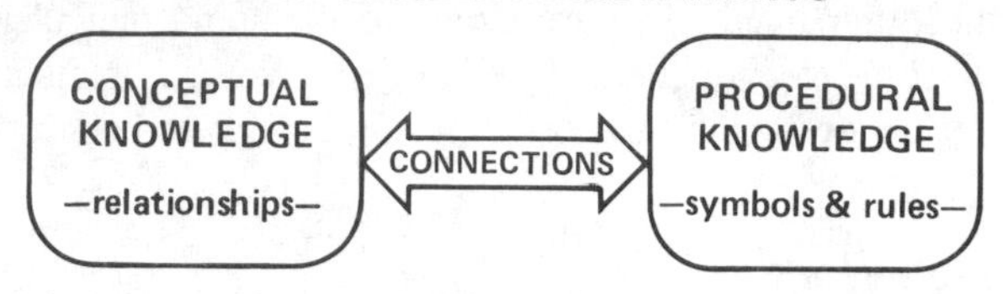

Figure 22.2
The goal of teaching mathematics developmentally.

Conceptual Knowledge

Conceptual knowledge of mathematics consists of relationships which are connected or integrated with other ideas that the learner possesses. Children can demonstrate their understanding of these relationships through the use of models. They can also talk about or describe relationships. In other words, in order to assess conceptual knowledge, some interaction with a model (drawings or materials) is almost required.

Procedural Knowledge

Procedural knowledge consists of the symbolism and procedures that are used in mathematics. This includes all numeration, the symbolism used for the various operations, the computational algorithms, the names of shapes and geometric properties, the identification of measurement units and the procedures of measuring, and so on. Procedural knowledge by itself is the easiest type of mathematics knowledge to assess. Students can use symbols, perform computations, identify shapes and properties, and make measurements. If they do these things correctly, they can be said to possess this procedural knowledge.

Connections

Procedural knowledge disconnected from concepts or without a conceptual basis is only instrumental understanding. We want students to not only have procedural knowledge but to connect it with underlying concepts, to have a rationale for the procedures. The assessment of this important connection involves having students make translations between conceptual models and the procedures they are using or to explain or justify why procedures work. Frequently this connection can be assessed along with conceptual knowledge.

Diagnostic Teaching

LEARN TO ASK YOURSELF "WHY?"

Awareness of Difficulties

The more you listen to children the more aware you will be of those times when they are having difficulties. You may notice problems or concerns during the course of a lesson or while you are going over a worksheet or quiz. Students may voice confusion, be unable to give a response, or may respond in a manner that indicates a lack of understanding. A more worrisome difficulty occurs when a child who you thought was following along shows up failing on a chapter or cumulative test. Once presented, difficulties or deficits with material you have worked on require some response, some form of remediation on the part of a good teacher.

Remediation Is Not Reteaching

One of the most common mistakes that teachers make when they observe a student difficulty is to reteach, to repeat what has already been done, provide more exercises, or cajole the student to "try harder." If a child has trouble with a particular exercise, there is a tendency to "show them" one more time how to do it. Unfortunately, reteaching, or simply showing the child how, is treating the symptom without any thought about the possible cause. Using a medical analogy, this is like giving aspirin for a recurring headache but ignoring the source of the headache. If infection, stress, eye trouble, and so on, remain, then the symptom will reappear.

Perhaps the most important thing you can take from this chapter is this: When a child demonstrates some difficulty with the material you are teaching, the first question you should ask yourself is, "Why? What is the cause of this child's problem?" This may require you to pause, reflect, and consider a wide variety of possible hypotheses instead of what may seem a more direct, immediate solution to the problem. Solving the immediate difficulty is always a short-term Band-Aid approach that may not address the real cause.

Considering Possible Causes

Trying to determine what is at the root of a child's difficulty is challenging, interesting, and highly rewarding. Where should you look? In Figure 22.3, three broad categories of factors are illustrated, each of which interacts and contributes to learning. The cause of a child's problem can usually be found in one of these areas or some combination of them.

Conceptual or Procedural Knowledge Deficits. This area may be the first to consider. What concepts, what procedures or symbols are necessary for this material to be understood? Perhaps the child lacked a prerequisite concept that was necessary to developing the idea you were working on. For example, if a fourth-grade child has difficulty with a whole

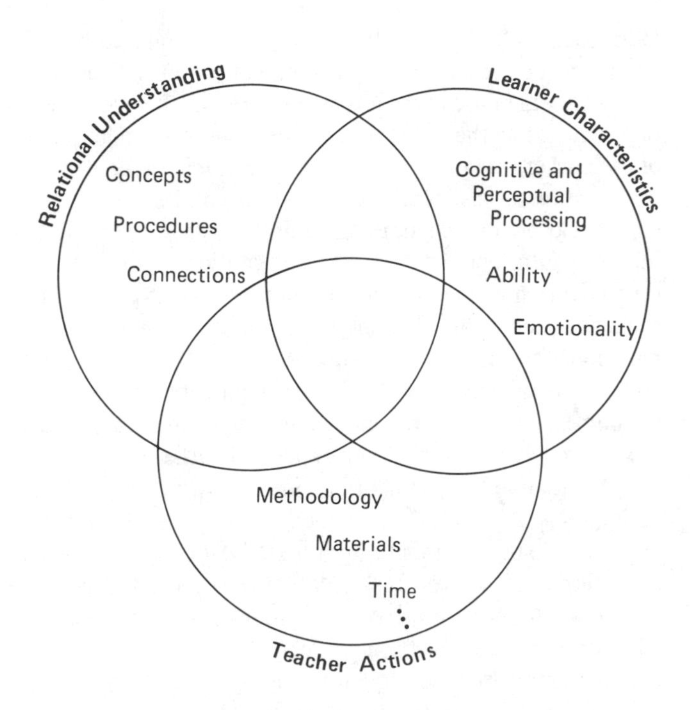

Figure 22.3
Factors contributing to learning are also sources of learning problems.

number long division, the trading ideas would not be understood if there was a deficit in place-value understanding. Or perhaps the concepts are there, but the connection with how they are to be written down is missing. Perhaps the problem is with basic multiplication facts or lack of subtraction skills. For most topics there is a broad list of concepts, procedures, and connections that may be missing or mislearned.

The best reaction, the best remediation, is based on accurately determining the cause. If simple basic fact mastery is the major difficulty in long division, place-value remediation is not very helpful. If the child has difficulty with multiplication, that should be attended to rather than the entire algorithm. If confusion is traced to place-value knowledge, then base ten pieces should be incorporated in the remediation. Simply showing the child or providing additional practice is the least effective use of your time and effort.

Learner Characteristics. These characteristics include a wide range of concerns that you should always keep in mind as possibly contributing to learning problems. The entire range of learning disabilities, or cognitive and perceptual processing problems, are included here. These are discussed in the next chapter. Emotional problems are of a different nature and can frequently interfere with learning. And, finally, many children learn at a slower rate or lag developmentally behind their peers. Each of these types of learner problems is found more and more frequently in the regular classroom, requiring care and attention from the classroom teacher.

When a problem can be traced to a particular learner trait, then the remediation must either confront that trait effectively or overtly sidestep it so that the child can learn in a manner that capitalizes on strengths and avoids weaknesses. For example, suppose that a child failed to learn some recently developed material due to an attention difficulty rather than a concept deficit. Remediation might include reteaching or practice that is designed in very short segments. Another idea would be to have the child work with a partner or work with a tape recording you make that focuses his attention periodically as he works. Attending to a child's emotional stress may produce greater gains than direct efforts at remediation.

Teaching Actions. Actions that you have provided must also be considered as a source of difficulty. None of us will ever be flawless in our methods of teaching. Not only is teaching a very imprecise science, but so many options are available that we can never be certain we have used the "best" approach for the situation. "Did I use a good model or manipulative to develop this idea?" "Did I provide enough practice with this procedure before I moved forward?" "Did I present this material orally when a drawing or a model might have been better?" In most instances, choices of teaching strategies will interact with the areas of learner knowledge and learner characteristics. Recognition that we do make choices and that some choices may not be the best for some children is a giant step toward effective remediation.

LISTEN, REACT, LISTEN, REACT, . . .

Listening, in all of the forms described earlier in this chapter, is a significant part of diagnostic teaching. The more actively and carefully you listen, the earlier you can detect problems. Reacting, by attending to the cause of the problem (rather than the problem itself), is the second part of diagnostic teaching. Treat the problem, not the symptom! And then continue to listen. "Was the treatment effective?" "Did I figure out the correct cause?" "Do I need more information?" Never assume that the first attempt at remediation is the correct one or is all that can be done. Diagnostic teaching is a continuous process of listening carefully and attending to causes.

CONDUCTING DIAGNOSTIC INTERVIEWS

Reasons to Consider an Interview

It takes extra time and effort to sit down one-on-one with a child and conduct an interview. There are several good reasons why the effort may be worthwhile. The most obvious reason is that you need more information concerning a particular child and how he or she is thinking. You may have already made several attempts at remediation and require more insight before continuing.

A second reason is to get information about the effectiveness of your instruction; to determine how well children are actually constructing the ideas that you want. For example, are you sure that they have a good understanding of equivalent fractions or are they just doing the exercises according to rote rules? To check, you could select three or four representative students to interview and base additional class instruction on what you learn from them.

A third reason is to grow as a teacher, to learn more about the way children learn and think. You might decide to conduct at least one interview a week, changing the topic and the children you interview. In addition to growth in your knowledge of children, you will also grow in your ability to conduct a good interview.

Planning an Interview

There is no magic right way to plan or structure an interview. In fact, flexibility is a key ingredient. You should, however, have some overall game plan before you begin, and be prepared with key questions and materials. Figure 22.4 is a basic skeleton of one interview strategy. Depending on the topic and situation, some elements may be omitted or others added.

Begin an interview with questions that are easy or closest to what the child is likely able to do, usually some form of procedural exercise. For numeration or computation topics, begin with a pencil-and-paper task such as doing a computation, writing or comparing numerals, or solving a simple translation problem. When the opening task has been completed, ask the child to explain what was done. "How would you explain this to a second grader (or your younger sister)?" "What does this (point to something on the paper) stand for?" "Tell me about why you do it that way." At this point you may try a similar task but with a different feature; for example, after doing 372 − 54 try 403 − 37. The second problem has a zero in the tens place, a possible source of difficulty.

The next phase of the interview involves the use of models or drawings that the child can use to demonstrate understanding of the earlier procedural task. Computations can be done with base ten materials, blocks or counters can be used, number lines explored, grid paper used for drawing, and so on. Be careful not to interject or teach. The temptation to do so is sometimes overwhelming. Watch and listen. Next, explore connections between what was done with models and what was done with pencil and paper. Many children will do the very same task and get two different answers. Does it matter to the child? How do they explain the discrepancy? Can they connect actions using models to what they wrote or explained earlier?

Alternative beginnings to an interview include making an estimate of the answer to either a computation or a word problem, doing a computation mentally, or trying to predict the solution to a given task. Notice that the interview does not generally proceed the way instruction does. That is, in an interview, the conceptual explanations and discussions in general come after the procedural activity. Your goal is not to use the interview to teach but to find out where the child is in terms of concepts and procedures at this time.

Suggestions for Effective Interviews

The following suggestions have been adapted from excellent discussions of interviewing children by Labinowicz (1985; 1987), Scheer (1980), and Liedtke (1988).

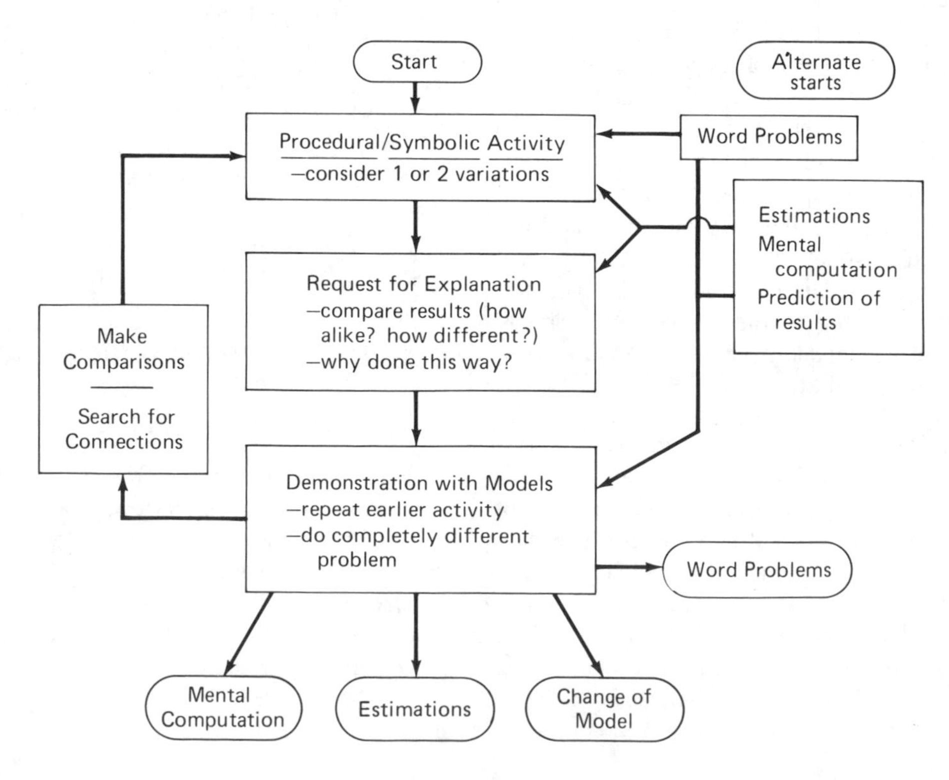

Figure 22.4
A skeleton plan for a diagnostic interview.

1. *Be accepting and neutral as you listen to the child.* Smiles, frowns, and other body language can make a child think that the answer he gave is right or wrong. Develop neutral responses such as, "Uhuh," "I see," or even a silent nod of the head.
2. *Avoid cueing or leading the child.* "Are you sure about that?" "Now look again closely at what you just did." "Wait. Is that really what you mean?" These responses will indicate to children that they have made some mistake and cause them to change their responses. This can mask what they really think and understand. A similar form of leading is a series of easily answered questions that direct the student to a correct response. That is teaching, not interviewing.
3. *Do wait silently.* Wait after you ask a question. Give the student plenty of time before you try a different question or probe. After the child makes a response, wait again! This second wait time is even more important because it allows and encourages the child to elaborate on the initial thought and provide you with more information. Wait even when the response is correct. Waiting also can be relaxing and give you a bit more time to think about the direction you want the interview to take. Your wait time will almost never be as long as you imagine it is.
4. *Do not interrupt.* Let children's thoughts flow freely. Encourage children to use their own words and ways of writing things down. Intervening with questions or by correcting language can be distracting to the child's thinking.
5. *Phrase questions in a directive manner to prevent avoidance.* Use "Show me . . .," "Do . . .," or "Try . . .," rather than "Can you . . .," or "Will you . . . ." In the latter form, the child can simply say "no," leaving you in a vacuum.
6. *Avoid confirming a request for validation.* Students frequently follow answers or actions with, "Is that right?" This query can easily be answered with a neutral, "That's fine," or "You're doing OK," regardless of whether the answer is right or wrong.

One final thought. Interviewing is not an easy thing to do well. Many teachers are timid about it and fail to take the time. But not much damage is possible, and the rewards of listening to children, both for you and your students, are so great you really do not want to pass it up.

For Discussion and Exploration

1. Select one chapter from the teacher's edition of any standard textbook at a grade level of interest. Compare the chapter test with the objectives that are in the front of the chapter and with the types of instruction and activities provided in the student pages and in the teacher notes. Which items test conceptual knowledge? Which items test procedural knowledge? Do any items test the connection between the two? Are the objectives assessed completely? If you were teaching this chapter, what additional assessments, if any, would you feel you might need?
2. Examine a commercial diagnostic achievement test. Select a particular grade level of interest. Consider the information on a broad area of the curriculum such as geometry or computation and compare the information the test provides with what is usually taught at that grade level. Use a current textbook as a guide to curriculum. After looking at the test as a whole, what do you think the score or subscores of the test would tell you as a teacher? What else could you learn from using this test?
3. Select a chapter from a textbook and examine the content and objectives. Make up a test that you could give to a full class that you think adequately assesses the conceptual knowledge and connections with procedural knowledge for the content of the chapter. Your test can include physical models if it is reasonable to use them with a full class. How is your test different than the book test for that chapter?
4. Read "Assessing for Learning: The Interview Method," by Labinowicz in the November 1987 *Arithmetic Teacher.* Try conducting all or part of the diagrammed interview that accompanies that short article with a second- or third-grade child or role-play the interview with a peer.
5. Select a topic for a particular grade level and design a rough outline (major questions and materials) for a diagnostic interview. Number relationships, operations, place value and computation, fractions and fraction computation, decimals, ratio, measurement, or geometry are suggested broad areas to work on. These should be narrowed further by the grade level. You might want to read the short article, "Diagnosis in Mathematics: The Advantages of an Interview," by Liedtke in the November 1988 *Arithmetic Teacher.*
6. Find one or more students and try your interview (see 5). Try to make either a video or audio tape of the interview. Discuss your interview technique and what you learned about the child. An alternative is to role-play your interview with a peer.

Suggested Readings

Clark, H. C. (1986). How to check elementary mathematics papers. *Arithmetic Teacher, 34*(1), 37–38.

Driscol, M. J. (1980). Diagnosis: Taking the mathematical pulse. In *Research within reach: Elementary school mathematics.* St. Louis, MO: CEMREL, Inc.

Driscol, M. J. (1980). Evaluation in mathematics education part one: Looking beneath and beyond the tests. In *Research within reach: Elementary school mathematics.* St. Louis, MO: CEMREL, Inc.

Driscol, M. J. (1980). Unlocking the mind of a child: Teaching for remediation in mathematics. In *Research within reach: Elementary school mathematics.* St. Louis, MO: CEMREL, Inc.

Ginsburg, H. P. (1987). *Assessing the arithmetic abilities and instructional needs of students.* Austin, TX: PRO-ED.

Howden, H. (1983). An alternative to conventional methods of evaluation. In L. G. Shufelt (Ed.), *The agenda in action.* Reston, VA: National Council of Teachers of Mathematics.

Inskeep, J. E. (1978). Diagnosing computational difficulty in the classroom. In M. N. Suydam (Ed.), *Developing computational skills.* Reston, VA: National Council of Teachers of Mathematics.

Labinowicz, E. (1987). Assessing for learning. *Arithmetic Teacher, 35*(3), 22–25.

Liedtke, W. (1988). Diagnosis in mathematics: The advantages of an interview. *Arithmetic Teacher, 36*(3), 26–29.

Lindquist, M. M. (1988). Assessing through questioning. *Arithmetic Teacher, 35*(5), 16–19.

Richardson, K. (1988). Assessing understanding. *Arithmetic Teacher, 35*(6), 39–41.

Schoen, H. L. (1979). Using the individual interview to assess mathematics learning. *Arithmetic Teacher, 27*(3), 34–37.

Underhill, R. (1988). Mathematical evaluation and remediation. In T. R. Post (Ed.), *Teaching mathematics in grades K–8: Research based methods.* Boston: Allyn and Bacon.

CHAPTER 23

MATHEMATICS AND CHILDREN WITH SPECIAL NEEDS

Exceptional Children

The Education for All Handicapped Children Act, or Public Law 94-142, was passed in 1975. The law mandates the most appropriate education for all handicapped children in the "least restrictive environment." Since the passage of P.L. 94-142, more and more children with some form of handicapping condition are being placed in regular education classrooms for all or part of the school day. The trend to accommodate these exceptional children within regular educational settings appears to be increasing.

Another group of exceptional children are the talented and gifted. These students also require special consideration. They represent a significant national resource and should be stimulated and provided with opportunities to reach their full potential.

Both regular and special education teachers should understand that the basic principles, strategies, and materials appropriate for any sound developmental instruction are also the principles, strategies, and materials appropriate for exceptional children.

Children with Perceptual and Cognitive Processing Deficits

CHARACTERISTICS OF SPECIFIC PROCESSING DEFICITS

A variety of terms are commonly used to label children who have difficulties processing information. Frequently these special children are referred to as *learning disabled* (LD) or *specific learning disabled* (SLD). Others are said to have *attention deficit disorders* (ADD). According to Glennon and Cruickshank (1981), *perceptual and cognitive processing deficit* would be a more accurate term than learning disability. These processing deficits are thought to be the result of some neurological dysfunction. Children with a processing deficit may be quite intelligent and by most definitions are not retarded. Here is a list of specific perceptual and cognitive processing deficits, which has been adapted from one developed by Bley and Thornton (1981).

PERCEPTUAL AND COGNITIVE PROCESSING DEFICITS

Perceptual Deficits
- Figure-ground
- Discrimination
- Spatial organization

Memory Deficits
- Short-term memory
- Long-term memory
- Sequential memory

Integrative Deficits
- Closure
- Receptive language
- Expressive language
- Abstract reasoning

Attention Disorders
- Attention
- Impulsivity
- Compulsiveness

When teaching mathematics to a child with a processing deficit, your focus should be on the child's pattern of learning strengths and weaknesses and not exclusively on failures in mathematics. Processing deficits span content areas. They are not unique to particular disciplines.

It is important that the teacher have as much detailed information as possible about children with cognitive or perceptual deficits so that instruction can be modified in the most effective manner. The school psychologist is the principal source of help to the teacher in identifying a child's specific learning problem. The teacher can work with the psychologist to aid in an accurate diagnosis. The psychologist can help the teacher better understand the nature of the dysfunction and how it affects learning. Both can work together to design the most effective instructional plan. Without a clear diagnosis, there can be a tendency to treat all children with "learning disabilities" the same. This global approach fails to capitalize on individual strengths and deal directly with specific weaknesses.

Perceptual Deficits

Perceptual deficits may be visual or auditory. Children with *figure-ground* problems have difficulty sorting out or recognizing component parts of what they see or hear. Separate steps of computational algorithms may be confused with others. For example, the multiplication and subtraction procedures within long division may be confused or missed completely. When listening to directions with several parts, these children may have difficulty understanding the complete procedure. An entire page of exercises can be an incomprehensible maze.

Discrimination difficulties refer to the inability to discern differences in things that are seen or heard. Numbers or problems on the page become confused. The word "forty" may be heard as "fourteen." Numbers are frequently written backward. Coins are frequently confused.

Spatial organization deficits are observed in children who cannot organize their work on paper, cannot understand pictures of three-dimensional solids, or are confused with orientations of up, down, left, right, under, and so on. These children frequently have difficulty writing numbers or drawing figures.

Memory Deficits

Memory deficits can also be specifically visual or auditory. Some children may have more difficulty recalling things seen than things heard, or vice versa. Children with *short-term memory* deficits can have trouble recalling things for even a few seconds. This causes difficulty copying from the board or the book or in following directions. From the time the child's eyes leave the board to the time he begins to write, the idea may already be forgotten. When he or she reads or listens to a verbal problem, the first parts of the problem may not be retained. After the reading, the entire problem may be mysterious.

Long-term memory deficits are manifested quite differently. Children with this problem will show less trouble with new material on the day presented but will have difficulty retaining material over days or weeks. Mastery of basic facts is a hallmark problem for children with long-term memory problems.

Sequential memory problems refer to the inability to retain an order to a sequence of events, or a series of steps in a procedure. These children frequently will ask, "What do I do next?" when doing a computational procedure such as adding fractions with unlike denominators. They may know how to do the steps but not know which comes next.

Integrative Disorders

Integrative disorders refer to some dysfunction that makes it difficult for children to make connections or associations between ideas, to see similarities, to build one idea upon another.

Children with *closure* difficulties do not see or hear things in the same "obvious" groupings or clusters that they are presented in. They fail to see simple repeating patterns, to group the digits in large numbers that they are reading in a logical order, or to make connections between two similar problems or procedures. Counting on may be difficult. To solve word problems that require identifying key numbers, relationships, or actions and then seeing how these are the same as one of the four basic operations may be exceedingly difficult for these children.

A *receptive language* deficit is exhibited by the child who frequently asks, "What do you mean?" or gives a blank expression when a question is asked. If a problem is explained a bit differently than what was presented earlier, the child may not connect this different approach with the previous learning. The same can be true in visual presentations. If the model or diagram for a concept is different superficially although still representing the same idea, these children may not see the common relationships.

Expressive language deficits refer to the inability to verbalize what is known. The difficulty is in retrieving the specific words or symbols that are used to express their idea. These children would be able to identify a procedure that was done incorrectly but be unable to explain the error.

Children with *abstract reasoning* deficits have much more difficulty constructing relationships and connecting abstract ideas with symbolism than do other children. Their reliance on models is much greater than most children. Memory, language, and proficiency with the models may not present a difficulty. However, since it is difficult for these children to connect ideas, to integrate new thoughts with existing ones, it is also difficult for them to abstract a relationship from their work with materials.

Attention Disorders

Many children with processing deficits are very easily distracted, unable to focus for any length of time on one task at a time. They are distracted by the slightest sounds, touches, visual stimuli, smells, hunger, and so forth, that would likely go unnoticed by the average student. Another feature of these *attention disturbances* is an overattention to unimportant details such as a variation in color of two essentially alike base ten models, or a wrinkle in a sheet of paper. They may repeatedly erase numbers or drawings to fix unimportant features.

Impulsivity is observed in children who tend to respond almost immediately to questions but give little thought to their answers. Consequently they frequently make what seem to be strange or even totally unrelated responses either orally or in written format.

Compulsiveness refers to a tendency to continue on with a train of thought without attending to changes in the task or without sensing when a task is completed. On a page of exercises with mixed operations these children may do all problems using the same operation as the first.

ADAPTING INSTRUCTION FOR CHILDREN WITH PROCESSING DEFICITS

Instructional methods for children with perceptual and cognitive processing deficits must be designed by taking into consideration the specific deficit or deficits involved. Children will not outgrow or overcome their problems with learning, but they can be helped to cope with them. Three general principles can be kept in mind: (1) Make an effort to add structure to your instruction, avoiding free and open discovery approaches. (2) Rely more heavily on models, spending a longer time and with less variation. (3) Teach to strengths while avoiding weaknesses.

Teaching to Perceptual Deficits

With perceptual deficits the general rule is to focus on the mode of learning that is a strength for the child and avoid or provide special helping techniques for the mode of learning that causes difficulties.

The following specific suggestions may be useful:

Seat the child near to you and/or near to the chalkboard.

Keep the child's desk or workspace clear of clutter. One task at a time.

Provide directions in written form for children with auditory problems and in verbal form for those with visual difficulties.

Keep your voice moderate and at an even level. Repeat main ideas.

Structure the page for the child. Use a marker to block out all but one row of exercises or cut out a tagboard template to mask all but one problem at a time (Figure 23.1). Cut up worksheets into single problems or rows of exercises or prepare separate exercise sheets with tasks arranged in separate boxes to avoid confusion.

Help the child with methods for organizing written work. Have computations done on centimeter grid paper with one numeral per square. Provide paper with columns drawn or templates in which to record computations (Figure 23.2).

Use a tape recorder (and headphones) with instruction on the tape to help students who have difficulty learning from visual materials.

Perceptual Deficits
Figure-ground
Discrimination
Spatial organization

Memory Deficits
Short-term memory
Long-term memory
Sequential memory

Integrative Deficits
Closure
Receptive language
Expressive language
Abstract reasoning

Attention Disorders
Attention
Impulsivity
Compulsiveness

Figure 23.1
Template frames problem to decrease perceptual problems.

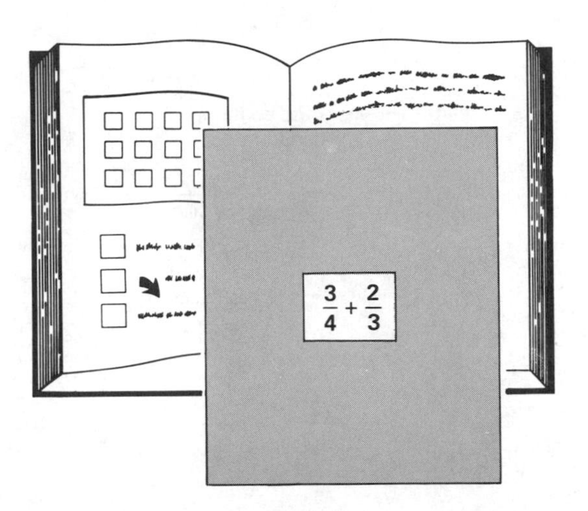

Figure 23.2
Centimeter grid paper and prepared recording templates can help with organization.

Teaching to Memory Deficits

Diminish the load on short-term memory by breaking tasks and directions into very small steps. Long-term memory problems require overlearning, frequent practice, and as many associations with other ideas as possible. The following more specific suggestions may be useful:

For short-term auditory memory problems, make an effort to give only one instruction at a time rather than a series of steps. If possible, write the steps of a task down on paper.

Provide a model exercise. Make each step clear (Figure 23.3)

For basic facts use strategies and relationships as in Chapter 7. Provide fact charts (Figure 23.4) and cross out facts that are mastered.

Permit the regular use of calculators. (Good for all students!)

The ideas in Figures 23.1–23.4 have been adapted from Bley and Thornton, *Teaching mathematics to the learning disabled* (1981).

Teaching to Integrative Disorders

The best general principle for helping children with integrative disorders is to emphasize the same methods of a conceptual, developmental approach to learning.

Stay with the use of models much longer than usual.

Simplify instruction. Use familiar models with simple directions.

Avoid any symbolism that is unfamiliar and not useful to the child.

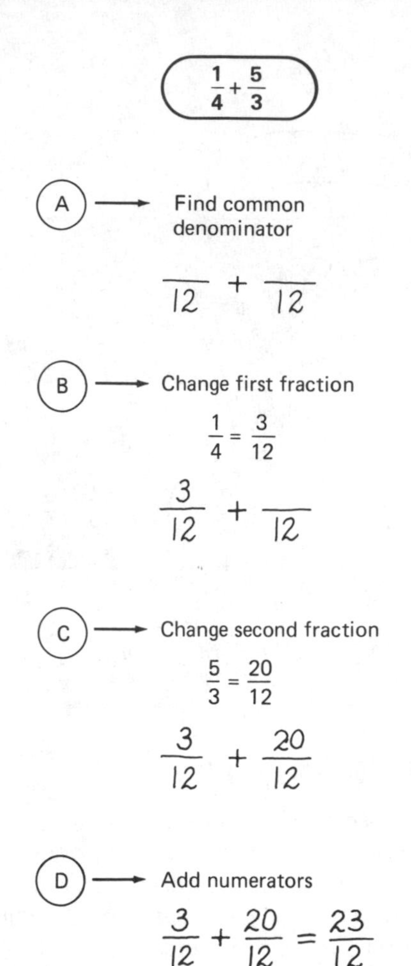

Figure 23.3
An example problem in step-by-step form. Each step can be put on a separate card.

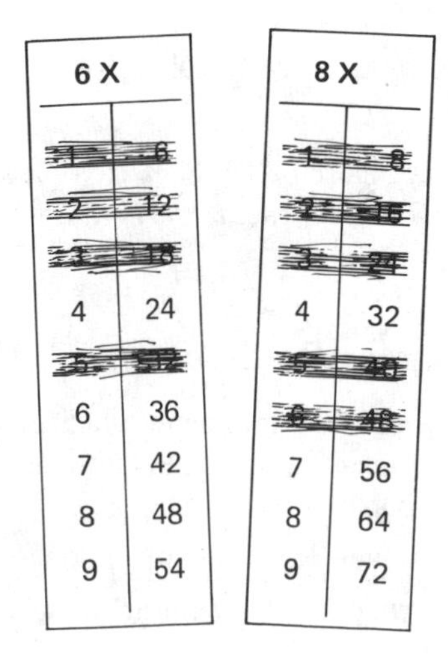

Figure 23.4
Permit the use of fact charts but encourage memorization.

Provide repetitive practice of a newly learned idea. Children will feel comfortable with the success provided by doing a mastered task and the overlearning will help make connections.

Teaching to Attention Disorders

For children with attention problems try to diminish the sources of distraction and decrease the demand for long-term attention.

Seat the child facing a bare wall. Isolate him from potential disturbances in the room.

Make assignments very short. Provide only one or two exercises to be done and then give one or two more when those are complete.

Have the child work with a "buddy."

Play a blank tape on which a tone or bell sounds intermittently every 2 to 5 minutes. The tone reminds the child to refocus his attention to the task. Similarly, set a timer to "buzz" in 3 minutes. After each buzz, the child resets the timer and returns to work.

Slow Learners and the Mildly Mentally Handicapped

CHARACTERISTICS OF SLOW LEARNERS AND MILDLY MENTALLY HANDICAPPED CHILDREN

The term *mentally handicapped* is used to describe those children with general intellectual functioning that is significantly below average. Frequently the term *educable mentally retarded* (EMR) is used instead. These children have measured intelligence scores (IQ scores) between 50 and 70 or between two and three standard deviations below the mean. The *slow learner* also exhibits general below average intelligence, usually having an IQ score between 70 and 90 (Callahan & MacMillan, 1981). These children are usually well below their expected grade level in all areas of the curriculum.

Many mildly handicapped children will spend all or part of their school day in a regular education classroom. Most slow learners do not qualify for special education classes and are almost always taught by the regular education teacher.

ADAPTING INSTRUCTION AND CURRICULUM

Modifications in Instruction

Callahan and MacMillan (1981) contend that mildly mentally handicapped children "seem to learn in the very same fashion as their nonretarded peers do, albeit a little less efficiently" (p. 156). The most significant difference in their learning is in the time that is required. The implication of this conclusion, they explain, is that there is no need for some special set of materials or techniques for these children.

There are some additional suggestions that teachers will find useful. First, in selecting models and materials, simplicity and clarity of purpose should be the major criterion. Use the one "best" model over and over rather than attempt to use a variety of models for the same concept. For example, these children may not be able to see how pie sections, fraction strips, and sets of counters all model the same idea since they are very slow to construct relationships. Second, make your instruction much more intense and explicit than is necessary with regular children. Repeat and model introductory ideas several times. Provide very simple tasks using models that the students can easily repeat until they become comfortable with the idea. Third, since many of these special children have difficulty attending to task, instruction should be in small increments with no more than one new idea at a time.

Care should be taken that instruction in the classroom and instruction provided by any supplementary program be carefully coordinated. Vocabulary, methods, and materials should all match as closely as possible. Without this coordination, the added diversity of two different programs can inadvertently cause confusion rather than help.

Modifications in Curriculum

Since mentally handicapped children learn much more slowly than regular children, it follows that less content can be learned during the years they are in school. It makes sense to focus the instructional time on those areas that are going to be of the most value to these students as adults. For example, while computational skills take up much of the regular curriculum, they are very complex and difficult for the mentally handicapped. These "skills" are accessible to them with the use of the calculator. At the same time, it is important that these students develop a conceptual knowledge of numbers, especially in terms of real-world referents. Therefore, the curriculum should be strong in the development of number concepts and should assist the student in learning numbers in the context of money and familiar measurements.

Mathematics for the Gifted and Talented

DESCRIPTION OF THE MATHEMATICALLY GIFTED

Renzulli (1978) proposed a definition of *gifted and talented* that requires the presence and interaction of three distinct characteristics: (1) above average general intelligence (not necessarily superior), (2) task commitment, and (3) creativity. In addition to the traits described by the Renzulli definition, the *mathematically gifted* student will exhibit specific characteristics in mathematics. These will include a clearly demonstrated interest in things mathematical, mastery of mathematical skills at an early age, an ability to reason analytically, and an ability to perceive mathematical patterns and generalizations (Ridge & Renzulli, 1981; House, 1987).

INSTRUCTION FOR THE MATHEMATICALLY GIFTED

Acceleration or Enrichment

Debate is common over the relative merits of vertical acceleration versus horizontal acceleration or enrichment. *Vertical acceleration* refers to movement through the regular curriculum at a more rapid pace. *Enrichment* is the expansion of the regular curriculum to include topics not generally encountered and the study of many standard topics in greater depth. The NCTM position on acceleration was first articulated in the *Agenda for action,* which noted that "programs for the gifted student should be based on a sequential program of enrichment through ingenious problem solving opportunities rather than acceleration alone" (NCTM, 1980). In 1983 the council issued a position statement on vertical acceleration, noting that "vertical acceleration should be implemented *only* for the extremely talented and productive student" and then, only in consultation with the student, parents, teachers, and counselors. Later, in 1986 the council issued a position on providing for the mathematically talented in which a preference for "a broad and enriched view of mathematics in a context of higher expectation" was clearly articulated. (See box on next page.)

Gifted Students in the Regular Classroom

The classroom teacher who has only one or two gifted students has a difficult struggle meeting their needs. To accommodate various extensions and projects, the classroom teacher has several options to consider: (1) Allow advanced students to work independently or in a small group while directing the rest of the class as a group. (2) Group students by ability within the class, providing some instruction with each group as is frequently done in reading. (3) Group gifted students from several classrooms or grade levels to create a larger group of students who can work together and be provided with planned, directed instruction. Each of these options has advantages and disadvantages. Approaches which permit talented students to interact or work together provide a more productive environment for enrichment but require extra work and preparation. Separation and special classes also raise difficult questions: How will the students be selected? What will the program content be? Will these students miss material in the regular class?

A Triad Model for Planning Activities

Too frequently the highly capable student is provided with clever puzzles, computer games or explorations, or a variety of independent study projects that fail to fit together in any cohesive manner to create a program. While certainly better than nothing, these projects can become superficial pastimes

Provisions for Mathematically Talented and Gifted Students

The following statement is an official NCTM position, developed by the Council's Instructional Issues Advisory Committee and adopted by the Board of Directors.

All students deserve the opportunity to achieve their full potential; talented and gifted students in mathematics deserve no less. It is a fundamental responsibility of all school districts to identify mathematically talented and gifted students and to design and implement programs that meet their needs. Further, it is the responsibility of mathematics educators to provide appropriate instruction for such students.

The identification of mathematically talented and gifted students should be based on multiple assessment measures and should involve teachers, counselors, administrators, and other professional staff. In determining admission to talented and gifted programs, the evaluators must consider the student's total educational development as well as his or her mathematical ability, achievement, and aspirations. Eligible students and their parents should fully understand the nature and demands of the program before making a commitment to participate. Unqualified students should not be admitted for any reason.

The needs of mathematically talented and gifted students cannot be met by programs of study that only accelerate these students through the standard school curriculum, nor can they be met by programs that allow students to terminate their study of mathematics before their graduation from high school. The curriculum should provide for all mathematically talented and gifted students every year they are in school. These students need enriched and expanded curricula that emphasize higher-order thinking skills, nontraditional topics, and applications of skills and concepts in a variety of contexts.

Therefore, the National Council of Teachers of Mathematics recommends that all mathematically talented and gifted students should be enrolled in a program that provides a broad and enriched view of mathematics in a context of higher expectation. Acceleration within such a program is recommended only for those students whose interests, attitudes, and participation clearly reflect the ability to persevere and excel throughout the entire program.

(October 1986)

or busy work and even turn students off of the idea of a serious investigation.

Renzulli (1977) suggested a model for the design of enrichment programs consisting of three types of activities. Type I activities are *general explorations to stimulate interest.* Type II activities are called *group training activities* in which processes and skills related to a particular interest are developed and refined. Type III activities are *investigations of real problems.*

The following example illustrates the Renzulli model. A brief introduction to the Logo computer language is provided. Students are given an assortment of task cards from which to select and work on. The open exploration of Logo might go on for several sessions with challenges coming from the teacher, from books or task cards, or from the students themselves. These Type I activities are designed to explore and develop interest. If interest is there, the students are in need of Type II activities to develop their problem-solving skills and their knowledge of some of the procedural aspects of the Logo language. The teacher can plan and assist these activities by providing guided exploration with any number of available books. Students need not become Logo experts. However, they should realize that there is a knowledge base that they can develop and apply to their explorations. A Type III activity might then consist of a real project that will sustain their interest and draw on the processes of the Type II experiences. This might be a Logo contest to be entered at the school or system level, the design or modification of a game that other students in the class can then play, or perhaps the application of Logo to assist the teacher with planning or grading. An important feature of the Type III activity is the sharing of the results of the efforts with other students, teachers, or groups outside the school. This exposition removes the project from the realm of superficiality and provides a sense of real worth.

Ideas and Beginnings

The start of good explorations for the gifted is not quite as difficult as it may seem on the surface. Type I activities are not all "high tech" or even highly advanced. Many common materials and teacher resource books provide good beginnings from which further study into Type II and III activities can easily develop.

Geometric models such as pattern blocks, Mira, kaleidoscopes, string designs, construction of geometric solids from sticks or poster board, the design of unusual tessellations, geoboards, and grid paper explorations all can be used to pose interesting tasks and explorations. Many commercial geometric materials come with resource books to assist teachers.

Calculators, even simple ones, can provide interesting challenges. For example, how can the calculator be used to multiply two large numbers if the product is more than eight digits causing an overload? How many times do you have to multiply 0.99 by itself before the result is less than 0.01? Graphics calculators open up an even larger door of explorations.

Probability problems can be explored at a wide range of levels. Computers or tables of random numbers can be used to create simulations of interesting phenomena such as batting averages, rainfall, lottery chances, traffic patterns, or even airplane crashes. Theoretical probabilities can be compared with actual experiments, and discrepancies in the results can be discussed.

Many computer software packages provide opportunity for exploration without special computer knowledge. Examples include *Rockey's Boots* (Learning Company), *Factory* and *Super Factory* (Sunburst), and more open-ended explorations such as those provided by the *Geometric Supposer* packages (Sunburst).

Most any collection of problem-solving task cards, strategy games or puzzles, or number games offer good and challenging explorations.

Every teacher should make it a habit to collect a resource of ideas, materials, and activities to provide challenges for able students and to enrich their own knowledge of mathematics as well. There is no need to wait until a gifted child appears in the classroom. Examine commercial materials that are available, read the *Student math notes* provided quarterly to NCTM members, and read the *Arithmetic Teacher* and/or the *Mathematics Teacher* regularly. With a little effort it is easy to get "hooked" on activities that go beyond the regular curriculum. You will find yourself using many ideas with your entire class.

For Discussion and Exploration

1. Select a basal textbook for a grade level of your interest and a chapter on any topic within the book. Consider an introductory lesson in the chapter and discuss how the lesson would be modified for
   a. A student with a visual perceptual processing deficit.
   b. A student with an integrative processing deficit.
   Or select any particular processing disorder of your interest.
2. What physical models would you select for teaching early basic number concepts to slow learners? Design a series of lessons for these children to develop number concepts. How do your lessons differ from what you might have planned for regular students?
3. Repeat question two for one of the following topics:
   Base ten numeration
   Fraction concepts
   Decimal and/or percent concepts
4. What broad topics in the fourth- to eighth-grade curriculum do you think are the most important for the mildly mentally handicapped? Which topics do you think could be eliminated? Defend your selections and compare them with others.
5. Refer to one of the resources at the end of this chapter or to one suggested by your instructor. Find an activity that could serve as a Type I exploration for talented and gifted students in a grade level of your choice. Explore the activity yourself. What additional related study would be appropriate as a Type II activity? Can you design a "real problem" or exploration that would be a Type III activity?
6. Examine your school's software library and explore a piece of software that might be appropriate for gifted elementary students. Describe why you think it would be appropriate for special students.
7. Visit a special education classroom. Observe the mathematics instruction of a single student. Ask the teacher about the child's particular learning deficit or degree of mental handicap. How is the instruction you observe meeting the particular needs of the child?

Suggested Readings

Bley, N. S., & Thornton, C. A. (1982). Help for learning disabled students in the mainstream. In L. Silvey (Ed.), *Mathematics for the middle grades (5–9)*. Reston, VA: National Council of Teachers of Mathematics.

Bley, N. S., & Thornton, C. A. (1989). *Teaching mathematics to the learning disabled* (2nd ed.). Austin, TX: PRO-ED.

Brown, S. I., & Walter, M. I. (1983). *The art of problem posing.* Philadelphia: Franklin Institute Press.

Cawley, J. F. (1985). *Cognitive strategies and mathematics for the learning disabled.* Austin, TX: PRO-ED.

Glennon, V. J. (Ed.) (1981). *The mathematical education of exceptional children and youth: An interdisciplinary approach.* Reston, VA: National Council of Teachers of Mathematics.

Haag, V., Kaufman, B., Martin, E., & Rising, G. (1986). *Challenge: A program for the mathematically talented.* Menlo Park, CA: Addison-Wesley.

House, P. A. (Ed.). (1987). *Providing opportunities for the mathematically gifted, K–12.* Reston, VA: National Council of Teachers of Mathematics.

Johnson, S. W. (1979). *Arithmetic and learning disabilities: Guidelines for identification and remediation.* Boston: Allyn and Bacon.

Jones, S. M. (1982). Don't forget math for special students: Activities to identify and use modality strengths of learning disabled children. *School Science and Mathematics, 82,* 118–126.

Lamon, W. E. (Ed.). (1984 Summer). Educating mathematically gifted and talented children. Special issue of *Focus on Learning Problems in Mathematics, 6.*

Meyers, M. J., & Burton, G. (1989). Yes you can . . . plan appropriate instruction for learning disabled students. *Arithmetic Teacher, 36*(7), 46–50.

Moyer, M. B., & Moyer, J. C. (1985). Ensuring that practice makes perfect: Implications for children with learning disabilities. *Arithmetic Teacher, 33*(1), 40–42.

Vance, J. H. (1986). The low achiever in mathematics: Readings from the *Arithmetic Teacher. Arithmetic Teacher, 33*(5), 20–23.

Wilmot, B., & Thornton, C. A. (1989). Mathematics teaching and learning: Meeting the needs of special learners. In P. R. Trafton (Ed.), *New directions for elementary school mathematics.* Reston, VA: National Council of Teachers of Mathematics.

APPENDIX A

CURRICULUM AND EVALUATION STANDARDS FOR SCHOOL MATHEMATICS*

Summaries of Changes in Content and Emphasis in Mathematics K–4 and 5–8[†]

INCREASED ATTENTION

NUMBER

- Number sense
- Place-value concepts
- Meaning of fractions and decimals
- Estimation of quantities

OPERATIONS AND COMPUTATION

- Meaning of operations
- Operation sense
- Mental computation
- Estimation and the reasonableness of answers
- Selection of an appropriate computational method
- Use of calculators for complex computation
- Thinking strategies for basic facts

GEOMETRY AND MEASUREMENT

- Properties of geometric figures
- Geometric relationships
- Spatial sense
- Process of measuring
- Concepts related to units of measurement
- Actual measuring
- Estimation of measurements
- Use of measurement and geometry ideas throughout the curriculum

PROBABILITY AND STATISTICS

- Collection and organization of data
- Exploration of chance

DECREASED ATTENTION

NUMBER

- Early attention to reading, writing, and ordering numbers symbolically

OPERATIONS AND COMPUTATION

- Complex paper-and-pencil computations
- Isolated treatment of paper-and-pencil computations
- Addition and subtraction without renaming
- Isolated treatment of division facts
- Long division
- Long division without remainders
- Paper-and-pencil fraction computation
- Use of rounding to estimate

GEOMETRY AND MEASUREMENT

- Primary focus on naming geometric figures
- Memorization of equivalencies between units of measurement

*NCTM Commission on Standards for School Mathematics (1989). *Curriculum and Evaluation Standards for School Mathematics.* Reston, VA: National Council of Teachers of Mathematics. Reprinted by permission.

[†](pages 18, 74, and 75)

INCREASED ATTENTION

PATTERNS AND RELATIONSHIPS
- Pattern recognition and description
- Use of variables to express relationships

PROBLEM SOLVING
- Word problems with a variety of structures
- Use of everyday problems
- Applications
- Study of patterns and relationships
- Problem-solving strategies

INSTRUCTIONAL PRACTICES
- Use of manipulative materials
- Cooperative work
- Discussion of mathematics
- Questioning
- Justification of thinking
- Writing about mathematics
- Problem-solving approach to instruction
- Content integration
- Use of calculators and computers

DECREASED ATTENTION

PROBLEM SOLVING
- Use of clue words to determine which operation to use

INSTRUCTIONAL PRACTICES
- Rote practice
- Rote memorization of rules
- One answer and one method
- Use of worksheets
- Written practice
- Teaching by telling

5–8 Mathematics

INCREASED ATTENTION

PROBLEM SOLVING
- Pursuing open-ended problems and extended problem-solving projects
- Investigating and formulating questions from problem situations
- Representing situations verbally, numerically, graphically, geometrically, or symbolically

COMMUNICATION
- Discussing, writing, reading, and listening to mathematical ideas

REASONING
- Reasoning in spatial contexts
- Reasoning with proportions
- Reasoning from graphs
- Reasoning inductively and deductively

CONNECTIONS
- Connecting mathematics to other subjects and to the world outside the classroom
- Connecting topics within mathematics
- Applying mathematics

NUMBER/OPERATIONS/COMPUTATION
- Developing number sense
- Developing operation sense
- Creating algorithms and procedures
- Using estimation both in solving problems and in checking the reasonableness of results
- Exploring relationships among representations of, and operations on, whole numbers, fractions, decimals, integers, and rational numbers
- Developing an understanding of ratio, proportion, and percent

DECREASED ATTENTION

PROBLEM SOLVING
- Practicing routine, one-step problems
- Practicing problems categorized by types (e.g., coin problems, age problems)

COMMUNICATION
- Doing fill-in-the-blank worksheets
- Answering questions that require only yes, no, or a number as responses

REASONING
- Relying on outside authority (teacher or an answer key)

CONNECTIONS
- Learning isolated topics
- Developing skills out of context

NUMBER/OPERATIONS/COMPUTATION
- Memorizing rules and algorithms
- Practicing tedious paper-and-pencil computations
- Finding exact forms of answers
- Memorizing procedures, such as cross multiplication, without understanding
- Practicing rounding numbers out of context

INCREASED ATTENTION

PATTERNS AND FUNCTIONS

- Identifying and using functional relationships
- Developing and using tables, graphs, and rules to describe situations
- Interpreting among different mathematical representations

ALGEBRA

- Developing an understanding of variables, expressions, and equations
- Using a variety of methods to solve linear equations and informally investigate inequalities and nonlinear equations

STATISTICS

- Using statistical methods to describe, analyze, evaluate, and make decisions

PROBABILITY

- Creating experimental and theoretical models of situations involving probabilities

GEOMETRY

- Developing an understanding of geometric objects and relationships
- Using geometry in solving problems

MEASUREMENT

- Estimating and using measurement to solve problems

INSTRUCTIONAL PRACTICES

- Actively involving students individually and in groups in exploring, conjecturing, analyzing, and applying mathematics in both a mathematical and a real-world context
- Using appropriate technology for computation and exploration
- Using concrete materials
- Being a facilitator of learning
- Assessing learning as an integral part of instruction

DECREASED ATTENTION

PATTERNS AND FUNCTIONS

- Topics seldom in the current curriculum

ALGEBRA

- Manipulating symbols
- Memorizing procedures and drilling on equation solving

STATISTICS

- Memorizing formulas

PROBABILITY

- Memorizing formulas

GEOMETRY

- Memorizing geometric vocabulary
- Memorizing facts and relationships

MEASUREMENT

- Memorizing and manipulating formulas
- Converting within and between measurement systems

INSTRUCTIONAL PRACTICES

- Teaching computations out of context
- Drilling on paper-and-pencil algorithms
- Teaching topics in isolation
- Stressing memorization
- Being the dispenser of knowledge
- Testing for the sole purpose of assigning grades

A GUIDE TO LEARNING ABOUT TEACHING DEVELOPMENTALLY

Reflecting on Student Activities

Each of the activities in this book, in every student textbook, in every article that you will ever read about teaching mathematics, and in every talk you will ever hear about teaching mathematics has been designed by someone to help children learn mathematics—to construct a concept or relationship or to develop a procedure. To evaluate these activities as candidates for use in your own classroom is to understand how they contribute to student growth or development. This means to go beyond what the child *does that is observable* and reflect on what the child *does mentally* during the activity. This four-step guide for reflecting on student activities is suggested as a good way not only to evaluate activities but to learn about teaching mathematics.

The only really effective way to use the guide is to actually do the activity. Get a friend and do it together. Talk about it. Think about how it would work in a classroom. Better yet, *do it with kids!*

Notice that the four steps are placed in order from the most superficial knowledge of an activity, *how to do it,* to the most important and most challenging understanding that we can have about an activity, *how it helps children grow mathematically*. Until you have reached step 4 and reflected on how an activity contributes to the growth of the child, you have not really learned about that activity as a teacher. It is step 4

TEACHER FOUR-STEP REFLECTIVE THOUGHT GUIDE FOR MATHEMATICS LEARNING ACTIVITIES

STEP 1: **HOW IS THE ACTIVITY DONE?**

Actually do the activity. Learn as much about the actual doing aspect of the activity as possible.

How do *you* do it?

How would *children* do it? (They don't know what you do!)

What materials are needed?

What are the steps?

What is written down or recorded?

STEP 2: **WHAT DOES THE TEACHER HAVE TO DO?**

Focus on what you need to do to conduct the activity in your class.

What directions would you need to give to students so that they could do the activity? How should they be given? Oral? Written? Demonstration?

How would you have to prepare students for the activity? What would they need to know before hand to do it?

STEP 3: **WHAT IS THE PURPOSE OF THE ACTIVITY?**

What is the activity designed to develop?

Concepts and relationships?

Connections between concepts and procedures?

Procedural knowledge (practice, skill, drill)?

STEP 4: **HOW DOES THE ACTIVITY ACCOMPLISH THE PURPOSE?**

How does the activity promote the required reflective thought for its specific purpose? What must children reflect on to do the activity? Will it necessarily promote reflective thought or is it possible that it may be done mindlessly?

For conceptual and connecting activities: How will the activity help children create the concept or relationship in their own minds?

For procedural activities: Is the activity effective in providing the necessary practice? Does it capitalize on or contribute to conceptual development?

that puts you "inside" children's minds and causes you to think about the concepts involved, how they create them, how valuable the activity is.

Teaching is much more than guiding children to *do activities*. It is helping them *construct ideas*. Every time we reflect on the activities we have children do, both before and after we try them with children, we learn more about teaching. The little four-step guide will help us learn to teach and continue to learn to teach as we do so.

Other Considerations

The essence or most significant feature of any activity is embodied in the response to step 4 of the guide. However, there are some additional things that we want to consider if we plan to use the activity with our class.

1. How could the activity be varied? Are there other materials that could be used? Could it be made more (or less) challenging? Can it be adapted to a range of abilities?
2. How will the activity sustain itself? If I am not there to direct each step of the way, how will the children know what to do next? Can I make it independent so that I do not have to be there?
3. What feedback will I get? If I do not actually watch children do the activity, how will I know what has gone on? Can some form of recording show more than just answers?
4. What other activities are related to this one? Does this fit into a broader collection of activities that have similar or related goals? How will the activity fit with my student textbook pages?
5. What are some ways I could use this activity? basic development? homework? enrichment? remediation? out of class project?

Teacher Learning Kits

Teachers and future teachers reading this text should consider making a "teacher learning kit" for themselves to use when studying (learning about teaching). A learning kit is a small collection of materials that you can use to investigate the activities in the text as you read about them. They might be put in a large plastic storage bag and kept with the book.

The materials you will want in the kit will vary a bit from chapter to chapter. There is no need to have a huge bag of "stuff" to carry. Change the kit as you progress through the book. But most of all, *do the activities* and *reflect on them.*

MATERIALS FOR LEARNING KITS

For All Chapters:

A copy of the reflective thought guide.

A calculator with an automatic constant (see Chapter 20).

Two kinds of counters equivalent to the following:

a. Squares. About thirty or forty 1-inch squares cut from poster board.

b. "Generic" counters, such as a small bag of dry lima beans or elbow macaroni, or pennies. One type should be two-sided. For example, spray-paint the limas on one side or use two-sided matboard to cut the squares from.

Several sheets of plain tagboard (you will need to replenish periodically).

Tools: felt tip marker, scissors, ruler.

Special Materials for Specific Chapters

Each chapter from 5 through 19 requires some special materials. Most of these can be made from tagboard or copied from the masters in Appendix C. When you need something of that type, make a simple set for your kit and use it. There is no need to be elaborate or fancy. You do not need to laminate or color or decorate. Use quickly made paper versions of things you would make from tagboard for students. Exceptions to this rule are base ten materials and fraction materials. For these topics it is important that you experience actually using the same materials that you would expect children to use and to consider how small hands would deal with them. You should plan on making base ten strips and squares and fraction models ahead of time.

APPENDIX C

BLACK-LINE MASTERS AND MATERIALS CONSTRUCTION TIPS

BLACK-LINE MASTERS

Index

SUGGESTIONS FOR USE AND CONSTRUCTION OF MATERIALS

Permission is given to reproduce any of the pages in this appendix for classroom use.

Card Stock

A good way to have many materials made quickly and adequately for students is to have them duplicated on *card stock* at a printing store that does photocopying. Card stock is a heavy paper, not quite as heavy as tagboard, that print shops all have in stock. It comes in a wide assortment of colors. If materials are printed on card stock and laminated before being cut apart, the result is quite good. Use this method for making the dot cards, fraction pieces, tangrams, or tiles.

Making Base Ten Strips and Squares

For making base ten strips and squares (p. 421), the master can first be duplicated on card stock (or even paper) and then mounted on white poster board. This makes the individual pieces a bit thicker and easier to handle. The most effective and easiest method of mounting the copies to the poster board is with spray adhesive. Spray adhesive can be purchased in art supply stores. The next best approach is dry mounting. Rubber cement is a last alternative. To make ones and tens pieces even heavier and therefore handled even more easily, mount the grid onto *mount board* or *illustration board.* This material is about the thickness of picture matting and can be cut into strips with a good paper cutter. Look for it in art supply stores. Hundreds squares need not be that heavy.

This method of making thicker materials can also be used to make tiles or tangrams much heavier.

(Notice that if you are using beans glued to popsicle sticks for tens, the grids on p. 423 can be duplicated for an adequate hundreds model.)

Fraction Pieces

You will notice that the black-line masters for fraction pie pieces have two different fractional parts on each page. Before duplicating these onto colored card stock, make two copies of each page. From these copies cut and paste a new master for each fractional part. Each page should have two disks of the same fractional part. With these new masters you will be able to duplicate each fractional part on a different color of card stock.

It takes a lot of time to cut out fraction disks because they are circular and because there is some need for accuracy. If you plan to do it yourself, you should laminate the tagboard before you cut out the pieces. Make fraction kits consisting of two complete disks for each fractional part and place each complete kit in a plastic bag. These will be surprisingly durable and can be replenished each year as individual pieces get lost.

Another alternative is to have students cut out the disks and make their own kits to keep. Charge each student the small duplication cost and forgo the lamination. The problem with this is that many students, even in the fifth and sixth grade, either cannot or will not take the time to cut out the pieces accurately.

Transparencies

Many of the black lines are useful as overhead transparencies. Using a copy of one of the pages, use a thermal transparency process to make overhead transparencies. Check with your media experts.

Of special value as transparencies are the large 10 x 10 bean chart, the 100 square, the multiplication array, the base ten grid paper, the 10,000 chart, and the fraction sieve.

Making Geoboards

It is possible to mass produce geoboards so that large numbers of them can be made quickly and quite cheaply. The master on page 445 is for a 7½-inch board. Seventy-two boards this size can be made for under $15.00. Get one or two other teachers to do this with you so that the mass production technique and cost savings can best be appreciated.

Use ⅝-inch particle board. This can be purchased at lumber stores or home supply stores and is very inexpensive, especially if purchased in a 4-foot by 8-foot sheet. Have someone with a table saw cut the board into 7½-inch squares. These squares should be cut fairly accurately. (Go to the shop teacher at the junior or senior high school for help.) Purchase 1-inch #16 wire brads. These have no heads. The #16 refers to the thickness. You want the nails as thick as possible so that they will not bend. Buy nails in bulk at a hardware store. You will want about a pound of nails to make 25 boards.

Make a "geoboard maker" out of a piece of ½-inch thick plywood or other half-inch thick lumber. You will need about 1-foot square. Tape a copy of the geoboard master in the center of the board. Nail strips of wood (about 1 inch by 6 inches) around all four sides of the master. These strips should be about 1/16 of an inch outside of the outline of the geoboard. This will allow the squares of particle board to fit snugly but allow for minor errors in cutting them out. Now drill small holes through each dot on the master. The holes should be just barely large enough for one of the brads to be pulled through. Be certain that the holes are drilled perpendicular to the board.

Now you are ready to make geoboards. Put the geoboard maker over one of the geoboards. Place a nail in each hole and hammer it flush. Then with a screw driver, gently pry the geoboard maker off of the board. The nails will each be a uniform half-inch height and all perfectly arranged. Try not to bang on the geoboard maker because it eventually suffers from overuse. However, one maker should serve the production of several hundred geoboards.

You may want to paint your geoboards a dark color before nailing in the nails (or spray paint them afterward). This makes it easier to see brown rubber bands on the boards.

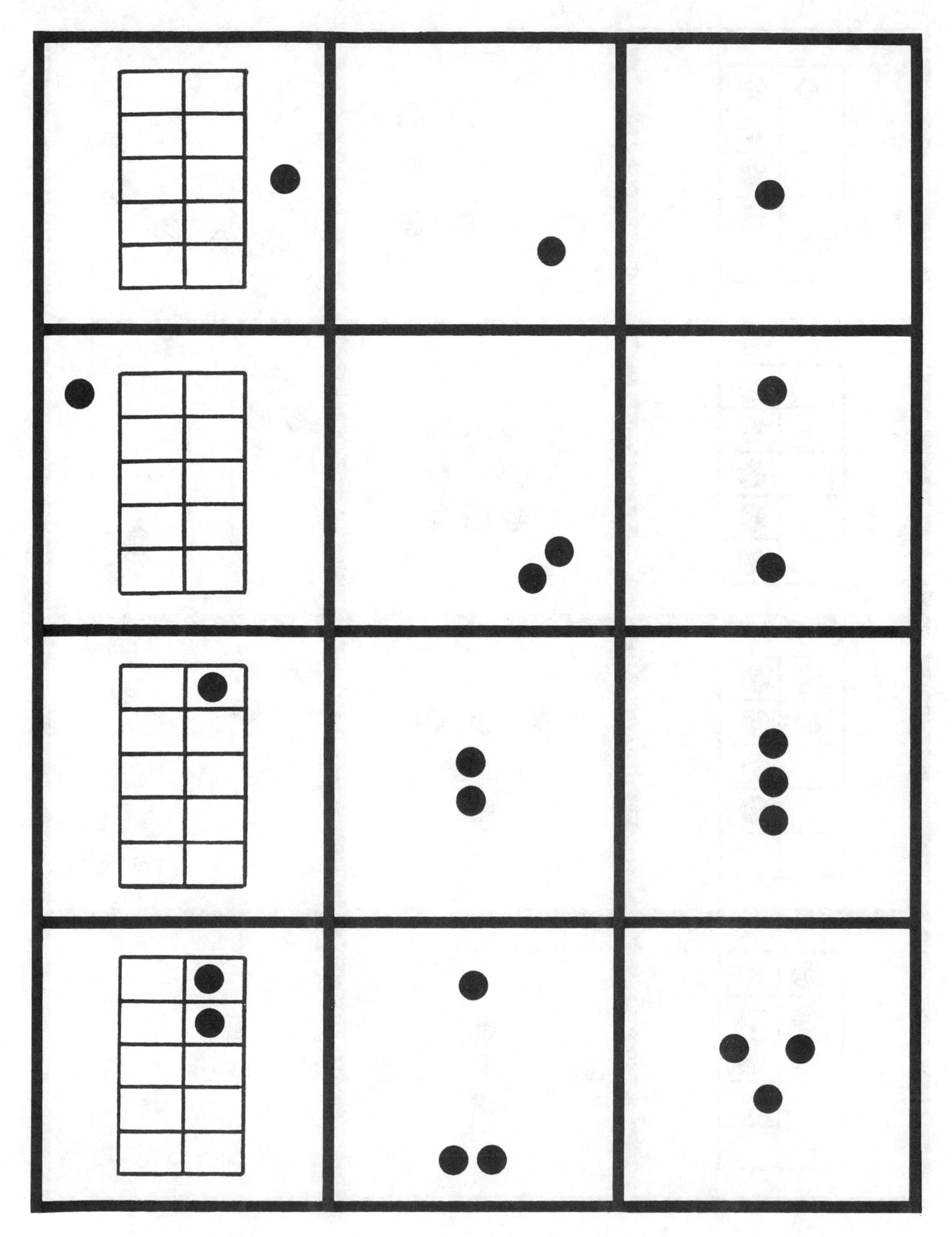

C1
Dot cards.

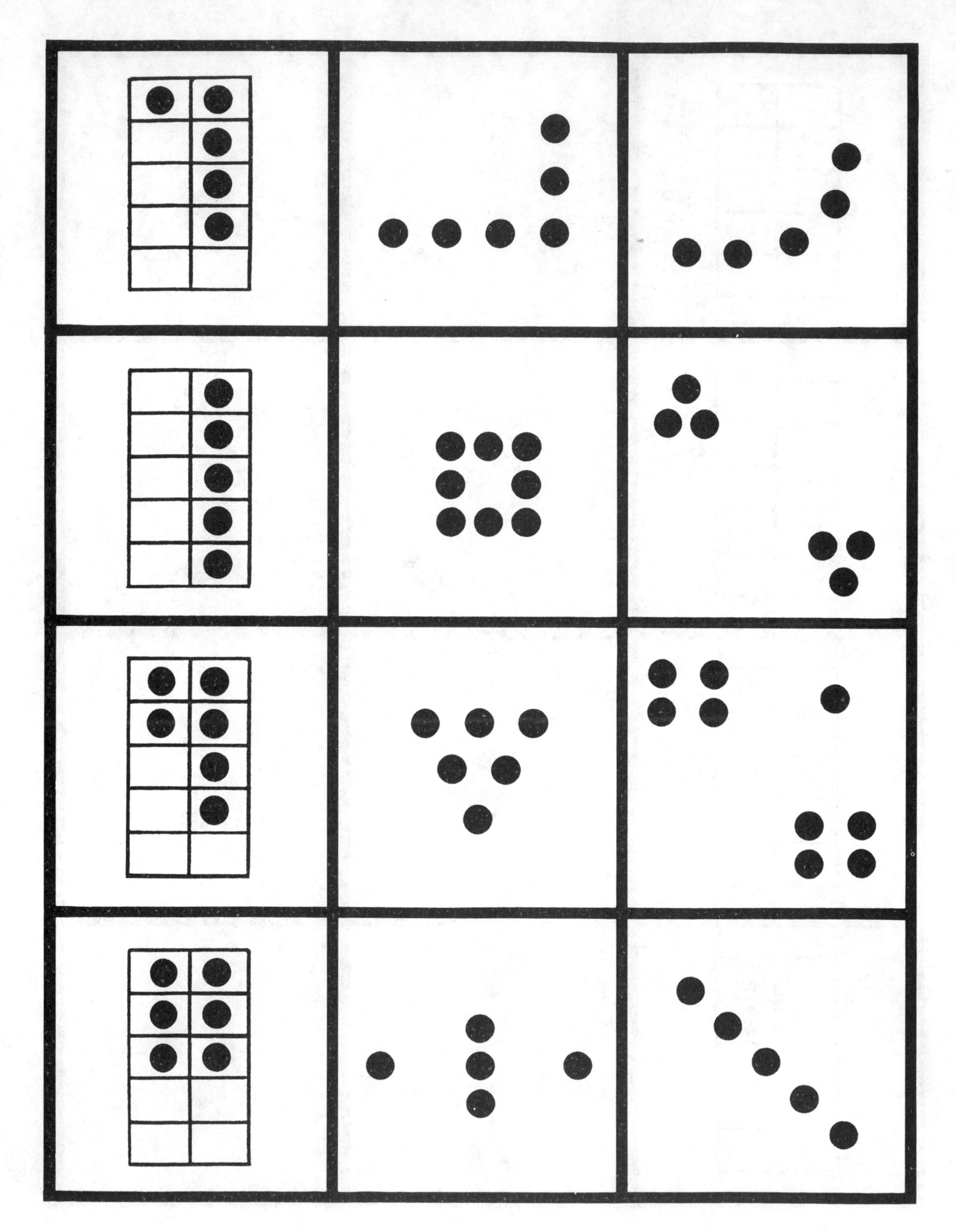

C1
Dot cards.

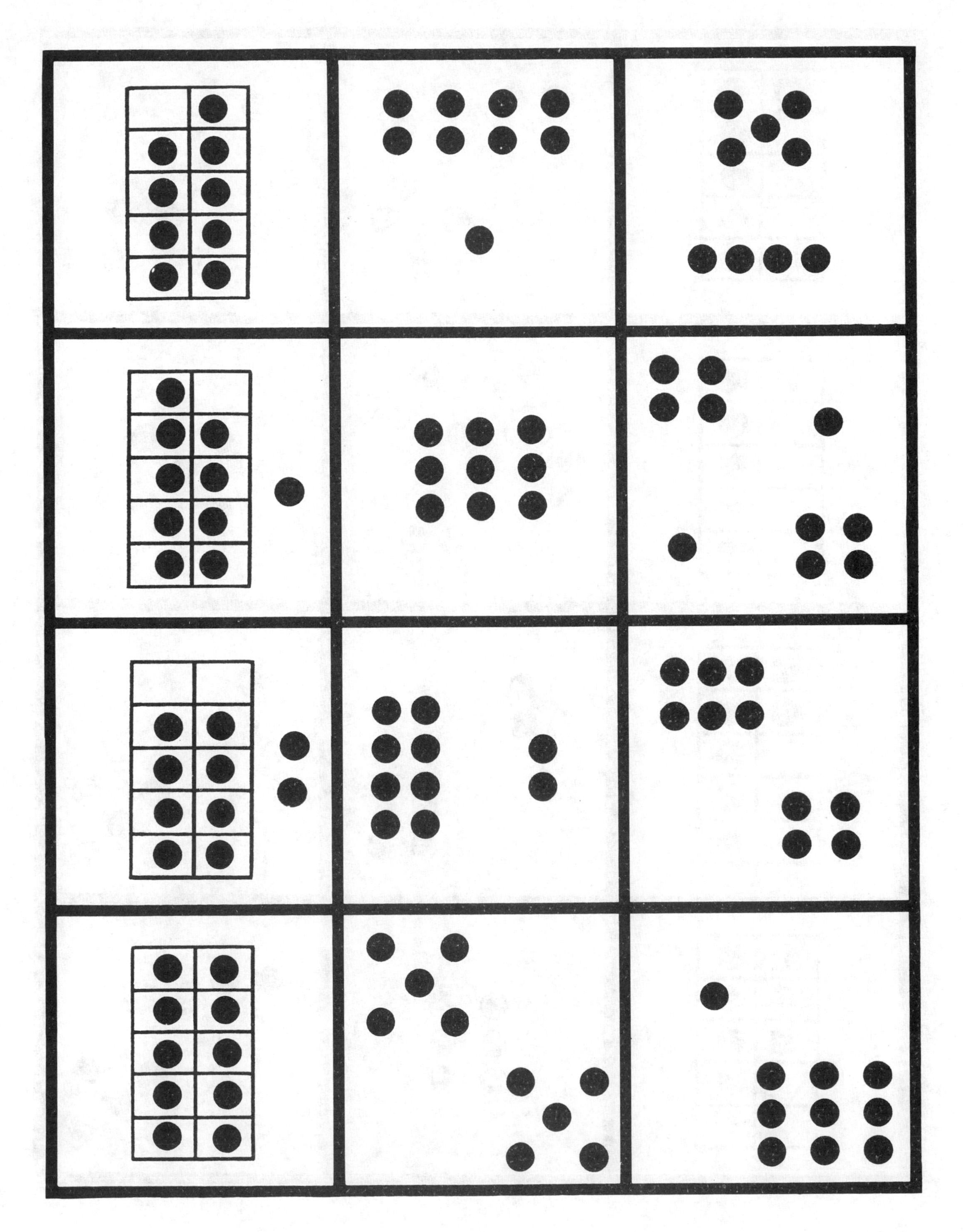

C1
Dot cards.

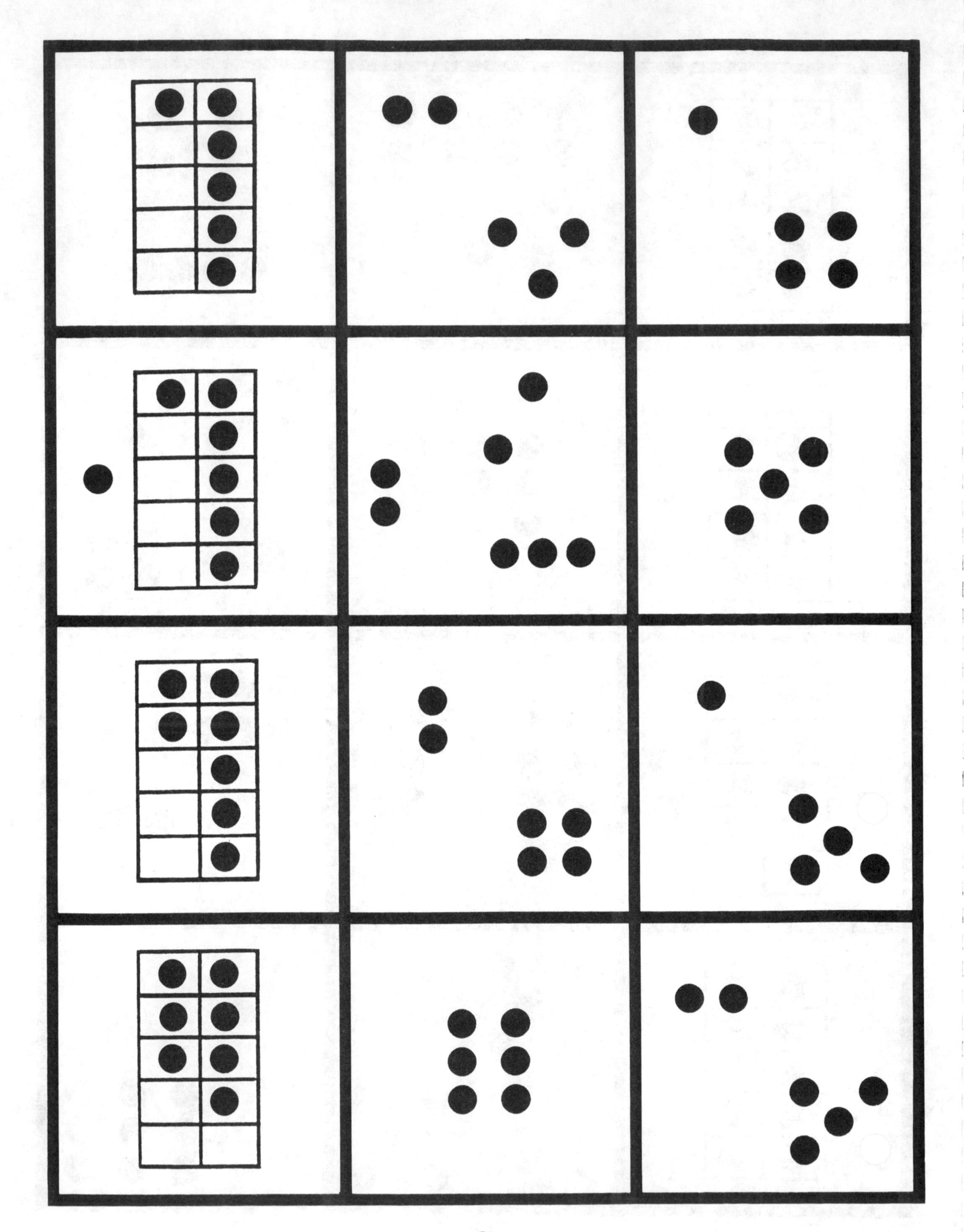

C1
Dot cards.

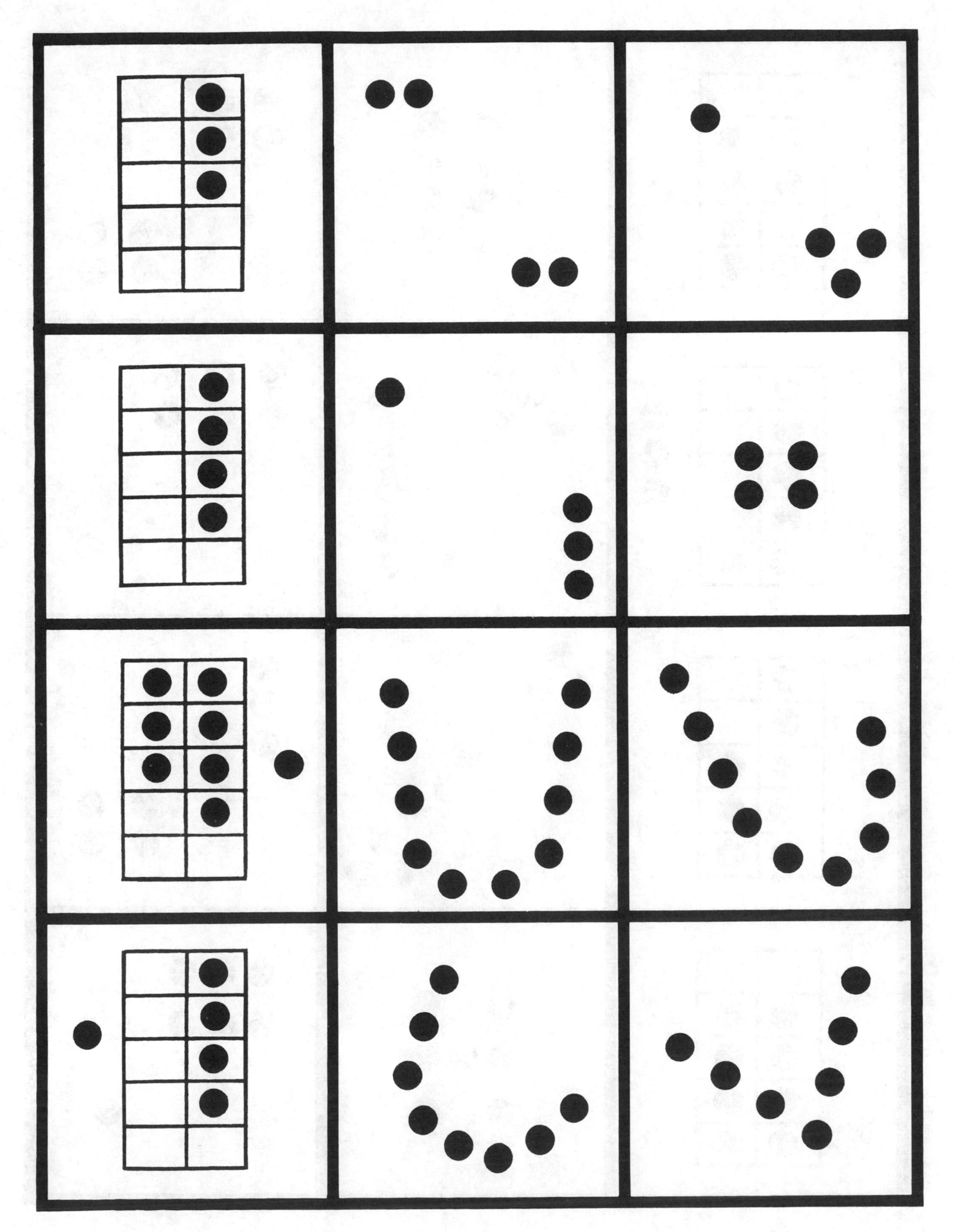

C1
Dot cards.

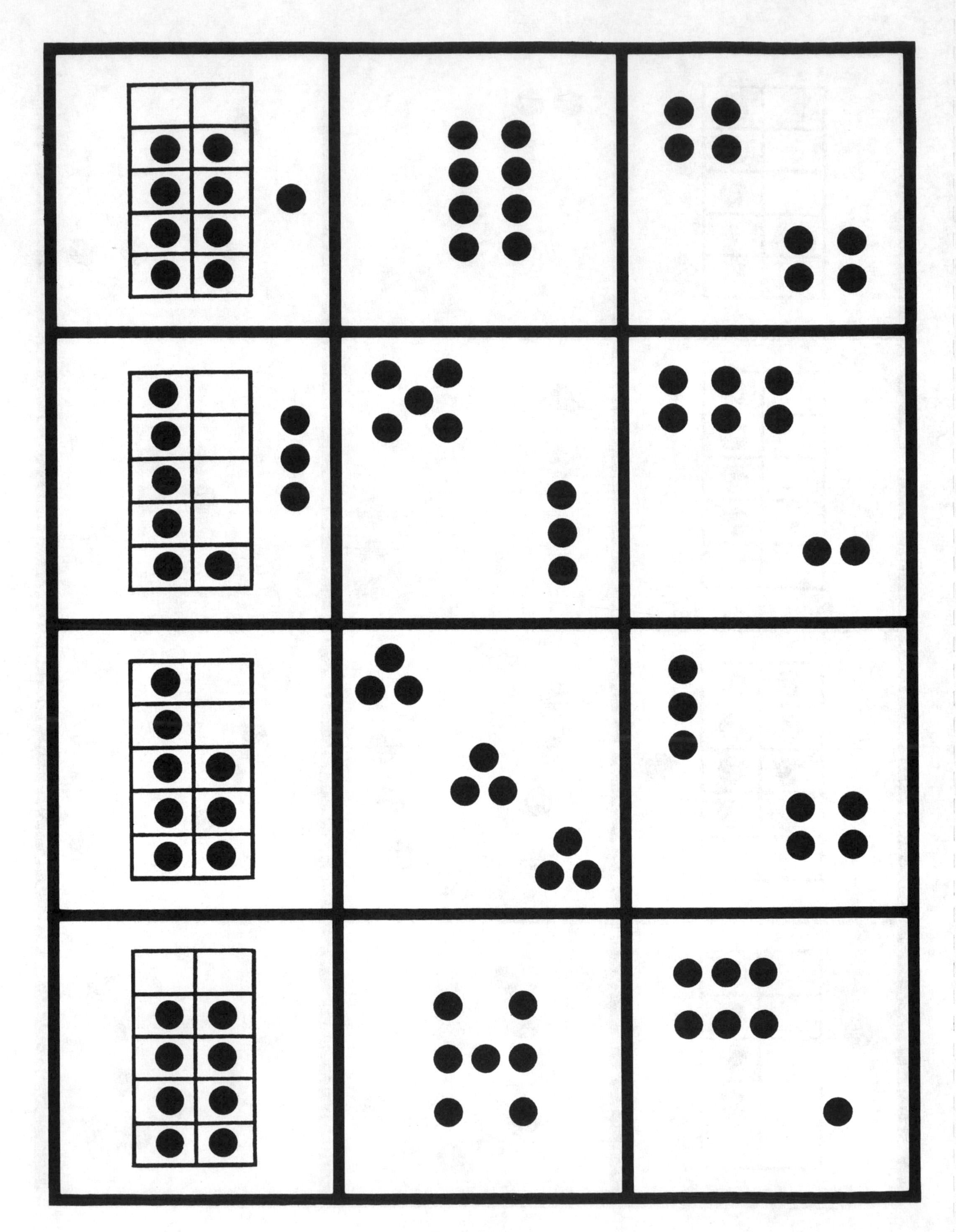

C1
Dot cards.

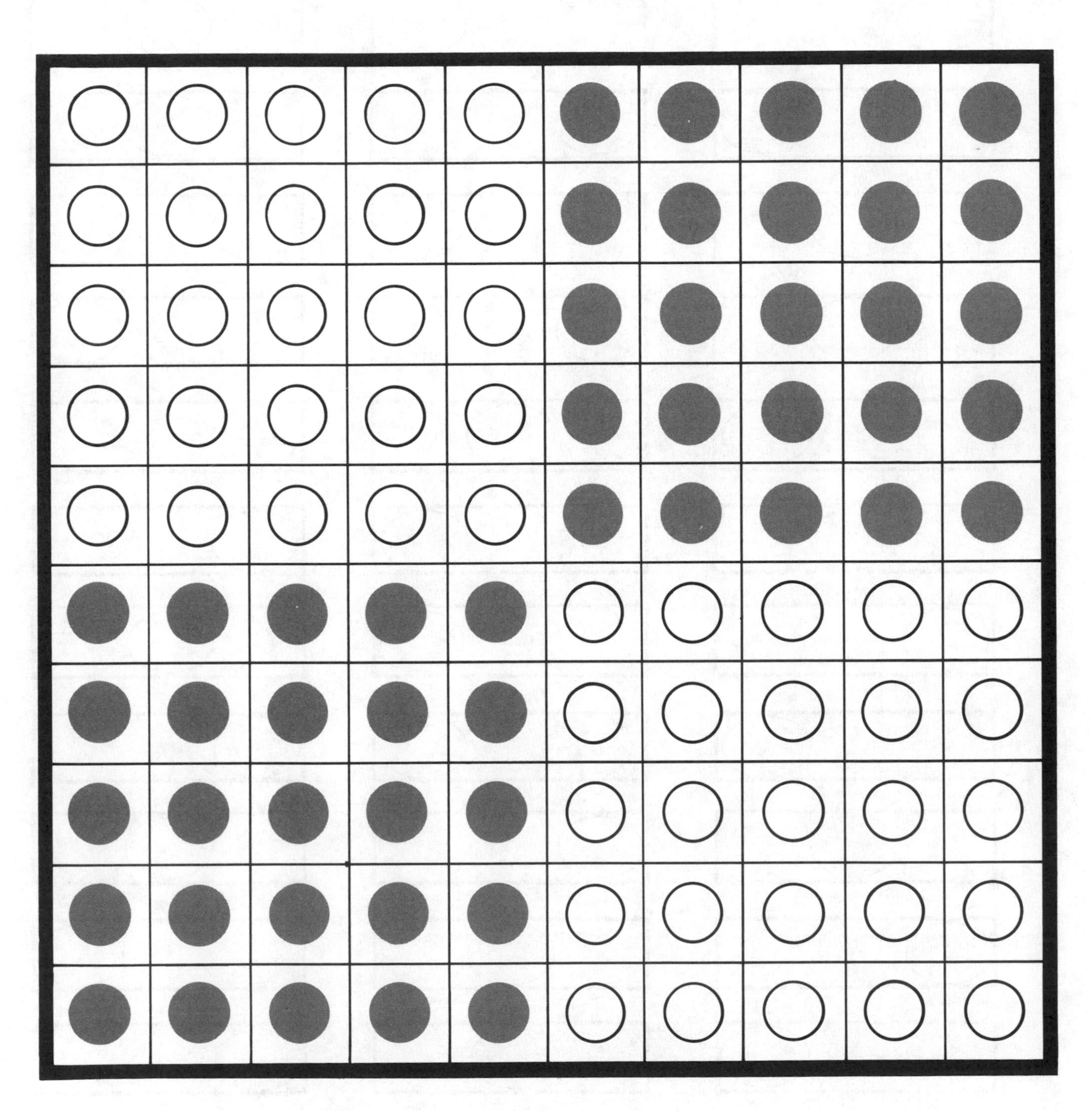

C2
10 × 10 multiplication array.

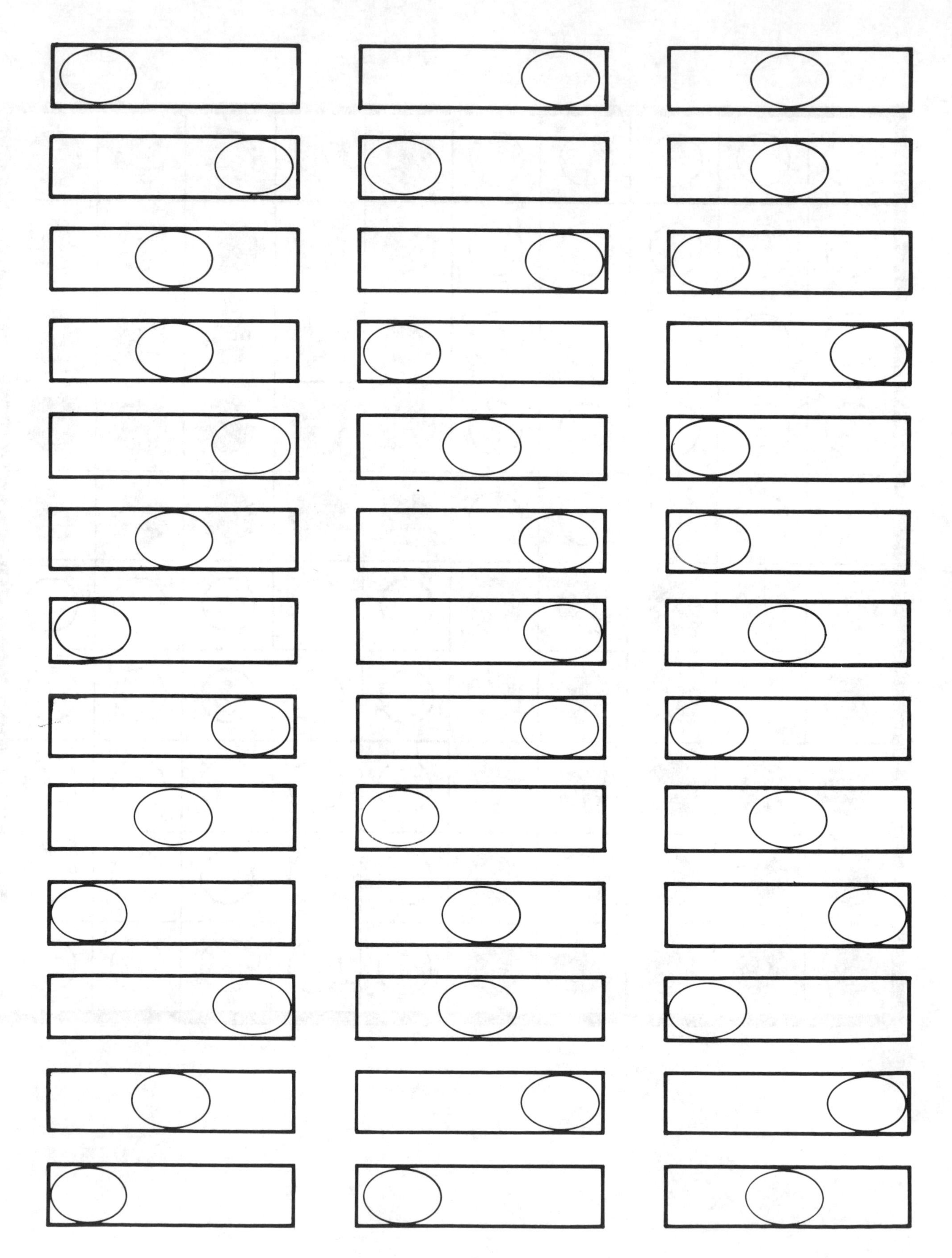

C3
Missing-part blanks.

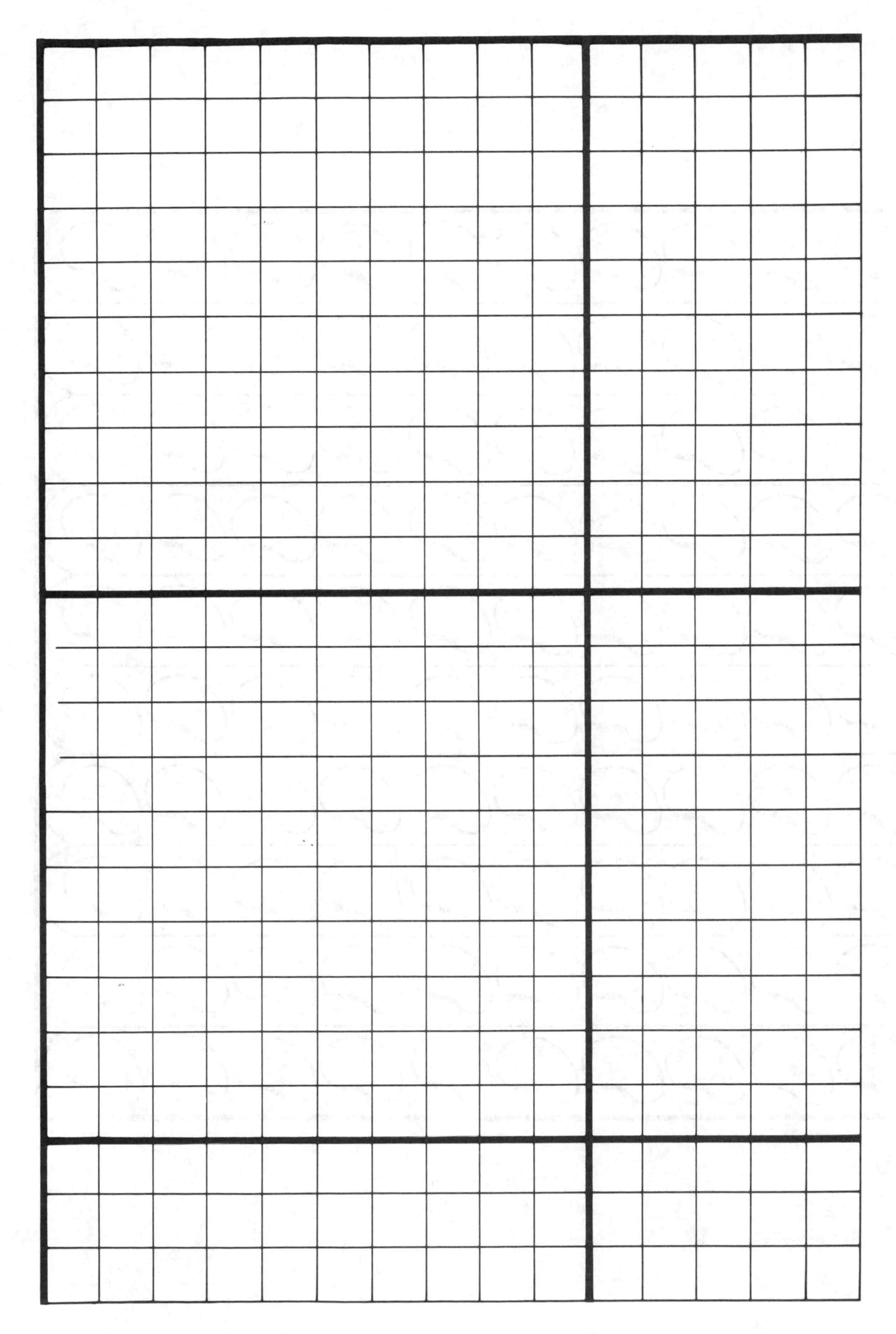

C4
Base ten materials grids.

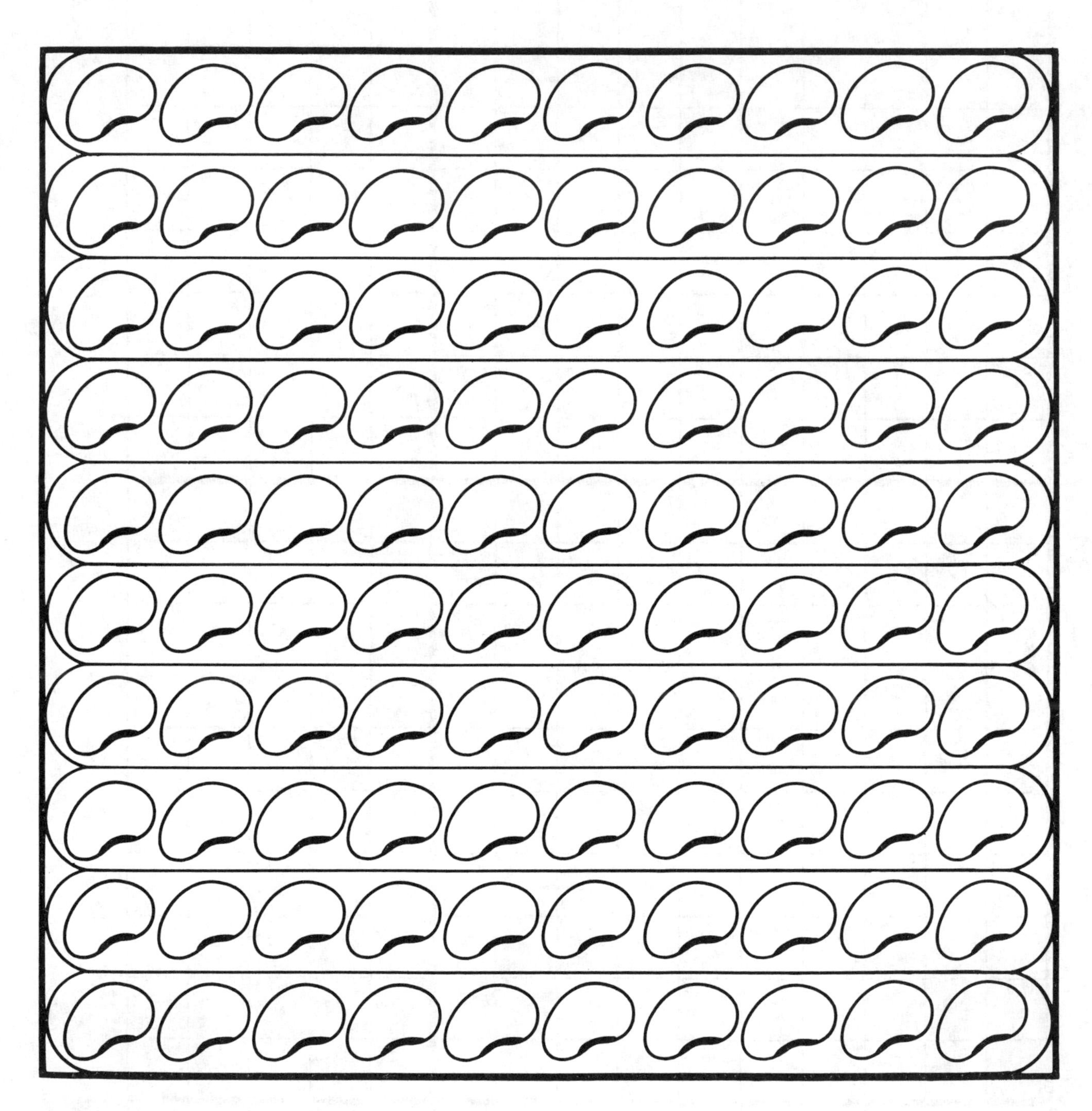

C5
10 × 10 bean chart.

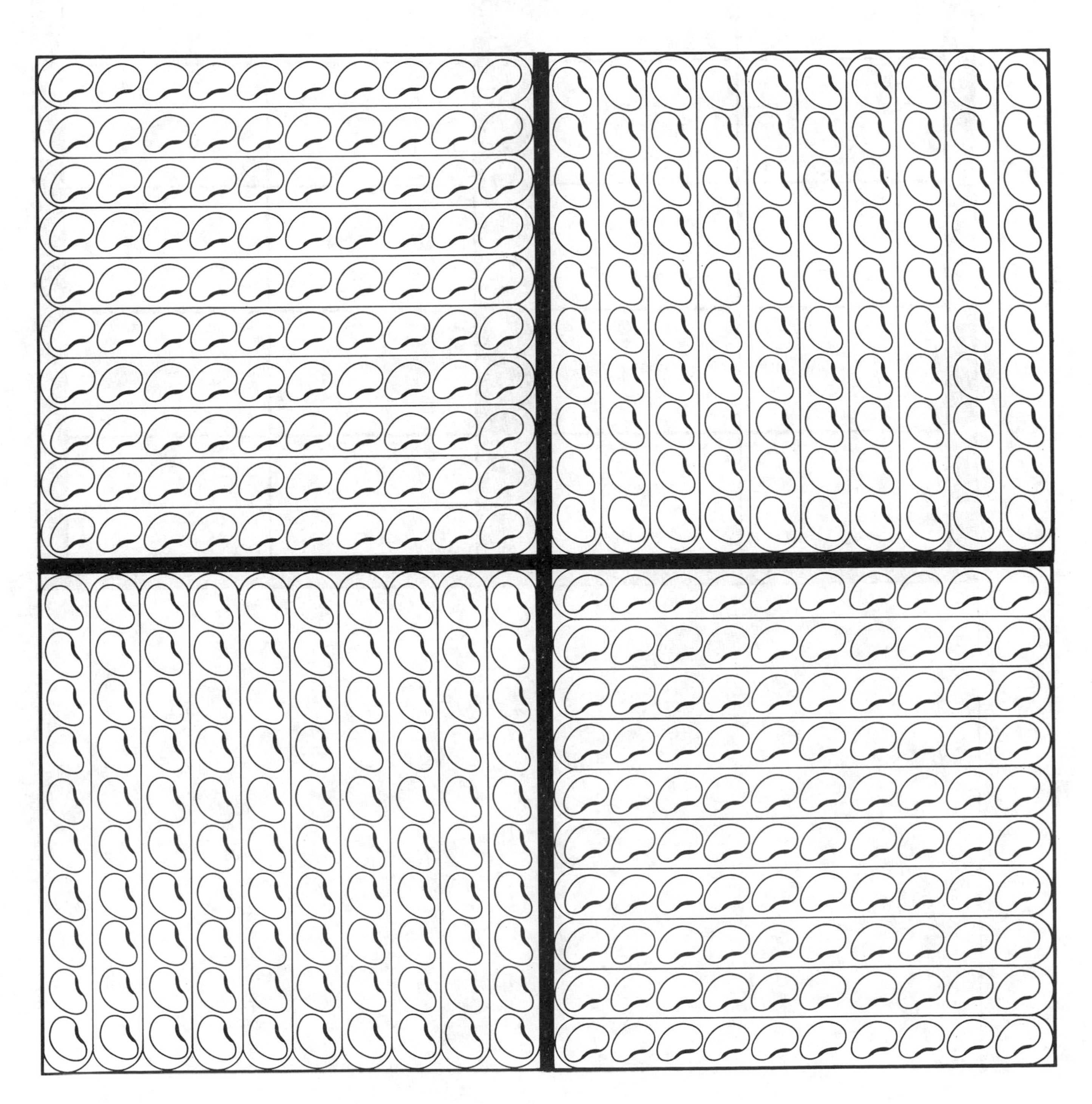

C6
100's Model for bean stick base ten pieces.

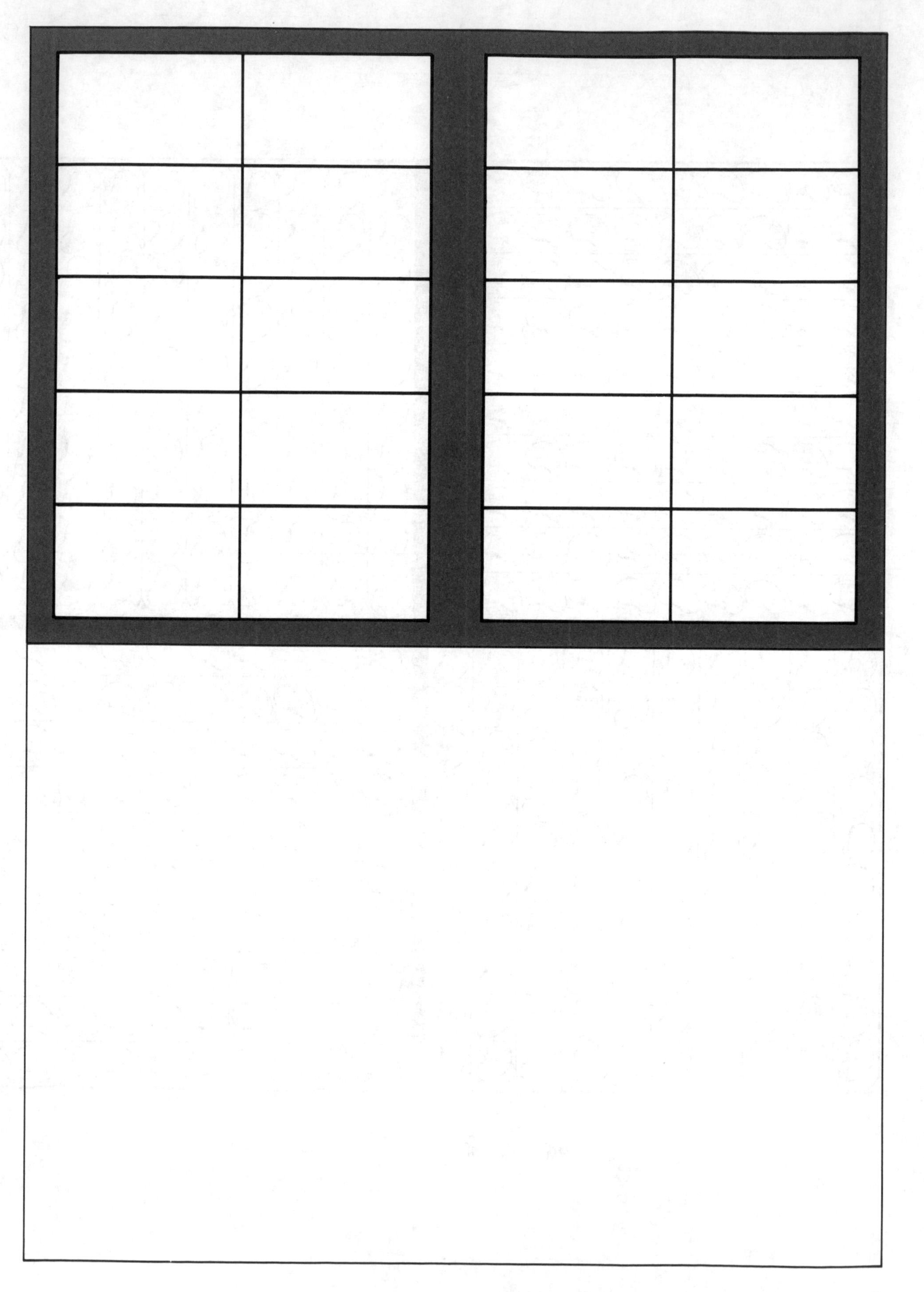

C7
Place-value mat (with ten-frames).

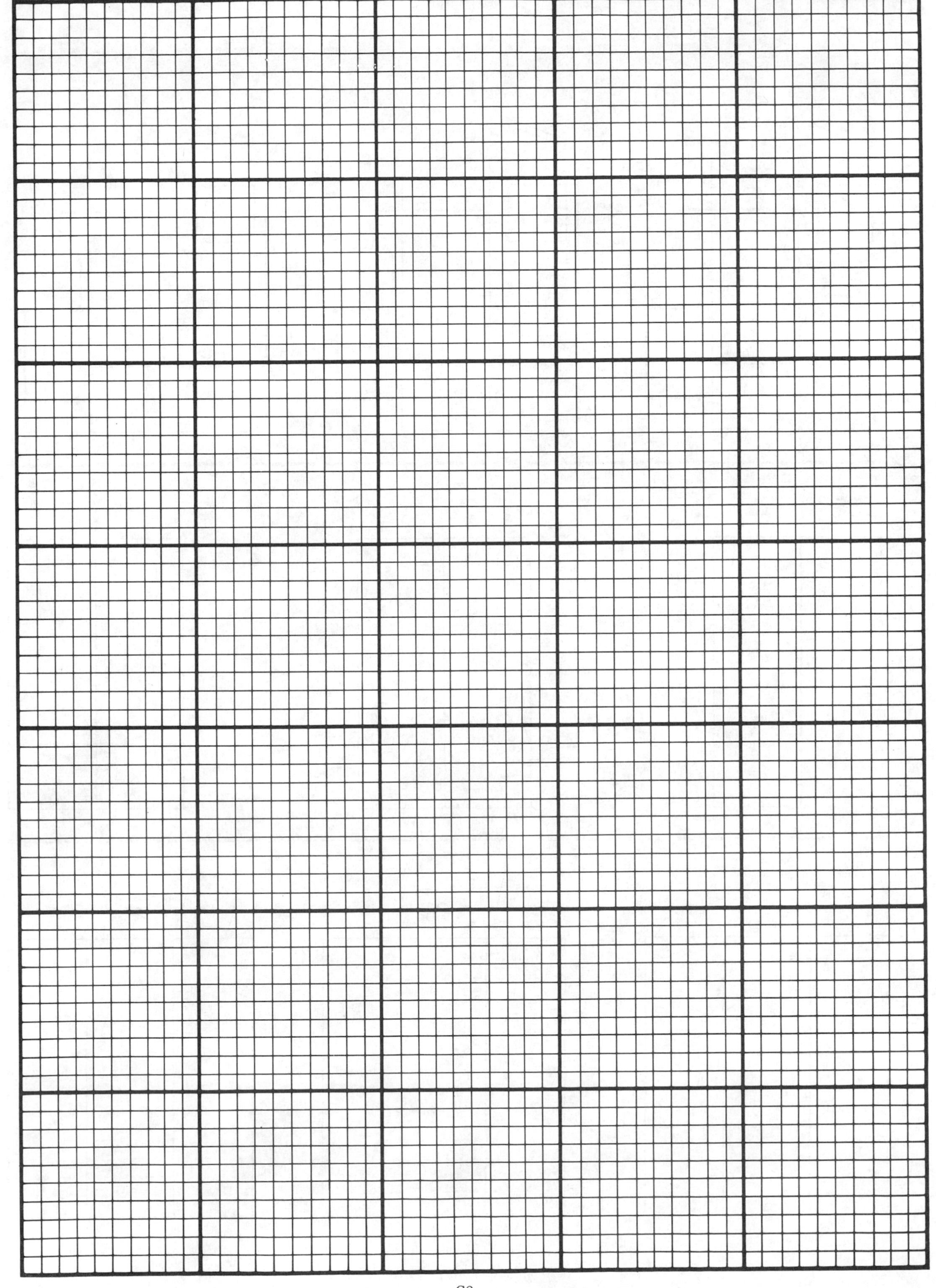

C8

Base ten grid paper.

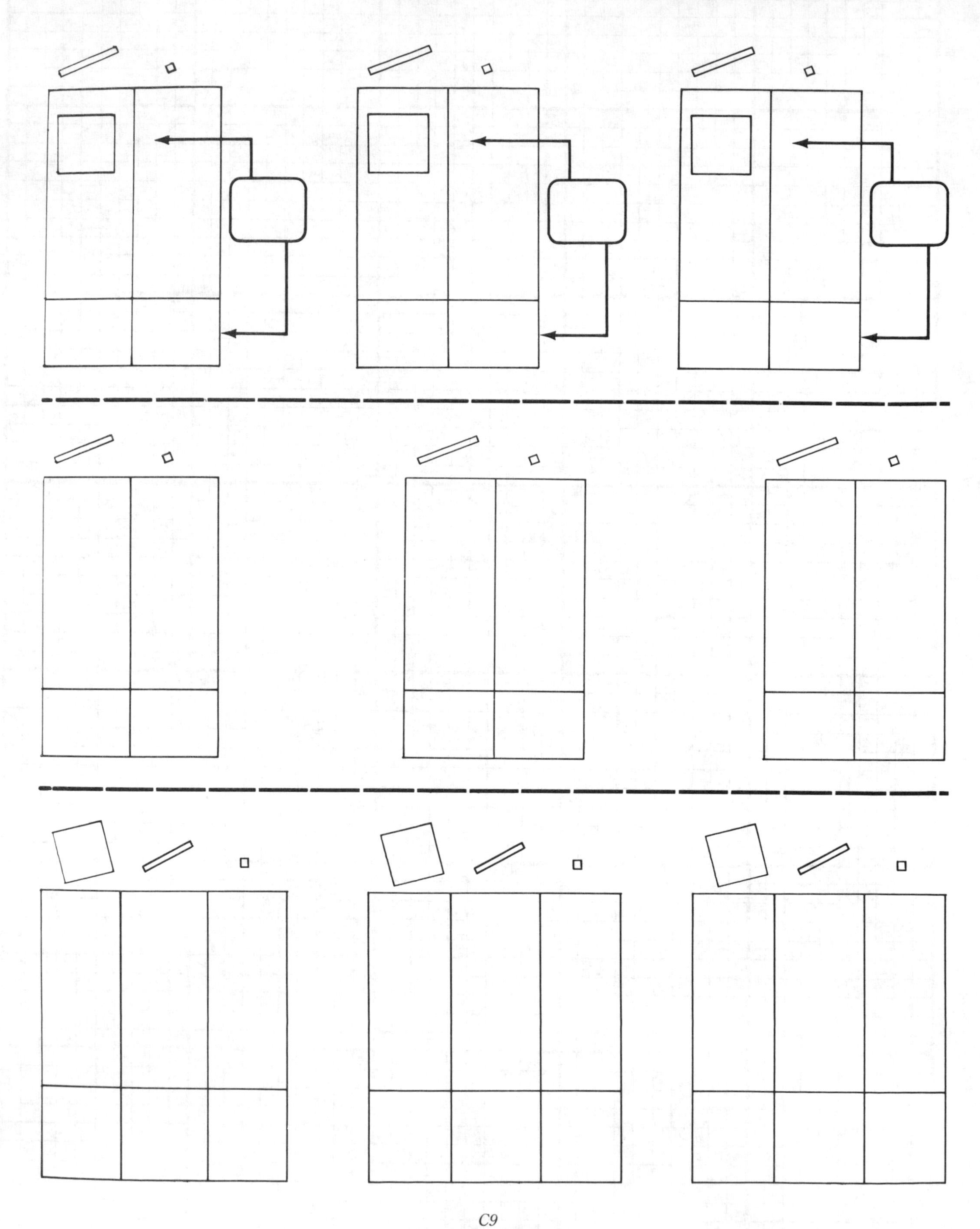

C9
Addition and subtraction record blanks.

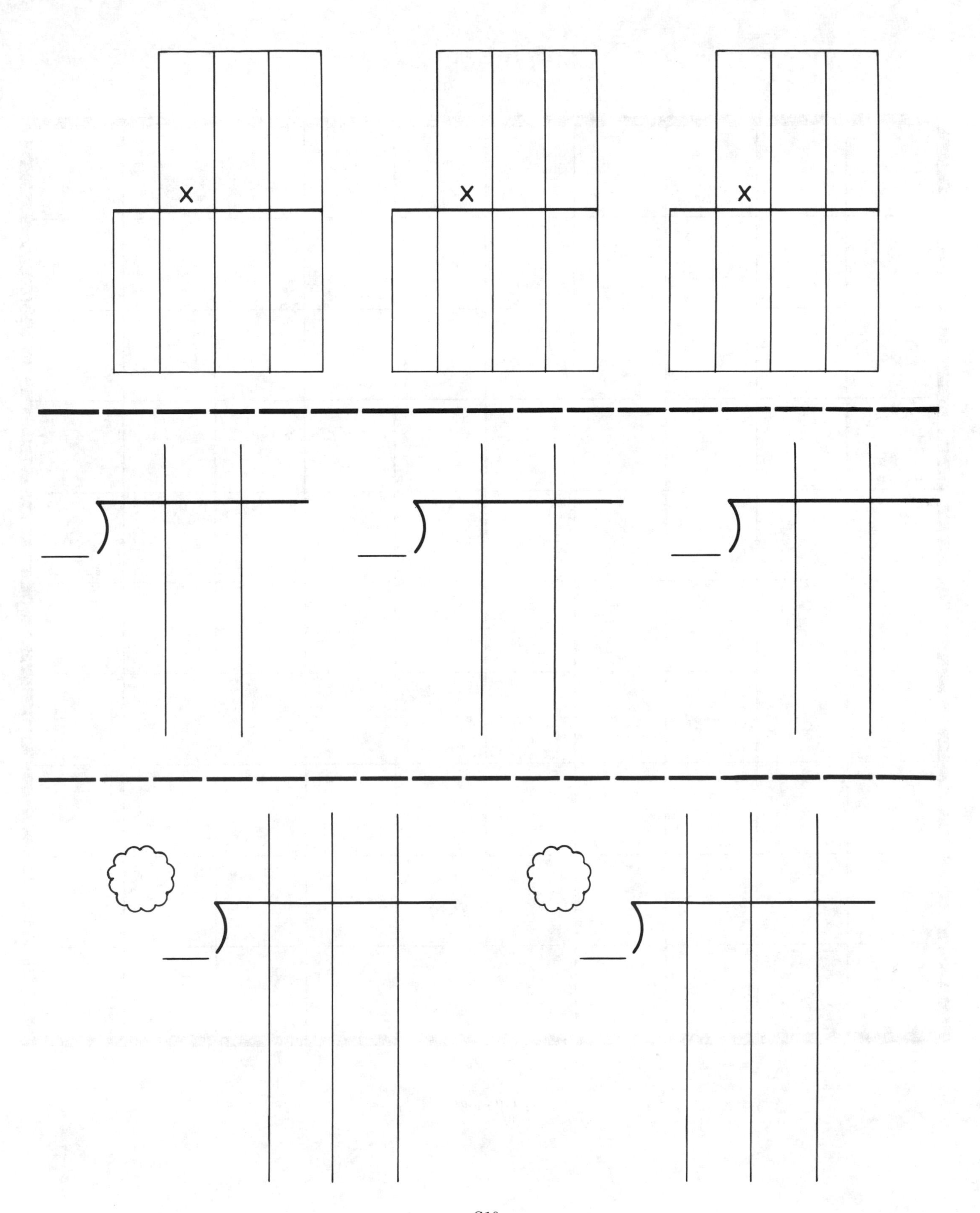

C10
Multiplication and division record blanks.

C11
A 100 square.

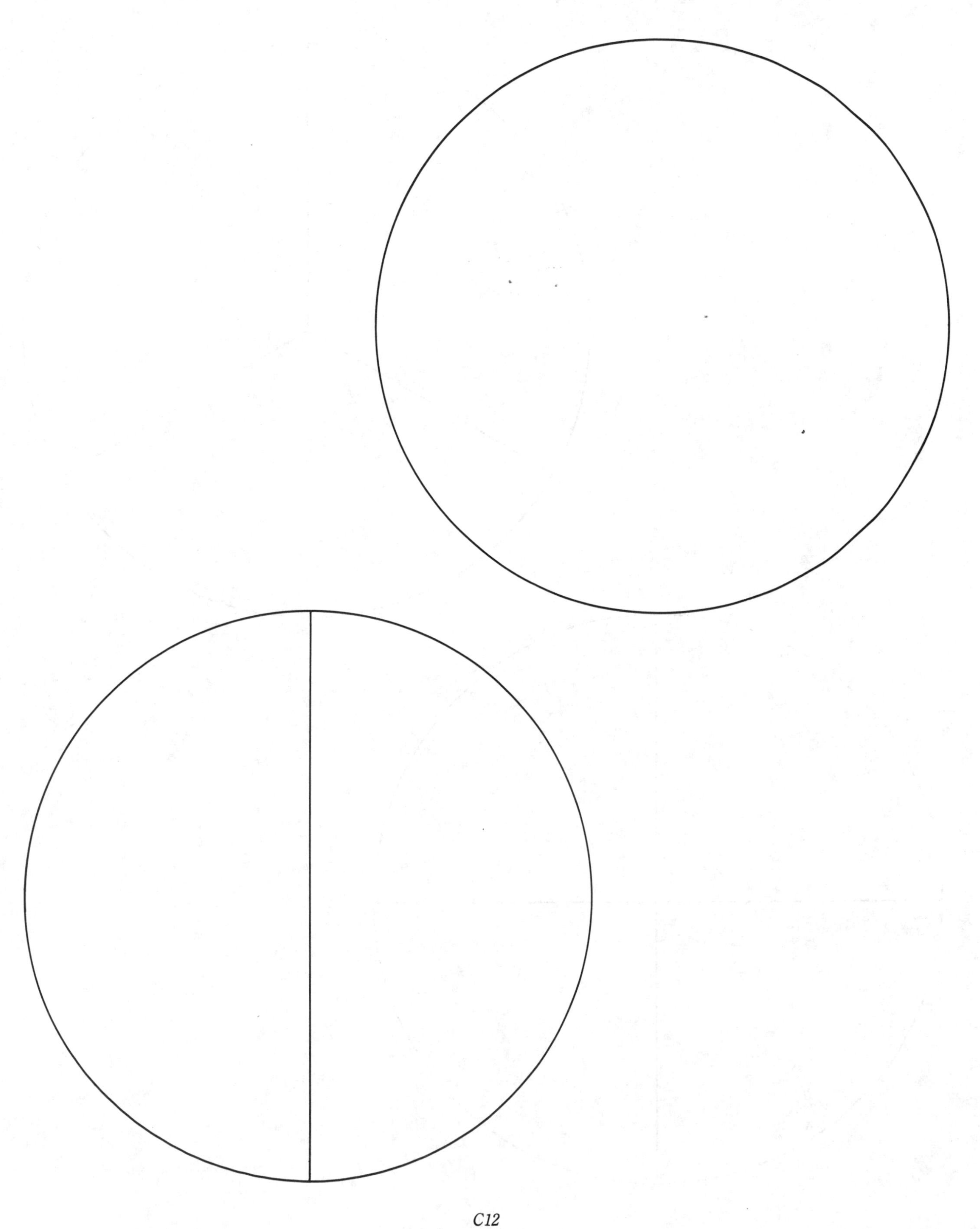

C12
Circular fraction pieces.

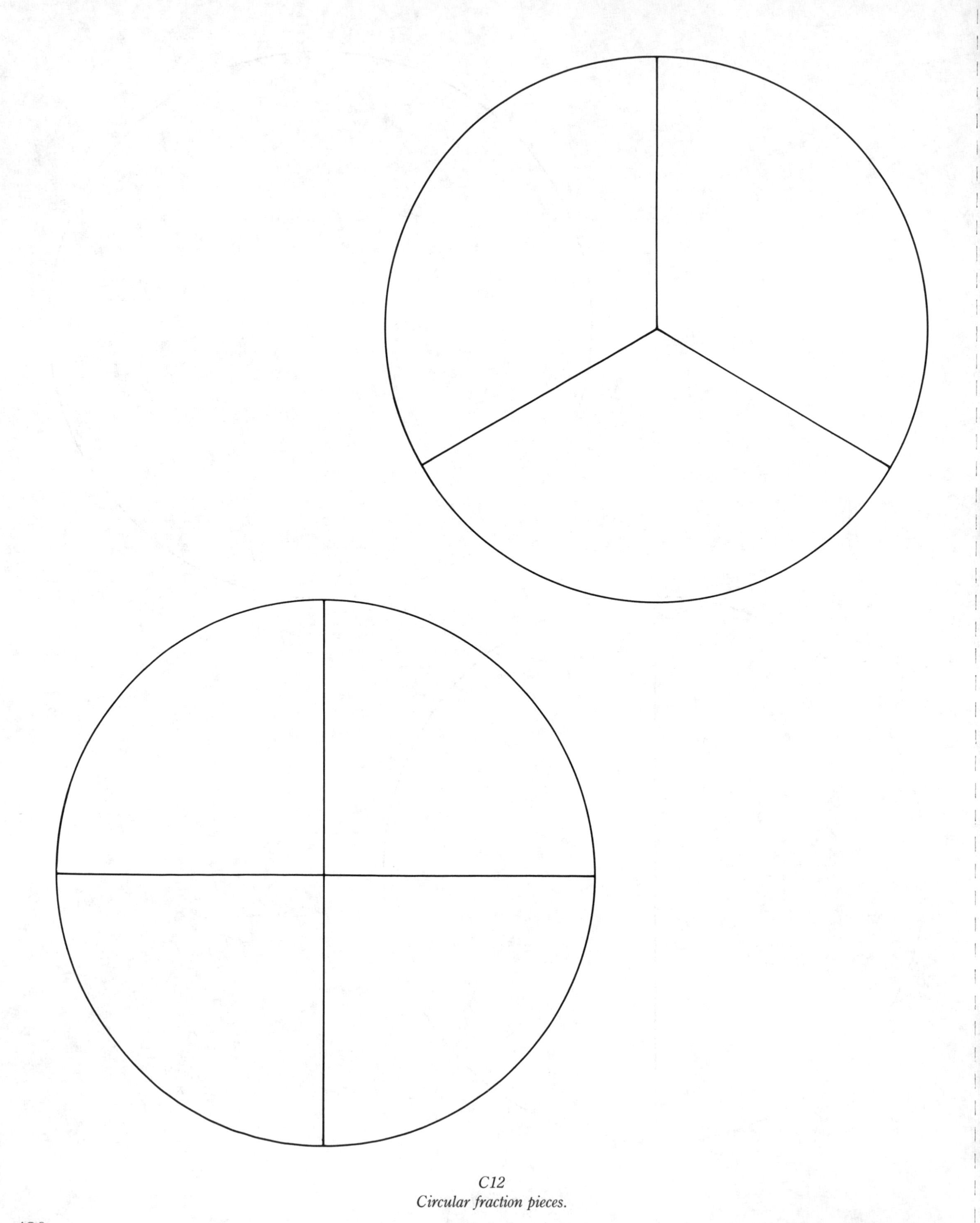

C12
Circular fraction pieces.

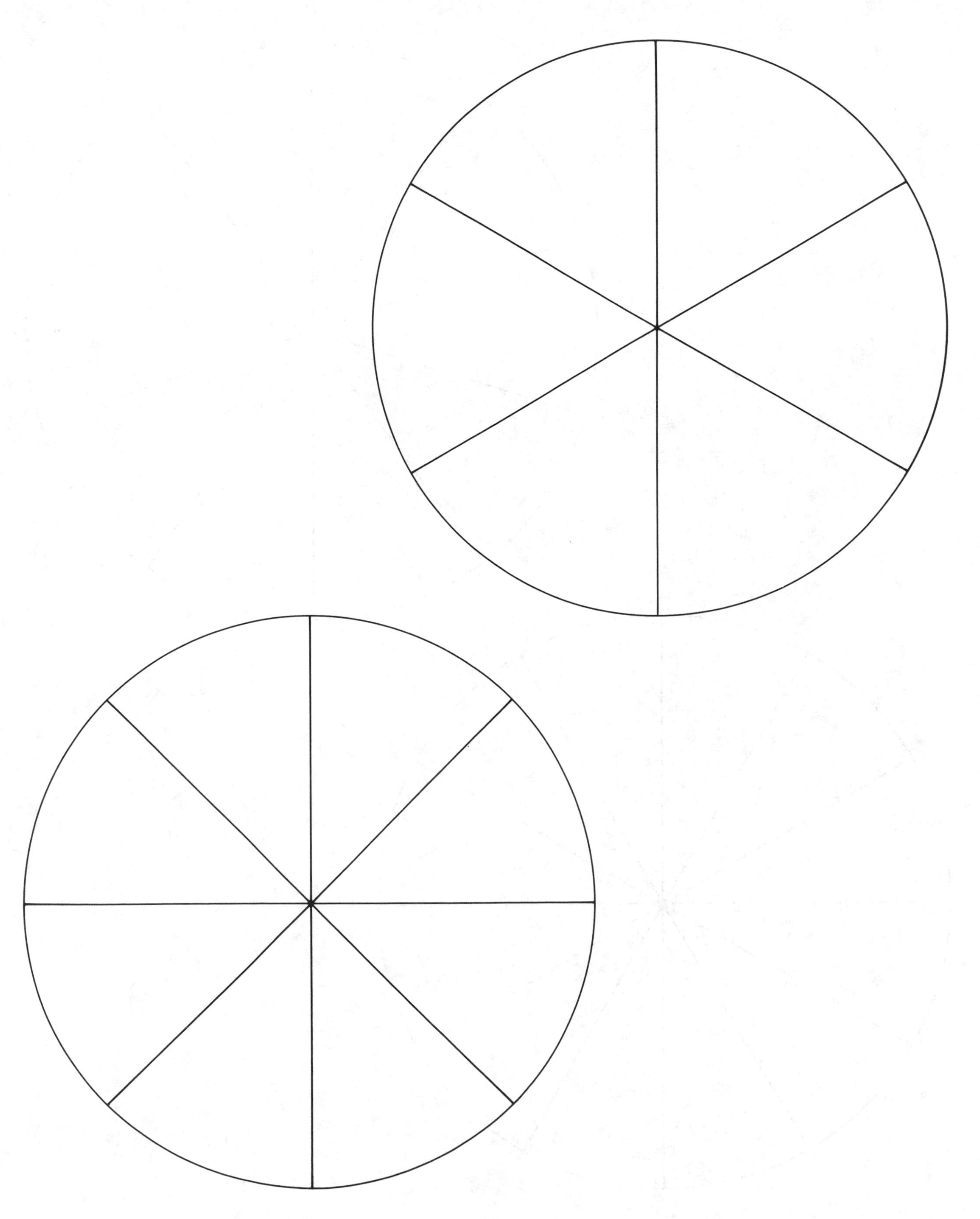

C12
Circular fraction pieces.

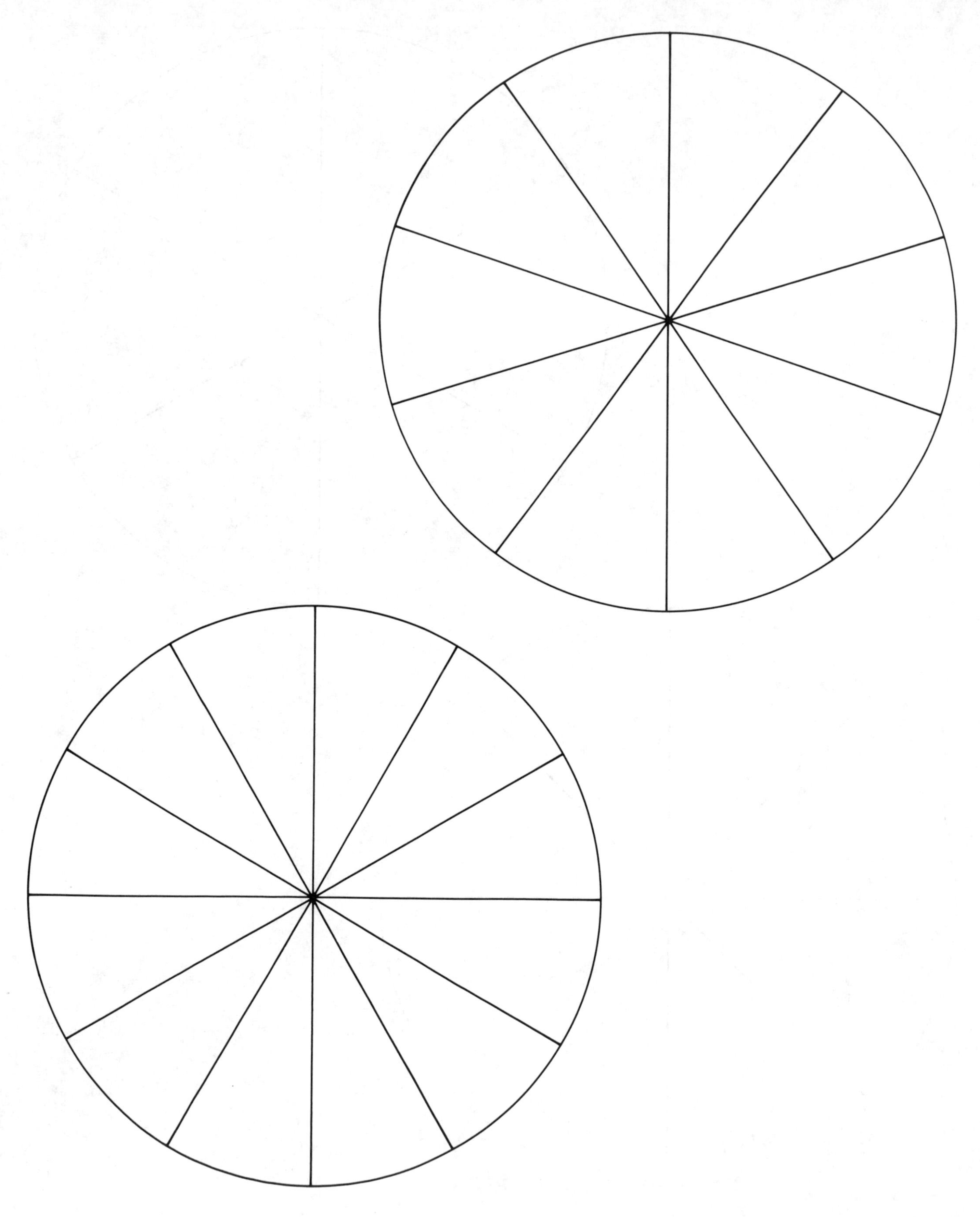

C12
Circular fraction pieces.

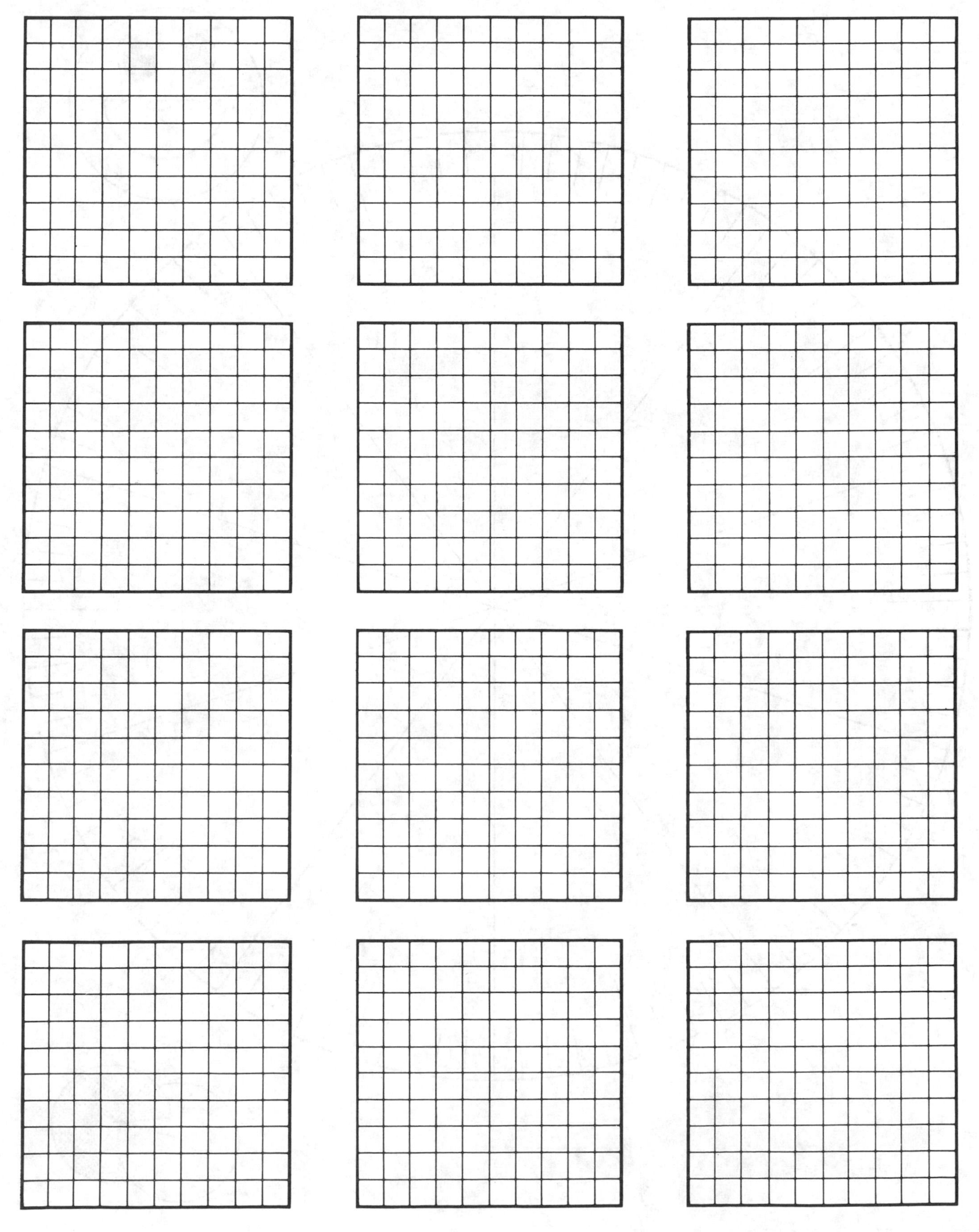

C13
10 × 10 grids.

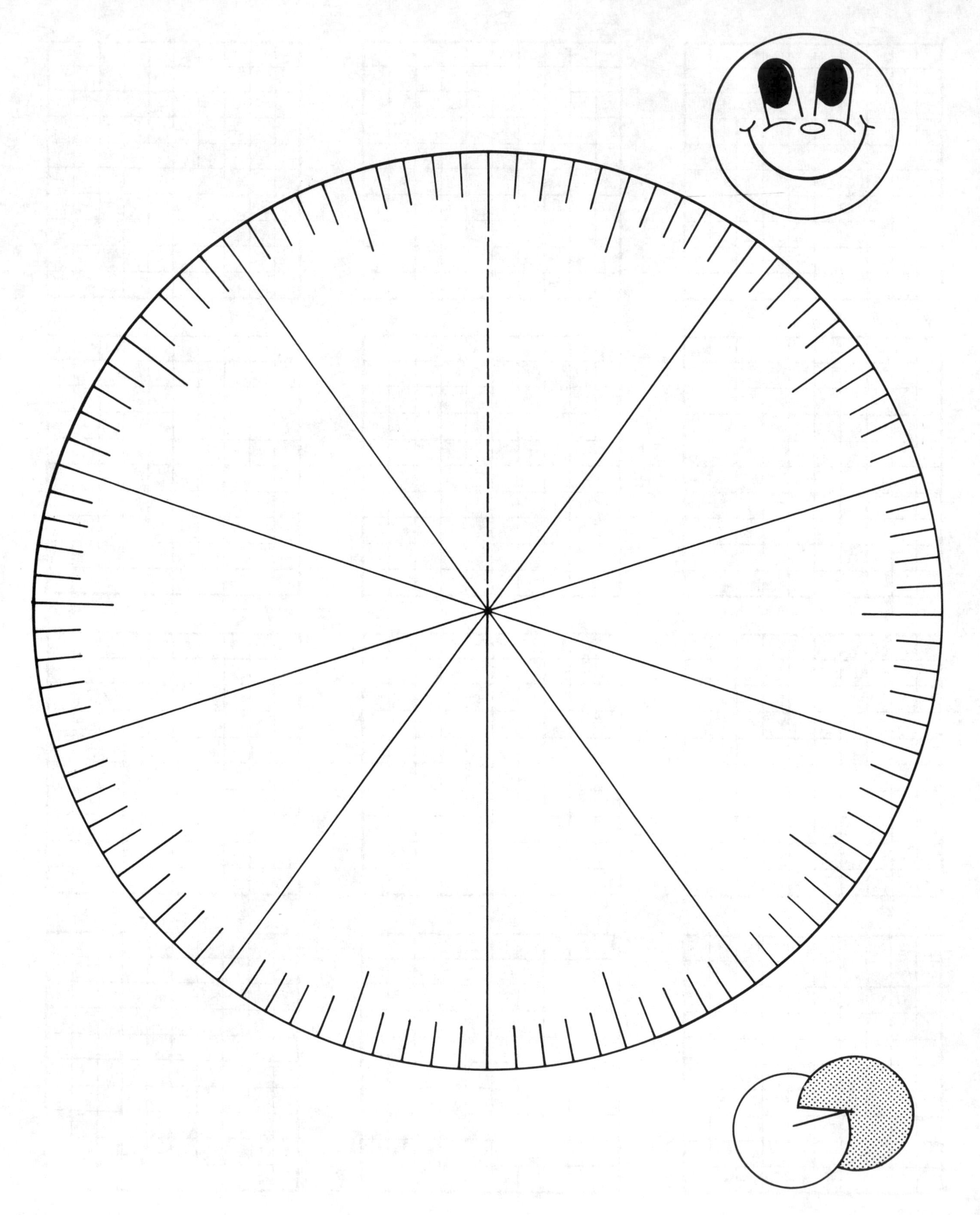

C14
Hundredth's disk.

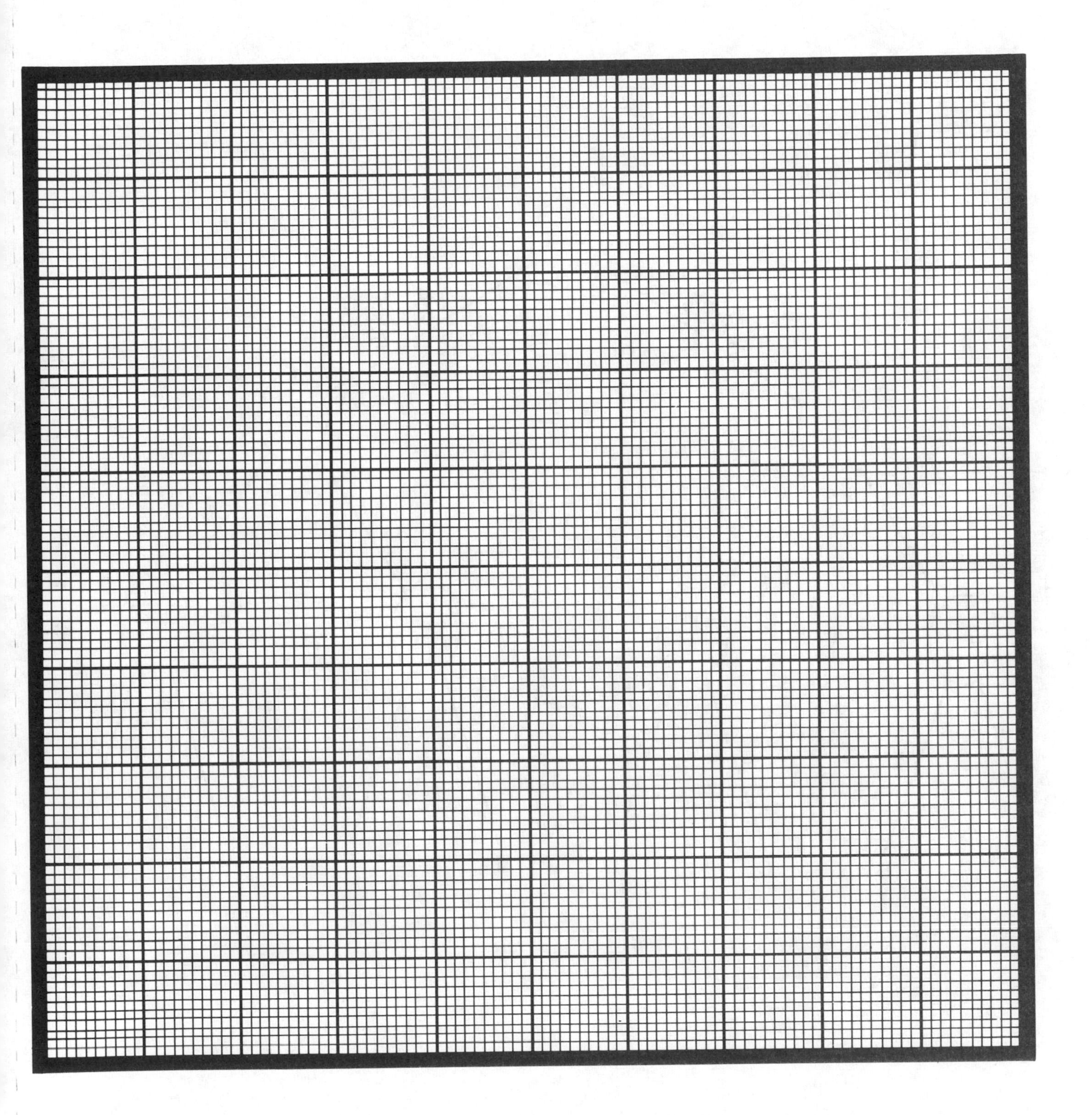

C15
10,000 grid.

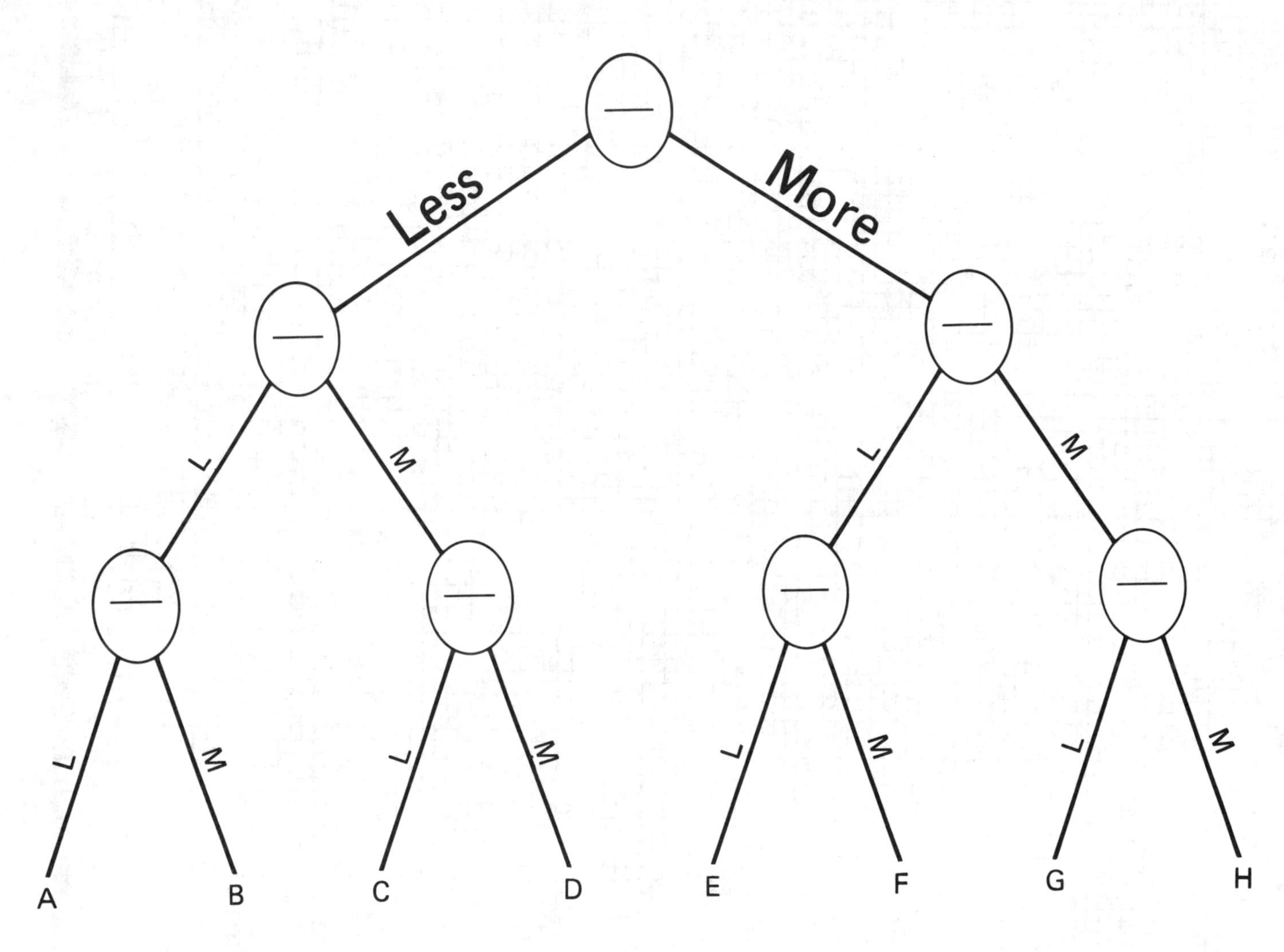

C16
Decimal/fraction sieve.

C17
2-cm square grid.

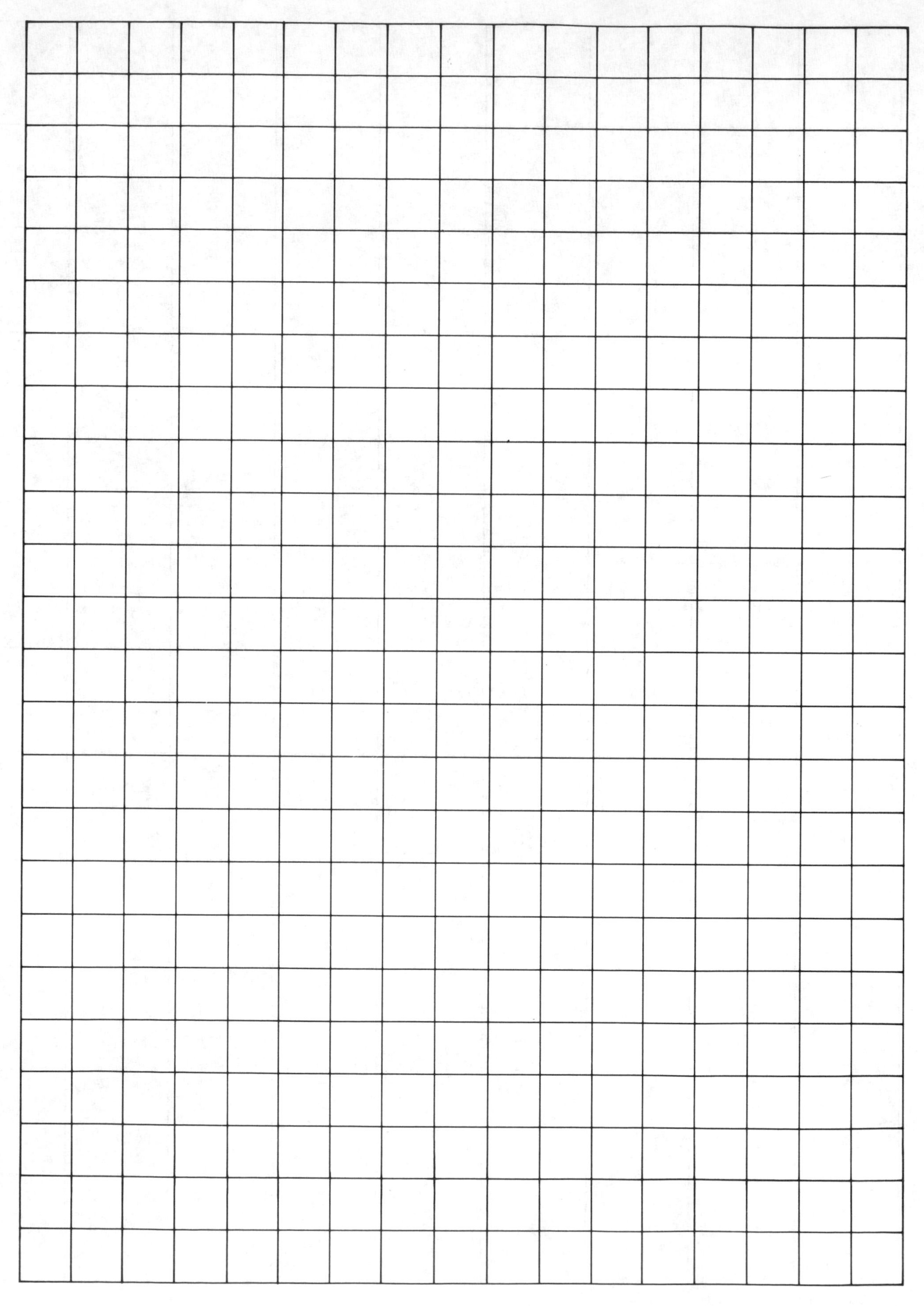

C17
1-cm square grid.

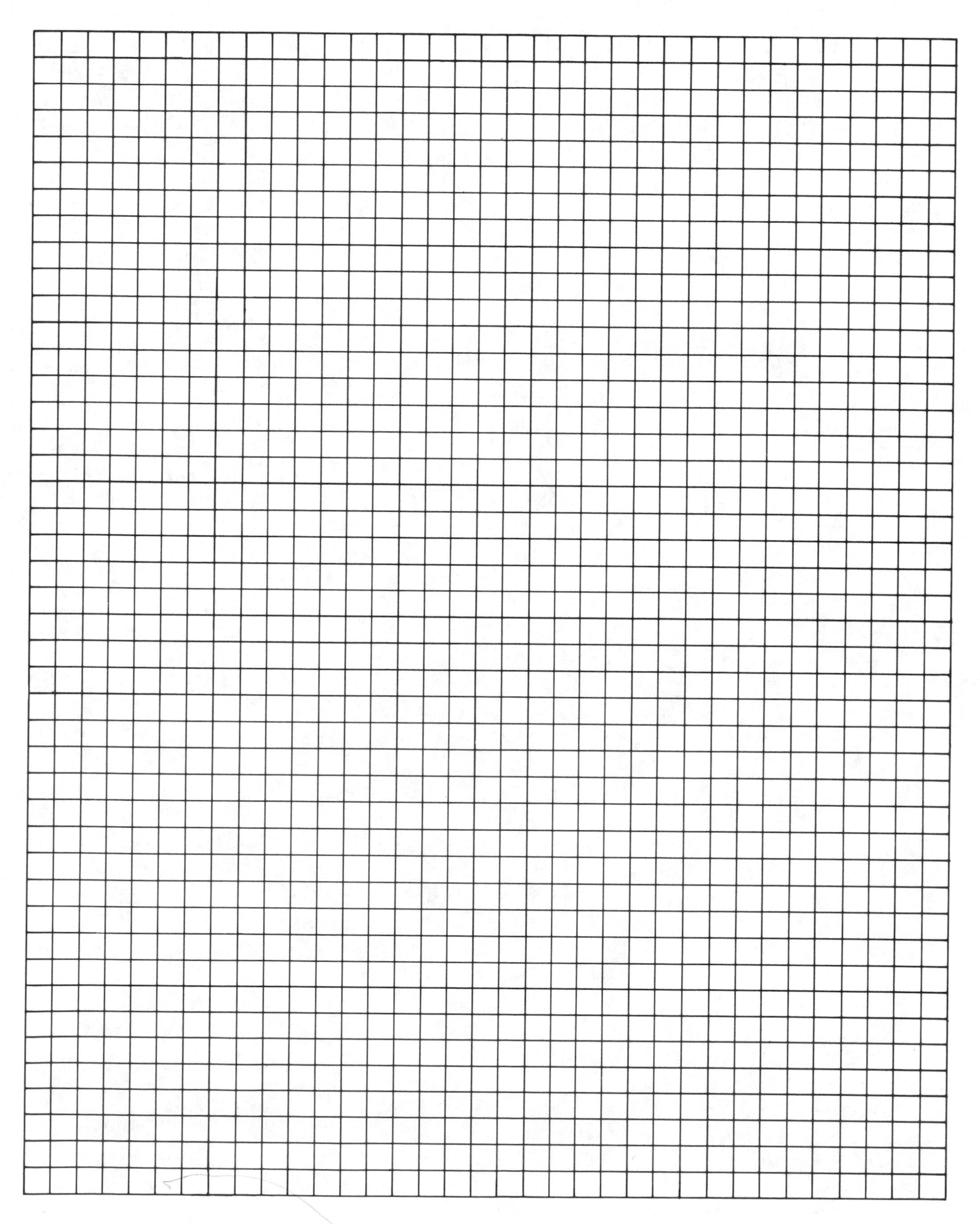

C17
0.5-cm square grid.

C17
1-cm square dot grid.

C17
2-cm isometric grid.

C17
1-cm isometric dot grid.

C17
1-cm square/diagonal grid.

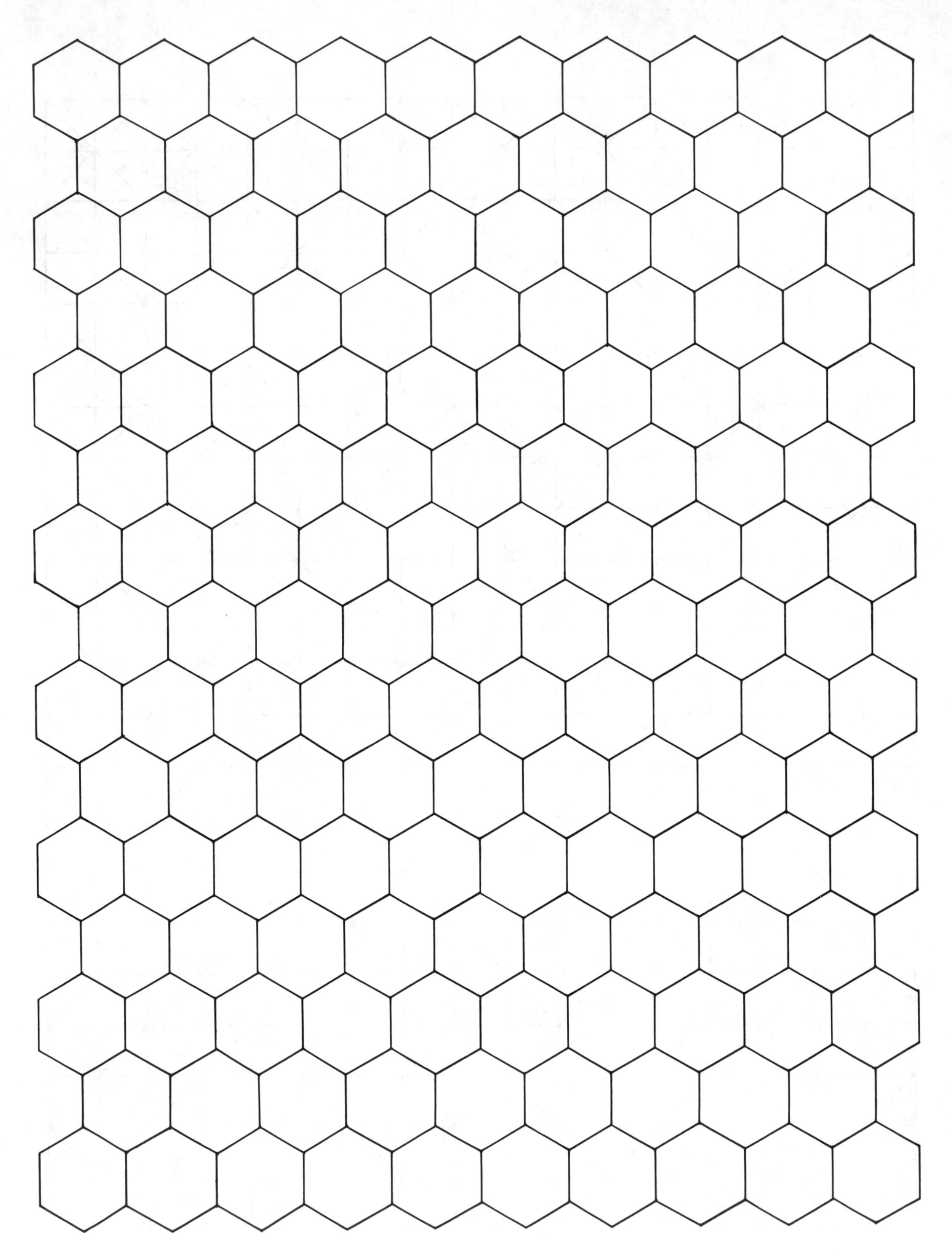

C17
1-cm hex grid.

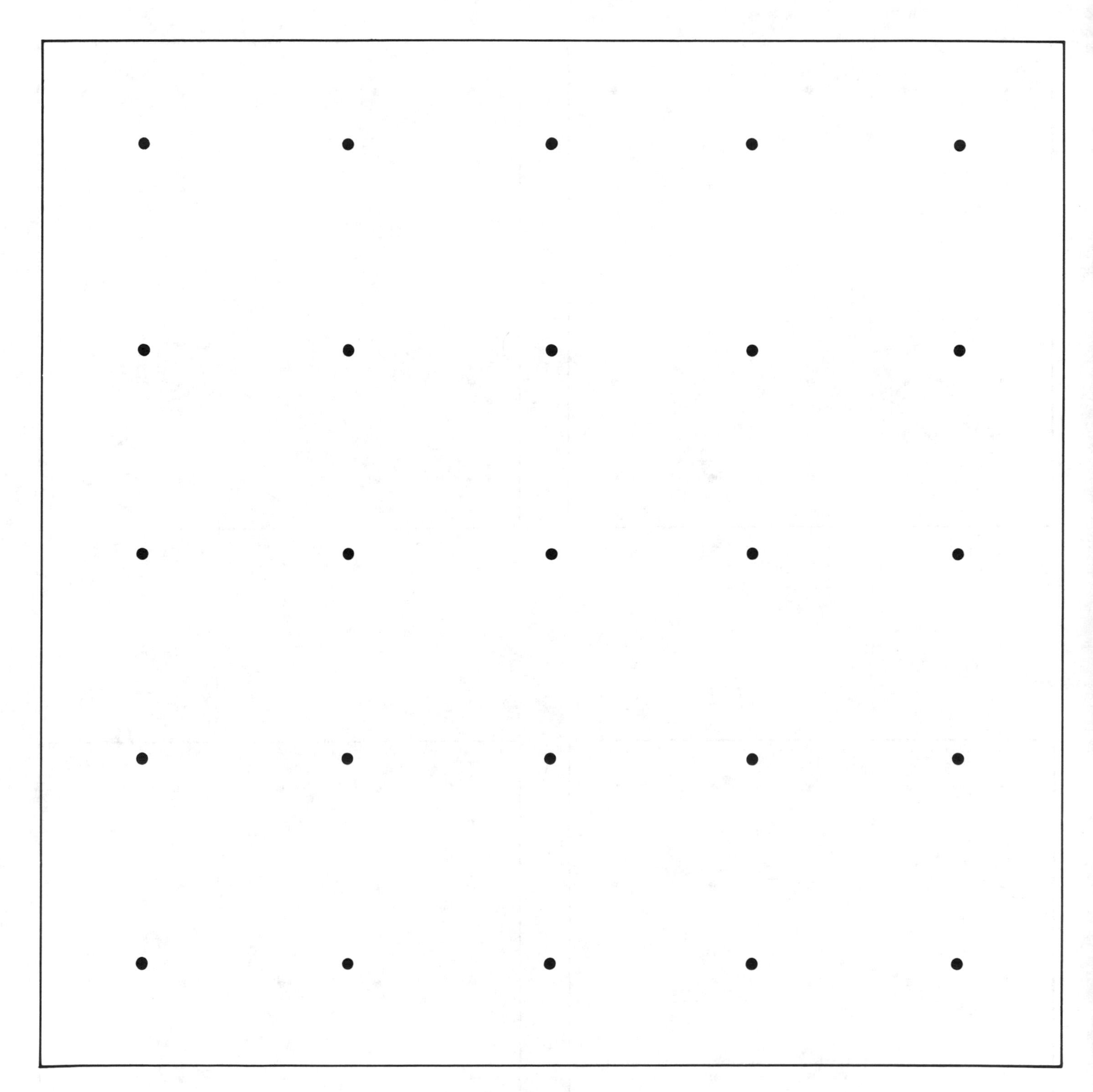

C18
Geoboard pattern.

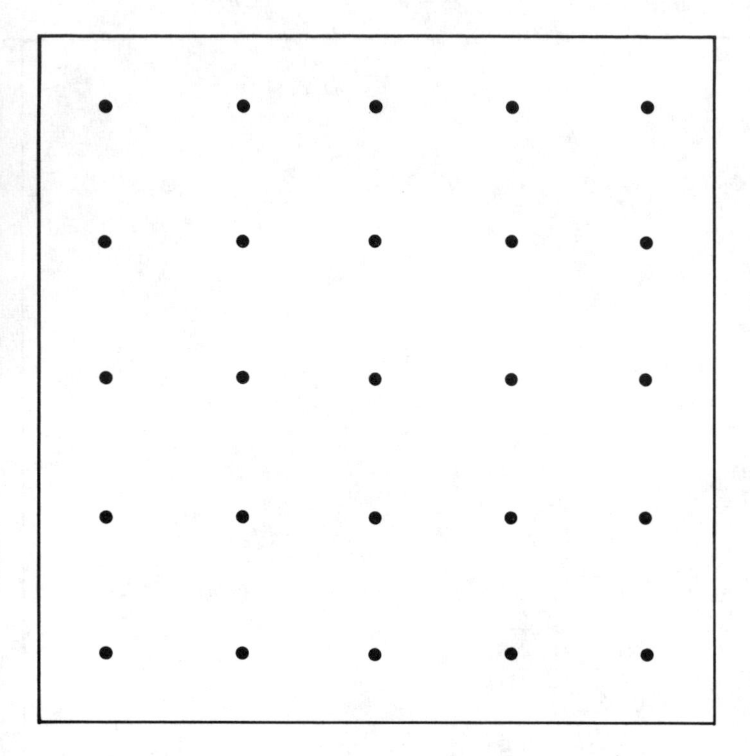

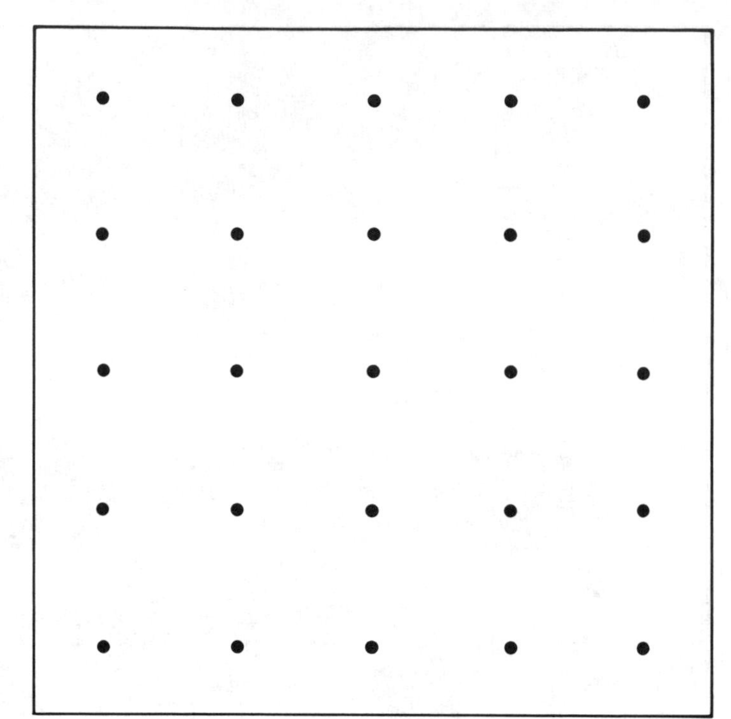

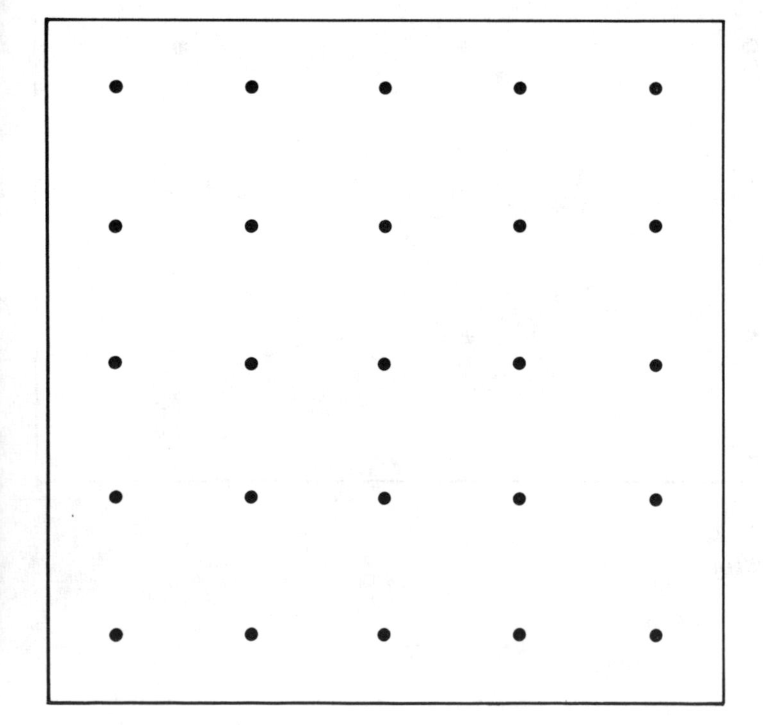

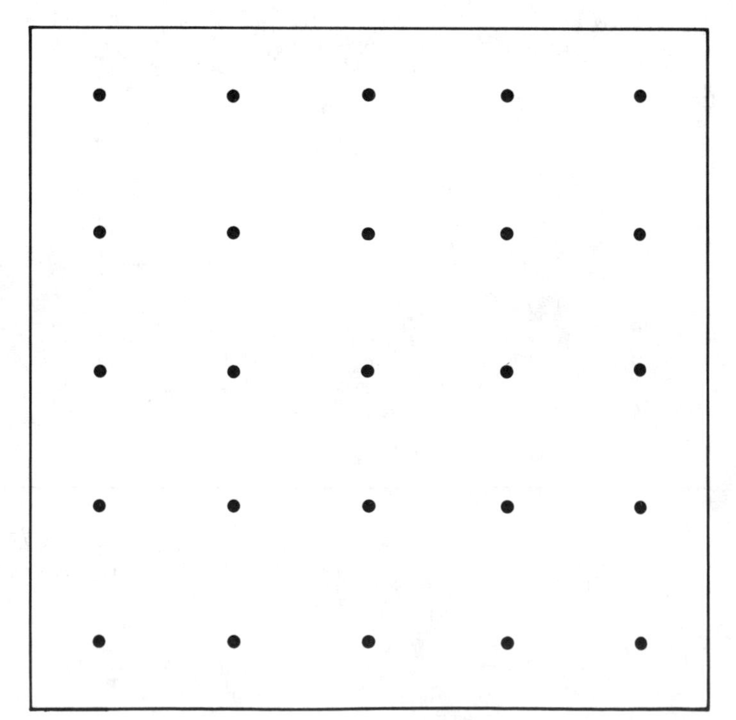

C19
Geoboard recording sheets.

C20
Right-triangle tiles.

C20
5-cm triangle tiles.

C20

Curvey tiles.

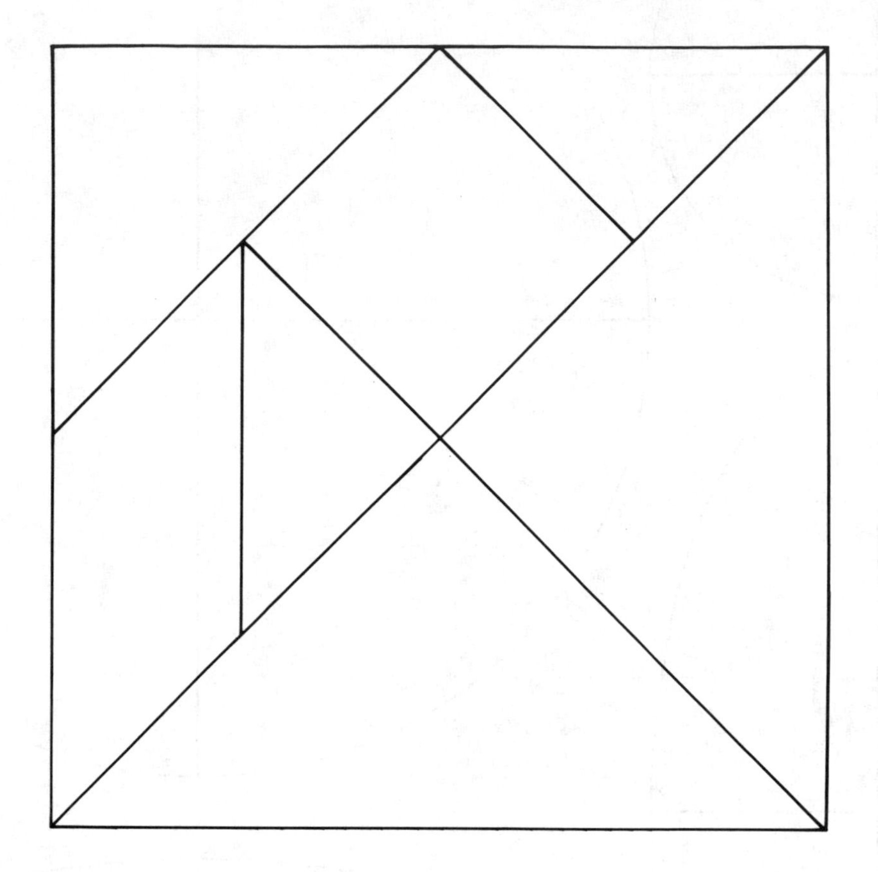

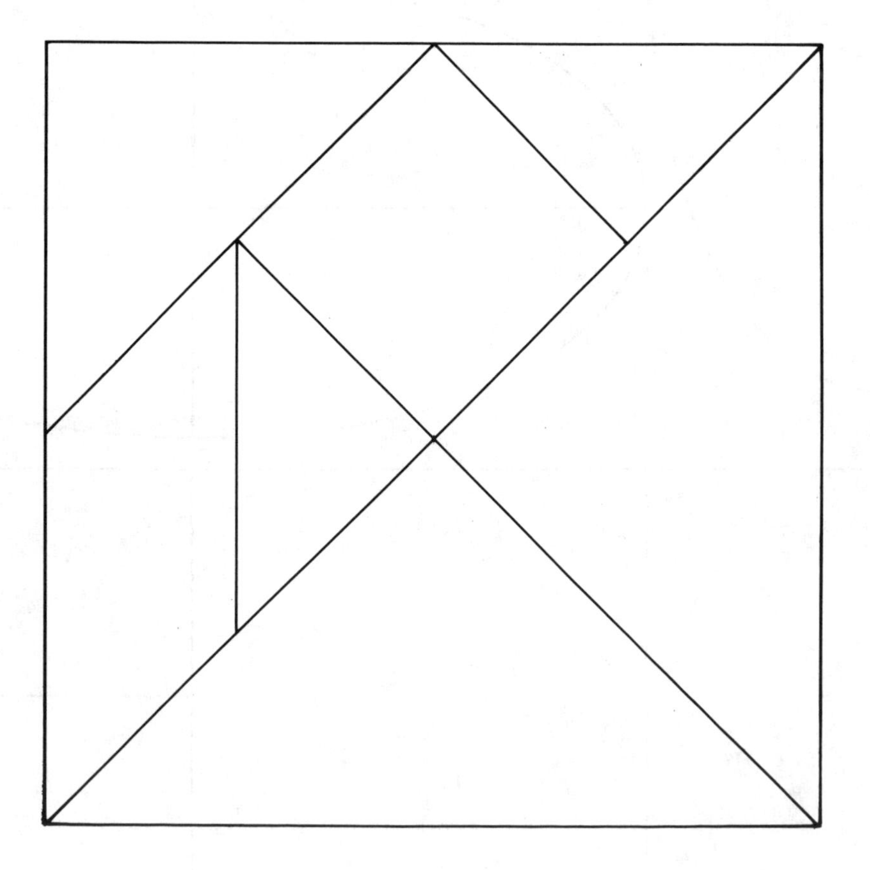

C21
Tangrams.

C22
Woozle cards.

REFERENCES

Armstrong, R. D. (1981). An area model for solving probability problems. In A. P. Schulte (Ed.), *Teaching probability and statistics*. Reston, VA: National Council of Teachers of Mathematics.

Au, W. K., Horton, J., & Ryba, K. (1987). Logo, teacher intervention, and the development of thinking skills. *The Computing Teacher, 15*(3), 12–16.

Baratta-Lorton, M. (1976). *Mathematics their way*. Menlo Park, CA: Addison-Wesley.

Baratta-Lorton, M. (1979). *Work jobs II*. Menlo Park, CA: Addison-Wesley.

Baroody, A. J. (1985). Mastery of the basic number combinations: Internalization of relationships or facts? *Journal for Research in Mathematics Education, 16,* 83–98.

Baroody, A. J. (1987). *Children's mathematical thinking: A developmental framework for preschool, primary, and special education teachers*. New York: Teachers College Press.

Beatty, L. S., Madden, R., Gardner, E. F., & Karlsen, B. (1984). *Stanford diagnostic mathematics test (3rd ed.)*. San Antonio, TX: The Psychological Corporation.

Behr, M. J., Erlwanger, S., & Nichols, E. (1976). How children view equality sentences. *PMDC Technical Report No. 3.* Tallahassee: Florida State University.

Behr, M. J., Lesh, R., Post, T. R., & Silver, E. A. (1983). Rational-number concepts. In R. Lesh & M. Landau (Eds.), *Acquisition of mathematics concepts and processes*. New York: Academic Press.

Bennett, A. (1982). *Decimal squares: Step by step teachers guide, Readiness to advanced levels in decimals*. Fort Collins, CO: Scott Scientific, Inc.

Bezuszka, S., Kenney, M., & Silvey, L. (1977). *Tessellations: The geometry of patterns*. Palo Alto, CA: Creative Publications.

Bley, N. S., & Thornton, C. A. (1981). *Teaching mathematics to the learning disabled*. Rockville, MD: Aspen Systems Corp.

Blume, G. W., & Schoen, H. L. (1988). Mathematical problem-solving performance of eighth-grade programmers and nonprogrammers. *Journal for Research in Mathematics Education, 19,* 142–156.

Bolster, L. C., et al. (1988). *Invitation to mathematics: 2*. Glenview, IL: Scott, Foresman.

Booth, L. R. (1988). Children's difficulties in beginning algebra. In A. F. Coxford (Ed.), *The ideas of algebra, K–12*. Reston, VA: National Council of Teachers of Mathematics.

Bright, G. W., Behr, M. J., Post, T. R., & Wachsmuth, I. (1988). Identifying fractions on number lines. *Journal for Research in Mathematics Education, 19,* 215–232.

Brownell, W., & Chazal, C. (1935). The effects of premature drill in third grade arithmetic. *Journal of Educational Research, 29,* 17–28.

Bruner, J. S. (1963). *The process of education*. New York: Vintage Books.

Bruni, J. V. (1977). *Experiencing geometry*. Belmont, CA: Wadsworth.

Burger, W. F. (1985). Geometry. *Arithmetic Teacher, 32,*(6), 52–56.

Burger, W. F., & Shaughnessy, J. M. (1986). Characterizing the van Hiele levels of development in geometry. *Journal for Research in Mathematics Education, 17,* 13–48.

Burns, M. (1987). *A collection of math lessons from grades 3 through 6.* (The Math Solutions Publications), New Rochelle, NY: Cuisenaire Company of America (distributor).

Burns, M., & Tank, B. (1988). *A collection of math lessons from grades 1 through 3.* (The Math Solutions Publications), New Rochelle, NY: Cuisenaire Company of America (distributor).

California State Department of Education. (1985). *Mathematics framework for California Public Schools kindergarten through grade twelve*. Sacramento, CA: The Department.

Callahan, L. G., & Glennon, V. (1975). *Elementary school mathematics: A guide to current research (4th ed.)*. Washington, DC: Association for Supervision and Curriculum Development.

Callahan, L. G., & MacMillan, D. L. (1981). Teaching mathematics to slow-learning and mentally retarded children. In V. J. Glennon (Ed.), *The mathematical education of exceptional children and youth: An interdisciplinary approach*. Reston, VA: National Council for Teachers of Mathematics.

Campbell, P. F. (1984). Using a problem-solving approach in the primary grades. *Arithmetic Teacher, 32*(4), 11–14.

Campbell, P. F. (1988). Microcomputers in the primary mathematics classroom. *Arithmetic Teacher, 35*(6), 22–30.

Carrier, C., Post, T. R., & Heck, W. (1985). Using microcomputers with fourth-grade students to reinforce arithmetic skills. *Journal for Research in Mathematics Education, 16,* 45–51.

Carroll, W. M. (1988). Cross sections of clay solids. *Arithmetic Teacher, 35*(7), 6–11.

Carpenter, T. P. (1986). Conceptual knowledge as a foundation for procedural knowledge: Implications from research on the initial learning. In J. Hiebert (Ed.), *Conceptual and procedural knowledge: The case of mathematics.* Hillsdale, NJ: Lawrence Erlbaum.

Carpenter, T. P., & Moser, J. M. (1983). The acquisition of addition and subtraction concepts. In R. Lesh & M. Landau (Eds.), *Acquisition of mathematics concepts and processes.* New York: Academic Press.

Carpenter, T. P., & Moser, J. M. (1982). The development of addition and subtraction problem-solving skills. In T. P. Carpenter (Ed.), *Addition and subtraction: A cognitive perspective.* Hillsdale, NJ: Lawrence Erlbaum.

Cauley, K. M. (1988). Construction of logical knowledge: Study of borrowing in subtraction. *Journal of Educational Psychology, 80,* 202–205.

Center for Innovation in Education. (1977). *Mathematics their way newsletter V.* Saratoga, CA.

Chalouh, L., & Herscovics, N. (1988). Teaching algebraic expressions in a meaningful way. In A. F. Coxford (Ed.), *The ideas of algebra, K–12.* Reston, VA: National Council of Teachers of Mathematics.

Charles, R., et al. (1985). *Problem-solving experiences in mathematics (grades 1 to 8).* Menlo Park, CA: Addison-Wesley.

Charles, R., & Lester, F. (1982). *Teaching problem solving: What, why & how.* Palo Alto, CA: Dale Seymour.

Charles, R., Lester, F., & O'Daffer, P. (1987). *How to evaluate progress in problem solving.* Reston, VA: National Council of Teachers of Mathematics.

Clements, D. H. (1984). Training effects on the development and generalization of Piagetian logical operations and knowledge of number. *Journal of Educational Psychology, 76,* 766–776.

Clements, D. H. (1985a). *Computers in early and primary education.* Englewood Cliffs, NJ: Prentice-Hall.

Clements, D. H. (1985b). Logo programming: Can it change how children think?. *Electronic Learning, 4*(4), 28, 74–75.

Clements, D. H. (1985c). Research on Logo in education: Is the turtle slow but steady, or not even in the race? *Logo in the Schools, 2*(2/3), 55–71.

Clements, D. H. (1986). Early studies on Logo and problem solving. *Logo Exchange, 5*(2), 23–25.

Coburn, T. (1987). *How to teach mathematics using a calculator.* Reston, VA: National Council of Teachers of Mathematics.

Connolly, A. J. (1988). *Key math—revised: A diagnostic inventory of essential mathematics.* Circle Pines, MN: American Guidance Service.

Connolly, A. J., Nachtman, W., & E. M. Pritchett. (1976). *Key math diagnostic arithmetic test.* Circle Pines, MN: American Guidance Service.

Coombs, B., & Harcourt, L. (1986). *Explorations 1.* Don Mills, Ontario: Addison-Wesley.

Crowley, M. L. (1987). The van Hiele model of the development of geometric thought. In M. M. Lindquist (Ed.), *Learning and teaching geometry, K–12.* Reston, VA: National Council of Teachers of Mathematics.

Dalbey, J., & Linn, M. C. (1985). The demands and requirements of computer programming: A literature review. *Journal of Educational Computing Research, 1,* 253–274.

Davis, E. J. (1978). Suggestions for teaching the basic facts of arithmetic. In M. N. Suydam (Ed.)., *Developing computational skills.* Reston, VA: National Council of Teachers of Mathematics.

Davidson, P. S. (1975). *Chip trading activities: Teacher's guide.* Fort Collins, CO: Scott Scientific, Inc.

Demana, F., & Osborne, A. (1988). Choosing a calculator: Four-function foul-ups. *Arithmetic Teacher, 35*(7), 2–3.

Dossey, J. A., Mullis, I. V. S., Lindquist, M. M., & Chambers, D. L. (1988). *The mathematics report card: Are we measuring up?* Princeton, NJ: Educational Testing Service.

Elementary Science Study. (1966). *Teacher's guide for attribute games and problems.* Nashua, NH: Delta Education.

Erlwanger, S. (1975), Case studies of children's conceptions of mathematics: 1. *Journal of Children's Mathematical Behavior, 1,* 157–183.

Fair, J. (1987). *Tangram treasury (Books A, B, and C).* New Rochelle, NY: Cuisenaire Company of America.

Fitch, D. M. (1987). *Logo data tool kit.* Cambridge, MA: Terrapin, Inc.

Flanders, J. R. (1987) How much of the content in mathematics textbooks is new? *Arithmetic Teacher, 35*(1), 18–23.

Fuson, K. C. (1984). More complexities in subtraction. *Journal for Research in Mathematics Education, 15,* 214–225.

Fuson, K. C., & Brinko, K. T. (1985). The comparative effectiveness of microcomputers and flash cards in the drill and practice of basic mathematics facts. *Journal for Research in Mathematics Education, 16,* 225–232.

Fuson, K. C., & Hall, J. W. (1983). The acquisition of early number word meanings: A conceptual analysis and review. In H. P. Ginsburg (Ed.), *The development of mathematical thinking.* New York: Academic Press.

Fuson, K. C., Secada, W. G., & Hall, J. W. (1983). Matching, counting, and conservation of numerical equivalence. *Child Development, 54,* 91–97.

Garofalo, J. (1987). Metacognition and school mathematics. *Arithmetic Teacher 34*(9), 22–23.

Gelman, R., & Gallistel, C. R. (1978). *The child's understanding of number.* Cambridge, MA: Harvard University Press.

Gelman, R., & Meck, E. (1986). The notion of principle: The

case of counting. In J. Hiebert (Ed.), *Conceptual and procedural knowledge: The case of mathematics.* Hillsdale, NJ: Lawrence Erlbaum.

Ginsburg, H. P. (1977). *Children's arithmetic: The learning process.* New York: Van Nostrand.

Ginsburg, H. P., & Baroody, A. J. (1983). *The test of early mathematics ability.* Austin, TX: PRO-ED.

Glennon, V. J., & Cruickshank, W. M. (1981). Teaching mathematics to children and youth with perceptual and cognitive processing deficits. In V. J. Glennon (Ed.), *The mathematical education of exceptional children and youth: An interdisciplinary approach.* Reston, VA: National Council of Teachers of Mathematics.

Gnanadesikan, M., Schaeffer, R., & Swift, J. (1987). *The art and techniques of simulation: Quantitative literacy series.* Palo Alto, CA: Dale Seymour.

Goldenberg, E. P. (1986). Learning to think algebraically: Word problems and data flow. *Logo Exchange, 5*(2), 16–20.

Good, T. L., & Grouws, D. A. (1979). The Missouri mathematics effectiveness project: An experimental study in fourth-grade classrooms. *Journal of Educational Psychology, 71,* 355–362.

Good, T. L., Grouws, D. A., & Ebmeier, H. (1983). *Active mathematics teaching.* New York: Longmans.

Goodnow, J., Hoogeboom, S., Moretti, G., Stephens, M., & Scanlin, A. (1987). *The problem solver: Activities for learning problem solving strategies.* Palo Alto, CA: Creative Publications.

Hamrick, K. (1979). Oral language and readiness for the written symbolization of addition and subtraction. *Journal for Research in Mathematics Education, 10,* 188–194.

Hatano, G. (1982). Learning to add and subtract: A Japanese perspective. In T. P. Carpenter (Ed.), *Addition and subtraction: A cognitive perspective.* Hillsdale, NJ: Lawrence Erlbaum.

Hazekamp, D. W. (1986). Components of mental multiplying. In H. Schoen (Ed.), *Estimation and mental computation.* Reston, VA: National Council of Teachers of Mathematics.

Hemre, R., & Dessart, D. D. (1986). Effects of hand-held calculators in precollege mathematics education: A meta-analysis. *Journal for Research in Mathematics Education, 17,* 83–99.

Hendrickson, A. D. (1986). Verbal multiplication and division problems: Some difficulties and some solutions. *Arithmetic Teacher, 33*(8), 26–33.

Herscovics, N., & Kieran, C. (1980). Constructing meaning for the concept of equation. *Mathematics Teacher, 73,* 573–580.

Hiebert, J., & Lefevre, P. (1986). Conceptual and procedural knowledge in mathematics: An introductory analysis. In J. Hiebert (Ed.), *Conceptual and procedural knowledge: The case of mathematics.* Hillsdale, NJ: Lawrence Erlbaum.

Hiebert, J., & Wearne, D. (1986). Procedures over concepts: The acquisition of decimal number knowledge. In J. Hiebert (Ed.), *Conceptual and procedural knowledge: The case of mathematics.* Hillsdale, NJ: Lawrence Erlbaum.

Hoffer, A. (1983). Van Hiele-based research. In R. Lesh & M. Landau (Eds.), *Acquisition of mathematics concepts and processes.* New York: Academic Press.

Hoffer, A. (1988a). Geometry and visual thinking. In T. R. Post (Ed.), *Teaching mathematics in grades K–8: Research based methods.* Boston: Allyn and Bacon.

Hoffer, A. (1988b). Ratios and proportional thinking. In T. R. Post (Ed.), *Teaching mathematics in grades K–8: Research based methods.* Boston: Allyn and Bacon.

Holden, L. (1986). *Fraction factory.* Palo Alto, CA: Creative Publications.

Holden, L., & Roper, A. (1980). *The pattern factory.* Palo Alto, CA: Creative Publications.

Hope, J. A. (1986). Mental calculation: Anachronism or basic skill. In H. Schoen (Ed.), *Estimation and mental computation.* Reston, VA: National Council of Teachers of Mathematics.

Hope, J. A., Leutzinger, L., Reys, B. J., & Reys, R. R. (1988). *Mental math in the primary grades.* Palo Alto, CA: Dale Seymour.

Hope, J. A., Reys, B. J., & Reys, R. (1987). *Mental math in the middle grades.* Palo Alto, CA: Dale Seymour.

Hope, J. A., Reys, B. J., & Reys, R. (1988). *Mental math in the junior high school.* Palo Alto, CA: Dale Seymour.

Houde, R., & Yerushalmy, M. (1988). *Geometry problems and projects: Blackline masters for use with the geometric supposers.* Pleasantville, NY: Sunburst Communications.

House, P. A. (Ed.) (1987). *Providing opportunities for the mathematically gifted, K–12.* Reston, VA: National Council of Teachers of Mathematics.

Institute for Research on Teaching. (1988). Reinventing the meaning of "knowing" in mathematics. *Communication Quarterly, 11*(1).

Kamii, C. K. (1985). *Young children reinvent arithmetic.* New York: Teachers College Press.

Karplus, E. F., Karplus, R., & Wollman W. (1974). Intellectual development beyond elementary school (Vol. IV): Ratio, the influence of cognitive style. *School Science and Mathematics, 74,* 476–482.

Karplus, R., Pulos, S., & Stage, E. K. (1983). Proportional reasoning of early adolescents. In R. Lesh & M. Landau (Eds.), *Acquisition of mathematics concepts and processes.* New York: Academic Press.

Keedy, M. L. (1989). Textbooks and curriculum—Whose dilemma? *Arithmetic Teacher, 36*(7), 6.

Kieran, C. (1988). Two different approaches among algebra learners. In A. F. Coxford & A. P. Shulte (Eds.), *The ideas of algebra, K–12.* Reston, VA: National Council of Teachers of Mathematics.

Kouba, V. L., Brown, C. A., Carpenter, T. P., Lindquist, M. M., Silver, E. A., & Swafford, J. O. (1988a). Results of the fourth NAEP assessment of mathematics: Number, operations, and word problems. *Arithmetic Teacher, 35*(8), 14–19.

Kouba, V. L., Brown, C. A., Carpenter, T. P., Lindquist, M. M., Silver, E. A., & Swafford, J. O. (1988b). Results of the fourth NAEP assessment of mathematics: Measurement, geometry, data interpretation, attitudes, and other topics. *Arithmetic Teacher, 35*(9), 10–16.

Kovatch, N. (1984). Functions of output. *National Logo Exchange, 3*(4), 1–2.

Labinowicz, E. (1980). *Piaget primer: Thinking, learning, teaching.* Menlo Park, CA: Addison-Wesley.

Labinowicz, E. (1985). *Learning from children: New beginnings for teaching numerical thinking.* Menlo Park, CA: Addison-Wesley.

Labinowicz, E. (1987). Assessing for learning: The interview method. *Arithmetic Teacher, 35*(3), 22–24.

Landwehr, J. M., Swift, J., & Watkins, A. E. (1987). *Exploring surveys and information from samples: Quantitative literacy series.* Palo Alto, CA: Dale Seymour.

Landwehr, J. M. & Watkins, A. E. (1987). *Exploring data: Quantitative literacy series.* Palo Alto, CA: Dale Seymour.

Leutzinger, L. P. (1979). The effects of counting-on on the acquisition of addition facts in grade one (Doctoral dissertation. University of Iowa, 1979). *Dissertation Abstracts International,* 40, 3765A.

Leutzinger, L. P., Rathmell, E. C., & Urbatsch, T. D. (1986). Developing estimation skills in the primary grades. In H. Schoen (Ed.), *Estimation and mental computation.* Reston, VA: National Council of Teachers of Mathematics.

Liedtke, W. (1988). Diagnosis in mathematics: The advantages of an interview. *Arithmetic Teacher, 36*(3), 26–29.

Lindquist, M. M. (1987). Estimation and mental computation: Measurement. *Arithmetic Teacher, 34*(5), 16–17.

Madell, R. (1985). Children's natural processes. *Arithmetic Teacher, 32*(7), 20–22.

Madell, R., & Larkin, E. (1977). *Picturing numeration.* Palo Alto, CA: Creative Publications.

Markovits, Z., Eylon, B. S., & Bruckheimer, M. (1988). Difficulties students have with the function concept. In A. F. Coxford (Ed.), *The ideas of algebra, K–12.* Reston, VA: National Council of Teachers of Mathematics.

Mason, J., Burton, L., & Stacey, K. (1982). *Thinking mathematically.* London: Addison-Wesley.

McKnight, C. C., Crosswhite, F. J., Dossey, J. A., Kifer, E., Swafford, J. O., Travers, K. J., & Cooney, T. J. (1987). *The underachieving curriculum: Assessing U.S. school mathematics from an international perspective.* Champaign, IL: Stipes.

Michigan State Board of Education. (1988). *Michigan essential goals and objectives for mathematics education.* Lansing, MI: Michigan Department of Education.

Milauskas, G. A. (1987). Creative geometry problems can lead to creative problem solvers. In M. M. Lindquist (Ed.), *Learning and teaching geometry, K–12.* Reston, VA: National Council of Teachers of Mathematics.

Minnesota Educational Computing Consortium. (1984). *Estimation* (computer software). St. Paul, MN: MECC.

Morris, J. (1981). *How to develop problem solving using a calculator.* Reston, VA: National Council of Teachers of Mathematics.

National Commission on Excellence in Education. (1983). *A nation at risk: The imperative for educational reform.* Washington, DC: U.S. Government Printing Office.

National Council of Supervisors of Mathematics. (1977). Position paper on basic mathematical skills. *Arithmetic Teacher 25*(2), 18–22.

National Council of Supervisors of Mathematics. (1988). Essential mathematics for the 21st century. NCSM.

National Council of Teachers of Mathematics. (1980). *An agenda for action: Recommendations for school mathematics of the 1980s.* Reston, VA: The Council.

National Council of Teachers of Mathematics. (1983). *A position statement: Vertical acceleration.* Reston, VA: The Council.

National Council of Teachers of Mathematics. (1986a). *A position statement: Calculators in the mathematics classroom.* Reston, VA: The Council.

National Council of Teachers of Mathematics. (1986b). *A position statement: Provisions for mathematically talented and gifted students.* Reston, VA: The Council.

National Council of Teachers of Mathematics: Commission on Standards for School Mathematics. (1989). *Curriculum and evaluation standards for school mathematics.* Reston, VA: The Council.

National Research Council. (1989). *Everybody counts: A report to the nation on the future of mathematics education.* Washington, DC: National Academy of Sciences.

Noelting, G. (1980). The development of proportional reasoning and the ratio concept: 1. Differentiation of stages. *Educational Studies in Mathematics, 11,* 217–253.

Papert, S. (1980). *Mindstorms: Children, computers, and powerful ideas.* New York: Basic Books.

Payne, J. N. (1976). Review of research on fractions. In R. A. Lesh & D. A. Bradbard (Eds.), *Number and measurement: Papers from a research workshop* (pp. 145–187). Columbus Ohio: ERIC/SMEAC.

Polya, G. (1957). *How to solve it: A new aspect of mathematical method.* Princeton, NJ: Princeton University Press.

Post, T. (1980). The role of manipulative materials in the learning of mathematics. In M. M. Lindquist (Ed.), *Selected issues in mathematics education.* Chicago: National Society for the Study of Education.

Post, T. (1981). Fractions: Results and implications from the national assessment. *Arithmetic Teacher, 28*(9), 26–31.

Post, T. R., Behr, M. J., & Lesh, R. (1988). Proportionality and development of prealgebra understandings. In A. Coxford (Ed.), *The ideas of algebra, K–12.* Reston, VA: National Council of Teachers of Mathematics.

Post, T. R., Wachsmuth, I., Lesh, R., & Behr, M. J. (1985). Order and equivalence of rational numbers: A cognitive analysis. *Journal for Research in Mathematics Education, 16,* 18–36.

Pothier, Y., & Sawada, D. (1983). Partitioning: The emergence of rational number ideas in young children. *Journal for Research in Mathematics Education, 14,* 307–317.

Quintero, A. H. (1986). Children's conceptual understanding of situations involving multiplication. *Arithmetic Teacher, 33*(5), 34–37.

Ranucci, E. R., & Teeters, J. L. (1977). *Creating Escher-type drawings.* Palo Alto, CA: Creative Publications.

Rathmell, E. C. (1978). Using thinking strategies to teach the basic facts. In M. N. Suydam (Ed.), *Developing computational skills.* Reston, VA: National Council of Teachers of Mathematics.

Reader's Digest Services. (1983). *Mathematics assessment/prescriptive edu-disks.* Pleasantville, NY: Reader's Digest Services.

Renzulli, J. S. (1977). *The enrichment triad model: A guide for developing defensible programs for the gifted and talented.* Wethersfield, CT: Creative Learning Press.

Renzulli, J. S. (1978). What makes giftedness? Reexamining a definition. *Phi Delta Kappan, 60,* 180–184, 261.

Resnick, L. (1983). A developmental theory of number understanding. In H. P. Ginsburg (Ed.), *The development of mathematical thinking.* New York: Academic Press.

Reys, R. E., & Reys, B. J. (1983). *Guide to using estimation skills and strategies (GUESS) Box I & II.* Palo Alto, CA: Dale Seymour.

Reys, R. E., & Reys, B. J. (1987). Calculators in the classroom: How can we make it happen? *Arithmetic Teacher. 34*(6), 12–14.

Reys, R., Trafton, P., Reys, B., & Zawojewski, J. (1987). *Computational estimation: (Grades 6, 7, 8).* Palo Alto, CA: Dale Seymour.

Ridge, H., & Renzulli, J. S. (1981). Teaching mathematics to the talented and gifted. In V. J. Glennon (Ed.), *The mathematical education of exceptional children and youth: An interdisciplinary approach.* Reston, VA: National Council of Teachers of Mathematics.

Ross, S. H. (1986). The development of children's place-value numeration concepts in grades two through five. Paper presented at the annual meeting of the American Educational Research Association, San Francisco. *ERIC Document Reproduction Service no. ED 273 482.*

Ross, S. H. (1989). Parts, wholes, and place value: A developmental perspective. *Arithmetic Teacher, 36*(6), 47–51.

Scheer, J. K. (1980). The etiquette of diagnosis. *Arithmetic Teacher, 27*(9), 18–19.

Schoenfeld, A. H. (1985). *Mathematical problem solving.* New York: Academic Press.

Schoenfeld, A. H. (1988). What's all the fuss about metacognition? In A. H. Schoenfeld (Ed.), *Cognitive science and mathematics education.* Hillsdale, NJ: Lawrence Erlbaum.

Seymour, D. (1971). *Tangramath.* Palo Alto, CA: Creative Publications.

Shumway, R. J. (1987). *101 ways to learn mathematics using BASIC (K–8).* Englewood Cliffs, NJ: Prentice-Hall.

Shumway, R. J. (1988). Calculators and computers. In T. R. Post (Ed.), *Teaching mathematics in grades K–8.* Boston: Allyn and Bacon.

Silver, E. A. (1986). Using conceptual and procedural knowledge: A focus on relationships. In J. Hiebert (Ed.), *Conceptual and procedural knowledge: The case of mathematics.* Hillsdale, NJ: Lawrence Erlbaum.

Skemp, R. (1978). Relational understanding and instrumental understanding. *Arithmetic Teacher, 26*(3), 9–15.

Skemp, R. (1979). Goals of learning and qualities of understanding. *Mathematics Teaching, 88,* 44–49.

Smith, S. (1984). Microcomputers in the middle school. In V. P. Hansen & M. J. Zweng (Eds.), *Computers in mathematics education.* Reston, VA: National Council of Teachers of Mathematics.

Steinberg, R. M. (1985). Instruction on derived facts strategies in addition and subtraction. *Journal for Research in Mathematics Education, 16,* 337–355.

Suydam, M. N. (1984). Microcomputers in mathematics instruction. *Arithmetic Teacher, 32*(2), 35.

Suydam, M., & Higgins, J. L. (1977). *Activity based learning in elementary school mathematics: Recommendations from research,* Columbus, OH: ERIC/SMEAC.

Thompson, C. S., & Hendrickson, A. D. (1986). Verbal addition and subtraction problems: Some difficulties and some solutions. *Arithmetic Teacher, 33*(7), 21–25.

Thompson, C. S., & Van de Walle, J. A. (1984a). Modeling subtraction situations. *Arithmetic Teacher, 32*(2), 8–12.

Thompson, C. S., & Van de Walle, J. A. (1984b). The power of 10. *Arithmetic Teacher, 32*(3), 6–11.

Thornton, C. A. (1982). Doubles up—Easy! *Arithmetic Teacher 29*(8), 20.

Thornton, C. A., & Noxon, C. (1977). *Look into the facts: (Addition, subtraction, multiplication, division).* Palo Alto, CA: Creative Publications.

Thornton, C. A., & Toohey, M. A. (1984). *A matter of facts: (Addition, subtraction, multiplication, division).* Palo Alto, CA: Creative Publications.

Thornton, C. A., Tucker, B. F., Dossey, J. A., & Bazik, E. F. (1983). *Teaching mathematics to children with special needs.* Menlo Park, CA: Addison-Wesley.

Trafton, Paul R. (1984). Toward more effective, efficient instruction in mathematics. *The Elementary School Journal, 84,* 514–528.

Usiskin, Z. (1987). Why elementary algebra can, should, and must be an eighth-grade course for average students. *Mathematics Teacher, 80,* 428–438.

Usiskin, Z. (1988). Conceptions of school algebra and uses of variables. In A. F. Coxford (Ed.), *The ideas of algebra, K–12.* Reston, VA: National Council of Teachers of Mathematics.

Van de Walle, J. A. (1983). Focus on the connections between concepts and symbolism. *Focus on Learning Problems in Mathematics, 5,* 5–13.

Van de Walle, J. A., & Thompson, C. S. (1981). A posterboard balance helps write equations. *Arithmetic Teacher, 28*(9), 4–8.

VanLehn, K. (1983). On the representation of procedures in repair theory. In H. P. Ginsburg (Ed.), *The development of mathematical thinking.* New York: Academic Press.

Wagner, S. (1981). Conservation of equation and function under transformation of variable. *Journal for Research in Mathematics Education, 12,* 107–118.

Watt, D. (1983). *Learning with Logo.* New York: McGraw Hill.

Wells, G. W. (1981). The relationship between the processes involved in problem solving and the processes involved in computer programming. *Dissertation Abstracts International,* 42, 2009A-2010A. (University Microfilms No. 81-23,791).

Wheatley, G. H., & Hersberger, J. (1986). A calculator estimation activity. In H. Schoen (Ed.), *Estimation and mental computation.* Reston, VA: National Council of Teachers of Mathematics.

Winter, M. J., Lappan, G. Phillips, E., & Fitzgerald, W. (1986). *Middle grades mathematics project: Spatial visualization.* Menlo Park, CA: Addison-Wesley.

Wirtz, R. (1974). *Mathematics for everyone.* Washington, DC: Curriculum Development Associates.

Wirtz, R. (1978). *An elementary mathematics curriculum for all children.* Paper presented at the National Council of Supervisors of Mathematics annual meeting, San Diego, CA.

Woodcock, R. W., & Johnson, M. B. (1989). *Woodcock-Johnson revised tests of achievement.* Allen, TX: DLM Teaching Resources.

Worth, J. (1987). *Developing primary pupils' problem solving attitudes and abilities.* Paper presented at NCTM Annual Meeting, Anaheim, CA.

Zawojewski, J. S. (1988). Teaching statistics: Mean, median, and mode. *Arithmetic Teacher, 35*(7), 25–26.

INDEX